Current Topics in Membranes, Volume 40

Cell Lipids

Current Topics in Membranes, Volume 40

Series Editors

Arnost Kleinzeller
Department of Physiology
University of Pennsylvania
School of Medicine
Philadelphia, Pennsylvania

Douglas M. Fambrough
Department of Biology
The Johns Hopkins University
Baltimore, Maryland

Yale Series Editors

Joseph F. Hoffman and Gerhard Giebisch
Department of Cellular and Molecular Physiology
Yale University School of Medicine
New Haven, Connecticut

Murdoch Ritchie
Department of Pharmacology
Yale University School of Medicine
New Haven, Connecticut

Current Topics in Membranes, Volume 40

Cell Lipids

Guest Editor

Dick Hoekstra

Department of Physiological Chemistry
University of Groningen
Groningen, The Netherlands

ACADEMIC PRESS
An Imprint of Elsevier

San Diego   New York   Boston   London   Sydney   Tokyo   Toronto

Academic Press, Inc.

An Imprint of Elsevier

525 B Street, Suite 1900, San Diego, California 92101-4495

United Kingdom Edition published by
Academic Press Limited
24–28 Oval Road, London NW1 7DX

International Standard Serial Number: 0070-2161

ISBN-13: 978-0-12-153340-3
ISBN-10: 0-12-153340-9

Printed and bound in the United Kingdom
Transferred to Digital Printing, 2011

Contents

Contents

PART V Summary and Perspectives

Contributors

Numbers in parentheses indicate the pages on which the authors' contributions begin.

Alain Bienvenüe (319), Laboratoire de Biologie Physico-chimique, CNRS URA 530 "interactions membranaires", Université Montpellier II, 34095 Montpellier, France

Torsten Börchers (261), Department of Biochemistry, University of Münster, D-48149 Münster, Germany

Eric G. Bremer (387), Department of Immunology/Microbiology, Rush University, Rush-Presbyterian-St. Luke's Medical Center, Chicago, Illinois 60612

Thomas Eichholtz (439), Division of Cellular Biochemistry, The Netherlands Cancer Institute, 1066 CX Amsterdam, The Netherlands

Anthony H. Futerman (93), Department of Membrane Research and Biophysics, Weizmann Institute of Science, Rehovot 76100, Israel

Ann-Sofi Härmälä (483), Department of Biochemistry and Pharmacy, Åbo Akademi University, 20520 Turku, Finland

Helmut Hauser (167), Laboratorium für Biochemie, Eidgenössische Technische Hochschule, CH-8092 Zürich, Switzerland

Dick Hoekstra (503), Department of Physiological Chemistry, University of Groningen, 9712 KZ Groningen, The Netherlands

Peter L. Hordijk (439), Division of Cellular Biochemistry, The Netherlands Cancer Institute, 1066 CX Amsterdam, The Netherlands

Ken Jacobson (111), Department of Cell Biology and Anatomy, University of North Carolina at Chapel Hill, Chapel Hill, North Carolina 27599

Kees Jalink (439), Division of Cellular Biochemistry, The Netherlands Cancer Institute, 1066 CX Amsterdam, The Netherlands

Jan Willem Kok (503), Department of Physiological Chemistry, University of Groningen, 9712 KZ Groningen, The Netherlands

Greta M. Lee (111), Department of Cell Biology and Anatomy, University of North Carolina at Chapel Hill, Chapel Hill, North Carolina 27599

Gert Lipka (167), Laboratorium für Biochemie, Eidgenössische Technische Hochschule, CH-8092 Zürich, Switzerland

María Paz Marzolo (579), Departamento de Gastroenterología, Facultad de Medicina, Pontificia Universidad Católica, Santiago, Chile

Jonathan C. McIntyre (453), Department of Molecular Genetics, Biochemistry, and Microbiology, University of Cincinnati College of Medicine, Cincinnati, Ohio 45267

Alfred H. Merrill, Jr. (361), Department of Biochemistry, Emory University School of Medicine, Atlanta, Georgia 30322

Wouter H. Moolenaar (439), Division of Cellular Biochemistry, The Netherlands Cancer Institute, 1066 CX Amsterdam, The Netherlands

Flavio Nervi (579), Departamento de Gastroenterología, Pontificia Universidad Católica, Santiago, Chile

Jos A. F. Op den Kamp (7), Department of Lipid Biochemistry, Centre for Biomembranes and Lipid Enzymology, University of Utrecht, 3584 CH Utrecht, The Netherlands

Bernadette C. Ossendorp (217), Centre for Biomembranes and Lipid Enzymology, University of Utrecht, 3584 CH Utrecht, The Netherlands

M. Isabella Pörn (483), Department of Biochemistry and Pharmacy, Åbo Akademi University, 20520 Turku, Finland

Attilio Rigotti (579), Departamento de Gastroenterología, Facultad de Medicina, Pontificia Universidad Católica, Santiago, Chile

Enrique Rodriguez-Boulan (295), Department of Cell Biology and Anatomy, Cornell University Medical College, New York, New York 10021

Ben Roelofsen (7), Department of Lipid Biochemistry, Centre for Biomembranes and Lipid Enzymology, University of Utrecht, 3584 CH Utrecht, The Netherlands

Josette Sainte Marie (319), Laboratoire de Biologie Physico-chimique, CNRS URA 530 "interactions membranaires", Université Montpellier II, 34095 Montpellier, France

Konrad Sandhoff (75), Institut für Organische Chemie und Biochemie der Universität Bonn, D-53121 Bonn 1, Germany

Dick Schaap (413), Division of Cellular Biochemistry, The Netherlands Cancer Institute, 1066 CX Amsterdam, The Netherlands

Alan J. Schroit (47), Department of Cell Biology, University of Texas M. D. Anderson Cancer Center, Houston, Texas 77030

Kai Simons (619), European Molecular Biology Laboratory, 6900 Heidelberg, Germany

Richard G. Sleight (453), Department of Molecular Genetics, Biochemistry, and Microbiology, University of Cincinnati College of Medicine, Cincinnati, Ohio 45267

J. Peter Slotte (483), Department of Biochemistry and Pharmacy, Åbo Akademi University, 20520 Turku, Finland

Gerry T. Snoek (217), Centre for Biomembranes and Lipid Enzymology, University of Utrecht, 3584 CH Utrecht, The Netherlands

Friedrich Spener (261), Department of Biochemistry, University of Münster, D-48149 Münster, Germany

Charles C. Sweeley (357), Department of Biochemistry, Michigan State University, East Lansing, Michigan 48824

Wim J. van Blitterswijk (413, 439), Division of Cellular Biochemistry, The Netherlands Cancer Institute, 1066 CX Amsterdam, The Netherlands

Emile van Corven (439), Division of Cellular Biochemistry, The Netherlands Cancer Institute, 1066 CX Amsterdam, The Netherlands

Laurens L. M. van Deenen (1), Centre for Biomembranes and Lipid Enzymology, University of Utrecht, 3584 CH Utrecht, The Netherlands

Rob van der Bend (413, 439), Division of Cellular Biochemistry, The Netherlands Cancer Institute, 1066 CX Amsterdam, The Netherlands

Gerhild van Echten (75), Institut für Organische Chemie und Biochemie der Universität Bonn, D-53121 Bonn 1, Germany

Gerrit van Meer (539), Department of Cell Biology, Medical School, University of Utrecht, 3584 CH Utrecht, The Netherlands

Wouter van 't Hof (539), Department of Cell Biology, Medical School, University of Utrecht, 3584 CH Utrecht, The Netherlands

Binks W. Wattenberg (565), Cell Biology Unit, The Upjohn Company, Kalamazoo, Michigan 49007

Karel W. A. Wirtz (217), Centre for Biomembranes and Lipid Enzymology, University of Utrecht, 3584 CH Utrecht, The Netherlands

David E. Wolf (143), Worcester Foundation for Experimental Biology, Shrewsbury, Massachusetts 01545

Philip L. Yeagle (197), Department of Biochemistry, University at Buffalo School of Medicine, Buffalo, New York 14214

Chiara Zurzolo (295), Department of Cell Biology and Anatomy, Cornell University Medical College, New York, New York 10021

Previous Volumes in Series

Current Topics in Membranes and Transport

Volume 20   Molecular Approaches to Epithelial Transport* (1984)
Edited by James B. Wade and Simon A. Lewis

Volume 21   Ion Channels: Molecular and Physiological Aspects (1984)
Edited by Wilfred D. Stein

Volume 22   The Squid Axon (1984)
Edited by Peter F. Baker

Volume 23   Genes and Membranes: Transport Proteins and Receptors*
(1985)
Edited by Edward A. Adelberg and Carolyn W. Slayman

Volume 24   Membrane Protein Biosynthesis and Turnover (1985)
Edited by Philip A. Knauf and John S. Cook

Volume 25   Regulation of Calcium Transport across Muscle
Membranes (1985)
Edited by Adil E. Shamoo

Volume 26 Na^+-H^+ Exchange, Intracellular pH, and Cell Function*
(1986)
Edited by Peter S. Aronson and Walter F. Boron

Volume 27   The Role of Membranes in Cell Growth and Differentiation
(1986)
Edited by Lazaro J. Mandel and Dale J. Benos

Volume 28   Potassium Transport: Physiology and Pathophysiology*
(1987)
Edited by Gerhard Giebisch

Volume 29   Membrane Structure and Function (1987)
Edited by Richard D. Klausner, Christoph Kempf and Jos van
Renswoude

* Part of the series from the Yale Department of Cellular and Molecular Physiology

Introduction

Laurens L. M. van Deenen
Centre for Biomembranes and Lipid Enzymology, University of Utrecht, 3584 CH
Utrecht, The Netherlands

The emergence of lipid dynamics as one of the most exciting new directions of biomembrane research is highlighted by this volume, which integrates lipid biochemistry with cell biology. During the past decade, several important links have been made between classical techniques of lipid research and modern approaches in cell biology. As a result, the new membranology now is complemented by a vast knowledge of the intriguing pathways of transmembrane and intracellular signaling that involve lipid mediators and modulators, as well as by new insights on lipid traffic and sorting.

The understanding of the diversity of functions of membrane lipids has progressed rather slowly with time. The recognition of lipids as the components that account for the osmotic and electrical barrier of cells dates back well over a century. The concept that the permeability barrier is made of a continuous lipid bilayer originated in 1925, but progress in X-ray analysis and electron microscopy was required before the problem of membrane structure could be tackled by direct experimentation. This problem turned out to be very complex because of the possible production of artifacts and because of pitfalls in the interpretation of results in terms of molecular organization of proteins and lipids. However, one of the outstanding features in the more recent history of studies of membrane structure is the change from the strongly conflicting views that existed in the 1960s to a general agreement, during the 1970s, that all biological interfaces indeed share a basic structural architecture. The flourishing research on artificial membranes (lipid bilayer membranes and liposomes) also contributed to reconciling some of the opposing views on membrane organization. Since that time, research on membranes became one of the most rapidly expanding areas in the life sciences.

Natural and reconstituted membranes have been investigated using a great variety of sophisticated chemical and physical techniques. In an astonishingly short period of time, key information has been obtained on structure and topography of many membrane proteins including receptors, ion pumps and channels, transporters, and photosynthetic reaction centers. Seeing the massive amount of information that has been obtained on the amino acid sequences and the detailed molecular orientations of these proteins within the lipid bilayer is amazing. Currently, researchers in the field are on the brink of major discoveries: the understanding of the molecular mechanism of action of a diversity of membrane proteins with distinct functions.

Meanwhile, physical studies on synthetic phospholipids have demonstrated that intermediary nonbilayer structures may contribute to the dynamic state of membranes. A modulation of bilayer–nonbilayer transitions by a variety of effectors has been proposed to be involved in biological processes such as membrane fusion and protein translocation. Indeed, physical effectors of lipid organization can be visualized to contribute to the dynamic behavior of membranes, but metabolic reactions of the lipid constituents have been demonstrated unequivocally to play a role in some primary functions of membranes. The discovery of the crucial role of lipid mediators and modulators in transmembrane and intracellular signaling represents one of the most exciting events of cellular biochemistry in the past decade. However, these new ideas, produced by the merging of cell biology and lipid biochemistry, had a long incubation time. For example, in the 1960s a diacyl–monoacylphosphoglyceride cycle (involving phospholipases A_1 and A_2 and transacylation of 1-acyl and 2-acyl phosphoglyceride isomers) was formulated. This cycle was considered to be responsible for repair of damaged membrane lipids and also was argued to contribute to a dynamic state of membrane organization because of the marked differences in physical properties between diacyl and monoacyl analogs. Only some decades later did researchers determine that the release of arachidonic acid by phospholipase A_2 represents the onset of a complex signaling network with a wide range of cellular responses. Another, even more striking, example is offered by the history of the action of hormones and neurotransmitters on phospholipase C-mediated hydrolysis of polyphosphoinositides. In the early 1950s, good evidence existed that various agonists stimulated the incorporation of ^{32}P into phospholipids of their target cells. At an early stage, these effects were shown to be caused by turnover of phosphoryl units on preformed lipids rather than by synthesis *de novo*. The turnover was found to occur in both phosphatidic acid and inositol phospholipids, a discovery that led to the formulation of a cycle relating these two lipids. Although the subsequent elucidation of the structure of the complex components, such as triphosphoinositides, gave further impetus to hypotheses and investigations on the physiological meaning of the cycle, only in the 1980s did eye-opening experiments demonstrate that the phospholipase C-catalyzed hydrolysis of polyphosphoinositides (with the formation of inositol triphosphate and diglyceride) is an intermediary step in the mechanism leading from agonist stimulation of receptors to the formation of second messengers. The demonstration of the central role of lipids in various networks of signal transduction is one of the most important recent achievements in membrane research, but several other avenues of inquiry have been opened that lead to different aspects of lipid dynamics.

The elucidation of enzymatic pathways and the intriguing topology of lipid biosynthesis has provoked new studies on transbilayer movement and intracellular transport of different classes of membrane components. Investigations on the biogenesis of surface polarity generated new approaches to establish the mechanism involved in the sorting of lipids and proteins, to account for the different composition of distinct domains of cell membranes. Particular support has been given for sphingolipids and glycolipids to reside in the outer membrane region so they can participate in surface recognition. Observing that various topics led to the exploration of the pathology of biomembranes has been rewarding.

The study of biomembranes has matured to the point that teaching, at an advanced level, requires several (text)books that discuss different specialized subareas. This volume in the "Current Topics in Membranes" series provides an authorative overview of the dynamics of membrane lipids, while achieving a high degree of integration of cell biology and biochemistry. The book will help readers of various disciplines obtain clear cut information on the present status of an important aspect of biomembranes. At the same time, the volume may have a catalyzing impact on the research of scientists working in one of the most exciting and promising areas of current molecular cell biology.

PART I

Lipid Metabolism and Dynamics

CHAPTER 1

Plasma Membrane Phospholipid Asymmetry and Its Maintenance: The Human Erythrocyte as a Model

Ben Roelofsen and Jos A. F. Op den Kamp
Department of Lipid Biochemistry, Centre for Biomembranes and Lipid Enzymology,
University of Utrecht, 3584 CH Utrecht, The Netherlands

I. INTRODUCTION

Inspired by an observation made by Maddy as early as 1964, as well as by personal experiences, in the early 1970s Bretscher proposed an "asymmetrical lipid bilayer structure for biological membranes" (Bretscher, 1972a,b). This original proposal was based on the observation that exposure of intact bovine (Maddy, 1964) or human (Bretscher, 1972a,b) erythrocytes to exogenous NH_2-group-specific chemical reagents

Current Topics in Membranes, Volume 40
Copyright © 1994 by Academic Press, Inc. All rights of reproduction in any form reserved.

did not result in the labeling of either of the two aminophospholipids—phosphatidylethanolamine (PE) and phosphatidylserine (PS)—to any appreciable extent, whereas Bretscher observed extensive labeling—particularly of PE—when open ghost membranes were subjected to such treatment. Soon thereafter, the concept of the asymmetric distribution of the four major phospholipid classes in the human red cell membrane could be completed by results of studies involving highly purified phospholipases (Verkleij *et al.*, 1973; Zwaal *et al.*, 1975). The subsequent development of other tools and techniques, in addition to chemical reagents and phospholipases, as well as the extension of such studies to cells other than erythrocytes, provided a firm basis for the concept of transbilayer phospholipid asymmetry as a characteristic feature of plasma membranes (Op den Kamp, 1979; Etemadi, 1980; van Deenen, 1981; Roelofsen, 1982). Although phospholipid asymmetry has been reported also to exist in a number of subcellular membranes, the experimental data are not as conclusive in those cases as in the case of plasma membranes (Op den Kamp, 1979; Roelofsen, 1982; Devaux, 1991,1992). Regarding the four major phospholipid classes found in plasma membranes, researchers generally accept that the choline-containing sphingomyelin (SM) and phosphatidylcholine (PC) dominate the exofacial leaflet, whereas both the aminophospholipids (PE and PS) typically are concentrated in the cytoplasmic side of the lipid bilayer.

Shortly after the idea of an asymmetric phospholipid bilayer had become an established fact, questions regarding its generation and maintenance were posed. Those questions regarding maintenance have received particular attention during the last 15 years. Enhanced accessibility of the aminophospholipids in intact human erythrocytes to purified phospholipases A_2, observed after mild oxidative cross-linking of their membrane skeletal proteins, led Haest and colleagues (1978) to propose an important role for this protein network—and more specifically for the component spectrin—in maintaining phospholipid asymmetry in the red cell membrane. Six years later, a novel and intriguing mechanism for the maintenance of phospholipid asymmetry was discovered by Seigneuret and Devaux (1984), who had observed an ATP-dependent rapid inward translocation of spin-labeled analogs of both aminophospholipids that previously had been inserted into the outer membrane leaflet of intact human erythrocytes. Subsequent studies by Devaux and others provided substantial evidence for the existence of this system in a number of different (plasma) membranes, unraveled quite a number of its characteristics (for reviews, see Zachowski and Devaux, 1990; Devaux, 1991,1992; Schroit and Zwaal, 1991), and demonstrated its (almost) absolute specificity for diacyl-PE and -PS as suitable candidates for translocation in the human erythrocyte membrane (Morrot *et al.*, 1989).

No doubt, this so-called "flippase"—the major topic of Chapter 2—plays a prominent role in maintaining phospholipid asymmetry in the red cell membrane. However, and in contrast to ideas suggested in a number of papers (Calvez *et al.*, 1988; Gudi *et al.*, 1990; Zachowski and Devaux, 1990; Devaux, 1991,1992; Pradhan *et al.*, 1991), we feel confident that the membrane skeleton also makes a major contribution to this phenomenon, as becomes evident from a variety of studies on model systems, native and modified normal, as well as pathologic, erythrocytes. Another point of controversy is the possible loss of phospholipid asymmetry, either artificially induced in normal erythrocytes or existing *in situ* in this cell type under pathologic conditions. Careful analysis of the conditions under which phospholipid asymmetry in the red cell membrane is either maintained or lost provides strong support for the concept that, indeed, the flippase and the membrane skeleton both are involved in maintaining proper transbilayer phospholipid asymmetry in this membrane.

II. PLASMA MEMBRANE PHOSPHOLIPID ASYMMETRY

Since the discovery of plasma membrane structure, many various techniques have been developed to assess the transbilayer distribution of membrane phospholipids. The tools used, the conditions of their application, and the membrane studied all must meet several strictly determined prerequisites to gain reliable and conclusive results (Op den Kamp, 1979; Etemadi, 1980; Roelofsen, 1982). Not many—if any at all—approaches fulfill all these essential requirements. Additionally, the eventual presence of, for such studies, unfavorable characteristics of a particular membrane—bilayer instability, rapid flip-flop of phospholipids, and so on—may induce additional problems that are not easy to solve. Such experimental imperfections are undoubtedly responsible for some uncertainties concerning the transbilayer phospholipid distribution in a number of subcellular membranes. However, even studies on some plasma membranes, particularly those in which remarkably high fractions of PS have been assigned to the outer membrane leaflet (Fontaine *et al.*, 1980; Sessions and Horwitz, 1983; Record *et al.*, 1984; Hinkovska *et al.*, 1986; Pelletier *et al.*, 1987; Lipka *et al.*, 1991), may arouse some suspicion. The fact that the aminophospholipid translocase plays an important role in maintaining phospholipid asymmetry adds a new prerequisite to the list of requirements for phospholipid "sidedness" experiments, that is, a careful control on the presence of sufficient Mg^{2+}-ATP—the fuel of the flippase—during the experiment. For example, the fact that malaria infection of erythrocytes has been claimed to cause the appearance of vast amounts of PS in the exofacial leaflet of the host cell plasma membrane may be ascribed simply to overlooking this point (see Section V,B,7).

The transbilayer distribution of the four major phospholipid classes in the normal human erythrocyte was established in the early 1970s using highly purified phospholipases (Verkleij et al., 1973; Zwaal et al., 1975), and has been confirmed in many studies that have been published since that time. The two choline-containing phospholipids dominate the outer leaflet, in which we find 76% of the PC and 82% of the SM. Both aminophospholipids, on the other hand, are for the most part (80% of the PE) or even exclusively (PS) located in the cytoplasmic half of the bilayer. More recent studies (Bütikofer et al., 1990; Gascard et al., 1991) showed that approximately 20% of each of the following phospholipids is present in the outer membrane leaflet of the human red cell membrane: phosphatidic acid (PA), phosphatidylinositol (PI), and phosphatidylinositol 4,5-bisphosphate (PIP_2). Interestingly, phosphatidylinositol 4-monophosphate (PIP) failed to be detected, suggesting its exclusive localization in the cytoplasmic leaflet, a feature that PIP would share with PS. However, after two decades of investigations, some researchers are still uncertain whether in normal healthy (human) erythrocytes all the PS is, indeed, located in the inner membrane leaflet (Devaux, 1992). This challenge seems to be based on no more than two references to previously published papers (van Meer et al., 1981; Dressler et al., 1984). The first paper reports the detection of some 8% of the PS in the outer leaflet of intact human erythrocytes. The second study, however, for which it was suggested that more than 20% of the PS was assigned to the outer half of the membrane bilayer, in fact showed that, not only in control cells but also in those cells that remained after a heat-induced pinching off of membrane vesicles, no PS could be detected in this particular leaflet. Note, however, that in an overwhelming number of studies the virtually complete absence of PS in the exofacial leaflet of the normal human erythrocyte was shown. These studies involved a variety of chemical reagents and phospholipases, as well as more sophisticated techniques such as the prothrombinase assay (Bevers et al., 1982) or the application of a chemically modified pig pancreatic phospholipase A_2 (N^ε-palmitoyl-Lys 116 ε-amidinated pig pancreatic phospholipase A_2, Pal-116-AMPA), which possesses superior membrane-penetrating capacities (van der Wiele et al., 1988). Hence, not much room seems to exist for any serious doubts about the exclusive localization of PS in the cytoplasmic leaflet. In this context, also note that the small amount of PS that is found routinely in the outer leaflet of the plasma membrane of resting platelets has been suggested to originate from a small fraction of activated platelets (Schroit and Zwaal, 1991). Further, many studies involving sphingomyelinase C demonstrated that 80–85% of the SM is located in the outer membrane leaflet. Nevertheless, this conclusion has been questioned (Allan and Walklin, 1988). These authors observed a considerable extent of endovesiculation when intact human erythrocytes were exposed to the action of a commercial preparation of *Staphylococcus aureus* sphingomyelinase C. Hence, these researchers argued that those endocytic vesicles may have withdrawn their membranes from any further attack by the enzyme, resulting in an incomplete hydrolysis (by about 15%) of the SM in the outer leaflet which, as a consequence, was proposed to accommodate all the SM present in the membrane. However, realize that, not only will the complete incubation medium be trapped inside those vesicles, but such enzymes usually exhibit preferential interactions with their target membranes, making the event that the sphingomyelinase C would not be able to complete its action very unlikely, even under those conditions. Moreover, the preparation of sphingomyelinase C that originally had been used to assess the transbilayer distribution of SM in the human red cell membrane (Verkleij *et al.*, 1973; Zwaal *et al.*, 1975) is known to have an extremely high membrane-penetrating capacity (Demel *et al.*, 1975). This enzyme easily degrades up to 82% of the SM in human red cells pretreated with *Naja naja* (or bee venom) phospholipase A_2, in contrast to the commercial preparation used by Allan and Walklin (1988) which almost completely fails to do so under these conditions, which do not cause endovesiculation.

Each of the phospholipid classes comprises a great number of molecular species, differing in the composition of their fatty acyl constituents. PC was the first class for which the transbilayer distribution was determined (Renooy *et al.*, 1974); the inner and outer pools appeared to have an essentially identical composition. This finding could be explained by the existence of a continuous transbilayer movement—so-called flip–flop—of those molecules. The rate of this passive diffusion process is highly dependent on the molecular species (van Meer and Op den Kamp, 1982) and ranges from a half-time value of 3 hr for 1-palmitoyl, 2-linoleoyl PC to 27 hr for dipalmitoyl PC at 37°C (Middelkoop *et al.*, 1986).

Note that the behavior of SM, the other choline-containing phospholipid of the red cell membrane, is essentially different from that of PC. For SM, we found that the molecular species composition of the outer monolayer pool differed considerably from that of the fraction in the inner leaflet (Boegheim *et al.*, 1983). The inner layer pool appeared to be characterized by a relatively high content (73%) of fatty acids with less than 20 carbon atoms, whereas these account for only 31% of the total fatty acids in the SM in the outer leaflet; the ratio of saturated to unsaturated fatty acids in the two pools was similar. This marked difference in molecular species composition of the two SM pools on either side of the bilayer was the very first indication that, in contrast to PC, SM might not be subject to any appreciable transbilayer movement. More recent studies (Zachowski *et al.*, 1985; Tilley *et al.*, 1986) provided direct evidence for this phenomenon. Indeed, the highly static transbilayer distribution of SM in the (human) red cell membrane is most intriguing, particularly since no conditions have been found to date under which this distribution can be shifted to any appreciable extent, not even conditions that cause considerable alterations in the transbilayer stability and orientation of the glycerophospholipids (see Section V).

Although subject to relatively fast transbilayer movements, revealing a half-time value for inward translocation of about 30 min, a certain asymmetry in the transbilayer distribution of molecular species is observed for PE also (Middelkoop, 1989; Hullin *et al.*, 1991), with respect to the diacyl as well as the 1-alkenyl, 2-acyl (plasmalogen) subclass of this phospholipid. The study by Hullin *et al.* (1991) revealed an enrichment of the monoene species in the outer leaflet and a preferential localization of the polyunsaturates in the inner one. These results are at variance with the observations of Middelkoop (1989), who did not observe any particular pattern in the asymmetric distribution of the various molecular species over both halves of the bilayer. Note, however, that Middelkoop discriminated the two pools of PE from one another by treating the intact cells with phospholipase A_2 plus sphingomyelinase C, conditions known to degrade the outer monolayer pool to completion, whereas Hullin *et al.*, (1991) applied trinitrobenzenesulfonic acid (TNBS) at 0°C, resulting in the labeling of no more than 4–8% of the PE in the outer leaflet. Hence, despite the precautions taken, a selective labeling of particular (domains of) molecular species of PE in the outer membrane layer cannot be excluded with certainty. Preliminary studies by Middlekoop (1989) showed that the rate, as well as the ultimate extent, of inward translocation of three individual (radiolabeled) molecular species of diacyl PE decreased in the order 16:0/18:2 > 18:1/18:1 > 16:0/16:0. Interestingly, this sequence appears to be the same as that of the preference by which these species of endogenous PE are located in the inner leaflet, suggesting that the unequal distribution of (particular) species of PE over the inner and outer monolayer pools is generated and maintained by a corresponding preference of the aminophospholipid translocase. Finally, the relative amounts of the plasmalogen (1-alkenyl, 2-acyl) subclass of PE, constituting approximately 40% of the total PE, are essentially identical in both pools (Middelkoop, 1989).

Since, in the normal human erythrocyte, virtually all the PS is confined to the cytoplasmic leaflet of the membrane, questions concerning its transbilayer mobility and the molecular species composition in the outer leaflet are irrelevant.

III. BIOGENESIS OF RED CELL PHOSPHOLIPID ASYMMETRY

In a first attempt to gain insight into the constitution of plasma membrane phospholipid asymmetry during red cell biogenesis, comparative studies were performed on mature murine erythrocytes and Friend erythroleukemic cells (FELCs). FELCs are erythroid cells, derived rom susceptible mouse spleens infected with the Friend virus complex, that are blocked at an early stage of their differentiation pathway, presumably between the burst-forming unit erythroid and the colony-forming unit erythroid stages. These cells can be induced to differentiate, and may reach the reticulocyte stage under appropriate conditions (Patel and Lodisch, 1987). FELCs are assumed to resemble pro-erythroblasts closely (Marks and Rifkind, 1978). The studies done involved the application of three independent techniques—treatment of intact cells with phospholipases A_2 and C, fluorescamine labeling, and a PC-transfer protein-mediated exchange procedure (Rawyler *et al.*, 1984,1985)—and led to several observations. The same extent of asymmetry in transbilayer distribution of SM in the mature erythrocyte membrane is observed already in the plasma membrane of the FELC: 79–85% in the outer leaflet. Although not as absolute as in the erythrocyte (none of the PS in the outer leaflet), a highly asymmetric distribution of PS (10–20% in the outer leaflet) also is observed in the plasma membrane of the precursor cells. The latter finding agrees well with observations by Schroit's group (Connor *et al.*, 1989). PC, PE, and PI, on the other hand, still are distributed randomly over both plasma membrane leaflets of the pro-erythroblast. Additional studies (Nijhof *et al.*, 1986; van der Schaft *et al.*, 1987a) showed that this condition is maintained up to the normoblast stage, the last stage in development before extrusion of the nucleus. Since the asymmetric distribution of phospholipids in the reticulocyte membrane is already identical to that in the mature erythrocyte (van der Schaft *et al.*, 1987a), evidently abrupt changes in *glycero*phospholipid distribution take place during the enucleation of the normoblast, or immediately thereafter. Indeed, researchers have suggested that the ultimate phospholipid asymmetry has not yet been established in the plasma membrane that surrounds the extruding nucleus, in marked contrast to the situation in the developing reticulocyte (Schlegel *et al.*, 1982).

Although major components of the membrane skeleton, such as α and β spectrin and protein 4.1, are synthesized already (and degraded again) during the early stages of erythroid differentiation (Woods et al., 1986; Lehnert and Lodish, 1988), the formation of a stable membrane skeleton, which coincides with the synthesis of the Band 3 protein (Lazarides and Woods, 1989; Hanspal and Palek, 1992), does not occur before the late erythroblast stage is reached and may be completed only during the process of enucleation (Nijhof et al., 1986; Nijhof and Wierenga, 1988) or shortly thereafter.

Comparative studies on the transbilayer equilibration of radiolabeled analogs of the three major glycerophospholipids, previously inserted into the outer leaflet of the plasma membrane of mature murine erythrocytes and FELCs, revealed some interesting results (Middelkoop et al., 1989a). Rapid reorientations of the newly introduced aminophospholipids (PE and PS) in favor of the inner membrane leaflet were observed in fresh mouse erythrocytes; the inward translocation of PC in this membrane proceeded relatively slowly. In FELCs, on the other hand, all three glycerophospholipids equilibrated over both halves of the plasma membrane very rapidly, that is, within 1 hr, resulting in an entirely random distribution of PC and PE; an asymmetric distribution in favor of the inner monolayer was observed only for PS. Experiments by Devaux (1991) also showed that the transmembrane movement of spin-labeled PS and PE is faster in erythroblasts than in erythrocytes. Lowering the ATP level in the FELCs caused a reduction in the rate of inward translocation of both aminophospholipids, but not of PC, indicating that this translocation of PS and PE is clearly ATP dependent (Middelkoop et al., 1989a). Hence, the situation in the plasma membrane of the FELC is unique in the sense that, although an ATP-dependent translocase (cf. Section IV) is present and active for both PS and PE, its activity results in an asymmetric distribution of PS, but not of PE. This remarkable situation might be a consequence of the fact that, in contrast to the mature red cell, this precursor cell still lacks a completely assembled membrane skeletal network and, therefore, its contribution to stabilizing the proper asymmetric phospholipid distribution that is known to exist in the mature red cell.

IV. MAINTENANCE OF PHOSPHOLIPID ASYMMETRY

A. ATP-Dependent Aminophospholipid Translocation

The presence, in the erythrocyte membrane, of a system that mediates an active transport of both aminophospholipids from the outer to the inner membrane leaflet, has been well documented in a variety of studies using different exogenous probes, including spin-labeled (Seigneuret and Devaux, 1984; Zachowski *et al.*, 1985,1986) or fluorescently labeled (Connor and Schroit, 1988,1989) phospholipid analogs, short-chain diacyl glycerophospholipids (Daleke and Huestis, 1985,1989), and radiolabeled long-chain diacyl glycerophospholipids and SM (Tilley *et al.*, 1986). In addition to the red cell membrane, this cytosolic Mg^{2+}-ATP-requiring system also has been found in the plasma membranes of platelets (Sune *et al.*, 1987; Tilly *et al.*, 1990), lymphocytes (Zachowski *et al.*, 1987), fibroblasts (Martin and Pagano, 1987), and synaptosomes (Zachowski *et al.*, 1990), and even in the membrane of a subcellular organelle, the chromaffin granule (Zachowski *et al.*, 1989).

Although the exact nature of this flippase has not been established—whether it is the 120-kDa Mg^{2+}-ATPase (Morrot *et al.*, 1990) or a 31-kDa Rh antigen-like polypeptide (Connor and Schroit, 1988,1991; Schroit *et al.*, 1990), eventually in concerted action with an endofacial protein (Connor and Schroit, 1990)—several of its features have been characterized in great detail. The system exhibits a markedly high degree of specificity, not only regarding the chemical structure of the molecules it can accept for translocation (Morrot *et al.*, 1989) but also regarding discrimination of the naturally occurring L stereoisomers from the unnatural D isomers, the latter not being translocated (Martin and Pagano, 1987).

Inward translocation of both aminophospholipids has been reported to be inhibited by extracellularly added vanadate (50 μM) (Seigneuret and Devaux, 1984; Zachowski *et al.*, 1986). The intracellular presence of Mg^{2+} (2 mM) is most essential for flippase functioning, whereas cytosolic Ca^{2+} (1 μM) appears to be strongly inhibitory to the translocation process. The extracellular presence of Ca^{2+}, as well as that of a series of other divalent cations, at concentrations of up to 1 mM has no effect (Bitbol *et al.*, 1987). Finally, the translocase has been shown to be inhibited by thiol reagents such as N-ethylmaleimide (NEM), pyridyldithioethylamine (PDA), and diamide (Connor and Schroit, 1988,1989,1990,1991; Daleke and Huestis, 1985,1989). Note, however, that this information has been derived from studies involving short-chain diacyl PE and PS or their fluorescently labeled analogs. Particularly for diamide, considerable inward translocation of radiolabeled long-chain diacyl PE and -PS still was observed when these compounds were inserted into the outer membrane leaflet of (fresh) human erythrocytes previously treated with the thiol reagent (Middelkoop *et al.*, 1989b). This apparent discrepancy may raise some doubts about the reliability of the unnatural probe molecules in their application as reporter molecules. Will they, indeed, show a behavior that is identical to that of their native counterparts under *all* circumstances?

More detailed information on the specific features of the aminophospho-lipid translocase can be found in several reviews (Zachowski and Devaux, 1990; Devaux, 1991,1992; Schroit and Zwaal, 1991) and, of course, in Chapter 2.

B. Membrane Skeleton–Aminophospholipid Interaction

The involvement of the membrane skeleton, a two-dimensional protein network that underlies—and interacts with—the cytoplasmic side of the red cell membrane (Haest, 1982; Cohen and Gascard, 1992; Hanspal and Palek, 1992; Liu and Derick, 1992; Luna and Hitt, 1992; Pinder *et al.*, 1992), in maintaining its phospholipid asymmetry, was suggested for the first time by Haest and colleagues (1978). These investigators observed a marked increase in the hydrolysis of both aminophospholipids when dia-mide (or tetrathionate)-treated intact human erythrocytes were exposed subsequently to pure phospholipases A_2 under nonlytic conditions (Haest *et al.*, 1978). Although this observation has been argued to reflect a trans-verse destabilization of the lipid bilayer resulting in enhanced lipid translo-cation rates (Franck *et al.*, 1982; Mohandas *et al.*, 1982; Bergmann *et al.*, 1984) rather than an *in situ* change in phospholipid asymmetry (Franck *et al.*, 1986), it clearly indicates that a perturbation of the membrane skeleton, caused by the diamide/tetrathionate-induced mild oxidative cross-linking of its components, strongly impairs its interaction with the inner monolayer phospholipids. Indeed, numerous studies involving model systems such as lipid monolayers and liposomes provided evidence for possible interactions of the aminophospholipids—most specifically PS—with components of the membrane skeleton, such as spectrin (Sweet and Zull, 1970; Mombers *et al.*, 1979,1980; Bonnet and Begard, 1984; Subbarao *et al.*, 1991) and protein 4.1 (Sato and Ohnishi, 1983; Cohen *et al.*, 1988; Rybicki *et al.*, 1988; Shiffer *et al.*, 1988), as well as with the cytoskeleton isolated from human platelets (Comfurius *et al.*, 1989). Other studies have shown, however, that the interaction between PS and spectrin is rather weak (Maksymiw *et al.*, 1987; Bitbol *et al.*, 1989), leading to the conclusion that this protein cannot be responsible for retaining PS in the cytoplasmic leaflet of the red cell membrane. Why should a strong interaction between PS and the membrane skeleton be required? Even a weak interaction may be sufficient to prevent a flipping back of PS mole-cules to the outer leaflet. The behavior of PS depends entirely on the actual conditions and their mutual influences. A strong, let alone a permanent, interaction between PS and the membrane skeleton may not be required to shift an equilibrium distribution of PS entirely in favor of the inner leaflet. The fact that PS, like the other phospholipids, exhibits a rapid lateral mobility (Morrot *et al.*, 1986) indicates already that a strong and permanent fixation of PS in that leaflet does not occur.

C. Concerted Action of Aminophospholipid Translocase and Membrane Skeleton

In studies of normal erythrocytes in which the membrane skeleton had been perturbed by diamide or heat treatment, of reversibly sicklable cells (RSCs), and of pro-erythroblasts (FELCs; see Section III), observations have been made that provide strong indications for the involvement of the membrane skeleton in the maintenance of phospholipid asymmetry in the red cell membrane.

Although mild oxidative cross-linking of membrane skeletal proteins by treatment of intact cells with diamide or tetrathionate causes a destabilization of the lipid bilayer, as reflected by a considerably enhanced flip–flop of PC molecules (Franck *et al.*, 1982), it appears—in contrast to earlier observations (Haest *et al.*, 1978; Williamson *et al.*, 1982)—not to give rise to an alteration in phospholipid asymmetry and an appearance of PS in the outer monolayer (Franck *et al.*, 1986; Middelkoop *et al.*, 1989b). This latter observation, which is of considerable interest, was demonstated convincingly by the negative response that diamide-treated fresh erythrocytes exhibited in the prothrombinase assay, as well as by the absence of any hydrolysis of PS when those cells were exposed to Pal-116-AMPA (Middelkoop *et al.*, 1989b). Given the specificity and high sensitivity of both these techniques, treatment of fresh intact human erythrocytes with those oxidative reagents could be concluded not to give rise to an outward translocation of more than 1 mol% of the PS, if any. However, endogenous PS began to appear in the outer membrane leaflet after diamide-treated cells had been incubated for 4 hr at 37°C, a process that progressively increased during additional incubation (Middelkoop *et al.*, 1989b). Interestingly, the very first appearance of PS in the exofacial leaflet of these cells coincided markedly with a drop in cellular ATP to 10% of the original level. This result agrees well with the immediate appearance of exofacial PS in diamide-treated cells that have been depleted of their ATP previously, a condition that by itself does not give rise to any appreciable change in the transbilayer distribution of PS (Middelkoop *et al.*, 1988,1989b). The observations just discussed suggest that, in diamide-treated (fresh) erythrocytes, the aminophospholipid translocase is still in operation, a suggestion that appears to be in conflict with the reported inhibitory effect that this treatment supposedly exerts on the flippase

(Daleke and Huestis, 1985,1989; Connor and Schroit, 1988,1990,1991). Although this inhibition of an inward translocation of aminophospholipids in diamide-treated red cells may provide some information about the suitability of the applied probe molecules under these conditions (Connor *et al.*, 1990; Colleau *et al.*, 1991), rather than truthfully reporting the condition of the flippase, the actual reason for this discrepancy is not immediately obvious. Whatever the reason, however, the application of radiolabeled long-chain diacyl glycerophospholipids as exogenous probe molecules unambiguously demonstrated that the aminophospholipids, particularly PS, still are translocated in favor of the inner membrane leaflet of fresh human erythrocytes previously exposed to diamide, although the efficiency of the process seems to be impaired compared with that in control cells (Fig. 1). Of particular interest, however, is the observation that, after 95% of the labeled PS accumulated in the inner leaflet within the first 10 hr of incubation, a fraction of it reappears in the outer leaflet during the subsequent 10–15 hr. At the same time point, the distribution of the labeled PE seems to reach an equilibrium at which less than the normal 80% of this lipid is present in the inner leaflet. The reverse is observed for the labeled PC since, in the diamide-treated cell, more than the normal 25% accumulates in the inner monolayer and does so at a

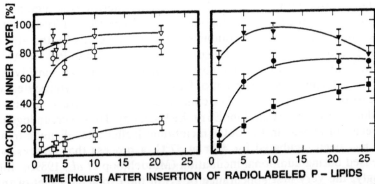

FIGURE 1 Transbilayer migration of radiolabeled glycerophospholipids in control (*left*) and diamide-treated (*right*) human erythrocytes at 37°C. Trace amounts of radiolabeled phospholipids were inserted into the outer membrane leaflet of intact cells using nonspecific lipid transfer protein from bovine liver. After a 30-min incubation, which started at zero time, cells were washed thoroughly and subsequently incubated at 37°C to facilitate a re-equilibration of the probe molecules. At the indicated time points, cells were treated with Pal-116-AMPA (see text). The fractions of the radiolabeled phospholipids that could be degraded this way were assumed to be present in the exofacial membrane leaflet. The results shown are the means of 8 independent experiments; bars indicate corresponding SD values. PS, Phosphatidylserine (∇, $\blacktriangledown$); PE, phosphatidylethanolamine ($\bigcirc$, $\bullet$); PC, phosphatidylcholine ($\square$, $\blacksquare$).

higher flip rate (Fig. 1). Note that, as mentioned earlier with respect to endogenous PS, the (re)appearance of some of the labeled PS in the outer leaflet coincides with an essentially complete deprivation of cellular ATP. Also, treatment of the cells with Pal-116-AMPA 20–25 hr after their exposure to diamide causes a degradation of the endogenous phospholipids, under still nonlytic conditions, to the following extents: 10% of the PS, 30% of the PE, and only 50% of the PC (Middelkoop et al., 1989b). Indeed, the resemblance between the ultimate distribution of the probe molecules and that of the endogenous glycerophospholipids in the membrane of the diamide-treated, ATP-depleted erythrocyte is most striking.

Another similarity that is apparent concerns the behavior of the radiolabeled long-chain diacyl glycerophospholipids in the plasma membrane of the pro-erythroblast (Section III) compared with that in diamide-treated cells (this section). Both cases show that, under conditions of sufficiently high cytosolic ATP levels, the flippase is capable of maintaining (a certain extent of) phospholipid asymmetry in the plasma membrane, despite an incomplete (Friend cell) or perturbed (diamide-treated cell) membrane skeleton. However, as soon as ATP is lacking, a time-dependent increase in the net migration of PS to the exofacial membrane leaflet is observed.

Identical observations have been made regarding yet another system, the sickled erythrocyte. Deoxygenation of RSCs induces them to sickle, thereby causing a local (mechanical) uncoupling of the lipid bilayer from the membrane skeleton by protrusion of bundles of polymerized hemoglobin S through gaps in this protein network (Allan et al., 1982; Franck et al., 1985a; Liu et al., 1991). Although this process by itself does not cause a marked change in phospholipid asymmetry (Raval and Allan, 1984; Franck et al., 1985a; Middelkoop et al., 1988), it results—as in the diamide-treated red cell—in an acceleration of phospholipid flip–flop (Franck et al., 1983,1985a; Mohandas et al., 1985). However, changes in phospholipid asymmetry, that is, an exposure of PS in the outer leaflet, are observed when sickling is induced in RSCs previously depleted of their ATP (Middelkoop et al., 1988). Energy deprivation alone does not affect the normal transbilayer distribution in oxygenated (discoid) RSCs. As in the case of diamide-treated cells, disturbing the interaction between membrane skeleton and bilayer, which in sickled erythrocytes occurs in the characteristic spicules that protrude from the cells, is not sufficient to cause changes in the transbilayer orientation of the phospholipids. Again, as long as the cell contains sufficiently high levels of ATP, the aminophospholipid translocase appears to be able to retain all the endogenous PS in the inner leaflet, despite a possibly reduced efficiency (Blumenfeld et al., 1991).

The prothrombinase assay that also was used in the studies on RSCs detected appreciable amounts of PS in the outer leaflet of the spectrin-free spicules that had been released from the RSCs by repetitive sickling and unsickling (Franck *et al.*, 1985a). The isolation of these spicules is rather time consuming, making a considerable loss of ATP quite likely. Similarly, low cytosolic ATP levels also might have been responsible for the altered phospholipid asymmetry that has been observed in spectrin-deficient vesicles isolated from heat (50°C)-treated human erythrocytes (Dressler *et al.*, 1984); asymmetry is maintained fully in the remnant cell, despite a heat-denatured membrane skeleton. The latter result agrees well with studies showing that heating red cells for 15 min at 50°C does not affect phospholipid asymmetry in their membranes (Gudi *et al.*, 1990) and does not impair the activity of the flippase (Calvez *et al.*, 1988). Even the spectrin-free vesicles thus produced are still capable of translocating PE and PS, provided the vesicles contained ATP. Subjecting Rhesus monkey erythrocytes to this heat treatment, on the other hand, does induce a partial loss of phospholipid asymmetry, probably because heat treatment of these cells causes not only a perturbation of the membrane skeleton but also a malfunctioning of the flippase (Kumar *et al.*, 1990).

A rather remarkable observation that should be mentioned here concerns vesicles that are released during *in vitro* maturation of guinea pig reticulocytes (Vidal *et al.*, 1989). Those vesicles have been found to be essentially devoid of a number of membrane proteins, most specifically spectrin, whereas experiments with spin-labeled phospholipids have demonstrated the absence of an active aminophospholipid translocase. Nevertheless, the vesicles are claimed to have a normal phospholipid asymmetry. Researchers have suggested (Vidal *et al.*, 1989) that the maintenance of phospholipid asymmetry in those vesicles can be ascribed to the high curvature of their bilayer or to the interaction of phospholipids with other membrane constituents.

With the possible exception of the final study, all the independent studies described lead to the conclusion that an alteration ("loss") of phospholipid asymmetry in the erythrocyte membrane may occur only when both of two prerequisites are fulfilled: (1) a malfunctioning or inhibition of the aminophospholipid translocase and (2) a perturbation, destruction, or removal of the membrane skeleton. The observations discussed in Section V regarding the eventual change in phospholipid asymmetry in the red cell membrane—either artificially induced or existing *in situ* under pathological conditions—will be considered in light of these two prerequisites.

D. Bidirectional or Outward Phospholipid Pumping?

Several studies involving spin-labeled or fluorescently labeled phospholipid analogs have attempted to address the issue of whether phospholipid asymmetry in the red cell membrane is maintained (1) solely by the ATP-dependent inward transport of PS and PE, (2) by a bidirectional functioning of this system in which the membrane skeleton plays a decisive role in retaining the aminophospholipids in the inner leaflet (Williamson *et al.*, 1987), or (3) by a bidirectional functioning of the flippase with different efficiencies for the inward and outward translocations (Bitbol and Devaux, 1988; Devaux, 1988). These three models all result in an accumulation of PS and PE in the cytoplasmic leaflet, leaving no other options for the two choline-containing phospholipids than to accumulate passively in the outer leaflet. However, the possible existence of a selective outward transloca-tion also has been proposed for PC (and SM?), a process that may (Connor *et al.*, 1992) or may not (Devaux, 1991) be ATP dependent.

Those experiments in which labeled phospholipid analogs have been applied clearly are intended to gain information about the kinetics of flip–flop of the endogenous phospholipids in the red cell membrane. How-ever, on this very point the results of those studies may question the suitability of the analogs. Consider the situation of glycerophospholipids in the native, normal, human erythrocyte. From the fact that, in this particular membrane, PC is distributed over outer and inner leaflet in a ratio of 3 to 1 (Zwaal *et al.*, 1975), we immediately conclude that the rate constant for the outward movement (k_o) of these molecules is 3 times higher than that of the inward movement (k_i), at least under (normal) steady state conditions. The actual translocation rates, on the other hand, differ considerably from one molecular species to another (Middelkoop *et al.*, 1986). However, since the molecular species compositions of the outer and inner PC pools are the same, the ratio $k_o : k_i$ is the same for each individual species despite the fact that, for instance, the actual values of k_o and k_i for 16:0/18:2 PC are more than 9 times higher than those for 16:0/16:0 PC. However, also in the case of PE, which is pumped continously from outer to inner leaflet by the ATP-driven flippase, steady state conditions dictate that the number of molecules that are translocated from outer to inner layer is compensated for by the migration of exactly the same number of PE molecules in the opposite direction. If not, its 20 to 80 distribution over outer and inner leaflet, respectively, would experience dramatic changes during the 120-day life-span of the cell. Since such changes do not occur, the (average) value of the $k_i : k_o$ ratio for PE is 4.

Consequent to the discussion in Section II, PS flip–flop in the human red cell membrane is considered to be nonexistent since this lipid is believed to be confined exclusively to the cytoplasmic leaflet. Let us nevertheless assume that 1 mol% of the PS, which is the estimated detection limit of the most advanced techniques (see Section II), might be present in the outer leaflet; the corresponding $k_i : k_o$ ratio should be 99. Clearly, this value, as well as those for the endogenous PE and PC fractions (see previous discussion), is at considerable variance with the ones that can be calculated from k_i and k_o (or the corresponding half-time) values determined using the lipid analogs (Herrmann and Müller, 1986; Williamson *et al.*, 1987; Bitbol and Devaux, 1988). Those differences indicate that the probe molecules do not genuinely reflect the behavior of the endogenous phospholipids, but merely create their own history. Also, the (re)appearance in the outer leaflet of 20% of the fluorescently labeled PS does not seem to correspond to the behavior of natural PS in the red cell membrane (Connor *et al.*, 1990); neither does the possibility of removing up to 80% of this analog (originally present in the inner leaflet) during a 6-hr incubation of the cells in the presence of 1% bovine serum albumin (BSA) (Connor *et al.*, 1992). The fact that the spin- or fluorescently labeled PS analogs, which *re*appear in the outer leaflet, can be extracted so easily with BSA suggests that those probe molecules have a higher affinity for BSA than for the flippase. This result again points to an abnormal behavior, since the flippase should be expected to be in charge of an immediate translocation of those molecules back to the inner leaflet. Yet another remarkable observation is the rapid translocation that spin-labeled SM appears to experience in guinea pig erythrocytes (Sune *et al.*, 1988). The inward translocation of SM reveals a half-time value of 32 min, indicating a rate that is 6 times higher than that for the fastest PC species (1-palmitoyl, 2-linoleoyl) in the human erythrocyte (Middelkoop *et al.*, 1986).

The inside-to-outside translocation (flop) of PE molecules, as well as of the spin- and fluorescently labeled PS analogs, is the immediate consequence of the continuous flippase-mediated active transport of those molecules from the outer to the inner layer. Thus, inhibition of the flippase by whatever means (vanadate, sulfhydryl reagents, Mg^{2+} and/or ATP depletion) should be expected to cause an inhibition of the inside-to-outside migration of the aminophospholipids also, which is not necessarily indicative of a protein—and, eventually, ATP—dependency of this process. On the other hand, the transbilayer movement of phospholipids, which is known to be almost immeasurably slow in pure lipid bilayers, is considerably enhanced on the incorporation of integral membrane proteins. This enhancement has been demonstrated specifically for glycophorin and protein band 3 incorporated into PC vesicles (Gerritsen *et al.*, 1980). Thus, one can, indeed, speak in terms of a protein-mediated bidirectional transbilayer movement of glycerophospholipids (including PC) in the red cell membrane, a mechanism that is, however, not likely to be specific.

In summary, we feel that the observations that led to the proposal of a protein-mediated, ATP-dependent, selective or nonselective, bidirectional transbilayer movement of all three glycerophospholipids (Connor *et al.*, 1992), or of a selective protein carrier in charge of the outward translocation of aminophospholipids (Bitbol and Devaux, 1988), can be explained easily by the continuous ATP-fired pumping of PE molecules (and PS analogs) from the outer to the inner monolayer pool, in conjunction with the (probably) aspecific facilities for phospholipid flip–flop that are created by other integral membrane proteins.

V. CHANGES IN RED CELL PHOSPHOLIPID ASYMMETRY

Alterations in red cell phospholipid asymmetry can be induced artificially *in vitro* or may occur *in situ* under certain pathological conditions. Both classes of alterations have been studied intensively to gain a better insight into the conditions and mechanisms that govern the transbilayer distribution and dynamics of the red cell phospholipids.

Changes in asymmetry usually are referred to in terms such as "loss of asymmetry" or "scrambling of phospholipids." This terminology implicitly refers to a complete randomization of all phospholipid classes over both leaflets, that is, the creation of a symmetric phospholipid membrane. However, whether such a situation can ever be reached can be strongly doubted. Not only is a highly static asymmetry in the transbilayer distribution of SM characteristic of the normal red cell membrane (see Section II), but many studies have shown that this situation is maintained fully under a variety of conditions that give rise to accelerated flip–flop of the glycerophospholipids or even to changes in their transbilayer distribution (Haest *et al.*, 1978; Lubin *et al.*, 1981; Kuypers *et al.*, 1984; Rawyler *et al.*, 1985; Schwartz *et al.*, 1987; van der Schaft *et al.*, 1987a). This peculiar behavior of SM perhaps is best illustrated by the fact that SM asymmetry is maintained completely in vesicles that have been pinched off by heat treatment of erythrocytes, whereas asymmetry is lost for PE and PS (Dressler *et al.*, 1984). Hence, a shift of aminophospholipids from inner to outer monolayer, as a consequence of whatever conditions, can be compensated for only by a shift of some of the PC in the opposite direction. In principle, two possibilities must be considered for the ultimate (passive) redistribution of the three glycerophospholipids that can occur under conditions of a complete failure of the mechanisms that normally generate and maintain phospholipid asymmetry (Fig. 2). In both situations, I and II, the outward migration of aminophospholipids is compensated for by an inward migration of PC. Either PC re-equilibrates over both layers in a 1 : 1 ratio (Fig. 2, I) or all three glycerophospholipids mix and re-equilibrate on the basis of entire equality (Fig. 2, II). In the situation II, one

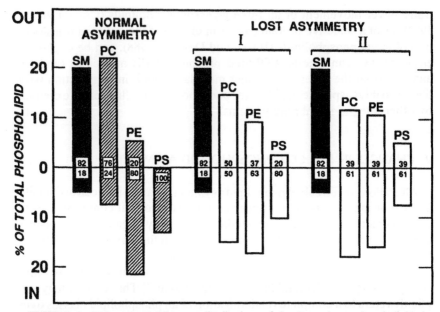

FIGURE 2 Ultimate transbilayer redistributions of the three glycerophospholipids in the human erythrocyte membrane that may be expected to occur under complete failure of the mechanisms that normally maintain their asymmetric distribution. In either of the two possibilities shown, sphingomyelin (SM) is not subject to transbilayer redistributions (see text). (I) Proportional outward migration of phosphatidylethanolamine (PE) and phosphatidylserine (PS), which is limited by a compensatory inward migration of phosphatidylcholine (PC) to reach an equal distribution over both leaflets. (II) Identical transbilayer distributions of the three glycerophospholipids that will be established when the system does not discriminate among the different subclasses.

cannot expect to have more than 40% of each of the aminophospholipids in the outer leaflet, whereas in situation I this amount is as little as 20% in the case of PS. Hence, particularly when also considering SM, a complete loss of asymmetry or scrambling of phospholipids may not occur easily. These considerations also imply that reports that claim the appearance of more than 20%—not to mention over 40%—of the PS in the exofacial leaflet, under whatever conditions, should be considered with some reservation.

A. Artificially Induced Changes in Red Cell Phospholipid Asymmetry

1. Intracellular Ca^{2+}

Loading intact erythrocytes with Ca^{2+} (up to 1 mM) by incubating the cells in the presence of ionophore A23187 causes a considerable alteration in transbilayer phospholipid asymmetry and dynamics, as expressed by the appearance of endogenous PS and an increased amount of endogenous PE in the outer leaflet (Schwartz et al., 1985; Chandra et al., 1987; Henseleit et al., 1990), a redistribution of spin-labeled phospholipid analogs (Williamson et al., 1992), and an accelerated transbilayer reorientation of probes such as lyso-PC and palmitoylcarnitine (Henseleit et al., 1990) or spin-labeled analogs of PC and SM (Williamson et al., 1992). The effects that enhanced intracellular Ca^{2+} concentrations may exert are manifold and include activation of an endogenous phospholipase C to degrade polyphosphoinositides (Allan and Thomas, 1981); direct effects on membrane skeletal proteins (Bennet, 1989) and their interaction with the lipid bilayer (Mombers et al., 1980); activation of transglutaminase to produce cross-linking of spectrin (Lorand et al., 1987); activation of endogenous proteases that also can affect components of the membrane skeleton (Grasso et al., 1986; Whatmore et al., 1992); activation of the Ca^{2+} pump, eventually leading to ATP depletion (Chandra et al., 1987); inhibition of the aminophospholipid translocase (Bitbol et al., 1987); and induction of nonbilayer structures (Cullis and de Kruijff, 1979). Hence, enhanced intracellular Ca^{2+} concentrations create conditions that inevitably will lead to a considerable alteration in phospholipid asymmetry, that is, a perturbation (or even loss) of membrane skeleton–bilayer interactions and an inhibition of the flippase. The induction of nonbilayer structures may facilitate the redistribution process further, eventually including the reorientation of a previously inserted spin-labeled analog of SM (Williamson et al., 1992). A decreased exposure of endogenous SM, to compensate for the Ca^{2+}/ionophore-induced increase in the exposure of PS (and PE), has been claimed to occur in platelets (Schroit and Zwaal, 1991); however, this suggestion is not supported by the actual experimental data (Bevers et al., 1983). The most dramatic effects that enhanced intracellular Ca^{2+} levels can exert are illustrated particularly by the shedding of vesicles from the plasma membrane, not only in erythrocytes but also, and even more pronouncedly, in platelets (Comfurius et al., 1990).

Several of these effects of Ca^{2+} will be equally responsible for the marked changes that have been observed in ghosts prepared in the presence of this divalent cation (Williamson et al., 1985). Normal asymmetry is reported to be restored, however, when the ghosts are resealed in the presence of ATP (Verhoven et al., 1992). Since the aminophospholipid translocase is inhibited strongly by even low intracellular concentrations of Ca^{2+}, these ions must be pumped out first, a task for which the Ca^{2+}-ATPase is very well equipped. Note, however, that particularly the Verhoven study exclusively involved the use of merocyanine 540 to monitor transbilayer phospholipid distributions, and that the suitability of this probe for this particular purpose has been questioned (Allan et al., 1989).

2. Hypotonic Lysis

Hypotonic lysis also has been claimed to cause changes in red cell phospholipid asymmetry (Connor et al., 1990; Schrier et al., 1992). Those changes are reported to be either prevented by including Mg^{2+} in the lysis buffer (Connor et al., 1990) or undone by resealing the ghosts in the presence of Mg^{2+}-ATP. Because of the high dilution factors that have been applied for ghost preparation in both these studies, some of the membrane skeletal proteins might have been removed (Clark and Ralston, 1990), or their interactions with the bilayer transiently disturbed. In conjunction with the loss of ATP that accompanies the lysis of the cells, the requirements for changes in phospholipid asymmetry seem to have been fulfilled again. Note, however, that the study by Schrier et al. (1992) showed a complete randomization to occur only for (all) short-chain spin-labeled phospholipid analogs, whereas a previously established asymmetry of long-chain spin-labeled PS is affected only marginally by the lysis step. Regarding the other studies (Connor et al., 1990), how the preservation of phospholipid asymmetry reported to occur during hemolysis in the presence of either Mg^{2+} or Mg^{2+}-ATP could have been determined in ghosts that remained unsealed under these conditions is quite puzzling.

3. Oxidative Stress

Several studies have appeared on the effects that oxidative reagents such as H_2O_2 (Wali et al., 1987; Pradhan et al., 1990), phenylhydrazine (Jain and Hochstein, 1980; Jain, 1985; Arduini et al., 1986,1989), and diamide (Wali et al., 1987; Section III,B) might exert on red cell phospholipid organization. Although such treatments invariably affect the membrane skeletal proteins, causing a perturbation of this network, reports on changes in phospholipid asymmetry (whether or not they occur and, if so, to what extent) are highly controversial. Except for a decrease in the inward translocation rate of spin-labeled PE in H_2O_2-treated human erythrocytes (Herrmann and Devaux, 1990), not much is known about the fate of the flippase under those conditions, making a detailed and critical evaluation of such conflicting results not possible.

4. Local Anesthetics

The results obtained by different groups regarding the effects that local anesthetics may exert on transbilayer movement and distribution of phospholipids in the intact erythrocyte are not easily comparable because of the use of different techniques. Schrier and colleagues (Schrier et al., 1983) found that no alteration in phospholipid asymmetry occurred during stomatocytosis as induced by the cationic amphipathic drugs vinblastine, primaquine, and chlorpromazine (CPZ). Such changes were observed, however, during subsequent endocytosis; an enhanced hydrolysis of PE; when those cells were exposed to bee venom phospholipase A_2 alone, whereas the combined action of this enzyme with *S. aureus* sphingomyelinase C, conditions known to degrade the outer monolayer to completion (Zwaal *et al.*, 1975), revealed an only marginal increase in exofacial PE, but still no hydrolysis of PS. The extent to which SM and PC could be degraded under those conditions tended to decrease slightly.

An appreciably greater increase in the hydrolysis of PE, but not of PS, by exogenous bee venom phospholipase A_2 was observed in erythrocytes treated with tetracaine, dibucaine, or amphotericin B (Schneider *et al.*, 1986). These drugs also caused a marked acceleration in the flip of exogenous lyso-PC, which appeared to be reversible since this flip reverted to its normal rate on subsequent removal of the perturbing agents by washing. Interestingly, the enhanced accessibility of PE appeared to be irreversible, since it was maintained on removal of the drugs. However, whether the action of the flippase was impaired by those treatments is not known.

In contrast to the accelerated flip of lyso-PC, normal (slow) transverse mobility of PC and the absence of mobility of SM were observed for their spin-labeled analogs in human erythrocytes previously treated with sublytic concentrations of CPZ (Rosso *et al.*, 1988). Moreover, this treatment was found to cause a decrease in the rate of flippase-mediated inward translocation of spin-labeled PE and PS, suggesting a direct effect of CPZ on the translocase. Further, CPZ was noted to affect a pre-established transbilayer distribution of the spin-labeled analogs. Fractions of 10–15% of both spin-labeled aminophospholipids flipped back instantaneously to the outer leaflet on addition of CPZ, a redistribution that was compensated for by a shift of comparable quantities of PC and SM in the opposite direction. Devaux and colleagues (Rosso *et al.*, 1988) interpreted these observations to support the view that phospholipid asymmetry in the red cell membrane is controlled and, when needed, corrected by the flippase. The authors admitted, however, that these experiments could not rule out the hypothesis that the membrane skeleton also could be partially responsible for the maintenance of aminophospholipid transmembrane asymmetry.

B. Phospholipid Asymmetry under Pathological Conditions

1. Membrane Skeletal Defects

To consider the role the membrane skeleton plays in maintaining phospholipid asymmetry, phospholipid distributions have been studied in erythrocytes that have a notorious defect in this skeletal network and/or in its interaction with the membrane bilayer.

Hereditary pyropoikilocytosis (HPP) is a congenital hemolytic anemia caused by an abnormal membrane skeleton (Hanspal and Palek, 1992), and is characterized by a partial deficiency in spectrin and an impaired capacity of the $\alpha\beta$ spectrin dimers to self-associate into tetramers. This inability causes HPP cells to fragment dramatically at 46°C, producing strangely shaped poikilocytes and microspherocytes (Franck et al., 1985b), whereas normal erythrocytes undergo such changes at 49°C. Already at 37°C, HPP cells exhibit a PC flip–flop in their membranes that is twice as fast as that in normal erythrocytes. In both cell types, PC flip–flop increases slowly with increasing temperature, but is accelerated suddenly at 44°C and 46°C in HPP and control cells, respectively (Frank et al., 1985b). Although these temperatures are a few degrees below the respective transition temperatures of spectrin in those cells, the marked acceleration in PC flip–flop has been interpreted to reflect early major changes in the structure of this skeletal protein. Despite the enhanced flip–flop of PC in HPP cells, which is already manifest at 37°C and which is supposedly indicative of an abnormality in the membrane skeleton of those cells, bee venom phospholipase A_2 fails to detect any abnormality in their phospholipid asymmetry at this temperature.

Bee venom phospholipase A_2, as well as TNB, detects an enhanced accessibility of both PE and PS in the red blood cells of a patient with hereditary spherocytosis (HS); these cells had only approximately one-third the normal complement of spectrin (Schwartz et al., 1985). In spectrin-deficient spherocytic mouse erythrocytes, on the other hand, bee venom phospholipase A_2 degrades one-third of the PE, compared with none in the control cells (Lubin and Chiu, 1982). PS remains inaccessible in the spectrin-deficient mouse cells. In human and in mouse cells, the enhanced accessibility of aminophospholipids to the exogenous probe is accompanied by a decrease in the hydrolysis of PC.

Most interestingly, 30% of the PS can be degraded by the combined action of bee venom phospholipase A_2 and S. aureus sphingomyelinase C on intact human erythrocytes deficient in protein 4.1 (Schwartz et al., 1985), whereas this treatment does not give rise to an enhanced hydrolysis of PE. No change is seen in the extent to which SM is degraded, but the hydrolysis of PC appears to be decreased. The specific bilayer translocation of PS, as observed in the 4.1-deficient cells, is rather intriguing considered in light of the observations of Sato and Ohnishi (1983) of a high affinity binding of purified protein 4.1 to PS-containing membranes.

Although these studies clearly support the importance of the membrane skeleton in stabilizing phospholipid asymmetry and transbilayer mobility in the red cell membrane, unfortunately no information is available about the status of the aminophospholipid translocase in these cells.

2. Antigenic Abnormalities

Studies of the organization of the phospholipid bilayer also include studies of red cells with an established antigenic abnormality. The En(a−) phenotype, in which the MN antigens and glycophorin A are absent (van Meer *et al.*, 1981), and the Leach phenotype, missing the Gerbich antigen and minor sialoglycoproteins (Kuypers *et al.*, 1985), show no abnormalities in phospholipid transbilayer organization when probed with highly purified phospholipases. The acanthocytic McLeod erythrocyte, which exhibits a weak expression of the Kell antigens, also is found to have a normal phospholipid distribution (Kuypers *et al.*, 1985) and an active aminophospholipid translocase (Redman *et al.*, 1989), but has enhanced transbilayer mobility of PC (Kuypers *et al.*, 1985). This characteristic also is observed in the stomatocytic cells of the Rh_{null} phenotype, which are missing—among others—the Rh(D) polypeptide, an integral protein that appears to be associated with the membrane skeleton (Kuypers *et al.*, 1984). Interestingly, the combined action of phospholipase A_2 and sphingomyelinase C degrades over 40% of the PE in those cells, a value twice as high as that in normal erythrocytes. Although no information is available about the status of the flippase, which was still awaiting discovery at the time of these experiments, the presence of an active aminophospholipid transport in Rh_{null} cells that, despite enhanced (PC) flip–flop, is able to generate and maintain PS asymmetry has been demonstrated (Smith and Daleke, 1990).

3. Neonatal Erythrocytes

Although not within the definition of pathology, neonatal erythrocytes have been shown to exhibit subtle changes in their transbilayer phospholipid distribution. Using the combined action of phospholipase A_2 and sphingomyelinase C, enhanced hydrolysis of PE and decreased degradation of PC have been observed in those cells (Matovcik *et al.*, 1986). No degradation of PS could be detected in this study, which contrasts with the observations of Jain (1986) which showed the accessibility of 8–9% of the PS for both bee venom phospholipase A_2 and TNBS as well as reduction in Russell's viper venom (RVV) clotting time, another indication of PS being present in the exofacial leaflet. The reason for those changes remains unknown, but they may explain the enhanced spontaneous endocytosis and the decrease in lifetime that are observed for these cells.

4. Diabetes Mellitus

No changes in RVV clotting time are observed for red cells from patients with diabetes mellitus type II, although 12–18% of the PS in those cells appears to be accessible to phospholipase A_2 and TNBS (Wali et al., 1988). The authors add to this apparent controversy that the enhanced susceptibility of PE and PS to enzymatic and chemical modification may not represent a static increase in the outer leaflet content of these phospholipids, but an increase in their transbilayer dynamics in the erythrocytes. Whatever its nature, some abnormality in the organization of the lipid bilayer seems to be evident in those cells, as is indicated by their enhanced adherence to endothelial cells as well (Wali et al., 1988).

5. Chronic Myeloid Leukemia

A rather excessive high degree of hydrolysis of PS (up to 30%) is observed, not only when intact erythrocytes from chronic myeloid leukemia (CML) patients are treated with N. naja snake venom phospholipase A_2 but also when they are exposed to the phospholipase A_2 from hog pancreas (Kumar and Gupta, 1983). The latter result is of particular interest since pig pancreatic phospholipase A_2 normally cannot attack its substrates in the intact human erythrocyte because of the weak membrane-penetrating capacity of the enzyme in relation to the tight packing of the (phospho)lipids in the outer leaflet of the cell membrane (Demel et al., 1975). Decreased lipid packing in the exofacial leaflet of the CML erythrocytes also is demonstrated by merocyanine 540 staining (Reed et al., 1985; Kumar et al., 1987). The accessibility of not less than 30% of the PS in the outer leaflet of the CML erythrocytes is confirmed by TNBS labeling (Kumar et al., 1987), which also shows an only slightly enhanced accessibility of PE, that is, from 20% in normal cells to 26% in the CML erythrocytes. Exposure of PS on the surface of these cells conforms with the observation that they are more adherent to endothelial cells and more readily phagocytosed by macrophages in vitro than are normal erythrocytes (Schlegel et al., 1990). CML erythrocytes have been shown to possess some abnormalities in the architecture of the membrane skeleton, that is, enhanced cross-linking of spectrin (Kumar and Gupta, 1983) and decreased tetramer formation of these proteins that—as in the case of the HPP erythrocytes—is accompanied by an abnormal heat sensitivity of the cells (Basu et al., 1988). However, a perturbation of the membrane skeleton–lipid bilayer interaction can result in the appearance of vast amounts of PS in the outer membrane leaflet only when accompanied by impaired functioning of the flippase. Whether or not this active translocation system can be affected by the recently discovered new class of Ca^{2+}-sensitive proteins, of which a 67-kDa annexin has been found to be present in CML erythrocytes in abnormal amounts and intracellular distribution (Fujimagari *et al.*, 1990), remains unknown. Apart from such considerations, one may wonder whether an erythrocyte in which not less than 30% of the PS is exposed at the outer surface will have any chance of survival *in vivo*. In other words, could the dramatic change in PS asymmetry detected in those abnormal cells have occurred during handling of the cells after the blood was drawn? Experiments in our laboratory that involved the combined action of *N. naja* phospholipase A_2 and *S. aureus* sphingomyelinase C failed to detect any abnormality in the phospholipid distribution in red cells from three individuals who suffered from CML (M. van Linde-Sibenius Trip and B. Roelofsen, unpublished observations).

6. Sickle Cells

Sickle cells are likely to be the best characterized of the various abnormal erythrocytes, particularly with respect to changes in the organization of their lipid bilayer. As already discussed (Section IV,C), studies on these cells have provided a major contribution to our understanding of the mechanisms involved in the maintenance of phospholipid asymmetry in plasma membranes. Similarly, as in the case of the diamide-treated erythrocyte, these cells also have taught us that a disturbance in the interaction between membrane skeleton and lipid bilayer causes a considerable destabilization of the latter, as reflected by an accelerated PC flip–flop (Franck *et al.*, 1982, 1983, 1985a) and that, as the probable consequence of this event, several otherwise straightforward techniques for assessing phospholipid topology easily can give rise to misleading results. This warning specifically applies to the use of phospholipases, which detected a considerable outward migration of PE and PS in the experiments described earlier (Haest *et al.*, 1978; Lubin *et al.*, 1981), since later studies have shown that neither in the diamide-treated erythrocyte (Franck *et al.*, 1986; Middelkoop *et al.*, 1989b) nor in the deoxygenated (sickled) RSC (Franck *et al.*, 1985a; Middelkoop *et al.*, 1988) do such changes occur *in situ* to any appreciable extent. Only when the uncoupling of the membrane skeleton from the lipid bilayer is accompanied by a dis- or nonfunctioning flippase, as, for example, under energy depriving conditions, does an appreciable fraction of PS (and an increased amount of PE) emerge in the exofacial leaflet (Middelkoop *et al.*, 1988, 1989b).

7. Malaria

Finally, the malaria-infected erythrocyte, for which highly contradictory results have been reported regarding possible changes in phospholipid topology in the host cell membrane, is worthy of consideration. Drastic changes in the asymmetric distribution of both PE and PS have been reported to occur during the intraerythrocytic development of *Plasmodium knowlesi* (Joshi *et al.*, 1987) and *Plasmodium falciparum* (Joshi and Gupta, 1988). Both studies involved the use of phospholipase A_2 from bee venom and pig pancreas, as well as merocyanine 540. Similar results were obtained by Schwartz *et al.* (1987) when probing *P. falciparum*-infected human erythrocytes with bee venom phospholipase A_2, sphingomyelinase C, and TNBS, although changes in PS distribution were observed only with TNBS. Unfortunately, the interpretation of the data is hampered by the fact that the marked increase in total lipid content of the cells, which occurs during parasite development, was not considered when making the calculations (Schwartz *et al.*, 1987).

The observations just discussed contrast with those obtained by van der Schaft *et al.* (1987b) and Beaumelle *et al.* (1988), who showed a normal asymmetry for SM and PS in *P. knowlesi*-infected monkey erythrocytes using phospholipases, fluorescamine, and merocyanine 540. However, since the action of conventional phospholipase A_2 on PC and PE did not plateau when used on the parasitized erythrocytes, nor did the fluorescamine labeling of PE (van der Schaft *et al.*, 1987b), the suitability of such probes under these conditions was doubted. These observations can be explained by the fact that the host cell membrane experiences a considerable extent of transbilayer destabilization that is dependent on the stage of parasite development. Accelerated transbilayer movements across the host cell membrane were observed for spin-labeled analogs of all three major glycerophospholipids, as well as for SM (Beaumelle *et al.*, 1988). Fluorescently (NBD-)labeled PC and PE, on the other hand, indicated that the behavior of PC was distinct from that of PE (Haldar *et al.*, 1989). Rapid transbilayer movements across the host cell membrane also were observed for radiolabeled long-chain diacyl PC (van der Schaft *et al.*, 1987b; Simões *et al.*, 1991), but not for radiolabeled SM (Simões *et al.*, 1991).

Despite these transbilayer instabilities, clear plateaus for the hydrolysis of PC and PE, but no degradation of PS, could be observed when *P. knowlesi*-infected monkey erythrocytes were exposed to Pal-116-AMPA (Moll *et al.*, 1990). The absence of PS in the outer leaflet of the host cell membrane of *P. falciparum*-infected human red cells was confirmed using the prothrombinase assay (Moll *et al.*, 1990). However, the same assay when used by Maguire *et al.* (1991) did detect surface-exposed PS in *P. falciparum*-infected cells. These authors also observed considerable changes in the transbilayer distribution of PC, PE, and SM in the host cell membrane when *P. falciparum*-infected cells were probed with bee venom phospholipase A_2 and sphingomyelinase C.

When critically evaluating the controversial results that have been generated by these studies, it is essential to consider the possible differences that may exist among the various studies with respect to the extent of parasitemia and to the stage of parasite development. Such a detailed discussion, which is beyond the scope of this chapter, can be found elsewhere (Vial et al., 1990; Simões et al., 1992). Whatever the importance of such differences, all studies in which a change in phospholipid asymmetry in the host cell membrane has been observed exhibit one common feature: the absence of glucose in the media in which the erythrocytes were kept during experimental handling (Joshi et al., 1987; Schwartz et al., 1987; Joshi and Gupta, 1988; Maguire et al., 1991). Withholding glucose from the cells may have a most dramatic effect, since the consumption of this fuel by the parasitized cells is known to increase up to two orders of magnitude (Roth et al., 1988). Therefore, starvation easily may cause a rapid and drastic drop in intracellular ATP and consequently an impaired functioning—or, eventually, a complete inhibition—of the aminophospholipid translocase. Since structural lesions in the membrane skeleton also are known to occur in the parasitized erythrocyte (Wallach and Conley, 1977; Deguercy et al., 1990), the two requirements for a change in phospholipid asymmetry are fulfilled again. Hence, from our studies (Moll et al., 1990) we infer that the plasma membranes of malaria-infected erythrocytes—even at the latest stage of parasite development, that is, the schizont—possess a normal phospholipid asymmetry, provided the cells are protected against energy deprivation.

C. Physiological Significance of Phospholipid Asymmetry and Changes Therein

Obviously nature has kept the exofacial leaflet of plasma membranes free of PS for good reasons, not only in erythrocytes and platelets, but probably also in all other cells for which the plasma membrane is exposed directly to the blood. The exposure of PS on the outer surface of intact erythrocytes (Tanaka and Schroit, 1983; McEvoy et al., 1986) or red cell ghosts (Schlegel et al., 1985) has been demonstrated to render them increasingly adherent to endothelial cells (Schlegel et al., 1985) and increasingly susceptible to recognition by macrophages (Tanaka and Schroit, 1983; McEvoy et al., 1986). Hence, the appearance of PS in the outer leaflet of the red cell membrane has been proposed to provide a signal for the sequestration of perturbed or senescent erythrocytes by the reticuloendothelial system. Indeed, Schroit et al. (1985) observed, in mice, a rapid *in vivo* clearance of erythrocytes containing fluorescent long-chain PS. Those cells appeared to accumulate in splenic macrophages and Kupffer cells. Although the efficiency of the aminophospholipid translocase has been shown to decrease on red cell aging (Herrmann and Devaux, 1990), no PS could be detected in the outer leaflet of old red cells (Shukla and Hanahan, 1982; Tanaka and Schroit, 1983). Although this failure could have been the result of the notorious impossibility of preparing a sample of really old erythrocytes by whatever centrifugation technique (Clark, 1988), it also could indicate that the reticuloendothelial system is so sensitive that erythrocytes already are sequestered when only a small percentage of the PS has migrated to the outer leaflet. If the latter situation occurred, again reasons exist to doubt the high amounts of PS that have been claimed to be present in the outer leaflet of some pathologic erythrocytes (see Section V,B), particularly when they are supposed to represent cells *in situ*.

Several studies indicate that the exposure of PS in the outer leaflet is a prerequisite for making a red cell competent for fusion processes induced by polyethyleneglycol (Baldwin *et al.*, 1990; Huang and Hui, 1990) or electrical breakdown (Song *et al.*, 1992). Fusion of erythrocytes with vesicular stomatitis virus, on the other hand, seems to depend on the packing characteristics of the outer membrane leaflet of the red cell rather than on the presence of any particular phospholipid (Herrmann *et al.*, 1990). The aminophospholipid translocase may play an important role in rendering a membrane fusion competent, as has been proposed for the chromaffin granule (Zachowski *et al.*, 1989). Further, Devaux emphasizes the role that the flippase may play in generating local invaginations in a membrane which, in combination with a fusion-competent membrane surface, create the conditions essential for endocytosis (Devaux, 1991,1992).

The most intriguing and important aspect of plasma membrane PS topology and changes therein, is related, no doubt, to its role in hemostasis and thrombosis. Activation of blood platelets is known to cause a rapid and dramatic increase in the exposure of PS on the outer surface. This phenomenon is accompanied, or possibly even caused, by membrane fusion events that supposedly take place during rapid transbilayer redistributions of (glycero)phospholipids. This process appears to be associated closely with the shedding of highly procoagulant microvesicles from the platelet surface. The exposure of PS at the exofacial surface of platelets and microvesicles is an (almost) absolute prerequisite for the formation of the so-called tenase and prothrombinase complexes that catalyze two consecutive reactions in the coagulation cascade. Although much progress has been made in the understanding of these phenomena, several questions remain unanswered. Since a more detailed discussion of this fascinating area is beyond the scope of this chapter, the interested reader is referred to two excellent reviews (Schroit and Zwaal, 1991; Zwaal *et al.*, 1992).

VI. CONCLUDING REMARKS

Many independent studies on human erythrocytes, in which abnormalities were either artificially induced or naturally occurring, provide ample evidence for the concept that the asymmetric distribution of glycerophospholipids is maintained by the concerted action of the ATP-dependent aminophospholipid translocase and the membrane skeleton. The latter seems of doubtless importance for the maintenance of an exclusive localization of PS in the cytoplasmic leaflet. The role of the membrane skeleton in keeping 80% of the PE in the inner monolayer may be less clear, however. Obviously, a continuous flippase-mediated pumping of PE molecules from outer to inner membrane leaflet contributes to the asymmetry. Hence, the asymmetric distribution of PE may be considered the result of a steady state dynamic process in which the membrane skeleton may play only a minor role. Although the rate and the extent of inward translocation of spin-labeled PE have been shown to depend on the intracellular ATP concentration (Zachowski *et al.*, 1986; Bitbol *et al.*, 1987), no change in the asymmetric distribution of endogenous PE can be observed in ATP-depleted cells. Indeed, other observations also indicate that an active aminophospholipid translocase is not required for the maintenance of asymmetry once it has been established (Vidal *et al.*, 1989; Schroit and Zwaal, 1991; and references therein). Since the pronounced asymmetric distribution of SM appears to be a very static and permanent one for which the mechanism and interactions remain unknown, PC has no other option than to occupy passively the space that is left open by SM and the aminophospholipids.

As was discussed in Section IV,D, to date no (conclusive) evidence exists that the maintenance of (glycero)phospholipid asymmetry in the red cell membrane is controlled by proteins other than the flippase and some of those that constitute the membrane skeleton. The concerted action of both these systems is certainly of physiological significance. First, clearly preventing an appearance of PS on the outer surface of its membrane is in the immediate interest of cell survival. Further, during the *in vivo* circulation of even normal erythrocytes, a (local) uncoupling of the membrane skeleton from the lipid bilayer may occur when the cells are squeezed through very narrow passages. The flippase then would be in charge of an immediate back translocation of PS molecules which, consequent to the uncoupling, may have flopped from the inner to the outer membrane leaflet. The flippase thus provides a repair mechanism under such conditions. The contribution of the membrane skeleton to retaining aminophospholipids in the inner membrane leaflet, on the other hand, will save large amounts of energy (ATP) that otherwise would be consumed by the flippase, if it were the only system responsible for the maintenance of phospholipid asymmetry in the red cell membrane.

Acknowledgments

The authors are indebted to L. L. M. van Deenen for critically reading the manuscript. Albert C. Noorlandt is gratefully acknowledged for his expert assistance in preparing this manuscript.

References

Allan, D., and Thomas, P. (1981). Ca^{2+}-induced biochemical changes in human erythrocytes and their relation to microvesiculation. *Biochem. J.* **198**, 433–440.

Allan, D., and Walklin, C. M. (1988). Endovesiculation of human erythrocytes exposed to sphingomyelinase C: A possible explanation for the enzyme-resistant pool of sphingomyelin. *Biochim. Biophys. Acta* **938**, 403–410.

Allan, D., Limbrick, A. R., Thomas, P., and Westerman, M. P. (1982). Release of spectrin-free spicules on reoxygenation of sickled erythrocytes. *Nature (London)* **295**, 612–613.

Allan, D., Hagelberg, C., Kallen, K.-J., and Haest, C. W. M. (1989). Echinocytosis and microvesiculation of human erythrocytes induced by insertion of merocyanine 540 into the outer membrane leaflet. *Biochim. Biophys. Acta* **986**, 115–122.

Arduini, A., Chen, Z., and Stern, A. (1986). Phenylhydrazine-induced changes in erythrocyte membrane surface packing. *Biochim. Biophys. Acta* **862**, 65–71.

Arduini, A., Stern, A., Storto, S., Belfiglio, M., Mancinelli, G., Scurti, R., and Federici, G. (1989). Effect of oxidative stress on membrane phospholipid and protein organization in human erythrocytes. *Arch. Biochem. Biophys.* **273**, 112–120.

Baldwin, J. M., O'Reilly, R., Whitney, M., and Lucy, J. A. (1990). Surface exposure of phosphatidylserine is associated with the swelling and osmotically-induced fusion of human erythrocytes in the presence of Ca^{2+}. *Biochim. Biophys. Acta* **1028**, 14–20.

Basu, J., Kundu, M., Rakshit, M. M., and Chakrabarti, P. (1988). Abnormal erythrocyte membrane cytoskeletal structure in chronic myelogenous leukaemia. *Biochim. Biophys. Acta* **945**, 121–126.

Beaumelle, B. D., Vial, H. J., and Bienvenue, A. (1988). Enhanced transbilayer mobility of phospholipids in malaria-infected monkey erythrocytes: A spin-label study. *J. Cell. Physiol.* **135**, 94–100.

Bennett, V. (1989). The spectrin–actin junction of erythrocyte membrane skeletons. *Biochim. Biophys. Acta* **988**, 107–122.

Bergmann, W. L., Dressler, V., Haest, C. W. M., and Deuticke, B. (1984). Cross-linking of SH-groups in the erythrocyte membrane enhances transbilayer reorientation of phospholipids. Evidence for a limited access of phospholipids to the reorientation sites. *Biochim. Biophys. Acta* **769**, 390–398.

Bevers, E. M., Comfurius, P., van Rijn, J. M. L. M., Hemker, H. C., and Zwaal, R. F. A. (1982). Generation of prothrombin-converting activity and the exposure of phosphatidylserine at the outer surface of platelets. *Eur. J. Biochem.* **122**, 429–436.

Bevers, E. M., Comfurius, P., and Zwaal, R. F. A. (1983). Changes in membrane phospholipid distribution during platelet activation. *Biochim. Biophys. Acta* **736**, 57–66.

Bitbol, M., and Devaux, P. F. (1988). Measurement of outward translocation of phospholipids across human erythrocyte membrane. *Proc. Natl. Acad. Sci. U.S.A.* **85**, 6783–6787.

Bitbol, M., Fellmann, P., Zachowski, A., and Devaux, P. F. (1987). Ion regulation of phosphatidylserine and phosphatidylethanolamine outside–inside translocation in human erythrocytes. *Biochim. Biophys. Acta* **904**, 268–282.

Bitbol, M., Dempsey, C., Watts, A., and Devaux, P. F. (1989). Weak interaction of spectrin with phosphatidylcholine–phosphatidylserine multilayers: A ^{2}H and ^{31}P NMR study. *FEBS Lett.* **244**, 217–222.

Blumenfeld, N., Zachowski, A., Galacteros, F., Beuzard, Y., and Devaux, P. F. (1991). Transmembrane mobility of phospholipids in sickle erythrocytes: Effect of deoxygenation on diffusion and asymmetry. *Blood* **77**, 849–854.

Boegheim, J. P. J., Jr., van Linde, M., Op den Kamp, J. A. F., and Roelofsen, B. (1983). The sphingomyelin pools in the outer and inner layer of the human erythrocyte membrane are composed of different molecular species. *Biochim. Biophys. Acta* **735**, 438–442.

Bonnet, D., and Begard, E. (1984). Interaction of anilinonaphtyl labeled spectrin with fatty acids and phospholipids: A fluorescent study. *Biochem. Biophys. Res. Commun.* **120**, 344–350.

Bretscher, M. S. (1972a). Asymmetrical lipid bilayer structure for biological membranes. *Nature (New Biol.)* **236**, 11–12.

Bretscher, M. S. (1972b). Phosphatidylethanolamine: Differential labeling in intact cells and cell ghosts of human erythrocytes by a membrane-impermeable reagent. *J. Mol. Biol.* **71**, 523–528.

Bütikofer, P., Lin, Z. W., Chiu, D. T.-Y., Lubin, B., and Kuypers, F. A. (1990). Transbilayer distribution and mobility of phosphatidylinositol in human red blood cells. *J. Biol. Chem.* **265**, 16035–16038.

Calvez, J.-Y., Zachowski, A., Hermann, A., Morrot, G., and Devaux, P. F. (1988). Asymmetric distribution of phospholipids in spectrin-poor erythrocyte vesicles. *Biochemistry* **27**, 5666–5670.

Chandra, R., Joshi, P. C., Bajpai, V. K., and Gupta, C. M. (1987). Membrane phospholipid organization in calcium-loaded human erythrocytes. *Biochim. Biophys. Acta* **902**, 253–262.

Clark, M. R. (1988). Senescence of red blood cells: Progress and problems. *Physiol. Rev.* **68**, 508–554.

Clark, S. J., and Ralston, G. B. (1990). The dissociation of periferal proteins from erythrocyte membranes brought about by p-mercuribenzenesulfonate. *Biochim. Biophys. Acta* **1021**, 141–147.

Cohen, A. M., Liu, S. C., Lawler, J., Derick, L., and Palek, J. (1988). Identification of the protein 4.1 binding site to phosphatidylserine vesicles. *Biochemistry* **27**, 614–619.

Cohen, C. M., and Gascard, P. (1992). Regulation of post-translational modification of erythrocyte membrane and membrane-skeletal proteins. *Sem. Hematol.* **29**, 244–292.

Colleau, M., Hervé, P., Fellmann, P., and Devaux, P. F. (1991). Transmembrane diffusion of fluorescent phospholipids in human erythrocytes. *Chem. Phys. Lipids* **57**, 29–37.

Comfurius, P., Bevers, E. M., and Zwaal, R. F. A. (1989). Interaction between phosphatidylserine and the isolated cytoskeleton of human blood platelets. *Biochim. Biophys. Acta* **983**, 212–216.

Comfurius, P., Senden, J. M. G., Tilly, R. H. J., Schroit, A. J., Bevers, E. M., and Zwaal, R. F. A. (1990). Loss of membrane phospholipid asymmetry in platelets and red cells may be associated with calcium-induced shedding of plasma membrane and inhibition of aminophospholipid translocase. *Biochim. Biophys. Acta* **1026**, 153–160.

Connor, J., and Schroit, A. J. (1988). Transbilayer movement of phosphatidylserine in erythrocytes: Inhibition of transport and preferential labeling of a 31,000-Dalton protein by sulfhydryl reactive reagents. *Biochemistry* **27**, 848–851.

Connor, J., and Schroit, A. J. (1989). Transbilayer movement of phosphatidylserine in erythrocytes: Inhibitors of aminophospholipid transport block the association of photolabeled lipid to its transporter. *Biochim. Biophys. Acta* **1066**, 37–42.

Connor, J., and Schroit, A. J. (1990). Aminophospholipid translocation in erythrocytes: Evidence for the involvement of a specific transporter and an endofacial protein. *Biochemistry* **29**, 37–43.

Connor, J., and Schroit, A. J. (1991). Transbilayer movement of phosphatidylserine in erythrocytes. Evidence that the aminophospholipid transporter is a ubiquitous membrane protein. *Biochemistry* **28**, 9680–9685.

Connor, J., Bucana, C., Fidler, I. J., and Schroit, A. J. (1989). Differentiation-dependent expression of phosphatidylserine in mammalian plasma membranes: Quantitative assessment of outer-leaflet lipid by prothrombinase complex formation. *Proc. Natl. Acad. Sci. U.S.A.* **86**, 3184–3188.

Connor, J., Gillum, K., and Schroit, A. J. (1990). Maintenance of lipid asymmetry in red blood cells and ghosts: Effect of divalent cations and serum albumin on the transbilayer distribution of phosphatidylserine. *Biochim. Biophys. Acta* **1025**, 82–86.

Connor, J., Pak, C. H., Zwaal, R. F. A., and Schroit, A. J. (1992). Bidirectional transbilayer movement of phospholipid analogs in human red blood cells. Evidence for an ATP-dependent and protein-mediated process. *J. Biol. Chem.* **267**, 19412–19417.

Cullis, P. R., and de Kruijff, B. (1979). Lipid polymorphism and the functional roles of lipids in biological membranes. *Biochim. Biophys. Acta* **559**, 399–420.

Daleke, D. L., and Huestis, W. (1985). Incorporation and translocation of aminophospholipids in human erythrocytes. *Biochemistry* **24**, 5406–5416.

Daleke, D. L., and Huestis, W. (1989). Erythrocyte morphology reflects the transbilayer distribution of incorporated phospholipids. *J. Cell Biol.* **108**, 1375–1385.

Deguercy, A., Hommel, M., and Schrével, J. (1990). Purification and characterization of 37-kilodalton proteases from *Plasmodium falciparum* and *Plasmodium berghei* which cleave erythrocyte cytoskeletal proteins. *Mol. Biochem. Parasitol.* **38**, 233–244.

Demel, R. A., Geurts van Kessel, W. S. M., Zwaal, R. F. A., Roelofsen, B., and van Deenen, L. L. M. (1975). Relation between various phospholipid actions on human red cell membranes and the interfacial phospholipid pressure in monolayers. *Biochim. Biophys. Acta* **406**, 97–107.

Devaux, P. F. (1988). Phospholipid flippases. *FEBS Lett.* **234**, 8–12.

Devaux, P. F. (1991). Static and dynamic lipid asymmetry in cell membranes. *Biochemistry* **30**, 1163–1173.

Devaux, P. F. (1992). Protein involvement in transmembrane lipid asymmetry. *Annu. Rev. Biophys. Biomol. Struct.* **21**, 417–439.

Dressler, V., Haest, C. W. M., Plasa, G., Deuticke, B., and Erusalimsky, J. D. (1984). Stabilizing factors of phospholipid asymmetry in the erythrocyte membrane. *Biochim. Biophys. Acta* **775**, 189–196.

Etemadi, A. H. (1980). Membrane asymmetry. A survey and critical appraisal of the methodology. II. Methods for assessing the unequal distribution of lipids. *Biochim. Biophys. Acta* **604**, 423–475.

Fontaine, R. N., Harris, R. A., and Schroeder, F. (1980). Aminophospholipid asymmetry in murine synaptosomal plasma membrane. *J. Neurochem.* **34**, 269–277.

Franck, P. F. H., Roelofsen, B., and Op den Kamp, J. A. F. (1982). Complete exchange of phosphatidylcholine from intact erythrocytes after protein cross-linking. *Biochim. Biophys. Acta* **687**, 105–108.

Franck, P. F. H., Chiu, D. T.-Y., Op den Kamp, J. A. F., van Deenen, L. L. M., and Roelofsen, B. (1983). Accelerated transbilayer movement of phosphatidylcholine in sickled erythrocytes. A reversible process. *J. Biol. Chem.* **258**, 8435–8442.

Franck, P. F. H., Bevers, E. M., Lubin, B. H., Comfurius, P., Chiu, D. T.-Y., Op den Kamp, J. A. F., Zwaal, R. F. A., van Deenen, L. L. M., and Roelofsen, B. (1985a). Uncoupling of the membrane skeleton from the lipid bilayer. The cause of accelerated phospholipid flip-flop leading to an enhanced procoagulant activity of sickled cells. *J. Clin. Invest.* **75**, 183–190.

Franck, P. F. H., Op den Kamp, J. A. F., Lubin, B. H., Berentsen, W., Joosten, P., Briët, E., van Deenen, L. L. M., and Roelofsen, B. (1985b). Abnormal transbilayer mobility of phosphatidylcholine in hereditary pyropoikilocytosis reflects the increased heat sensitivity of the membrane skeleton. *Biochim. Biophys. Acta* **815**, 259–267.

Franck, P. F. H., Op den Kamp, J. A. F., Roelofsen, B., and van Deenen, L. L. M. (1986). Does diamide treatment of intact human erythrocytes cause a loss of phospholipid asymmetry? *Biochim. Biophys. Acta* **857**, 127–130.

Fujimagari, M., Williamson, P. L., and Schlegel, R. A. (1990). Ca^{2+}-dependent membrane-binding proteins in normal erythrocytes and erythrocytes from patients with chronic myeloid leukemia. *Blood* **75**, 1337–1345.

Gascard, P., Tran, D., Sauvage, M., Sulpice, J.-C., Fukami, K., Takenawa, T., Claret, M., and Giraud, F. (1991). Asymmetric distribution of phosphoinositides and phosphatidic acid in the human erythrocyte membrane. *Biochim. Biophys. Acta* **1069**, 27–36.

Grasso, M., Morelli, A., and De Flora, A. (1986). Ca^{2+}-induced alterations in the levels and subcellular distribution of proteolytic enzymes in human red blood cells. *Biochem. Biophys. Res. Commun.* **138**, 87–94.

Gerritsen, W. J., Henricks, P. A. J., de Kruijff, B., and van Deenen, L. L. M. (1980). The transbilayer movement of phosphatidylcholine in vesicles reconstituted with intrinsic proteins from the human erythrocyte membrane. *Biochim. Biophys. Acta* **600**, 607–619.

Gudi, S. R. P., Kumar, A., Bhakuni, V., Gokhale, S. M., and Gupta, C. M. (1990). Membrane skeleton–bilayer interaction is not the major determinant of membrane phospholipid asymmetry in human erythrocytes. *Biochim. Biophys. Acta* **1023**, 63–172.

Haest, C. W. M. (1982). Interactions between membrane skeleton proteins and the intrinsic domain of the erythrocyte membrane. *Biochim. Biophys. Acta* **694**, 331–352.

Haest, C. W. M., Plasa, G., Kamp, D., and Deuticke, B. (1978). Spectrin as a stabilizer of the phospholipid asymmetry in the human erythrocyte membrane. *Biochim. Biophys. Acta* **509**, 21–32.

Haldar, K., de Amorim, A. F., and Cross, G. A. M. (1989). Transport of fluorescent phospholipid analogues from the erythrocyte membrane to the parasite in *Plasmodium falciparum*-infected cells. *J. Cell Biol.* **108**, 2183–2192.

Hanspal, M., and Palek, J. (1992). Biogenesis of normal and abnormal red blood cell membrane skeleton. *Sem. Hematol.* **29**, 305–319.

Henseleit, U., Plasa, G., and Haest, C. W. M. (1990). Effects of divalent cations on lipid flip-flop in the human erythrocyte membrane. *Biochim. Biophys. Acta* **1029**, 127–135.

Herrmann, A., and Devaux, P. F. (1990). Alteration of the aminophospholipid translocase activity during *in vivo* and artificial aging of human erythrocytes. *Biochim. Biophys. Acta* **1027**, 41–46.

Herrmann, A., and Müller, P. (1986). A model for the asymmetric lipid distribution in the human erythrocyte membrane. *Biosci. Rep.* **6**, 185–191.

Herrmann, A., Clague, M. J., Puri, A., Morris, S. J., Blumenthal, R., and Grimaldi, S. (1990). Effect of erythrocyte transbilayer phospholipid distribution on fusion with vesicular stomatitis virus. *Biochemistry* **29**, 4054–4058.

Hinkovska, V. T., Dimitrov, G. P., and Koumanov, K. S. (1986). Phospholipid composition and phospholipid asymmetry of ram spermatozoa membranes. *Int. J. Biochem.* **18**, 1115–1121.

Huang, S. K., and Hui, S. W. (1990). Fluorescence measurements of fusion between human erythrocytes induced by poly(ethylene glycol). *Biophys. J.* **58**, 1109–1117.

Hullin, F., Bossant, M.-J., and Salem, N., Jr. (1991). Aminophospholipid molecular species asymmetry in the human erythrocyte plasma membrane. *Biochim. Biophys. Acta* **1061**, 15–25.

Jain, S. K. (1985). *In vivo* externalization of phosphatidylserine and phosphatidylethanolamine in the membrane bilayer and hypercoagulability by the lipid peroxidation of erythrocytes in rats. *J. Clin. Invest.* **76**, 281–286.

Jain, S. K. (1986). Presence of phosphatidylserine in the outer membrane bilayer of newborn human erythrocytes. *Biochem. Biophys. Res. Commun.* **136**, 914–920.

Jain, S. K., and Hochstein, P. (1980). Membrane alterations in phenylhydrazine-induced reticulocytes. *Arch. Biochem. Biophys.* **201**, 683–687.

Joshi, P., and Gupta, C. M. (1988). Abnormal membrane phospholipid organization in *Plasmodium falciparum*-infected human erythrocytes. *Brit. J. Haematol.* **68**, 255–259.

Joshi, P., Dutta, G. P., and Gupta, C. M. (1987). An intracellular simian malarial parasite (*Plasmodium knowlesi*) induces stage-dependent alterations in membrane phospholipid organization of its host erythrocyte. *Biochem. J.* **246**, 103–108.

Kumar, A., and Gupta, C. M. (1983). Red cell membrane abnormalities in chronic myeloid leukemia. *Nature (London)* **303**, 632–633.

Kumar, A., Daniel, S., Agarwal, S. S., and Gupta, C. M. (1987). Abnormal erythrocyte membrane phospholipid organisation in chronic myeloid leukemia. *J. Biosci.* **11**, 543–548.

Kumar, A., Gudi, S. R. P., Gokhale, S. M., Bhakuni, V., and Gupta, C. M. (1990). Heat-induced alterations in monkey erythrocyte membrane phospholipid organization and skeletal protein structure and interactions. *Biochim. Biophys. Acta* **1030**, 269–278.

Kuypers, F. A., van Linde-Sibenius Trip, M., Roelofsen, B., Tanner, M. J. A., Anstee, D. J., and Op den Kamp, J. A. F. (1984). Rh$_{null}$ human erythrocytes have an abnormal membrane phospholipid organization. *Biochem. J.* **221**, 931–934.

Kuypers, F. A., van Linde-Sibenius Trip, M., Roelofsen, B., Op den Kamp, J. A. F., Tanner, J. A., and Anstee, D. J. (1985). The phospholipid organisation in the membranes of McLeod and Leach phenotype erythrocytes. *FEBS Lett.* **184**, 20–24.

Lazarides, E., and Woods, C. (1989). Biogenesis of the red blood cell membrane-skeleton and the control of erythroid morphogenesis. *Annu. Rev. Cell Biol.* **5**, 427–452.

Lehnert, M. E., and Lodish, H. F. (1988). Unequal synthesis and differential degradation of α and β spectrin during murine erythroid differentiation. *J. Cell Biol.* **107**, 413–426.

Lipka, G., Op den Kamp, J. A. F., and Hauser, H. (1991). Lipid asymmetry in rabbit small intestinal brush border membrane as probed by an intrinsic phospholipid exchange protein. *Biochemistry* **30**, 11828–11836.

Liu, S.-C., and Derick, L. H. (1992). Molecular anatomy of the red blood cell membrane skeleton: Structure–function relationships. *Sem. Hematol.* **29**, 231–243.

Liu, S.-C., Derick, L. H., Agre, P., and Palek, J. (1990). Alteration of the erythrocyte membrane skeletal ultrastructure in hereditary spherocytosis, heridetary elliptocytosis, and pyropoikilocytosis. *Blood* **76**, 198–205.

Liu, S.-C., Derick, L. H., Zhai, S., and Palek, J. (1991). Uncoupling of the spectrin-based skeleton from the lipid bilayer in sickled red cells. *Science* **252**, 574–576.

Lorand, L., Barnes, N., Bruner-Lorand, J. A., Hawkins, M., and Michalska, M. (1987). Inhibition of protein cross-linking in Ca^{2+}-enriched human erythrocytes and activated platelets. *Biochemistry* **26**, 308–313.

Lubin, B., and Chiu, D. (1982). Membrane phospholipid organization in pathologic human erythrocytes. *In* "Membranes and Genetic Disease" (J. R. Shepperd, V. E. Anderson, and J. W. Eaton, eds.), pp. 137–150. Liss, New York.

Lubin, B., Chiu, D., Bastacky, J., Roelofsen, B., and van Deenen, L. L. M. (1981). Abnormalities in membrane phospholipid organization in sickled erythrocytes. *J. Clin. Invest.* **67**, 1643–1649.

Luna, E. J., and Hitt, A. L. (1992). Cytoskeleton–plasma membrane interactions. *Science* **258**, 955–964.

Maddy, A. H. (1964). Fluorescent label for the outer components of the plasma membrane. *Biochim. Biophys. Acta* **88**, 390–399.

Maguire, P. A., Prudhome, J., and Sherman, I. W. (1991). Alterations in erythrocyte membrane phospholipid organization due to the intracellular growth of the human malaria parasite *Plasmodium falciparum*. *Parasitol.* **102**, 179–186.

Maksymiw, R., Sui, S. F., Gaub, H., and Sackmann, E. (1987). Electrostatic coupling of spectrin dimers to phosphatidylserine containing lipid lamellae. *Biochemistry* **26**, 2983–2990.

Marks, P. A., and Rifkind, R. A. (1978). Erythroleukemic differentiation. *Ann. Rev. Biochem.* **47**, 419–448.

Martin, O. C., and Pagano, R. E. (1987). Transbilayer movement of fluorescent analogs of phosphatidylserine and phosphatidylethanolamine at the plasma membrane of cultured cells. Evidence for a protein-mediated and ATP-dependent process. *J. Biol. Chem.* **262**, 5890–5898.

Matovcik, L. M., Chiu, D., Lubin, B., Mentzer, W. C., Lane, P. A., Mohandas, N., and Schrier, S. L. (1986). The aging process of human neonatal erythrocytes. *Pediatr. Res.* **20**, 1091–1096.

McEvoy, L., Williamson, P., and Schlegel, R. A. (1986). Membrane phospholipid asymmetry as a determinant of erythrocyte recognition by macrophages. *Proc. Natl. Acad. Sci. U.S.A.* **83**, 3311–3315.

Middelkoop, E. (1989). "Transmembrane Phospholipid Asymmetry in Erythroid Cells: Mechanisms of Maintenance." Ph.D. Thesis, University of Utrecht, The Netherlands.

Middelkoop, E., Lubin, B. H., Op den Kamp, J. A. F., and Roelofsen, B. (1986). Flip–flop rates of individual molecular species of phosphatidylcholine in the human red cell membrane. *Biochim. Biophys. Acta* **855**, 421–424.

Middelkoop, E., Lubin, B. H., Bevers, E. M., Op den Kamp, J. A. F., Comfurius, P., Chiu, D. T.-Y., Zwaal, R. F. A., van Deenen, L. L. M., and Roelofsen, B. (1988). Studies on sickled erythrocytes provide evidence that the asymmetric distribution of phosphatidylserine in the red cell membrane is maintained by both ATP-dependent translocation and interaction with membrane skeletal proteins. *Biochim. Biophys. Acta* **937**, 281–288.

Middelkoop, E., Coppens, A., Llanillo, M., van der Hoek, E. E., Slotboom, A. J., Lubin, B. H., Op den Kamp, J. A. F., van Deenen, L. L. M., and Roelofsen, B. (1989a). Aminophospholipid translocase in the plasma membrane of Friend erythroleukemic cells can induce an asymmetric topology for phosphatidylserine but not for phosphatidylethanolamine. *Biochim. Biophys. Acta* **978**, 241–248.

Middelkoop, E., van der Hoek, E. E., Bevers, E. M., Comfurius, P., Slotboom, A. J., Op den Kamp, J. A. F., Lubin, B. H., Zwaal, R. F. A., and Roelofsen, B. (1989b). Involvement of ATP-dependent aminophospholipid translocation in maintaining phospholipid asymmetry in diamide-treated human erythrocytes. *Biochim. Biophys. Acta* **981**, 151–160.

Mohandas, N., Wyatt, J., Mel, S. F., Ross, M. E., and Shohet, S. B. (1982). Lipid translocation across the human erythrocyte membrane. Regulatory factors. *J. Biol. Chem.* **257**, 6537–6543.

Mohandas, N., Rossi, M., Bernstein, S., Ballas, S., Ravindranath, Y., Wyatt, J., and Mentzer, W. (1985). The structural organization of skeletal proteins influences lipid translocation across erythrocyte membrane. *J. Biol. Chem.* **260**, 14264–14268.

Moll, G. N., Vial, H. J., Bevers, E. M., Ancelin, M. L., Roelofsen, B., Comfurius, P., Slotboom, A. J., Zwaal, R. F. A., Op den Kamp, J. A. F., and van Deenen, L. L. M. (1990). Phospholipid asymmetry in the plasma membrane of malaria-infected erythrocytes. *Biochem. Cell Biol.* **68**, 579–585.

Mombers, C., Verkleij, A. J., de Gier, J., and van Deenen, L. L. M. (1979). The interaction of spectrin-actin with synthetic phospholipids. II. The interaction with phosphatidylserine. *Biochim. Biophys. Acta* **551**, 271–281.

Mombers, C., de Gier, J., Demel, R. A., and van Deenen, L. L. M. (1980). Spectrin–phospholipid interaction. A monolayer study. *Biochim. Biophys. Acta* **603**, 52–62.

Morrot, G., Cribier, S., Devaux, P. F., Geldwerth, D., Davoust, J., Bureau, J. F., Fellmann, P., Hervé, P., and Frilley, B. (1986). Asymmetric lateral mobility of phospholipids in the human erythrocyte membrane. *Proc. Natl. Acad. Sci. U.S.A.* **83**, 6863–6867.

Morrot, G., Hervé, P., Zachowski, A., Fellmann, P., and Devaux, P. F. (1989). Aminophospholipid translocase of human erythrocytes: Phospholipid substrate specificity and effect of cholesterol. *Biochemistry* **28**, 3456–3462.

Morrot, G., Zachowski, A., and Devaux, P. F. (1990). Partial purification and characterization of the human erythrocyte Mg^{2+}–ATPase. A candidate aminophospholipid translocase. *FEBS Lett.* **266**, 29–32.

Nijhof, W., and Wierenga, P. K. (1988). Biogenesis of red cell membrane and cytoskeletal proteins during erythropoiesis *in vitro*. *Exp. Cell Res.* **177**, 329–337.

Nijhof, W., van der Schaft, P. H., Wierenga, P. K., Roelofsen, B., Op den Kamp, J. A. F., and van Deenen, L. L. M. (1986). The transbilayer distribution of phosphatidylethanolamine in erythroid plasma membranes during erythropoiesis. *Biochim. Biophys. Acta* **862**, 273–277.

Op den Kamp, J. A. F. (1979). Lipid asymmetry in membranes. *Annu. Rev. Biochem.* **48**, 47–71.

Patel, V. P., and Lodish, H. F. (1987). A fibronectin matrix is required for differentiation of murine erythroleukemia cells into reticulocytes. *J. Cell Biol.* **105**, 3105–3118.

Pelletier, X., Mersel, M., Freysz, L., and Leray, C. (1987). Topological distribution of aminophospholipid fatty acids in trout intestinal brush-border membrane. *Biochim. Biophys. Acta* **902**, 223–228.

Pinder, J. C., Pekrun, A., Maggs, A. M., and Gratzer, W. B. (1992). Interaction of the red cell membrane skeleton with the membrane. *Biochem. Soc. Trans.* **20**, 774–776.

Pradhan, D., Weiser, M., Lumley-Sapanski, K., Frazier, D., Kemper, S., Williamson, P., and Schlegel, R. A. (1990). Peroxidation-induced perturbations of erythrocyte lipid organization. *Biochim. Biophys. Acta* **1023**, 398–404.

Pradhan, D., Williamson, P., and Schlegel, R. A. (1991). Bilayer/cytoskeleton interactions in lipid-symmetric erythrocytes assessed by a photoactivatable phospholipid analogue. *Biochemistry* **30**, 7754–7758.

Raval, P. J., and Allan, D. (1984). Sickling of sickle erythrocytes does not alter phospholipid asymmetry. *Biochem. J.* **223**, 555–557.

Rawyler, A., Roelofsen, B., and Op den Kamp, J. A. F. (1984). The use of fluorescamine as a permeant probe to localize phosphatidylethanolamine in intact Friend erythroleukaemic cells. *Biochim. Biophys. Acta* **769**, 330–336.

Rawyler, A., van der Schaft, P., Roelofsen, B., and Op den Kamp, J. A. F. (1985). Phospholipid localization in the plasma membrane of Friend erythroleukemic cells and mouse erythrocytes. *Biochemistry* **24**, 1777–1783.

Record, M., Tamer, A. El., Chap, H., and Douste-Blazy, L. (1984). Evidence for a highly asymmetric arrangement of ether- and diacyl-phospholipid subclasses in the plasma membrane of Krebs II ascites cells. *Biochim. Biophys. Acta* **778**, 449–456.

Redman, C. M., Huima, T., Robbins, E., Lee, S., and Marsh, W. L. (1989). Effect of phosphatidylserine on the shape of McLeod red cell acanthocytes. *Blood* **74**, 1826–1835.

Reed, J. A., Kough, R. H., Williamson, P., and Schlegel, R. A. (1985). Merocyanine 540 recognizes membrane abnormalities of erythrocytes in chronic myelogenous leukemia. *Cell Biol. Int. Rep.* **9**, 43–49.

Renooy, W., van Golde, L. M. G., Zwaal, R. F. A., Roelofsen, B., and van Deenen, L. L. M. (1974). Preferential incorporation of fatty acids at the inside of human erythrocyte membranes. *Biochim. Biophys. Acta* **363**, 287–292.

Roelofsen, B. (1982). Phospholipases as tools to study the localization of phospholipids in biological membranes. A critical review. *J. Toxicol. Toxin Rev.* **1**, 87–197.

Rosso, J., Zachowski, A., and Devaux, P. F. (1988). Influence of chlorpromazine on the transverse mobility of phospholipids in the human erythrocyte membrane: Relation to shape changes. *Biochim. Biophys. Acta* **942**, 271–279.

Roth, E. F., Jr., Calvin, M. L., Max-Audit, I., Rosa, J., and Rosa, R. (1988). The enzymes of the glycolytic pathway in erythrocytes infected with *Plasmodium falciparum* malaria parasites. *Blood* **76**, 1922, 1925.

Rybicki, A. C., Heath, R., Lubin, B., and Schwartz, R. S. (1988). Human erythrocyte protein 4.1 is a phosphatidylserine binding protein. *J. Clin. Invest.* **81**, 255–260.

Sato, S. B., and Ohnishi, S.-I. (1983). Interaction of a peripheral protein of the erythrocyte membrane, band 4.1, with phosphatidylserine-containing liposomes and erythrocyte inside-out vesicles. *Eur. J. Biochem.* **130**, 19–25.

Schlegel, R. A., Phelps, B. M., Cofer, G. P., and Williamson, P. (1982). Enucleation eliminates a differentiation-specific marker from normal and leukemic murine erythroid cells. *Exp. Cell Res.* **139**, 321–325.

Schlegel, R. A., Prendergast, T. W., and Williamson, P. (1985). Membrane phospholipid asymmetry as a factor in erythrocyte-endothelial cell interactions. *J. Cell. Physiol.* **123**, 215–218.

Schlegel, R. A., Kemper, S., and Williamson, P. (1990). Functional and pathological significance of phospholipid asymmetry in erythrocyte membranes. *J. Biosci.* **15**, 187–191.

Schneider, E., Haest, C. W. M., Plasa, G., and Deuticke, B. (1986). Bacterial cytotoxins, amphotericin B and local anesthetics enhance transbilayer mobility of phospholipids in erythrocyte membranes. Consequences for phospholipid asymmetry. *Biochim. Biophys. Acta* **855**, 325–336.

Schrier, S. L., Chiu, D. T.-Y., Yee, M., Sizer, K., and Lubin, B. (1983). Alteration of membrane phospholipid bilayer organization in human erythrocytes during drug-induced endocytosis. *J. Clin. Invest.* **72**, 1698–1705.

Schrier, S. L., Zachowski, A., Hervé, P., Kader, J.-C., and Devaux, P. F. (1992). Transmembrane redistribution of phospholipids of the human red cell membrane during hypotonic lysis. *Biochim. Biophys. Acta* **1105**, 170–176.

Schroit, A. J., and Zwaal, R. F. A. (1991). Transbilayer movements of phospholipids in red cell and platelet membranes. *Biochim. Biophys. Acta* **1071**, 313–329.

Schroit, A. J., Madsen, J. W., and Tanaka, Y. (1985). *In vivo* recognition and clearance of red blood cells containing phosphatidylserine in their plasma membrane. *J. Biol. Chem.* **260**, 5131–5138.

Schroit, A. J., Bloy, C., Connor, J., and Cartron, J. P. (1990). Involvement of Rh blood group polypeptides in the maintenance of aminophospholipid asymmetry. *Biochemistry* **29**, 10303–10306.

Schwartz, R. S., Chiu, D. T.-Y., and Lubin, B. (1985). Plasma membrane phospholipid organization in human erythrocytes. *Curr. Top. Hematol.* **4**, 63–112.

Schwartz, R. S., Olson, J. A., Raventos-Suarez, C., Yee, M., Heath, R. H., Lubin, B., and Nagel, R. L. (1987). Altered plasma membrane phospholipid organization in *Plasmodium falciparum*-infected human erythrocytes. *Blood* **69**, 401–407.

Seigneuret, M., and Devaux, P. F. (1984). ATP-dependent asymmetric distribution of spin-labeled phospholipids in the erythrocyte membrane: Relation to shape changes. *Proc. Natl. Acad. Sci. U.S.A.* **81**, 3751–3755.

Sessions, A., and Horwitz, A. F. (1983). Differentiation-related differences in the plasma membrane phospholipid asymmetry of myogenic and fibrogenic cells. *Biochim. Biophys. Acta* **728**, 103–111.

Shiffer, K. A., Goerke, J., Düzgünes, N., Fedor, J., and Shohet, S. B. (1988). Interaction of erythrocyte protein 4.1 with phospholipids. A monolayer and liposome study. *Biochim. Biophys. Acta* **937**, 269–280.

Shukla, S. D., and Hanahan, D. J. (1982). Membrane alterations in cellular aging: Susceptibility of phospholipids in density (age-)separated human erythrocytes to phospholipase A_2. *Arch. Biochem. Biophys.* **214**, 335–341.

Simões, A. P., Moll, G. N., Slotboom, A. J., Roelofsen, B., and Op den Kamp, J. A. F. (1991). Selective internalization of choline-phospholipids in *Plasmodium falciparum* parasitized human erythrocytes. *Biochim. Biophys. Acta* **1063**, 45–50.

Simões, A. P., Roelofsen, B., and Op den Kamp, J. A. F. (1992). Lipid compartmentalization in erythrocytes parasitized by *Plasmodium* spp. *Parasitol. Today* **8**, 18–21.

Smith, R. E., and Daleke, D. L. (1990). Phosphatidylserine transport in Rh_{null} erythrocytes. *Blood* **76**, 1021–1027.

Song, L. Y., Baldwin, J. M., O'Reilly, R., and Lucy, J. A. (1992). Relationships between the surface exposure of acidic phospholipids and cell fusion in erythrocytes subjected to electrical breakdown. *Biochim. Biophys. Acta* **1104**, 1–8.

Subbarao, N. K., MacDonald, R. J., Takeshita, K., and MacDonald, R. C. (1991). Characteristics of spectrin-induced leakage of extruded, phosphatidylserine vesicles. *Biochim. Biophys. Acta* **1063**, 147–154.

Sune, A., Bette-Bobillo, P., Bienvenue, A., Fellmann, P., and Devaux, P. F. (1987). Selective outside–inside translocation of aminophospholipids in human platelets. *Biochemistry* **26**, 2972–2977.

Sune, A., Vidal, M., Morin, P., Sainte-Marie, J., and Bienvenue, A. (1988). Evidence for bidirectional transverse diffusion of spin-labeled phospholipids in the plasma membrane of guinea pig blood cells. *Biochim. Biophys. Acta* **946**, 315–327.

Sweet, C., and Zull, J. E. (1970). Interaction of the erythrocyte-membrane protein, spectrin, with model systems. *Biochem. Biophys. Res. Commun.* **41**, 135–141.

Tanaka, Y., and Schroit, A. J. (1983). Insertion of fluorescent PS into the plasma membrane of red blood cells. Recognition by autologous macrophages. *J. Biol. Chem.* **258**, 11335–11343.

Tilley, L., Cribier, S., Roelofsen, B., Op den Kamp, J. A. F., and van Deenen, L. L. M.

(1986). ATP-dependent translocation of aminophospholipids across the human erythrocyte membrane. *FEBS Lett.* **194**, 21–27.

Tilly, R. H. J., Senden, J. M. G., Confurius, P., Bevers, E. M., and Zwaal, R. F. A. (1990). Increased aminophospholipid translocase activity in human platelets during secretion. *Biochim. Biophys. Acta* **1029**, 188–190.

Van Deenen, L. L. M. (1981). Topology and dynamics of phospholipids in membranes. *FEBS Lett.* **123**, 1–16.

Van der Schaft, P. H., Roelofsen, B., Op den Kamp, J. A. F., and van Deenen, L. L. M. (1987a). Phospholipid asymmetry during erythropoiesis. A study on Friend erythroleukemic cells and mouse reticulocytes. *Biochim. Biophys. Acta* **900**, 103–115.

Van der Schaft, P. H., Beaumelle, B., Vial, H., Roelofsen, B., Op den Kamp, J. A. F., and van Deenen, L. L. M. (1987b). Phospholipid organization in monkey erythrocytes upon *Plasmodium knowlesi* infection. *Biochim. Biophys. Acta* **901**, 1–14.

Van der Wiele, F., Chr., Atsma, W., Roelofsen, B., van Linde, M., van Binsbergen, J., Radvanyi, F., Raykova, D., Slotboom, A. J., and de Haas, G. H. (1988). Site-specific ε-NH$_2$ monoacylation of pancreatic phospholipase A$_2$. 2. Transformation of soluble phospholipase A$_2$ into a highly penetrating "membrane-bound" form. *Biochemistry* **27**, 1688–1694.

Van Meer, G., and Op den Kamp, J. A. F. (1982). Transbilayer movement of various phosphatidylcholine species in intact human erythrocytes. *J. Cell. Biochem.* **19**, 193–204.

Van Meer, G., Gahmberg, C. G., Op den Kamp, J. A. F., and van Deenen, L. L. M. (1981). Phospholipid distribution in human En(a-1) red cell membranes which lack the major sialoglycoprotein, glycophorin A. *FEBS Lett.* **135**, 53–55.

Verhoven, B., Schlegel, R. A., and Williamson, P. (1992). Rapid loss and restoration of lipid asymmetry by different pathways in resealed erythrocyte ghosts. *Biochim. Biophys. Acta* **1104**, 15–23.

Verkleij, A. J., Zwaal, R. F. A., Roelofsen, B., Comfurius, P., Kastelijn, D., and van Deenen, L. L. M. (1973). The asymmetric distribution of phospholipids in the human red cell membrane. A combined study using phospholipases and freeze-etch electron microscopy. *Biochim. Biophys. Acta* **323**, 178–193.

Vial, H. J., Ancelin, M.-L., Philippot, J. R., and Thuet, M. (1990). Biosynthesis and dynamics of lipids in *Plasmodium*-infected mature mammalian erythrocytes. *Blood Cells* **16**, 531–555.

Vidal, M., Sainte-Marie, J., Phillipot, J. R., and Bienvenue, A. (1989). Asymmetric distribution of phospholipids in the membrane of vesicles released during *in vitro* maturation of guinea pig reticulocytes: Evidence precluding a role for "aminophospholipid translocase". *J. Cell. Physiol.* **140**, 455–462.

Wali, R. K., Jaffe, S., Kumar, D., Sorgente, N., and Kalra, V. K. (1987). Increased adherence of oxidant-treated human and bovine erythrocytes to cultured endothelial cells. *J. Cell. Physiol.* **133**, 25–36.

Wali, R. K., Jaffe, S., Kumar, D., and Kalra, V. Y. (1988). Alterations in organization of phospholipids in erythrocytes as factor in adherence to endothelial cells in diabetes mellitus. *Diabetes* **37**, 104–111.

Wallach, D. F. H., and Conley, M. (1977). Altered membrane proteins of monkey erythrocytes infected with simian malaria. *J. Mol. Med.* **2**, 119–136.

Whatmore, J. L., Tang, E. K. Y., and Hickman, J. A. (1992). Cytoskeletal proteolysis during calcium-induced morphological transitions of human erythrocytes. *Exp. Cell Res.* **200**, 316–325.

Williamson, P., Bateman, J., Kozarsky, K., Mattocks, K., Hermanovicz, N., Choe,

H.-R., and Schlegel, R. A. (1982). Involvement of spectrin in the maintenance of phase-state asymmetry in the erythrocyte membrane. *Cell* **30**, 725–733.

Williamson, P., Algarin, L., Bateman, J., Choe, H.-R., and Schlegel, R. A. (1985). Phospholipid asymmetry in erythrocyte ghosts. *J. Cell. Physiol.* **123**, 209–214.

Williamson, P., Antia, R., and Schlegel, R. A. (1987). Maintenance of membrane phospholipid asymmetry. Lipid-cytoskeletal interactions or lipid pump? *FEBS Lett.* **219**, 316–320.

Williamson, P., Kulick, A., Zachowski, A., Schlegel, R. A., and Devaux, P. F. (1992). Ca^{2+} induces transbilayer redistribution of all major phospholipids in human erythrocytes. *Biochemistry* **31**, 6355–6360.

Woods, C. M., Boyer, B., Vogt, P. K., and Lazarides, E. (1986). Control of erythroid differentiation: Asynchronous expression of the anion transporter and the peripheral components of the membrane skeleton in AEV- and S_{13}-transformed cells. *J. Cell Biol.* **103**, 1789–1798.

Zachowski, A., and Devaux, P. F. (1990). Transmembrane movements of lipids. *Experientia* **46**, 644–656.

Zachowski, A., Fellmann, P., and Devaux, P. F. (1985). Absence of transbilayer diffusion of spin-labeled sphingomyelin in human erythrocytes. Comparison with the diffusion of several spin-labeled glycerophospholipids. *Biochim. Biophys. Acta* **815**, 510–514.

Zachowski, A., Favre, E., Cribier, S., Hervé, P., and Devaux, P. F. (1986). Outside–inside translocation of aminophospholipids in the human erythrocyte membrane is mediated by a specific enzyme. *Biochemistry* **25**, 2585–2590.

Zachowski, A., Herrmann, A., Paraf, A., and Devaux, P. F. (1987). Phospholipid outside–inside translocation in lymphocyte plasma membranes is a protein-mediated phenomenon. *Biochim. Biophys. Acta* **897**, 197–200.

Zachowski, A., Henry, J. P., and Devaux, P. F. (1989). Control of transmembrane lipid asymmetry in chromaffin granules by an ATP-dependent protein. *Nature (London)* **340**, 75–76.

Zachowski, A., and Morot Gaudry-Talarmain, Y. (1990). Phospholipid transverse diffusion in synaptosomes: Evidence for the involvement of the aminophospholipid translocase. *J. Neurochem.* **55**, 1352–1356.

Zwaal, R. F. A., Roelofsen, B., Comfurius, P., and van Deenen, L. L. M. (1975). Organization of phospholipids in human red cell membranes as detected by the action of various purified phospholipases. *Biochim. Biophys. Acta* **406**, 83–96.

Zwaal, R. F. A., Comfurius, P., and Bevers, E. M. (1992). Platelet procoagulant activity and microvesicle formation. Its putative role in hemostasis and thrombosis. *Biochim. Biophys. Acta* **1180**, 1–8.

CHAPTER 2

Protein-Mediated Phospholipid Movement in Red Blood Cells

Alan J. Schroit
Department of Cell Biology, University of Texas M. D. Anderson Cancer Center,
Houston, Texas 77030

I. INTRODUCTION

Erythrocyte membranes contain several major classes of lipids that are composed of different fatty acyl side chains or, in the case of sphingolipids, of different long-chain bases. Studies described over the last decade have established that these lipids are not distributed randomly in the plasma membrane; instead, certain species and lipids of a specific molecular composition are distributed asymmetrically across the membrane bilayer. This

Current Topics in Membranes, Volume 40

distribution is especially evident for the aminophospholipids phosphatidyl-serine (PS) and phosphatidylethanolamine (PE), which reside preferen-tially in the inner leaflet of the plasma membrane.

Although the distribution of phospholipids in red blood cells was found to be nonrandom more than 20 years ago (Bretscher, 1972), only recently have researchers determined that lipid sidedness plays a role in important cellular processes (see also Chapter 1). Many of these phenomena are associated directly with the "atypical" display of PS on the outer leaflet of the cell; for example various cell–cell interactions (Schlegel *et al.*, 1985; Schroit *et al.*, 1985), cell activation and hemostasis (Bevers *et al.*, 1983; Rosing *et al.*, 1985; Sims *et al.*, 1989), cell aging (Shukla and Hanahan, 1982; Herrmann and Devaux, 1990), fusion (Farooqui *et al.*, 1987; Schewe *et al.*, 1992; Song *et al.*, 1992), and apoptosis (Fadok *et al.*, 1992a,b) all require PS in the outer leaflet.

With the exception of diacylglycerol (Ganong and Bell, 1984) and cera-mide (Lipsky and Pagano, 1985), most lipids cannot move across the lipid bilayer of artificially generated vesicles (Pagano and Sleight, 1985), probably because of the charge and amphipathic nature of the lipids, properties that make passing through the hydrophobic membrane core energetically unfavorable. Because of this phenomenon, lipid asymmetry was, for the most part, believed to be a consequence of processes other than transbilayer lipid movement, particularly reorganization of the inner surface of the membranes by the action of phospholipases and acylases. Although these processes contribute to membrane lipid asymmetry, the discovery of an aminophospholipid-specific translocase (Seigneuret and Devaux, 1984) indicated that the nonrandom distribution of PS and PE is controlled by an active energy-requiring process. The existence of a "flippase" that shuttles PS and PE across the bilayer membrane suggested that a specific transmembrane distribution of these lipids is of major impor-tance in cell physiology.

Considerable progress has been made in understanding the molecular requirements of aminophospholipid movement in the human red blood cell. Many experiments have shown that the transbilayer movement of aminophospholipids is ATP and temperature dependent (Seigneuret and Devaux, 1984), stereospecific (Martin and Pagano, 1987), and sensitive to oxidation of membrane sulfhydryls (Daleke and Huestis, 1985; Connor and Schroit, 1988,1990). Although several candidate proteins for the transport function have been suggested (Morrot *et al.*, 1990; Schroit *et al.*, 1990), conclusive evidence of its identity awaits functional reconstitution in an artificially generated system. In this chapter, data that support the exis-tence of an aminophospholipid-specific translocase in the plasma mem-brane of red blood cells is reviewed. This information is summarized in the context of the mechanism or mechanisms likely to be responsible for generating and maintaining membrane phospholipid asymmetry.

II. PHOSPHOLIPID ASYMMETRY IN THE RED BLOOD CELL MEMBRANE

Extensive evidence supports the general concept that the two halves of the red cell membrane bilayer leaflet differ in phospholipid composition (see Chapter 1). This idea first was suggested by the pioneering "steady state" studies of the 1970s and, more recently, by studies that monitored the transbilayer movement of natural phospholipids and their analogs.

A. Distribution of Endogenous Lipids

Several techniques have been used to determine the transbilayer distribution of phospholipids in red blood cells (Etemadi, 1980), the most popular of which was based on the use of phospholipases to hydrolyze only lipids residing in the outer leaflet (Op den Kamp, 1979). After analysis by thin-layer chromatography (TLC), the fraction of hydrolyzed lipid was determined and was used to estimate the fraction of total lipid that was accessible to the enzymes. Other methods were based on chemical labeling of aminophospholipids with "nonpermeable" probes such as formylmethionyl (sulfone) methyl phosphate (Bretscher, 1972), trinitrobezenesulfonic acid (Gordesky et al., 1975), and fluorescamine (Rawyler et al., 1984).

Although these techniques directly modified the cell membrane and therefore were prone to generating artifacts, they clearly demonstrated that the phospholipid distribution of the red blood cell membrane is asymmetric. The studies showed that the outer leaflet is rich in the choline-containing phospholipids whereas the aminophospholipids preferentially occupy the inner leaflet (Fig. 1). Specifically, in human red cells, about 80% of the sphingomyelin (SM) and 75% of the phosphatidylcholine (PC) are located in the outer leaflet (Verkleij et al., 1973); 80% of the PE and most, if not all, of the PS is in the inner leaflet (Verkleij et al., 1973; Gordesky et al., 1975; Zwaal et al., 1975; Marinetti and Crain, 1978).

The different transbilayer distributions of the various phospholipid classes in cell membranes are likely to be regulated by different, yet interdependent, processes. For example, the identity of the molecular species of PC and PE that reside in the inner and outer leaflet lipid pools (Renooij et al., 1974; Marinetti and Crain, 1978), combined with the finding that lipid remodeling (acylation and deacylation) occurs at the cytoplasmic

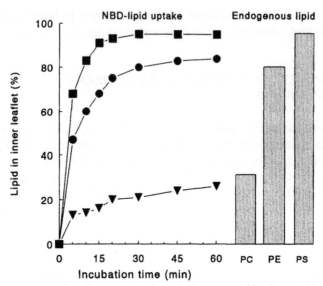

FIGURE 1 Aminophospholipid asymmetry in human red blood cells. (*Left*) Protein-mediated outside-to-inside movement (flip) of NBD-labeled phosphatidylserine (PS; ■), phosphatidylethanolamine (PE; ●), and phosphatidylcholine (PC; ▼) (Connor *et al.*, 1992a). (*Right*) Transmembrane distribution of endogenous phospholipids (Verkleij *et al.*, 1973).

surface of the membrane (Renooij *et al.*, 1974,1976; Marinetti and Cattieu, 1982), indicated that these lipids can undergo complete, albeit slow (van Meer and Op den Kamp, 1982; Middelkoop *et al.*, 1986), transbilayer equilibration. This phenomenon was observed for PC in rat (Renooij *et al.*, 1976; Bloj and Zilversmit, 1976; Kramer and Branton, 1979) and in human (van Meer and Op den Kamp, 1982; Middelkoop *et al.*, 1986) erythrocytes.

Other processes, however, also must be involved in generating and maintaining phospholipid asymmetry because passive diffusion processes cannot generate disproportionate phospholipid distributions. Although the unequal distribution of the choline-containing phospholipids might be the result of a compensatory mechanism for the strongly asymmetric localization of the aminophospholipids in the inner leaflet, data suggest that this is not the case (Connor *et al.*, 1992a). Aminophospholipid asymmetry is controlled, on the other hand, by specific energy-requiring mechanisms that involve an aminophospholipid-specific active transport system and possibly interactions between lipids in the inner leaflet and cytoskeletal proteins.

B. Transbilayer Movement of Aminophospholipid Analogs

1. Outside-to-Inside Movement (Flip)

Protein-mediated transport of aminophospholipids first was described in erythrocytes by Seigneuret and Devaux (1984). When spin-labeled PS and PE were added to red blood cells, a temperature- and time-dependent decrease was observed in the fraction of spin-labeled lipid that could be reduced by ascorbate. Since ascorbate cannot penetrate the cells' bilayer membrane, these findings indicated that the analogs were moved from their initial insertion site in the outer leaflet to the inner monolayer. Outside-to-inside movement of spin-labeled analogs was found to be slow for PC, fast for PE, and even faster for PS, which was transported with a half-time of about 5 min.

Similar observations were made when the movement of fluorescently labeled (Connor and Schroit, 1987) and isotopically labeled (Tilley *et al.*, 1986; Schroit *et al.*, 1987; Connor *et al.*, 1992a) lipids was monitored. With these probes, lipid movement can be monitored by removing the fraction of residual lipid not transported to the inner leaflet by the so-called "back-exchange" procedure (Struck and Pagano, 1980) or, similar to detecting spin-labeled lipids with ascorbate, by destroying the fraction of fluorescent lipid (in the outer leaflet) that is accessible to a reductant in the buffer (Mcintyre and Sleight, 1991).

Because PS does not translocate between leaflets of artificially generated vesicles (Tanaka and Schroit, 1986), the observation of its rapid and temperature-dependent transbilayer movement indicated that aminophospholipid transport is mediated by a facilitative transport mechanism that probably involves lipid-specific transport proteins (Table I). PS transport was shown to require ATP, since transport did not occur in ATP-depleted cells, was inhibited by vanadate, and could be reconstituted in erythrocyte ghosts resealed in the presence of Mg^{2+}-ATP but not ADP (Seigneuret and Devaux, 1984). Additional evidence that the transport of aminophospholipids depends on lipid-specific protein transporters came from experiments that showed transport to be inhibited by agents that react with membrane cysteines (Seigneuret and Devaux, 1984; Daleke and Huestis, 1985; Tilley *et al.*, 1986; Connor and Schroit, 1988,1990), as well as with arginines (Daleke, 1990) and histidines (Connor *et al.*, 1992a). Although these observations implicated the involvement of a specific protein, conclusive evidence of the specificity of the transporters was shown by the inability to transport D isomers of PS and PE (Martin and Pagano, 1987). Substrate specificity is dictated also, however, by the glyceride backbone and esterification of the *sn*-2 position. Substitution of glyceride with ceramide abolished translocation even when the polar head group contained

TABLE I

Inhibitors of Protein-Mediated Phosphatidylserine Transport in Red Blood Cells

	PS transport	
	Outside-to-inside (FLIP)	Inside-to-outside (FLOP)
Sulfhydryl reagents		
N-Ethylmaleimide	Zachowski et al. (1986)	
Iodoacetamide	Connor and Schroit (1988)	
Pyridyldithioethylamine	Connor and Schroit (1988)	Connor et al. (1992a)
Dithiopyridine	Connor and Schroit (1988)	
Diamide	Daleke and Huestis (1985)	
Ellman's reagent	Connor and Schroit (1990)	
Histidine reagents		
Phenacylbromide	Connor et al. (1992a)	Connor et al. (1992a)
Arginine reagents		
Phenylglyoxal	Daleke (1990)	
ATP depletion/competition		
Metabolic depletion	Seigneuret and Devaux (1984)	
NaN$_3$ and deoxyglucose	Connor and Schroit (1990)	Connor et al. (1992a)
Iodoacetate	Tilley et al. (1986)	
Iodoacetamide/inosine	Daleke and Huestis (1985)	
Vanadate	Seigneuret and Devaux (1984)	
Cations		
Ca^{2+}	Zachowski et al. (1986)	
Mg^{2+} depletion	Daleke and Huestis (1985)	Bitbol and Devaux (1988)

the phosphoserine moiety (Morrot et al., 1989); also, lysophosphatidyl-serine translocated across the bilayer membrane at very slow rates (Berg-mann et al., 1984). Although absolute rates of transport are influenced by the length of the fatty acid side chains and the presence of a particular reporter group (e.g., fluorescent, iodinated, or spin-labeled moieties), translocation rates are determined primarily by the chemical nature of the polar head group, as was shown in a series of elegant experiments in which progressive methylation of transportable PE to its monomethyl, dimethyl, and finally trimethyl (phosphatidylcholine) adducts resulted in a concomitant decrease of transport (Morrot et al., 1989).

Note that the conclusion of the existence of an ATP-dependent amino-phospholipid translocase in cell membranes includes the assumption that exogenously introduced phospholipid analogs are reliable reporters of endogenous phospholipid behavior. Although various lipid analogs clearly differ in their kinetics of transport, conclusions from these studies are supported by data showing similar rates of ATP-dependent transport of exogenously supplied "natural" lipids. The transport rates obtained by Daleke and Huestis (1985), who monitored cellular shape changes that depend on the leaflet localization of exogenously added (unlabeled) phospholipids, were similar to those obtained for various analogs. Similar data were obtained in studies of cells enriched with long-chain radiolabeled phospholipids monitored by phospholipase digestion (Tilley et al., 1986). Not only do these exogenously supplied "natural" lipids behave like the spin-labeled and fluorescent analogs, but their transport closely parallels the transport rate of endogenous PS exposed in the outer leaflet by treatment with Ca^{2+} ionophore (Comfurius et al., 1990) and monitored, after addition of EGTA, by the noninvasive prothrombinase complex assay (Bevers et al., 1982; Connor et al., 1989). However, some exceptions to these observations exist. NBD- and ^{125}I-Bolton–Hunter-labeled 1-oleoyl, 2-aminocaproyl PE (C_6-PE), for example, are not transported in human red blood cells (Colleau et al., 1991; Connor et al., 1992a). On the other hand, when the fluorescent and iodinated moieties are attached to a longer chain, aminododecanoyl lipid (C_{12}-PE), all analogs are transported at similar rates (Connor et al., 1992a). The reason for the inability of short-chain PE analogs to be transported is not known, but might be the distortion of the labeled acyl chain by the reporter group (Chattopadhyay and London, 1987).

2. Inside-to-Outside Movement (Flop)

Although protein-mediated inward movement of exogenously inserted aminophospholipid analogs proceeds to its appropriate equilibrium distribution, how other lipids, in particular the choline-containing phospholipids, are maintained preferentially in the outer leaflet is unclear. Although equilibrium distributions can, theoretically, be satisfied by immobilizing a fraction of the phospholipid in a particular leaflet, such a mechanism cannot explain how approximately 20% of exogenously supplied PC (Tilley et al., 1986) reaches the inner leaflet. Red blood cells may possess an active mechanism for moving lipids from the inner to the outer leaflet. If so, steady state equilibrium distributions may be satisfied by different rates of inward and outward movement. Indeed, Herrmann and Müller (1986) proposed that, for a simple two-compartment model, differences in rates of outside-to-inside movement and inside-to-outside movement are sufficient to sustain phospholipid asymmetry without requiring other interactive processes such as cytoskeletal involvement. This proposal suggests that the steady state equilibrium distribution of all membrane phospholipids could be determined exclusively by inward and outward translocation rates, irrespective of whether an active or passive process is responsible for movement in a particular direction.

Several studies have addressed the issue of whether active protein-mediated outward transport of lipid occurs. Using spin-labeled analogs, Bitbol and Devaux (1988) showed that the outward movement of aminophospholipids is faster than that of PC and requires Mg^{2+}. Although these data suggest some degree of lipid specificity, several experiments show that cells containing NBD-labeled lipids exclusively in the inner leaflet transport NBD-labeled PS, PE, and PC to the outer leaflet at similar rates, and only to the point at which normal membrane lipid asymmetry of the NBD-labeled probes is established (Connor et al., 1992a). When the experiment was initiated with 100% of the labeled probe in the inner leaflet, approximately 70% of the PC, 40% of the PE, and 15% of the PS redistributed to the outer leaflet. When the cells were incubated in the presence of bovine serum albumin (a lipid "acceptor"), the cells were depleted of the analogs with a half-time of about 90 min in a manner that was independent of the lipid species (Fig. 2).

Similar to the inward movement of aminophospholipids, the outward movement of PS, PE, and PC also seems to be energy dependent (Bitbol and Devaux, 1988; Connor et al., 1990,1992a). Both processes require ATP and are inhibited by low temperature, vanadate, sulfhydryl oxidants, and phenacylbromide (Table I).

III. CONTROL OF PHOSPHOLIPID ASYMMETRY IN RED BLOOD CELLS

The membrane bilayer with its embedded transmembrane proteins and the underlying network of peripheral membrane proteins, the cytoskeleton, are the two main structural components of the red blood cell. Although clearly an aminophospholipid translocase is required for transmembrane movement, whether the translocase is required to maintain phospholipid asymmetry once it has been established is unknown. Further, whether or not cytoskeletal proteins are required to establish membrane phospholipid asymmetry or to maintain that asymmetry once it has been established is still questionable (however, see Chapter 1).

A. Membrane Skeleton and Phospholipid Asymmetry

A role for the membrane skeleton in maintenance of membrane lipid asymmetry first was proposed by Haest and colleagues (1978; Haest, 1982), who observed that diamide treatment of red blood cells resulted not only in high molecular weight spectrin cross-links but also in a loss of membrane phospholipid asymmetry. This observation led to the conclu-

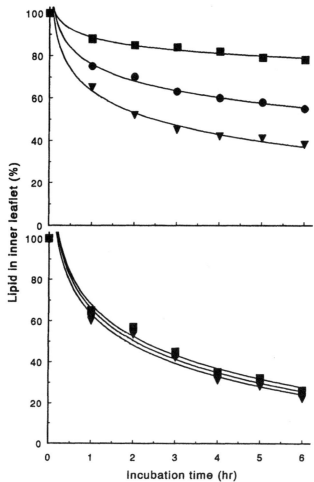

FIGURE 2 Protein-mediated inside-to-outside movement (flop) of NBD-labeled lipids. (*Top*) Spontaneous movement in the absence of a lipid acceptor. (*Bottom*) Lipid movement in the presence of bovine serum albumin. PS, Phosphatidylserine (■); PE, phosphatidylethanolamine (●); PC, phosphatidylcholine (▼).

sion that spectrin contributes to the maintenance of the inside orientation of aminophospholipids by a mechanism that involves selective cytoskeletal interactions that do not occur with the choline-containing phospholipids.

Other studies with liposomes and monolayer lipid films also supported the notion of cytoskeletal involvement in phospholipid asymmetry. In addition to spectrin (Haest *et al.*, 1977; Mombers *et al.*, 1979,1980; Bonnet and Begard, 1984; Cohen *et al.*, 1986; Sikorski *et al.*, 1987) for example, band 4.1 (Sato and Ohnishi, 1983; Cohen *et al.*, 1988; Shiffer *et al.*, 1988; Rybicki *et al.*, 1988) also interacted with PS in vesicles as well as in erythrocyte ghosts. Collectively, these data suggest that both spectrin and band 4.1 contribute to the maintenance of phospholipid asymmetry by their capacity to "fix" PS to the inner leaflet. These data led Williamson *et al.* (1987) to extend the fast flip/slow flop bidirectional model of Herrmann and Müller (1986; see Section II,B,2); they proposed that the translocase operates as an ATP-dependent bidirectional catalyst of equal transport efficiency in both directions. In contrast to Herrmann and Müller's (1986) cytoskeleton-independent theory, this model predicts that differences in flip and flop are caused by interaction of the aminophospholipids with the cytoskeleton, a mechanism that essentially shifts their equilibrium distribution toward the inner leaflet.

Although these studies suggested a major involvement of the membrane cytoskeleton in the maintenance of phospholipid asymmetry, other observations indicated that cytoskeletal proteins are not involved in this process. Raval and Allan (1984) observed that the distribution of phospholipids in spectrin-free microvesicles is unaltered from that in the remnant cells. In addition, heat-treated spectrin-denatured red cells (Gudi *et al.*, 1990) and spectrin-poor vesicles derived from these cells (Calvez *et al.*, 1988) also were found to maintain normal lipid asymmetry.

In summary, whether the cytoskeleton plays a role in maintaining membrane phospholipid asymmetry remains unclear. Based on data obtained with liposomes, however, one might assume that—although the cytoskeleton provides structural strength, resistance to deformation, and overall membrane stability—it has little to do with the sidedness of an individual lipid species.

B. Role of Translocase in Maintaining Phospholipid Asymmetry

The major function of the aminophospholipid translocase has been suggested to be maintaining phospholipid asymmetry during the life-span of the cell (Devaux, 1988). This concept is based on the notion that stable lipid asymmetry is incompatible with the spontaneous transverse diffusion rates of phospholipids in cells and in artificial membrane systems. In human red blood cells, for example, PC equilibrates between the two leaflets with half-times of 3–26 hr (van Meer and Op den Kamp, 1982; Middelkoop *et al.*, 1986), much shorter than the 120-day life-span of the cell. However, the transbilayer distribution of this lipid is preserved at roughly 75% outside and 25% inside (Gordesky and Marinetti, 1973; Verkleij *et al.*, 1973; Zwaal *et al.*, 1973,1975; Gordesky *et al.*, 1975).

Findings of an unequal PC distribution suggested that, like PS and PE asymmetry, some mechanism preserves PC asymmetry throughout the life-span of the cell. Perhaps the unequal distribution of the choline-containing phospholipids is a consequence of the preference of the amino-phospholipids for the inner leaflet, by a mechanism that is dependent on or independent of the mechanism responsible for aminophospholipid movement. This explanation does not suggest, however, that, once established, active lipid transport is required to maintain the asymmetry. In fact, ample evidence suggests that an active translocase is not required to maintain asymmetry (Haest, 1982; Tilley *et al.*, 1986; Comfurius *et al.*, 1990; Connor and Schroit, 1990; Henseleit *et al.*, 1990). For example, spectrin-free vesicles released from reticulocytes during *in vitro* maturation had the same asymmetric phospholipid distribution as the parent cell but no translocase activity (Vidal *et al.*, 1989). This observation was supported further by studies showing that inhibition of transporter-dependent inward movement also inhibits outward movement, a condition that presumably would result in a static membrane (Connor *et al.*, 1992a; see Table I). Although these observations suggest that the aminophospho-lipid translocase is not needed for maintaining asymmetry, these data do not exclude the participation of the cytoskeleton in this process (Section III,A). Indeed, Sugihara *et al.* (1991) showed that lipid asymmetry is maintained during elliptical deformation of red blood cells even when the aminophospholipid translocase is inhibited. Contrary to the evidence discussed in Section III,A, these data raise the possibility that maintenance of lipid asymmetry depends totally on the stabilizing effects of the membrane skeleton.

Since phospholipid transport is not apparent unless the lipids occupy the wrong half of the membrane bilayer, the actual role of the translocase may be to regenerate phospholipid asymmetry after it is disarranged. Although the translocase would correct for any backflow or leak of PS to the outer monolayer, its major function could be to restore phospholipid asymmetry when major structural rearrangements of the membrane occur. Work with activated platelets and Ca^{2+} ionophore-treated red cells showed that scrambling of membrane phospholipids during fusion events can be corrected only by an active aminophospholipid translocase (Comfurius *et al.*, 1990; Tilly *et al.*, 1990). This result suggests that the aminophospho-lipid translocase plays a principal role in preserving membrane asymmetry in cells that undergo endo- or exocytosis, cell division, and other membrane fusion events. This putative function predicts, therefore, a corrective role for the translocase in reestablishing membrane asymmetry following its deterioration.

In contrast, Devaux (1991) proposed that the aminophospholipid translocase triggers endocytosis by continuously forming membrane invaginations by accumulating aminophospholipids in the inner leaflet. This hypothesis predicts, therefore, that the ATP-dependent pump would be driven continuously and that endocytosis would be blocked by transporter-specific inhibitors. Although transbilayer movement of PS occurs when endocytosis is inhibited (Martin and Pagano, 1987), whether endocytosis depends on lipid transport is not known.

IV. CHARACTERIZATION OF THE AMINOPHOSPHOLIPID TRANSLOCASE

Without sufficient data to identify unequivocally the protein responsible for aminophospholipid transmembrane movement, designation of a specific transport protein and its mechanism of movement is still premature. Nevertheless, a significant amount of data is compatible with an amino-phospholipid transport role for two different red blood cell proteins: the 120-kDa Mg^{2+}-ATPase proposed by Zachowski and Devaux (Morrot *et al.*, 1990) and the 32-kDa Rh-expressing band-7 proteins suggested by Connor and Schroit (Schroit *et al.*, 1990; Connor *et al.*, 1992a).

In the absence of structural data, models for the mechanisms of these putative transport proteins must be based on several assumptions.

1. Specificity of aminophospholipid recognition is determined by a substrate recognition site on a specific membrane-associated transport protein.
2. Initial interaction between the substrate and the transporter occurs at a recognition site that is at or very close to the membrane–water interface.
3. The transmembrane transport protein forms a canalicular structure that "protects" the polar head group of the lipid from the hydrophobic membrane as the lipid moves through a hydrophilic channel.
4. Aminophospholipid movement requires hydrolyzable ATP.

Although the 120-kDa Mg^{2+}-ATPase and the 32-kDa band-7 proteins are known to have some of these characteristics, neither at the present time completely fulfills all the assumed requirements of a specific aminophospholipid translocase.

A. 120-kDa Mg^{2+}-ATPase

Based on corollary evidence for the biochemical requirements of amino-phospholipid transport and the activity of an ATPase from red blood cells,

Devaux and colleagues suggested that a 120-kDa Mg^{2+}-ATPase functions as the aminophospholipid transporter (Zachowski *et al.*, 1989; Morrot *et al.*, 1990). This hypothesis is based on observations that aminophospholipid transport, as well as a partially purified Mg^{2+}-ATPase from erythrocytes, is stimulated by PS, requires hydrolyzable ATP, and is inhibited by vanadate, fluoride, N-ethylmaleimide, and Ca^{2+} ions. That this enzyme may be the lipid transporter is supported by data showing that this Mg^{2+}-ATPase has no known function and, in contrast to the Ca^{2+}- and Na^+/K^+-ATPases, is not an ion pump (Forgac and Cantley, 1984).

In addition to triggering the formation of membrane invaginations (Devaux, 1991), the aminophospholipid translocase was proposed to be responsible for maintaining cell shape as well (Devaux, 1992; Farge and Devaux, 1992). Indeed, red blood cells require ATP for their normal discoid shape and undergo dramatic shape changes (Nakao *et al.*, 1960; Patel and Fairbanks, 1986; Xu *et al.*, 1991) and membrane fluctuations (Levine and Korenstein, 1991) when their ATP levels are manipulated and exogenous lipids are inserted (Daleke and Huestis, 1985,1989). These ATP-dependent shape changes depend on a "shape-change Mg^{2+}-ATPase" (Patel and Fairbanks, 1986) which, like the aminophospholipid transporter, is vanadate sensitive and is expressed in the presence of ouabain and EGTA. Since ATP-dependent lipid transport generates ATP-dependent echinocyte-to-discocyte transitions (Daleke and Huestis, 1985,1989), both of these activities may be associated with the same protein. However, although the properties of the shape-change Mg^{2+}-ATPase are similar to the properties proposed for the aminophospholipid translocase, this protein or group of proteins is not necessarily the translocase. The data are, in fact, consistent with the characteristics of certain protein kinases, enzymes that exhibit regulatory properties similar to those of the Mg^{2+}-ATPase (O'Brian and Ward, 1990).

B. 32-kDa Band-7 Proteins

The first indications that band-7 proteins (red blood cell polypeptides that migrate to the 30-kDa to 32-kDa region in SDS–polyacrylamide gels) might be involved in the transbilayer movement of aminophospholipids was obtained from experiments with photoactivatable ^{125}I-labeled N_3-PS and -PC (Schroit *et al.*, 1987). Photolysis of red blood cells incubated with these analogs at 37°C cross-linked the probes to membrane proteins. Analysis of SDS–PAGE gels by autoradiography revealed that, whereas the PC analog was distributed randomly among all membrane proteins, the

PS analog preferentially labeled proteins in the band-7 region. Additional evidence for the participation of these proteins in lipid transport is from similar studies in which lipid movement was inhibited (Connor and Schroit, 1991). Instead of preferentially labeling 32-kDa proteins, the photoactivatable PS became distributed randomly in a manner indistinguishable from that of the nontransportable PC analog.

Additional support for the involvement of the 32-kDa proteins was obtained by labeling red blood cells with the thiol disulfide exchange reagent ^{125}I-labeled pyridyldithioethylamine (PDA), a potent inhibitor of aminophospholipid transport (Connor and Schroit, 1988). Similar to the results obtained with photoactivatable PS, analysis of SDS-PAGE gels by autoradiography revealed that the probe preferentially labeled 32-kDa proteins and specifically and reversibly inhibited PS transport. The proteins labeled with photoactivatable PS and with ^{125}I-labeled PDA later were shown to be the same (Connor and Schroit, 1991).

Although these findings implicated the involvement of 32-kDa band-7 proteins in the inward movement of PS, these proteins also were found to be involved in the general maintenance of membrane phospholipid asymmetry (Connor et al., 1992a). This conclusion was reached based on data showing that the distribution of appropriately labeled PC, PE, and PS analogs that were localized exclusively in the inner leaflet were capable (1) of selectively labeling 32-kDa proteins on photolysis and (2) of outward movement that reestablished the normal membrane asymmetry of each lipid species (see Section II,B,2).

Findings that the same protein (or proteins) is involved in the inward movement of aminophospholipids as in the outward movement of the aminophospholipids and PC suggest that the cell slowly but continuously moves lipids to the outer leaflet and simultaneously keeps the aminophospholipids in the inner leaflet by rapidly transporting them back to the inside. Since the movement of a particular lipid species between bilayer leaflets is independent of the movement of another lipid species in the opposite direction (Connor et al., 1992a), the net effect of different rates of inward and outward movement is, in the case of PE and PS, dominant localization in the inner leaflet. Since PC, in contrast, is not transported actively to the inner leaflet, its equilibrium distribution favors the outside. Building a general model that is compatible with aminophospholipid-specific inward rates of movement and lipid species-independent rates of outward movement is difficult, but these results are compatible with the bidirectional transport model of Herrmann and Müller (1986; see Section II,B,2), which predicts that the translocase is a bidirectional pump with different rates of movement in each direction.

C. Is the "Transporter" More Than One Protein?

Although the band-7 polypeptides and the 120-kDa Mg^{2+}-ATPase are distinct proteins, the assignment of "transport activity" to one does not necessarily exclude the possible participation of the other. As discussed earlier, aminophospholipid transport not only requires hydrolyzable Mg^{2+}-ATP but also is likely to need a structure that forms a protective environment for the polar head group of the lipid to traverse the hydrophobic bilayer membrane. Some of the 32-kDa band-7 proteins have been shown to belong to a set of closely related isoforms that are associated with the Rh blood group system. Rh polypeptides may, therefore, be involved in the maintenance of membrane lipid asymmetry, especially since the Rh protein is a multispanning membrane polypeptide (Avent *et al.*, 1990; Cherif-Zahar *et al.*, 1990), a property common to other membrane proteins associated with transport and channel functions. Since Rh polypeptides do not seem to belong to the ABC (ATP-binding cassette) superfamily of active transporters, they may provide the protective pathway for transmembrane lipid movement while the Mg^{2+}-ATPase uses ATP and provides the driving force that enables transport to proceed. Thus, the functional "transporter" could be a complex of more than one protein. Indeed, the transport of fluorescently labeled PS analogs was shown to require the participation of a 32-kDa polypeptide and a presumably distinct protein located at the endofacial membrane surface (Connor and Schroit, 1990).

V. INVOLVEMENT OF Rh PROTEINS IN AMINOPHOSPHOLIPID MOVEMENT

A. Membrane Organization and Topology of Rh Polypeptides

The Rh blood system contains three immunologically defined primary epitopes—the D, c, and E antigens—that probably are formed by alternative splicing of a primary transcript (Colin *et al.*, 1991; Le van Kim *et al.*, 1992a) from two genes with significant homology (Avent *et al.*, 1990; Cherif-Zahar *et al.*, 1990; Le van Kim *et al.*, 1992b). These epitopes are carried by at least three distinct integral membrane proteins, approximately 32 kDa in size (Gahmberg, 1982,1983; Moore *et al.*, 1982; Blanchard *et al.*, 1988; Bloy *et al.*, 1988; Hughes-Jones *et al.*, 1988; Saboori *et al.*, 1988), that share well-conserved amino and carboxy termini and include 5 or 6 cysteine residues, some of which are palmitoylated (de Vetten and Agre, 1988; Hartel-Schenk and Agre, 1992). Most of the cysteines appear in the motif Cys–Leu–Pro, whereas one cysteine critical to the antigenicity of the protein (residue 285; Avent *et al.,* 1990) is present in the sequence motif Cys–His–Leu–Ile–Pro and is predicted to lie at the exofacial surface of the cell. Hydropathy analysis suggests that these polypeptides are hydrophobic, have 12 or 13 transmembrane regions, and have most of the polypeptide localized in the hydrophobic core of the membrane. Whether the carboxy (Krahmer and Prohaska, 1987; Bloy *et al.,* 1990; Suyama and Goldstein, 1992) or amino (Avent *et al.,* 1992) terminus is localized at the external surface is a topic of some controversy.

Several lines of evidence support the hypothesis that Rh proteins are assembled in the membrane in the form of a relatively large molecular complex (Hartel-Schenk and Agre, 1992) that contains, in addition to Rh polypeptides, Rh-related and -unrelated glycoproteins. For example, Rh-related glycoproteins that migrate as broad bands on SDS–PAGE have been shown to coprecipitate with polyclonal and monoclonal antibodies to Rh(D), Rh(E), and Rh(c) antigens (Moore and Green, 1987; Avent *et al.,* 1988a,b). Moreover, Ridgwell *et al.* (1992) isolated cDNA clones for a 50-kDa glycoprotein that is probably a member of the Rh complex. The full coding sequence of one of these clones predicts a 409-amino-acid *N*-glycosylated membrane protein with a sequence similar to that of the Rh polypeptides and hydropathy plots predicting, like the Rh polypeptides, up to 12 transmembrane domains. These data suggest that Rh and its related polypeptides are subunits of a large oligomeric complex. Indeed, hydrodynamic analysis of Triton X-100 solubilized membranes indicates that Rh polypeptides behave in a manner consistent with a protein or protein complex of 170 kDa (Hartel-Schenk and Agre, 1992). Such an oligomeric complex is compatible with a structure likely to have a transport or a channel function, and fits the expected properties of an aminophospholipid translocase.

B. Rh Proteins and Lipid Transport

Independent observations on the similarities between the biochemical properties of Rh polypeptides and the aminophospholipid transporter have led to the suggestion that Rh proteins might be involved in maintaining lipid asymmetry (de Vetten and Agre, 1988; Connor and Schroit, 1989). Both Rh polypeptides and the putative band-7 transporter are present in nonhuman erythrocytes, have similar molecular weights and sensitivity to sulfhydryl oxidants, are associated with membrane lipids, and are apparently nonglycosylated (Table II).

TABLE II

Common Properties of the Aminophospholipid Transporter and Rh Polypeptides

Characteristic	Transporter	References	Rh polypeptide	References
Apparent molecular mass on SDS gels	31–32 kDa	Schroit et al. (1987); Connor and Schroit (1988)	30–32 kDa	Gahmberg (1982); Moore et al. (1982)
Cysteine dependency	Transport abolished by all membrane-permeable sulfhydryl-reactive reagents[a]		Cysteine residue critical for Rh(D) antigenicity Antigenicity abolished by some sulfhydryl-reactive reagents	Ridgwell et al. (1983) Green (1967,1983); Abbott and Schachter (1976); A. J. Schroit (unpublished observations)
Histidine dependency	Transport abolished by phenacylbromide	Connor et al. (1992a)	Antigenicity abolished or decreased by phenacylbromide and diethylpyrocarbonate	Victoria et al. (1986)
Integral membrane protein	Transport activity resistant to proteases	Connor and Schroit (1989)	Rh polypeptides buried deeply in cell membrane	Gahmberg (1983); Bloy et al. (1990)
Presence in nonhuman erythrocytes	Transport activity of nonhuman red cells indistinguishable from that of human red cells	Connor and Schroit (1989)	Rh-related proteins present in nonhuman erythrocytes	Saboori et al. (1989)
Association with membrane lipids	Photolysis of ^{125}I-N$_3$-labeled lipid analogs labels 32-kDa proteins	Schroit et al. (1987); Connor and Schroit (1991); Connor et al. (1992a)	Rh antigenicity requires phospholipid Phospholipid has protective effect on Rh	Green (1972,1982); Green et al. (1984) Suyama and Goldstein (1990)

[a] See Table I.

A series of experiments has provided more direct evidence for the involvement of Rh polypeptides in membrane lipid asymmetry (Schroit *et al.*, 1990). After 32-kDa band-7 red blood cell proteins were labeled with the transport inhibitor ^{125}I-labeled PDA, or with the transportable substrate ^{125}I-labeled N_3-PS, the ability of monoclonal Rh antibodies to immunoprecipitate the labeled proteins was determined. Autoradiography of SDS–PAGE gels revealed that the immunoprecipitated Rh polypeptides were labeled with the iodinated probes, indicating that the labeled proteins were Rh polypeptides. Precipitation of the probe-labeled Rh protein was specific because immunoprecipitation occurred only when monoclonal antibody was incubated with cells of the appropriate Rh phenotype. Thus, anti-c, anti-D, and anti-E precipitate labeled 32-kDa polypeptides from cDE/cDE cells, but only monoclonal anti-c precipitates polypeptides from cde/cde cells because they do not express the D or E alleles. Similarly, only anti-D precipitates polypeptides from D− − cells because this phenotype lacks both the C/c and the E/e epitope.

Although photolabeling of Rh polypeptides with ^{125}I-labeled N_3-PS implicates involvement of the Rh complex in the inward movement of aminophospholipids, other studies suggest that these proteins also might be involved in the outward movement of lipid (Connor *et al.*, 1992a). Results from experiments using lipids occupy the inner leaflet (see Section II,B,2), photolysis results in labeling of 32-kDa band-7 proteins in a lipid species-independent manner. Immunoprecipitation with Rh antibodies reveals that the labeled band-7 proteins were Rh polypeptides. These results raise the possibility that the Rh blood group system might be involved in the maintenance of general phospholipid asymmetry.

VI. IS THE PS TRANSPORTER THE RH POLYPEPTIDE?

Conclusive evidence in support of the concept that Rh polypeptides (or the Rh complex) function as the aminophospholipid transporter is still lacking. Peptide mapping studies indicate, however, that protein regions that label with ^{125}I-labeled PDA (Connor *et al.*, 1992b) and with ^{125}I-labeled N_3-PS (J. Connor and A. J. Schroit, unpublished observations) share a significant degree of homology that is independent of the Rh phenotype. This observation is consistent with the existence of highly conserved regions in Rh polypeptides. Indeed, Rh(D), Rh(c), and Rh(E) all share a significant degree of homology in their amino acid sequences (Avent *et al.*, 1990; Cherif-Zahar *et al.*, 1990; Le van Kim *et al.*, 1992b). Additional support for the existence of homologous phenotype-independent Rh regions is from findings that polyclonal antiserum prepared against antibody-purified Rh(D) react on immunoblots not only with Rh(D) but also with Rh(c) and Rh(E) (Suyama and Goldstein, 1988). These data indicate that Rh polypeptides and proteins that label with ^{125}I-labeled PDA and ^{125}I-labeled N$_3$-PS possess highly conserved domains. Assuming that Rh and the transporter are the same, they may represent the region involved in aminophospholipid transport.

Since the ability of red blood cells to transport PS is independent of the Rh phenotype (Schroit et al., 1990), the epitopes responsible for Rh antigenicity are unlikely to play a role in lipid transport. This hypothesis is supported by results that show that transport is not inhibited by Rh antibodies (Schroit et al., 1990; Smith and Daleke, 1990) and that Rh$_{null}$ cells, cells that do not possess known Rh epitopes, transport PS normally (Schroit et al., 1990). Although the absence of Rh polypeptides in Rh$_{null}$ cells is, by definition, interpreted to mean that the aminophospholipid transporter cannot be Rh (Smith and Daleke, 1990), Rh$_{null}$ cells are defined exclusively by their inability to react with Rh antibodies. Because these cells contain ^{125}I-labeled PDA and ^{125}I-labeled N$_3$-PS labeled 32-kDa proteins that are indistinguishable from labeled proteins isolated from cells bearing Rh antigen (Schroit et al., 1990), Rh$_{null}$ cells may possess an antigenically silent "Rh-like" phenotype that cannot be detected by immunological techniques.

Comparative analysis of isotopically labeled 32-kDa polypeptides from Rh$_{null}$ cells and immunoprecipitated Rh(D) and Rh(c) does, in fact, show that Rh$_{null}$ cells possess a protein with structural properties very similar to those of antibody-binding Rh polypeptides. First, the M_r 31,900 Rh(D) and M_r 33,100 Rh(c)/Rh(E) polypeptides have been suggested to be replaced in Rh$_{null}$ by an M_r 33,800 Rh$_{null}$ polypeptide (Moore and Green, 1987). Second, size comparisons of labeled proteins in Triton X-100 solubilized membranes (Hartel-Schenk and Agre, 1992) from Rh-bearing and Rh$_{null}$ cells showed similar sedimentation coefficients and Stoke's radii (Connor et al., 1992b). Third, high performance liquid chromatography (HPLC) analysis and iodopeptide maps of ^{125}I-labeled PDA-labeled chymotryptic fragments from immunoprecipitated Rh(D), Rh(c), and 32-kDa Rh$_{null}$ proteins show that all labeled peptides are identical with the exception of one missing ^{125}I-labeled peptide from Rh$_{null}$ (Connor et al., 1992b). These data are consistent with the observation that Rh$_{null}$ cells lack an extracellular cysteine (Ridgwell et al., 1983) known to be present on the Rh(D) polypeptide and known to be important to its antigenicity (Green, 1967,1983; Abbott and Schachter, 1976). Note that these polymorphisms are not an exclusive property of Rh$_{null}$ cells; similar differences exist in the distribution of exofacial tyrosines between Rh(D) and Rh(c) (Blanchard et al., 1988; Saboori et al., 1988). Moreover, although many species have domains homologous to those of the human protein, they also exhibit divergent extracellular domains so, unlike Rh(D), many nonhuman cells cannot be surface-labeled with ^{125}I (Saboori et al., 1989). Collectively, these data suggest that the apparently nonimmunogenic Rh membrane-spanning region, the so-called "Rh core" (Saboori et al., 1988,1989; Agre and Cartron, 1991), might represent a constant region homologous among many species, including human Rh_{null}. On the other hand, the surface-accessible domain is probably a variable region that is responsible for the phenotypic and antigenic differences among cell types.

VII. CONCLUSION

Aminophospholipid asymmetry in eukaryotic cell membranes has been under intensive study since the first observations of ATP-dependent lipid movement across the red cell membrane by Seigneuret and Devaux (1984). Although real progress has been made in this field, many important details must be resolved to complete our understanding of the mechanisms and significance of active transbilayer lipid movement. Clearly, more experimentation is needed to identify and characterize all proteins that might be involved in lipid movement. Nevertheless, based on our current understanding of the biochemical requirements for PS translocation across the plasma membrane bilayer and on predictions of the protein structure from the nucleotide sequence of the putative transport protein (the Rh gene), we can envision a model of the aminophospholipid transporter.

Considering the low molecular weight of the aminophospholipids and the structural similarities of the Rh protein to other known transporters, an attractive hypothesis is that the polypeptide responsible for the movement of aminophospholipids across the bilayer membrane forms a pore that operates in a manner similar to that of other transporters of low molecular weight solutes. Although the specific amino acids that constitute the active site of the transporter are not known, one or more cysteine and histidine residues are likely to participate in the process because lipid movement is abolished by sulfhydryl oxidants (Connor and Schroit, 1988) and by histidine-specific reagents (Connor et al., 1992a), agents that also alter Rh immunoreactivity (Victoria et al., 1986). Interestingly, several of the Rh polypeptide cysteine residues are palmitoylated (de Vetten and Agre, 1988; Hartel-Schenk and Agre, 1992), suggesting that they may be involved in transport regulation, whereas the involvement of histidine residues suggests a relationship between a charge relay system and aminophospholipid movement.

Since lipid movement requires hydrolyzable ATP and the Rh nucleotide sequence does not encode a consensus ATP-binding site, the energy needs of active transport can reasonably be believed to be fulfilled by another protein such as an ATPase or protein kinase. This assumption implies that enzymes distinct from Rh are involved in lipid transport (Connor and Schroit, 1990; Schroit and Zwaal, 1990) and suggests that Rh polypeptides could be complexed to these enzymes, thereby fulfilling the energy requirements of the transporter. This possibility is supported by unrelated studies that have shown that the transporter/Rh protein does not exist in the red cell membrane as a single polypeptide but forms a complex (Hartel-Schenk and Agre, 1992) or "Rh cluster" (Bloy et al., 1988) with other membrane components, one of which may be an ATP-utilizing enzyme.

The detailed molecular events that control the functions of the aminophospholipid transporter are understood poorly and will remain so until all proteins involved in lipid movement are identified. Reconstitution of these proteins will set the stage for a better understanding of the processes at the molecular level and will provide a model system in which to study the precise molecular events that lead to transmembrane lipid movement.

Acknowledgments

My thanks to J. Connor and J. Killion for critical comments, helpful discussions, and suggestions. This work was supported in part by National Institutes of Health Grant DK-41714.

References

Abbott, R. E., and Schachter, D. (1976). Impermeant maleimides: Oriented probes of erythrocyte membrane proteins. J. Biol. Chem. 251, 7176–7183.

Agre, P., and Cartron, J. P. (1991). Molecular biology of the Rh antigens. Blood 78, 551–563.

Avent, N. D., Judson, P. A., Parsons, S. F., Mallinson, G., Anstee, D. J., Tanner, M. J., Evans, P. R., Hodges, E., Maciver, A. G., and Holmes, C. (1988a). Monoclonal antibodies that recognize different membrane proteins that are deficient in Rh_{null} human erythrocytes. One group of antibodies reacts with a variety of cells and tissues whereas the other group is erythroid specific. Biochem. J. 251, 499–505.

Avent, N. D., Ridgwell, K., Mawby, W. J., Tanner, M. J., Anstee, D. J., and Kumpel, B. (1988b). Protein sequence studies on Rh related polypeptides suggest the presence of at least two groups of proteins which associate in the human red cell membrane. Biochem. J. 256, 1043–1046.

Avent, N. D., Ridgwell, K., Tanner, M. J. A., and Anstee, D. J. (1990). cDNA cloning of a 30 kDa erythrocyte membrane protein associated with Rh (Rhesus)-blood-group-antigen expression. Biochem. J. 271, 821–825.

Avent, N. D., Butcher, S. K., Liu, W., Mawby, W. J., Mallinson, G., Parsons, S. F., Anstee, D. J., and Tanner, M. J. A. (1992). Localization of the C-termini of the Rh (Rhesus) polypeptides to the cytoplasmic face of the human erythrocyte membrane. J. Biol. Chem. 267, 15134–15139.

Bergmann, W. L., Dressler, V., Haest, C. W. M., and Deuticke, B. (1984). Reorientation rates and asymmetry of distribution of lysophospholipids between the inner and outer leaflet of the erythrocyte membrane. Biochim. Biophys. Acta 772, 328–336.

Bevers, E. M., Comfurius, P., van Rijn, J. L. M. L., Hemker, H. C., and Zwaal, R. F. A.

(1982). Generation of prothrombin-converting activity and the exposure of phosphatidyl-serine at the outer surface of platelets. *Eur. J. Biochem.* **122,** 429–436.

Bevers, E. M., Comfurius, P., and Zwaal, R. F. A. (1983). Changes in membrane phospho-lipid distribution during platelet activation. *Biochim. Biophys. Acta* **736,** 57–66.

Bitbol, M., and Devaux, P. (1988). Measurement of outward translocation of phospholipids across human erythrocyte membrane. *Proc. Natl. Acad. Sci. U.S.A.* **85,** 6783–6787.

Blanchard, D., Bloy, C., Hermand, P., Cartron, J. P., Saboori, A., Smith, B. L., and Agre, P. (1988). Two-dimensional iodopeptide mapping demonstrates erythrocyte Rh D, c, and E polypeptides are structurally homologous but nonidentical. *Blood* **72,** 1424–1427.

Bloj, B., and Zilversmit, D. B. (1976). Asymmetry and transposition rates of phosphatidyl-choline in rat erythrocyte ghosts. *Biochemistry* **15,** 1277–1283.

Bloy, C., Blanchard, D., Dahr, W., Beyreuther, K., Salmon, C., and Cartron, J. P. (1988). Determination of the N-terminal sequence of human red cell Rh(D) polypeptide and demonstration that the Rh(D), (c), and (E) antigens are carried by distinct polypeptide chains. *Blood* **72,** 661–666.

Bloy, C., Hermand, P., Blanchard, D., Cherif-Zahar, B., Goossens, D., and Cartron, J. P. (1990). Surface orientation and antigen properties of Rh and LW polypeptides of the human erythrocyte membrane. *J. Biol. Chem.* **265,** 21482–21487.

Bonnet, D., and Begard, E. (1984). Interaction of anilinonaphtyl labeled spectrin with fatty acids and phospholipids: A fluorescence study. *Biochem. Biophys. Res. Commun.* **120,** 344–350.

Bretscher, M. S. (1972). Asymmetrical lipid bilayer structure for biological membranes. *Nature (New Biol.)* **236,** 11–12.

Calvez, J. Y., Zachowski, A., Herrmann, A., Morrot, G., and Devaux, P. F. (1988). Asym-metric distribution of phospholipids in spectrin-poor erythrocyte vesicles. *Biochemistry* **27,** 5666–5670.

Chattopadhyay, A., and London, E. (1987). Parallax method for direct measurement of membrane penetration depth utilizing fluorescence quenching by spin-labeled phospho-lipids. *Biochemistry* **26,** 39–45.

Cherif-Zahar, B., Bloy, C., Le Van Kim, C., Blanchard, D., Bailly, P., Hermand, P., Salmon, C., Cartron, J. P., and Colin, Y. (1990). Molecular cloning and protein structure of a human blood group Rh polypeptide. *Proc. Natl. Acad. Sci. U.S.A.* **87,** 6243–6247.

Cohen, A. M., Liu, S. C., Derick, L. H., and Palek, J. (1986). Ultrastructural studies of the interaction of spectrin with phosphatidylserine liposomes. *Blood* **68,** 920–926.

Cohen, A. M., Liu, S. C., Lawler, J., Derick, L., and Palek, J. (1988). Identification of protein 4.1 binding site to Phosphatidylserine vesicles. *Biochemistry* **27,** 614–619.

Colin, Y., Cherif-Zahar, B., Le Van Kim, C. L., Raynal, V., Van Huffel, V., and Cartron, J. P. (1991). Genetic basis of the RhD-positive and RhD-negative blood group polymor-phism as determined by southern analysis. *Blood* **78,** 2747–2752.

Colleau, M., Herve, P., Fellmann, P., and Devaux, P. F. (1991). Transmembrane diffusion of fluorescent phospholipids in human erythrocytes. *Chem. Phys. Lipids* **57,** 29–37.

Comfurius, P., Senden, J. M. G., Tilly, R. H. J., Schroit, A. J., Bevers, E. M., and Zwaal, R. F. A. (1990). Loss of membrane phospholipid asymmetry in platelets and red cells may be associated with calcium-induced shedding of plasma membrane and inhibition of aminophospholipid translocase. *Biochim. Biophys. Acta* **1026,** 153–160.

Connor, J., and Schroit, A. J. (1987). Determination of lipid asymmetry in human red cells by resonance energy transfer. *Biochemistry* **26,** 5099–5105.

Connor, J., and Schroit, A. J. (1988). Transbilayer movement of phosphatidylserine in erythrocytes. Inhibition of transport and preferential labeling of a 31,000 Dalton protein by sulfhydryl reactive reagents. *Biochemistry* **27,** 848–851.

Connor, J., and Schroit, A. J. (1989). Transbilayer movement of phosphatidylserine in non-human erythrocytes: Evidence that the aminophospholipid transporter is a ubiquitous membrane protein. *Biochemistry* **28**, 9680–9685.

Connor, J., and Schroit, A. J. (1990). Aminophospholipid translocation in erythrocytes. Evidence for the involvement of a specific transporter and an endofacial protein. *Biochemistry* **29**, 37–43.

Connor, J., and Schroit, A. J. (1991). Transbilayer movement of phosphatidylserine in erythrocytes: Inhibitors of aminophospholipid transport block the association of photolabeled lipid to its transporter. *Biochim. Biophys. Acta* **1066**, 37–42.

Connor, J., Bucana, C., Fidler, I. J., and Schroit, A. J. (1989). Differentiation-dependent expression of phosphatidylserine in mammalian plasma membranes: Quantitative assessment of outer leaflet lipid by prothrombinase complex formation PS asymmetry. *Proc. Natl. Acad. Sci. U.S.A.* **86**, 3184–3188.

Connor, J., Gillum, K., and Schroit, A. J. (1990). Maintenance of lipid asymmetry in red blood cells and ghosts: Effect of divalent cations and serum albumin on the transbilayer distribution of phosphatidylserine. *Biochim. Biophys. Acta* **1025**, 82–86.

Connor, J., Pak, C. H., Zwaal, R. F. A., and Schroit, A. J. (1992a). Bidirectional transbilayer movement of phospholipid analogs in human red blood cells. *J. Biol. Chem.* **267**, 19412–19417.

Connor, J., Bar-Eli, M., Gillum, K. D., and Schroit, A. J. (1992b). Evidence for a structurally homologous Rh-like polypeptide in Rh$_{null}$ erythrocytes. *J. Biol. Chem.* **267**, 26050–26055.

Daleke, D. L. (1990). Inhibition of aminophospholipid transport in human erythrocytes by arginine reagents. *J. Cell Biol.* **111**, Abstr. 73A.

Daleke, D. L., and Huestis, W. H. (1985). Incorporation and translocation of aminophospholipids in human erythrocytes. *Biochemistry* **24**, 5406–5416.

Daleke, D. L., and Huestis, W. H. (1989). Erythrocyte morphology reflects the transbilayer distribution of incorporated phospholipids. *J. Cell Biol.* **108**, 1375–1385.

Devaux, P. F. (1988). Phospholipid flippases. *FEBS Lett.* **234**, 8–12.

Devaux, P. F. (1991). Static and dynamic lipid asymmetry in cell membranes. *Biochemistry* **30**, 1163–1173.

Devaux, P. F. (1992). Protein involvement in transmembrane lipid asymmetry. *Annu. Rev. Biophys. Biomol. Struct.* **21**, 417–439.

de Vetten, M. P., and Agre, P. (1988). The Rh polypeptide is a major fatty acid-acylated erythrocyte membrane protein. *J. Biol. Chem.* **263**, 18193–18196.

Etemadi, A.-H. (1980). Membrane asymmetry: A survey and critical appraisal of the methodology. II. Methods for assessing the unequal distribution of lipids. *Biochim. Biophys. Acta* **604**, 423–475.

Fadok, V. A., Voelker, D. R., Campbell, P. A., Cohen, J. J., Bratton, D. L., and Henson, P. M. (1992a). Exposure of phosphatidylserine on the surface of apoptotic lymphocytes triggers specific recognition and removal by macrophages. *J. Immunol.* **148**, 2207–2216.

Fadok, V. A., Savill, J. S., Haslett, C., Bratton, D. L., Doherty, D. E., Campbell, P. A., and Henson, P. M. (1992b). Different populations of macrophages use either the vitronectin receptor or the phosphatidylserine receptor to recognize and remove apoptotic cells. *J. Immunol.* **149**, 4029–4035.

Farge, E., and Devaux, P. F. (1992). Shape changes of giant liposomes induced by an asymmetric transmembrane distribution of phospholipids. *Biophys. J.* **61**, 347–357.

Farooqui, S. M., Wali, R. K., Baker, R. F., and Kalra, V. K. (1987). Effect of cell shape, membrane deformability and phospholipid organization on phosphate–calcium-induced fusion of erythrocytes. *Biochim. Biophys. Acta* **904**, 239–250.

Forgac, M., and Cantley, L. (1984). The plasma membrane (Mg^{2+})-dependent adenosine triphosphatase from the human erythrocyte is not an ion pump. *J. Membr. Biol.* **80**, 185–190.

Gahmberg, C. G. (1982). Molecular identification of the human Rho(D) antigen. *FEBS Lett.* **140**, 93–97.

Gahmberg, C. G. (1983). Molecular characterization of the human Rho(D) antigen. *EMBO J.* **2**, 223–228.

Ganong, B. R., and Bell, R. M. (1984). Transmembrane movement of phosphatidylglycerol and diacylglycerol. *Biochemistry* **23**, 4977–4983.

Gordesky, S. E., and Marinetti, G. V. (1973). The asymmetric arrangement of phospholipids in the human erythrocyte membrane. *Biochem. Biophys. Res. Commun.* **50**, 1027–1032.

Gordesky, S. E., Marinetti, G. V., and Love, R. (1975). The reaction of chemical probes with the erythrocyte membrane. *J. Membr. Biol.* **20**, 111–132.

Green, F. A. (1967). Erythrocyte membrane sulfhydryl groups and Rh antigen activity. *Immunochemistry* **4**, 247–257.

Green, F. A. (1972). Erythrocyte membrane lipids and Rh antigen activity. *J. Biol. Chem.* **247**, 881–887.

Green, F. A. (1982). Erythrocyte membrane phosphatidylcholine and Rh(D) cryolatency. *Immunol. Commun.* **11**, 25–32.

Green, F. A. (1983). The mode of attenuation of erythrocyte membrane Rho(D) antigen activity by 5,5'-dithiobis-(2-nitrobenzoic acid) and protection against loss of activity by bound anti-Rho(D) antibody. *Mol. Immunol.* **20**, 769–775.

Green, F. A., Hui, H. L., Green, L. A., Heubusch, P., and Pudlak, W. (1984). The phospholipid requirement for Rho(D) antigen activity: Mode of inactivation by phospholipases and of protection by anti-Rho(D) antibody. *Mol. Immunol.* **21**, 433–438.

Gudi, S. R. P., Kumar, A., Bhakuni, V., Gokhale, S. M., and Gupta C. M. (1990). Membrane skeleton-bilayer interaction is not the major determinant of membrane phospholipid asymmetry in human erythrocytes. *Biochim. Biophys. Acta* **1023**, 63–72.

Haest, C. W. M. (1982). Interactions between membrane skeleton proteins and the intrinsic domain of the erythrocyte. *Biochim. Biophys. Acta* **694**, 331–352.

Haest, C. W. M., Kamp, D., Plasa, G., and Deuticke, D. (1977). Intra- and intermolecular cross-linking of membrane proteins in intact erythrocytes and ghosts by SH-oxidizing agents. *Biochim. Biophys. Acta* **469**, 226–230.

Haest, C. W. M., Plasa, G., Kamp, D., and Deuticke, B. (1978). Spectrin as stabilizer of the phospholipid asymmetry in the erythrocyte membrane. *Biochim. Biophys. Acta* **509**, 21–32.

Hartel-Schenk, S., and Agre, P. (1992). Mammalian red cell membrane Rh polypeptides are selectively palmitoylated subunits of a macromolecular complex. *J. Biol. Chem.* **267**, 5569–5574.

Henseleit, U., Plasa, G., and Haest, C. W. M. (1990). Effects of divalent cations on lipid flip–flop in the human erythrocyte membrane. *Biochim. Biophys. Acta* **1029**, 127–135.

Herrmann, A., and Devaux, P. F. (1990). Alteration of the aminophospholipid translocase activity during *in vivo* and artificial aging in human erythrocytes. *Biochim. Biophys. Acta* **1027**, 41–46.

Herrmann, A., and Müller, P. (1986). A model for the asymmetric lipid distribution in the human erythrocyte membrane. *Biosci. Rep.* **6**, 185–191.

Hughes-Jones, N. C., Bloy, C., Gorick, B., Blanchard, D., Doinel, C., Rouger, P., and Cartron, J. P. (1988). Evidence that the c, D and E epitopes of the human Rh blood group system are on separate polypeptide molecules. *Mol. Immunol.* **25**, 931–936.

Krahmer, M., and Prohaska, R. (1987). Characterization of the human Rh (Rhesus)-specific polypeptides by limited proteolysis. *FEBS Lett.* **226**, 105–108.

Kramer, R. M., and Branton, D. (1979). Retention of lipid asymmetry in membranes on polylysine-coated polyacrylamide beads. *Biochim. Biophys. Acta* **556**, 219–232.

Le van Kim, C., Cherif-Zahar, B., Raynal, V., Mouro, I., Lopez, M., Cartron, J. P., and Colin, Y. (1992a). Multiple Rh messenger RNA isoforms are produced by alternative splicing. *Blood* **80**, 1074–1078.

Le van Kim, C., Mouro, I., Cherif-Zahar, B., Raynal, V., Cherrier, C., Cartron, J. P., and Colin, Y. (1992b). Molecular cloning and primary structure of the human blood group RhD polypeptide. *Proc. Natl. Acad. Sci. U.S.A.* **89**, 10925–10929.

Levine, S., and Korenstein, R. (1991). Membrane fluctuations in erythrocytes are linked to MgATP-dependent dynamic assembly of the membrane skeleton. *Biophys. J.* **60**, 733–737.

Lipsky, N. G., and Pagano, R. E. (1985). A vital stain for the Golgi apparatus. *Science* **228**, 745–747.

Marinetti, G. V., and Cattieu, K. (1982). Asymmetric metabolism of phosphatidylethanolamine in the human red cell membrane. *J. Biol. Chem.* **257**, 245–248.

Marinetti, G. V., and Crain, R. C. (1978). Topology of amino-phospholipids in the red cell membrane. *J. Supramol. Struct.* **8**, 191–213.

Martin, O. C., and Pagano, R. E. (1987). Transbilayer movement of fluorescent analogs of phosphatidylserine and phosphatidylethanolamine at the plasma membrane of cultured cells. *J. Biol. Chem.* **262**, 5890–5898.

Mcintyre, J. C., and Sleight, R. G. (1991). Fluorescence assay for phospholipid membrane asymmetry. *Biochemistry* **30**, 11819–11827.

Middelkoop, E., Lubin, B. H., Op den Kamp, J. A. F., and Roelofsen, B. (1986). Flip flop rates of individual molecular species of phosphatidylcholine in the human red cell membrane. *Biochim. Biophys. Acta* **855**, 421–424.

Mombers, C. A. M., Verkleij, A. J., de Gier, J., and van Deenen, L. L. M. (1979). The interaction of spectrin-actin and synthetic phospholipids. II. The interaction with phosphatidylserine. *Biochim. Biophys. Acta* **551**, 271–281.

Mombers, C. A. M., de Gier, J., Demel, R. A., and van Deenen, L. L. M. (1980). Spectrin-phospholipid interaction. A monolayer study. *Biochim. Biophys. Acta* **603**, 52–62.

Moore, S., and Green, C. (1987). The identification of specific Rhesus polypeptide blood group ABH active glycoprotein complexes in the human red-cell membrane. *Biochem. J.* **244**, 735–741.

Moore, S., Woodrow, C. F., and McClelland, D. B. L. (1982). Isolation of membrane components associated with human red cell antigens Rho(D), (c) (E), and Fya. *Nature (London)* **295**, 529–531.

Morrot, G., Hervé, P., Zachowski, A., Fellmann, P., and Devaux, P. F. (1989). Aminophospholipid translocase of human erythrocytes: Phospholipid substrate specificity and effect of cholesterol. *Biochemistry* **28**, 3456–3462.

Morrot, G., Zachowski, A., and Devaux, P. F. (1990). Partial purification and characterization of the human erythrocyte Mg^{2+}–ATPase. *FEBS Lett.* **266**, 29–32.

Nakao, M., Nakao, T., Tatibana, M., and Yoshikawa, H. (1960). Shape transformation of erythrocyte ghosts on addition of adenosine triphosphate to the medium. *J. Biochem.* **47**, 694–695.

O'Brian, C. A. O., and Ward, N. E. (1990). Characterization of a Ca^{2+}- and phospholipid-dependent ATPase reaction catalyzed by rat brain protein kinase C. *Biochemistry* **29**, 4278–4282.

Op den Kamp, J. A. F. (1979). Lipid asymmetry in membranes. *Annu. Rev. Biochem.* **48**, 47–71.

Pagano, R. E., and Sleight, R. G. (1985). Defining lipid transport pathways in animal cells. *Science* **229**, 1051–1057.

Patel, V. P., and Fairbanks, G. (1986). Relationship of major phosphorylation reactions and Mg^{2+}–ATPase activities to ATP-dependent shape change of human erythrocyte membranes. *J. Biol. Chem.* **261,** 3170–3177.

Raval, P. J., and Allan, D. (1984). Phospholipid asymmetry in the membranes of intact human erythrocytes and in spectrin-free microvesicles derived from them. *Biochim. Biophys. Acta* **772,** 192–196.

Rawyler, A., Roelofsen, B., and Op den Kamp, J. A. F. (1984). The use of fluorescamine as a permeant probe to localize phosphatidylethanolamine in intact Friend erythroleukemic cells. *Biochim. Biophys. Acta* **769,** 330–336.

Renooij, W., van Golde, L. M. G., Zwaal, R. F. A., Roelofsen, B., and van Deenen, L. L. M. (1974). Preferential incorporation of fatty acids at the inside of the human erythrocyte membranes. *Biochim. Biophys. Acta* **363,** 287–292.

Renooij, W., van Golde, L. M. G., Zwaal, R. F. A., Roelofsen, B., and van Deenen, L. L. M. (1974). Preferential incorporation of fatty acids at the inside of the human erythrocyte membranes. *Biochim. Biophys. Acta* **363,** 287–292.

Ridgwell, K., Roberts, S. J., Tanner, M. J. A., and Anstee, D. J. (1983). Absence of two membrane proteins containing extracellular thiol groups on Rh_{null} human erythrocytes. *Biochem. J.* **213,** 267–269.

Ridgwell, K., Spurr, N. K., Laguda, B., Macgeoch, C., Avent, N. D., and Tanner, M. J. A. (1992). Isolation of cDNA clones for a 50-kDa glycoprotein of the human erythrocyte membrane associated with Rh (Rhesus) blood-group antigen expression. *Biochem. J.* **287,** 223–228.

Rosing, J., Bevers, E. M., Comfurius, P., Hemker, H. C., van Dieijen, G., Weiss, H. J., and Zwaal, R. F. A. (1985). Impaired factor X and prothrombin activation associated with decreased phospholipid exposure in platelets from a patient with a bleeding disorder. *Blood* **65,** 1557–1561.

Rybicki, A. C., Heath, R., Lubin, B. H., and Schwartz, R. S. (1988). Human erythrocyte protein 4.1 is a phosphatidylserine binding protein. *J. Clin. Invest.* **81,** 255–260.

Saboori, A. M., Smith, B. L., and Agre, P. (1988). Polymorphism in the M_r 32,000 Rh protein purified from Rh(D) positive and negative erythrocytes. *Proc. Natl. Acad. Sci. U.S.A.* **85,** 4042–4045.

Saboori, A. M., Denker, B. M., and Agre, P. (1989). Isolation of proteins related to the RH polypeptides from nonhuman erythrocytes. *J. Clin. Invest.* **83,** 187–190.

Sato, S. B., and Ohnishi, S. I. (1983). Interaction of a peripheral protein of the erythrocyte membrane, band 4.1, with phosphatidylserine-containing liposomes and erythrocyte inside-out vesicles. *Eur. J. Biochem.* **130,** 19–25.

Schewe, M., Müller, P., Korte, T., and Herrmann, A. (1992). The role of phospholipid asymmetry in calcium-phosphate-induced fusion of human erythrocytes. *J. Biol. Chem.* **267,** 5910–5915.

Schlegel, R. A., Prendergast, T. W., and Williamson, P. (1985). Membrane phospholipid asymmetry as a factor in erythrocyte–endothelial cell interactions. *J. Cell. Physiol.* **1123,** 215–218.

Schroit, A. J., and Zwaal, R. F. A. (1990). Transbilayer movement of phospholipids in red cell and platelet membranes. *Biochim. Biophys. Acta* **1071,** 313–329.

Schroit, A. J., Madsen, J. W., and Tanaka, Y. (1985). *In vivo* recognition and clearance of red blood cells containing phosphatidylserine in their plasma membranes. *J. Biol. Chem.* **260,** 5131–5138.

Schroit, A. J., Madsen, J., and Ruoho, A. E. (1987). Radioiodinated, photoaffinity-labeled phosphatidylcholine and phosphatidylserine: Transfer properties and differential photoreactive reaction with human erythrocyte membrane proteins. *Biochemistry* **26,** 1812–1819.

Schroit, A. J., Bloy, C., Connor, J., and Cartron, J.-P. (1990). Involvement of Rh blood group polypeptides in the maintenance of aminophospholipid asymmetry. *Biochemistry* **29**, 10303–10306.

Seigneuret, M., and Devaux, P. F. (1984). ATP-dependent asymmetric distribution of spin-labeled phospholipids in the erythrocyte membrane: Relation to shape changes. *Proc. Natl. Acad. Sci. U.S.A.* **81**, 3751–3755.

Shiffer, K. A., Goerke, J., Duzgunes, N., Fedor, J., and Shohet, S. B. (1988). Interaction of erythrocyte protein 4.1 with phospholipids. A monolayer and liposome study. *Biochim. Biophys. Acta* **937**, 269–280.

Shukla, S. D., and Hanahan, D. J. (1982). Membrane alteration in cellular aging: Susceptibility of phospholipids in density (age)-related human erythrocytes to phospholipase A_2. *Arch. Biochem. Biophys.* **214**, 335–341.

Sikorski, A. F., Michalak, K., and Bobrowska, M. (1987). Interaction of spectrin with phospholipids. Quenching of spectrin intrinsic fluorescence by phospholipid suspensions. *Biochim. Biophys. Acta* **904**, 55–60.

Sims, P. J., Wiedmer, T., Esmon, C. T., Weiss, H. J., and Shattil, S. J. (1989). Assembly of the platelet prothrombinase complex is linked to vesiculation of the platelet plasma membrane. Studies in Scott syndrome: An isolated defect in platelet procoagulant activity. *J. Biol. Chem.* **264**, 17049–17057.

Smith, R. E., and Daleke, D. L. (1990). Phosphatidylserine transport in Rh null erythrocytes. *Blood* **76**, 1021–2027.

Song, L. Y., Baldwin, J. M., O'Reilly, R., and Lucy, J. A. (1992). Relationship between surface exposure of acidic phospholipids and cell fusion in erythrocytes subjected to electrical breakdown. *Biochim. Biophys. Acta* **1104**, 1–8.

Struck, D. K., and Pagano, R. E. (1980). Insertion of fluorescent phospholipids into the plasma membrane of a mammalian cell. *J. Biol. Chem.* **255**, 5404–5419.

Sugihara, T., Sugihara, K., and Hebbel, R. P. (1991). Phospholipid asymmetry during erythrocyte deformation: Maintenance of the unit membrane. *Biochim. Biophys. Acta* **1103**, 303–306.

Suyama, K., and Goldstein, J. (1988). Antibody produced against isolated Rh(D) polypeptide reacts with other Rh-related antigens. *Blood* **72**, 1622–1626.

Suyama, K., and Goldstein, J. (1990). Enzymatic evidence for differences in the placement of Rh antigens within the red cell membrane. *Blood* **75**, 255–260.

Suyama, K., and Goldstein, J. (1992). Membrane orientation of Rh(D) polypeptide and partial localization of its epitope-containing domain. *Blood* **79**, 808–812.

Tanaka, Y., and Schroit, A. J. (1986). Calcium phosphate-induced immobilization of fluorescent phosphatidylserine in synthetic bilayer membranes: Inhibition of lipid transfer between vesicles. *Biochemistry* **25**, 2141–2148.

Tilley, L., Cribier, S., Roelofson, B., Op den Kamp, J. A. F., and van Deenen, L. L. M. (1986). ATP-dependent translocation of amino phospholipid across the human erythrocyte membrane. *FEBS Lett.* **194**, 21–27.

Tilly, R. H. J., Senden, J. M. G., Comfurius, P., Bevers, E. M., and Zwaal, R. F. A. (1990). Increased aminophospholipid translocase activity in human platelets during secretion. *Biochim. Biophys. Acta* **1029**, 188–190.

Van Meer, G., and Op den Kamp, J. A. F. (1982). Transbilayer movement of various phosphatidylcholine species in intact human erythrocytes. *J. Cell. Biochem.* **19**, 193–204.

Verkleij, A. J., Zwaal, R. F. A., Roelofsen, B., Comfurius, P., Kastelijn, D., and van Deenen, L. L. M. (1973). The asymmetric distribution of phospholipids in the human red cell membrane. A combined study using phospholipases and freeze-etching electron microscopy. *Biochim. Biophys. Acta* **323**, 178–193.

Victoria, E. J., Branks, M. J., and Masouredis, S. P. (1986). Rh antigen immunoreactivity after histidine modification. *Mol. Immunol.* **23,** 1039–1044.

Vidal, M., Sainte-Marie, J., Phillipot, J. R., and Bienvenue, A. (1989). Asymmetric distribution of phospholipids in the membrane of vesicles released during in vitro maturation of guinea pig reticulocytes: Evidence precluding a role for "aminophospholipid translocase." *J. Cell. Physiol.* **140,** 455–462.

Williamson, P., Antia, R., and Schlegel, R. A. (1987). Maintenance of membrane phospholipid asymmetry. *FEBS Lett.* **219,** 316–320.

Xu, Y. H., Lu, Z. Y., Conigrave, A. D., Auland, M. E., and Roufogalis, B. D. (1991). Association of vanadate-sensitive Mg^{2+}–ATPase and shape change in intact red blood cells. *J. Cell. Biochem.* **46,** 284–290.

Zachowski, A., Favre, E., Cribier, S., Herve, P., and Devaux, P. F. (1986). Outside-inside translocation of aminophospholipids in the human erythrocyte membrane is mediated by a specific enzyme. *Biochemistry* **25,** 2585–2590.

Zachowski, A., Henry, J. P., and Devaux, P. F. (1989). Control of transmembrane lipid asymmetry in chromaffin granules by an ATP-dependent protein. *Nature (London)* **340,** 75–76.

Zwaal, R. F. A., Roelofsen, B., and Colley, C. M. (1973). Localization of red cell membrane constituents. *Biochim. Biophys. Acta* **300,** 159–182.

Zwaal, R. F. A., Roelofsen, B., Comfurius, P., and van Deenen, L. L. M. (1975). Organization of phospholipids in human red cell membranes as detected by the action of various purified phospholipases. *Biochim. Biophys. Acta* **406,** 83–96.

CHAPTER 3

Metabolism of Gangliosides: Topology, Pathobiochemistry, and Sphingolipid Activator Proteins

Konrad Sandhoff and Gerhild van Echten
Institut für Organische Chemie und Biochemie der Universität Bonn, D-53121 Bonn 1, Germany

I. INTRODUCTION

Glycosphingolipids (GSL) constitute a family of complex membrane components that are assembled in cell- and species-specific patterns in all eukaryotic cells. Gangliosides are sialic acid-containing GSLs. They are highly enriched in nervous tissue, in which their more complex derivatives—namely, di-, tri-, and tetrasialogangliosides—are particularly prevalent. Except for G_{M4}, all gangliosides are derived from lactosylceramide (LacCer) and contain glucosylceramide (GlcCer) as a backbone (see Fig. 1).

The importance of gangliosides and their implications in biological functions have been reviewed in detail (Zeller and Marchase, 1992). The possi-

GP1c(NeuAcα2 → 8NeuAcα2 → 3Galβ1 → 3GalNAcβ1 → 4(NeuAcα2 → 8NeuAcα2 → 8NeuAcα2 → 3)Galβ1 → 4Glcβ1 → 1Cer)

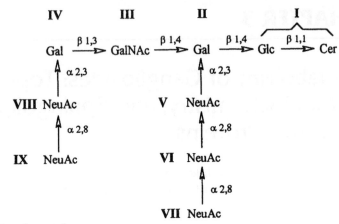

II   - inner galactose
IV   - outer galactose

GlcCer	glucosylceramide,	**I**
LacCer	lactosylceramide,	**I, II**
GM3	**I, II, V**	
GD3	**I, II, V, VI**	
GT3	**I, II, V - VII**	
GA2	**I - III**	
GM2	**I - III, V**	
GD2	**I - III, V, VI**	
GT2	**I - III, V - VII**	
GA1	**I - IV**	
GM1a	**I - V**	
GD1b	**I - VI**	
GT1c	**I - VII**	
GM1b	**I - IV, VIII**	
GD1a	**I - V, VIII**	
GT1b	**I - VI, VIII**	
GQ1c	**I - VIII**	
GD1c	**I - IV, VIII, IX**	
GT1a	**I - V, VIII, IX**	
GQ1b	**I - VI, VIII, IX**	
GP1c	**I - IX**	

FIGURE 1   The terminology used for gangliosides is that recommended by Svennerholm (1963). The structures of glycosphingolipids (GSLs) derived from LacCer are part of the G_{P1c} structure, containing the sugar residues shown on the bottom.

ble role in signal transduction of their biosynthetic intermediates, for example, ceramide (Cer), or of their breakdown products, for example, sphingosine and sphingosine 1-phosphate, is a provocative hypothesis, substantially discussed in the literature (for review, see Merrill, 1991; Kolesnick, 1992; Olivera and Spiegel, 1992).

II. GLYCOSPHINGOLIPID BIOSYNTHESIS

A. Subcellular Localization and Topology

GSL biosynthesis starts in the endoplasmic reticulum (ER) with the formation of ceramide, which then is glycosylated stepwise in the Golgi apparatus. The mechanism by which the growing molecules are transported from the ER through the Golgi cisternae to the plasma membrane has not yet been defined (for review, see Schwarzmann and Sandhoff, 1990). Although a vesicle-bound exocytic membrane flow has been considered, the involvement of glycolipid binding and/or transfer proteins cannot be excluded at this time (for review, see Sandhoff and van Echten, 1993).

The enzymes catalyzing the formation of ceramide via 3-dehydrosphinganine, sphinganine, and dihydroceramide are bound to the cytosolic face of the ER membrane (Mandon *et al.*, 1992). The 4-*trans* double bond of the sphingoid base is introduced after acylation of sphinganine at the level of dihydroceramide (Rother *et al.*, 1992). The localization and topology of the enzyme catalyzing the desaturation of dihydroceramide has not yet been studied.

All the glycosylation steps presented in Fig. 2 are localized in the Golgi apparatus. Information about the localization of the individual glycosyltransferases within the Golgi stack was derived from metabolic studies in cultured cells using transport inhibitors (Miller-Prodraza and Fishman, 1984; Saito *et al.*, 1984; Hogan *et al.*, 1988; van Echten and Sandhoff, 1989; van Echten *et al.*, 1990a). The most impressive uncoupling of ganglioside biosynthesis was observed in primary cultured neurons in the presence of brefeldin A (van Echten *et al.*, 1990a). The results suggested that G_{M3} and G_{D3} are synthesized in early Golgi compartments, whereas complex gangliosides such as G_{M1a}, G_{D1a}, G_{D1b}, G_{T1b}, and G_{Q1b} are formed in a late Golgi compartment or even in the *trans* Golgi network, beyond the brefeldin A-induced block. These findings later were confirmed in Chinese hamster ovary (CHO) cells (Young *et al.*, 1990).

These indirect pieces of evidence obtained using drugs in cultured cells were supported by Golgi subfractionation studies in rat liver (Trinchera and Ghidoni, 1989; Trinchera *et al.*, 1990) as well as in primary cultured

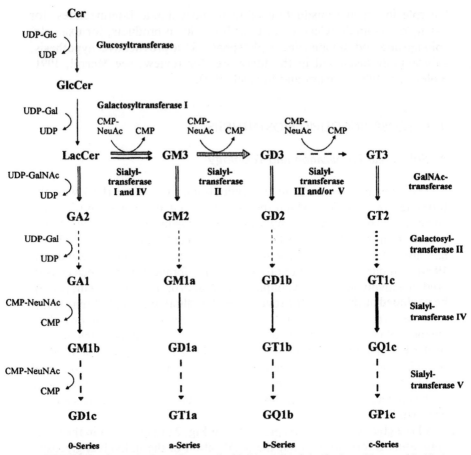

FIGURE 2 General scheme for ganglioside biosynthesis beginning with ceramide (Cer). All steps are catalyzed by glycosyltransferases of Golgi membranes. All glycosylation steps (except the formation of GlcCer and possibly LacCer) have a luminal topology.

neurons (Iber *et al.,* 1992a). However, data obtained in different cell systems should be interpreted with caution, especially when unnatural and truncated sphingolipid precursors such as NBD-C_6-ceramide [*N*{-6-(7-nitrobenzo-2-oxa-1,3-diazol-4-yl)-aminohexanoyl}-D-*erythro* sphingosine] are used that undergo transbilayer movement (flip–flop) in membranes (reviewed by Sandhoff and van Echten, 1993).

With respect to the topology of the glycosyltransferases involved in GSL biosynthesis, many studies were done during the last decade. Glucosyltransferase and galactosyltransferase I, which catalyze the formation of GlcCer and LacCer, respectively, are accessible from the cytosolic side of the Golgi (Trinchera *et al.*, 1991). The precise localization of ceramide glucosyltransferase is, however, not yet clear and could be on the Golgi and/or in a pre-Golgi compartment (Coste *et al.*, 1985,1986; Futerman and Pagano, 1991; Trinchera *et al.*, 1991; Jeckel *et al.*, 1992; Chapter 4). Also, the cytosolic topology of galactosyltransferase I is not unambiguously clear. The finding that mutant CHO cells, with an intact enzyme but lacking the translocator for UDP-Gal into the Golgi lumen, have strongly reduced levels of LacCer (Deutscher and Hirschberg, 1986) suggests a luminal topology for galactosyltransferase I.

The sequential addition of additional monosaccharide and sialic acid residues to the growing oligosaccharide chain, yielding G_{M3}, G_{D3}, and more complex gangliosides, is catalyzed by membrane-bound glycosyltransferases that are restricted to the luminal face of the Golgi membranes (Carey and Hirschberg, 1981; Fleischer, 1981; Yusuf *et al.*, 1983; Fig. 2). However, how LacCer and GlcCer, the starting materials for the synthesis of complex gangliosides, are transferred from the cytosolic to the luminal side of the Golgi membranes remains unknown. Similarly, how ceramide formed at the cytosolic surface of ER membranes reaches the luminal side of early Golgi membranes for sphingomyelin biosynthesis is unclear (Futerman *et al.*, 1990; Jeckel *et al.*, 1990; Chapter 4).

Only a few glycosyltransferases are involved in the elongation of the oligosaccharide chains of LacCer, G_{M3}, G_{D3}, and G_{T3}, thereby generating the manifold members of the different ganglioside series (Fig. 2). Brady and co-workers suggested very early that the same glycosyltransferases might be involved in mono- and disialoganglioside synthesis (Cumar *et al.*, 1972; Pacuszka *et al.*, 1978). Years later, our group proved for the first time that almost all glycosyltransferases involved in ganglioside biosynthesis are rather unspecific for their acceptors. Competition experiments performed in rat liver Golgi demonstrated that these enzymes catalyze the transfer of the same sugar molecule to analogous glycolipid acceptors that differ in the number of sialic acid residues bound to the inner galactose of the oligosaccharide chain (Pohlentz *et al.*, 1988; Iber and Sandhoff, 1989; Iber *et al.*, 1989,1991,1992b); these enzymes also use neoglycolipids with changed hydrophobic anchors (Pohlentz *et al.*, 1992). For example, sialyltransferase V *in vitro* generates G_{D1c} from G_{M1b}, G_{T1a} from G_{D1a}, G_{Q1b} from G_{T1b}, G_{P1c} from G_{Q1c}, and even G_{T3} from G_{D3} (Iber *et al.*, 1992b).

Although many studies on GSL biosynthesis have been published during the last two decades, many questions concerning the topology of some enzymes involved in sphingolipid biosynthesis (e.g., dihydroceramide desaturase) and the mechanism of intracellular transfer of sphingolipids remain unanswered.

B. Regulation

Even more poorly understood is the regulation of GSL biosynthesis. However, during embryogenesis, ontogenesis, and cell differentiation a good correlation has been reported between the expression of one ganglioside and the relative activity of the respective glycosyltransferase catalyzing its formation from its GSL precursor, when sufficiently available (Yu et al., 1988; Daniotti et al., 1991; Nakamura et al., 1992). Earlier studies by Hashimoto et al. (1983) and by Nagai and colleagues (Nakakuma et al., 1984; Nagai et al., 1986) also suggested that the formation of specific GSL patterns is under the control of expression of respective glycosyltransferase activities. This concept is in agreement with observations in transformed cells that alterations in the glycosylation of different glycoconjugates correspond to respective changes in the expression of the relevant glycosyltransferases (Coleman et al., 1975; Nakaishi et al., 1988a,b; Matsuura et al., 1989; Ruan and Lloyd, 1992). Regulation of ganglioside biosynthesis at the transcriptional level also was suggested by an increase of G_{M3}-synthase activity (sialyltransferase I) in HeLa cells after addition of butyrate (as well as propionate and pentanoate) to the culture medium, a process that required RNA and protein synthesis (Fishman et al., 1976). Similar observations were made on treatment of neuroblastoma cells with retinoic acid (Moskal et al., 1987), which also stimulated the biosynthesis of other membrane components such as phospholipids (Li and Ladisch, 1992).

Evidence for an epigenetic regulation of GSL biosynthesis is also available. Dawson and co-workers demonstrated a direct relationship between cyclic AMP-dependent phosphorylation and G_{M2}-synthase activity (McLawhon et al., 1981; Scheideler and Dawson, 1986). The importance of protein phosphorylation for the regulation of ganglioside biosynthesis also became clear from experiments with okadaic acid, a potent inhibitor of protein phosphatases (Hattori and Horiuchi, 1992).

To date, no evidence exists for a feedback control of GSL biosynthesis by exogenous gangliosides in cell culture or in vivo. Indirect evidence for such control has been obtained only in vitro when very high concentrations of gangliosides (100–500 μM) were used (Richardson et al., 1977; Nores and Caputto, 1984; Yusuf et al., 1987; Shukla et al., 1991). However, in cultured neurons, we observed that sphingosine (a catabolic product of sphingolipids) down-regulates serine palmitoyltransferase, the enzyme that catalyzes the first step of sphingolipid biosynthesis (van Echten et al., 1990b; Mandon et al., 1991). The key regulatory role of this enzyme also was suggested by Holleran et al. (1991), who studied the regulation of epidermal sphingolipid biosynthesis by changes in the permeability barrier of the skin.

For the regulation of ganglioside biosynthesis, we assume that the sequence LacCer $\rightarrow$ G_{M3} $\rightarrow$ G_{D3} (Fig. 2) might be essential (Pohlentz *et al.*, 1988; Ruan and Lloyd, 1992), G_{M3} being the branching point at which ganglioside biosynthesis can be diverted either to G_{M2} ("a" series) or to G_{D3} ("b" series). We have shown that lowering the pH from 7.4 to 6.2 in the culture medium of murine cerebellar cells causes a reversible shift from a to b series gangliosides (Iber *et al.*, 1990). This effect could be explained easily by the different pH profiles of the two key regulatory glycosyltransferases involved; GalNAc-transferase and sialyltransferase II (Fig. 2; Iber *et al.*, 1990). However, the physiological relevance of this phenomenon is unclear.

Extracellular proteins inhibiting certain glycosyltransferases also may be involved in the regulation of GSL biosynthesis, as proposed by Caputto and co-workers (Quiroga and Caputto, 1988; Quiroga *et al.*, 1991).

Despite several experimental approaches, regulation of GSL biosynthesis is far from understood. However, purification of several glycosyltransferases (Gu *et al.*, 1990; Melkerson-Watson and Sweeley, 1991; Chatterjee *et al.*, 1992; Nagata *et al.*, 1992), cloning of the respective genes, and analysis of the properties of the enzymes are in progress, and probably will provide additional information about this topic.

III. LYSOSOMAL DEGRADATION OF GLYCOSPHINGOLIPIDS

A. Topology and Mechanism

After endocytosis, possibly via noncoated invaginations, GSLs reach the same endosomes through which ligands internalized via coated pits are channeled, and accumulate in a time-dependent manner in multivesicular bodies (Tran *et al.*, 1987). Endosomes are important sorting centers from which materials may be directed to the lysosomes, to the Golgi, or back to the plasma membrane (Kok *et al.*, 1991; Chapter 20). Whether the components of the plasma membrane finally become part of the lysosomal membrane is not clear. We think this assumption is quite unlikely, considering the massive luminal glycocalyx coat of lysosomal membranes that makes selective degradation by lysosomal enzymes impossible. Alternatively, the occurrence of multivesicular bodies suggested the idea that parts of the endosomal membranes—possibly those enriched in components derived from the plasma membrane—bud off into the endosomal lumen und thus form intraendosomal vesicles. These vesicles could be delivered by successive processes of membrane fission and fusion directly into the lysosol for final degradation of their components (reviewed by Fürst and Sandhoff, 1992).

B. Role of Sphingolipid Activator Proteins

In the lysosomes, GSLs are degraded stepwise by the action of specific acid exohydrolases. Degradation of glycolipids with short hydrophilic head groups also requires the assistance of small glycoprotein cofactors called sphingolipid activator proteins (SAPs) to attack their lipid substrates. Some of these SAPs have been identified on GSL binding proteins. They complex membrane-bound GSLs, lift them from the plane of the membrane, and present them as substrates to the lysosomal hydrolases.

The five SAPs known to date are encoded on two genes. One gene carries the genetic information for the G_{M2}-activator and the second for the *sap* precursor (prosaposin), which is processed to four homologous proteins—saposins A–D (*sap* A to D; Fürst *et al.*, 1988; O'Brien *et al.*, 1988; Nakano *et al.*, 1989). The function, specificity, physicochemical properties, mechanism of action, and biosynthesis of SAPs have been reviewed in detail by Fürst and Sandhoff (1992) and therefore are not discussed further in this chapter.

C. Pathobiochemistry of Inherited Enzyme and Activator Protein Deficiencies

Almost all the lysosomal exohydrolases involved in degradation of GSLs can be deficient and, thus, lead to a lipid storage disease. The inherited deficiency of one of these ubiquitously occurring enzymes causes the lysosomal storage of its substrates. The biochemical as well as the clinical characteristics of the diseases resulting from these enzyme defects have been reviewed in detail (Barranger and Ginns, 1989; Moser *et al.*, 1989; O'Brien, 1989; Sandhoff *et al.*, 1989). The analysis of sphingolipid storage diseases without detectable hydrolase deficiency resulted in the identification of several point mutations in the G_{M2}-activator gene (reviewed by Fürst and Sandhoff, 1992) and in the *sap* precursor gene (Fig. 3). Mutations affecting the *sap* C domain (Gaucher factor) resulted in variant forms of Gaucher disease, whereas mutations affecting the *sap* B domain (sulfatide activator) led to different forms of metachromatic leukodystrophy. A mutation in the initiation codon ATG of the *sap* precursor gene resulted in defects of several activator proteins and, accordingly, in the storage of the respective sphingolipids—ceramide, glucosylceramide, lactosylceramide,

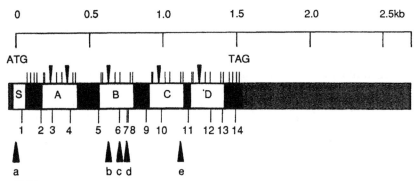

FIGURE 3 cDNA of the *sap* precursor encodes a sequence of 524 (or 527) amino acids (Holtschmidt *et al.*, 1991), including a signal peptide of 16 amino acids (termed S) for entry into the ER (Nakano *et al.*, 1989). The four domains of the precursor (termed saposins A–D) correspond to the mature proteins found in human tissues: A, *sap* A; B, *sap* B or sulfatide activator; C, *sap* C or glucosylceramidase activator; D, *sap* D. The positions of cysteine residues are marked by vertical bars; those of *N*-glycosylation sites are marked by arrow heads. The positions of the 14 introns are numbered. The positions of the known mutations leading to disease are given as (a) A1 → T (Met 1 → Leu) (Schnabel *et al.*, 1992), (b) C650 → T (Thr 217 → Ile) (Kretz *et al.*, 1990; Rafi *et al.*, 1990), (c) 33-bp insertion after G777 (11 additional amino acids after Met 259) (Zhang *et al.*, 1990,1991), (d) G722 → C (Cys 241 → Ser) (Holtschmidt *et al.*, 1991), and (e) G1154 → T (Cys 385 → Phe) (Schnabel *et al.*, 1991).

and ganglioside G_{M3}—in the tissue of the patient (Harzer *et al.*, 1989; Schmidt *et al.*, 1992; Schnabel *et al.*, 1992).

The inherited defects of the G_{M2}-activator lead to the fatal AB variant of G_{M2}-gangliosidosis, causing primarily neuronal storage of ganglioside G_{M2} (Conzelmann and Sandhoff, 1978). The metabolic defects of cultured fibroblasts from an AB variant patient could be corrected when purified G_{M2}-activator was added to the culture medium, using G_{M2}-activator purified from post mortem human tissues (Sonderfeld *et al.*, 1985) as well as a carbohydrate-free human G_{M2}-activator fusion protein obtained by expression of an appropriate cDNA construct in *Escherichia coli* (Klima *et al.*, 1993).

D. Degree of Enzyme Deficiency and Development of Different Clinical Diseases

Inherited deficiencies of the same enzyme may have quite different clinical consequences. Lysosomal storage diseases constitute a particularly striking example of this general phenomenon. The syndromes caused by allelic mutations, for example, of a lysosomal sphingolipid hydrolase, may range from most severe infantile diseases that lead to death in early childhood; to late-infantile, juvenile, and adult forms; to late-adult variants with onset of symptoms in adulthood, slow progression of the disease, and almost normal life-span. Clinically healthy probands with severely reduced activity (10–20% of normal) of a lysosomal enzyme (e.g., arylsulfatase A) also were identified ("pseudodeficiency"). In addition to this temporal variability, the symptomatology also may vary widely.

An obvious choice for the major factor that determines the severity and course of a disease is the residual activity of the affected enzyme. In the case of lysosomal storage diseases, however, the residual activities found in all variants are generally very low and only small differences (if any) are found among variants. We have suggested a simple kinetic model that describes the correlation between the residual activity of an enzyme and the turnover rate of its substrate in a limited compartment such as the lysosome (Conzelmann and Sandhoff, 1983/84). In essence, the model states that substrate concentration in the lysosome is not regulated but will build up until the degrading enzyme becomes saturated to such an extent that the degradation rate equals the influx rate (dynamic equilibrium). Since V_{max} of almost all lysosomal enzymes is sufficiently high to keep this steady state substrate concentration of their substrates ($[S]_{eq}$) far below K_m, a reduction of the activity of one enzyme can, within wide limits, be compensated by an increase of $[S]_{eq}$ so that, due to the higher degree of saturation of the remaining enzyme, the turnover rate will still be equal to the influx rate (Fig. 4). Only when the residual activity falls below a critical threshold (i.e., when $[S]_{eq}$ mathematically becomes infinite), the turnover rate will be reduced to a fraction of substrate influx. A small variation of the residual activity in this critical region has a large influence on the velocity of the storage process and, hence, on the development of disease. The model also is able to explain the occurrence of pseudodeficient probands (persons with a residual activity still above the critical threshold).

Model studies verified the postulated correlation between residual activity and the turnover of the substrate in cell culture (Leinekugel et al., 1992). The radiolabeled substrates ganglioside G_{M2} and sulfatide were added to cultures of skin fibroblasts with different activities of β-hexosaminidase A (originating from patients with G_{M2}-gangliosidoses) or arylsulfatase A (originating from patients with metachromatic leukodystrophy). The uptake and turnover of the substrates were measured. In both series of experiments, the correlation between residual enzyme activity and turnover rate of the substrate was essentially as predicted: degradation rate increased steeply with the residual activity to reach the control level

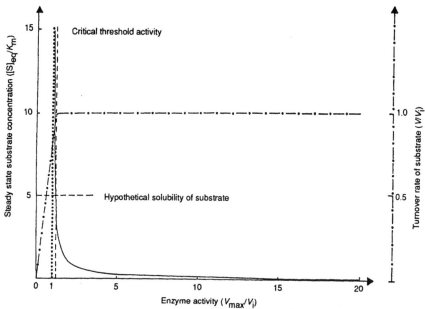

FIGURE 4 Theoretical correlation between the residual activity of an enzyme and the steady state concentration and turnover rate of its substrate in a defined compartment (Conzelmann and Sandhoff, 1983/84). The substrate concentration is expressed as multiples of K_M, turnover rate and enzyme activity (V_{max}) as multiples of the influx rate V_i. Above the critical threshold activity ($V_{max} > V_i$), the turnover rate is limited by the influx rate; below this threshold ($V_{max} < V_i$), it is limited by the remaining capacity of the deficient enzyme. In the latter case, the substrate accumulates at a rate of $V_{acc} = V_i - V_{max}$. Steady state substrate concentration $[S]_{eq}$ (—); turnover (flux) rate (–·–); critical threshold activity (. . .); critical threshold activity, taking limited solubility of substrate into account (–––).

at a residual activity of approximately 10–15% of normal. All cells with an activity above this critical threshold had a normal turnover.

On the basis of their residual enzyme activity below this critical threshold, as determined *in vitro* and *in situ*, different groups of G_{M2}-gangliosidoses corresponding to infantile, juvenile, and adult variants could be distinguished with only marginal overlap, and were well separated from the heterozygotes.

References

Barranger, J. A., and Ginns, E. I. (1989). Glucosylceramide lipidosis: Gaucher disease. *In* "The Metabolic Basis of Inherited Disease" (C. R. Scriver, A. L. Beaudet, W. S. Sly, and D. Valle, eds.), 6th Ed., Vol. II, pp. 1677–1698. McGraw–Hill, New York.

Carey, D. J., and Hirschberg, C. B. (1981). Topography of sialoglycoproteins and sialyltransferases in mouse and rat liver Golgi. *J. Biol. Chem.* **256**, 989–993.

Chatterjee, S., Ghosh, N., and Khurana, S. (1992). Purification of uridine diphosphate-galactose: Glucosyl ceramide, $\beta 1 \rightarrow 4$ galactosyltransferase from human kidney. *J. Biol. Chem.* **267**, 7148–7153.

Coleman, P. L., Fishman, P. H., Brady, R. O., and Todaro, G. J. (1975). Altered ganglioside biosynthesis in mouse cell cultures following transformation with chemical carcinogens and X irradiation. *J. Biol. Chem.* **250**, 55–60.

Conzelmann, E., and Sandhoff, K. (1978). AB variant of infantile G_{M2}-gangliosidosis: Deficiency of a factor necessary for stimulation of hexosaminidase A-catalyzed degradation of ganglioside G_{M2} and glycolipid G_{A2}. *Proc. Natl. Acad. Sci. U.S.A.* **75**, 3979–3983.

Conzelmann, E., and Sandhoff, K. (1983/84). Partial enzyme deficiencies: Residual activities and the development of neurological disorders. *Dev. Neurosci.* **6**, 58–71.

Coste, H., Martel, M.-B., Azzar, G., and Got, R. (1985). UDP glucose-ceramide glycosyltransferase from porcine submaxillary glands is associated with the Golgi apparatus. *Biochim. Biophys. Acta* **814**, 1–7.

Coste, H., Martel, M.-B., and Got, R. (1986). Topology of glucosylceramide synthesis in Golgi membranes from porcine submaxillary glands. *Biochim. Biophys. Acta* **858**, 6–12.

Cumar, F. A., Talman, J. F., and Brady, R. O. (1972). The biosynthesis of a disialoganglioside by galactosyltransferase from rat brain tissue. *J. Biol. Chem.* **247**, 2322–2327.

Daniotti, J. L., Landa, C. A., Rösner, H., and Maccioni, H. J. F. (1991). GD3 prevalence in adult rat retina correlates with the maintenance of a high GD3-/GM2-synthase activity ratio throughout development. *J. Neurochem.* **57**, 2054–2058.

Deutscher, S. L., and Hirschberg, C. B. (1986). Mechanism of galactosylation in the Golgi apparatus. *J. Biol. Chem.* **261**, 96–100.

Fishman, P. H., Bradley, R. M., and Henneberry, R. C. (1976). Butyrate-induced glycolipid biosynthesis in HeLa cells: Properties of the induced sialyltransferase. *Arch. Biochem. Biophys.* **172**, 618–626.

Fleischer, B. (1981). Orientation of glycoprotein galactosyltransferase and sialyltransferase enzymes in vesicles derived from rat liver Golgi apparatus. *J. Cell Biol.* **89**, 246–255.

Fürst, W., and Sandhoff, K. (1992). Activator proteins and topology of lysosomal sphingolipid catabolism. *Biochim. Biophys. Acta* **1126**, 1–16.

Fürst, W., Machleidt, W., and Sandhoff, K. (1988). The precursor of sulfatide activator protein is processed to three different proteins. *Hoppe Seyler's Z. Physiol. Chem.* **369**, 317–328.

Futerman, A. H., and Pagano, R. E. (1991). Determination of the intracellular sites and topology of glucosylceramide synthesis in rat liver. *Biochem. J.* **280**, 295–302.

Futerman, A. H., Stieger, B., Hubbard, A. L., and Pagano, R. E. (1990). Sphingomyelin synthesis in rat liver occurs predominantly at the *cis* and medial cisternae of the Golgi apparatus. *J. Biol. Chem.* **265**, 8650–8657.

Gu, X.-B., Gu, T.-J., and Yu, R. K. (1990). Purification to homogeneity of GD3 synthase and partial purification of GM3 synthase from rat brain. *Biochem. Biophys. Res. Commun.* **166**, 387–393.

Harzer, K., Paton, B. C., Poulos, A., Kustermann-Kuhn, B., Roggendorf, W., Grisar, T., and Popp, M. (1989). Sphingolipid activator protein (SAP) deficiency in a 16-week-old atypical Gaucher disease patient and his fetal sibling: Biochemical signs of combined sphingolipidoses. *Eur. J. Pediatr.* **149**, 31–39.

Hashimoto, Y., Suzuki, A., Yamakawa, T., Miyashita, N., and Moriwaki, K. (1983). Expression of GM1 and GD1a in mouse liver is linked to the H_2 complex on chromosome 17. *J. Biochem.* **94**, 2043–2048.

Hattori, M. A., and Horiuchi, R. (1992). Enhancement of ganglioside GM3 synthesis in okadaik-acid-treated granulosa cells. *Biochim. Biophys. Acta* **1137**, 101–106.

Hogan, M. V., Saito, M., and Rosenberg, A. (1988). Influence of monensin on ganglioside anabolism and neurite stability in cultured chick neurons. *J. Neurosci. Res.* **20**, 390–394.

Holleran, W. M., Feingold, K. R., Mao-Qiang, M., Gao, W. N., Lee, J. M., and Elias, P. M. (1991). Regulation of epidermal sphingolipid synthesis by permeability barrier function. *J. Lipid Res.* **32**, 1151–1158.

Holtschmidt, H., Sandhoff, K., Kwon, H. Y., Harzer, K., Nakano, T., and Suzuki, K. (1991). Sulfatide activator protein: Alternative splicing generates three messenger RNAs and a newly found mutation responsible for a clinical disease. *J. Biol. Chem.* **266**, 7556–7560.

Iber, H., and Sandhoff, K. (1989). Identity of GD1c, GT1a and GQ1b synthase in Golgi vesicles from rat liver. *FEBS Lett.* **254**, 124–128.

Iber, H., Kaufmann, R., Pohlentz, G., Schwarzmann, G., and Sandhoff, K. (1989). Identity of GA1-, GM1a-, and GD1b synthase in Golgi vesicles from rat liver. *FEBS Lett.* **248**, 18–22.

Iber, H., van Echten, G., Klein, R. A., and Sandhoff, K. (1990). pH-Dependent changes of ganglioside biosynthesis in neuronal cell culture. *Eur. J. Cell Biol.* **52**, 236–240.

Iber, H., van Echten, G., and Sandhoff, K. (1991). Substrate specificity of $\alpha2\rightarrow3$ sialyltransferases in ganglioside biosynthesis of rat liver Golgi. *Eur. J. Biochem.* **195**, 115–120.

Iber, H., van Echten, G., and Sandhoff, K. (1992a). Fractionation of primary cultured neurons: Distribution of sialyltransferases involved in ganglioside biosynthesis. *J. Neurochem.* **58**, 1533–1537.

Iber, H., Zacharias, C., and Sandhoff, K. (1992b). The c-series gangliosides GT3, GT2, and GP1c are formed in rat liver Golgi by the same set of glycosyltransferases that catalyze the biosynthesis of asialo-, a-, and b-series gangliosides. *Glycobiology* **2**, 137–142.

Jeckel, D., Karrenbauer, A., Birk, R., Schmidt, R. R., and Wieland, F. (1990). Sphingomyelin is synthesized in the *cis* Golgi. *FEBS Lett.* **261**, 155–157.

Jeckel, D., Karrenbauer, A., Burger, K. N. J., van Meer, G., and Wieland, F. (1992). Glycosylceramide is synthesized at the cytosolic surface of various Golgi subfractions. *J. Cell Biol.* **117**, 259–267.

Klima, H., Klein, A., van Echten, G., Schwarzmann, G., Suzuki, K., and Sandhoff, K. (1993). Overexpression of a functionally active human GM2-activator protein in *Escherichia coli*. *Biochem. J.* **292**, 571–576.

Kok, J. W., Babia, T., and Hoekstra, D. (1991). Sorting of sphingolipids in the endocytic pathway of HT29 cells. *J. Cell Biol.* **114**, 231–239.

Kolesnick, R. (1992). Ceramide: a novel second messenger. *Trends Cell Biol.* **2**, 232–236.

Kretz, K. A., Carson, G. S., Morimoto, S., Kishimoto, Y., Fluharty, A. L., and O'Brien, J. S. (1990). Characterization of a mutation in a family with saposin B deficiency: A glucosylation site defect. *Proc. Natl. Acad. Sci. U.S.A.* **87**, 2541–2544.

Leinekugel, P., Michel, S., Conzelmann, E., and Sandhoff, K. (1992). Quantitative correlation between the residual activity of β-hexosaminidase A and arylsulfatase A and the severity of the resulting lysosomal storage disease. *Hum. Genet.* **88**, 513–523.

Li, R., and Ladisch, S. (1992). Alteration of neuroblastoma ganglioside metabolism by retinoic acid. *J. Neurochem.* **59**, 2297–2303.

Mandon, E. C., van Echten, G., Birk, R., Schmidt, R. R., and Sandhoff, K. (1991). Sphingolipid biosynthesis in cultured neurons. Down-regulation of serine palmitoyltransferase by sphingoid bases. *Eur. J. Biochem.* **198**, 667–674.

Mandon, E., Ehses, I., Rother, J., van Echten, G., and Sandhoff, K. (1992). Subcellular localization and membrane topology of serine palmitoyltransferase, 3-dehydrosphinganine reductase, and sphinganine N-acyltransferase in mouse liver. *J. Biol. Chem.* **267**, 11144–11148.

Matsuura, H., Greene, T., and Hakomori, S.-I. (1989). An α-N-acetylgalactosaminylation at the threonine residue of a defined peptide sequence creates the oncofetal peptide epitope in human fibronectin. *J. Biol. Chem.* **264**, 10472–10476.

McLawhon, R. W., Schoon, G., and Dawson, G. (1981). Possible role of cyclic AMP in the receptor-mediated regulation of glycosyltransferase activities in neurotumor cell lines. *J. Neurochem.* **37**, 132–139.

Melkerson-Watson, L. J., and Sweeley, C. C. (1991). Purification to apparent homogeneity by immunoaffinity, chromatography and partial characterization of the GM3-ganglioside-forming enzyme, CMP-sialic acid : lactosylceramide α2,3-sialyltransferase (SAT-1), from rat-liver Golgi. *J. Biol. Chem.* **266**, 4448–4457.

Merrill, A. M. (1991). Cell regulation by sphingosine and more complex sphingolipids. *J. Bioenerg. Biomembr.* **23**, 83–104.

Miller-Prodraza, H., and Fishman, P. H. (1984). Effects of drugs and temperature on biosynthesis and transport of glycosphingolipids in cultured neurotumor cells. *Biochim. Biophys. Acta* **804**, 44–51.

Moser, H. W., Moser, A. B., Chen, W. W., and Schram, A. W. (1989). Ceramidase deficiency: Farber lipogranulomatosis. *In* "The Metabolic Basis of Inherited Disease" (C. R. Scriver, A. L. Beaudet, W. S. Sly, and D. Valle, eds.), 6th Ed., Vol. II, pp. 1645–1654. McGraw-Hill, New York.

Moskal, J. R., Lockney, M. W., Marvel, C. C., Trosko, J. E., and Sweeley, C. C. (1987). Effect of retinoic acid and phorbol-12-myristate-13-acetate on glycosyltransferase activities in normal and transformed cells. *Cancer Res.* **47**, 787–790.

Nagai, Y., Nakaishi, H., and Sanai, Y. (1986). Gene transfer as a novel approach to the gene-controlled mechanism of the cellular expression of glycosphingolipids. *Chem. Phys. Lipids* **42**, 91–103.

Nagata, Y., Yamashiro, S., Yodai, J., Lloyd, K. O., Shiku, H., and Furukawa, K. (1992). Expression cloning of β1,4 N-acetylgalactosaminyltransferase cDNAs that determine the expression of GM2 and GD2 gangliosides. *J. Biol. Chem.* **267**, 12082–12089.

Nakaishi, H., Sanai, Y., Shiroki, K., and Nagai, Y. (1988a). Analysis of cellular expression of gangliosides by gene transfection I: GD3 expression in *myc*-transfected and transformed 3Y1 correlates with anchorage-independent growth activity. *Biochem. Biophys. Res. Commun.* **150**, 760–765.

Nakaishi, H., Sanai, Y., Shibuya, M., and Nagai, Y. (1988b). Analysis of cellular expression of gangliosides by gene transfection. II. Rat 3Y1 cells transformed with several DNAs containing oncogenes (*fes, fps, ras,* and *src*) invariably express sialosylparagloboside. *Biochem. Biophys. Res. Commun.* **150**, 766–774.

Nakakuma, H., Sanai, Y., Shiroki, Y., and Nagai, Y. (1984). Gene-regulated expression of glycolipids: Appearance of GD3 ganglioside in rat cells on transfection with transforming gene E1 of human adenovirus type 12 DNA and its transcriptional subunits. *J. Biochem.* **96**, 1471–1480.

Nakamura, M., Tsunoda, A., Sakoe, K., Gu, J., Nishikawa, A., Taniguchi, N., and Saito, M. (1992). Total metabolic flow of glycosphingolipid biosynthesis is regulated by UDP-GlcNAc : lactosylceramide β1→3N-acetylglucosaminyltransferase and CMP-NeuAc : lactosylceramide α2→3 sialyltransferase in human hematopoietic cell line HL-60 during differentiation. *J. Biol. Chem.* **267**, 23507–23514.

Nakano, T., Sandhoff, K., Stümper, J., Christomanou, H., and Suzuki, K. (1989). Structure of full-length cDNA coding for sulfatide activator, a co-β-glucosidase and two other homologous proteins: Two alternate forms of the sulfatide activator. *J. Biochem. (Tokyo)* **105**, 152–154.

Nores, G. A., and Caputto, R. (1984). Inhibition of the UDP-*N*-acetylgalactosamine: GM3, *N*-acetylgalactosaminyltransferase by gangliosides. *J. Neurochem.* **42**, 1205–1211.

O'Brien, J. S. (1989). β-Galactosidase deficiency (GM1, gangliosidosis, galactosialidosis, and Morquio syndrome type B); Ganglioside sialidase deficiency (mucolipidosis IV). *In* "The Metabolic Basis of Inherited Disease" (C. R. Scriver, A. L. Beaudet, W. S. Sly, and D. Valle, eds.), 6th Ed., Vol. II, pp. 1797–1806. McGraw-Hill, New York.

O'Brien, J. S., Kretz, K. A., Dewji, N., Wenger, D. A., Esch, F., and Fluharty, A. L. (1988). Coding of two sphingolipid activator proteins (SAP-1 and SAP-2) by same genetic locus. *Science* **241**, 1098–1101.

Olivera, A., and Spiegel, S. (1992). Ganglioside GM1 and sphingolipid breakdown products in cell proliferation and signal transduction pathways. *Glycoconjugate J.* **9**, 110–117.

Pacuszka, T., Duffard, R. O., Nishimura, R. N., Brady, R. O., and Fishman, P. H. (1978). Biosynthesis of bovine thyroid gangliosides. *J. Biol. Chem.* **253**, 5839–5846.

Pohlentz, G., Klein, D., Schwarzmann, G., Schmitz, D., and Sandhoff, K. (1988). Both GA2, GM2, and GD2 synthases and GM1b, GD1a, and GT1b synthases are single enzymes in Golgi vesicles from rat liver. *Proc. Natl. Acad. Sci. U.S.A.* **85**, 7044–7048.

Pohlentz, G., Schlemm, S., and Egge, H. (1992). 1-Deoxy-1-phosphatidylethanolamino-lactitol-type neoglycolipids serve as acceptors for sialyltransferases from rat liver Golgi vesicles. *Eur. J. Biochem.* **203**, 387–392.

Quiroga, S., and Caputto, R. (1988). An inhibitor of the UDP-acetylgalactosamine: GM3, *N*-acetylgalactosaminyltransferase: Purification and properties, and preparation of an antibody to this inhibitor. *J. Neurochem.* **50**, 1695–1700.

Quiroga, S., Panzetta, P., and Caputto, R. (1991). Internalization of the inhibitor of the *N*-acetylgalactosaminyltransferase by chicken embryonic retina cells—Reversibility of the inhibitor effects. *J. Neurosci. Res.* **30**, 414–420.

Rafi, M. A., Zhang, X.-L., De Gala, G., and Wenger, D. A. (1990). Detection of a point mutation in sphingolipid activator protein-1 mRNA in patients with a variant form of metachromatic leukodystrophy. *Biochem. Biophys. Res. Commun.* **166**, 1017–1023.

Richardson, C. L., Keenan, T. W., and Morré, D. J. (1977). Ganglioside biosynthesis. Characterization of CMP-*N*-acetylneuraminic acid: lactosylceramide sialyltransferase in Golgi apparatus from rat liver. *Biochim. Biophys. Acta* **488**, 88–96.

Rother, J., van Echten, G., Schwarzmann, G., and Sandhoff, K. (1992). Biosynthesis of sphingolipids: Dihydroceramide and not sphinganine is desaturated by cultured cells. *Biochem. Biophys. Res. Commun.* **189**, 14–20.

Ruan, S., and Lloyd, K. O. (1992). Glycosylation pathways in the biosynthesis of gangliosides in melanoma and neuroblastoma cells: Relative glycosyltransferase levels determine ganglioside patterns. *Cancer Res.* **52**, 5725–5731.

Saito, M., Saito, M., and Rosenberg, A. (1984). Action of monensin, a monovalent cationo-phore, on cultured human fibroblasts: Evidence that it induces high cellular accumulation of glucosyl- and lactosylceramide. *Biochemistry* **23**, 1043–1046.

Sandhoff, K., and van Echten, G. (1993). Ganglioside metabolism. Topology and regulation. *In* "Advances in Lipid Research" (R. M. Bell, Y. A. Hannun, and A. H. Merrill, Jr., eds.). Vol. 26, pp. 119–142. Academic Press.

Sandhoff, K., Conzelmann, E., Neufeld, E. F., Kaback, M. M., and Suzuki, K. (1989). The GM2 gangliosidosis. *In* "The Metabolic Basis of Inherited Disease" (C. R. Scriver, A. L. Beaudet, W. S. Sly, and D. Valle, eds.), 6th Ed., Vol. II, pp. 1808–1839. McGraw–Hill, New York.

Scheideler, M., and Dawson, G. (1986). Direct demonstration of the activation of UDP-N-acetylgalactosamine : GM3-N-acetylgalactosaminyltransferase by cyclic AMP. *J. Neurochem.* **46**, 1639–1643.

Schmidt, B., Paton, B. C., Sandhoff, K., and Harzer, K. (1992). Metabolism of GM1 ganglioside in cultured skin fibroblasts: Anomalies in gangliosidoses, sialidoses, and sphingolipid activator protein (SAP, saposin) 1 and prosaposin deficiencies. *Hum. Genet.* **89**, 513–518.

Schnabel, D., Schröder, M., and Sandhoff, K. (1991). Mutation in the sphingolipid activator protein 2 in a patient with a variant of Gaucher disease. *FEBS Lett.* **284**, 57–59.

Schnabel, D., Schröder, M., Fürst, W., Klein, A., Hurwitz, R., Zenk, T., Weber, G., Harzer, K., Paton, B., Poulos, A., Suzuki, K., and Sandhoff, K. (1992). Simultaneous deficiency of sphingolipid activator proteins 1 and 2 is caused by a mutation in the initiation codon of their common gene. *J. Biol. Chem.* **267**, 3312–3315.

Schwarzmann, G., and Sandhoff, K. (1990). Metabolism and intracellular transport of glycosphingolipids. *Biochemistry* **29**, 10865–10871.

Shukla, G. S., Shukla, A., and Radin, N. S. (1991). Gangliosides inhibit glucosylceramide synthase: A possible role in ganglioside therapy. *J. Neurochem.* **56**, 2125–2132.

Sonderfeld, S., Conzelmann, E., Schwarzmann, G., Burg, J., Hinrichs, U., and Sandhoff, K. (1985). Incorporation and metabolism of ganglioside G_{M2} in skin fibroblasts from normal and G_{M2} gangliosidosis subjects. *Eur. J. Biochem.* **149**, 247–255.

Svennerholm, L. (1963). Chromatografic separation of human brain gangliosides. *J. Neurochem.* **10**, 613–623.

Tran, D., Carpentier, J.-L., Sawano, I., Gorden, P., and Orci, L. (1987). Ligands internalized through coated or noncoated invaginations follow a common intracellular pathway. *Proc. Natl. Acad. Sci. U.S.A.* **84**, 7957–7961.

Trinchera, M., and Ghidoni, R. (1989). The glycosphingolipid sialyltransferases are localized in different sub-Golgi compartments in rat liver. *J. Biol. Chem.* **264**, 15766–15769.

Trinchera, M., Pirovano, B., and Ghidoni, R. (1990). Sub-Golgi distribution in rat liver of CMP-NeuAc : GM3- and CMP-NeuAc : GT1b $\alpha 2 \rightarrow 8$ sialyltransferases and comparison with the distribution of the other glycosyltransferase activities involved in ganglioside biosynthesis. *J. Biol. Chem.* **265**, 18242–18247.

Trinchera, M., Fabbri, M., and Ghidoni, R. (1991). Topography of glycosyltransferases involved in the initial glycosylations of gangliosides. *J. Biol. Chem.* **266**, 20907–20912.

van Echten, G., and Sandhoff, K. (1989). Modulation of ganglioside biosynthesis in primary cultured neurons. *J. Neurochem.* **52**, 207–214.

van Echten, G., Iber, H., Stotz, H., Takatsuki, A., and Sandhoff, K. (1990a). Uncoupling of ganglioside biosynthesis by Brefeldin A. *Eur. J. Cell Biol.* **51**, 135–139.

van Echten, G., Birk, R., Brenner-Weiss, G., Schmidt, R. R., and Sandhoff, K. (1990b). Modulation of sphingolipid biosynthesis in primary cultured neurons by long-chain bases. *J. Biol. Chem.* **265**, 9333–9339.

Young, W. W., Jr., Lutz, M. S., Mills, S. E., and Lechler-Osborn, S. (1990). Use of Brefeldin A to define sites of glycosphingolipid synthesis GA2/GM2/GD2 synthase is *trans* to the BFA block. *Proc. Natl. Acad. Sci. U.S.A.* **87**, 6838–6842.

Yu, R. K., Macala, L. J., Taki, T., Weinfeld, H. M., and Yu, F. S. (1988). Developmental changes in ganglioside composition and synthesis in embryonic rat brain. *J. Neurochem.* **50**, 1825–1829.

Yusuf, H. K. M., Pohlentz, G., and Sandhoff, K. (1983). Tunicamycin inhibits ganglioside biosynthesis in rat liver Golgi apparatus by blocking sugar nucleotide transport across the membrane vesicles. *Proc. Natl. Acad. Sci. U.S.A.* **80**, 7075–7079.

Yusuf, H. K. M., Schwarzmann, G., Pohlentz, G., and Sandhoff, K. (1987). Oligosialoganglisides inhibit GM2- and GD3-synthesis in isolated Golgi vesicles from rat liver. *Hoppe Seyler's Z. Physiol. Chem.* **368,** 455–462.

Zeller, C. B., and Marchase, R. B. (1992). Gangliosides as modulators of cell function. *Am. J. Physiol. Cell Physiol.* **262,** C1341–C1355.

Zhang, X.-L., Rafi, M. A., De Gala, G., and Wenger, D. A. (1990). Insertion in the mRNA of a metachromatic leukodystrophy patient with sphingolipid activator protein-1 deficiency. *Proc. Natl. Acad. Sci. U.S.A.* **87,** 1426–1430.

Zhang, X.-L., Rafi, M. A., De Gala, G., and Wenger, D. A. (1991). The mechanism for a 33-nucleotide insertion in messenger RNA causing sphingolipid activator protein (SAP-1)-deficient metachromatic leukodystrophy. *Hum. Genet.* **87,** 211–215.

CHAPTER 4

Ceramide Metabolism Compartmentalized in the Endoplasmic Reticulum and Golgi Apparatus

Anthony H. Futerman
Department of Membrane Research and Biophysics, Weizmann Institute of Science, Rehovot 76100, Israel

I. INTRODUCTION

Two suggestions have generated significant interest in the cell biology of sphingolipids. First, work in the mid-1980s led to the proposal that the long-chain sphingoid base of sphingolipids is involved in signal transduction (Hannun and Bell, 1989). More recently, researchers have proposed that ceramide and ceramide phosphate also act as second messengers (Kolesnick, 1992). These proposals are being widely tested and are the subject of a number of reviews (see for instance, Chapters 13, 14, and 15). Second, sphingolipids have been implied to play important roles in the functioning of the secretory pathway, in which they may be involved in regulating protein transport and retention in the endoplasmic reticulum (ER) and Golgi apparatus (Machamer, 1991; Rosenwald et al., 1992) and in directing transport from the *trans*-Golgi network to the plasma membrane (PM) of polarized cells (reviewed by Nelson, 1992; van Meer and Burger, 1992; Chapter 21). If sphingolipids are indeed involved in the functioning of the secretory pathway, their metabolism in and transport between the various compartments of the pathway must be controlled carefully.

In this chapter, I review studies published during the last several years that have demonstrated that the early steps of sphingolipid synthesis are compartmentalized in the ER and Golgi apparatus. In particular, I discuss the sites and membrane topology of three enzymes that reside in these compartments—acyl CoA:sphinganine *N*-acyltransferase (dihydroceramide synthase), UDP-glucose:ceramide glucosyltransferase [glucosylceramide (GlcCer) synthase], and phosphatidylcholine (PC):ceramide choline phosphotransferase [sphingomyelin (SM) synthase]. Although the sites and topology of these enzymes have now been elucidated, based mainly on subcellular fractionation studies, little is known about the regulation of synthesis or mechanisms of intracellular transport of sphingolipids in the ER and Golgi apparatus. Ultimately, understanding the regulation of sphingolipid synthesis will depend on isolating these enzymes and subsequently analyzing them at the molecular level. None of the three enzymes has been isolated to date, but I hope the next volume in this series devoted to the cell biology of lipids will contain reports of the isolation and purification of a number of enzymes involved in sphingolipid metabolism.

II. SITES OF CERAMIDE METABOLISM

The central molecule in sphingolipid metabolism is ceramide. Ceramide is formed by the condensation of serine and palmitoyl CoA, followed by reduction of 3-ketosphinganine, acylation of the amide group of sphinganine, and finally dehydrogenation of dihydroceramide. Ceramide subsequently is modified by addition of various polar (e.g., phosphocholine) or nonpolar (e.g., glucose, galactose) moieties at carbon-1 of the sphingoid base. Some of the reactions in which ceramide is involved are illustrated in Fig. 1 and are discussed in detail in the following section.

A. Ceramide Synthesis

The predominant long-chain base of sphingolipids is sphingosine (*trans*-D-*erythro*-2-amino-4-octadecene-1,3-diol), although sphinganine (D,L-

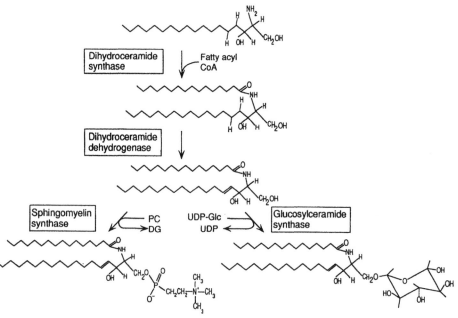

FIGURE 1 Scheme for ceramide metabolism. Additional details of the reactions are given in the text. DG, diacylglycerol; PC, phosphatidylcholine; Glc, glucose.

erythro-1,3-dihydroxy-2-aminooctadecane) is found also, but in much lower amounts. Only recently have researchers accepted that the 4-*trans* double bond of sphingolipids is added after acylation of the sphingoid base (Fig. 1), despite observations published from *in vivo* studies in the 1970s that strongly suggested the same idea (Ong and Brady, 1973; reviewed by Merrill and Jones, 1990). Thus, some early studies examining the site of acylation of sphingoid bases used sphingosine, not sphinganine, as the substrate (see, for instance, Walter *et al.*, 1983). This is unlikely to have been of any consequence in localizing the site of acylation of the sphingoid long-chain base, since dihydroceramide synthase is likely to be able to utilize both sphingosine and sphinganine as substrate, as demonstrated by the observation that fumonisin, an inhibitor of sphingoid base acylation, inhibits acylation of both sphingosine and sphinganine (Wang *et al.*, 1991; Hirschberg *et al.*, 1993).

In early subcellular fractionation studies, ceramide synthesis was observed in microsomal fractions from either mouse brain (Morell and Radin, 1970) or rat liver (Walter *et al.*, 1983). Useful as these and other studies were in demonstrating that ceramide synthesis occurred early in the secretory pathway, the studies were limited since, in one case, no attempts were made to distinguish between the various membranes that copurify in a microsomal fraction (Morell and Radin, 1970) and, in the other, although ER and Golgi apparatus fractions were separated, no attempts were made to quantify the levels of ceramide synthesis in any of the separated fractions.

The localization of dihydroceramide synthase to the ER now has been demonstrated unambiguously using subcellular fractions from mouse (Mandon et al., 1992) or rat (Hirschberg et al., 1993) liver. In the former study, two other enzymes of sphingolipid synthesis—serine dihydropalmitoyltransferase and 3-sphinganine reductase—also were localized to the ER. Both these studies are characterized by careful separation of the ER from the Golgi apparatus, and by analysis of contamination with Golgi apparatus membranes in the ER fractions and vice versa using marker enzymes for each compartment (Fig. 2). This type of analysis permits calculation of the levels of contamination of isolated membrane fractions and is essential if definitive conclusions are to be reached concerning the precise intracellular localization of an enzyme activity. In the case of ER and Golgi apparatus fractions, analysis is relatively easy since established enzyme markers are readily available for each compartment (i.e., glucose 6-phosphatase for the ER and galactosyltransferase for the Golgi apparatus). Problems arise, however, in attempting to separate subfractions of the Golgi apparatus from each other or from the pool of membranes known collectively as the "intermediate compartment" (see subsequent discussion) and, consequently, in analyzing the contamination of these fractions.

Another prerequisite for comparing enzyme activities in different subcellular fractions is that the substrates are present at saturating concentrations so the rate of the reaction is directly proportional to the amount of enzyme present in the fractions and is not limited in any way by availability of substrates. This requirement is particularly important in studies of membrane-bound enzymes for which the substrates are often not soluble in aqueous media. In our analysis of the distribution of dihydroceramide synthase in rat liver fractions (Hirschberg et al., 1993), care was taken to ensure that sphinganine quantitatively transferred into the membranes of the fractions so the rate of formation of dihydroceramide was not limited by availability of sphinganine or palmitoyl CoA. In addition, the rate of the reaction was found to be critically dependent on the ratio of palmitoyl CoA and bovine serum albumin (BSA) in the reaction mixture. At low ratios, palmitoyl CoA bound tightly to BSA and the apparent concentration of palmitoyl CoA was reduced; at high ratios, palmitoyl CoA acted as a detergent and severely reduced dihydroceramide synthesis, presumably by disrupting membrane integrity or enzyme structure. Analysis of sphin-

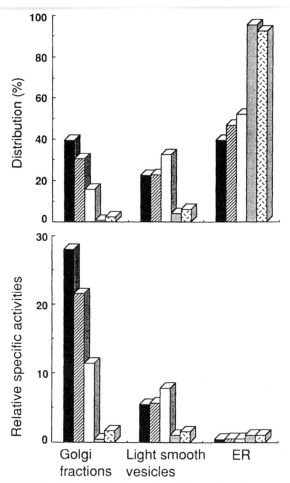

FIGURE 2 Analysis of the distribution and specific activities of enzymes in microsomal subfractions. Subcellular fractions were prepared as described (Futerman *et al.*, 1990; Futerman and Pagano, 1991; Hirschberg *et al.*, 1993). The Golgi apparatus fraction consists of pooled *trans* and *cis*/medial fractions, and the ER fraction of pooled smooth and rough microsomal fractions. Percentage distribution is calculated as the ratio of the activity in the fraction divided by the total recovered activity. Relative specific activity is given as the ratio of the specific activity in the fraction compared with the specific activity in a microsomal suspension. Note that both parameters must be taken into account to determine the location of an enzyme activity. Thus, for dihydroceramide synthesis (cross-hatch), no increase in relative specific activity is observed between the microsomal suspension and the ER; the highest specific activity is observed in the Golgi fractions and in light smooth vesicles. However, more than 90% of the recovered activity is found in the ER fraction; the distribution of dihydroceramide synthase exactly parallels that of the ER marker glucose 6-phosphatase (shaded). For SM synthase (black), significant amounts of activity are found in the ER fraction, but this activity corresponds to contaminating Golgi apparatus membranes (Futerman *et al.*, 1990) since a similar amount of the Golgi apparatus marker galactosyl transferase (hatched) is found in the ER fraction. However, the highest specific activity of SM synthase is found in the Golgi apparatus fractions. Note also the difference in relative specific activities and distribution between SM synthase and GlcCer synthase (white), particularly between the Golgi and light smooth vesicle fractions.

ganine N-acyltransferase in the absence of BSA using high concentrations of palmitoyl CoA therefore may lead to an underestimation of enzyme activity.

A key step in the regulation of sphingolipid metabolism may be the dehydrogenation of dihydroceramide to ceramide (Fig. 1). This reaction has been observed *in vivo* by incubating cerebellar neurons or B104 neuroblastoma cells with dihydroceramide (Rother *et al.*, 1992), but has yet to be observed *in vitro*. The double bond of sphingolipids has been postulated to be added in a subcellular compartment different from that in which dihydroceramide is synthesized (Merrill and Wang, 1986; Merrill and Jones, 1990). Isolation of both these enzymes will resolve this issue. The availability of fumonisin, an inhibitor of dihydroceramide synthase, presumably will permit the purification of dihydroceramide synthase by affinity chromatography, although solubilizing the enzyme in an active form, which may prove difficult, will be necessary first.

B. Glucosylceramide Synthesis

Synthesis of GlcCer by the transfer of glucose from UDP-glucose to ceramide is the first step in the formation of most complex glycosphingolipids (Fig. 1). Since the sequential glycosylation of sphingolipids occurs in the Golgi apparatus, GlcCer initially was assumed also to be synthesized in this organelle; studies performed in the mid-1980s seemed to support this idea. Isolation of a Golgi-rich fraction from porcine submaxillary glands (Coste *et al.*, 1985) revealed a 20-fold enrichment in the specific activity of GlcCer synthesis relative to a postnuclear supernatant, but a 6-fold increase in specific activity also was found in a smooth ER fraction that was not characterized further. No analysis of the distribution of various marker enzymes in the subcellular fractions was provided, so the distribution and recoveries of GlcCer synthesis in subcellular compartments of porcine submaxillary glands could not be determined. Other studies demonstrated that GlcCer synthesis occurred in microsomal fractions of BHK-21 cells (Suzuki *et al.*, 1984) and of rat brain (reviewed by Sasaki, 1990), but these fractions contained membranes derived from both the ER and Golgi apparatus. Use of fluorescently labeled analogs of ceramide demonstrated that GlcCer was synthesized in cultured cells before being transported out of the Golgi apparatus (Lipsky and Pagano, 1983; van Meer *et al.*, 1987).

Three studies have been performed that sought to determine the precise localization of GlcCer synthesis in the ER or Golgi apparatus. First, highly enriched and well-characterized subcellular fractions were obtained from rat liver (Futerman and Pagano, 1991). More than 40% of the recovered GlcCer synthesis was associated with an enriched Golgi apparatus fraction. However, between 65 and 75% of a Golgi marker enzyme was found in this fraction, suggesting that only a portion of the total GlcCer synthesis was being recovered. Nearly 30% of the recovered GlcCer synthase was observed in another fraction, which contained 23% of an ER marker but only 14% of a Golgi marker (see Table 1 in Futerman and Pagano, 1991). Subsequent analysis of Golgi apparatus subfractions showed that substantial amounts of GlcCer were synthesized in a heavy (*cis*/medial) Golgi apparatus subfraction, a light smooth vesicle fraction that was almost devoid of an ER marker enzyme (glucose 6-phosphatase), and a heavy vesicle fraction (smooth microsomes) (see Table 2 in Futerman & Pagano, 1991). These results demonstrated that GlcCer synthesis was distributed fairly widely among the various subcompartments that constitute the ER and Golgi apparatus or, alternatively, that significant amounts of GlcCer were synthesized in a pre-Golgi apparatus compartment that consisted of light smooth vesicles and copurified with an intact Golgi apparatus fraction, a *cis*/medial Golgi apparatus subfraction, and a smooth microsome fraction. The light smooth vesicle fraction was not characterized further, but the density and profile of marker enzymes in the fraction suggested that it contained membranes derived from a compartment intermediate in density between the *cis*/medial Golgi apparatus and the smooth ER. The possibility that this compartment corresponds to the intermediate compartment seen, for example, in Vero cells (Schweizer *et al.*, 1991) has not been tested to date.

A second study also used subcellular fractions from rat liver (Trinchera *et al.*, 1991). GlcCer synthesis was observed in a Golgi apparatus fraction in which a 19% yield and 159-fold enrichment of galactosyltransferase was obtained. No GlcCer was synthesized in an ER fraction, although these data must be viewed in light of the fact that no activity was observed in the whole rat liver homogenate, suggesting that the detergent-based assay of GlcCer synthesis used in this study was not sufficiently sensitive to detect low levels of activity. Because of the inability to detect GlcCer synthesis in any other fraction, no analysis of the distribution of GlcCer synthase in rat liver was provided. Note that a high specific activity of an enzyme in one particular subcellular fraction does not necessarily imply that the major portion of the activity is found in that fraction. Both specific activities and distribution of recovered activities must be analyzed if unambiguous conclusions are to be drawn about the intracellular localization of an enzyme. In addition, an assay of enzyme activity in which membrane integrity is not destroyed is highly desirable, particularly for enzymes that act on substrates that are membrane associated. Such an assay has been developed for both GlcCer and SM synthase in which a short-chain radioactive fatty acid [N-([1-^{14}C]hexanoyl-D-*erythro*-ceramide)] replaces the naturally occurring fatty acid (Futerman and Pagano, 1992). This lipid analog rapidly transfers between donor and acceptor membranes in the absence of detergent.

An assay based on this approach was used to analyze the site of GlcCer synthesis in the third study, using both rat liver and HepG2 fractions (Jeckel *et al.*, 1992). Sucrose gradients were used to separate fractions containing GlcNAc-phosphotransferase from fractions containing galactosyltransferase (presumed proximal and distal Golgi apparatus markers, respectively), although the gradient used was only successful in separating the marker enzymes in 4 of the 12 experiments performed. Significant amounts of GlcCer were synthesized in the fraction containing GlcNAc-phosphotransferase, but, in addition, activity was observed in the fraction containing the distal marker galactosyltransferase. No quantification of the distribution was provided. The use of GlcNAc-phosphotransferase as a proximal Golgi marker may be somewhat problematic, since this protein has never been demonstrated unequivocally by immunocytochemical methods to be located in the proximal Golgi apparatus. Two peaks of GlcCer synthase activity were observed in fractions from HepG2 cells but, since no distribution of Golgi marker enzymes was provided, assessing the significance of the separation of GlcCer synthase into two peaks is difficult.

The subtle differences in localization of GlcCer synthesis observed in these three studies emphasize the need to isolate GlcCer synthase and produce antibodies that would permit immunolocalization of the enzyme. Integral membrane enzymes are notoriously difficult to solubilize and purify, but preliminary steps toward isolating GlcCer synthase have been reported with its solubilization from porcine submaxillary glands using zwitterionic detergents such as CHAPS and Zwittergent 3-14 (Durieux *et al.*, 1990a). Enzyme activity was found to be highly dependent on its lipid environment, and reconstitution studies indicated a requirement for PC to maintain full activity (Durieux *et al.*, 1990b).

C. Sphingomyelin Synthesis

Only within the past decade has the donor of the phosphocholine moiety of SM been recognized to be PC and not CDP-choline (Voelker and Kennedy, 1982; Fig. 1). The site of transfer of phosphocholine from PC to ceramide, however, still was disputed until a few years ago; the ER, Golgi apparatus, and PM were all candidates. Notable among these was the

PM, which was shown in many studies to synthesize significant amounts of SM (see references in Futerman *et al.*, 1990). However, although some SM is synthesized in the PM (see subsequent discussion), the major site of SM synthesis is now believed to be the Golgi apparatus. Early subcellular fractionation studies were flawed, both by the low purity of the subcellular fractions obtained from tissue culture cells and by inadequate characterization of the fractions. The Golgi apparatus appeared likely to be a major site of SM synthesis from observations made by Pagano and colleagues, who synthesized a fluorescent derivative of ceramide in which the naturally occurring fatty acid was replaced by hexanoic acid to which an NBD group was attached (C_6-NBD-ceramide). This derivative permitted direct observation by fluorescence microscopy of the translocation and sequestering of the labeled precursor and its metabolites in living cells (Lipsky and Pagano, 1983). Under conditions in which C_6-NBD-ceramide was trapped in the Golgi apparatus, these investigators noted that it still was metabolized to C_6-NBD-SM (as well as C_6-NBD-GlcCer). Fractionation studies with Chinese hamster ovary (CHO) cells supported the suggestion that significant amounts of SM were synthesized in the Golgi apparatus (Lipsky and Pagano, 1985).

Because of the controversy surrounding the site of SM synthesis, we undertook a study to determine unambiguously the site of SM synthesis using subcellular fractions from rat liver and using N-([1-^{14}C]hexanoyl-D-*erythro*-ceramide as substrate. Care was taken to ensure that all assays of SM synthesis were carried out under conditions in which the rate of formation of SM was linear with respect to time and protein and was not limited by availability of substrate (Futerman *et al.*, 1990). Of the SM synthase activity, 87% was found in a Golgi apparatus fraction; the remaining 13% of the activity was found in a PM fraction. In addition, activity was enriched significantly in a *cis*/medial Golgi apparatus fraction that was characterized by Western blot analysis using an antibody against mannosidase II. Similar conclusions using rat liver fractions were reached in another study using a truncated ceramide analog with eight carbon atoms in both the long-chain base and the fatty acid. Synthesis of SM was found in an early (*cis*) compartment of the Golgi apparatus (Jeckel *et al.*, 1990).

A subtle but consistent difference has been observed in the distribution of GlcCer and SM synthesis (Futerman and Pagano, 1991; Jeckel *et al.*, 1992), as indicated in Fig. 2, in which the distribution and relative specific activities of SM and GlcCer synthase are compared in microsomal subfractions. Particularly striking is the difference in their relative specific activities in the Golgi apparatus fraction (see preceding discussion). The differences in subcellular localization of GlcCer and SM synthesis may be important for metabolic regulation (see subsequent discussion), and the understanding of their regulation will be advanced once the enzymes have been purified. Encouraging progress in this direction has been made by the solubilization and functional reconstitution of SM synthase from CHO cells (Hanada *et al.*, 1991). Interestingly, PC was required for full enzyme activity, as was the case for GlcCer synthase, although PC is also a substrate for SM synthase. Moreover, diacylglycerol, the product of the removal of phosphocholine from PC (Fig. 1), appeared to act as an inhibitory regulator, suggesting that SM synthase may be regulated by product inhibition.

III. TOPOLOGY OF CERAMIDE METABOLISM

The topology of the three enzymes discussed earlier has been determined. Proteases and membrane impermeable inhibitors have been used to demonstrate that dihydroceramide synthase is oriented with its active site facing the cytosol (Mandon *et al.*, 1992; Hirschberg *et al.*, 1993). Once dihydroceramide is synthesized, it may be able to undergo transbilayer movement to the luminal leaflet. Short acyl chain analogs of ceramide undergo rapid transbilayer movement (Pagano, 1989), although no direct demonstration has been made that naturally occuring (dihydro) ceramide also is able to undergo transbilayer movement.

SM, GlcCer, and complex glycosphingolipids are located primarily at the external leaflet of the PM of animal cells (Pagano, 1989; Schwarzmann and Sandhoff, 1990; Koval and Pagano, 1991). Since these sphingolipids are believed to travel from the Golgi apparatus to the PM by vesicle-mediated transport (Pagano and Sleight, 1985; Schwarzmann and Sandhoff, 1990) and do not undergo transbilayer movement, one might expect that SM and GlcCer would be synthesized at the luminal leaflet of the Golgi apparatus or pre-Golgi apparatus compartment, both of which are topographically equivalent to the external leaflet of the PM. Indeed, for SM, controlled protease digestion showed this to be the case. In intact Golgi apparatus-derived vesicles, SM synthase was totally inaccessible to added proteases but, on disruption of the vesicles, SM synthase was destroyed rapidly (Futerman *et al.*, 1990). Somewhat surprisingly, similar protease digestion experiments suggested that GlcCer synthase was located at the cytoplasmic surface of the Golgi apparatus in porcine submaxillary glands (Coste *et al.*, 1986) and in rat or rabbit liver (Coste *et al.*, 1986; Futerman and Pagano, 1991; Trinchera *et al.*, 1991; Jeckel *et al.*, 1992).

Protease digestion experiments sometimes can produce spurious results. Therefore, the topology of GlcCer synthase also has been examined by other methods. First, membrane impermeable compounds were shown to inhibit GlcCer synthase (Coste *et al.*, 1986; Trinchera *et al.*, 1991; Jeckel *et al.*, 1992). Second, the extent of UDP-glucose translocation into the Golgi apparatus was shown to be insufficient to account for the amount of GlcCer synthesized (Futerman and Pagano, 1991). Third, sepharose-immobilized ceramide was found to be a substrate for GlcCer synthase (Trinchera *et al.*, 1991). Finally, in semi-intact CHO cells, C_6-NBD-GlcCer but not C_6-NBD-SM was released into the medium under conditions in which vesicular flow to the PM was inhibited, suggesting that GlcCer is not "trapped" within the lumen of the Golgi apparatus as is SM (Jeckel *et al.*, 1992). Collectively, these results strongly suggest that GlcCer is synthesized at the cytosolic surface of intracellular membranes, so GlcCer must be translocated across the membranes of the Golgi apparatus to be glycosylated further to form higher order glycosphingolipids (Hoekstra and Kok, 1992; Chapter 20). No evidence for a GlcCer translocase exists at present, but phospholipid translocases exist in both the ER and the PM (Pagano, 1990; Zachowski and Devaux, 1990). Alternatively, lactosylceramide (LacCer) and not GlcCer may be translocated across the membrane, since protease digestion experiments have suggested that LacCer synthase (UDP-Gal:glucosylceramide $\beta1\rightarrow4$galactosyltransferase) also is located on the cytosolic surface of the Golgi apparatus (Trinchera *et al.*, 1991). This idea is supported by observations that both LacCer and GlcCer synthase can be released selectively from intact Golgi cisternae by cathepsin D (Trinchera *et al.*, 1991), but is contradicted by results using mutant cell lines that lacked the transporter of UDP-Gal into the lumen of the Golgi apparatus. In these mutant cells, isolated Golgi apparatus vesicles were able to translocate UDP-Gal at only 2% of the rate observed for vesicles from wild-type cells and, as a result, a strong reduction in galactose-containing glycosphingolipids was observed; GlcCer constituted over 90% of the glycosphingolipids synthesized (Brandli *et al.*, 1988). Additional work clearly is needed to resolve the issue of the topology of LacCer synthase and to determine which glycosphingolipid is translocated across the membrane bilayer.

IV. TRANSPORT OF CERAMIDE BETWEEN THE ENDOPLASMIC RETICULUM AND GOLGI APPARATUS

After synthesis at the cytosolic surface of the ER, dihydroceramide is dehydrogenated to ceramide. Since the intracellular localization of dihydroceramide dehydrogenase has not been determined, whether dehydrogenation occurs before or after transport out of the ER is not clear. The mechanism by which (dihydro)ceramide moves out of the ER is not known (Hoekstra and Kok, 1992) but two potential mechanisms exist, namely, transport by the same vesicles that transport proteins out of the ER along the secretory pathway or, alternatively, transport as lipid monomers through the cytosol by a protein-facilitated mechanism involving a ceramide transfer protein.

If this latter mechanism is correct, then treatments that affect vesicle traffic in the secretory pathway would not be expected to disrupt ceramide transport. Several studies have addressed this issue using mitotic and interphase HeLa cells (Collins and Warren, 1992). Interphase cells first were labeled with [^{3}H]serine and shown to synthesize ceramide, GlcCer, LacCer, and SM, as well as some gangliosides. In contrast, the only sphingolipids obtained after labeling mitotic cells with [^{3}H]serine were GlcCer, SM, and, to a much lesser extent, LacCer. These results demonstrate that ceramide is delivered to the pre- or early Golgi apparatus where it is metabolized to GlcCer or SM although vesicle traffic from the ER to Golgi apparatus is inhibited, strongly supporting the idea that ceramide moves out of the ER by a facilitated mechanism, but that subsequent steps require vesicle transport through the Golgi apparatus. However, note that the locations of ceramide, GlcCer, and SM synthesis have not been determined in HeLa cells, although no reason exists to suspect that they will be different from those observed in liver fractions and other cultured cells.

Additional evidence supporting a facilitated mechanism comes from studies in which all four stereoisomers of sphinganine were shown to be acylated *in vivo,* but only the D-*erythro* and, to some extent, the L-*threo* isomers subsequently are converted to SM and GlcCer (Stoffel and Bister, 1973), perhaps implying that D-*erythro*-ceramide is transported out of the ER by a transfer protein that is stereospecific to the long-chain base. The inability of GlcCer and SM synthases to metabolize all the sphinganine stereoisomers is not a result of the specificity of the enzymes themselves since, when C$_6$-NBD-Cer containing different stereoisomers of sphingosine was added to cultured cells, SM synthase was able to metabolize all four stereoisomers and GlcCer synthase was able to metabolize D-*erythro*- and L-*erythro*-sphingosine (Pagano and Martin, 1988).

Two other novel approaches have been used to analyze the transport of lipids between the ER and the Golgi apparatus. The synthesis of mannosylated forms of inositolceramides has been shown to be dependent on genes that control the flow of secretory vesicles from the ER to the Golgi apparatus in *Saccharomyces cerevisiae*. Using *sec23*, a temperature-sensitive mutant in which newly synthesized secretory proteins remain blocked in the ER at 37°C, a significant change in the profile of sphingolipids synthesized was observed compared with wild-type. Similar changes were observed with other *sec* mutants. Although this study has no direct bearing on the transport of ceramide from the ER to the Golgi apparatus, since the transport of an inositolphosphoceramide is proposed to be inhibited in the *sec* mutants, this type of approach might be used to determine whether ceramide is transported between the ER and the Golgi apparatus by incubating cells with a precursor of ceramide such as radioactive serine. Again, note that the conclusions from such studies would be easier to interpret if the precise sites of synthesis of sphingolipids in the ER and the Golgi apparatus were known.

The second approach involves isolation and analysis of the lipid composition of the "transition vesicles" that are responsible for transporting material from the ER to the Golgi apparatus. A cell-free system has been developed in which the lipid composition of transitional ER was shown to be similar in almost all respects to that of transition vesicles, which were not enriched in cerebrosides or SM (Moreau *et al.*, 1991). This system must be characterized further before definitive conclusions can be reached about whether certain lipids are included or excluded from transition vesicles.

A provocative paper published by Wieland and colleagues (1987) suggested that *de novo* lipid synthesis could not account for the large amount of lipid that was proposed to be transported to the Golgi apparatus by "bulk flow" of vesicles. Based on this suggestion, the concept of retrograde flow between the Golgi apparatus and the ER first was promoted, and subsequently has been strengthened by studies with brefeldin A (Klausner *et al.*, 1992). Are sphingolipids involved in this anterograde/retrograde pathway? Presumably not, if ceramide moves by a facilitated mechanism between the ER and the Golgi apparatus. How much ceramide is transported between the ER and the Golgi apparatus? Could a putative ceramide transport protein be efficient enough to supply the Golgi apparatus with enough ceramide to account for total synthesis GlcCer and SM? To address these questions, the rate of synthesis of GlcCer and SM must be determined *in vivo*. Interestingly, when CHO cells were labeled with [^{3}H]palmitate, no detectable lag in the synthesis of either ceramide or GlcCer was observed following the addition of the precursor, although 5–6 min passed before LacCer was detected (Young *et al.*, 1992). Since the conversion of GlcCer to LacCer appears to depend on vesicle transport (according to studies with mitotic cells; see preceding discussion), these data could be interpreted to suggest that protein-mediated lipid transport of sphingolipids is much more rapid than vesicular transport, although other explanations (such as partially overlapping sites of synthesis of these two lipids) should be considered. If such a ceramide transfer protein exists, it is to be hoped that it will soon be isolated and characterized. The use of photoactivatable analogs of ceramide might prove useful for isolation.

V. CONCLUSIONS AND PERSPECTIVES

In this chapter, I have summarized work performed within the past few years that conclusively demonstrates that the early steps of sphingolipid synthesis are compartmentalized in the ER and the Golgi apparatus (Fig. 3). However, the reason sphingolipid synthesis is compartmentalized remains obscure. Perhaps compartmentalization allows cells to regulate synthesis. If ceramide is transported by a facilitated mechanism between the ER and the Golgi apparatus, then synthesis of SM and GlcCer (and, consequently, of glycosphingolipids) could be regulated by controlling the activity of the putative ceramide transfer protein. Alternatively, the synthesis of glycosphingolipids could be regulated by controlling the activity of the putative GlcCer or LacCer translocase. SM synthesis could be regulated by delivery of ceramide to the *luminal* leaflet of the Golgi apparatus. Determining why cells might need to regulate sphingolipid synthesis at multiple sites, and whether sphingolipid synthesis is regulated

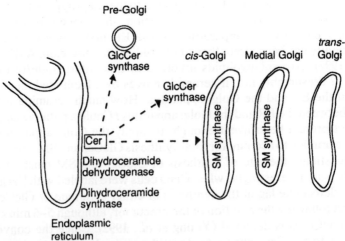

FIGURE 3 Sites and topology of enzymes of ceramide metabolism in the ER and Golgi apparatus. Note that the location of dihydroceramide dehydrogenase is not known. See the text for a discussion of the possible mechanisms by which ceramide (Cer) may be transported between the ER and Golgi apparatus.

by any of these mechanisms, awaits the availability of tools that would permit modification of the enzymes, transfer proteins, and translocases at the molecular level. Such studies are entirely dependent on prior isolation and purification of these proteins.

Another possible reason for localizing different enzymes to different compartments might be to generate specific lipid compositions in particular compartments. Since vesicle fission, sorting, and fusion may depend in part on the topology and composition of lipids within intracellular vesicles, the synthesis of sphingolipids in discrete compartments may provide a means of regulating lipid composition and, perhaps, the ability of a compartment to mediate vesicular events. Lipid composition has been suggested to regulate the closure of fusion pores in mast cells (Oberhauser et al., 1992). A phosphatidylinositol transfer protein in yeast has been shown to elevate the phosphatidylinositol : PC ratio of the Golgi apparatus above that of the ER, leading to activation of secretion in a late Golgi apparatus compartment (Bankaitis et al., 1990). No evidence exists to suggest that sphingolipid transport or transfer is involved in similar processes, but the possibility that sphingolipids are involved in such events is worth considering.

Alternatively, the generation of specific sphingolipid compositions in the ER and the Golgi apparatus could be involved in the retention of proteins in specific cisternae. According to this hypothesis, hydrophobic sequences of integral membrane proteins buried deep in the lipid bilayer may interact with specific lipid microenvironments in specific compartments (Machamer, 1991). This theory should be relatively easy to test by examining the distribution of ER or Golgi apparatus resident proteins after manipulating sphingolipid composition using inhibitors of sphingolipid synthesis.

Two inhibitors already have proved useful in examining the role of sphingolipids in the secretory pathway. Inhibition of GlcCer synthase by D-threo-1-phenyl-2-decanoylamino-3-morpholino-1-propanol (PDMP) was shown to affect the transport of a viral protein through the Golgi apparatus of infected cells (Rosenwald et al., 1992). In addition, inhibition of sphingolipid synthesis by fumonisin has been shown to retard axonal outgrowth severely in cultured hippocampal neurons (R. Harel and A. H. Futerman, 1993). If sphingolipids are essential components of vesicles that deliver newly synthesized material to the growing axon, then inhibition of synthesis would affect axonal outgrowth since vesicle transport to the axon would be disrupted. Currently we are testing this hypothesis by examining whether protein transport to the cell surface is inhibited in the presence of fumonisin and, if so, at what stage in the secretory pathway protein transport is inhibited.

In summary, the demonstration that sphingolipid synthesis is compartmentalized within the ER and the Golgi apparatus should make possible
the design of experiments that will address the important issues of how and
why the early steps of sphingolipid synthesis and transport are regulated.

Acknowledgments

The data presented in Fig. 2 concerning the site of GlcCer and SM synthesis were obtained
while the author was in the laboratory of Richard Pagano, Carnegie Institution of Washington,
Baltimore, Maryland. Work from the author's laboratory is supported by the Josef Cohn
Center for Biomembrane Research at the Weizmann Institute; by the Minerva Foundation,
Munich, Germany; and by the Basic Research Foundation of the Israel Academy of Sciences
and Humanities. The author is the incumbent of the Recanati Career Development Chair
in Cancer Research.

References

Bankaitis, V. A., Aitken, J. R., Cleves, A. E., and Dowhan, W. (1990). An essential
    role for a phospholipid transfer protein in yeast Golgi function. *Nature (London)* **347**,
    561–562.
Brandli, A. W., Hansson, G. C., Rodriguez-Boulan, E., and Simons, K. (1988). A polarized
    epithelial cell mutant deficient in translocation of UDP-galactose into the Golgi complex.
    J. Biol. Chem. **263**, 16283–16290.
Collins, R. N., and Warren, G. (1992). Sphingolipid transport in mitotic HeLa cells. *J. Biol.
    Chem.* **267**, 24906–24911.
Coste, H., Martel, M.-B., Azzar, G., and Got, R. (1985). UDP-glucose–ceramide glucosyl-
    transferase from porcine submaxillary glands is associated with the Golgi apparatus.
    Biochim. Biophys. Acta **814**, 1–7.
Coste, H., Martel, M.-B., and Got, R. (1986). Topology of glucosylceramide synthesis in
    Golgi membranes from porcine submaxillary glands. *Biochim. Biophys. Acta* **858**, 6–12.
Durieux, I., Martel, M.-B., and Got, R. (1990a). Solubilization of UDP-glucose–ceramide:
    glucosyltransferase from the Golgi apparatus. *Biochim. Biophys. Acta* **1024**, 263–266.
Durieux, I., Martel, M.-B., and Got, R. (1990b). Effect of phospholipids on UDP-glucose:
    ceramide glucosyltransferase from Golgi membranes. *Int. J. Biochem.* **22**, 709–715.
Futerman, A. H., and Pagano, R. E. (1991). Determination of the intracellular sites and
    topology of glucosylceramide synthesis in rat liver. *Biochem. J.* **280**, 295–302.
Futerman, A. H., and Pagano, R. E. (1992). Use of N-([1-^{14}C]hexanoyl) sphingolipids to
    assay sphingolipid metabolism. *Meth. Enzymol.* **209**, 437–446.
Futerman, A. H., Stieger, B., Hubbard, A. L., and Pagano, R. E. (1990). Sphingomyelin
    synthesis in rat liver occurs predominantly at the cis and medial cisternae of the Golgi
    apparatus. *J. Biol. Chem.* **265**, 8650–8657.
Hanada, K., Horii, M., and Akamatsu, Y. (1991). Functional reconstitution of sphingomyelin
    synthase in Chinese hamster ovary cell membranes. *Biochim. Biophys. Acta* **1086**,
    151–156.
Hannun, Y. A., and Bell, R. M. (1989). Functions of sphingolipids and sphingolipid break-
    down products in cellular regulation. *Science* **243**, 500–507.
Harel, R., and Futerman, A. H. (1993). Inhibition of sphingolipid synthesis affects axonal
    outgrowth in cultured hippocampal neurons. *J. Biol. Chem.* **268**, 14476–14481.
Hirschberg, K., Rodger, J., and Futerman, A. H. (1993). The long chain base of sphingolipids

is acylated at the cytosolic surface of the endoplasmic reticulum in rat liver. *Biochem. J.* **290**, 751–757.

Hoekstra, D., and Kok, J. W. (1992). Trafficking of glycosphingolipids in eukaryotic cells; sorting and recycling of lipids. *Biochim. Biophys. Acta* **1113**, 277–294.

Jeckel, D., Karrenbauer, A., Birk, R., Schmidt, R. R., and Wieland, F. T. (1990). Sphingomyelin is synthesized in the cis Golgi. *FEBS Lett.* **261**, 155–157.

Jeckel, D., Karrenbauer, A., Burger, K. N. J., Van-Meer, G., and Wieland, F. (1992). Glucosylceramide is synthesized at the cytosolic surface of various Golgi subfractions. *J. Cell Biol.* **117**, 259–267.

Klausner, R. D., Donaldson, J. G., and Lippincott-Schwartz, J. (1992). Brefeldin-A—Insights into the control of membrane traffic and organelle structure. *J. Cell Biol.* **116**, 1071–1080.

Kolesnick, R. (1992). Ceramide: A novel second messenger. *Trends Cell Biol.* **2**, 232–236.

Koval, M., and Pagano, R. E. (1991). Intracellular transport and metabolism of sphingomyelin. *Biochim. Biophys. Acta* **1082**, 113–125.

Lipsky, N. G., and Pagano, R. E. (1983). Sphingolipid metabolism in cultured fibroblasts: Microscopic and biochemical studies employing a fluorescent ceramide analogue. *Proc. Natl. Acad. Sci. U.S.A.* **80**, 2608–2612.

Lipsky, N. G., and Pagano, R. E. (1985). Intracellular translocation of fluorescent sphingolipids in cultured fibroblasts. Endogenously synthesized sphingomyelin and glucosylcerebroside analogues pass through the Golgi apparatus en route to the plasma membrane. *J. Cell Biol.* **100**, 27–34.

Machamer, C. E. (1991). Golgi retention signals: Do membranes hold the key. *Trends Cell Biol.* **1**, 141–144.

Mandon, E. C., Ehses, I., Rother, J., Van Echten, G., and Sandhoff, K. (1992). Subcellular localization and membrane topology of serine palmitoyltransferase, 3-dehydrosphinganine reductase, and sphinganine *N*-acyltransferase in mouse liver. *J. Biol. Chem.* **267**, 11144–11148.

Merrill, A. H., and Jones, D. D. (1990). An update of the enzymology of sphingomyelin metabolism. *Biochim. Biophys. Acta* **1044**, 1–12.

Merrill, A. H., and Wang, E. (1986). Biosynthesis of long chain (sphingoid) bases from serine by LM cells. *J. Biol. Chem.* **261**, 3764–3769.

Moreau, P., Rodriguez, M., Casagne, C., Morre, D., and Morre, D. J. (1991). Trafficking of lipids from the endoplasmic reticulum to the Golgi apparatus in a cell-free system from rat liver. *J. Biol. Chem.* **266**, 4322–4328.

Morell, P., and Radin, N. S. (1970). Specificity in ceramide biosynthesis from long chain bases and various fatty acyl coenzyme A's by brain microsomes. *J. Biol. Chem.* **245**, 342–350.

Nelson, W. J. (1992). Regulation of cell surface polarity from bacteria to animals. *Science* **258**, 948–955.

Oberhauser, A. F., Monck, J. R., and Fernandez, J. M. (1992). Events leading to the opening and closing of the exocytic fusion pore have markedly different temperature dependencies: Kinetic analysis of single fusion events in patch-clamped mouse mast cells. *Biophys. J.* **61**, 800–809.

Ong, D. E., and Brady, R. N. (1973). *In vivo* studies on the introduction of the 4-t-double bond of the sphingenine moiety of rat brain ceramides. *J. Biol. Chem.* **248**, 3884–3888.

Pagano, R. E. (1989). A fluorescent derivative of ceramide: Physical properties and use in studying the Golgi apparatus of animal cells. *Meth. Cell Biol.* **29**, 75–85.

Pagano, R. E. (1990). Lipid traffic in eukaryotic cells: Mechanisms for intracellular transport and organelle-specific enrichment of lipids. *Curr. Opin. Cell Biol.* **2**, 652–663.

Pagano, R. E., and Martin, O. C. (1988). A series of fluorescent N-acylsphingosines: Synthesis, physical properties and studies in cultured cells. *Biochemistry* **27**, 4439–4445.

Pagano, R. E., and Sleight, R. G. (1985). Defining lipid transport pathways in animal cells. *Science* **229**, 1051–1057.

Rosenwald, A. G., Machamer, C. E., and Pagano, R. E. (1992). Effects of a sphingolipid synthesis inhibitor on membrane transport through the secretory pathway. *Biochemistry* **31**, 3581–3590.

Rother, Y., van Echten, G., Schwarzmann, G., and Sandhoff, K. (1992). Biosynthesis of sphingolipids: dihydroceramide and not sphinganine is desaturated by cultured cells. *Biochem. Biophys. Res. Commun.* **189**, 14–20.

Sasaki, T. (1990). Glycolipid transfer protein and intracellular transport of glucosylceramide. *Experientia* **46**, 611–616.

Schwarzmann, G., and Sandhoff, K. (1990). Metabolism and intracellular transport of glycosphingolipids. *Biochemistry* **29**, 10865–10871.

Schweizer, A., Matter, K., Ketcham, C. M., and Hauri, H.-P. (1991). The isolated ER-Golgi intermediate compartment exhibits properties that are different from the ER and cis-Golgi. *J. Cell Biol.* **113**, 45–54.

Stoffel, W., and Bister, K. (1973). Stereospecificities in the metabolic reactions of the four isomeric sphinganines (dihydrosphingosines) in rat liver. *Hoppe-Seyler's Z. Physiol. Chem.* **354**, 169–181.

Suzuki, Y., Ecker, C. P., and Blough, H. A. (1984). Enzymatic glucosylation of dolichol monophosphate and transfer of glucose from isolated dolichol-D-glucose phosphate to ceramide by BHK-21 cell microsomes. *Eur. J. Biochem.* **143**, 447–453.

Trinchera, M., Fabbri, M., and Ghidoni, R. (1991). Topography of glycosyltransferases involved in the initial glycosylation of gangliosides. *J. Biol. Chem.* **266**, 20907–20912.

van Meer, G., and Burger, K. N. J. (1992). Sphingolipid trafficking - sorted out. *Trends Cell Biol.* **2**, 332–337.

van Meer, G., Stelzer, E. H. K., Wijnaendts-van-Resandt, R. W., and Simons, K. (1987). Sorting of sphingolipids in epithelial (Madine-Darby canine kidney) cells. *J. Cell Biol.* **105**, 1623–1635.

Voelker, D. R., and Kennedy, E. P. (1982). Cellular and enzymic synthesis of sphingomyelin. *Biochemistry* **21**, 2753–2759.

Walter, V. P., Sweeney, K., and Morre, J. (1983). Neutral lipid precursors for gangliosides are not formed by rat liver homogenates or by purified cell fractions. *Biochim. Biophys. Acta* **750**, 346–352.

Wang, E., Norred, W. P., Bacon, C. W., Riley, R. T., and Merrill, A. H. (1991). Inhibition of sphingolipid biosynthesis by fumonosins. Implications for diseases associated with *Fusarium moniliforme*. *J. Biol. Chem.* **266**, 14486–14490.

Wieland, F. T., Gleason, M. L., Serafini, T. A., and Rothman, J. E. (1987). The rate of bulk flow from the endoplasmic reticulum to the cell surface. *Cell* **50**, 289–300.

Young, W. W., Lutz, M. S., and Blackburn, W. A. (1992). Endogenous glycosphingolipids move to the cell surface at a rate consistent with bulk flow estimates. *J. Biol. Chem.* **267**, 12011–12015.

Zachowski, A., and Devaux, P. F. (1990). Transmembrane movement of lipids. *Experientia* **466**, 644–656.

CHAPTER 5

Lateral Mobility of Lipids in Membranes

Greta M. Lee and Ken Jacobson
Department of Cell Biology and Anatomy, University of North Carolina at Chapel Hill,
Chapel Hill, North Carolina 27599

I. INTRODUCTION

The ability of lipids to move freely within the plane of the membrane serves several purposes in addition to being a solvent for membrane proteins. Lipids, as second messengers, move from the site of cleavage to their target molecule. Lipids interact with membrane proteins to modulate their function, as occurs with integrin modulating factor 1 which increases the binding avidity of integrin molecules (Hermanowski-Vosatka *et al.*,

1992). The free mobility of lipids also allows membranes to be flexible, readily accommodating shape changes such as ruffling and extension of filopodia. Membrane fusion, which facilitates recycling and renewal of membrane components, also requires the lateral mobility of lipids.

The measurement of lipid lateral mobility yields a diffusion coefficient (D) and, depending on the method, a mobile fraction. In general, membrane lipids have a higher D and mobile fraction than membrane proteins. The D of lipids is also less manipulatable than that of membrane proteins (Koppel *et al.*, 1981; Boullier *et al.*, 1992). However, the lateral diffusion of membrane lipids can be altered by a variety of factors (see Table I), can be limited to specific regions in some cells (see Chapter 6), and may be restricted selectively by associated cytoplasmic proteins (Bazzi and Nelsestuen, 1991).

II. MEASUREMENT OF LATERAL MOBILITY

Fluorescence recovery after photobleaching (FRAP) and single particle tracking (SPT) methods are presented briefly. Magnetic resonance methods for measuring lateral mobility are not discussed.

A. Fluorescent Lipid Analogs

In fluorescence measurements, the mobility of luminescent lipid analogs usually is used as an indicator of lipid diffusion. The lipid analogs employed have a fluorescent label attached covalently (see Fig. 1 for structures of some representative examples) and can be classified as phospholipid and sphingolipid analogs or as anionic and cationic membrane probes (see Haugland, 1992, for review). The probes most useful for FRAP studies will have appreciable absorption at one or more visible laser lines (typically, for the argon or argon–krypton ion lasers). Phospholipid analogs may be labeled covalently on their head groups with fluorescein, BODIPY, rhodamine, Texas Red, or NBD (Fig. 1A) or on their acyl chains with NBD (Fig. 1B) or BODIPY. Anionic membrane probes include fatty acids labeled with fluorescein conjugated to the carboxyl group (Fig. 1C) or NBD conjugated to the terminal methyl group (Fig. 1D) or to other positions along the chain. The cationic membrane probes include the popular carbocyanine dyes, alkylated with two chains with 12 or more carbon atoms to anchor them in the membrane. For FRAP studies, the most widely used members of this family are the alkylated indocarbocyanines (Fig. 1E), known by the generic acronym DiI. For a complete review of the structure and membrane properties of DiI, including its use for domain detection, see Wolf (1988; Chapter 6).

Fluorescence measurements of short-range lateral diffusion use pyrene excimer emission and usually employ phospholipid or fatty acid analogs with hydrophobic pyrene conjugated to the hydrocarbon chain.

In fluid phase bilayers, different probes seem to report similar mobility values (Derzko and Jacobson, 1980; El Hage Chahine *et al.*, 1993), presumably because the motion of the probe depends on diffusion of the host bilayer lipids. However, in phase-separated systems (Klausner and Wolf, 1980; Vaz, 1992), the mobility reported will, in general, depend critically on probe structure because probes of different structures will partition differently into the various phases present.

B. Photobleaching Methods

The simplest version of the family of photobleaching techniques (for reviews, see Kapitza and Jacobson, 1986; Petersen *et al.*, 1986; Wolf, 1989) is called spot photobleaching. This version of the method is based on photobleaching a small circular region on the surface of a membrane bearing fluorescent molecules, thus destroying the emission from that region. The recovery of fluorescence because of the diffusion and/or flow of unbleached fluorophores from the surrounding area into the irradiated area is measured. If only isotropic lateral diffusion occurs, the recovery kinetics, characterized by the time $(\tau_{1/2})$ to obtain 50% of full recovery, are related to D, the diffusion coefficient by the equation $D = W_s^2 \gamma / 4\tau_{1/2}$ where W_s is the $1/e^2$ radius of the Gaussian profile laser beam used for both photobleaching and measuring fluorescence and γ is a parameter that depends on the extent of photobleaching (Axelrod *et al.*, 1976). The mobile fraction of the measured population of probe molecules is obtained from the degree to which the final fluorescence level approaches the prebleach fluorescence value. In this simple version, typically an ion laser is used to excite the fluorescence, which is observed in a fluorescence microscope equipped for incident light excitation. The laser beam is focused to waist on the secondary image plane of the microscope by a weak biconvex lens; the objective further focuses the light on a small spot on the specimen (Jacobson *et al.*, 1977). The $1/e^2$ spot diameter ($2 W_s$) is a function of the objective power and can be adjusted, typically, from $\leq 1 \mu m$ to as large as $50 \mu m$.

Variations on the spot photobleaching technique have been developed. Koppel (1979) introduced a multipoint analysis of the basic FRAP experiment by monitoring recovery at several points within the bleached and

TABLE I

Diffusion of Lipids in Red Blood Cell Membranes as Measured by FRAP

Species	Analog[a]	Diffusion coefficient ($\times 10^{-9}$)	Mobile fraction	Temperature[b] (°C)	Comments	References
Human						
normal	DiI	8.1 ± 0.3	95	37		Boullier et al. (1986)
reversible sickle	DiI	7.7 ± 0.2	90	37		
irreversible sickle	DiI	7.3 ± 0.2	94	37		
Human						
normal ghosts	NBD-PE	2 ± 0.1	81	20	External only	Rimon et al. (1984)
normal ghosts	NBD-PE	2.7 ± 0.1	79	20	Both sides of bilayer	
spherocyte ghosts	NBD-PE	2.3 ± 0.1	83	20	External only	
spherocyte ghosts	NBD-PE	3.0 ± 0.1	71	20	Both sides of bilayer	
normal ghosts	NBD-PE	4.2 ± 0.2	82	35	External only	
normal ghosts	NBD-PE	5.5 ± 0.2	78	35	Both sides of bilayer	
spherocyte ghosts	NBD-PE	6.3 ± 0.3	86	35	External only	
spherocyte ghosts	NBD-PE	7.7 ± 0.3	86	35	Both sides of bilayer	
Human ghosts	DiIC$_{18}$	1.9	—	30		Thompson and Axelrod (1980)
Human						
intact cells	DiIC$_{18}$(5)	8.2 ± 1.2	—	25		Bloom and Webb (1983)
intact cells	DiIC$_{18}$(5)	21 ± 4.6	—	37		
resealed ghosts	DiIC$_{18}$(5)	7.8 ± 2.3	—	25		
resealed ghosts	DiIC$_{18}$(5)	19 ± 4.4	—	37		
dilution ghosts	DiIC$_{18}$(5)	6.3 ± 2.1	—	25		
dilution ghosts	DiIC$_{18}$(5)	15 ± 3.1	—	37		
Human						
intact cells	NBD-PC	1.1 ± 0.4	—	20	Slow comp. (majority of the signal)	Morrot et al. (1986)

	Lipid	D		Temp.[b]	Comment	Reference
	NBD-PE	5.4 ± 1.7	—	20	Fast comp. (majority of the signal)	
	NBD-PS	5.0 ± 0.8	—	20	Fast comp. (majority of the signal)	
	NBD-PC	1.4 ± 0.2	—	35	Slow comp. (majority of the signal)	
	NBD-PE	7.8 ± 3.1	—	35	Fast comp. (majority of the signal)	
	NBD-PS	8.5 ± 2.0	—	35	Fast comp. (majority of the signal)	
ghosts (sealed in the presence of ATP)	NBD-PC	1.2 ± 0.2	—	20	Slow comp. (majority of the signal)	
	NBD-PE	4.7 ± 0.9	—	20	Fast comp. (majority of the signal)	
	NBD-PS	5.9 ± 0.6	—	20	Fast comp. (majority of the signal)	
	NBD-PC	1.9 ± 0.3	—	35	Slow comp. (majority of the signal)	
	NBD-PE	8.8 ± 2.3	—	35	Fast comp. (majority of the signal)	
	NBD-PS	7.8 ± 1.2	—	35	Fast comp. (majority of the signal)	Koppel et al. (1981)
Mouse						
normal (ghosts)	NBD-PE	14 ± 5	—	23		
spherocytic (ghosts)	NBD-PE	15 ± 5	—	23		
Turkey						
(ghosts)	NBD-PE	4.5	82	24		
	NBD-PE	1.4	47	9		Henis et al. (1982)

[a] Abbreviations: DiI, tetramethyl indocarbocyanine perchlorate; DiIC$_{18}$, 1,1'-dioctadecyl-3,3,3',3'-tetramethylindocarbocyanine perchlorate; DiIC$_{18}(S)$, 1,1'-dioctadecyl-3,3,3',3'-tetramethylindodicarbocyanine perchlorate; NBD-PC, 1-acyl-2-[N-(4-nitrobenzo-2-oxa-1,3-diazoyl) amino]caproyl-sn-glycero-3-phosphocholine; NBD-PE, phosphatidyl-N-(NBD)ethanolamine; NBD-PS, 1-acyl-2-[6-(NBD-aminocaproyl)]phosphatidylserine.
[b] Temperature at which FRAP measurement was performed.

115

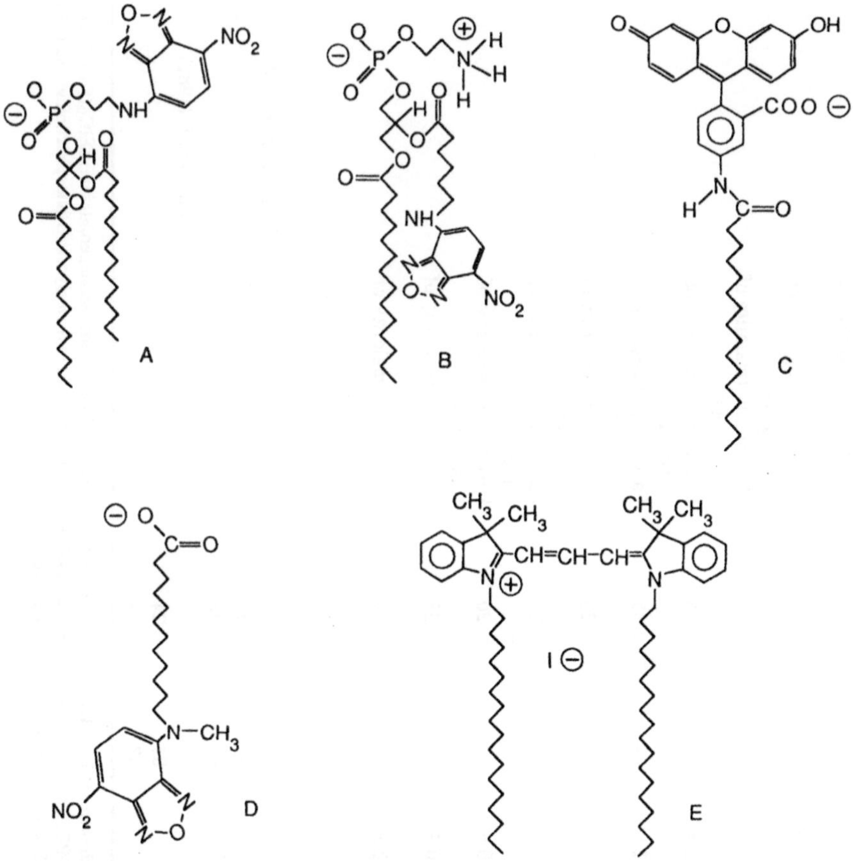

FIGURE 1 Structures of lipid probes frequently used in FRAP experiments. (A) NBD phosphatidylethanolamine, NBD fluorochrome on the head group (NBD-PE). (B) NBD phosphatidylcholine, NBD fluorochrome on an acyl chain (C_6-NBD-PC). (C) Fluorescein conjugated to the head group of a fatty acid. (D) Fatty acid with NBD on an acyl chain. (E) DilC$_{18}$, 3′,3′-dioctadecylindocarbocyanine.

surrounding area, allowing diffusion and flow to be recognized readily. Pattern photobleaching involves the bleaching of periodic patterns on the specimen (Smith and McConnell, 1978). The lateral transport of the observed fluorophores causes the gradual exponential decay of these bleached patterns over time. The evaluation of this decay yields the diffusion coefficient. Patterns usually have been produced by inserting a Ronchi ruling in the path of the laser at the rear focal plane of the microscope, but also have been generated by the interference of two coherent laser beams for fringe pattern photobleaching (Davoust et al., 1982).

Imaging methods combined with photobleaching can yield detailed information on more complicated lateral transport processes. In spot photobleaching, fluorescence detection normally is done by a photomultiplier so, unless the laser beam or stage is scanned, no spatial information on the recovery process can be obtained directly. In the "video-FRAP" technique, the photomultiplier is replaced by a very sensitive video camera that is calibrated to operate as a time-resolved (30-msec resolution) spatial photometer (Kapitza et al., 1985). Thus, spatial information on the fluorescence redistribution process is obtained.

C. Single Particle Tracking

Colloidal gold (Lee et al., 1991,1993) or fluorescent latex (Fein et al., 1993) 30-nm beads are conjugated to antibodies against a fluorochrome on the lipid head group in the SPT technique. The colloidal gold is detected using video-enhanced microscopy. The images, usually a series of 200 or more, are recorded on a rapid storage medium such as an optical disk recorder. Additional image processing is used to locate the centroids of each particle or bead in each succeeding image. A tracking program then is used to produce trajectories for each particle by locating the nearest centroid in each succeeding image. From the trajectory, the mean square displacement for each time interval (see Fig. 2) is computed. The slope of the plot of the mean square displacement against the time interval is used to compute the diffusion coefficient.

Single particle tracking, as the name implies, allows for the movements of individual or small clusters of lipid molecules to be followed. The clusters occur because a properly stabilized, single 30-nm gold particle can be conjugated to as many as 20–60 antibodies (DeRoe et al., 1987). Alternatively, the antibody against the fluorochrome can be mixed with an irrelevant antibody before conjugation so the gold is stabilized but will bind only one or a few lipid molecules (Lee et al., 1991). In the case of artificial planar bilayers, the paucivalent gold yields a higher average diffusion coefficient than the multivalent gold. With plasma membranes, the valency of the gold has little effect on diffusion because the pericellular matrix on the extracellular surface of the plasma membrane produces sufficient viscous drag to mask any effect of valency (Lee et al., 1993). An apparent value for this matrix viscosity of 0.5–0.9 poise was calculated by comparing D obtained by FRAP with D obtained by SPT.

Tracking single or small clusters of lipids allows greater spatial resolution than can be obtained with FRAP. This spatial resolution can be very useful in detecting domains. However, because of the stochastic nature

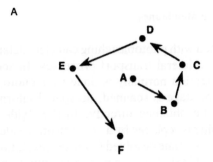

A

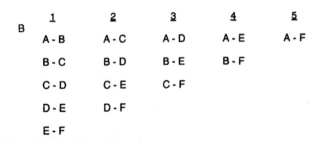

	1	2	3	4	5
B	A - B	A - C	A - D	A - E	A - F
	B - C	B - D	B - E	B - F	
	C - D	C - E	C - F		
	D - E	D - F			
	E - F				

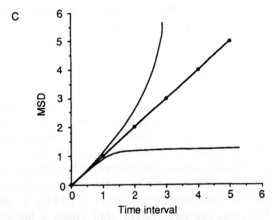

C

FIGURE 2 The computation of D (diffusion coefficient) from a trajectory and interpretation of the resulting MSD plot. (A) In the trajectory shown, each point represents the centroid of a gold particle in a series of images recorded at equally spaced times. (B) Displacements that are used in computing the mean square displacement (MSD) for each time interval. As the length of the time interval increases, fewer displacements are included in the average. (C) MSD plotted against time interval. For two-dimensional diffusion, the equation of the plot is MSD = 4Dt where the slope equals 4D. The theoretical plot for random diffusion is a straight line. In reality, for a single trajectory of 200 points, the line is linear only for the shorter time intervals in which many displacements have been averaged for each data point. The rest of the plot reflects the shape of the trajectory, curving up when running and tumbling occurs or curving down when the trajectory loops back on itself. If many more

of Brownian motion, the number of particles tracked and the time span of the observation are important considerations (Qian *et al.*, 1991; Lee *et al.*, 1993). The operational definition of the mobile fraction also becomes important. For a given experiment, a range of diffusion coefficients will be obtained; and some gold particles will be completely stationary. The immobile fraction can be considered the number of stationary gold particles or, more analogous to the FRAP experiments, the number of particles with a D below a certain value.

In general, single particle tracking of lipids has yielded average D values $(1.1–7.3 \times 10^{-9} \text{ cm}^2/\text{sec})$ slightly lower than those measured by FRAP $(5.4–13 \times 10^{-9} \text{ cm}^2/\text{sec})$ (Lee *et al.*, 1991,1993).

D. Shorter Range Measurements

The FRAP measurements discussed in this chapter are micron-scale measurements that derive from equilibration of a concentration gradient in fluorescent molecules imposed by photobleaching. Such experiments measure diffusion over distances ranging from 1 to tens of microns.

However, other measurements rely in some way on collisions or interactions between probe molecules. These measurements give information on (nanometer scale) mobility with distances on the order of molecular diameters (1–15 nm). Such measurements involve pyrene excimer formation (Mueller and Galla, 1987; Sassaroli *et al.*, 1990), fluorescence quenching (Blackwell and Whitmarsh, 1990), or Heisenberg spin exchange in magnetic resonance (see, for example, Nageem *et al.*, 1989); the derived diffusion coefficient depends heavily on the dynamic model for the relevant collisional interactions. In general, the calculated nanometer-scale microscopic diffusion coefficients are somewhat larger than the micron-scale macroscopic diffusion coefficients obtained by photobleaching (see, for example, Blackwell and Whitmarsh, 1990; Sassaroli *et al.*, 1990; Shin *et al.*, 1991). In the case of the intermediate distance range (~100 nm) provided by ^{31}P-NMR measurements of phospholipid mobility in 150-nm diameter erythrocyte membrane microvesicles, diffusion coefficients larger than FRAP values in ghosts also are obtained (Gawrisch *et al.*, 1986); however, the authors suggest that the microvesicles probably are depleted of membrane proteins, leading to larger D values.

---

data points are included or if the MSDs of several trajectories are averaged together, one can detect overall patterns such as flow or directed transport (upward curvature) and restriction to a domain or percolation (downward curvature followed by leveling).

III. ARTIFICIAL MEMBRANES

A. Multibilayers

Early FRAP studies examined the lateral mobility of lipid analogs in multibilayer systems that are formed by hydrating phospholipids after removing organic solvent by evaporation (Wu et al., 1977; Fahey and Webb, 1978; Smith and McConnell, 1978; Derzko and Jacobson, 1980; McCown et al., 1981). These investigators found lipid diffusion coefficients of $<10^{-10}$ to 10^{-7} cm²/sec, depending on the temperature, phase, lipid bilayer composition and hydration state. These studies have been reviewed carefully elsewhere (Vaz et al., 1982).

Multibilayer systems also can be used to study the lateral diffusion of biologically important lipoidal compounds. For example, Chazotte et al. (1991) studied the diffusion of a fluorescently labeled ubiquinone analog and concluded that it behaved substantially as a phospholipid with respect to its lateral mobility and orientation within the bilayer. These results are consistent with the role of this molecule as a mobile electron carrier that functions in mitochondrial electron transport and whose rate of diffusion can be rate limiting in maximum (uncoupled) electron transport. Balcom and Petersen (1993) investigated the diffusion of isoprenoid alcohols of various length including dolichol, concluding similarly that these molecules diffuse at the same rate as phospholipids with little dependence of D on the length of the probe molecule. These results support the free area description of bilayer lateral diffusion (Vaz et al., 1984).

More recently, measurements have been made on planar bilayer systems and in phase separating bilayer systems. These studies are discussed in the two following sections.

B. Supported Planar Bilayers

Planar bilayers, in which the membranes are formed or deposited on a substratum, have become useful in studying binding of ligands to membranes and, in modeling cell–cell interactions, as a controllable membrane for one partner (for reviews, see Thompson and Palmer, 1988; Thompson et al., 1988; Tamm and Kalb, 1993). When lateral diffusion of lipid analogs in these bilayers was measured by the FRAP technique, diffusion coefficients were in the range of 10^{-8}–10^{-7} cm²/sec for liquid crystalline phase bilayers. D dropped by over two orders of magnitude in gel phase bilayers (for example, see Tamm and McConnell, 1985). The finding that lipid analog probes diffuse similarly in both monolayers of planar supported bilayers (Tamm, 1988) indicates that a water layer exists between the substratum and the bilayer; indeed, neutron scattering measurements indicate that this layer is, on average, about 30 Å thick (Johnson et al., 1991).

The influence of coupling between monolayers on diffusion has been investigated theoretically (Evans and Sackmann, 1988) and experimentally in planar supported bilayers (Merkel et al., 1989). The lipid analog diffusion coefficient in the distal monolayer (furthest from the substratum) was greater when the proximal monolayer (closest to the substratum) was coupled covalently to the substratum but in a fluid state than when the proximal monolayer was composed of solid cadmium arachidate linked via salt bridges to the substratum. In the latter situation, the probe experiences additional interfacial drag as a result of the apposition of the fluid mono-layer to a solid monolayer. Such measurements allowed estimates of the coefficients that characterize the molecular friction between apposed monolayers.

Lee et al. (1991) used SPT (nanovid microscopy) to follow the move-ments of lipid molecules in planar supported bilayers containing fluoresce-in-conjugated phosphatidylethanolamine (PE) labeled with 30-nm gold anti-fluorescein (anti-Fl). Multivalant gold probes were prepared by conju-gating only anti-Fl to the gold, whereas paucivalent probes were made by mixing an irrelevant antibody with the anti-Fl prior to conjugation. The multivalent gold probes had an average lateral diffusion coefficient of 0.26×10^{-8} cm^2/sec; paucivalent probes had an average D of 0.73×10^{-8} cm^2/sec compared with a D of 1.3×10^{-8} cm^2/sec measured by FRAP of Fl–PE in planar bilayers. The authors suggested that the multivalent gold binds several lipids, forming a disk up to 30–40 nm in diameter, resulting in reduced diffusion with respect to the paucivalent gold, which binds one or a very few lipids.

C. Phase-Separated Systems

After initial studies by Rubenstein et al. (1979) and Klausner and Wolf (1980), Vaz, Thompson, and co-workers (Vaz et al., 1989,1990; Bultmann et al., 1991; Almeida et al., 1993) have provided the most comprehensive description of lateral diffusion in two- and three-component phase-sepa-rated lipid bilayer membranes. The FRAP technique gives information on long-range diffusion (over several micrometers), which is convenient since translational diffusion ought to be measured over a scale that is large compared with the typical domain size. Knowing the degree to which the fluorescently labeled tracer molecule partitions into each phase and its diffusion coefficient in each phase is also crucial. Given this situation and knowledge, such FRAP measurements give information about domains in these membranes in terms of their shape, size, and connectivity. These studies have been reviewed elsewhere (Vaz, 1992). The sophistication that these studies have reached is demonstrated by a study in which Almeida *et al.* (1992), using a membrane-spanning probe, were able to show that, in a binary mixture of two saturated phosphatidylcholines (PCs), the solid phase domains in either monolayer were superimposed exactly. In contrast, in a particular glycolipid-saturated PC mixture, the solid domains in one monolayer were independent of the distribution of solid domains in the apposed monolayer.

IV. CELLULAR MEMBRANES

A. *Overview of the Tables*

Lipid lateral mobility has been measured in both plant and animal cells under a variety of conditions. Table I gives a summary of measurements made on red blood cell membranes using FRAP. Table II gives a summary for other cell types. The tables are not exhaustive but are intended to overview the values for D and the mobile fraction that have been found for lipid lateral mobility under a variety of conditions. In some cases, the temperature at which the FRAP measurements were done is not listed because no mention of it was made in the respective article. When multiple components to diffusion were found, either the average or the major component is given. The structures for some of these probes are shown in Fig. 1. Because of space limitations, the full results of many studies are not presented; thus, the reader may not fully appreciate the purpose of the study from the table.

Table II is organized loosely into overlapping categories based on the comparisons made in each report. The categories are presented in the following order:

1. possible measurement variables (size of photobleached spot and objective

2. effect of temperature

3. comparison of different lipid probes

4. effects of different treatments such as drugs and growth conditions

5. morphology including normal versus blebbed membranes

6. developmental stage and the effect of transformation.

TABLE II
Lateral Mobility of Lipids in a Variety of Cell Types as Measured by FRAP

Cell type[a]	Analog[b]	Diffusion coefficient ($\times10^{-9}$)	Mobile fraction	Temperature[c]	Comments[d]	References
Measurement variables						
Human skin fibroblast (CCD)						Yechiel and Edidin (1987)
40× objective	NBD-PC	0.9 ± 0.3	81	18–20	0.8 μm spot size	
63× objective	NBD-PC	0.9 ± 0.5	88	18–20	0.4 μm spot size	
90× objective	NBD-PC	0.9 ± 0.5	89	18–20	0.35 μm spot size	
Mouse hepatoma cells (HEPA-OVA)						Edidin and Stroynowski (1991)
0.6 μm spot size	DiIC$_{16}$	0.6	66	—		
1.1 μm spot size	DiIC$_{16}$	1.0	60	—		
Temperature						
Human fibroblasts (BG-9)	DiIC$_{16}$	3.5	—	5		Jacobson et al. (1981)
		8.0	—	20		
		16.0	—	37		
Different probes compared						
Soybean protoplasts	NBD-PE	2.8 ± 1.3	35	18	1st component	Metcalf et al. (1986)
	NBD-PE	11 ± 12	62	37	1st component	
	DiIC$_{18}$	3.4 ± 1.9	42	18	1st component	
	DiIC$_{18}$	12 ± 16	56	37	1st component	
	NBD-PC	1.3 ± 0.6	27	18	1st component	
	DiIC$_{14}$	5.7 ± 3.0	39	18	1st component	
	C$_{12}$NFl	25 ± 1.4	69	18	1st component	
Swiss 3T3 fibroblasts	NBD-PE	14 ± 13	68	18		Metcalf et al. (1986)
	DiIC$_{18}$	11 ± 9	47	18		

(continued)

TABLE II (*Continued*)

Cell type[a]	Analog[b]	Diffusion coefficient ($\times 10^{-9}$)	Mobile fraction[c]	Temperature[c]	Comments[d]	References
Chinese hamster lung fibroblasts (V79)	NBD-PC	0.4 ± 0.1	—	22		El Hage Chahine *et al.* (1993)
	NBD-SPC	0.4 ± 0.1	—	22		
	NBD-PS	0.3 ± 0.2	—	22		
	NBD-SPC	0.6 ± 0.2	—	37		
3T3 mouse fibroblast	DiIC$_{18}$	6 ± 3	82	rt		Wolf *et al.* (1980)
	AcT$_5$RSD	0.4 ± 0.2	72	rt		
	AcRSD	0.3 ± 0.2	62	rt		
	TRSD	0.3 ± 0.2	76	rt		
CHO cells	HEDAF	0.9	100	—	Inserted	Dupou *et al.* (1988)
	ASte	2.3	97	—	Inserted	
	ANno	2.0	97	—	Inserted	
	ANno	1.6	75	—	Metabolic incorporation labels variety of membrane lipids	

Effectors or treatments

Effectors or treatments					
Aplysia neurons					Treistman *et al.* (1987)
control	C_6-NBD-PC	1.8 ± 0.2	58		
5% EtOH	C_6-NBD-PC	1.7 ± 0.2	60		
control	Rh-PE	1.7 ± 0.2	75		
5% EtOH	Rh-PE	3.4 ± 0.6	71		
Maize protoplasts					Furtula *et al.* (1990)
0.45 *M* mannitol	LY-Chol	1.3 ± 0.3	36	Low mobile fraction has	
0.90 *M* mannitol	LY-Chol	0.3 ± 0.1	50	been reported for	
0.45 *M* mannitol	LY-DC-PE	2.2 ± 0.9	52	other plant cells (see	
0.90 *M* mannitol	LY-DC-PE	2.3 ± 1.2	46	Metcalf *et al.*, 1986)	
Saccharomyces cerevisiae					Greenberg and Axelrod (1993)
inol mutant	$DiIC_{18}$	0.8	56	—	Averaged fast and slow components
inol+inositol	$DiIC_{18}$	1.5	58	—	Averaged fast and slow components
inol+inositol, trypsinized	$DiIC_{18}$	108.0	96	—	Averaged fast and slow components
inol mutant	Rh-PE	3.4	58	—	Averaged fast and slow components
inol+inositol	Rh-PE	3.2	34	—	Averaged fast and slow components
opi3 mutant	Rh-PE	4.3	38	—	Averaged fast and slow components

(continued)

125

TABLE II (Continued)

Cell type[a]	Analog[b]	Diffusion coefficient ($\times 10^{-9}$)[b]	Mobile fraction	Temperature[c]	Comments[d]	References
opi3+choline	Rh-PE	2.2	100	—	Averaged fast and slow components	
Rat smooth muscle cells	DiIC$_{18}$	38	100	—	Averaged fast and slow components	Greenberg and Axelrod (1993)
Human colon adenocarcinoma (HT29)	NBD-PE	3.3 ± 0.2	50	12	Trypsin did not affect diffusion coefficient	Cohen et al. (1990)
V79-UF (Chinese hamster) +cis-vaccenic acid	DiIC$_{18}$	3.5 ± 1.0	—	—	Cells incubated with sodium azide and 2-deoxyglucose	Hayashi et al. (1987)
+arachidonic acid	DiIC$_{18}$	4.1 ± 0.8	—	—		
BCL$_1$ (murine B-cell leukemia) Control	DiI	3.8	—	37		Keating et al. (1988)
50 µg 1 ml lipopolysaccharide for 24 hr	DiI	2.2	—	37		
+testosterone for 6 hr	DiI	2.0	—	37		
+10^{-6}M glucocorticoid for 6 hr	DiI	3.5	—	37		
Keratinocytes in normal Ca^{2+}	DiIC$_{14}$	5.2 ± 0.4	—	—		Tertoolen et al. (1988)
in low Ca^{2+}	DiIC$_{14}$	3.2 ± 0.3	—	—		
7 hr after switch from low to normal	DiIC$_{14}$	7.1 ± 0.5	—	—		
Human endothelial cells untreated	Fl-PE	8.2 ± 2.6	>80	37	100 U/ml for 3–4 days	Stolpen et al. (1988)
TNF	Fl-PE	7.0 ± 1.7	>80	37	200 U/ml for 3–4 days	
IFN-γ	Fl-PE	7.1 ± 2.2	>80	37	100 U/ml + 200 U/ml	
TFN + IFN-γ	Fl-PE	5.1 ± 1.5	>80	37		

C3H mouse fibroblast	FI-PE	5.4 ± 2.7	69	rt		Lee et al. (1993)
PtK$_1$	FI-PE	6.3 ± 1.3	71	rt		Lee et al. (1993)
Fish scale fibroblast	FI-PE	9.5 ± 1.7	74	rt		Lee et al. (1993)
Goldfish keratocyte	FI-PE	6.2 ± 2.4	64	rt		Lee et al. (1993)
Bovine endothelial cells						Nakache et al. (1985)
on secreted ECM	NBD-PC	1.0 ± 0.3	68	10	Both cell surfaces	
on secreted ECM	NBD-PC	8.0 ± 0.2	70	10	Apical surface	
on collagen type I	NBD-PC	7.0 ± 0.2	56	10	Both surfaces	
on collagen type I	NBD-PC	8.2 ± 0.2	60	10	Apical surface	
on fibronectin	NBD-PC	7.5 ± 0.2	48	10	Both surfaces	
on fibronectin	NBD-PC	7.5 ± 0.2	64	10	Apical surface	
Morphology (including blebbing)						
Mouse leukemic cells (RDM$_4$)						
normal cell	DiIC$_{16}$	11 ± 2	92	rt		Wu et al. (1982)
bulbous cell	DiIC$_{16}$	40 ± 2	96	rt	Treated with concanavalin A	
Mouse spleen cells						
normal cell	DiIC$_{16}$	15 ± 6	90	rt		Wu et al. (1982)
bulbous cell	DiIC$_{16}$	32 ± 15	92	rt	Treated with concanavalin A	
Mouse spleen lymphocytes	DiIC$_{18}$	17 ± 3	95	—		Dragsten et al. (1979)
Human lymphocytes						
flat	DiIC$_{18}$	9.5 ± 2.2	96	—	No microvilli	Dragsten et al. (1979)
round	DiIC$_{18}$	8.6 ± 2.1	91	—	Microvilli present	
Rat hepatocytes						
unblebbed region	DiIC$_{16}$	8.1	80	25	Hypoxia for 20 min	Wang et al. (1992)
blebbed region	DiIC$_{16}$	12.3	99	25	Hypoxia for 20 min	
unblebbed region	DiIC$_{16}$	2.0	30	25	Hypoxia for 60 min	
blebbed region	DiIC$_{16}$	24.6	85	25	Hypoxia for 60 min	

(continued)

TABLE II (*Continued*)

Cell type[a]	Analog[b]	Diffusion coefficient ($\times 10^{-9}$)	Mobile fraction	Temperature[c]	Comments[d]	References
Mouse muscle fibers						Tank *et al.* (1982)
normal cells	NBD-PC	1.0 ± 0.5	92	22		
blebs	NBD-PC	15.0 ± 6.0	100	22		
L6 rat embryo myoblasts						Tank *et al.* (1982)
normal cells	NBD-PC	4.4 ± 1.5	100	22		
blebs	NBD-PC	12.0 ± 4.0	99	22		
Schistosoma mansoni						Foley *et al.* (1986)
normal	Fl-PE	0.4 ± 0.09	29	—	Surface membrane	
normal	DiIC$_{14}$	3.1 ± 2.4	25	—	Surface membrane	
normal	DiIC$_{18}$	3.2 ± 2.0	20	—	Surface membrane	
bleb	DiIC$_{18}$	3.9 ± 0.8	83	—	Surface membrane	
normal	C$_{18}$-Fl	3.6 ± 1.9	79	—	Surface membrane	
bleb	C$_{18}$-Fl	2.3 ± 0.5	84	—	Surface membrane	
normal	C$_{18}$-Rh	0.4 ± 0.07	79	—	Surface membrane	
Development (including transformation						
3T3 mouse fibroblasts						Eldridge *et al.* (1980)
normal	DiI	9 ± 0.8	95	—		
transformed	DiI	8.7 ± 1.3	94	—		
Chick embryo fibroblasts						Boullier *et al.* (1992)
normal	DiIC$_{18}$	1.4 ± 0.03	85	35		
RSV transformed	DiIC$_{18}$	1.4 ± 0.1	85	35		
RSV-NY68	DiIC$_{18}$	0.7 ± 0.2	85	35	Transient decrease 5 hr after transformation	
MDCK cells						Jesaitis and Yguerabide (1986)
subconfluent	HEDAF	0.6	78	rt		
confluent	HEDAF	2.0	92	rt		

Sample	Probe	Value	%	Temp	Conditions	Reference
Mouse sperm						Wolf et al. (1986a)
control head	DiC_{16}	1.2 ± 0.2	60	—	Whittingham's medium + 0.4% BSA	
control tail	DiC_{16}	1.6 ± 0.3	69	—		
hyperactivated head	DiC_{16}	2.6 ± 0.3	45	—	Whittingham's plus 3% BSA	
hyperactivated tail	DiC_{16}	7.8 ± 1.5	53	—		Wolf et al. (1986b)
Mouse germ cells						
pachytene spermatocyte	DiC_{16}	1.8 ± 0.2	87	—		
round spermatid	DiC_{16}	2.2 ± 0.2	81	—		
testicular sperm	DiC_{16}	2.8 ± 0.3	65	—	Anterior head	
cauda epididymal sperm	DiC_{16}	2.1 ± 0.2	61	—	Anterior head	
Urodelean (*Pleurodeles watl*) ectodermal cells						Dupou et al. (1987)
blastula	HEDAF	3.5	100	—		
mid-gastrula	HEDAF	2.2	100	—		
Xenopus laevis cells						Gadenne et al. (1984)
early blastula	HEDAF	2.4 ± 0.1	72	rt		
late blastula	HEDAF	3.6 ± 0.2	82	rt		
mid-gastrula	HEDAF	3.9 ± 0.2	80	rt		

[a]Abbreviations: ECM, extracellular matrix; IFN-γ, interferon; MDCK, Madin–Darby canine kidney; RSV, Rous sarcoma virus; TNF, tumor necrosis factor.

[b]Abbreviations of lipid analogs: AcT$_3$RSD, acetyl-5-trinitrophenyl-rhodamine-stearoyldextran; AcRSD, acetylrhodamine-stearoyldextran; ANno, 9-(2-anthryl)-nonanoic acid (anthracene); ASte, 12-(9-anthroyloxy)-stearic acid; C_{12}NFl, 5-(N-dodecanoyl)-aminofluorescein; C_{18}Fl, 5-(N-octadecanoyl)-aminofluorescein; C_{18}Rh, octadecylrhodamine B chloride; DiC_{14}, 1,1′-ditetradecyl-3,3,3′,3′-tetramethylindocarbocyanine perchlorate; DiC_{16}, 1,1′-dihexadecyl-3,3,3′,3′-tetramethylindocarbocyanine perchlorate; DiC_{18}, 1,1′-dioctadecyl-3,3,3′,3′-tetramethylindocarbocyanine perchlorate; DiC_{18}(5), 1,1′-dioctadecyl-3,3,3′,3′-tetramethylindodicarbocyanine perchlorate; Fl-PE, fluorescein dihexadecanoyl phosphoethanolamine (fluorescein on the head group); HEDAF, 5-(N-hexadecanoyl)-aminofluorescein; LY-DC-PE, dilithium 4-amino-N-(3-[β-(dilauroyl-sn-glycero-3-phosphoethanolamino) ethylsulfonyl]phenyl)-1,8-naphthalimide-3,6-disulfonate; LY-Chol, dilithium 4-amino-N-{[β-(carbo(5-cholesten-3β-yl)oxy) hydrazino-carbonyl]amino}-1,8-naphthalimide-3,6-disulfonate; C_6NBD-PE, NBD on acyl chain via amino linkage; NBD-PC, 1-acyl-2-[N-(4-nitrobenzo-2-oxa-1,3-diazoyl)-amino]caproyl-sn-glycero-3-phosphocholine; NBD-PE, phosphatidyl-N-(4-nitrobenzo-2-oxa-1,3-diazoyl)ethanolamine; NBD-PS, 1-acyl-2-[6-(NBD-amino)caproyl]phosphatidylserine; NBD-SPC, N-[6-(NBD-amino)caproyl]sphingosylphosphocholine; Rh-PE, lissamine rhodamine B on head group amino terminus; TRSD, trinitrophenyl-rhodamine-stearoyldextran.

[c]Temperature at which FRAP measurement was performed. rt, Room temperature.

[d]BSA, Bovine serum albumin.

Studies using similar lipid analogs have been placed near each other within the confines of these categories.

In most of the studies in which the same probe and cells were measured at different temperatures (Jacobson *et al.*, 1981; Bloom and Webb, 1983; Metcalf *et al.*, 1986), temperature can be seen to have a major effect on lipid mobility (Tables I, II) and, thus, should be considered when comparing results from different studies. Additional variability may be caused by several other factors. First, bilayer composition differences lead to different probe mobilities even in a purely fluid membrane. However, awareness is growing of the possibility of domain structures in biological membranes (see Chapter 6). Different probes may partition differently into domains (see subsequent discussion), which would lead to variability in the reported diffusivities (see Jacobson and Vaz, 1992). Finally, systematic instrument errors, particularly uncertainties in the laser beam diameter, have large effects on the reported mobilities (see next section).

Deliberate manipulation of the membrane by factors such as ethanol, drugs, interferon, Ca^{2+}, osmolarity, and the substratum for cell growth also has an effect, frequently producing statistically significant changes in D by a factor of 2 or less (Tables I,II). In the case of bovine endothelial cells grown on secreted extracellular matrix, the NBD-PC mobility in the substratum-attached surface was reduced by at least an order of magnitude compared with that in cells grown on Type I collagen or fibronectin alone (Nakache *et al.*, 1985). This result suggests a role for the extracellular matrix in determining the lateral organization of plasma membrane apposed to the substratum. The effect of morphology is most dramatic in the case of membrane blebs in mouse muscle cells and myoblasts (Tank *et al.*, 1982). In some cases, an order of magnitude increase in the NBD-PC diffusion coefficient was measured, perhaps as a result of protein depletion in the bleb membrane.

Lipid diffusion can be modulated slightly by developmental stage (Gadenne *et al.*, 1984) but oncogenic transformation has no significant effect on DiI diffusion (Eldridge *et al.*, 1980), apart from a transient decrease in mobility (Boullier *et al.*, 1992).

B. Call for the Use of Standards

Perusal of the tables reveals that values for D and the mobile fraction are quite comparable within a study but do not always compare well between experiments although the same types of cells, lipid probe, and temperature have been used. For instance, with human red blood cells, DiI yields similar values for intact cells and ghosts as measured by Bloom and Webb (1983). However, the D for DiI in human red blood cell ghosts is much lower as measured by Thompson and Axelrod (1980); the value reported for intact cells at 37°C by Boullier *et al.* (1986) is about one-third the value reported by Bloom and Webb (1983) for the same temperature. A similar situation can be found for the diffusion of DiI in fibroblasts at 35–37°C. Apart from the biological variables mentioned earlier, a major source of the variation is due to uncertainties in spot size from laboratory to laboratory (Greenberg and Axelrod, 1993); lesser variation is due to various curve fitting procedures used to extract the D values. A solution to this problem involves the adoption of a universal standard that can be used by each laboratory to calibrate their instruments or to normalize their measurements. One standard preparation uses rhodamine-labeled bovine serum albumin (BSA) in phosphate-buffered glycerol solution (90% glycerol by weight) (Thompson and Axelrod, 1980; Foley *et al.*, 1986). Alternatively, Edidin and Stroynowski (1991) used 2% fluorescent IgG in water/98% anhydrous glycerol. By varying the concentration of glycerol and, thus, the viscosity of the solution, a standard curve at 25°C could be prepared by measuring both the viscosity of the solution and the diffusion coefficient. Since diffusion scales with inverse viscosity (η), a plot of D vs $1/\eta$ should be linear. To give a thin film of even depth, the solution simply is sandwiched between a slide and a coverslip, which can be sealed with valap ($1:1:1$ vaseline, lanolin, paraffin) to prevent drying. The diffusion of these proteins in water and, for reference, the viscosity of glycerol solutions at various temperatures and concentrations have been published. Measured values on the standard specimen then can be adjusted to the published values and to the expected viscosity dependence, giving a correction factor. However, large departures of the measured from the published values should be investigated because the FRAP technique is capable of measuring protein diffusion in solution quite accurately (see, for example, Jacobson *et al.*, 1976; Barisas and Leuther, 1979).

C. Lipid Domains

Several studies have reported unusually low values for the diffusion of one or more lipid analogs in the plasma membrane of a variety of cell types (Metcalf *et al.*, 1986; Morrot *et al.*, 1986; Yechiel and Edidin, 1987; Edidin and Stroynowski, 1991; El Hage Chahine *et al.*, 1993; Greenberg and Axelrod, 1993). With the exception of the studies by Edidin and co-workers, these studies identified more than one component in the analysis of the recovery curve. The finding that D varied with spot size (Edidin and Stroynowski, 1991), or that the recovery was multicomponent (Greenberg and Axelrod, 1993), has been used to infer the presence of lipid domains in the membrane. Although lipid domains will be discussed much more extensively in Chapter 6, the elegant work of Davoust and co-workers in detecting lipid mobility differences between inner and outer leaflets deserves mention. In the erythrocyte membrane, Morrot *et al.* (1986) found, using the ATP-dependent aminophospholipid translocase, that NBD-labeled aminophospholipids [phosphatidylserine (PS); PE] exhibited ≥ 5-fold or greater in D values in the inner monolayer compared with NBD-PC diffusing primarily in the outer monolayer, suggesting a difference in the effective fluidity and/or domain structure of the inner and outer monolayers.

In a study analogous to that performed with red cells, El Hage Chahine *et al.* (1993) report that, in hamster lung fibroblasts, an NBD-PC derivative as well as an NBD-sphingolipid derivative also showed very low mobility ($\sim 4 \times 10^{-10}$ cm^2/sec) at 22°C; both derivatives preferentially incorporated into the outer monolayer of the plasma membrane. However, an NBD-PS analog displayed two diffusing components: a slow one associated with outer monolayer diffusion and a fast component ($D \sim 3 \times 10^{-9}$ cm^2/sec), presumably due to flip–flop of the aminophospholipid probe to the inner monolayer by an ATP-requiring aminophospholipid translocase (see Chapter 2). Thus, the two monolayers of the bilayer can be considered separate domains. These authors suggest that early work showing large D values for lipid probes may have overlooked the superposition of two different lipid mobilities in the inner and outer monolayer. However, with SPT of gold-labeled lipids, D for Fl-PE was found to be $\sim 1 \times 10^{-9}$ cm^2/sec for a variety of cell types (Lee *et al.*, 1993). SPT only measures the diffusion of lipids in the outer leaflet of the bilayer since the gold tag is limited to the cell surface. The D measured by SPT is a low estimate because of the drag on the 30-nm gold particle produced by the pericellular matrix and because each gold probe may be binding several lipids (Lee *et al.*, 1993).

D. Putative Bulk Lipid Flows

The retrograde lipid flow (RLF) hypothesis, articulated by Bretscher (1984), postulated that membrane insertion at the leading edge is accompanied by endocytosis over the entire cell surface; this phenomenon would set up a front to rear flow within the lipid bilayer as lipid moves from source (insertion) to sink (endocytosis). This intriguing hypothesis accounted for cell movement since membrane insertion essentially deposited a new leading edge. The hypothesis also accounted for capping because the lipid flow would sweep slowly diffusing aggregates of membrane proteins (patches) to the rear pole of the cell in a "cap". However, several explicit predictions concerning protein movement that were derived from the RLF model could not be confirmed experimentally.

First, cell membrane proteins (or protein aggregates) with vastly different lateral diffusion coefficients would be expected to distribute with markedly different gradients along the length of the cell; the slower moving protein (or aggregate) would be predicted to accumulate at the trailing edge of the cell in a much steeper gradient. In fact, Holifield *et al.* (1990) showed, in a test using the murine cell membrane protein pgp-1, that this does not occur. pgp-1, a member of the CD44 family, is unusual because it can be capped in fibroblasts using only a monoclonal antibody, in addition to being capped in the conventional way using the monoclonal antibody followed by a secondary polyclonal antibody. pgp-1 is cleared off the lamella to the trailing edge of the cell to the same extent in both forms of capping, in clear contradiction to the prediction of the RLF model. In fact, Holifield and Jacobson (1991) showed that capping in fibroblasts only results in clearance of protein from the leading lamella to a zone known as the "null border" ahead of the nucleus. Behind the null border, aggregated patches of the protein actually move forward as the trailing edge retracts.

Conversely, if two proteins have similar D values, the RLF model would predict that both should distribute with a similar gradient along the length of the cell. The GP4F NIH 3T3 cell line permanently expresses the viral hemagglutinin, HA, as well as endogenous pgp-1. Both molecules have similar D values (2–3×10^{-10} cm^2/sec) as measured by FRAP after tagging with antibody. However, pgp-1 was capped but HA was not, instead maintaining a uniform distribution over the cell surface (Holifield *et al.*, 1990). The different behaviors of these two proteins contradict the predictions of the RLF model.

If proteins were capped by an RLF, they would be expected as they randomly diffuse, to drift simultaneously toward the trailing edge as they experience the putative rearward-directed lipid flow. In fact, concanavalin A (Con A) receptors, identified by Con A-coated gold beads and observed by SPT, only move rearward when they *do not* diffuse (Sheetz *et al.*, 1989; Kucik *et al.*, 1990). Periods of rearward-directed movement are interspersed with periods of random diffusion in which no detectable rearward drift occurs. This result is consistent with the idea that rearward movement of Con A receptors is coupled tightly to a rearward-directed cortical cytoskeletal movement. Occasionally, the Con A receptors disengage from this mechanism and undergo free lateral diffusion.

Finally, in a direct test of the RLF model, Lee *et al.* (1990) photobleached a line mark in a DiI plasma membrane stain; the mark was perpendicular

to the direction of motion of rapidly moving polymorphonuclear leukocytes. The expectation from the RLF model was that the line would move
rearward with respect to the cell, in a displacement that should have been
detectable given the speed of the cell. In fact, the line did not move with
respect to the leading edge (it did move forward with respect to the
substratum), suggesting that the bilayer moves forward passively as the
cell moves.

These experiments establish that bulk rearward flow of bilayer does not
occur. However, membrane insertion has been shown to sustain a bulk
forward flow of the lipid bilayer. In rapidly growing *Xenopus* neurites,
stained locally with $DiIC_{12}$, the labeled region moves in an anterograde
manner consistent with the addition of new membrane both at the cell
body and along the neurite, but not at the growth cone itself (Popov *et
al.*, 1993). In the context of cell locomotion, local membrane insertion of
cell substratum receptors (e.g., integrins) and other adhesive molecules
could be crucial to providing the membrane–substratum coupling essential
for traction, as Bretscher has pointed out. When insertion occurs close
to the site of endocytosis, a coupling of the two processes may exist and
local flows may occur.

V. BIOLOGICAL IMPLICATIONS

The lateral mobility of lipids can have an effect on multiple cellular
processes because of lipid involvement in signal transduction (Shukla and
Halenda, 1991; Mathias *et al.*, 1993; see also chapters in Part V). In fact,
cytoplasmic proteins such as protein kinase C appear to restrict the lateral
mobility of plasma membrane phospholipids (Bazzi and Nelsestuen, 1991).
This ability may be especially important when the stimulus is localized
to a small region of the cell. For example, when phosphoinositol 4,5-
bisphosphate (PIP_2) is cleaved in response to localized stimulation by a
growth factor, the PIP_2 in that area of the membrane could be depleted
or lateral diffusion could maintain a continual fresh supply of PIP_2. The
temporary depletion of the substrate phospholipid could limit the duration
of the signal elicited by a given stimulatory event. However, diffusion
could result in a steady but limited supply of substrate lipid and thus limit
the magnitude of the signal rather than the duration. In addition, lateral
mobility may determine the area affected by the stimulation by limiting
the molecules to which the lipid messenger has access. (For a discussion
of diffusion limited reactions, see Berg and von Hippel, 1985; McCloskey
and Poo, 1986).

The lateral mobility of lipids in terms of diffusion coefficients has been
discussed. Putting the diffusion of lipids into the context of movement

over the cell surface will help the reader appreciate the time frame in which the lipid replenishment could occur and the distance a lipid second messenger can travel in a relatively short time. In PtK cells at 37°C, gold-labeled lipids with an average D of 2.3×10^{-9} cm^2/sec have an average net displacement of 2.1 ± 1.3 μm (average $\pm$ SD for 27 gold particles) in 6.6 sec (G. Lee and K. Jacobson, unpublished results). The range of displacement in this time was 0.1–5.5 μm. One gold-tagged lipid had a net displacement of 8.9 μm in 40 sec. A fibroblast or PtK cell attached to a substrate is typically 50–70 μm in its longest dimension. Thus, a lipid second messenger could move one-sixth of the way across the cell in 40 sec. For a longer time frame, consider the fusion of DiI-labeled erythrocytes with a fibroblast. Within 1–2 min of fusion, the cell surface had become uniformly labeled (Wojcieszyn et al., 1983). Since the erythrocytes were as much as 20 μm apart on the cell surface, the lipids had to diffuse over a 10 μm distance in this 1–2 min time interval. Within 10 min of fusion of liposomes to the apical surface of Madin-Darby canine kidney (MDCK) cells, radioactively labeled dipalmitoyl PC from the fused liposomes had diffused to the basal surface of the cell (Knoll et al., 1988). The radioactive PC had passed the tight junction, but the authors did not indicate a distinction between leaflets of the bilayer. In contrast, rhodamine PE will pass the tight junction only in the cytoplasmic leaflet of the bilayer (van Meer and Simons, 1986).

The examples just given indicate that, in some situations, random diffusion can result in the mixing of lipids over the whole cell surface. In other instances, regional restrictions prevent the free lateral mobility of lipids, resulting in enrichment of specific lipids into areas known as domains. The significance of these restrictions with respect to biological function is not clear at this time. A striking example of regional restriction is the inability of labeled lipids located on either monolayer to cross from the axon through the axon hillock to the cell body of hippocampal neurons, indicating the presence of a barrier to lipid diffusion between the initial segment and the axon hillock (Kobayashi et al., 1992). This restriction is similar, although not identical, to the inhibition of movement of fluorescently labeled outer monolayer lipids from the apical to the basolateral surfaces of epithelial cells with tight junctions. These results provide evidence that specialized structures in the plasma membrane can restrict lipid diffusion. Additional evidence for lipid regionalization has been found in other cell types such as sperm, and is discussed in Chapter 6.

Lipids also may play a role in limiting lateral mobility of certain membrane proteins. Cholesterol is required for the formation of caveolae, structurally unique invaginations in the plasma membrane (Rothberg et al., 1990). These structures also are enriched in glycosyl-phosphatidylinositol (GPI)-linked proteins (Anderson et al., 1992) and may account for the high (50%) immobile fraction observed for these proteins (Ishihara *et al.*, 1987; Edidin and Stroynowski, 1991; Zhang *et al.*, 1991,1992). Although microdomains of GPI-linked proteins and specific lipids have not been demonstrated definitively, Triton X-100-insoluble complexes of cholesterol, glycerophospholipids, sphingolipids, and GPI-linked proteins have been isolated from the Golgi network and the plasma membrane (Brown, 1992). In addition to an enrichment of GPI-linked proteins in cholesterol-maintained caveolae, the association of some GPI-linked proteins with sphingolipids may lead to the temporary immobilization of these proteins on delivery to the plasma membrane (Hannan *et al.*, 1993; See also Chapter 11).

VI. SUMMARY

Lipid lateral mobility is important in cell signaling, membrane flexibility, fusion, and modulation of protein function. Lipid lateral mobility, expressed as a diffusion coefficient and a mobile fraction, can be measured by a variety of techniques including FRAP, SPT, magnetic resonance, pyrene excimer formation, and fluorescence quenching. Lateral diffusion of lipids has been measured in both artificial and biological membranes. Many factors can affect lipid mobility, but the effects are small compared with modulations of the mobility of membrane proteins. Large-scale directed mobility, or lipid flow, has not been detected in cell membranes except in the case of growing neurites. Restrictions to lipid lateral mobility have been found to occur and may have important biological functions.

Acknowledgments

We thank Brad Chazotte and Fen Zhang for their helpful comments on the manuscript, and Jennifer Lipfert and April Humphrey for their help in preparing the manuscript.

References

Almeida, P. F. F., Vaz, W. L. C., and Thompson, T. E. (1992). Lateral diffusion and percolation in two-phase, two-component lipid bilayers. Topology of the solid-phase domains in-plane and across the lipid bilayer. *Biochemistry* **31**, 7198–7210.

Almeida, P. F. F., Vaz, W. L. C., and Thompson, T. E. (1993). Percolation and diffusion in three-component lipid bilayers: Effect of cholesterol on an equimolar mixture of two phosphatidylcholines. *Biophys. J.* **64**, 1–14.

Anderson, R. G. W., Kamen, B. A., Rothberg, K. G., and Lacey, S. W. (1992). Potocytosis: Sequestration and transport of small molecules by caveolae. *Science* **255**, 410–411.

Axelrod, D., Koppel, D. E., Schlessinger, J., Elson, E., and Webb, W. W. (1976). Mobility measurement by analysis of fluorescence photobleaching recovery kinetics. *Biophys. J.* **16**, 1055–1069.

Balcom, B. J., and Petersen, N. O. (1993). Lateral diffusion in model membranes is independent of the size of the hydrophobic region of molecules. *Biophys. J.* **65**, 630–637.

Barisas, B. G., and Leuther, M. D. (1979). Fluorescence photobleaching recovery measurement of protein absolute diffusion constants. *Biophys. Chem.* **10**, 221–229.

Bazzi, M. D., and Nelsestuen, G. L. (1991). Extensive segregation of acidic phospholipids in membranes induced by protein kinase-C and related proteins. *Biochemistry* **30**, 7961–7969.

Berg, O. G., and von Hippel, P. H. (1985). Diffusion-controlled macromolecular interactions. *Annu. Rev. Biophys. Biophys. Chem.* **14**, 131–160.

Blackwell, M. F., and Whitmarsh, J. (1990). Effect of integral membrane proteins on the lateral mobility of plastiquinone in phosphatidylcholine proteoliposomes. *Biophys. J.* **58**, 1259–1271.

Bloom, J. A., and Webb, W. W. (1983). Lipid diffusibility in the intact erythrocyte membrane. *Biophys. J.* **42**, 295–305.

Boullier, J. A., Brown, B. A., Bush, J. C., Jr., and Barisas, B. G. (1986). Lateral mobility of a lipid analog in the membrane of irreversible sickle erythrocytes. *Biochim. Biophys. Acta* **856**, 301–309.

Boullier, J. A., Peacock, J. S., Roess, D. A., and Barisas, B. G. (1992). Protein and lipid lateral diffusion in normal and Rous sarcoma virus transformed chick embryo fibroblasts. *Biochim. Biophys. Acta* **1107**, 193–199.

Bretscher, M. S. (1984). Endocytosis: Relation to capping and cell locomotion. *Science* **224**, 681–686.

Brown, D. A. (1992). Interactions between GPI-anchored proteins and membrane lipids. *Trends Cell Biol.* **2**, 338–343.

Bultmann, T., Vaz, W. L. C., Melo, E., Sisk, R. B., and Thompson, T. E. (1991). Fluid-phase connectivity and translational diffusion in a eutectic, two-component, two-phase phosphatidylcholine bilayer. *Biochemistry* **30**, 5573–5579.

Chazotte, B., Wu, E.-S., and Hackenbrock, C. R. (1991). The mobility of a fluorescent ubiquinone in model lipid membranes. Relevance to mitochondrial electron transport. *Biochim. Biophys. Acta* **1058**, 400–409.

Cohen, E., Ophir, I., Henis, Y. I., Bacher, A., and Ben Shaul, Y. (1990). Effect of temperature on the assembly of tight junctions and on the mobility of lipids in membranes of HT29 cells. *J. Cell Sci.* **97**, 119–125.

Davoust, J., Devaux, P., and Leger, L. (1982). Fringe pattern photobleaching, a new method for the measurements of transport coefficients of biological macromolecules. *EMBO J.* **1**, 1233–1238.

DeRoe, C., Courtoy, P. J., and Baudhuin, P. (1987). A model of protein-colloidal gold interactions. *J. Histochem. Cytochem.* **35**, 1191–1198.

Derzko, Z., and Jacobson, K. (1980). Comparative lateral diffusion of fluorescent lipid analogues in phospholipid multibilayers. *Biochemistry* **19**, 6050–6057.

Dragsten, P., Henkert, P., Blumenthal, R., Weinstein, J., and Schlessinger, J. (1979). Lateral diffusion of surface immunoglobulin, Thy-1 antigen, and a lipid probe in lymphocyte plasma membranes. *Proc. Natl. Acad. Sci. U.S.A.* **76**, 5163–5167.

Dupou, L., Gualandris, L., Lopez, A., Duprat, A. M., and Tocanne, J. F. (1987). Alterations in lateral lipid mobility in the plasma membrane of urodelean ectodermal cells during gastrulation. *Exp. Cell Res.* **169**, 502–513.

Dupou, L., Lopez, A., and Tocanne, J. F. (1988). Comparative study of the lateral motion of extrinsic probes and anthracene-labelled constitutive phospholipids in the plasma membrane of Chinese hamster ovary cells. *Eur. J. Biochem.* **171**, 669–674.

Edidin, M., and Stroynowski, I. (1991). Differences between the lateral organization of conventional and inositol phospholipid-anchored membrane proteins. A further definition of micrometer scale membrane domains. *J. Cell Biol.* **112**, 1143–1150.

Eldridge, C. A., Elson, E. L., and Webb, W. W. (1980). Fluorescence photobleaching recovery measurements of surface lateral mobilities on normal and SV-40 transformed mouse fibroblasts. *Biochemistry* **19**, 2075–2079.

El Hage Chahine, J. M., Cribier, S., and Devaux, P. (1993). Phospholipid transmembrane domains and lateral diffusion in fibroblasts. *Proc. Natl. Acad. Sci. U.S.A.* **90**, 447–451.

Evans, E., and Sackmann, E. (1988). Translational and rotational drag coefficients for a disk moving in a liquid membrane associated with a rigid substrate. *J. Fluid Mech.* **194**, 553–561.

Fahey, P. F., and Webb, W. W. (1978). Lateral diffusion in phospholipid bilayer membranes and multilamellar liquid crystals. *Biochemistry* **17**, 3046–3053.

Fein, M., Unkeless, J., Chuang, F. Y. S., Sassaroli, M., da Costa, R., Vaananen, H., and Eisinger, J. (1993). Lateral mobility of lipid analogs and GPI-anchored proteins in supported bilayers determined by fluorescent bead tracking. *J. Memb. Biol.* **135**, 83–92.

Foley, M., MacGregor, A. N., Kusel, J. R., Garland, P. B., Downie, T., and Moore, I. (1986). The lateral diffusion of lipid probes in the surface membrane of *Schistosoma mansoni*. *J. Cell Biol.* **103**, 807–818.

Furtula, V., Khan, I. A., and Nothnagel, E. A. (1990). Selective osmotic effect on diffusion of plasma membrane lipids in maize protoplasts. *Proc. Natl. Acad. Sci. U.S.A.* **87**, 6532–6536.

Gadenne, M., van Zoelen, E. J. J., Tencer, R., and de Laat, S. W. (1984). Increased rate of capping of concanavalin A receptors during early *Xenopus* development is related to changes in protein and lipid mobility. *Dev. Biol.* **104**, 461–468.

Gawrisch, K., Stibenz, D., Mops, A., Arnold, K., Linss, W., and Halbnuber, K.-J. (1986). The rate of lateral diffusion of phospholipids in erythrocyte microvesicles. *Biochim. Biophys. Acta* **856**, 443–447.

Greenberg, M., and Axelrod, D. (1993). Anomalously slow mobility of fluorescent lipid probes in the plasma membrane of the yeast *Saccharomyces cerevisiae*. *J. Memb. Biol.* **131**, 115–127.

Hannan, L. A., Lisanti, M. P., Rodriguez-Boulan, E., and Edidin, M. (1993). Correctly sorted molecules of a GPI-anchored protein are clustered and immobile when they arrive at the apical surface of MDCK cells. *J. Cell Biol.* **120**, 353–358.

Haugland, R. (1992). *In* "Handbook of Fluorescent Probes and Research Chemicals" (K. D. Larison, ed.) 5th Ed. Part VI, pp. 235–274. Molecular Probes, Eugene, Oregon.

Hayashi, Y., Urade, R., and Kito, M. (1987). Distribution of phospholipid molecular species containing arachidonic acid and cholesterol in V79-UF cells. *Biochim. Biophys. Acta* **918**, 267–273.

Henis, Y. I., Rimon, G., and Felder, S. (1982). Lateral mobility of phospholipids in turkey erythrocytes. *J. Biol. Chem.* **257**, 1407–1411.

Hermanowski-Vosatka, A., Van Strijp, J. A. G., Swiggard, W. J., and Wright, S. D. (1992). Integrin modulating factor-1: A lipid that alters the function of leukocyte integrins. *Cell* **68**, 341–352.

Holifield, B. F., and Jacobson, K. (1991). Mapping trajectories of pgp-1 membrane protein patches on surfaces of motile fibroblasts reveals a distinct boundary separating capping on the lamella and forward transport on the retracting tail. *J. Cell Sci.* **98**, 191–203.

Holifield, B. F., Ishihara, A., and Jacobson, K. (1990). Comparative behavior of membrane protein-antibody complexes on motile fibroblasts: Implications for a mechanism of capping. *J. Cell Biol.* **111**, 2499–2512.

Ishihara, A., Hou, Y., and Jacobson, K. (1987). The Thy-1 antigen exhibits rapid lateral diffusion in the plasma membrane of rodent lymphoid cells and fibroblasts. *Proc. Natl. Acad. Sci. U.S.A.* **84**, 1290–1293.

Jacobson, K., and Vaz, W., Eds. (1992). "Domains in Biological Membranes." *Comments Mol. Cell Biophys.* **8**, 1–114.

Jacobson, K., Wu, E.-S., and Poste, G. (1976). Measurement of the translational mobility of concanavalin A in glycerol-saline solutions and on the cell surface by fluorescence recovery after photobleaching. *Biochim. Biophys. Acta* **433**, 215–222.

Jacobson, K., Derzko, Z., Wu, E. S., Hou, Y., and Poste, G. (1977). Measurement of the lateral mobility of cell surface components in single, living cells by fluorescence recovery after photobleaching. *J. Supramol. Struct.* **5**, 565–576.

Jacobson, K., Hou, Y., Derzko, Z., Wojcieszyn, J., and Organisciak, D. (1981). Lipid lateral diffusion in the surface membrane of cells and in multibilayers formed from plasma membrane lipids. *Biochemistry* **20**, 5268–5275.

Jesaitis, A. J., and Yguerabide, J. (1986). The lateral mobility of the (Na +, K +)-dependent ATPase in Madin–Darby canine kidney cells. *J. Cell Biol.* **102**, 1256–1263.

Johnson, S. J, Bayerl, T. M., McDermott, D. C., Adam, G. W., Rennie, A. R., Thomas, R. K., and Sachmann, E. (1991). Structure of an adsorbed dimyristoylphosphatidylcholine bilayer measured with specular reflection of neutrons. *Biophys. J.* **59**, 289–294.

Kapitza, H.-G., and Jacobson, K. (1986). Lateral motion of membrane proteins. *In* "Techniques for the Analysis of Membrane Proteins" (I. Ragan and R. J. Cherry, eds.), pp. 345–376. Chapman and Hall, London.

Kapitza, H. G., McGregor, G., and Jacobson, K. A. (1985). Direct measurement of lateral transport in membranes by using time-resolved spatial photometry. *Proc. Natl. Acad. Sci. U.S.A.* **82**, 4122–4126.

Keating, K. M., Barisas, B. G., and Roess, D. A. (1988). Glucocorticoid effects on lipid lateral diffusion and membrane composition in lipopolysaccharide-activated B-cell leukemia 1 cells. *Cancer Res.* **48**, 59–63.

Klausner, R. D., and Wolf, D. E. (1980). Selectivity of fluorescent lipid analogues for lipid domains. *Biochemistry* **19**, 6199–6203.

Knoll, G., Burger, K. N., Bron, R., van Meer, G., and Verkleij, A. J. (1988). Fusion of liposomes with the plasma membrane of epithelial cells: Fate of incorporated lipids as followed by freeze fracture and autoradiography of plastic sections. *J. Cell Biol.* **107**, 2511–2521.

Kobayashi, T., Storrie, B., Simons, K., and Dotti, C. G. (1992). A functional barrier to movement of lipids in polarized neurons. *Nature (London)* **359**, 647–650.

Koppel, D. E. (1979). Fluorescence redistribution after photobleaching. A new multipoint analysis of membrane translational dynamics. *Biophys. J.* **28**, 281–292.

Koppel, D. E., Sheetz, M. P., and Schindler, M. (1981). Matrix control of protein diffusion in biological membranes. *Proc. Natl. Acad. Sci. U.S.A.* **78**, 3576–3580.

Kucik, D. F., Elson, E. L., and Sheetz, M. P. (1990). Cell migration does not produce membrane flow. *J. Cell Biol.* **111**, 1617–1622.

Lee, G. M., Ishihara, A., and Jacobson, K. A. (1991). Direct observation of Brownian motion of lipids in a membrane. *Proc. Natl. Acad. Sci. U.S.A.* **88**, 6274–6278.

Lee, G. M., Zhang, F., Ishihara, A., McNeil, C. L., and Jacobson, K. A. (1993). Unconfined lateral diffusion and an estimate of pericellular matrix viscosity revealed by measuring the mobility of gold-tagged lipids. *J. Cell Biol.* **120**, 25–35.

Lee, J., Gustafsson, M., Magnusson, K.-E., and Jacobson, K. (1990). The direction of membrane lipid flow in locomoting polymorphonuclear leukocytes. *Science* **247**, 1229–1233.

Mathias, S. A., Younes, A., Kan, C.-C., Orlow, I., Joseph, C., and Kolesnick, R. N. (1993). Activation of the sphingomyelin signaling pathway in intact EL4 cells and in a cell-free system by 1L-1β. *Science* **259**, 519–522.

McCloskey, M. A., and Poo, M.-M. (1986). Rates of membrane-associated reactions: Reduction of dimensionality revisited. *J. Cell Biol.* **102,** 88–96.

McCown, J. T., Evans, E., Diehl, S. E., and Wiles, H. C. (1981). Degree of hydration and lateral diffusion in phospholipid multibilayers. *Biochemistry* **20,** 3134–3138.

Merkel, R., Sackmann, E., and Evans, E. (1989). Molecular friction and epitactic coupling between monolayers in supported bilayers. *J. Physiol. (France)* **50,** 1535–1555.

Metcalf, T. N., III, Wang, J. L., and Schindler, M. (1986). Lateral diffusion of phospholipids in the plasma membrane of soybean protoplasts: evidence for membrane lipid domains. *Proc. Natl. Acad. Sci. U.S.A.* **83,** 95–99.

Morrot, G., Cribier, S., Devaux, P. F., Geldwerth, D., Davoust, J., Bureau, J. F., Fellmann, P., Herve, P., and Frilley, B. (1986). Asymmetric mobility of phospholipids in the human erythrocyte membrane. *Proc. Natl. Acad. Sci. U.S.A.* **83,** 6863–6867.

Mueller, H.-J., and Galla, H.-J. (1987). Chain length and pressure dependence of lipid translational diffusion. *Eur. Biophys. J.* **14,** 485–491.

Nageem, A., Rananavare, S. B., Sastry, V. S. S., and Freed, J. H. (1989). Heisenberg spin exchange and molecular diffusion in liquid crystals. *J. Chem. Phys.* **91,** 6887–6905.

Nakache, M., Schreiber, A. B., Gaub, H., and McConnell, H. M. (1985). Heterogeneity of membrane phospholipid mobility in endothelial cells depends on cell substrate. *Nature (London)* **317,** 75–77.

Petersen, N. O., Felder, S., and Elson, E. L. (1986). Measurement of lateral diffusion by fluorescence photobleaching recovery. *In* "Handbook of Experimental Immunology, 1. Immunochemistry" (D. M. Weir, ed.) pp. 24.1–24.23. Blackwell Scientific, London.

Popov, S., Brown, A., and Poo, M.-M. (1993). Forward plasma membrane flow in growing nerve processes. *Science* **259,** 244–246.

Qian, H., Sheetz, M. P., and Elson, E. L. (1991). Single particle tracking analysis of diffusion and flow in two-dimensional systems. *Biophys. J.* **55,** 21–28.

Rimon, G., Meyerstein, N., and Henis, Y. I. (1984). Lateral mobility of phospholipids in the external and internal leaflets of normal and hereditary spherocytic human erythrocytes. *Biochim. Biophys. Acta* **775,** 283–290.

Rothberg, K. G., Ying, Y.-S., Kamen, B. A., and Anderson, R. G. W. (1990). Cholesterol controls the clustering of the glycophospholid-anchored membrane receptor for 5-methyltetrahydrofolate. *J. Cell Biol.* **111,** 2931–2938.

Rubenstein, J. L., Smith, B. A., and McConnell, H. M. (1979). Lateral diffusion in binary mixtures of cholesterol and phosphatidylcholines. *Proc. Natl. Acad. Sci. U.S.A.* **76,** 15–18.

Sassaroli, M., Vanhkoneu, M., Perry, D., and Eisinger, J. (1990). Lateral diffusivity of lipid analogue excimeric probes in dimyristoyl phosphatidylcholine bilayers. *Biophys. J.* **57,** 281–290.

Sheetz, M. P., Turner, S., Qian, H., and Elson, E. L. (1989). Nanometer-level analysis demonstrates that lipid flow does not drive membrane glycoprotein movements. *Nature (London)* **340,** 284–288.

Shin, Y.-K., Ewert, U., Budil, D. E., and Freed, J. H. (1991). Microscopic versus macroscopic diffusion in model membranes by electron spin resonance spectral-spatial imaging. *Biophys. J.* **59,** 950–957.

Shukla, S. D., and Halenda, S. P. (1991). Phospholipase D in cell signalling and its relationship to phospholipase C. *Life Sci.* **48,** 851–866.

Smith, B. A., and McConnell, H. M. (1978). Determination of molecular motion in membranes using periodic pattern photobleaching. *Proc. Natl. Acad. Sci. U.S.A.* **75,** 2759–2763.

Stolpen, A. H., Golan, D. E., and Pober, J. S. (1988). Tumor necrosis factor and immune interferon act in concert to slow the lateral diffusion of proteins and lipids in human endothelial cell membranes. *J. Cell Biol.* **107**, 781–789.

Tamm, L. K. (1988). Lateral diffusion and fluorescence microscope studies on a monoclonal antibody specifically bound to supported phospholipid bilayers. *Biochemistry* **27**, 1450–1457.

Tamm, L. K., and Kalb, E. (1993). Microspectrofluorometry on supported planar membranes. *In* "Molecular Luminescence Spectroscopy: Methods and Applications" (R. Shulman, ed.) pp. 139–173. John Wiley and Sons, New York.

Tamm, L. K., and McConnell, H. M. (1985). Supported phospholipid bilayers. *Biophys. J.* **47**, 105–113.

Tank, D. W., Wu, E.-S., and Webb, W. W. (1982). Enhanced molecular diffusibility in muscle membrane blebs: Release of lateral constraints. *J. Cell Biol.* **92**, 207–212.

Tertoolen, L. G., Kempenaar, J., Boonstra, J., de Laat, S. W., and Ponec, M. (1988). Lateral mobility of plasma membrane lipids in normal and transformed keratinocytes. *Biochem. Biophys. Res. Commun.* **152**, 491–496.

Thompson, N. L., and Axelrod, D. (1980). Reduced lateral mobility of a fluorescent lipid probe in cholesterol-depleted erythrocyte membrane. *Biochim. Biophys. Acta* **597**, 155–165.

Thompson, N. L., and Palmer, A. G. (1988). Model cell membranes on planar substrates. *Comments Mol. Cell Biophys.* **5**, 39–56.

Thompson, N. L., Palmer, A. G., Wright, L., and Scarborough, P. F. (1988). Fluorescence techniques for supported planar model membranes. *Comments Mol. Cell Biophys.* **5**, 109–131.

Treistman, S. N., Moynihan, M. M., and Wolf, D. E. (1987). Influence of alcohols, temperature, and region on the mobility of lipids in neuronal membrane. *Biochim. Biophys. Acta* **898**, 109–120.

van Meer, G., and Simons, K. (1986). The function of tight junctions in maintaining differences in lipid composition between the apical and basolateral cell surface domains of MDCK cells. *EMBO J.* **5**, 1455–1464.

Vaz, W. L. C. (1992). Translational diffusion in phase-separated lipid bilayer membranes. *Comments Mol. Cell Biophys.* **8**, 17–36.

Vaz, W. L. C., Derzko, Z. I., and Jacobson, K. A. (1982). Photobleaching measurements of the lateral diffusion of lipids and proteins in artificial phospholipid bilayer membranes. *Cell Surface Rev.* **8**, 83–135.

Vaz, W. L., Goodsaid-Zalduondo, F., and Jacobson, K. (1984). Lateral diffusion of lipids and proteins in bilayer membranes. *FEBS Lett.* **174**, 199–207.

Vaz, W. L. C., Melo, E., and Thompson, T. E. (1989). Translational diffusion and fluid domain connectivity in a two-component, two-phase phospholipid bilayer. *Biophys J.* **56**, 869–876.

Vaz, W. L. C., Melo, E., and Thompson, T. E. (1990). Fluid phase connectivity in an isomorphous two-component, two-phase phosphatidylcholine bilayer. *Biophys. J.* **58**, 273–275.

Wang, X. F., Kuo, S. C., Lemasters, J. J., and Herman, B. (1992). Measurements of plasma membrane architecture during hypoxia using multiple fluorescent spectroscopic techniques. *SPIE Proc. 1640 on Time-Resolved Laser Spectroscopy in Biochemistry* **3**, 301–308.

Wojcieszyn, J. W., Schlegel, R. A., Lumley-Sapanski, K., and Jacobson, K. (1983). Studies on the mechanism of polyethylene glycol-mediated cell fusion using fluorescent membrane and cytoplasmic probes. *J. Cell Biol.* **96**, 151–159.

Wolf, D. E. (1988). Probing the lateral organization and dynamics of membranes. *In* "Spectro-scopic Membrane Probes" (L. M. Loew, ed.), Vol. II, pp. 193–220. CRC Press, Boca Raton, Florida.

Wolf, D. E. (1989). Designing, building, and using a fluorescence recovery after photobleach-ing instrument. *Meth. Cell Biol.* **30**, 271–306.

Wolf, D. E., Henkart, P., and Webb, W. W. (1980). Diffusion, patching, and capping of stearoylated dextrans on 3T3 cell plasma membranes. *Biochemistry* **19**, 3893–3904.

Wolf, D. E., Hagopian, S. S., and Ishijima, S. (1986a). Changes in sperm plasma membrane lipid diffusibility after hyperactivation during in vitro capacitation in the mouse. *J. Cell Biol.* **102**, 1372–1377.

Wolf, D. E., Scott, B. K., and Millette, C. F. (1986b). The development of regionalized lipid diffusibility in the germ cell plasma membrane during spermatogenesis in the mouse. *J. Cell Biol.* **103**, 1745–1750.

Wu, E.-S., Jacobson, K., and Papahadjopoulos, D. (1977). Lateral diffusion in phospholipid multibilayers measured by fluorescence recovery after photobleaching. *Biochemistry* **16**, 3936–3941.

Wu, E.-S., Tank, D. W., and Webb, W. W. (1982). Unconstrained lateral diffusion of concanavalin A receptors on bulbous lymphocytes. *Proc. Natl. Acad. Sci. U.S.A.* **79**, 4962–4966.

Yechiel, E., and Edidin, M. (1987). Micrometer-scale domains in fibroblast plasma mem-branes. *J. Cell Biol.* **105**, 755–760.

Zhang, F., Crise, B., Su, B., Hou, Y., Rose, J. K., Bothwell, A., and Jacobson, K. (1991). Lateral diffusion of membrane-spanning and glycosylphosphatidylinositol-linked pro-teins: Toward establishing rules governing the lateral mobility of membrane proteins. *J. Cell Biol.* **115**, 75–84.

Zhang, F., Schmidt, W. G., Hou, Y., Williams, A. F., and Jackson, K. (1992). Spontaneous incorporation of the glycosyl-phosphatidylinositol-linked protein Thy-1 into cell mem-branes. *Proc. Natl. Acad. Sci. U.S.A.* **89**, 5231–5235.

CHAPTER 6

Microheterogeneity in Biological Membranes

David E. Wolf
Worcester Foundation for Experimental Biology, Shrewsbury, Massachusetts 01545

Protoplasm,[1] simple or nucleated, is the formal basis of all life. It is the clay of the potter: which, bake it and paint it as he will, remains clay, separated by artifice, and not by nature from the commonest brick or sun-dried clod. (Huxley, 1866, p. 142)

[1] In this chapter, which deals as much with what we do not know as with what we know about lipid domains, it is appropriate to excuse the use of as archaic a term as "protoplasm." Dr. Huxley would have been the first to admit to the limitations of this term in 1866 and to have expressed his surety that, by the present day, its meaning would have yielded considerably, but never completely, to the systematic pursuit of the scientific method.

Current Topics in Membranes, Volume 40

I. INTRODUCTION—THE FUNDAMENTAL QUESTION

Do lipid domains exist in biological membranes? Unfortunately, at our current state of knowledge, this fundamental question must be answered with another question, or perhaps equivocation. What is meant by lipid domains? If the question is whether or not biological membranes can be modeled accurately by a homogeneous fluid lipid bilayer, the answer is certainly not! If the question is whether compositional heterogeneity exists, the answer is yes. If, however, the question is whether this compositional heterogeneity results from lateral phase separation due to lipid–lipid interaction, the answer, in most cases, remains uncertain.

The goals of this chapter are to review the meaning of lateral phase separation in the context of binary lipid bilayer membranes and to describe some of the experimental evidence that exists for compositional heterogeneity and possible lateral phase separations in biological membranes. The homogeneous bilayer certainly is not an adequate model for most biomembranes. The multiphasic lipid bilayer may or may not be.

II. LATERAL PHASE SEPARATIONS—COEXISTENT LIPID DOMAINS

Considering only the lipids, biological membranes are compositionally complex. Two diametrically opposed hypotheses have evolved from this complexity. The first point of view is so much compositional complexity exists that it averages out spatial heterogeneity, so the bilayer acts as a homogeneous fluid medium. The second point of view is that, if the sole purpose of the lipid bilayer were to form a fluid matrix, why do membranes contain so many different lipids? In other words, only one or two lipids would suffice to make a fluid bilayer. Neither of these hypotheses leads very far. However, an important starting point is realizing that, even in three dimensions, homogeneous mixing is more the exception than the rule.

A striking graphic example is given by Hildebrand and Scott (1962), who showed 10 coexistent phases in a test tube. Different molecular species have different interaction energies with one another, which generally tends the system to spatial heterogeneity. This point is illustrated by Monte Carlo studies by Freire and Snyder (1980) of lateral lipid interactions in two-component lipid bilayers. The assumption of different interaction energies for homo and hetero interactions leads to lateral phase separations and lipid domains. Essentially, bilayer organization is analogous to the organization of metal alloys (for instance, see Moore, 1972).

A. What Are Lateral Phase Separations?

Reviewing what these lateral phase separations are will be useful. If one prepares lipid bilayer membranes from a single lipid component, these bilayers are well known to have a phase transition below which they are essentially crystalline or gel and above which they are essentially fluid. Figure 1 shows differential scanning calorimetry (DSC) thermograms (differential heat flow as a function of temperature) for dioleoyl, distearoyl, and dimyristoyl phosphatidylcholine (DOPC, DSPC, and DMPC, respectively). DOPC has a sharp phase transition (latent heat of melt) at − 12°C, DSPC at 55°C, and DMPC at 23°C. If one prepares membranes from binary mixtures of two lipids, and if they mix homoegeneously, one expects a broad transition spanning the two transition temperatures of the individual lipids. However, as shown in Fig. 1, this does not actually occur. Membranes formed from a 1:1 binary mixture of DOPC:DSPC, rather than showing a broad transition, exhibit two sharp transitions, one near − 12°C, and the other near 53°C, because the membrane lipids, rather than mixing homogeneously to form a single phase at all temperatures, laterally separate at temperatures intermediate between the two melts into two domains—one fluid and rich in DOPC, the other gel and rich in DSPC. We know that these phases are not purely one form of lipid because the lower

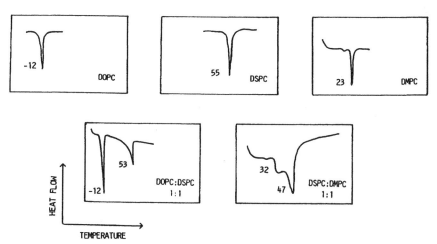

FIGURE 1 Differential scanning calorimetry (DSC) thermograms of one- and two-component bilayer lipid membranes of dioleoyl, distearoyl, and dimyristoyl phosphatidylcholine (DOPC, DSPC, DMPC). Reprinted from Wolf (1992) with permission.

phase transition is slightly elevated due to doping of this phase with DSPC and the higher phase transition is depressed due to doping with DOPC. As also shown in Fig. 1, even mixing of two similar saturated lipids (1:1 DSPC:DMPC) does not lead to a homogeneous bilayer.

Additional supporting evidence for coexistent gel and fluid lipid domains in binary lipid membranes is from a wide variety of biophysical techniques (Ladbrooke and Chapman, 1969; Steim *et al.*, 1969; Shimshick and McConnell, 1973; Hui and Parsons, 1974, 1975; Mabrey and Sturtevant, 1976; Stewart *et al.*, 1979; Hui *et al.*, 1983; Huang and Mason, 1986; Melchior, 1986). Notably, such domains can be demonstrated using X-ray diffraction, in which the fluid state appears as a broad diffuse band and the crystalline gel as a sharp 4.2-Å band in the powder patterns (Hui and Parson, 1974; Janiak *et al.*, 1976; Hui *et al.*, 1983). The domain structure of such binary membranes has, in fact, been visualized directly using rapid freeze-fracture electron microscopy (Stewart *et al.*, 1979). The reader also is referred to an interesting wet-stage electron microscopic approach that involves contrast enhancement by occulting the 4.2-Å band in the diffraction plane of the microscope (Hui and Parsons, 1975).

B. Lipid Phase Diagrams

Studies such as those described in the preceding section, when done as a function of temperature and composition, lead to phase diagrams for these membranes. An idealized phase diagram for such a binary lipid membrane is shown in Fig. 2. We assume that the membrane is composed of two lipids, A and B. Lipid A has a characteristic melting temperature T_A and lipid B has a higher characteristic melting temperature T_B. The phase diagram displays temperature vs. mole fraction of lipid B, X_B. The melting behavior for a given mole fraction X is defined by moving along a vertical (constant composition) line from X on the x axis. Thus, if $X = 0$ we have a pure lipid A membrane which is solid for $T < T_A$ and fluid for $T > T_A$. If $X = 1$ we have a pure lipid B membrane which is solid for $T < T_B$ and fluid for $T > T_B$. At intermediate compositions such as a 1:1 A:B membrane where $X = 0.5$, the membrane is solid below a temperature defined by the solidus line and fluid above temperatures defined by the liquidus line. At intermediate temperatures, we have coexistent fluid and gel phases. The composition and proportion of these two phases is defined by the "tie" line at that temperature, as illustrated in Fig. 2. The composition of the gel phase is given by the intersection of the tie line with the solidus line, X_G, whereas that of the fluid is given by the intersection of the tie line with the liquidus line, X_F. The proportion of the two phases,

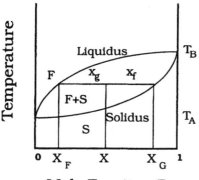

FIGURE 2   Idealized phase diagram of a binary lipid mixture showing coexistent gel (S) and fluid (F) phases. At composition X, fluid phases of composition X_F coexist with gel phases of composition X_G. The proportion of gel to fluid phase is x_g/x_f.

fluid to gel, is defined by the intersection of the tie line with the vertical line at that composition, that is, x_f/x_g.

A relatively simple phase diagram like that in Fig. 2 occurs for membranes composed of DMPC and dipalmitoyl phosphatidylcholine (DPPC). However, even more complex phase behavior can occur for binary lipid mixtures (Shimshick and McConnell, 1973) such as mixtures of DMPC and DSPC, for which the mole fraction of DSPC <0.6 exhibits two coexistent gel phases below the solidus line (Shimshick and McConnell, 1973). Figure 3 shows an idealized phase diagram exhibiting two coexistent fluid

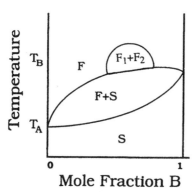

FIGURE 3   Idealized phase diagram of a binary lipid mixture that, in addition to having coexistent fluid (F) and gel (S) phases, has a region in which two fluid phases ($F_1 + F_2$) coexist.

phases in a region above the liquidus line ($F_1 + F_2$). Such behavior has been reported for membranes composed of mixtures of dielaidoyl phosphatidylchioline and dipalmitoyl phosphatidylethanolamine (Wu and McConnell, 1975).

Phase behavior becomes extremely complex as the number of components increases. For an interesting example of a three-component system, the reader is referred to Small and Shipley (1974), who consider the isothermal phase behavior of atherosclerotic plaques. The number of possible phases ultimately is limited by Gibbs' phase rule (for instance, see Moore, 1972), which states that the number of possible phases equals the number of components minus the number of degrees of freedom plus two. The degrees of freedom equal the number of intensive variables needed to define the system, less the number that cannot be varied independently. For a comprehensive review of the physical chemistry of lipids, the reader is referred to Small (1986).

C. Effects of Cholesterol

However, the effects of cholesterol are worth briefly considering. At high concentrations, cholesterol acts as a "plasticizer" causing the membrane to be in an intermediate gel–fluid state (Demel and de Kruyff, 1976). Nonetheless, cholesterol preferentially associates with different lipid classes: association for sphingomyelin > phosphatidylcholine > phosphatidylethanolamine (Demel et al., 1977). Cholesterol forms associations with phospholipids in stoichiometric ratios of 1:4, 1:2, and 1:1 (Engleman and Rothman, 1972; Philips and Finer, 1974; Verkleij et al., 1974; Hui and Parsons, 1975; Copland and McConnell, 1980; Melchior et al., 1980; Brulet and McConnell, 1976; Demel and de Kruyff, 1976; Opella et al., 1976; Haberkorn et al., 1977; Huang, 1977; Estep et al., 1978, 1979; Gershfeld, 1978; Mabrey et al., 1978; Calhoun and Shipley, 1979; Lange et al., 1979; Rubenstein et al., 1979; Blume, 1980; Recktenwald and McConnell, 1981). The phase diagram for cholesterol DPPC bilayers has been studied extensively using a variety of techniques. These studies indicate organizational changes at cholesterol mole fractions of 0.20 and approximately 0.30 (Melchior et al., 1980; Recktenwald and McConnell, 1981). Below 0.20 mole fraction cholesterol and below the DPPC melt, these membranes appear to have coexistent fluid phases rich in cholesterol and gel phases rich in DPPC. These domains may be organized linearly and cause anisotropic membrane diffusion (Rubenstein et al., 1979).

D. Difficulties Extrapolating to Biological Membranes

As we have indicated already, extrapolation beyond relatively simple systems such as these binary mixtures to membranes with a pantheon of components is difficult, not just theoretically, but experimentally as well. The experimental problem is that techniques such as DSC measure the properties of the entire system and are not good at resolving changes confined to a small fraction of the membrane. This difficulty is complicated further by the fact that these approaches are best suited to measuring state changes in which large enthalpic changes occur. Detection of fluid–fluid immiscibility, for example, is problematic calorimetrically or by X-ray diffraction. An interesting approach to overcome this difficulty has been developed by Melchior (1986), who employed rapid freezing of the sample prior to calorimetry measurement. Rapid freezing forces the system into a metastable set of crystalline states that reflect the lateral organization of the bilayer at higher temperatures.

III. PROBING MEMBRANE ORGANIZATION

Because of the difficulty encountered using the techniques already discussed, much of our understanding of lateral organization in more complex bilayers and biological membranes has relied on the use of reporter probes (cf. Chapter 5). Indeed, much of the pioneering work by the McConnell laboratory (for instance, Shimshick and McConnell, 1973) was based on electron spin resonance (ESR) studies of the selective partitioning of the probe 2,2,6,6-tetramethylpiperidine-1-oxyl (TEMPO) between the aqueous and membrane phases. In this section, we describe several fluorescence probe approaches. In general, these techniques exploit either the selective partitioning of a probe between two phases or the distinct (i.e., nonaverage) physical properties of these probes in the two environments.

One complication is obvious. Whenever one uses probes, first one must be concerned about the possibility of perturbing the native system. Second, one must be concerned about the possibility that, on extrapolating behavior from a model system (e.g., a binary lipid bilayer to a biomembrane), one can never be certain that the same physical mechanism is at work. Nevertheless, focusing on the critical issue of whether the membrane can be modeled as a homogeneous fluid or whether some level of complexity and heterogeneity exists is important.

A. Changes in Lipid Diffusibility during Fertilization in the Sea Urchin Egg

As an example, we may consider changes in lipid diffusibility during fertilization in the sea urchin egg (Wolf *et al.*, 1981). If a membrane such as the egg plasmalemma is a homogeneous fluid and a physiological transformation such as fertilization occurs, one of three things can happen: fluidity can increase, decrease, or remain the same. We found, using fluorescence recovery aftery photobleaching (FRAP) and a series of fluorescent lipid analogs, principally the dialkylindocarbocyanines (diIC$_N$, where N is the alkyl chain length; see Fig. 1E in Chapter 5), that whether an increase, a decrease, or no change in fluidity occurred depended on the probe used. The underlying cause (mathematically) is seen in Fig. 4, in which we have plotted the diffusion coefficients of unfertilized and fertilized eggs as a function of alkyl chain length for the DiIC$_N$s. We see a relative minimum at $N = 12$ for the unfertilized eggs and a relative

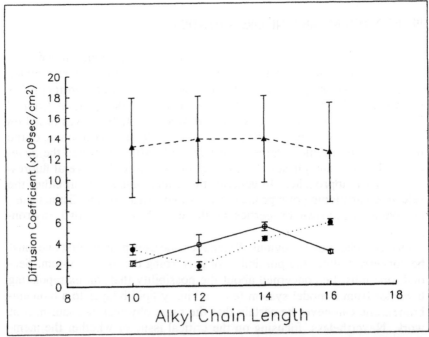

FIGURE 4 Diffusion coefficient of the DiIC$_N$s as a function of their alkyl chain length (N). Unfertilized sea urchin egg (●); fertilized sea urchin egg (○); homogeneous fluid phase in lipid bilayers (▲). Data are from Wolf *et al.* (1981) and Klausner and Wolf (1980).

maximum at $N = 14$ for the fertilized eggs. As a result, for $N = 16$ and 10, fertilization causes a decrease in diffusibility whereas, for $N = 12$ and 14, fertilization causes an increase in diffusibility.

What is the dependence of diffusibility on alkyl chain length in a homogeneous fluid bilayer? As shown in Fig. 4D, diffusibility in such a system is essentially independent of alkyl chain length (Klausner and Wolf, 1980). Thus, neither the unfertilized nor the fertilized sea urchin egg plasma membrane can be modeled by a homogeneous lipid bilayer. Klausner and Wolf (1980) have characterized the preference of the $DiIC_N$s as a function of their alkyl chain length for gel and fluid disaturated phosphatidylcholines (alkyl chain length M). This issue is pursued further by Ethier et al. (1983) and by Packard and Wolf (1985), who found that:

if $N < M$, the $DiIC_N$ prefers the fluid phase
if $N = M$, the $DiIC_N$ has no preference
if $N > M$, the $DiIC_N$ prefers the gel
if $N > > M$, the $DiIC_N$ prefers the fluid

Klausner and Wolf (1980) went on to show that the dependence of diffusion coefficient on $DiIC_N$, alkyl chain length is consistent with this preference rule, that is, in a mixed-phase gel–fluid bilayer, one will see an extreme in the dependence of D on N. In conclusion, the diffusion behavior of the $DiIC_N$s in the unfertilized and fertilized sea urchin egg plasma membrane is inconsistent with the membrane being a homogeneous fluid system and is more consistent with a mixed-phase system. However, as noted earlier, the data are insufficient to conclude that the heterogeneity within the egg plasmalemma is caused by the same lipid–lipid interactions that govern the binary lipid bilayers, namely, phase separations. Note that Weaver (1985) has continued this investigation and shown that if the dependence of D on temperature in the egg plasma membrane is studied for the different $DiIC_N$s, different characteristic transition temperatures dependent on N are observed. Also of interest are some provocative physiological experiments. Adair (1972) showed that a reversible block to fertilization exists in these ova that occurs between 25 and 30°C. Additionally, Glabe (1985a,b) showed that the sperm–egg adhesion protein, bindin, interacts only with bilayers that contain gel phases. This protein acts to agglutinate these membranes and, in so doing, promotes fusion.

B. Other Lipid Phase or Lipid Environmentally Selective Probes and Direct Visualization of Lipid Domains

Various environmentally selective fluorescent probes and light analogs other than the $DiIC_N$s have been exploited in a variety of techniques to probe lipid domains. Among the earliest examples of these probes were the *trans* and *cis* isomers of paranaric acid. Sklar *et al.* (1979) showed that the *trans* isomer of paranaric acid has a 3:1 preference for solid over fluid. In contrast, the *cis* isomer has a 1.7:1 preference for fluid over solid.

The fact that such selective probes exist raises the possibility that they could be exploited to visualize domains directly. This use would be possible only in special cases in which the domains were large enough to be resolved by the light microscope. A very dramatic example of such an application is found in the work of Peters and Beck (1983), who visualized large-scale phase separations in lipid monolayers using the fluorescent lipid *N*-4-nitrobenzo-2-oxa-1,3-diazole egg phosphatidylethanolamine (NBD-PE). Another example is the monolayer study by Lösche *et al.* (1983), who used the lipid analog 3,3'-diotadecyloxocarbocyanine. A similar approach can be used on certain lipid bilayer membranes, such as mixtures of dielaidoyl phosphatidylcholine and dipalmitoyl phosphatidyl ethanolamine (4:1; Wolf, 1987). In this case, the probe used was phosphatidylcholine acyl chain labeled with NBD (NBD-PC). A very promising approach is that of enhancing the differences in fluorescence of different environments using low light level video image processing (Haverstick and Glaser, 1987, 1989; Rodgers and Glaser, 1990). Yechiel and Edidin (1987) have visualized NBD-PC-selective domains on human skin fibroblasts by sequentially photographing the cell as the NBD-PC photobleaches. Larger scale regionalizations on cells are visualized readily using environment-selective probes (Williamson *et al.*, 1981; Wolf and Voglmayr, 1984; Wolf *et al.*, 1985). For example, Wolf and Voglmayr (1984) have shown that $DiIC_{16}$ stains the entire sperm plasma membrane, but stains the anterior region of the sperm head most strongly. This labeling reflects the segregation of native anionic lipids (Forrester, 1980; Bearer and Friend, 1982).

C. Lifetime Heterogeneity Analysis

The alternative approach of using a probe that partitions into all environments, but has a property that is different in different phases and can be separated experimentally, is epitomized by lifetime heterogeneity studies performed by the Kleinfeld laboratory (Klausner *et al.*, 1980; Pjura, 1985). These studies used the probe diphenylhexatriene (DPH). DPH does not show a selective partitioning between gel and fluid domains; however, its fluorescence lifetime is highly sensitive to phase environment. In fluid-phase bilayers, DPH fluorescence decay is nearly monoexponential with a characteristic lifetime of 6 to 8 nsec. In gel-phase bilayers, DPH again shows nearly monoexponential decay with a characteristic lifetime of 9 to 10 nsec. In mixed gel–fluid bilayers, and in certain cell membranes, DPH exhibits lifetime heterogeneity that manifests as multiexponential decay, with three characteristic lifetimes of ~7 nsec, ~9 nsec, and ~2 nsec. The 2-nsec lifetime is faster than that reported for DPH in any single-phase homogeneous system and may reflect a population of DPH at domain interfaces in the bilayer. Pjura (1985) has studied the lifetime heterogeneity of DPH in CH1 lymphoma cell plasma membrane vesicles and in protein-free reconstituted lipid vesicles from these membranes. He found lifetime heterogeneity below 39°C, but monoexponential decay above this temperature. The longer ~9-nsec lifetime, possibly reflecting gel domains, is lost.

Note that recent advances in instrumentation and analysis are revolutionizing fluorescence lifetime measurements (Knutson *et al.*, 1983; Parasassi *et al.*, 1984). Significantly, the decay observed in cell membranes is inconsistent with the decay observed in homogeneous fluid bilayers. The homogeneous bilayer is not a complete model for these membranes, indicating that heterogeneity is functioning at some level.

As another example of how lifetime can be used to study domains, consider the studies by Packard and Wolf (1985) of the lifetimes of the $DiIC_N$s in homogeneous and mixed-phase bilayers. As already described, in this case the probe exhibits selectivity for gel vs. fluid domains. Packard and Wolf (1985) measured the fluorescence decay of $DiIC_{12}$, $DiIC_{18}$ and $DiIC_{22}$ at room temperature in fluid DOPC, gel DSPC, and mixed-phase DOPC : DSPC bilayers. These investigators found double exponential decay in all phases with a long lifetime ranging from 0.84 to 1.42 nsec and a short lifetime ranging from 0.30 to 0.42 nsec. To overcome this complication, they calculated the average lifetime $<t>$ in each system as

$$<t> = (f_1t_1^2 + f_2t_2^2)/(f_1t_1 + f_2t_2)$$

where f and t are the amplitude and lifetime of the individual components of the double exponential decay. These average lifetimes are shown in Fig. 5. By comparing the average lifetime for the mixed-phase system with that of the two single-phase systems, one can obtain the fraction of the signal from the gel state: 0.01 for $DiIC_{12}$, 0.93 for $DiIC_{18}$ and 0.22 for $DiIC_{22}$. These preferences are seen to be consistent with the rules given earlier.

IV. FLUORESCENCE RECOVERY AFTER PHOTOBLEACHING MEASUREMENTS OVER DIFFERENT DISTANCE SCALES

A very different approach to studying "micrometer" scale domains in biological membranes has been developed by Yechiel and Edidin (1987). As expected, if the diffusibility of a fluorescent lipid analog in a homoge-

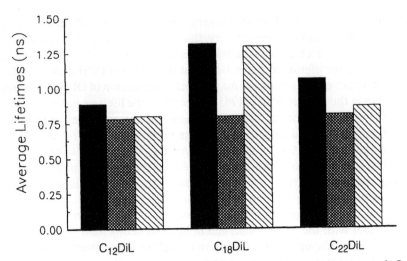

FIGURE 5 Average lifetimes of $DiIC_{12}$, $DiIC_{18}$, and $DiIC_{22}$ at room temperature in fluid-phase DOPC (solid), gel-phase DSPC (cross-hatched), and mixed-phase 1:1 DOPC:DSPC (hatched). The average lifetime can be used to determine the phase partition properties of these analogs. Data are from Packard and Wolf (1985).

neous fluid membrane is measured by FRAP, no dependence on the exp(-2) beam radius of the laser beam is seen. The mobile fraction is always 100% and the diffusion coefficient is constant. However, when Yechiel and Edidin measured the diffusibility of NBD-PC in human skin fibroblasts, they found that the mobile fraction decreased radically for beam radii above about 0.5 μm. Differences in the diffusion coefficient were observed also. Once again, the membrane cannot be modeled by a homogeneous fluid bilayer. Yechiel and Edidin (1987) suggested that the situation is one in which diffusants are corralled into domains within which diffusion is relatively free, but diffusion between these domains is difficult. Some level of heterogeneity is at play. The cause of these domains may be lateral phase separations or may result from some other cell-specific structure. As already described, Yechiel and Edidin (1987) also observed sequestration of the NBD-PC into some sort of domain.

Several points should be made. First, although accurate measurements of the beam radius is difficult, the general trend of increasing radius with decreasing magnification should hold. Second, although the system may deviate from pure Gaussian profile, which might affect to the second order the calculations of diffusion coefficients, the measurements of mobile fraction are independent of beam size and profile.

The concept of scales of diffusion is supported by pyrene quenching data (Eisinger *et al.*, 1986) and ESR data (Shin *et al.*, 1991). Note that applicability of the diffusion equation, Fick's second law, does not allow this kind of scaling since invariance under translation and rotation is fundamental to this equation (Sommerfeld, 1949). Nagle (1992) has sought to explain these dependencies in terms of long tail kinetic theory (Scher *et al.*, 1991). This theory is a random walk theory that seeks to describe the variability of the molecular interaction energies encountered by a diffusant, that is, the theory assumes that the membrane has a fundamental microheterogeneity.

V. IMMOBILE LIPID IN THE MAMMALIAN SPERM PLASMA MEMBRANE

As indicated at the outset, many of the techniques described indicate that the plasma membrane cannot be treated as a homogeneous fluid bilayer and behaves in a manner more consistent with mixed-phase lipid bilayers. The difficulty is determining whether lipid phase separations do or do not exist in biomembranes. For instance, we ask whether the domains observed by Yechiel and Edidin (1987) are caused by lipid separations or by some other cell-specific structure. The same question can be asked of the long fluorescence lifetimes observed for DPH in CH1 lymphoma cell plasma membranes or of the dependence of $DiIC_N$ diffusibility on chain length in sea urchin egg plasma membranes.

A. Lipid Diffusion and Differential Scanning Calorimetry of Sperm Membranes

One case in which the situation is clearer, although not "air tight," is the mammalian sperm plasma membrane. We have studied lipid analog diffusion in mammalian sperm plasma membranes extensively (for instance, see Wolf and Voglmayr, 1984; Wolf *et al.*, 1985). We already have described the fact that large-scale regionalizations of lipids occur in the sperm plasma membrane and find that, despite the physical continuity of the plasmalemma, lipid diffusibility is regionalized as well. Interestingly, we have found that, in ram, human, mouse, and hamster sperm, not all the lipid in the outer leaflet of the plasma membrane is free to diffuse. A nondiffusing fraction develops during spermatogenesis (Wolf *et al.*, 1986) and continues to evolve during epididymal maturation (Wolf and Voglmayer, 1984) and capacitation (Wolf *et al.*, 1985). These fractions are reminiscent of the immobile fractions observed in binary lipid bilayers containing gel phases (Klausner and Wolf, 1980) and contrast with the 100% recoveries observed in homogeneous fluid-phase bilayer membranes. When lipid bilayers are prepared from lipid extracts of the anterior region of sperm head plasma membranes and lipid diffusion in these bilayers is measured, again immobile fractions are found (Wolf *et al.*, 1988), indicating that lipid–lipid interactions may be sufficient to account for the immobile lipid in native sperm plasma membranes. These results suggest that, in sperm plasma membranes, sufficient gel-phase domains may exist to be detected by DSC. Figure 6 shows the results of such a DSC study (Wolf *et al.*, 1990). Figure 6A shows the thermogram of bilayer membranes reconstituted from a lipid extract of a plasma membrane fraction from the anterior region of the ram sperm head. We observe at least two major

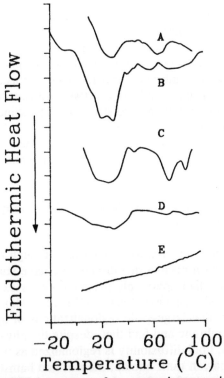

FIGURE 6 (A–D) DSC thermograms of ram sperm plasma membrane. (A) Rehydrated lipid extract of a plasma membrane fraction from the anterior region of the head; (B) sample similar to A, rehydrated in 1:1 ethylene glycol:deionized water; (C) plasma membrane fraction from the anterior region of the head; (D) sample similar to C, rehydrated in 1:1 ethylene glycol:deionized water. (E) Human erythrocyte lipids. Thermograms B and C have had a constant slope subtracted from them to correct for pan imbalance. Reprinted from Wolf *et al.* (1990) with permission.

transitions (one beginning below 12°C and ending near 39°C; the other beginning near 48°C and ending near 74°C). These results may be contrasted with thermograms from a similar lipid extract of erythrocyte ghost plasma membranes (Fig. 6E), in which no evidence exists of any transitions. Significantly, the first transition ends at ~39°C, the core temperature of the ewe. Thus, at physiological temperature, the ram sperm plasma membrane appears to be in a state of coexistent domains. The transitions in Fig. 6 are broader than those observed in single and binary phase lipid bilayers (see Fig. 1). We cannot conclude definitively that the transitions observed in the ram sperm plasma membrane represent gel-to-fluid phase transitions. However, the enthalpies of these melts, determined from the areas of these transitions, are consistent with such transitions. Figure 6B shows reconstituted ram sperm membranes similar to those in Fig. 6A, but here the measurement was made in 1:1 ethylene glycol:water to enable us to see the onset of the lower transition, which is below the freezing point of water. We observe that the first transition begins near −8°C. Unfortunately, this treatment seems to distort and split the individual transitions. This distortion effect may result from the presence of significant amounts of glycolipids in these membranes. Sperm plasma membranes are well known to contain unique glycolipids (Ishizuka *et al.*, 1973; Selivonchick *et al.*, 1980; Nickolopoulou *et al.*, 1985, 1986a,b; Mack *et al.*, 1987; Parks *et al.*, 1987). Thompson and Tillack (1985) suggested that these glycolipids play a dominant role in defining bilayer lipid organization. Figure 6C and D shows thermograms of intact plasma membrane vesicles (i.e., not reconstituted lipid extracts) in water and in 1:1 ethylene glycol:water, respectively. Here, we first denatured any protein transitions by cycling the system up and down in temperature until reproducible transitions were observed. Again, at least two major transitions were observed similar to those observed in Fig. 6A. The first transition ends near 40°C. Second and third transitions extend from approximately 60 to 75°C and 78 to 88°C, respectively.

Thus, although these transitions are not as sharp as those observed in model binary lipid membranes, lipid phase transitions do occur. At physiological temperature, the sperm plasma membrane is in a state between two transitions and, thus, has coexistent lipid phases.

B. Mammalian Sperm Compared with Mammalian Somatic Cells

Speculating about the differences between the sperm plasma membrane and the plasma membranes of mammalian somatic cells that result in these relatively dramatic transitions is of interest. The mol% of cholesterol is

43%, which is comparable to that in other cell types (Wolf *et al.*, 1990). Figure 7 shows the lipid composition (Wolf *et al.*, 1990). Similar results have been obtained by Parks and Hammerstedt (1985). We see that the ram sperm plasma membrane has an unusually high amount of ether-linked phospholipids. The major plasma membrane lipid species is plasmalogen phosphatidylcholine. Also of significance is the fact that Parks and Hammerstedt (1985) have shown that these membranes contain large amounts of polyunsaturated fatty acid chains, particularly docosahexaenoyl (22:6). These polyunsaturates are concentrated on the ether-linked phosphatidylcholine. The other dominant fatty acid chains are the saturates C_{14}, C_{16}, and C_{18}. The results of this work agree with numerous studies on whole sperm lipids (Quinn and White, 1967; Scott *et al.*, 1967; Poulos *et al.*, 1973). The major lipid species in ruminant sperm is a choline ether-linked

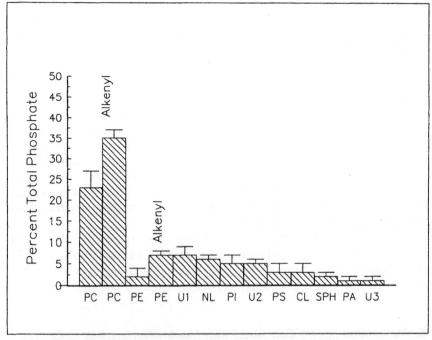

FIGURE 7 Phospholipid composition of the anterior region of the head plasma membrane of ejaculated ram spermatozoa. PC, Phosphatidylcholine; PE, phosphatidylethanolamine; U1, U2, U3, unidentified lipids; NL, neutral lipid; PI, phosphatidylinositol; PS, phosphatidylserine; CL, cardiolipin; SPH, sphingomyelin; PA, phosphatidic acid. Data are from Wolf *et al.* (1990).

glycerolphospholipid with a polyunsaturate chain (22:5 or 22:6) at the 2 position and a saturate (i.e., 16:0) ether-linked (either as alk-l'-enyl or as akyl) at the 1 position (Horrocks and Sharma, 1982).

Note that the thermal phase properties of the gel-to-fluid transition for ester- and ether-linked phospholipids are quite similar. However, ether-linked lipids exhibit polymorphisms in the gel state and appear to form interdigitated gel phases (Kim *et al.*, 1987a,b). ^{2}H-NMR studies suggest that 1-palmitoyl, 2-docosahexaenoyl phosphatidylcholine also forms inter-digitated phases in the fluid state (Salmon *et al.*, 1987). Also note that bovine rod outer segments also contain large amounts of 1-palmitoyl, 2-docosahexaenoyl phosphatidylcholine. Evidence also exists of phase separation in these rod membranes, based on fluorescence depolarization studies using paranaric acid (Sklar *et al.*, 1979).

Although sperm do not have unusual amounts of glycolipids, several glycolipids do appear to be unique to sperm (e.g., monogalactosyldiacyl-glycerol sulfate.) As noted earlier, glycolipids have been suggested to play a controlling role in defining membrane lipid organization. The alterations and splitting of the transitions observed in the ram sperm plasma membrane in the presence of ethylene glycol are consistent with such a role for glycolipids.

Drawing any conclusions about the cause of domains in the ram sperm plasma membrane is certainly premature. Are these domains created by glycolipids, polyunsaturated chains, ether-linked lipids, or some other factor? Clearly these membranes exhibit both large- and small-scale levels of membrane mosaicism. The same phase separations that cause small-scale sequestration of the membrane into lipid domains also may result in the large-scale regionalization of the membrane. Lipid regionalization also may, in turn, lead to protein regionalization by virtue of the preferential solubility (chemical potential) of the proteins in different places. Membrane regionalization ultimately may prove to be an important element in the control of mammalian fertilization (for instance, see Wolf *et al.*, 1992).

VI. CONCLUSION

As indicated at the outset of this chapter, we cannot yet state definitely the cause and physical nature of lipid domains in biological membranes. Most mammalian cell plasma membranes do not exhibit sufficient long-range order and cooperativity for domains to be detected by techniques such as DSC. A notable exception to this rule is the ram sperm plasmalemma, in which transitions can be observed calorimetrically. Nevertheless, clearly biological membranes cannot, in most cases, be modeled as homogeneous single-phase fluid bilayers. Evidence from a broad spectrum of techniques demonstrates nonideal behavior and supports the hypothesis of structural and compositional heterogeneity in some ensemble of membrane domains. In many instances, this nonideal behavior can be mimicked partially in mixed-phase binary lipid membranes. The organization of lipid bilayers into multiple phases or domains is analogous in two dimensions to the organization of metal alloys in three dimensions. Biological membranes are compositionally complex. This complexity is augmented by the association of the membrane with the cyto- and exoskeletons. All this complexity functions to compound the heterogeneity of membranes and to confound the observer. Several levels of domain organization and causes of this heterogeneity may coexist. Nevertheless, nothing is uniquely or metaphysically intrinsic to biological systems. As Huxley taught us many years ago, a fundamental physical basis to biological structure and function does exist (Huxley, 1866).

> But, the man of science, who, forgetting the limits of philosophical inquiry, slides from these formulae and symbols into what is commonly understood by materialism, seems to me to place himself on a level of the mathematician, who should mistake the x's and y's with which he works his problems, for real entities and with this further disadvantage, as compared with the mathematician, that the blunders of the latter are of no practical consequence, while the errors of systematic materialism may paralise the energies and destroy the beauty of a life. (Huxley, 1866, p. 165)

Acknowledgments

The author thanks the Worcester Foundation for Experimental Biology for its continuing support of his research, and Christine McKinnon for her invaluable assistance in the preparation of this chapter.

References

Adair, W. S. (1972). "Effects of Temperature on Fertilization in the Purple Sea Urchin, *S. purpuratus*." Masters Thesis. University of California at San Diego.

Bearer, E. L., and Friend, D. S. (1982). Modifications of anionic-lipid domains preceding membrane fusion in guinea pig sperm. *J. Cell Biol.* **92**, 604–615.

Blume, A. (1980). Thermotropic behavior of phosphatidylethanolamine–cholesterol and phosphatidylethanolamine–phosphatidylcholine–cholesterol mixtures. *Biochemistry* **19**, 4908–4912.

Brulet, P., and McConnell, H. M. (1976). Kinetics of phase equilibrium in a binary mixture of phospholipids. *J. Am. Chem. Soc.* **98**, 1314–1318.

Calhoun, W. I., and Shipley, G. G. (1979). Sphingomyelin–lecithin bilayers and their interaction with cholesterol. *Biochemistry* **18**, 1717–1722.

Copland, B. R., and McConnell, B. (1980). The rippled structure in bilayer membranes of phosphatidylcholine and binary mixtures of phosphatidycholine and cholesterol. *Biochim. Biophys. Acta* **599**, 95–109.

Demel, R. A., and de Kruyff, B. (1976). The function of sterols in membranes. *Biochim. Biophys. Acta* **457**, 109–132.

Demel, R. A., Jansen, J. W. C. M., van Dijck, P. W. M., and van Deenen, L. L. M. (1977). The preferential interaction of cholesterol with different classes of phospholipids. *Biochim. Biophys. Acta* **465**, 1–10.

Eisinger, J., Flores, J., and Petersen, W. P. (1986). A milling crowd model for local and long-range obstructed lateral diffusion. Mobility of excimeric probes in the membrane of intact erythrocytes. *Biophys. J.* **49**, 987–1001.

Engleman, D. M., and Rothman, J. E. (1972). The planar organization of lecithin-cholesterol bilayers. *J. Biol. Chem.* **447**, 3694–3697.

Estep, T. N., Mountcastle, D. B., Biltonen, R. L., and Thompson, T. E. (1978). Studies on the anomalous thermotropic behavior of aqueous dispersion of dipalmitoylphosphatidylcholine–cholesterol mixtures. *Biochemistry* **17**, 1984–1989.

Estep, T. N., Mountcastle, D. B., Barenholz, Y., Biltonen, R. L., and Thompson, T. E. (1979). Thermal behavior of synthetic sphingomyelin–cholesterol dispersion. *Biochemistry* **18**, 2112–2117.

Ethier, M. F., Wolf, D. E., and Melchior, D. L. (1983). Calorimetric investigation of the phase partitioning of the fluorescent carbocyanine probes in phosphatidylcholine bilayers. *Biochemistry* **22**, 1178–1182.

Forrester, I. (1980). Effects of digitonin and polymixin B on plasma membrane of ram spermatozoa—An EM study. *Arch. Androl.* **4**, 195–204.

Freire, E., and Snyder, B. (1980). Monte Carlo studies of the lateral organization of molecules in two-component lipid bilayers. *Biochim. Biophys. Acta* **600**, 643–654.

Gershfeld, N. L. (1978). Equilibrium studies of lecithin-cholesterol interactions. I. Stoichiometry of lecithin-cholesterol complexes in bulk systems. *Biophys. J.* **22**, 469–488.

Glabe, C. G. (1985a). Interaction of the sperm adhesive protein binding with phospholipid vesicles. I. Specific association of binding with gel-phase phospholipid vesicles. *J. Cell Biol.* **100**, 794–799.

Glabe, C. G. (1985b). Interaction of the sperm adhesive protein binding with phospholipid vesicles. II. Binding induces the fusion of mixed-phase vesicles that contain phosphatidylcholine and phosphatidylserine *in vitro*. *J. Cell Biol.* **100**, 800–806.

Haberkorn, R. A., Griffin, R. A., Meadows, M. D., and Olfield, E. (1977). Deuterium nuclear magnetic resonance investigation of the dipalmitoyl lecithin-cholesterol water system. *J. Am. Chem. Soc.* **99**, 7353–7355.

Haverstick, D. M., and Glaser, M. (1987). Visualization of Ca^{2+}-induced phospholipid domains. *Proc. Natl. Acad. Sci. U.S.A.* **84**, 4475–4479.

Haverstick, D. M., and Glaser, M. (1989). Influence of proteins on the reorganization of phosphoilipid bilayers into large domains. *Biophys. J.* **55**, 677–682.

Hildebrand, J. H., and Scott, R. L. (1962). "Regular Solutions." Prentice Hall, Englewood Cliffs, New Jersey.

Horrocks, L. A., and Sharma, M. (1982). Plasmalogens and O-alkyl glycerophospholipids. *In* "Phospholipids" (J. N. Hawthorne and G. B. Ansell, eds.), pp. 51–93. Elsevier Biomedical Press, Amsterdam.

Huang, C-H. (1977). A structural model for the cholesterol–phosphatidylcholine complexes in bilayer membranes. *Lipids* **12**, 348–356.

Huang, C-H., and Mason, J. T. (1986). Structure and properties of mixed-chain phospholipid assemblies. *Biochim. Biophys. Acta* **864**, 423–470.

Hui, S. W., and Parsons, P. F. (1974). Electron diffraction of wet biological membranes. *Science* **184**, 77–78.

162   David E. Wolf

Hui, S. W., and Parsons, D. (1975). Direct observation of domains in wet lipid bilayers. *Science* **190**, 383–384.

Hui, S. W., Boni, L. T., Stewart, T. P., and Isac, T. (1983). Identification of phosphatidylserine and phosphatidylcholine in calcium-induced phase separated domains. *Biochemistry* **22**, 3511–3516.

Huxley, T. H. (1866). On the physical basis of life. *Reprinted* in "Method and Results Essays." D. Appleton and Company, New York (1901).

Ishizuka, F., Suzuki, M., and Yamakawa, L. (1973). Isolation and characterization of a novel sulfoglucolipid, seminolipid, from boar testis and spermatozoa. *J. Biochem.* **73**, 77–87.

Janiak, M. J., Small, D. M., and Shipley, G. G. (1976). Nature of the thermal pretransition of synthetic phospholipids. Dimyristoyl- and dipalmitoyl lecithin. *Biochemistry* **15**, 4575–4580.

Kim, J. T., Mattai, J., and Shipley, G. G. (1987a). Gel-phase polymorphism in ether-linked dihexadecylphosphatidylcholine bilayers. *Biochemistry* **26**, 6592–6598.

Kim, J. T., Mattai, J., and G. G. Shipley (1987b). Bilayer interactions of ether- and ester-linked phospholipids: Dihexadecyl- and dipalmitoylphosphatidylcholines. *Biochemistry* **26**, 6599–6603.

Klausner, R. D., and Wolf, D. E. (1980). Selectivity of fluorescent lipid analogues for lipid domains. *Biochemistry* **19**, 6199–6203.

Klausner, R. D., Kleinfeld, A. M., Hoover, R. L., and Karnovsky, M. J. (1980). Lipid domains in membranes: Evidence derived from structural perturbations induced by free fatty acids and lifetime heterogeneity analysis. *J. Biol. Chem.* **255**, 1286–1295.

Knutson, J. R., Beechem, J., and Brand, L. (1983). Simultaneous analysis of multiple fluorescence decay curves: A global approach. *Chem. Phys. Lett.* **102**, 501–507.

Ladbrooke, B. D., and Chapman, D. (1969). Thermal analysis of lipids, proteins and biological membranes. A review and summary of some recent studies. *Chem. Phys. Lipids* **3**, 304–367.

Lange, Y., D'Alesandro, J. S., and Small, D. M. (1979). The affinity of cholesterol for phosphatidylcholine and sphingomyelin. *Biochim. Biophys. Acta* **556**, 388–398.

Lösche, M., Sackmann, E., and Möshwald, H. (1983). A fluorescnece microscopic study concerning the phase diagram of phospholipids. *Ber. Bunsenges Phys. Chem.* **87**, 848–852.

Mabrey, S., and Sturtevant (1976). Investigations of phase transitions of lipids and lipid mixtures by high sensitivity differential scanning calorimetry. *Proc. Natl. Acad. Sci. U.S.A.* **73**, 3862–3866.

Mabrey, S., Mateo, P. L., and Sturtevant, J. M. (1978). High-sensitivity scanning calorimetric study of mixtures of cholesterol with dimyristoyl- and dipalmitoylphosphatidylcholines. *Biochemistry* **17**, 2464–2468.

Mack, S. R., Zaneveld, L. J. D., Peterson, R. N., Hunt, W., and Russell, L. D. (1987). Characterization of human sperm plasma membrane: Glycolipids and polypeptides. *J. Exp. Zool.* **243**, 339–346.

Melchior, D. L. (1986). Lipid domains in fluid membranes: A quick-freeze differential scanning calorimetry study. *Science* **234**, 1577–1580.

Melchior, D. L., Scavitto, F. J., and Steim, J. M. (1980). Dilatometry of dipalmitoyllecithin–cholesterol bilayers. *Biochemistry* **19**, 4828–4834.

Moore, W. J. (1972). "Physical Chemistry," 4th Ed. Prentice-Hall, Englewood Cliffs, New Jersey.

Nagle, J. F. (1992). Long tail kinetics in biophysics? *Biophys. J.* **63**, 366–370.

Nikolopoulou, M., Soucek, D. A., and Vary, J. C. (1985). Changes in the lipid content of boar sperm plasma membranes during epidydymal maturation. *Biochim. Biophys. Acta* **851,** 486–498.

Nikolopoulou, M., Soucek, D. A., and Vary, J. C. (1986a). Lipid composition of the membrane released after an *in vitro* acrosome reaction of epidydymal boar sperm. *Lipids* **21,** 566–570.

Nikolopoulou, M., Soucek, D. A., and Vary, J. D. (1986b). Modulation of the lipid composition of boar sperm plasma membranes during an acrosome reaction *in vitro. Arch. Biochem. Biophys.* **250,** 30–37.

Opella, S. J., Yesinowski, J. P., and Waugh, J. S. (1976). Nuclear magnetic resonance description of molecular motion and phase separation of cholesterol in lecithin dispersions. *Proc. Natl. Acad. Sci. U.S.A.* **73,** 3812–3815.

Packard, B. S., and Wolf, D. E. (1985). Fluorescence properties of dialkyl indocarbocyanine dyes in phospholipid membranes. *Biochemistry* **24,** 5176–5181.

Parasassi, T., Conti, F., and Gratton, E. (1984). Study of heterogeneous emission of parinaric acid isomers using multifrequency phase fluorometry. *Biochemistry* **23,** 5660–5664.

Parks, J. E., and Hammerstedt, R. H. (1985). Developmental changes occurring in the lipids of ram epididymal spermatozoa plasma membrane. *Biol. Reprod.* **32,** 653–668.

Parks, J. E., Arion, J. W., and Fotte, R. H. (1987). Lipids of plasma membrane and outer acrosomal membrane from bovine spermatozoa. *Biol. Reprod.* **37,** 1249–1258.

Peters, R., and Beck, K. (1983). Translational diffusion in phospholipid monolayers measured by fluorescence microphotolysis. *Proc. Natl. Acad. Sci. U.S.A.* **80,** 7183–7187.

Philips, M. C., and Finer, E. G. (1974). The stoichiometry and dynamics of lecithin-cholesterol clusters in bilayer membranes. *Biochim. Biophys. Acta* **356,** 199–206.

Pjura, W. J. (1985). "Lipid Phase Heterogeneity in Membranes: A Fluorescence Spectroscopy Study." Ph.D. Thesis. Harvard University, Cambridge, Massachusetts.

Poulos, A., Voglmayr, J. K., and White, I. G. (1973). Phospholipid changes in spermatozoa during passage through the genital tract of the bull. *Biochim. Biophys. Acta* **306,** 194–202.

Quinn, P. J., and White, I. G. (1967). Phospholipid and cholesterol content of epididymal and ejaculated ram spermatozoa and seminal plasma in relation to cold shock. *Austral. J. Biol. Sci.* **20,** 1205–1215.

Recktenwald, D. J., and McConnell, H. M. (1981). Phase equilibria in binary mixtures of phosphatidylcholine and cholesterol. *Biochemistry* **20,** 4505–4510.

Rodgers, W., and Glaser, M. (1991). Characterization of lipid domains in erytrocyte membranes. *Proc. Natl. Acad. Sci. U.S.A.* **88,** 1364–1368.

Rubenstein, J. L. R., Smith, B. A., and McConnell, H. M. (1979). Lateral diffusion in binary mixture of cholesterol and phosphatidylcholines. *Proc. Natl. Acad. Sci. U.S.A.* **76,** 15–18.

Salmon, A., Dodd, J. W., Williams, G. D., Beach, J., and Brown, M. F. (1987). Configurational statistics of acyl chains in polyunsaturated lipid bilayers from ^{2}H NMR. *J. Am. Chem. Soc.* **109,** 2600–2609.

Scher, H., Shlessinger, M. F., and Bendler, J. T. (1991). Time-scale invariance in transport and relaxation. *Physics Today* 26–34.

Scott, J. W., Voglmayr, J. K., and Setchell, B. P. (1967). Lipid composition and metabolism in testicular and ejaculated ram spermatozoa. *Biochem. J.* **102,** 456–461.

Selivonchick, D. P., Schmid, P. C. Natarajan, V., and Schmid, H. H. O. (1980). Structure and metabolism of phospholipids in bovine epididymal spermatozoa. *Biochim. Biophys. Acta* **618,** 242–254.

Shimshick, E. J., and McConnell, H. M. (1973). Lateral phase separation in phospholipid membranes. *Biochemistry* **12,** 2351–2360.

Shin, Y. K., Ewert, U., Budil, D. E., and Freed, J. H. (1991). Microscopic versus macroscopic diffusion in model membranes by electron spin resonance spectral-spatial imaging. *Biophys. J.* **9**, 950–957.

Sklar, L. A., Miljanich, G. P., and Dratz, E. A. (1979). Phospholipid lateral phase separation and the partition of *cis*-parinaric acid and *trans*-parinaric acid among aqueous solid lipid and fluid lipid phases. *Biochemistry* **18**, 1707–1716.

Small, D. M. (1986). The physical chemistry of lipids from alkanes to phospholipids. *In* "Handbook of Lipid Research Series" (D. Hanahan, ed.), Vol. 3. Plenum Press, New York.

Small, D. M., and Shipley, G. G. (1974). Physical–chemical basis of lipid deposition in atherosclerosis. *Science* **185**, 222–229.

Sommerfeld, A. (1949). "Lectures on Theoretical Physics," Academic Press, New York.

Steim, J. M., Tourtellotte, M. E., Reinert, J. C., McElhaney, A. N., and Rader, R. L. (1969). Calorimetric evidence for the liquid-crystalline state of lipids in a biomembrane. *Proc. Natl. Acad. Sci. U.S.A.* **63**, 104–109.

Stewart, T. P., Hui, S. W., Portis, A. R., Jr., and Papahadjopoulos, D. (1979). Complex phase mixing of phosphatidyl choline and phosphatidyl serine in multi-lamellar membrane vesicles. *Biochim. Biophys. Acta* **556**, 1–16.

Thompson, T. E., and Tillack, T. W. (1985). Organization of glycosphingolipids in bilayers and plasma membranes of mammalian cells. *Annu. Rev. Biophys. Biophys. Chem.* **14**, 361–386.

Verkleij, A. J., Ververgaert, P. H. J. Th., deKruyff, B., and Van Deenen, L. L. M. (1974). The distribution of cholesterol in bilayers of phosphatidylcholines as visualized by freeze factoring. *Biochim. Biophys. Acta* **373**, 495–501.

Weaver, F. (1985). Studies on the Effects of Temperature and Membrane Composition on the Organization of Eukaryotic Cell Membranes." Ph.D. Thesis. The Johns Hopkins University, Baltimore, Maryland.

Williamson, P. L., Massey, W. A., Phelps, B. M., and Schlegel, R. A. (1981). Membrane phase state and the rearrangement of hematopoietic cell surface receptors. *Mol. Cell. Biol.* **1**, 128–135.

Wolf, D. E. (1987). Probing the lateral organization and dynamics of membranes. *In* "Spectroscopic Membrane Probes" (L. Loew, ed.), pp. 193–220. CRC Press, Boca Raton, Florida.

Wolf, D. E. (1992). Lipid domains: The parable of the blind men and the elephant. *Comments Biol. Cell. Biophys.* **8**, 83–95.

Wolf, D. E., Kinsey, W., Lennarz, W., and Edidin, M. (1981). Changes in the organization of the sea urchin egg plasma membrane upon fertilization: Indications from the lateral diffusion rates of lipid-soluble fluorescent dyes. *Dev. Biol.* **81**, 133–138.

Wolf, D. E., and Voglmayr, J. K. (1984). Diffusion and regionalization in membranes of maturing ram spermatozoa. *J. Cell Biol.* **98**, 1678–1684.

Wolf, D. E., Hagopian, S. S., and Ishijima, S. (1985). Changes in sperm plasma membrane lipid diffusibility after hyperactivation during *in vitro* capacitation in the mouse. *J. Cell Biol.* **102**, 1372–1377.

Wolf, D. E., Scott, B. K., and Millette, C. F. (1986). The development of regionalized diffusibility in the germ cell plasma membrane during spermatogenesis in the mouse. *J. Cell Biol.* **103**, 1745–1750.

Wolf, D. E., Lipscomb, A. C., and Maynard, V. M. (1988). Causes of nondiffusing lipid in the plasma membrane of mammalian spermatozoa. *Biochemistry* **27**, 860–865.

Wolf, D. E., Maynard, V. M., McKinnon, C. A., and Melchior, D. L. (1990). Lipid domains in the ram sperm plasma membrane demonstrated by differential scanning calorimetry. *Proc. Natl. Acad. Sci. U.S.A.* **87,** 6893–6896.

Wolf, D. E., McKinnon, C. A., Leyton, L., Lakoski-Loveland, K., and Saling, P. M. (1992). Protein dynamics in sperm membranes: Implications for sperm function during gamete interaction. *Mol. Reprod. Dev.* **33,** 228–234.

Wu, S. H.-W., and McConnell, H. M. (1975). Phase separation in phospholipid membranes. *Biochemistry* **14,** 847–854.

Yechiel, E., and Edidin, M. (1987). Micrometer-scale domains in fibroblast plasma membranes. *J. Cell Biol.* **105,** 755–760.

in the two specific plasma membrane demonstrated by differential scanning calorimetry. *Proc. Natl. Acad. Sci. U.S.A.* **87,** 843–856.

Wolf, D. E., McKinnon, C. A., Leyton, L., Lakoski Loveland, K., and Saling, P. M. (1992). Protein dynamics in sperm membranes: Implications for sperm function during gamete interaction. *Mol. Reprod. Dev.* **33,** 228–234.

Wu, S. H.-W., and McConnell, H. M. (1975). Phase separation in phospholipid membranes. *Biochemistry* **14,** 847–854.

Yechiel, E., and Edidin, M. (1987). Micrometer-scale domains in fibroblast plasma membranes. *J. Cell Biol.* **105,** 755–760.

CHAPTER 7

Lipid Dynamics in Brush Border Membrane

Helmut Hauser and Gert Lipka
Laboratorium für Biochemie, Eidgenössische Technische Hochschule, CH-8092 Zürich, Switzerland

I. INTRODUCTION

This chapter addresses the lipid dynamics of the small intestinal brush border membrane (BBM), sometimes also referred to as the microvillus plasma membrane. The BBM is the apical part of the plasma membrane of enterocytes, the cells lining the small intestines. The plasma membrane of these cells is highly polarized, both structurally and functionally, the apical part (BBM) is specialized in the digestion and absorption of nutrients and compounds present in the chyme of the small intestine.

The purpose of this chapter is to summarize our understanding of the lipid dynamics in BBM with special reference to lipid absorption. Lipid absorption refers to the first step of the absorptive process, that is, the insertion of exogenous lipid into the external monolayer of the lipid bilayer of BBM. Small intestinal BBM appears to be specially equipped for the uptake of exogenous lipids such as cholesterol, particularly from bile salt micelles. This fact apparently has not been placed on record to date.

II. PREPARATION OF BRUSH BORDER MEMBRANE VESICLES

Unless stated otherwise, results pertaining to the lipid dynamics of BBM discussed here have been obtained with brush border membrane vesicles (BBMV) routinely prepared in our laboratory from frozen small intestines of rabbit. The method we use for the preparation of BBMVs is based on a procedure originally developed by Schmitz et al. (1973) and later modified by Kessler et al. (1978). An essential step in the preparation of BBMVs is the Ca^{2+} precipitation of contaminating membranes. We reported that BBMVs prepared by Ca^{2+} precipitation have an exceptionally high content of free fatty acids and lysophospholipids (Hauser et al., 1980), a result that is not too surprising since the intestinal mucosa is among the richest sources of phospholipases. Further, phospholipase A_2 has been shown to have the highest specific activity in BBM. Phospholipases A_2 and B can be activated by fatty acids, so phospholipid degradation may become autocatalytic if fatty acids are produced during the preparation. Another disadvantage of the preparation is that the phospholipase activity in mucosal homogenates is retained at low temperatures. To minimize phospholipid degradation, we introduced the following modification (Hauser et al., 1980): all the buffers contain EGTA (at 5 mM), insuring that the concentration of free Ca^{2+} is low, thus preventing the possible activation of intrinsic phospholipases. Further, in the presence of EGTA, Mg^{2+} appears to replace Ca^{2+} effectively in the selective precipitation of contaminating membranes.

The preparation used by Hauser et al. (1980) yields BBMVs with lysophosphatidylcholine and lysophosphatidylethanolamine contents of less than 2–3% each. The resulting BBMVs were shown to be relatively homogenous with respect to size, with an average hydrodynamic radius of 100 nm (Kessler et al., 1978; Perevucnik et al., 1985). These vesicles are essentially free of basolateral plasma membrane and nuclear, mitochondrial, microsomal, and cytosolic contaminants (Hauser et al., 1980); Thurnhofer and Hauser, 1990b). BBMVs satisfy another criterion that is essential to the study of lipid dynamics: more than 90% are oriented right side out (Klip et al., 1979).

III. CHARACTERIZATION OF BRUSH BORDER MEMBRANE VESICLES

A. Lipid Composition

The characterization of BBMVs containing a number of specific hydrolases and transport proteins usually is carried out by monitoring the activity of these marker enzymes. For instance, in our laboratory the specific activities of sucrase-isomaltase, of K^+-stimulated phosphatase (which serves as a marker enzyme for contamination of BBMV with basolateral plasma membrane), and of D-glucose uptake are measured routinely. Although researchers reported that not all BBMVs are sealed (Gains and Hauser, 1984), BBMVs accumulate D-glucose transiently against a concentration gradient in the presence of a Na^+ gradient. During active D-glucose transport, the concentration of D-glucose in BBMV may exceed the equilibrium concentration approximately 30–fold.

Another property used to standardize BBM preparations in our laboratory is the electrophoretic mobility of BBMVs, which is related directly to the surface charge density and, hence, to the surface potential (Hauser et al., 1980).

The lipid composition of small intestinal BBM prepared from different species by different preparative methods is now available (for a summary, see Proulx, 1991). Although significant differences exist among species, some common features have evolved from these analyses: the BBM is characterized by a relatively high cholesterol:phospholid ratio, a low lipid:-protein ratio, and a high glycosphingolipid content (Table I). Cholesterol accounts for more than half the neutral lipids, the other major neutral lipids are free fatty acids. The relatively high fatty acid content is probably the result of intrinsic phospholipase A activity associated with this membrane. Phosphatidylcholine and phosphatidylethanolamine are the major phospholipids, amounting to 60–70% of the total phospholipids. The major acidic (negatively charged) phospholipids are phosphatidylinositol and phosphatidylserine, amounting to 15–30% of the total phospholipids (Table II; Proulx, 1991). BBM is rich in glycosphingolipids; note that the glycosphingolipid content of the apical part of the plasma membrane is 1.5 to 2 times higher than that of the basolateral membrane.

BBMVs prepared from total rabbit small intestines are used routinely in our laboratory. Unless stated otherwise, our discussion of the lipid dynamics is based on measurements carried out with this kind of preparation. These BBMVs are characterized by a lipid:protein weight ratio of 0.5 ± 0.2, a phospholipid content of 0.19 ± 0.5 mg/mg protein, and a cholesterol content of 50 ± 5 μg/mg protein yielding a cholesterol:phos-

TABLE I

Characterization of Brush Border Membrane Vesicles[a]

Total lipid (mg/mg protein)	0.5 ± 0.15
Lipid phosphorus[b] (μg/mg protein)	8 ± 2 (range 6–12)
Cholesterol (μg/mg protein)	50 ± 5
Neutral lipid/phospholipid/glycolipid (molar ratio)	1:1.2:1.1
Phospholipid/monohexoside (molar ratio)	2.2 ± 0.2
Specific activity sucrase	1.7 ± 0.7 units/mg protein (range 0.8 – 2.5 units/mg)
K^+-Stimulated phosphatase	8 ± 3 mU/mg protein (range 6–12 mU/mg)
Electrophoretic mobility (cm^2/V-sec)	−1.6 ± 0.15 × 10^{-4} (range −1.4 to −1.8 × 10^{-4})

[a]Values are the mean ± SD of 10–12 experiments.

[b]From the average phospholipid composition (Table II) and the fatty acid composition (see Hauser et al., 1980), the effective phospholipid molecular weight was calculated as 720. Using this value, the total phospholipid content is 0.19 ± 0.05 mg/mg protein (range 0.14–0.28 mg/mg protein). The large standard deviation and range obtained for the lipid content when referenced to protein are due to variations in the protein content of different preparations.

pholipid mole ratio of 0.5 ± 0.05. The lipid composition and some properties of rabbit small intestinal BBMVs are summarized in Tables I and II. In comparison to BBMVs from other species, the cholesterol content is apparently lower (cf. Proulx, 1991), amounting to ~10% of the total lipid (Hauser et al., 1980). The molar ratio of neutral lipids to phospholipids to glycolipids is 1:1.2:1.1 (Table I). As mentioned, all BBM analyzed to date are characterized by a high glycosphingolipid content. In rabbit as well as in mouse and rat, the major glycosphingolipid is monohexosylceramide (Proulx, 1991). The major fatty acids of the lipids of rabbit BBM are palmitic and stearic acid, accounting for 40–60%; the remainder consists mainly of oleic, linoleic, and polyunsaturated fatty acids (Hauser et al., 1980; Proulx, 1991).

B. Stability

One inherent problem of BBMVs is their instability. On incubation of BBMVs at temperatures > 0°C, membrane proteins are released. The problem is serious because incubation at 37°C for 15 hr was shown to liberate ~30% of the total membrane protein (Thurnhofer and Hauser,

TABLE II

Phospholipid Composition of Rabbit Intestinal Brush Border Membranes[a]

Phospholipid	Two-dimensional TLC[b]		Two-dimensional TLC[c] (after correction)	One-dimensional TLC[d]
Phosphatidic acid	1.2 ± 0.5		1.2	nd[e]
Phosphatidylethanolamine	23.3 ± 1.4 ⎤			
Ethanolamine plasmalogen	8.5 ± 1.0 ⎬ 35.0		35.6	35.2 ± 3.0
Alkylacylglycerophospho-ethanolamine	3.2 ± 1.5 ⎦			
Phosphatidylserine	7.5 ± 0.8 ⎤		7.4	
	⎬ 11.1			9.8 ± 2.0
Lysophosphatidylethanolamine	3.6 ± 1.0 ⎦		2.3	(2.3)
Phosphatidylcholine	27.3 ± 1.7 ⎤			
	⎬ 30.3		33.3	32.1 ± 3.0
Alkyl and alkenylglycerophosphocholine	3.0 ± 1.7 ⎦			
Lysophosphatidylcholine	3.1 ± 1.5		1.8	1.8 ± 1.0
Phosphatidylinositol	8.3 ± 1.0		8.2	9.5 ± 1.0
Sphingomyelin	10.5 ± 0.4		10.3	11.2 ± 1.6

[a]Results are expressed as percentage of total lipid phosphorus and are presented as the mean ± SD of 12 experiments.

[b]A possible explanation for the higher lysophospholipid content found by two-dimensional TLC is that in this case the lipid extracts in $CHCl_3:CH_3OH$ (2:1, v/v) were subjected to a Folch wash with 0.2 vol 0.025 M $CaCl_2$.

[c]Data were corrected assuming that the inherent lysophospholipid content is that determined immediately by one-dimensional TLC.

[d]Samples analyzed by one-dimensional TLC were Folch-washed with 0.2 M NaCl.

[e]nd, Not determined.

1990b). The protein loss is probably caused by switching on intrinsic or membrane-bound proteinases in the course of the preparation of BBMVs. The mechanism by which the activity of these proteinases is controlled is still unknown. Various proteinase inhibitors including phenylmethylsulfonyl fluoride, EDTA, aprotinin, pepstatin, and bacitracin, applied singly and in combination, failed to stop the release of membrane proteins. Detergent solubilization was shown to enhance proteolysis (Gains and Hauser, 1981), leading to massive degradation of integral membrane proteins. This finding lends additional support to the notion that intrinsic proteinases are responsible for the cleavage and release of membrane proteins. The release of membrane protein observed on incubation of BBM at temperatures > 0°C is accompanied by significant phospholipid hydrolysis. After incubation at 37°C for 15 hr, ~60% of the total membrane phospholipid was found to be degraded (Thurnhofer and Hauser, 1990b). Lipid analysis of BBMVs after incubation revealed that the aminophospholipids are affected primarily: half the phosphatidylethanolamine and all the phosphatidylserine were degraded under these conditions. The presence of EDTA in the buffer markedly slowed the degradation of phospholipids, indicating that phospholipid degradation probably is caused by intrinsic phospholipases. The instability of BBMVs at temperatures > 0°C and the difficulties in controlling and inhibiting the intrinsic hydrolases is a serious drawback of this BBM preparation. Self-digestion has to be borne in mind when working with BBMVs, and necessitates the conduction of appropriate, sometimes tedious, control experiments.

IV. LIPID PACKING AND MOTION IN RABBIT SMALL INTESTINAL BRUSH BORDER MEMBRANE

A. ^{31}P Nuclear Magnetic Resonance

Lipids of BBMVs pack as bilayers, as can be demonstrated by ^{31}P-NMR for lipids present in the BBM and for lipids extracted from BBM and dispersed in an aqueous medium (pH $\approx$ 7). The line shape of proton-decoupled ^{31}P-NMR spectra of BBMVs and the total lipids extracted from BBM is typical of lipid bilayers (Fig. 1). Over the temperature range shown, symmetric powder spectra are obtained, indicating that the lipid molecules undergo rapid motional averaging about the bilayer normal. For BBMVs as well as for the liposomes made from the lipid extract, the chemical shielding anisotrophy $\Delta\sigma$ increased linearly with decreasing temperature between 40°C and 0°C (data not shown). No change in slope or break in the $\Delta\sigma$ − temperature relationship occurs, indicating that the presence of a possible lipid order–disorder transition is not reflected in the temperature dependence of the polar group motion, at least not in the temperature range between 0 and 40°C. Over the temperature range of 0 to 40°C, similar $\Delta\sigma$ values were measured for BBMVs and liposomes, indicating that the presence of proteins has little effect on the motional averaging of the phospholipid polar group, at least not within the error of the measurement. The chemical shielding anisotropy increased from $|\Delta\sigma| = 38 \pm 1$ ppm at 37°C to $|\Delta\sigma| = 41 \pm 1.5$ ppm at 0°C. At temperatures well below −20°C, the motion of the phospholipid polar group ceases and axially asymmetric ^{31}P powder NMR spectra are observed for BBMVs and for the lipid extracts of BBMVs (cf. Fig. 1; Mütsch *et al.*, 1983). The conclusions derived from ^{31}P-NMR are corroborated by electron spin resonance (ESR) studies discussed in a subsequent section.

B. Differential Scanning Calorimetry

The thermal behavior of BBMVs and liposomes made from the lipid extract were studied by differential scanning calorimetry (DSC) (Mütsch *et al.*, 1983). Figure 2 shows DSC thermograms of BBMVs and of the total lipid extract. The first heating run of BBMVs yields, reproducibly, a broad endothermic transition between 10 and 30°C with a peak at ~25°C and several endothermic transitions in the temperature range 50–80°C. On cooling this suspension of BBMVs, a broad reversible exothermic transition occurs between 25 and 12°C with a peak at ~20°C. On subsequent heating–cooling cycles, the first endothermic transition between 10 and 30°C is observed reproducibly on heating, as is the exothermic transition at ~20°C on cooling, whereas the high temperature endothermic transitions are detected only in the first heating run. On heating the lipid extract of BBM, a broad reversible endothermic transition between 10 and 30°C is obtained with a peak at ~20°C, similar to the first endothermic transition of BBMVs. On cooling the lipid extract, a broad reversible exothermic transition centered at ~14°C is observed reproducibly (Fig. 2). Based on the similar thermal behavior of BBMVs and the lipid extract, the reversible endothermic transition at ~25°C is assigned to the order–disorder (gel-to-liquid crystal) transition of the lipid bilayer of BBMVs. Based on the irreversible nature of the high-temperature transitions of BBMVs, these transitions are proposed to be caused by the irreversible denaturation of membrane proteins. The thermal behavior of BBMVs after proteolytic treatment with papain or alkaline treatment at pH 11 was shown to be identical to that of untreated BBMVs (Mütsch *et al.*, 1983). Therefore, the excessive loss of up to ~70% of peripheral proteins produced by the proteolytic and/or alkaline treatment of BBMVs has no effect on the thermal behavior of the lipid, indicating that lipid packing is not affected by this type of protein.

The broad lipid phase transition of both BBMVs and the total lipid extract is of low enthalpy, indicating that the lipid cooperativity of the transition is low. Similar thermal behavior was reported for rat BBM (Brasitus *et al.*, 1980). Generally, plasma membranes that characteristically have a high cholesterol content exhibit either no order–disorder transition or a broad low-enthalpy transition, indicative of low lipid coop-

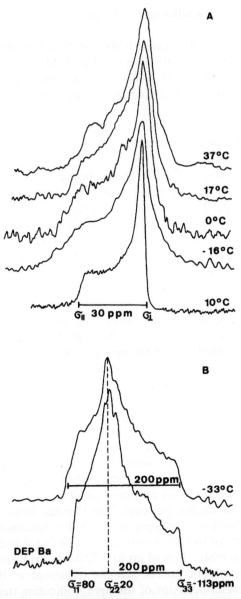

FIGURE 1 Proton-decoupled ^{31}P-NMR spectra recorded at 121 MHz on a Bruker CXP 300 Fourier-transform spectrometer (Bruker Instruments, Karlsruhe, Germany). Chemical shielding was measured relative to 85% orthophosphoric acid. (A) The lipids (50–100 mg) extracted from brush border membrane vesicles (BBMV) were dispersed in 1 ml buffer (10 mM HEPES/Tris, pH 7.0, 0.3 M D-mannitol, 5 mM EDTA, 0.02% NaN$_3$) and ^{31}P-NMR spectra were recorded at different temperatures. For comparison, a spectrum from a 10% unsonicated 1-palmitoyl-2-oleoyl-3-sn-phosphatidylcholine dispersion in H$_2$O is included (bottom, 10°C). The shielding anisotropy of this compound is $\Delta\sigma = \sigma_\parallel - \sigma_\perp \approx -30$ ppm obtained

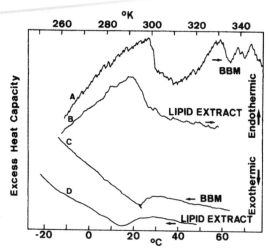

FIGURE 2 Differential scanning calorimetry (DSC) thermograms of brush border membrane vesicles (BBMV) dispersed in 10 mM HEPES/Tris buffer, pH 7.6, containing 0.3 M D-mannitol. To enrich BBM, the dispersion was centrifuged at 100,000 g for 1 hr at 4°C; the resulting pellet of BBM was filled into the DSC pan (for details, see Mütsch et al., 1983). Heating curves are shown for BBMV (A, first heating run) and for the lipid extract of BBMV dispersed in the same buffer (B). Cooling curves recorded at 5°C/min are shown for BBMV (C) and for the lipid extract (D). Reproduced from Mütsch et al. (1983) with permission.

erativity. Lipid–protein interactions cannot be responsible for the low enthalpy and lack of cooperativity, since the protein-free membrane lipids also give broad low-enthalpy phase transitions (cf. Fig. 2). The high cholesterol content of the BBM, in conjunction with the lipid heterogeneity, is likely to be the main reason for the low enthalpy of the lipid phase transitions shown in Fig. 2. The cholesterol content was shown to vary in BBM from species to species, ranging from 10 to 25% of the total lipid content (Proulx, 1991). Chapman and co-workers showed that the effect of

---

from the computer simulation of the axially symmetric powder spectrum. (B) Axially asymmetric powder spectrum obtained from the lipid extract of BBMV dispersed in buffer (*top*). An almost superimposable spectrum was obtained from a pellet of BBMV centrifuged at 60,000 g for 30 min (data not shown). For comparison, the axially asymmetric ^{31}P powder spectrum obtained from barium diethyl phosphate at room temperature is included (*bottom*). The principal values of the tensor components σ_{11}, σ_{22}, and σ_{33} of this compound agreed within 1 ppm with published values and within 3 ppm with the tensor components derived from the ^{31}P-NMR spectrum of a dispersion of the lipid extract. For details, see Mütsch et al. (1983). Reproduced from Mütsch et al. (1983) with permission.

increasing quantities of cholesterol in dipalmitoyl phosphatidylcholine–cholesterol mixtures is to reduce the enthalpy of the phase transition of the phospholipid and that, at a phospholipid : cholesterol mole ratio of 1, the phase transition is no longer detectable (Ladbrooke *et al.*, 1968; Ladbrooke and Chapman, 1969). Consistent with this finding, human red blood cells containing ~24% cholesterol (with respect to the total lipid content) exhibit an order–disorder transition only after removal of the cholesterol (Ladbrooke *et al.*, 1968).

The main conclusion of the DSC study (Fig. 2), however, is that small intestinal BBM functions in the liquid crystalline state at a temperature well above the lipid phase transition. Therefore, the thermotropic lipid phase transition itself is unlikely to play a physiological role, since the transition temperature is well separated from the body temperature. The thermal behavior of BBMVs as determined by DSC is similar to that of aqueous dispersions of the lipids extracted from BBMVs. This result is evidence that the thermotropic properties of BBMVs are determined primarily by the lipid composition of this membrane. Lipid–protein interactions seem to play a minor role in this context.

C. Electron Spin Resonance Spin-Labeling

BBMVs were labeled with a variety of spin-labels with the aim of probing the membrane fluidity or microviscosity. For this purpose, fatty acid spin-labels were used: stearic acid with the 4,4'-dimethyl-3-oxazolidinyloxy group attached to carbon atoms 5, 12 or 16; a spin-labeled cholesterol analog, 4,4'-dimethylspiro[5α-cholestane-3,2'-oxazolidin]-3'-yloxyl (3-doxyl-5α-cholestane); spin-labeled phosphatidylcholines such as 1-palmitoyl-2-(5-doxylstearoyl)-3-*sn*-phosphatidylcholine (5-doxyl-PC) and 1-palmitoyl-2-(8-doxylpalmitoyl)-3-phosphatidylcholine (8-doxyl-PC); and 2,2,6,6-tetramethylpiperidinyloxy (TEMPO). Details pertaining to the methods of labeling BBMVs are discussed in Section V. In parallel experiments, the lipids extracted from BBMVs were labeled with the same spin-labeled molecules with the aim of probing the properties of these lipids dispersed in aqueous media and comparing the properties of the protein-free membrane lipids with those of the lipids present in BBM.

Figure 3 shows ESR spectra of 5-, 12-, and 16-doxylstearate incorporated into BBMVs. Similar ESR spectra were obtained when these spin-labels were incorporated in the total lipid extract of BBM dispersed in the same buffer. For the two series of experiments, ESR spectra were recorded under exactly the same conditions. The lipid extract dispersed in aqueous medium (pH ~ 7) yields liposomes that are primarily large

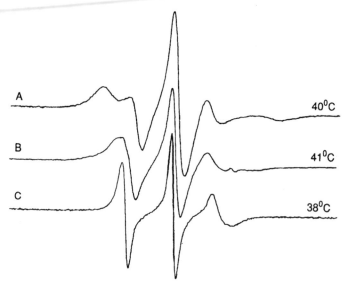

FIGURE 3   Electron spin resonance (ESR) spectra of 5-, 12- and 16-doxylstearate (spectra A, B, and C, respectively) incorporated into brush border membrane vesicles (BBMV) dispersed in 5 mM HEPES buffer, pH 7.6, containing 0.3 M D-mannitol and 5 mM EGTA. The protein concentration was (A) 15 mg/ml (~8.5 mg lipid/ml; lipid : spin-label mole ratio = 200) and (B, C) 27 mg/ml (~15 mg/lipid ml; lipid : spin-label ratio = 100). For details, see Hauser *et al.* (1982). Reproduced from Hauser *et al.* (1982) with permission.

unilamellar vesicles with a diameter $\geq$100 nm. The line shapes of the spectra of 5- and 12-doxylstearate are typical of the label present in a lipid bilayer and undergoing rapid but anisotropic motion. Line shapes similar to that shown in Fig. 3 were obtained with BBMVs labeled with 5- and 8-doxyl-PC. Figure 4A shows the temperature dependence of the ESR spectra of 5-doxyl-PC-labeled BBMVs. The maximum hyperfine splitting $2T_\parallel$ derived from the anisotropic ESR spectra (Fig. 4A) is a measure of the anisotropy of motion. The $2T_\parallel$ values obtained with the labels in BBMVs and liposome are summarized in Table III. The data show that the $2T_\parallel$ values of liposomes generally are smaller than those obtained with BBMVs, indicating that the anisotropy of motion is less in the protein-free bilayer.

In Fig. 4B and C, evidence is presented that the spin-labeled molecules are, indeed, incorporated into the lipid bilayer of the BBM. The solid lines in Fig. 4B and C give the temperature dependence of the hyperfine splitting $2T_\parallel$ of BBMVs determined with 5-doxyl-PC and 8-doxyl-PC, respectively. In these experiments, a minimum of spin-label was incorporated into the BBM, yielding mole ratios of lipid : spin-label of $\geq$200. Note

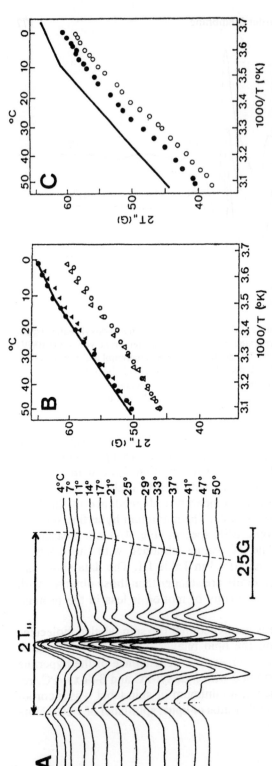

FIGURE 4 4(A) Temperature dependence of the ESR spectra of 5–doxyl–phosphatidylcholine (5–doxyl-PC) incorporated into brush border membrane vesicles (BBMV; 10 mg lipid/ml) dispersed in 10 mM HEPES/Tris buffer, pH 7.6, containing 0.3 M D-mannitol, 5 mM EDTA, and 0.2% NaN$_3$. (B) Maximum hyperfine splittings $2T_{\parallel}$ (G) of 5-doxyl-PC in BBMV and phospholipid small unilamellar vesicles (SUV) as a function of 1/T. The solid line represents BBMV (2.5 mg total lipid/ml) labeled with 5-doxyl-PC. The closed symbols represent BBMV labeled with 5-doxyl-PC by incubation with spin-labeled SUV of egg phosphatidylcholine (EPC) (●) and dioleoyl PC (▲) in the absence of PC exchange protein; open symbols represent SUV of EPC (○) and dioleoyl PC (△) labeled with 5-doxyl-PC at a lipid : spin-label ratio of 140. (C) Maximum hyperfine splittings $2T_{\parallel}$ (G) of 8-doxyl-PC in BBMV and phospholipid SUV as a function of 1/T. The solid line represents BBMV (2.5 mg total lipid/ml) labeled with 8-doxyl-PC using PC exchange protein according to Barsukov et al. (1980). The closed circles represent BBMV labeled with 8-doxyl-PC by incubation with an about 5-fold excess of spin-labeled SUV (14 mg lipid/ml) of dioleoyl PC. Open circles represent SUV of dioleoyl PC labeled with 8-doxyl-PC at a lipid : spin-label mole ratio of 65. For details, see Mütsch et al. (1986). Reproduced from Mütsch et al. (1986) with permission.

TABLE III

Hyperfine Splitting Constants $2T_\parallel$ and $2T_\perp$ of Various Spin Labels in Brush Border Membrane Vesicles and Liposomes Made from Extracted Lipids

Spin label	Temperature (°C)	BBMV $2T_\parallel$ (G)	BBMV $2T_\perp$ (G)	Liposomes $2T_\parallel$ (G)	Liposomes $2T_\perp$ (G)
5-Doxylstearic acid	4	64.3	nd[a]	61.0	nd
	25	57.3	18.3	54.0	19.0
	40	53.3	19.4	51.0	20.2
5-Doxyl-phosphatidylcholine	4	64.5	nd	64.2	nd
	25	58.6	16.8		
	40	51.5	18.5	52.5	19.1
12-Doxylstearic acid	4	60.0	nd	53.0	nd
	25	45.8	19.8	44.2	20.2
	40	41.0	21.3	40.0	21.5

[a] nd, Not determined.

that BBMVs described in Fig. 4B and C were spin-labeled by incubating them with small unilamellar vesicles (SUV) made of EPC (egg PC) or dioleoyl PC containing the spin-label at a lipid:spin-label mole ratio of 10. The open symbols in Fig. 4B and C represent hyperfine splittings obtained with the spin-labels present in EPC and dioleoyl PC bilayers. Clearly the anisotropy of motion of the spin probes in PC bilayers is significantly smaller than that in bilayers of BBMVs (Fig. 4B and C). The actual hyperfine splitting values measured for BBMVs (filled symbols in Fig. 4B and C) depend on the experimental conditions of the incubation, more precisely on the lipid mole ratio of donor (SUV) to acceptor (BBMV). In Fig. 4B, BBMVs were in excess whereas, in Fig. 4C, a fivefold excess (referred to BBM lipid) of SUVs made of dioleoyl PC labeled with 8-doxyl-PC was incubated with BBMVs. In the latter case, over the total temperature range measured, the $2T_\parallel$ values (filled circles) of the spin-label present in BBM are still greater than those for the label present in dioleoyl PC bilayers. However, these values are significantly smaller than those characteristic of BBM (cf. filled circles and solid line in Fig. 4C). This result is interpreted to mean that, on incubation of BBMVs with an excess of spin-labeled dioleoyl PC, not only the spin-label but also substantial amounts of dioleoyl PC are incorporated into the BBM. As a result, the membrane fluidity increases and, hence, the hyperfine splitting values decrease significantly. Density gradient centrifugation of the incubation mixture lends support to this interpretation. This method shows that, after incubation of BBMVs with radiolabeled EPC SUVs, part of the liposomal EPC is associated with BBM protein, indicating that liposomal phospholipid is, indeed, incorporated into the BBM. This evidence and the evidence presented in Fig. 4 collectively support the notion that, on incubation of BBMVs with SUVs made of PC, PC is transferred from SUVs to BBM and eventually is inserted into the lipid bilayer of BBMVs (Mütsch *et al.*, 1986).

The temperature dependence of the order parameter S of spin-labeled BBMVs is shown in Fig. 5. The order parameter is another ESR parameter related to the hyperfine splitting $2T_\parallel$ that is, like $2T_\parallel$, a measure of the anisotropy of the motion. Under favorable conditions, order parameters may be derived directly from anisotropic ESR spectra (Berliner, 1976), as is the case for spin-labeled BBM. Order parameters were measured for BBMVs and compared with those of liposomes made of the total lipid extract of the BBMVs. Over the total temperature range shown in Fig. 5,

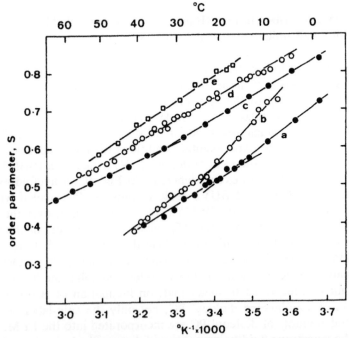

FIGURE 5 Order parameter S as a function of $1/T$ (K^{-1}) for 12-doxylstearate incorporated into BBMV (O, b) and liposomes made from total lipids extracted from BBMV (●, a); for 5-doxylstearate in BBMV (O, d) and liposomes (●, c); and 5-doxyl-PC incorporated into BBMV (□, e). For experimental conditions see Hauser *et al.* (1982).

higher S values are obtained for the spin-label in BBMVs than for the same label in liposomes, indicating that the presence of integral membrane proteins increases the order of the lipid packing or decreases the microviscosity. Similar results were reported for other cell membranes (Rottem et al., 1970; Tourtelotte et al., 1970).

The values obtained for the order parameter S of BBMVs may be compared with those of other mammalian cell membranes and bacterial membranes (cf. Hauser et al., 1982). Such a comparison of order parameters of various mammalian cell membranes labeled with 5-doxylstearate at 37°C (or body temperature) reveals that the order parameter of BBMVs (S ≈ 0.60) is at the high end of the range of values (0.50–0.60). However, this value is still lower than the order parameter obtained with highly ordered bacterial membranes, for example, of *Halobacterium cutirubrum* (S = 0.71 at 37°C). One main conclusion of the spin-label study of BBMVs is that, compared with other mammalian cell membranes, the lipid bilayer of BBM is characterized by a relatively high order of lipid packing or a low microviscosity. This conclusion, drawn from ESR spin-labeling, is qualitatively in good agreement with results obtained with BBM from other species using different methods such as fluorescence polarization (Schachter and Shinitzky, 1977; Brasitus et al., 1980).

For the temperature dependence of the order parameters shown in Fig. 5, note that the S vs. 1/T relationship is linear for BBMVs probed with 5-doxylstearate and 5-doxyl-PC; in contrast, these relationships are characterized by a break if BBMVs or liposomes are probed with 12-doxylstearate. The same is true for liposomes probed with 5-doxylstearate. The break points in the linear relationships lie between 18 and 30°C, which is within the range of the order–disorder lipid phase transition observed by DSC (cf. Figs. 2 and 5). The lack of a break point in the S vs. 1/T function is observed for BBMVs with spin-labels with doxyl groups that are close to the polar interface of the lipid bilayer. Such a result suggests that the fluidity of this region is low and that this region is not undergoing the lipid order–disorder transition. In other words, the melting of the hydrocarbon chains during the phase transition appears to be restricted to the central region of the lipid bilayer.

The ESR spectra of BBMVs and liposomes made from the lipid extract of BBMVs, both labeled with 16-doxylstearate, are relatively simple three-line spectra (Fig. 3C), indicating that the central region of the lipid bilayer is fairly fluid and the spin group tumbles almost isotropically. Assuming, to a first approximation, that the motion of the doxyl group of 16-doxylstearate is isotropic, the ESR spectra may be evaluated in terms of rotational correlation times τ. Since several assumptions are involved in the derivation of τ, the absolute values may be subject to criticism. However, the temperature dependence of the spectra reveals break points

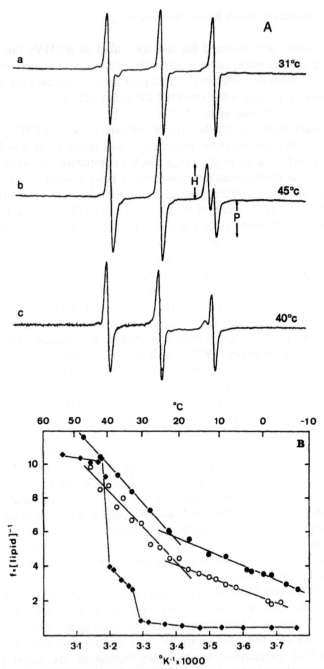

FIGURE 6 (A) Electron spin resonance (ESR) spectra of (a,b) TEMPO (10 μM) added to dipalmitoyl PC (46 mg/ml, 60 mM) dispersed in H$_2$O and (c) TEMPO (100 μM) added to brush border membrane vesicles (BBMV; 39 mg protein/ml, ~21 mg lipid/ml) dispersed in 5 mM HEPES buffer, pH 7.6, containing 0.3 M D-mannitol and 5 mM EGTA. The

at ~13°C for BBMVs and liposomes, again within the range of the lipid order–disorder transition.

As seen in lipid bilayers and other biological membranes, a motional gradient apparently exists along the bilayer normal in the sense that the lipid motion increases significantly along the bilayer normal toward the center line of the lipid bilayer. More specifically, the segmental motion of a lipid molecule in both BBMVs and lipid bilayers made from the lipid extract is highly anisotropic at the polar/apolar interface; the anisotropy of motion decreases, probably in a nonlinear fashion, toward the center of the lipid bilayer.[1] Experimental data are consistent with a steeper flexibility gradient in BBMVs than in bilayers made from the lipid extract. Qualitatively similar fluidity (microviscosity) gradients were reported for lipid bilayers and for biological membranes using primarily spin-labeling and NMR methods (Rottem et al., 1970; Hubbell and McConnell, 1971; Jost et al., 1971; Seelig, 1971; Levine et al., 1972; Devaux et al., 1975).

Information concerning the microviscosity of BBMVs also can be derived from partitioning experiments using the water-soluble spin-label TEMPO. This label has been shown to partition between water and lipid bilayers, provided the lipid bilayer is in the liquid crystalline state (Shimshick and McConnell, 1973; Marsh and Watts, 1981; Hauser et al., 1982). This phenomenon is demonstrated in Fig. 6, which also shows the temperature dependence of TEMPO partitioning into BBMVs and liposomes made from the lipid extract. The data of Figs. 5 and 6 support the conclusion that the fluidity (microviscosity) of the protein-free lipid bilayer is greater (lower) than that of BBMVs. The temperature dependence of TEMPO partitioning into BBMVs and liposomes exhibits break points at ~20°C (Fig. 6). The results obtained with the water-soluble spin-label TEMPO are entirely consistent with the other spin-labeling results presented earlier.

Similar ESR spectra were obtained when 3-doxyl-5α-cholestane was incorporated into BBMVs and liposomes made from the lipid extract of BBMVs (cf. Hauser et al., 1982; Mütsch et al., 1983). This label confirms

---

[1] Such conclusions derived from ESR spin-labeling are based on the assumption that the motion of the spin probe is representative of the average motion of the lipid molecules, that is, that the probe molecule reflects the average motion of the lipid molecules. This assumption is reasonable and generally accepted.

---

temperature in a is below and that in b above the gel-to-liquid crystal transition temperature of dipalmitoyl PC. (B) TEMPO partitioning into BBMV (O), into liposomes made from the lipid extract of BBMV (●), and into liposomes made from dipalmitoyl PC (◆) as a function of 1/T. Lipids extracted from BBMV were dispersed at 9.2 mg/ml in the same buffer as BBMV containing 100 μM TEMPO. Other experimental details are as in A. The spectral parameter $f = H/(H + P)$ was divided by the lipid concentration in g/ml. For details, see Hauser et al. (1982). Reproduced from Hauser et al. (1982) with permission.

the conclusions drawn from the spin-label studies discussed earlier and provides additional information pertaining to the lipid dynamics in BBM. The temperature dependence of the ESR spectra of the cholestane spin-label incorporated into the lipid extract of BBM is shown in Fig. 7A. Also shown is the temperature dependence of the maximum hyperfine splitting $2T_\perp$ (Fig. 7B). The line shapes of the spectra are typical of anisotropic motion of the steroid nucleus. The maximum hyperfine splitting $2T_\perp$ increases from ~40 G at 50°C to 43 G at room temperature. A value of about 40 G is interpreted to result from the rapid rotation of the 3-doxyl-5α-cholestane molecule about its long axis, which averages out the hyperfine splitting tensor components T_{zz} and T_{xx} to yield $2T_\perp = T_{zz} + T_{xx} \approx 40G$. The temperature dependence of the hyperfine splitting $2T_\perp$ shows a discontinuity between 0 and 20°C, indicated by the hatched area of Fig. 7B. Below this region, that is, at temperatures <0°C, the spectral line shape approaches that of a powder pattern with $2T_\perp$ values close to the T_{zz} tensor component, indicating that the rotation about the long axis of the steroid nucleus is frozen at these temperatures. In the temperature region 0–20°C (hatched area of Fig. 7B), a transition occurs between the motionally averaged spectra observed above ~20°C and the almost immobilized spectra observed below 0°C. This transition region partially overlaps the temperature range of the lipid order–disorder transition of BBMVs. The ESR spectra obtained with the 3-doxyl-5α-cholestane spin-label in this temperature range are composite, consisting of at least two component spectra. One possible explanation is that the phase transition is accompanied by two-dimensional phase separations of the lipids. Such a process would give rise to two or even more environments for the spin-label, provided the exchange rate between these different environments is slow on the ESR time scale.

The temperature dependence of both 5-doxyl-PC (cf. Fig. 4) and 3-doxyl-5α-cholestane (cf. Fig. 7) incorporated in BBMVs and liposomes made from the lipid extract of BBMVs indicates that the molecular rotation about the long axis of these spin labels is frozen at temperatures below ~0°C. The ESR spectra obtained with the two labels under these conditions are almost immobilized, probably because the lipids crystallize and form a gel phase below the lipid phase transition. At first sight, the results obtained with 5-doxyl-PC and the cholestane spin-label at temperatures between 0 and −20°C (cf. Fig. 7) seem to be at variance with the ^{31}P-NMR data (Fig. 1). The ^{31}P-NMR spectra indicate that the motional averaging of the phospholipid molecules about the axis parallel to the bilayer normal is maintained at temperatures below the lipid phase transition, down to ~−20°C (Fig. 1). One explanation of this apparent discrepancy is that more than one mechanism is responsible for the averaging of the ^{31}P-NMR chemical shielding tensor. Motional averaging of this tensor could be due

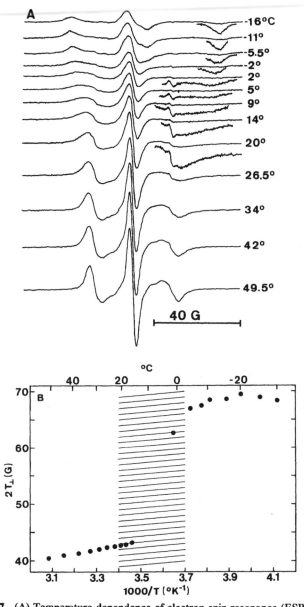

FIGURE 7 (A) Temperature dependence of electron spin resonance (ESR) spectra of 3-doxyl-5α-cholestane incorporated into total lipids extracted from brush border membrane vesicles (BBMV). The lipids were dispersed in 10 m*M* HEPES/Tris buffer, pH 7.5, containing 0.3 *M* D-mannitol, 5 m*M* EDTA, and 0.02% NaN₃. For temperatures < 20°C, vertical expansions of the high-field part of the spectra are shown (Mütsch *et al.*, 1983). (B) Maximum hyperfine splittings 2T$_\perp$ (G) as a function of 1/T (°K⁻¹). 2T$_\perp$ values were derived from the ESR spectra shown in A. The hatched region between ~20 and −2°C represents the transition from motional-averaged to immobilized spectra (Mütsch *et al.*, 1983). Reproduced from Mütsch *et al.* (1983) with permission.

to (1) rotation of the phospholipid molecule as a whole about its molecular long-axis and (2) rotation of the phospholipid polar group about one of the C–C bonds of glycerol. For instance, the first glycerol bond HC–CH$_2$OP has been shown to be aligned almost parallel to the bilayer normal; rotation of the polar group about this bond has been postulated (Hauser, 1981; Hauser et al., 1981). Clearly fast rotation of the polar group about this bond would lead to an axially symmetric shielding tensor. The explanation, then, is that above the lipid phase transition temperature of BBM both mechanisms of motional averaging are effective, whereas below the phase transition between ~0 and 20°C single-bond rotation prevails, still effectively averaging the chemical shielding tensor (Mütsch et al., 1983).

The results discussed so far are relevant to the basic lipid dynamics in BBM. Summarizing these results, we conclude that ^{31}P-NMR and ESR spin-labeling are consistent with the lipids of BBM forming a bilayer structure. When total lipids of BBM are extracted and dispersed in an aqueous medium, the preferred arrangement is also the bilayer structure, at least under physiological conditions (Hauser et al., 1982; Mütsch et al., 1983). Several lines of independent evidence show that the lipids present in BBM undergo an order–disorder (gel-to-liquid crystal) transition, as do those in liposomal dispersions of lipids extracted from BBMV. Evidence from DSC (Fig. 2) and from the temperature dependence of different ESR parameters discussed earlier (Figs. 5–7) supports this concept. A third line of evidence not discussed here is provided by the temperature dependence of the activity of various integral proteins of BBM such as the D-glucose transport protein. This evidence was discussed in detail elsewhere (Brasitus et al., 1980; Mütsch et al., 1983). Further, ESR spin-labeling provides evidence that the fluidity (microviscosity) of the lipid bilayer of BBM is generally low (high) with respect to other mammalian cell membranes, a result that is consistent with tight packing of the lipids in BBM. As in lipid bilayers and other biological membranes, a flexibility gradient occurs along the bilayer normal that appears to be pronounced in BBMVs. This gradient is characterized by highly anisotropic motion of the hydrocarbon chain segments close to the polar/apolar interface and by almost isotropic motion of these segments and, hence, an almost liquid environment at the center of the BBM.

V. LIPID TRANSFER AND LIPID EXCHANGE INTERACTIONS BETWEEN BRUSH BORDER MEMBRANE VESICLES AND LIPID PARTICLES

In discussing the lipid packing and dynamics of BBMVs, we noted that these properties are not exceptional but are typical of plasma membranes, except for the tight and ordered packing of the lipid bilayer and the rather steep fluidity gradient. This section is addressed to a unique property of the BBM that may be physiologically important and related to functional properties of this membrane. The unique feature of BBM is the protein-mediated lipid absorption from various donor particles, as first described by Thurnhofer and Hauser (1990a,b). To introduce radiolabeled PC or spin-labeled PC into BBM, we originally used water-soluble PC exchange protein isolated from beef liver (Barsukov *et al.*, 1980; Hauser *et al.*, 1982). The experiment carried out by Barsukov *et al.* (1980) indicated that more PC was exchanged between EPC SUVs and BBMVs in the presence of PC exchange protein than theoretically predicted on the basis of the total PC content of BBMVs. Another anomaly of the Barsukov experiment was that, in the control experiment carried out in the absence of beef liver PC exchange protein, an unexpectedly high amount of PC was transferred from EPC SUVs to BBM. This experiment clearly shows that a significant quantity of radiolabeled PC is transferred somehow from EPC SUVs and is inserted into the lipid bilayer of BBM in the absence of exogenous PC exchange protein. Good use has been made of this finding and, as mentioned in Section IV,C, BBMVs were spin-labeled with 3-doxyl-5α-cholestane and 5-doxyl-PC by incubating BBMV with PC SUVs containing the spin-label (Hauser *et al.*, 1982) in the absence of exchange protein.

These observations prompted a systematic study of the lipid exchange between EPC SUVs and BBMVs. The main conclusion resulting from this study was that the uptake of cholesterol and certain phospholipids such as PC and phosphatidylinositol is protein mediated (Thurnhofer and Hauser, 1990a,b). Our primary interest has been focused ever since on cholesterol absorption by BBMVs. This process is facilitated by (an) integral membrane protein(s) of the BBM with its active center(s) exposed on the external or luminal side of the membrane. This is true for cholesterol absorption from micelles as well as from SUVs as donor particles. However, a significant difference exists between the two kinds of donor particles: with micelles as donor particles, net transfer of cholesterol from donor to acceptor occurs (Thurnhofer and Hauser, 1990a); in contrast, with SUVs as donor particles, true mass exchange occurs, that is, at equilibrium cholesterol or PC is distributed evenly between the lipid pools of the donor and acceptor particles (Thurnhofer and Hauser, 1990a,b). That lipid exchange indeed occurs between SUVs and BBMVs was shown by a double-labeling experiment. BBMVs labeled with [³H]dipalmitoyl PC were incubated with EPC SUVs labeled with [¹⁴C]dipalmitoyl PC; pseudo-first-order rate constants (k_1) were determined for the forward and backward reactions of PC exchange. Within the error of the measurement, the same k_1 values were measured for the forward and backward reactions (Thurnhofer and Hauser, 1990b). The exchange reaction with SUVs as well as the net lipid transfer from micelles as the donor are second order reactions. The mechanisms are collision-induced lipid exchange and transfer, respectively (Thurnhofer and Hauser, 1990a,b), that are catalyzed by integral membrane proteins. If one EPC molecule is transferred from EPC SUVs to BBMVs, the question arises whether one EPC molecule is strictly exchanged for one PC molecule of BBM or whether the exchange involves other lipids of BBM. After incubation of BBMVs with EPC SUVs at room temperature for ~24 hr, the lipids of both SUVs and BBM were extracted and subjected to thin-layer chromatography (TLC) analysis (Lipka et al., 1991). The main conclusion from this experiment is that not only the phospholipids of BBM but also cholesterol and glycolipids participate in the lipid exchange. This exchange reaction between EPC SUVs and BBMVs, although protein mediated and interesting per se, is probably not relevant physiologically. The physiologically important donor particles, from which lipid absorption takes place in the upper small intestines, are mixed bile salt micelles and not SUVs. Nevertheless, the exchange reaction in the model system described here provides a great deal of information about the nature of the protein(s) that catalyze(s) it. The protein is selective with respect to the lipid acquired when interacting with the donor membrane. However, it is nonspecific with respect to the selection of the lipid it moves in the reverse direction from the acceptor to the donor membrane. This feature of the lipid exchange reaction is remarkable. An explanation on a molecular level must await the elucidation of the structure of the protein(s) involved.

The kinetics of the lipid uptake by BBMVs from various donor particles, mainly EPC SUVs and cholate mixed micelles, was the subject of previous publications (Thurnhofer and Hauser, 1990a,b; Thurnhofer et al., 1991). Psuedo-first-order rate constants for cholesterol absorption by BBMVs from various donor particles are summarized in Table IV. Evidence is presented that, after proteolytic treatment of BBMVs with papain, the pseudo-first-order rate constants are reduced significantly. The values obtained after papain treatment were comparable to or smaller than the first-order rate constants measured for cholesterol exchange between two populations of SUVs in the absence of protein (Thurnhofer and Hauser, 1990a). The rate of cholesterol absorption from taurocholate mixed micelles is much higher (by a factor of ~10^3) than that from EPC SUVs or lyso-PC (Table IV; Thurnhofer and Hauser, 1990b). In another double-labeling experiment, using as donor particles taurocholate mixed micelles containing trace quantities of both [^{3}H]cholesterol and sodium [^{14}C]cholate, researchers showed that the absorption of BBMVs of sodium cholate is negligible compared with that of cholesterol. This result is convincing evidence that the protein catalyzing cholesterol absorption can differenti-

TABLE IV

Pseudo-First-Order Rate Constants for Lipid Absorption in Brush Border Membranes[a]

Substrate	Donor	Acceptor	k_1 (hr^{-1})	$t_{1/2}$ (hr)	BBM lipid/ donor lipid
Cholesterol	SUV of EPC	BBMV	2.65	0.26	3.33
EPC	SUV of EPC	BBMV	0.601	1.15	3.33
Cholesterol	SUV of EPC	BBMV (papain treated)	0.080	8.7	3.33
Cholesterol	Lyso-EPC/ EPC, mixed micelles	BBMV	5.3	0.13	3.33
Cholesterol	Taurocholate mixed micelles	BBMV	1×10^4	6.9×10^{-5} ($= 0.25$ sec)	3.33
[^{3}H]DPPC	Taurocholate mixed micelles	BBMV	0.714×10^4	0.35 sec	3.33
Cholesterol	Taurocholate mixed micelles	BBMV (proteinase K treated)	0.63	1.1	3.33

[a]Abbreviations: EPC, egg phosphatidylcholine; SUV, small unilamellar vesicles; BBMV, brush border membrane vesicles; DPPC, dipalmitoyl phosphatidylcholine.

ate between the ring system of cholesterol and that of bile salts. After treatment of BBMVs with proteinase K, cholesterol absorption from taurocholate mixed micelles is not only much reduced, but is a true first-order reaction characterized by a half-time of ~1 hr (Table IV). Similar rate constants are measured for cholesterol transfer from taurocholate mixed micelles to EPC SUVs in the absence of any protein, which is clearly a passive process. The data summarized in Table IV provide clear cut evidence that cholesterol absorption by BBMVs is protein mediated. These data also emphasize the special role bile salt micelles play as donor particles in the process of lipid absorption in the small intestines.

The absorption of PC from various donor particles by BBMVs is also protein mediated; we have shown that the same integral membrane protein is likely to be responsible for this effect. The protein involved was shown to have a lipid binding site with properties typical of a nonspecific lipid binding protein (Thurnhofer and Hauser), 1990b; Thurnhofer et al., 1991). Studying the PC exchange between EPC SUVs and BBMVs, the phospho-

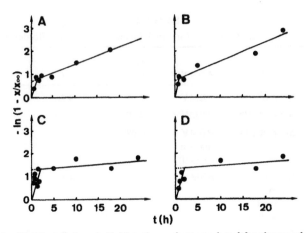

FIGURE 8 Kinetics of phospholipid exchange between brush border membrane vesicles (BBMV) and small unilamellar vesicles (SUV) of egg phosphatidylcholine (EPC). BBMV as donors were suspended in 10 mM HEPES/Tris buffer, pH 7.3, containing 0.3 M D-mannitol and 1 mM EDTA to a final concentration of 6–10 mg total lipid/ml and incubated with SUV of EPC as the acceptors at 4–7 mg lipid/ml at room temperature. The weight ratio of total lipid of acceptor to donor was kept constant at 0.7. The time course of phospholipid exchange was linearized according to the formula $-\ln(1 - x/x_\infty) = k_1[(a + b)/b]t$, where x and x_∞ represent the fractional transfer of phospholipid from BBMV to EPC-SUV at time t and at equilibrium, respectively, and k_1 is the pseudo-first-order rate constant. (A) Sphingomyelin, (B) phosphatidylethanolamine, (C) phosphatidylserine, (D) phosphatidylinositol. The solid lines were fitted to the experimental data points by linear regression analysis. Each data point represents the average of two to three measurements. Reproduced from Lipka *et al.* (1991) with permission.

lipid exchange can be shown to be biphasic (see Fig. 8; Lipka *et al.*, 1991). In the initial fast phase, the exchange of exogenous EPC for different BBM phospholipids (sphingomyelin, phosphatidylethanolamine, phosphatidylserine, phosphatidylinositol) is described by pseudo-first-order rate constants (k_1) that agree within the error of the measurement (cf. Fig. 8; Table V; Lipka *et al.*, 1991). The average k_1 value obtained for the transfer of the four phospholipids from BBMVs to EPC SUVs in exchange for exogenous PC is 0.38 ± 0.03, corresponding to a half-time of 1.8 hr. This k_1 value agrees well with the pseudo-first-order rate constant k_1 obtained for the transfer of EPC from SUVs to BBMVs. From the fact that forward and backward reactions are characterized by the same rate constants, and from the balance of lipid movement in and out of BBM, we conclude that all phospholipids and cholesterol present in the external layer of BBM have equal probability of participating in the lipid exchange and that this process is characterized by a 1:1 stoichiometry (Lipka *et al.*, 1991).

TABLE V

Pseudo-First-Order Rate Constants k_1 and Half Times $t_{1/2}$ for Phospholipid Exchange

Phospholipid	Fast exchange[a]		Slow exchange	
	k_1 (hr^{-1})	$t_{1/2}$ (hr)	k_1 (hr^{-1})	$t_{1/2}$ (hr)
Sphingomyelin	0.36 ± 0.07	1.9 ± 0.4	0.073 ± 0.005	9 ± 0.6
Phosphatidylethanolamine	0.39 ± 0.05	1.8 ± 0.2	0.085 ± 0.006	8 ± 0.6
Phosphatidylserine	0.39 ± 0.08	1.8 ± 0.4	0.009 ± 0.006	75 ± 50
Phosphatidylinositol	0.39 ± 0.03	1.8 ± 0.2	0.008 ± 0.006	85 ± 60

[a]The exchange of phospholipids between brush border membrane vesicles as the donors and small unilamellar egg phosphatidylcholine vesicles as the acceptors is biphasic, that is, an initial fast phase of exchange is followed by a slow phase of lipid exchange. The data are derived from Fig. 8.

The biphasic character of the phospholipid exchange between EPC SUVs as donors and BBMVs (shown in Fig. 8) is interpreted to mean that BBM phospholipids are present in two pools. These two pools not only differ in their rate of phospholipid exchange as shown in Fig. 8, but also differ in their accessibility to exogenously added phospholipases (Lipka et al., 1991). The k_1 values for the slow exchange of the two isoelectric phospholipids phosphatidylethanolamine and sphingomyelin are smaller by a factor of 5 than the k_1 value for the fast phase; those of the two negatively charged phospholipids are reduced even further and are smaller by a factor of 40–50 (Table V). The two lipid pools revealed by phospholipid exchange and by digestion with phospholipases were assigned tentatively to phospholipid molecules located on the outer and inner layer of the BBM (Lipka et al., 1991). Information concerning the pool size and, hence, the asymmetric or tranverse distribution of phospholipids in BBM can be derived from the data in Fig. 8, as discussed in detail by Lipka et al. (1991). The percentages of phospholipids present in the outer and inner monolayers of the BBM are summarized in Table VI. One aspect of this table that is conspicuous is the good agreement of the results obtained with the two independent methods. The exception is sphingomyelin; possible explanations of this discrepancy were discussed by Lipka et al. (1991). As seen in Table VI, the neutral (isoelectric) phospholipids such as PC and phosphatidylethanolamine are located preferentially on the inner (cytoplasmic) side (to ~70%). In contrast, sphingomyelin as the third isoelectric lipid may have a preference for the outer layer of BBM, as indicated by the phospholipase treatment (Table VI). In contrast, the negatively charged phospholipids phosphatidylserine and phosphatidylinositol exhibit a more even distribution between outer and inner monolayer; 40–45% is located on the outer and 55–60% is located

TABLE VI

Asymmetric Distribution of Phospholipids in the Brush Border Membrane[a]

Phospholipid	Phospholipid exchange		Digestion with lipases	
	Outer layer (%)	Inner layer (%)	Outer layer (%)	Inner layer (%)
Sphingomyelin	31 ± 2	69 ± 2	63 ± 3	37 ± 3
Phosphatidylcholine	nd[b]	nd	32 ± 2	68 ± 2
Phosphatidylethanolamine	28 ± 2	72 ± 2	34 ± 3	66 ± 3
Phosphatidylserine	42 ± 3	58 ± 3	44 ± 9	56 ± 9
Phosphatidylinositol	40 ± 4	60 ± 4	40 ± 10	60 ± 10

[a]Values are presented as the mean ± standard deviation.
[b]nd, Not determined.

on the inner bilayer of BBM. Compared with other plasma membranes, the asymmetric distribution of PC with the majority on the cytoplasmic side is unusual. A possible explanation is that BBM is a plasma membrane rich in glycosphingolipids that are known to be located exclusively on the outer layer of the membrane (see Chapter 20). The abundance of glycosphingolipids in the outer layer may be balanced by the accumulation of PC in the inner layer of BBM.

If the assignment of the two lipid pools to the outer and inner monolayer of BBM is correct, then the pseudo-first-order rate constants of the second slow phase of phospholipid exchange represent the rate constants for transverse or flip–flop motion of phospholipids in BBM. For the isoelectric phospholipids phosphatidylethanolamine and sphingomyelin, the half-times of this motion are ~8 hr, and for the negatively charged phospholipids the values for the half-time are about 10 times larger.

VI. CONCLUSIONS

The lipid dynamics of BBM were probed with different techniques including ^{31}P-NMR, ESR spin-labeling, DSC, and kinetic methods using radiolabeled and spin-labeled molecules. The general consensus emerging from the application of these techniques is that lipids of BBM form a typical bilayer structure. The molecular and segmental motion of the lipids within the bilayer are characteristic of plasma membranes, except that the lipid fluidity (microviscosity) of BBM is high (low) in comparison with other plasma membranes. A rather steep flexibility (fluidity) gradient exists in BBMVs with almost crystalline hydrocarbon chain packing close to the lipid–water interface and almost fluid hydrocarbon chain in the center of the lipid bilayer. The tight lipid packing and relatively high order in the lipid bilayer of BBM is determined primarily by the lipid composition of this membrane. Saturated fatty acids prevailing in the hydrocarbon chain region of the bilayer may account in part for this observation. Lipid packing is modulated only little by the presence of integral membrane proteins.

In addition to the straightforward in-plane lipid dynamics, BBM has the unique capacity of transporting lipids in a transverse direction (i.e., parallel to the bilayer normal). By this we do not mean in the usual transverse or flip–flop motion of lipid molecules from the outer to the inner monolayer of the bilayer and vice versa, but as the insertion into the BBM of exogenous lipids from donor particles and the exchange of lipids between BBM and donor particles. Donor particles studied to date are micelles or small unilamellar vesicles. A prerequisite for this type of transverse motion is the collision-induced contact of micelles or single bilayer vesicles with the BBM. The capacity of BBM to take up exogenous lipids or to exchange lipids with donor particles is more remarkable considering the tight lipid packing of BBM. This property of BBM is very likely to be related to functional properties and adds a special note to the lipid dynamics of this membrane. The ability of BBM to insert dietary lipid molecules into its lipid bilayer and/or to exchange exogenous lipids for endogenous lipids is associated with integral membrane protein(s), the active sites of which are exposed on the external luminal side of BBM. Although protein-mediated lipid absorption or exchange has been demonstrated for dietary lipids such as cholesterol and diacyl PC, it may, however, not occur for other dietary lipids such as fatty acids and monoacylglycerols.

Good use has been made of this unique property of BBM. Not only was it utilized to incorporate labeled lipid molecules readily to probe the lipid dynamics of BBM, but it also provided information on the transverse distribution of phospholipids and on the rate of the transverse or flip–flop motion of phospholipids in BBM.

References

Barsukov, L. I., Hauser, H., Hasselbach, H.-J., and Semenza, G. (1980). Phosphatidylcholine exchange between brush border membrane vesicles and sonicated liposomes. *FEBS Lett.* **115**, 189–192.

Berliner, L. J. (1976). "Spin Labeling. Theory and Applications." Academic Press, London.

Brasitus, T. A., Tall, A. R., and Schachter, D. (1980). Thermotropic transitions in rat intestinal plasma membranes studied by differential scanning calorimetry and fluorescence polarization. *Biochemistry* **19**, 1256–1261.

Devaux, P. F., Bienvenüe, A., Lanquin, G., Brisson, A. D., Vignais, P. M., and Vignais, P. V. (1975). Interaction between spin-labeled acyl-coenzyme A and the mitochondrial adenosine diphosphate carrier. *Biochemistry* **14**, 1272–1280.

Gains, N., and Hauser, H. (1981). Detergent-induced proteolysis of rabbit intestinal brush border vesicles. *Biochim. Biophys. Acta* **646**, 211–217.

Gains, N., and Hauser, H. (1984). Leakiness of brush-border vesicles. *Biochim. Biophys. Acta* **722**, 161–166.

Hauser, H. (1981). The polar group conformation of 1,2-dialkylphosphatidylcholines. An NMR study. *Biochim. Biophys. Acta* **646**, 203–210.

Hauser, H., Howell, K., Dawson, R. M. C., and Bowyer, D. E. (1980). Rabbit small intestinal brush border membrane preparation and lipid composition. *Biochim. Biophys. Acta* **602**, 567–577.

Hauser, H., Pascher, I., Pearson, R. H., and Sundell, S. (1981). Preferred conformation and molecular packing of phosphatidylethanolamine and phosphatidylcholine. *Biochim. Biophys. Acta* **650**, 21–51.

Hauser, H., Gains, N., Semenza, G., and Spiess, M. (1982). Orientation and motion of spin-labels in rabbit small intestinal brush border vesicle membranes. *Biochemistry* **21**, 5621–5628.

Hubbell, W. L., and McConnell, H. M. (1971). Molecular motion in spin-labeled phospholipids and membranes. *J. Am. Chem. Soc.* **93**, 314–326.

Jost, P. C., Libertini, L. J., Herbert, V. C., and Griffith, O. H. (1971). Lipid spin labels in lecithin multilayers. Motions along fatty acid chains. *J. Mol. Biol.* **59**, 77–99.

Kessler, M., Acuto, O., Storelli, C., Murer, H., Müller, M., and Semenza, G. (1978). A modified procedure for the rapid preparation of efficiently transporting vesicles from small intestinal brush border membranes. *Biochim. Biophys. Acta* **506**, 136–154.

Klip, A., Grinstein, S., and Semenza, G. (1979). Transmembrane disposition of the phlorizin binding protein of intestinal brush borders. *FEBS Lett.* **99**, 91–96.

Ladbrooke, B. D., and Chapman, D. (1969). Thermal analysis of lipids, proteins and biological membranes. A review and summary of some recent studies. *Chem. Phys. Lipids* **3**, 304–367.

Ladbrooke, B. D., Williams, R. M., and Chapman, D. (1968). Studies on lecithin–cholesterol–water interactions by differential scanning calorimetry and X-ray diffraction. *Biochim. Biophys. Acta* **150**, 333–340.

Levine, Y. K., Birdsall, N. J. M., Lee, A. G., and Metcalfe, J. C. (1972). Carbon-13 nuclear magnetic resonance relaxation measurements of synthetic lecithins and the effect of spin-labeled lipids. *Biochemistry* **11**, 1416–1421.

Lipka, G., Op den Kamp, J. A. F., and Hauser, H. (1991). Lipid asymmetry in rabbit small intestinal brush border membrane as probed by an intrinsic phospholipid exchange protein. *Biochemistry* **30**, 11828–11836.

Marsh, and Watts, A. (1981). ESR spin label studies in liposomes. *In* "Liposomes: From Physical Structure to Therapeutic Applications" (C. G. Knight, ed.), pp. 139–188. Elsevier, Amsterdam.

Mütsch, B., Gains, N., and Hauser, H. (1983). Order-disorder phase transition and lipid dynamics in rabbit small intestinal brush border membranes. Effect of proteins. *Biochemistry* **22**, 6326–6333.

Mütsch, B., Gains, N., and Hauser, H. (1986). Interaction of intestinal brush border membrane vesicles with small unilamellar phospholipid vesicles. Exchange of lipids between membranes is mediated by collisional contact. *Biochemistry* **25**, 2134–2140.

Perevucnik, G., Schurtenberger, P., Lasic, D. D., and Hauser, H. (1985). Size analysis of biological membrane vesicles by gel filtration, dynamic light scattering and electron microscopy. *Biochim. Biophys. Acta* **821**, 169–173.

Proulx, P. (1991). Structure-function relationships in intestinal brush border membranes. *Biochim. Biophys. Acta* **1071**, 255–271.

Rottem, S., Hubbell, W. L., Hayflick, L., and McConnell, H. M. (1970). Motion of fatty acid spin labels in the plasma membrane of mycoplasma. *Biochim. Biophys. Acta* **219**, 104–113.

Schachter, D., and Shinitzky, M. (1977). Fluorescence polarization studies of rat intestinal microvillus membranes. *J. Clin. Invest.* **59**, 536–548.

Schmitz, J., Preiser, H., Maestracci, D., Ghosh, B. K., Cerda, J. J., and Crane, R. K. (1973). Purification of the human intestinal brush border membrane. *Biochim. Biophys. Acta* **323**, 98–112.

Seelig, J. (1971). Flexibility of hydrocarbon chains in lipid bilayers. *J. Am. Chem. Soc.* **93**, 5017–5022.

Shimshick, E. J., and McConnell, H. M. (1973). Lateral phase separation in phospholipid membranes. *Biochemistry* **12**, 2351–2360.

Thurnhofer, H., and Hauser, H. (1990a). Uptake of cholesterol by small intestinal brush border membrane is protein-mediated. *Biochemistry* **29**, 2142–2148.

Thurnhofer, H., and Hauser, H. (1990b). The uptake of phosphatidylcholine by small intestinal brush border membrane is protein mediated. *Biochim. Biophys. Acta* **1024**, 249–262.

Thurnhofer, H., Schnabel, J., Betz, M., Lipka, G., Pidgeon, C., and Hauser, H. (1991). Cholesterol-transfer protein located in the intestinal brush border membrane. Partial purification and characterization. *Biochim. Biophys. Acta* **1064**, 275–286.

Tourtellotte, M. E., Branton, D., and Keith, A. (1970). Membrane structure: spin-labeling and freeze-etching of *Mycoplasma laidlawii*. *Proc. Natl. Acad. Sci. U.S.A.* **66**, 909–916.

CHAPTER 8

Lipids and Lipid-Intermediate Structures in the Fusion of Biological Membranes

Philip L. Yeagle
Department of Biochemistry, University at Buffalo School of Medicine, Buffalo, New York 14214

I. GENERAL CONSIDERATION OF MEMBRANE DEFORMATION NECESSARY FOR FUSION AND THE ROLE OF LIPIDS IN FUSION

Biological membranes contain the lipid bilayer as their fundamental structural element. This stable structure is formed from an assembly of amphipathic membrane lipids. The assembly of a lipid bilayer is a thermodynamically spontaneous process driven by the hydrophobic effect and the structure of the lipid molecules (for a detailed discussion, see Yeagle 1987). The resulting lipid bilayer normally is not only stable, but also is largely impermeable. The lipid bilayer thus is an impediment to the entry of material into a cell and the exit of material from a cell.

Membrane fusion, the joining of two membranes to form one membrane (or fission, the reverse of this process), is a means of overcoming the limitations to entry and exit of materials imposed by a stable lipid bilayer. The most difficult problem in understanding the membrane fusion process is the description of the pathway by which two membranes with stable lipid bilayers can mix their bilayer components and become one membrane. This process has been described as requiring at least three steps (Bentz and Ellens, 1988). (1) The two membranes that are going to fuse must approach each other closely. (2) The lipids of the two membranes must, at least transiently, populate higher energy structures that involve nonbilayer lipid assemblies. (3) The two membranes must mix their components and fuse into one membrane.

Conceptually, for two lipid bilayers to mix, some disruption of the bilayer structure must occur. Initially, the membranes must approach closely to fuse. Such a close approach may involve some removal of water from between the membranes. Thus dehydration of the membrane surface can have a role in membrane fusion. Surfaces of some bilayers can be relatively hydrophobic. For example, bilayers rich in phosphatidylethanolamine (PE) exhibit relatively hydrophobic surfaces that, consequently, interact poorly with water (Yeagle and Sen, 1986). Close approach of such surfaces is more energetically favorable than close approach of surfaces that interact extensively and more favorably with water, such as surfaces of bilayers enriched in phosphatidylcholine (PC). Therefore, at least in some systems, PE in the membrane enhances membrane fusion.

Close approach of two membranes is not sufficient to achieve membrane fusion. Two membranes can be in close proximity with each other and without membrane fusion taking place. The closeness of approach of two membrane surfaces will be governed by hydration forces (Rand and Parsegian, 1989) which, in turn, are determined in part by the lipid head group composition of the lipid bilayers. However, two membrane lipid bilayers that are close together can be stable without fusion, unless other factors are brought into play.

Successful membrane fusion necessitates transient disruption of otherwise stable bilayer structures. Such disruption may result from the intrinsic properties of the lipids that constitute the bilayer. Alternatively, such disruption may result from nonlipid components of the system. Included among these latter components is calcium, which can induce membrane fusion in some bilayers containing a large percentage of negatively charged lipids (Papahadjopoulos and Poste, 1975; Papahadjopoulos et al., 1977). Membrane fusion induced by calcium will not be discussed in detail here, although brief reference will be made to it later in the chapter.

II. NONLAMELLAR LIPID ASSEMBLIES

Phospholipids can form a variety of nonlamellar lipid assemblies. Most of these nonlamellar structures have, at one time or another, been pro-

posed to be involved as intermediates in membrane fusion because of the
need for some transient disruption of bilayer structure to permit the mixing
of the components of two (originally separate) membranes, thus leading
to membrane fusion. Therefore, reviewing some of the nonlamellar structures that lipid assemblies can adopt is appropriate (Luzzati and Husson,
1962; Reiss-Husson, 1967).

First to be considered is the hexagonal II structure, which can be
adopted by lipids such as PE (Cullis and de Kruijff, 1978b) and monogalactosyl diglyceride (Sen et al., 1981). The hexagonal II phase (H_{II}) consists
of tubes of lipids with the headgroups of the lipids directed toward the
interior of the tubes and the hydrocarbon chains extending toward the
outer surface of the tubes (see Fig. 1). The tubes then stack on each other
like pipes in storage (Turner and Gruner, 1992). Although still a matter
of some discussion, the hydrophobic exterior of the H_{II} tubes is likely to
be covered by a monolayer of lipid when the H_{II} phase is found in excess
water. Lipids such as PE and monogalactosyl diglyceride will form the
H_{II} phase from the lamellar phase in response to stress on the system
caused by an increase in temperature.

The H_{II} phase is characterized by a lipid surface with a negative radius
of curvature (Gruner, 1985). The facility (with respect to increased temperature or other stress on the system) with which the H_{II} phase is formed
is correlated with the intrinsic radius of curvature of the lipid (which can
be measured when the lipid is in a "relaxed" H_{II} phase). For example,
the radius of the water channel in the H_{II} phase of dioleoyl phosphatidylethanolamine decreases from about 22Å at 15°C to about 16 Å at 85°C (Tate
and Gruner, 1989). From this perspective, formation of the H_{II} phase is
a response of the system to the incorporation of lipids with a finite intrinsic
radius of curvature into a bilayer with a nearly infinite radius of curvature.

From a thermodynamic point of view, the tendency to form the H_{II}
phase can be related to the relative hydrophobicity of the membrane
surface (Yeagle and Sen, 1986). In the H_{II} phase, the head groups in the
interior of the tubes are less exposed to the aqueous phase (limited amount
of water in the tubes). The packing of lipid head groups in a surface of

Bilayer   Hexagonal II

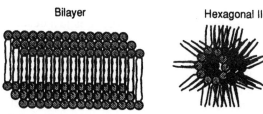

FIGURE 1   Schematic representation of the lamellar and H_{II} phase structures.

negative radius of curvature leads to a reduction in the hydration of the head groups, particularly with respect to hydration of the ester carbonyls of the lipid (Yeagle and Martin, 1976; Schmidt *et al.*, 1977). Therefore, the formation of the H_{II} phase can be seen as a response of the lipid system to a poor interaction of the bilayer surface with the aqueous phase (for a more complete discussion of this point, see Yeagle, 1987). Some of the lipids that promote membrane fusion also readily form H_{II} phases.

The cubic (Q) phase, or group of phases, is another nonlamellar structure that some phospholipids can form, as a side path on the way from lamellar to H_{II} (Lindblom and Rilfors, 1989). In other systems, Q can be induced by repeated cycling of the system above and below the transition temperature for the L_{α}–H_{II} phase transition. The cubic phase is a three-dimensional structure, resembling truncated H_{II} tubes connected at right angles. The symmetry of the cubic phase can be different depending on the components and conditions. This phase can be kinetically stable, so reversion to the lamellar phase can be very slow even when the system is incubated at temperatures well below the L_{α}–H_{II} phase transition temperature.

The concept of lipidic intramembranous particles was introduced by the observation of nonlamellar structures (in the form of "particles" in protein-free lipid systems) in freeze-fracture electron micrographs of lipid systems undergoing phase transitions from L_{α} to H_{II}. These structures were described variously as inverted micelles within a lipid bilayer (Verkleij *et al.*, 1979) or between two lipid bilayers (Hui *et al.*, 1981b). Combined freeze-fracture electron micrography, ^{31}P nuclear magnetic resonance (NMR), and X-ray diffraction studies revealed that these lipidic particles appeared in systems capable of undergoing the L_{α}–H_{II} phase transition at temperatures well below the L_{α}–H_{II} phase transition temperature (Hui *et al.*, 1981c). Thus, these nonlamellar structures could coexist with lipid bilayers.

Related to these structures is the interlamellar attachment (ILA) described by cryoelectron microscopy (Siegel *et al.*, 1989c). This structure contains connectivity between the opposing lipid monolayers of the two membranes and directly suggests a relationship to the cubic structure, to which the ILA was proposed to be a precursor. Similar structures previously were observed in freeze-fracture electron microscopy (Hui *et al.*, 1981a).

The formation of the nonlamellar structures just described results from some type of stress on the membrane structure. The existence of a balance between tendencies to form nonlamellar and lamellar structures has been termed "frustration" (Anderson *et al.*, 1988). The balance between lamellar and nonlamellar phases (such as H_{II}) is governed by several energy factors. One factor is the thermodynamics of the interaction of the membrane surface with the aqueous phase, as mentioned earlier. For example, in PE, the head groups are hydrogen bonded intermolecularly and lie approximately parallel to the bilayer surface (Yeagle *et al.*, 1976, 1977), which effectively neutralizes the head group dipole. Thus, this surface interacts poorly with water and PE will form the H_{II} phase readily (see preceding discussion).

One of the factors counteracting H_{II} phase formation is a problem in the packing of the lipids. Packing defects result from the stacking of the cylinders of the H_{II} phase. Energy is required for the hydrocarbon chains of the lipids to fill such voids between the cylinders of the H_{II} phase. Normal hydrocarbons can fill these voids and stabilize the H_{II} phase relative to the lamellar phase (Siegel *et al.*, 1989b; Turner and Gruner, 1992).

Another factor stabilizing the lamellar phase against transformation to nonlamellar configurations is protein. Some membrane proteins in biological membranes can stabilize bilayer structure. An example is rhodopsin. In the absence of rhodopsin, and in the presence of calcium, the lipids of the retinal rod outer segment disk membrane can form an H_{II} phase. However, in the native disk membrane in the presence of rhodopsin, the membrane is stable in the lamellar structure and has not been observed to form H_{II} phases under any conditions (deGrip *et al.*, 1979; Albert *et al.*, 1984; Mollevanger and deGrip, 1984). This observation is consistent with the requirement in cell biology to maintain stable lipid bilayers for cell viability.

Stress on the lipid system can alter the balance between factors such as those just described. Stress can come about in many ways. One form of stress is temperature. Increased temperature will increase the surface area per lipid in the membrane and, for those lipids that form a surface that interacts poorly with water, the system will be destabilized in favor of the H_{II} phase. Thus, increasing temperature will cause a phase change between the lamellar and the H_{II} phase.

Another form of stress is lipid composition. Membranes with higher levels of PE will form the H_{II} phase more readily than will systems low in PE (Hui *et al.*, 1981c). Likewise, membranes with higher levels of monogalactosyl diglycerides also will form the H_{II} phase more readily (Wieslander *et al.*, 1980). Membranes containing diacylglycerol, an intermediate in lipid metabolism, will form the H_{II} phase more readily than membranes of the (otherwise) same composition without diacylglycerol (Siegel *et al.*, 1989b).

A related issue is membrane asymmetry (Tullius *et al.*, 1989). For example, the inner leaflet of the erythrocyte membrane is enriched in PE compared with the outer leaflet. The other major lipid is phosphatidylserine

(PS), which could be phase-separated by calcium. Thus, the inner leaflet is more conducive to membrane fusion than the outer leaflet. If the stalk theory is correct (see subsequent discussion), membrane fusion initially requires the facing leaflets of lipids to form nonlamellar structures; thus, the lipid composition of one half of the membrane may be more important than that of the other half of the membrane in promoting membrane fusion.

The last form of stress to be discussed here is caused by protein. Some proteins can destabilize bilayers, in contrast to the example of rhodopsin given earlier. For example, signal peptides can be bilayer destabilizing in favor of nonlamellar structures (Killian *et al.*, 1990). In an observation directly related to membrane fusion, the "fusion peptide" of a viral fusion protein was found to destabilize the lamellar phase (Yeagle *et al.*, 1991). Thus, cell biology has proteins available that can disrupt the balance of forces controlling the lamellar–nonlamellar transitions that can be utilized selectively for regulation of vital biological membrane fusion events.

III. ROLE OF NONLAMELLAR INTERMEDIATES IN VESICLE FUSION

Several suggestions have been made for the possible role of nonlamellar structures in membrane fusion, based on the observations of nonlamellar assemblies that membrane lipids can form. The logic in these approaches to understanding fusion is that some transient disruption of bilayer structure apparently must occur for two membranes to fuse. Several of the nonlamellar structures reviewed in the previous section have been included in various schemes for membrane fusion.

Early suggestions were made that an H_{II} phase might be involved in fusion (Cullis and Hope, 1978). This suggestion was based on the observation that some compounds that promoted membrane fusion also promoted the formation of an H_{II} phase. However, the H_{II} phase is not a transient structure, but a stable lipid phase under appropriate conditions. Lipids in the H_{II} phase are in slow communication with lipids in the lamellar phase when the two phases coexist (Fenske and Cullis, 1992). The H_{II} phase involves complex and extensive membrane degradation. Further, close to the phase boundary, the kinetics of the L_α–H_{II} phase transition are very slow (Tate *et al.*, 1992). For these reasons, the H_{II} phase is not a likely candidate for the intermediate structure in the pathway of membrane fusion.

Correlations have been found between compounds that inhibit membrane fusion by incorporation into the lipid bilayer and the ability of the same compounds to increase the temperature of the L_α–H_{II} phase transition (Epand *et al.*, 1987). In this approach, nearness to the L_α–H_{II} phase boundary rather than formation of the H_{II} phase itself becomes important. This idea may be related to the concept of "frustration" advanced by Gruner, and suggests that a property of the lipid bilayer responds to the closeness of the L_α–H_{II} phase boundary while the system is lamellar. Although this idea does not lead to specific suggestions for the membrane fusion pathway, this approach is related to others described subsequently because the morphological flexibility of a membrane system, which is likely to be important to fusion, also is related to the nearness of the L_α–H_{II} phase boundary.

In another approach, Siegel used the inverted micelle intermediates (previously suggested to play a role in the L_α–H_{II} phase transition) to model the fusion pathway (Siegel, 1984). Figure 2 shows this structure schematically. These intermediates had, theoretically, the short lifetimes required to function as intermediates in a membrane fusion pathway. These intermediates also had the structural ability to mix the lipid compo-

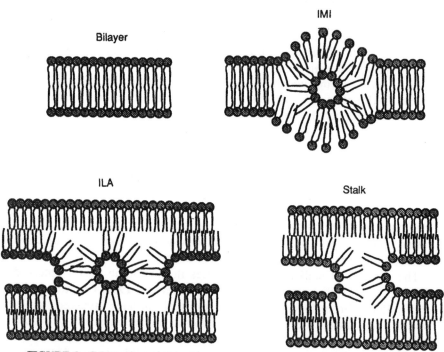

FIGURE 2 Schematic representation of nonlamellar structures discussed in the text. The open white areas within these structures represent packing defects, or voids, that can be filled only by perturbations of the lipid chains or by exogenous hydrophobic compounds.

nents of two opposing bilayers, as a prelude to the membrane fusion event. The next logical step was to identify whether such intermediates indeed played a role in membrane fusion.

Work on the fusion of large unilamellar vesicles (LUV) produced some interesting observations about the pathway of membrane fusion for membranes of N-methyldioleoyl phosphatidylethanolamine (N-methyl DOPE) and membranes consisting of mixtures of PE and PC. Initial rates of membrane fusion (as measured by contents mixing) were correlated directly with the appearance of nonlamellar phospholipid assemblies (Ellens et al., 1989). Although the system was mostly unilamellar, the initial rate of fusion was directly proportional to the percentage of the total lipids constituting the nonlamellar structures. The presence of the nonlamellar structures was detected by isotropic ^{31}P-NMR resonances (collapsed powder patterns) and by freeze-fracture electron microscopy. (X-Ray diffraction is not useful here because the nonlamellar structures are not found in a repeating lattice at low incidence.) Thus, this work suggested that the detected nonlamellar structures might be intermediates in the pathway of fusion, or might be formed utilizing an intermediate common to the membrane fusion pathway.

The latter suggestion is much more likely. The nonlamellar structures detected by ^{31}P-NMR could not be involved directly in fusion because they were not transient. The ^{31}P-NMR experiments took place over a long time interval, during which these structures were stable (Ellens et al., 1989). The kinetics of formation of the nonlamellar structures in N-methyl DOPE were reported to be slow relative to the initial rates of membrane fusion of the LUV (van Gorkom et al., 1992).

Specific structural suggestions for nonlamellar intermediates in the fusion pathway were made by Hui on the basis of freeze-fracture electron micrographs of a PC/PE system that underwent membrane fusion (Hui et al., 1981a). In this system, nonlamellar structures were observed that formed ILAs. These structures resembled stalks (see subsequent discussion) that provided connectivity between the two opposing leaflets of the fusing lipid bilayers. At some points, the fusion had gone to completion so fused bilayers rather than stalks were observed.

The concept of stalks as intermediates in membrane fusion received the attention of other investigators (Markin et al., 1984; Chernomordik et al., 1987). From theoretical calculations of energies involved, stalks were suggested as achievable intermediates in the fusion pathway. Figure 2 shows the structure of stalks schematically. Researchers suggested that very little activation energy was required to achieve this proposed intermediate state, and that the lifetime of such a structure could be sufficiently short to act as an intermediate in the fusion process. Stalks, therefore, were good candidates for the fusion pathway.

This approach has been elaborated on extensively by Siegel, who included energies involved in the packing voids that were not included in the previous calculations (Siegel, 1993). Whenever lipid surfaces exhibit a (relatively) short radius of curvature, a potential problem arises with packing the structure of that surface with the structure of the other lipid assemblies with which it must interact. This problem is apparent in the packing of the cylinders of the H_{II} phase, as discussed earlier. In the case of the stalk fusion intermediate, similar packing problems arise at the points of contact of the stalk with each of the bilayers from which it was formed (see Fig. 2). However, from an energy standpoint, these latter calculations again suggest that stalks would operate well as intermediates in membrane fusion.

From considerations such as these, considering the following hypothetical scheme that responds to the theoretical and experimental results already discussed becomes useful. Consider these fusion processes (as well as the development of other phases) to begin with a lamellar phase in which structural fluctuations can occur that permit the system to sample lipid assemblies with nonlamellar arrangements of the lipids (as least for a minority of the membrane lipids).

Step 1 represents the sampling of a nonlamellar intermediate structure I_α by a minority of the membrane lipid, which quickly relaxes to the bilayer structure L. Thus, under normal stable bilayer conditions, the forward rate constant for Step 1, k_{+1}, may be appreciable but the rate constant for the reverse step, k_{-1}, is large so the intermediate I_α does not accumulate. However, under appropriate conditions, the forward rate constant for Step 2, k_{+2}, can become appreciable and the process can go forward to a nonlamellar structure I_S. As will be discussed shortly, I_S is a structure that is stable under some conditions and can be observed, and essentially represents fusion, although not a completed fusion product.

Also in this scheme are pathways that may lead to H_{II} and to cubic phases (Q). For example, I_S has been observed before the formation of H_{II} phase, possibly as a precursor in some cases (Hui *et al.*, 1981c). However, the scheme also allows for the formation of H_{II} phase without the formation of I_S. Some L_α–H_{II} transitions appear as simple two state

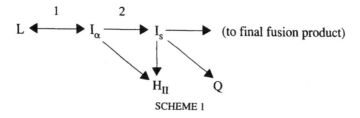

SCHEME 1

transitions with no evidence for (relatively) stable intermediates (Cullis and de Kruijff, 1978a). Q also may be a product that utilizes I_S as a precursor. Some systems clearly show the formation of I_S as a precursor to Q. In other systems that perhaps do not form I_S readily, extensive cycling of the system through the I_α–H_{II} transition temperature has been observed to induce the formation of Q, perhaps because of the presence of only a small proportion of I_S that is required as a precursor and because the temperature cycling sequentially produces more I_S as a function of time, which can lead to the formation of Q.

Information about the structure of I_S has been obtained through several lines of experimentation. I_S is manifest in ^{31}P-NMR spectra as an isotropic resonance (Hui *et al.*, 1981); Gagne *et al.*, 1985; Ellens *et al.*, 1989). In freeze-fracture electron micrographs, I_S appears as a "lipid particle" (Burnell *et al.*, 1980; Hui *et al.*, 1980; de Kruijff *et al.*, 1980). I_S is most likely an ILA, as seen in cross-fractures in freeze-fracture electron micrographs (Hui *et al.*, 1981b), although other models have been suggested (Verkleij *et al.*, 1979). Cryo-EM techniques have visualized the ILA more directly without fixation artifacts (Siegel *et al.*, 1989c).

I_S has been suggested to be an intermediate in membrane fusion (Verkleij *et al.*, 1980). The percentage of the membrane lipid in I_S was observed to be correlated directly with the initial rate of fusion of some phospholipid LUVs (Ellens *et al.*, 1989). Freeze-fracture electron micrographic evidence directly implicated I_S in membrane fusion (Hui *et al.*, 1981a). Diacylglycerols (Siegel *et al.*, 1989a), retinal, and retinol (Boesze-Battaglia *et al.*, 1992) all lower the energy barriers to the formation of I_S and enhance the initial rates of fusion of phospholipid vesicles. However, I_S is placed in Scheme 1 more as a product than as an intermediate for several reasons. I_S is a stable long-lived structure that can be observed with ^{31}P-NMR and freeze-fracture electron microscopy with no special efforts to trap short-lived intermediates. The lipids in I_S are in slow exchange with the lipids in the surrounding bilayer, indicating that interconversion between the ILA and the bilayer is relatively slow (on the seconds time scale; Fenske and Cullis, 1992). Development of I_S can be slow relative to the rates of membrane fusion, near the temperature at which it forms in N-methyl DOPE (van Gorkom *et al.*, 1992). Further, I_S is not found in all lipid vesicle systems that undergo membrane fusion. For these reasons, I_S is not likely to be a universal intermediate in membrane fusion reactions. Rather, it appears to be a product of sorts that represents the completion of most of the fusion event, stopping short of the expansion of the "fusion pore" (see subsequent discussion). This structure is stable in some fusion systems and is, therefore, observed, but in other fusion systems I_S apparently is not stable and is not detected by techniques that require substantial time for visualization.

Attention now is turned to a consideration of I_α, which plays the role of true intermediate in Scheme 1. I_α may be one intermediate or could represent several intermediates with several intervening steps. At this point, fusion simply requires at least one such intermediate. For the purposes of this discussion, the assumption will be made that this scheme is among the most simple and I_α represents one intermediate. This intermediate should have a short lifetime and should provide a pathway for the membrane fusion mechanism to proceed.

Evidence concerning the structure of I_α is, at present, indirect. The only experimental evidence to date is from studies of the inhibition of membrane fusion by certain small hydrophobic peptides. These antiviral peptides (Miller et al., 1968; Nicolaides et al., 1968; Richardson et al., 1980) were shown to inhibit viral infection by inhibiting viral fusion (Kelsey et al., 1990, 1991). The peptide ZfFG is a particular example of this class of peptides. More recently, information on the mechanism of action of these peptides has been obtained. These fusion-inhibitory peptides inhibit the formation of phospholipid assemblies with highly curved surfaces (Yeagle et al., 1992). In particular, surfaces with a negative small radius of curvature appear to be destabilized, including H_{II} phase (Epand, 1986) and the inside surface of small sonicated vesicles (Yeagle et al., 1992). Perhaps most interestingly, these antiviral peptides destabilize the highly curved surfaces of I_S. Although the formation of I_S is not inhibited directly, the resulting structure is forced into a larger radius of curvature. Therefore, the key effect of these inhibitors of membrane fusion is inhibiting the formation of highly curved (negative radius of curvature) lipid surfaces. These data suggest that I_α contains significant regions of negative radius of curvature in its structure. Data on inhibition of fusion of N-methyl DOPE LUVs by lyso-PC (Yeagle and Flanagan, 1991) also support the concept of the involvement of I_α (with negative radius of curvature) in membrane fusion.

The best current model for the structure of I_α is derived from the theoretical analysis of Siegel (1993). As described earlier, Siegel's approach builds on the stalk model. In this model, membrane fusion proceeds by a mixing of the outer monolayers of the two membranes to fuse through a stalk, formed of membrane lipids, that bridges the two membranes. This stalk has the features of low energy of formation and short lifetime. Further, the stalk fits the experimental data that suggest that I_α should have some surface(s) with (a) negative radius of curvature.

Therefore, at this time, the fusion mechanism modeled by Scheme 1 can be suggested to proceed structurally by the following pathway. Initially, stress on the membrane enhances the formation of transient nonlamellar structures such as stalks (although still likely to be relatively rare events), after the close approach of two membranes. These structural excursions to nonlamellar forms permit the mixing of the two outer mono-layers of the closely apposed membranes. The stalks then develop to form I_S or ILA, which represents the initial fusion product. From this structure, a pore can develop that allows mixing of the contents of the two compart-ments bounded by the two membranes in question. In some cases, I_S is sufficiently unstable that, kinetically, only a transition directly to a state in which a pore has formed would be observed. Finally, the pore can open and close in a manner similar to the behavior of ion channels (Ober-hauser *et al.*, 1992), until it either grows so a macroscopically observable fusion product results or closes so the fusion event fails.

Several studies have suggested that some of the same factors discussed in this section also may be important in divalent cation-induced fusion (Walter *et al.*, 1993). Normal hydrocarbons and diacylglycerols have been found to enhance the initial rate of fusion of PS (bovine brain) vesicles. This enhancement is analogous to the enhancement of fusion of *N*-methyl DOPE LUVs reported previously (Siegel *et al.*, 1989a). In the latter case, nonlamellar lipid intermediates are on the pathway of membrane fusion, as discussed earlier. Nonlamellar lipid intermediates also may be involved in divalent cation-induced fusion, thus expanding the applicability of Scheme 1 to membrane fusion between lipid vesicles.

IV. ROLE OF NONLAMELLAR INTERMEDIATES IN VIRAL FUSION

The discussion now focuses specifically on the role of nonlamellar inter-mediates in viral fusion (neutral pH fusion). Fusion of enveloped viruses is promoted by the viral fusion proteins. The structure of the viral fusion proteins, in general, contains a linear sequence of highly hydrophobic amino acids that is not involved in anchoring the fusion protein to the viral envelope membrane. (This anchoring is achieved by another hydrophobic sequence in the protein.) This hydrophobic sequence is required for the fusion event. Some homology among viral fusion proteins is observed in the sequence of this portion of the protein (Gallaher, 1987). The amino acid sequence required for membrane fusion is called the "fusion peptide" of the viral fusion protein. Figure 3 is a schematic representation of the fusion protein and its fusion peptide. In the case of Sendai virus, this fusion peptide (on the F protein of the Sendai virus) has been suggested to contact the target membrane bilayer during the fusion event (Novick and Hoekstra, 1988). Some contact with the target membrane also was suggested for the fusion peptide of the fusion protein (HA) of the influenza virus (Brunner, 1989).

Not only are the fusion peptides required for membrane fusion by the virion, but they also promote fusion of phospholipid vesicles independent

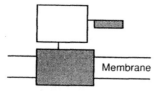

Fusion protein

FIGURE 3   Schematic representation of a fusion protein. One of the hydrophobic domains (shaded) anchors the fusion protein to the viral membrane. The other hydrophobic
domain (shaded) is the fusion peptide that is most likely to be exposed during membrane
fusion.

of the remainder of the fusion protein. The fusion peptides of measles
(Yeagle *et al.*, 1991), influenza (Wharton *et al.*, 1988), human immunodeficiency virus (HIV; Rafalski *et al.*, 1990), and simian immunodeficiency
virus (SIV; Martin *et al.*, 1991) all promote membrane fusion. This observation has encouraged a model of viral entry into cells in which the fusion
protein plays a role of insertion and destabilization of the target membrane
to facilitate the fusion of the viral envelope (membrane) and the membrane
of the target cell (Hoekstra and Kok, 1989).

Data suggest a mechanism for such promotion of membrane fusion by
these fusion peptides. The measles fusion peptide destabilizes lamellar
structures. In particular, in *N*-methyl DOPE, this fusion peptide stabilizes
nonlamellar structures with a short radius of curvature relative to the
lamellar phase (Yeagle *et al.*, 1991). Therefore, in the model of Scheme
1, the viral fusion peptide may facilitate membrane fusion by decreasing
the activation energy required for the formation of I_α, thus enhancing the
rate of the fusion reaction. A mechanism for this destabilization has been
suggested. Because of the distribution of hydrophobic and hydrophilic
amino acids on the surface of a helix constructed from the fusion peptides,
the fusion peptide has been proposed to insert into the bilayer at an oblique
angle (Voneche *et al.*, 1992). Such an angle of insertion might affect the
packing of the lipids in the membrane and could lead to bilayer destabilization, in contrast to the transmembrane segments of transmembrane proteins, which are predicted to insert perpendicular to the bilayer surface,
or to certain amphipathic peptides, which may orient parallel to the bilayer
surface and are not bilayer disrupting (Brasseru, 1991).

Interestingly, a peptide much shorter than those just described has been
found to be bilayer destabilizing and fusogenic (R. M. Epand and P. L.
Yeagle, unpublished observations, 1993). The peptide fFGOBz is similar to
the fusion inhibitory peptide ZfFG, except that the hydrophobic blocking
group has been removed from the amino terminus to the carboxy terminus.
In contrast to ZfFG, which stabilizes the lamellar phase and inhibits fusion,

fFGOBz destabilizes the lamellar phase (decreases the transition temperature for the L_α–H_{II} phase transition) and promotes fusion.

The other line of evidence for the role of nonlamellar intermediates in the fusion pathway of enveloped viruses is from inhibition studies. The antiviral peptides referred to earlier, which inhibit the fusion of N-methyl DOPE LUVs, also inhibit the fusion of Sendai virus with vesicle and cellular targets. This inhibition appears to occur by interference with the formation of highly curved surfaces, as described earlier. Therefore, the fusion pathway for at least some enveloped viruses is likely also to involve I_α structures with highly curved surfaces (likely surfaces with a negative radius of curvature). Some investigators have reported freeze-fracture electron micrograph data supporting the possible presence of nonlamellar structures at the fusion sites for influenza virus (Knoll *et al.*, 1988). This conclusion is supported by the observations of inhibition of viral fusion by lyso-PC (Yeagle and Flanagan, 1991; Vogel *et al.*, 1993). Incorporation of lyso-PC into membranes would be expected to destabilize lipid assemblies with a negative radius of curvature. Thus, inhibition of viral fusion by lyso-PC is consistent with the proposal of an intermediate such as I_α, with surfaces exhibiting a negative radius of curvature, in the fusion pathway.

Finally, note that Sendai virus can fuse with target membranes exhibiting I_S (and perhaps I_α) structures in the absence of a receptor (Kelsey *et al.*, 1990). In this case, the target membranes were LUVs of N-methyl DOPE and the initial rate of fusion of the Sendai was directly proportional to the percentage of the lipids in the target membrane in the I_S form. In the absence of a receptor, fusion normally does not occur. Thus, these data indicate that a nonlamellar structure could take the place of a receptor in viral fusion.

The possible role of nonlamellar intermediates in other biological membrane fusions has not been investigated to date. These events include the fusions involved in exocytosis, endocytosis, and intracellular vesicular transport. However, the observation of inhibition of exocytosis (Vogel *et al.*, 1993) by lyso-PC is interesting in this regard.

References

Albert, A. D., Sen, A., and Yeagle, P. L. (1984). The effect of calcium on the bilayer stability of lipids from bovine rod outer segment disk membranes. *Biochim. Biophys. Acta* **771**, 28–34.

Anderson, D. M., Gruner, S. M., and Leibler, S. (1988). Geometrical aspects of the frustration in the cubic phases of lyotropic liquid crystals. *Proc. Natl. Acad. Sci. U.S.A.* **85**, 5364–5368.

Bentz, J., and Ellens, H. (1988). Membrane fusion: Kinetics and mechanisms. *Colloids Surfaces* **30**, 65–112.

Boesze-Battaglia, K., Fliesler, S. J., Li, J., Young, J. E., and Yeagle, P. L. (1992). Retinal and retinol promote membrane fusion. *Biochim. Biophys. Acta* **1111**, 256–262.

Brasseru, R. (1991). Differentiation of lipid-associating helices by use of 3-dimensional molecular hydrophobicity potential calculations. *J. Biol. Chem.* **266**, 16120–16127.

Brunner, J. (1989). Testing topological models for the membrane penetration of the fusion peptide of influenza virus hemagluttinin. *FEBS Lett.* **257**, 369–372.

Burnell, E., van Aphen, L., Verkleij, A. J., and de Kruijff, B. (1980). ^{31}P-NMR and freeze fracture electron microscopy studies on *E. coli*. *Biochim Biophys. Acta* **863**, 213.

Chernomordik, L. V., Melikyan, G. B., and Chizmadzhev, Y. A. (1987). Biomembrane fusion: A new concept derived from model studies using two interacting planar lipid bilayers. *Biochim. Biophys. Acta* **906**, 309–352.

Cullis, P. R., and de Kruijff, B. (1978a). Lipid polymorphism and the functional roles of lipids in biological membranes. *Biochim. Biophys. Acta* **507**, 207–218.

Cullis, P. R., and de Kruijff, B. (1978b). The polymorphic phase behaviour of phosphatidylethanolamines of natural and synthetic origin. *Biochim. Biophys. Acta* **513**, 31–42.

Cullis, P. R., and Hope, M. J. (1978). Effects of fusogenic agent on membrane structure of erythrocyte ghosts and the mechanism of membrane fusion. *Nature (London)* **271**, 672–674.

de Grip, W. J., Drenthe, E. H. S., van Echteld, C. J. A., de Kruijff, B., and Verkleij, A. J. (1979). A possible role of rhodopsin in maintaining bilayer structure in the photoreceptor membrane. *Biochim. Biophys. Acta* **558**, 330.

de Kruijff, B., Cullis, P. R., and Verkleij, A. J. (1980). Non-bilayer lipid structure in model and biological membranes. *Trends Biochem. Sci.* March, 79–81.

Ellens, H., Siegel, D. P., Alford, D., Yeagle, P. L., Boni, L., Lis, L. J., Quinn, P. J., and Bentz, J. (1989). Membrane fusion and inverted phases. *Biochemistry* **28**, 3692–3703.

Epand, R. M. (1986). Virus replication inhibitory peptide inhibits the conversion of phospholipid bilayers to the hexagonal phase. *Biosci. Rep.* **6**, 647–653.

Epand, R. M., Epand, R. F., and McKenzie, R. C. (1987). Effects of viral chemotherapeutic agents on membrane properties—Studies of cyclosporin A, benzyloxycarbonyl-D-phe-L-phe-gly and amantadine. *J. Biol. Chem.* **262**, 1526–1529.

Fenske, D. B., and Cullis, P. R. (1992). Chemical exchange between lamellar and nonlamellar lipid phases. A one- and two-dimensional ^{31}P-NMR study, *Biochim. Biophys. Acta* **1108**, 201–209.

Gagne, J., Stamatatos, L., Diacovo, T., Hui, S. W., Yeagle, P. L., and Silvius, J. (1985). Physical properties and surface interactions of bilayer membranes containing N-methylated phosphatidylethanolamines. *Biochemistry* **24**, 4400–4408.

Gallaher, W. R. (1987). Detection of a fusion peptide sequence in the transmembrane protein of human immunodeficiency virus. *Cell* **50**, 327–328.

Gruner, S. M. (1985). Intrinsic curvature hypothesis for biomembrane lipid composition: A role for nonbilayer lipids, *Proc. Natl. Acad. Sci. U.S.A.* **82**, 3665–3669.

Hoekstra, D., and Kok, J. W. (1989). Entry mechanisms of enveloped viruses. Implications for fusion of intracellular membranes. *Biosci. Rep.* **9**, 273–305.

Hui, S. W., Stewart, T. P., and Yeagle, P. L. (1980). Temperature dependent morphological and phase behavior of sphingomyelin. *Biochim. Biophys. Acta* **601**, 271–281.

Hui, S. W., Stewart, T. P., Boni, L. T., and Yeagle, P. L. (1981a). Membrane fusion through a point defect in the bilayer. *Science* **212**, 921–923.

Hui, S. W., Stewart, T. P., Verkleij, A. J., and de Kruijff, B. (1981b). "Lipidic particles" are intermembrane attachment sites. *Nature (London)* **290**, 427–428.

Hui, S. W., Stewart, T. P., Yeagle, P. L., and Albert, A. D. (1981c). Bilayer to nonbilayer transition in mixtures of phosphatidylethanolamine and phosphatidylcholine: Implications for membrane properties. *Arch. Biochem. Biophys.* **207**, 227–240.

Kelsey, D. R., Flanagan, T. D., Young, J., and Yeagle, P. L. (1990). Peptide inhibitors of enveloped virus infection inhibit phospholipid vesicle fusion and Sendai virus with phospholipid vesicles. *J. Biol. Chem.* **265**, 12178–12183.

Kelsey, D. R., Flanagan, T. D., Young, J., and Yeagle, P. L. (1991). Inhibition of Sendai virus fusion with phospholipid vesicles and human erythrocyte membranes by hydrophobic peptides. *Virology* **182**, 690–702.

Killian, J. A., de Jong, A. M. P., Bijvelt, J., Verkleij, A. J., and de Kruijff, B. (1990). Induction of non-bilayer structures by functional signal peptides. *EMBO J.* **9**, 815–819.

Knoll, G., Burger, K. N., Bron, R., van Meer, G., and Verkleij, A. J. (1988). Fusion of liposomes with the plasma membrane of epithelial cells: Fate of incorporated lipids as followed by freeze fracture and autoradiography of plastic sections. *J. Cell Biol.* **107**, 2511–2521.

Lindblom, G., and Rilfors, L. (1989). Cubic phases and isotropic structures formed by membrane lipids—Possible biological relevance. *Biochim. Biophys. Acta* **988**, 221–256.

Luzzati, V., and Husson, F. (1962). The structure of the liquid-crystalline phases of lipid–water systems. *J. Cell Biol.* **12**, 207–210.

Markin, V. S., Kozolov, M. M., and Borovjagin, V. L. (1984). On the theory of membrane fusion: The stalk mechanism. *Gen. Physiol. Biophys.* **5**, 361.

Martin, I., Defrise-Quertain, F., Mandieau, V., Nielsen, N. M., Saermark, T., and Burny, A. (1991). Fusogenic activity of SIV peptides located in the GP32 amino-terminal domain. *Biochem. Biophys. Res. Commun.* **175**, 872–879.

Miller, F. A., Dixon, G. J., Arnett, G., Dice, J. R., Rightsel, W. A., Schabel, J. F. M., and McLean, J. I. W. (1968). Antiviral activity of carbobenzoxy di- and tripeptides on measles virus. *Appl. Microbiol.* **16**, 1489–1496.

Mollevanger, L. C. P. J., and de Grip, W. J. (1984). Phase behavior of isolated photoreceptor membrane lipids is modulated by bivalent cations. *FEBS Lett.* **169**, 256–260.

Nicolaides, E., de Wald, H., Westland, R., Lipnik, M., and Posler, J. (1968). Potential antiviral agents. Carbobenzoxy di- and tripeptides active against measles and herpes viruses. *J. Med. Chem.* **11**, 74–79.

Novick, S. L., and Hoekstra, D. (1988). Membrane penetration of Sendai virus glycoproteins during the early stages of fusion with liposomes as determined by hydrophobic photoaffinity labeling. *Proc. Natl. Acad. Sci. U.S.A.* **85**, 7433–7437.

Oberhauser, A. F., Monck, J. R., and Fernandez, J. M. (1992). Events leading to the opening and closing of the exocytoic fusion pore have markedly different temperature dependencies. Kinetic analysis of single fusion events in patch-clamped mouse mast cells. *Biophys. J.* **61**, 800–809.

Papahadjopoulos, D., and Poste, G. (1975). Calcium-induced phase separation and fusion in phospholipid membranes. *Biophys. J.* **15**, 945–948.

Papahadjopoulos, D., Vail, W. J., Newton, C., Nir, S., Jacobson, K., Poste, G., and Lazo, R. (1977). Studies on membrane fusion. III. The role of calcium-induced phase changes. *Biochim. Biophys. Acta* **465**, 579–598.

Rafalski, M., Lear, D., and DeGrado, W. F. (1990). Phospholipid interactions of synthetic peptides representing the N-terminus of HIV gp41. *Biochemistry* **29**, 7917–7922.

Rand, R. P., and Parsegian, V. A. (1989). Hydration forces between phospholipid bilayers. *Biochim. Biophys. Acta* **988**, 351–376.

Reiss-Husson, F. (1967). Structure des phases liquide-cristallines de differents phospholipides, monoglycerides, sphingolipides, anhydres ou en presence d'eau. *J. Mol. Biol.* **25**, 363–382.

Richardson, C. D., Scheid, A., and Choppin, P. W. (1980). Specific inhibition of paramyxovirus and myxovirus replication by oligopeptides with amino acid sequences similar to those at the N-termini of the F-1 or HA-2 viral polypeptides. *Virology* **105**, 205–222.

Schmidt, C. F., Barenholz, Y., Huang, C., and Thompson, T. E. (1977). Phosphatidylcholine C-13 labelled carbonyls as a probe of bilayer structure. *Biochemistry* **16**, 3948–3954.

Sen, A., Williams, W. P., Brain, A. P. R., Dickens, M. J., and Quinn, P. J. (1981). Formation of inverted micelles in dispersions of mixed galactolipids. *Nature (London)* **293**, 488–490.

Siegel, D. P. (1984). Inverted micellar structures in bilayer membranes. *Biophys. J.* **45**, 399–420.

Siegel, D. P. (1993). Modelling protein-induced fusion mechanisms: Insights from the relative stability of lipidic structures. *In* "Viral Fusion Mechanisms," (J. Bentz, ed.), pp. 475–511. CRC Press, Boca Raton, Florida.

Siegel, D. P., Banschbach, J., Alford, D., Ellens, H., Lis, L., Quinn, P. J., Yeagle, P. L., and Bentz, J. (1989a). Physiological levels of diacylglycerols in phospholipid membranes induce membrane fusion and stabilize inverted phases. *Biochemistry* **28**, 3703–3709.

Siegel, D. P., Banschbach, J., and Yeagle, P. L. (1989b). Stabilization of H-II phases by low levels of diglycerides and alkanes: An NMR, DSC, and X-ray diffraction study. *Biochemistry* **28**, 5010–5018.

Siegel, D. P., Burns, J. L., Chestnut, M. H., and Talmon, Y. (1989c). Intermediates in membrane fusion and bilayer/non-bilayer transitions imaged by time-resolved cryo-transmission electron microscopy. *Biophys. J.* **56**, 161–169.

Tate, M. W., and Gruner, S. M. (1989). Temperature dependence of the structural dimensions of the inverted hexagonal phase of PE-containing membranes. *Biochemistry* **28**, 4245–4253.

Tate, M. W., Shyamsunder, E., Gruner, S. M., and D'Amico, K. L. (1992). Kinetics of the lamellar-inverse hexagonal phase transition determined by time-resolved X-ray diffraction. *Biochemistry* **31**, 1081–1092.

Tullius, E. K., Williamson, P., and Schlegel, R. A. (1989). Effect of transbilayer phospholipid distribution on erythrocyte fusion. *Biosci. Rep.* **9**, 623–633.

Turner, D. C., and Gruner, S. M. (1992). X-ray diffraction reconstruction of the inverted hexagonal (II) phase in lipid-water systems. *Biochemistry* **31**, 1340–1355.

van Gorkom, L. C. M., Nie, S.-Q., and Epand, R. M. (1992). Hydrophobic lipid additives affect membrane stability and phase behavior of *N*-methyl DOPE. *Biochemistry* **31**, 671–677.

Verkleij, A. J., Mombers, C., Leunissen-Bijvelt, J., and Ververgaert, P. H. J. T. (1979). Lipidic intramembraneous particles. *Nature (London)* **279**, 162–163.

Verkleij, A. J., von Echteld, C. J. A., Gerritsen, W. J., Cullis, P. R., and de Kruijff, B. (1980). The lipidic particle as an intermediate structure in membrane fusion processes and bilayer to hexagonal H II transitions, *Biochim. Biophys. Acta* **600**, 620–624.

Vogel, S. S., Leikina, E. A., and Chernomordik, L. V. (1993). Pre-fusion intermediates of both pH and Ca-triggered biological fusion can be reversibly arrested with lysolipids. *Biophys. J.* **64**, A186.

Voneche, V., Portetelle, D., Kettmann, R., Willems, L., Limbach, K., Paoletti, E., Ruysschaert, J. M., Burny, A., and Brasseur, R. (1992). Fusogenic segments of bovine leukemia virus and simiain immunodeficiency virus are interchangeable and mediate fusion by means of oblique insertion in the lipid bilayer of their target cells. *Proc. Natl. Acad. Sci. U.S.A.* **89**, 3810–3814.

Walter, A., Yeagle, P. L., and Siegel, D. P. (1993). How do hexadecane and diglyceride increase divalent cation-induced lipid mixing rates in PS LUV?, *Biophys. J.* **64**, A187.

Wharton, S. A., Martin, S. R., Ruigrok, R. W. H., Skehel, J. J., and Wiley, D. C. (1988). Membrane fusion by peptide analogues of influenza virus haemagglutinin, *J. Gen. Virol.* **69**, 1847–1857.

Wieslander, A., Christiansson, A. L., Rilfors, L., and Lindblom, G. (1980). Lipid bilayer stability in membranes. Regulation of lipid composition in Acholeplasma laidlawii as governed by molecular shape. *Biochemistry* **19**, 3650–3655.

Yeagle, P. L. (1987). "The Membranes of Cells." Academic Press, Orlando, Florida.

Yeagle, P. L., and Flanagan, T. (1991). Inhibition of viral membrane fusion by amphipathic compounds. *Biophys. J.* **59**, 132a.

Yeagle, P. L., and Martin, R. B. (1976). Hydrogen-bonding of the ester carbonyls in phosphatidylcholine bilayers. *Biochem. Biophys. Res. Commun.* **69**, 775–780.

Yeagle, P. L., and Sen, A. (1986). Hydration and the lamellar to hexagonal (II) phase transition of phosphatidylethanolamine. *Biochemistry* **25**, 7518–7522.

Yeagle, P. L., Hutton, W. C., Huang, C., and Martin, R. B. (1976). Structure in the polar region of phospholipid bilayers. *Biochemistry* **15**, 2121–2124.

Yeagle, P. L., Hutton, W. C., Huang, C., and Martin, R. B. (1977). Phospholipid headgroup conformations; Intermolecular interactions and cholesterol effects, *Biochemistry* **16**, 4344–4349.

Yeagle, P. L., Epand, R. M., Richardson, C. D., and T. D. Flanagan (1991). Effects of the "fusion peptide" from measles virus on the structure of N-methyl dioleoylphosphatidylethanolalmine membranes and their fusion with Sendai virus. *Biochim. Biophys. Acta* **1065**, 49–53.

Yeagle, P. L., Young, J., Hui, S. W., and Epand, R. M. (1992). On the mechanism of inhibition of viral and vesicle membrane fusion by carbobenzoxy-D-phenylalanyl-L-phenylalanyl-glycine. *Biochemistry* **31**, 3177–3183.

PART II

Lipid–Protein Interactions

CHAPTER 9

Intracellular Phospholipid Transfer Proteins

Bernadette C. Ossendorp, Gerry T. Snoek, and Karel W. A. Wirtz
Centre for Biomembranes and Lipid Enzymology. University of Utrecht. 3584 CH
Utrecht, The Netherlands

I. INTRODUCTION

Membranes are essential to life because they separate cells from their environment and compartmentalize specific functions within the cell. One of the major components of cellular membranes is phospholipid. Phospholipids are synthesized predominantly in the endoplasmic reticulum (ER) from where they are transported to other subcellular membranes including the plasma membrane. Transfer of phospholipids may occur by several mechanisms among which are bulk lipid flow via membrane vesicles, lateral diffusion of lipids within the plane of the membrane through contact sites, spontaneous redistribution between membranes by lipid monomer equilibration, and redistribution by transfer proteins. The first evidence for the occurrence of intracellular lipid transfer proteins was presented by Wirtz and Zilversmit (1968), who observed that, *in vitro*, the membrane-free cytosol from rat liver stimulated the redistribution of radiolabeled phospholipids between mitochondria and microsomes. Since then, phospholipid transfer proteins have been detected in all mammalian tissues tested, as well as in plants, in yeast, and in other microorganisms. From these sources, several transfer proteins with different phospholipid specificity have been purified and characterized. For general reviews, refer to Kader *et al.* (1982), Helmkamp (1986), and Wirtz (1982,1991).

In mammalian tissues, at least three types of phospholipid transfer protein are present. One protein is highly specific for phosphatidylcholine (PC) and is designated PC transfer protein (PC-TP) (for review, see Wirtz *et al.*, 1986a). The second protein has a distinct preference for phosphatidylinositol (PI). However, this PI transfer protein (PI-TP) also is able to transfer PC (for review, see Helmkamp, 1990). The third protein mediates the transfer of a great variety of lipids, including glycolipids and cholesterol, and is designated nonspecific lipid transfer protein (nsL-TP). This protein is identical to sterol carrier protein 2 (for reviews, see Scallen *et al.*, 1985; Van Amerongen et al., 1985). Yeast contains a variety of lipid transfer proteins of which PI-TP is the best characterized (for review, see Paltauf and Daum, 1990). Plants contain lipid transfer proteins (LTP) that, in terms of specificity, strongly resemble mammalian nsL-TP (for review, see Kader, 1990). The strongly conserved primary structure of the mammalian LTPs and their ubiquity in nature are strong indications for an important physiological role for these proteins. At present, however, these proteins are much better known for their lipid transfer properties *in vitro* than for their functional properties *in vivo*.

During the last few years, considerable progress has been made in the study of nonspecific lipid transfer proteins from mammals (nsL-TP) and plants (LTP) as well as in our knowledge about PI-TP from mammalian sources and yeast. In this chapter, we discuss only these proteins and, where appropriate, make reference to other LTPs.

II. NONSPECIFIC LIPID TRANSFER PROTEINS

A. Purification and Biochemical Characterization

1. Mammalian nsL-TP

During the initial attempts to purify PC-TP from bovine liver, researchers soon determined that a phosphatidylethanolamine (PE) transfer activity also was present in the cytosolic fraction of this tissue. These PE and PC transfer activities could be separated easily by anion exchange chromatography (Wirtz *et al.*, 1972). Subsequently, the protein responsible for PE transfer activity was purified from rat (Bloj and Zilversmit, 1977) and bovine (Crain and Zilversmit, 1980a) liver. However, instead of being specific to or at least having a preference for PE, this protein (nsL-TP) was shown to transfer a great variety of lipids between membranes, including phospholipids, glycolipids, and cholesterol. In the meantime, extensive studies were carried out on proteins that nonenzymatically stimulated microsomal cholesterol biosynthesis. As a result, researchers found that rat liver cytosol contained two proteins required for the microsomal conversion of squalene to cholesterol (Scallen *et al.*, 1974). One of these proteins—sterol carrier protein 2 (SCP_2)—specifically activated the conversion of lanosterol to cholesterol. When, after many years of strenuous effort, SCP_2 from rat liver finally yielded to purification, this protein surprisingly appeared to be very similar to nsL-TP (Noland *et al.*, 1980). That these proteins were, indeed, identical was proven unambiguously by determining the amino acid sequences of bovine liver nsL-TP (Westerman and Wirtz, 1985) and rat liver SCP_2 (Pastuszyn *et al.*, 1987). Later, with the availability of cDNA clones, the primary structures of human and mouse nsL-TP also became known (Moncecchi *et al.*, 1991; Yamamoto *et al.*, 1991). In support of the importance of these proteins, sequence identity among the four species is more than 90% (see Fig. 1). Through cDNA analysis, researchers learned that nsL-TP is synthesized as a 15-kDa precursor with a leader sequence of 20 amino acid residues (Billheimer *et al.*, 1990; Ossendorp *et al.*, 1990; Moncecchi *et al.*, 1991; Yamamoto *et al.*, 1991). As shown in Fig. 1, this leader sequence is highly conserved.

A 1300- to 1500-fold purification is required to obtain a homogeneous preparation of nsL-TP from rat liver (Noland *et al.*, 1980; Poorthuis *et al.*, 1981). Glycerol (10%, v/v) and β-mercaptoethanol (5 mM) are used throughout the purification procedure to prevent inactivation of the protein. By protecting the single cysteine residue (Cys 71), β-mercaptoethanol may prevent the dimerization of nsL-TP (M_r of 14 kDa; pI of 8.6–9.6; Table I). This cysteine residue seems important since N-ethylmaleimide, p-hydroxymercurobenzoate, mercurichloride, and mersalyl inactivated rat and bovine nsL-TP (Poorthuis *et al.*, 1981; Van Amerongen *et al.*, 1985; Megli *et al.*, 1986). From the perspective of regulation, nsL-TP has two potential phosphorylation sites for protein kinase C; Ser 65 and Thr 85 (A. Van Amerongen and B. C. Ossendorp, unpublished results). In agreement with this observation, phosphorylation of nsL-TP by protein kinase C *in vitro* has been reported (Steinschneider *et al.*, 1989).

Some indications are that nsL-TPs with properties different from the ones just described do exist. From rat hepatoma 27, a form of nsL-TP

```
        -20                                           1
rat     M G F P E A A S S F R T H Q I S A A P T S S A G D G F K A N L I F
bovine  - - - - - - - - - - - - - - - - - . . S V . . . . . . . . V .
human   . . . . . . . . . . . . . . . E . V . . . . . S . . . . . . V .
mouse   . . . . . . . . . . . . . . . V . . . . . . . . . . . . . . V .

              20                                          40
rat     K E I E K K L E E E G E E F V K K I G G I F A F K V K D G P G G K
bovine  . . . . . . . D . . . Q . . . . . . . . . . . . . . . . . . .
human   . . . . . . . . . . . Q . . . . . . . . . . . . . . . . . . .
mouse   . . . . . . . . . . . Q . . . . . . . . . . . . . . . . . . .

                              60
rat     E A T W V V D V K N G K G S V L P D S D K K A D C T I T M A D S D
bovine  . . . . . . . . . . . . . . . . N . . . . . . . . . . . . . . .
human   . D . . . . . . . . . Q . . . . N . . . . . . . . . . . A . .
mouse   . . . . . . . A . . . . . . . . N . . . . . . . . . . . . . .

        80                                  100
rat     L L A L M T G K M N P Q S A F F Q G K L K I A G N M G L A M K L Q
bovine  . . . . . . . . . . . T . . . . . . . . . N . . . . . . . . . .
human   F . . . . . . . . . . . . . . . . P . . . T . . . . . . . . . .
mouse   . . . . . . . . . . . . . . . . . . . . . . . . . . . . . . .

            120
rat     S L Q L Q P D K A K L
bovine  N . . . . . G . . - -
human   N . . . . . G N . . .
mouse   N . . . . . G . . . .
```

FIGURE 1 Sequence identity among mammalian nsL-TPs. In the bovine, human, and mouse sequences, amino acid residues identical to those in the rat sequence are indicated with dots. Dashes in the bovine sequence indicate as yet unidentified residues.

was purified which, in contrast to the regular nsL-TP, has a pronounced activity toward Sphingomyelin (SM) (Dyatlovitskaya *et al.*, 1978). This nsL-TP (M_r of 11 kDa; pI of 5.2) was found in several carcinoma cell lines and in fetal rat liver, but not in adult rat liver (Dyatlovitskaya *et al.*, 1982). In chicken liver, three proteins of M_r 12, 30–36, and 55–60 kDa were found to react with the antibody raised against rat liver nsL-TP (Reinhart *et al.*, 1991). The 12-kDa protein was purified to homogeneity using immunoaffinity chromatography. Surprisingly, in agreement with the considerable differences in amino acid composition, the N-terminal 12 amino acid residues of the avian protein failed to show any homology with the N terminus of rat nsL-TP. In this connection, interestingly, a transfer protein with a preference for SM has been purified partially from chicken liver (Koumanov *et al.*, 1982).

2. Plant LTP

After the initial demonstration that cytosolic extracts from potato tubers contained proteins that facilitate the movement of phospholipids between membranes (Kader, 1975), these proteins were purified to homogeneity from maize seedlings (Douady et al., 1982), spinach leaves (Kader et al., 1984), and castor bean seedlings (Watanabe and Yamada, 1986). These LTPs are small (M_r of 9–10 kDa), basic (pI of 9–11), and nonspecific. They are very abundant, amounting to 1–4% of the cytosolic soluble protein. This high level puts LTP in the same class as fatty acid binding proteins (Paulussen and Veerkamp, 1990). In fact, in contrast to mammalian nsL-TP, LTP can bind fatty acids (Rickers et al., 1985). Since the discovery that a putative amylase/protease inhibitor from barley is actually an LTP (Bernhard and Somerville, 1989; Breu et al., 1989), the number of identified plant LTPs has increased dramatically.

Plant LTPs are much less conserved than mammalian nsL-TPs; sequence homologies range from 30 to 70%. Plant LTPs are synthesized as precursor molecules containing a leader sequence of 23–26 amino acid residues (see references in Table I). Another feature different from mammalian nsL-TP is the occurrence of several isoforms. For castor bean, four isoforms (LTP-A, -B, -C, and -D) have been characterized, with sequence homologies again in the range of 30–70% (Tsuboi et al., 1991). Amino acid sequences reported for maize, spinach, and barley LTPs are closest to that of LTP-D. Evidence for LTP isoforms also has come from studies on maize (Arondel et al., 1991), wheat (Dieryck et al., 1992), barley (Mundy and Rogers, 1986; Hughes et al., 1992; Molina et al., 1993), and tobacco (Fleming et al., 1992; Masuta et al., 1992). Although the significance of these isoforms is not clear, a tissue-specific distribution has been observed (see subsequent discussion).

B. DNA Analysis

1. Mammalian nsL-TP

Northern blotting experiments revealed two major and two minor nsL-TP related mRNAs (1.1 and 2.4 kb; 1.7 and 3.0 kb, respectively) in rat tissue (Mori et al., 1991; Ossendorp et al., 1991; Seedorf and Assmann, 1991) and in mouse tissue (Moncecchi et al., 1991). In Chinese hamster ovary (CHO) cells, the same mRNA species were found, with reversed intensities (Ossendorp et al., 1991). In human liver, 1.8-kb and 3.2-kb mRNA species were detected (Yamamoto et al., 1991). The rat 1.7- and 3.0-kb mRNA species were established to be generated by alternative polyadenylation of the 1.1- and 2.4-kb mRNA species, respectively (Seed-

TABLE I
Properties of Mammalian and Plant Nonspecific Lipid Transfer Proteins

Source	Molecular weight	Signal peptide	Isoelectric point	Substrate specificity[a]	References
Rat liver	13,500	20[b]	8.6	PC, PI, PE, PS, PG, SM, Chol	Bloj and Zilversmit (1977); Noland et al. (1980); Poorthuis et al. (1981); Traszkos and Gaylor (1983)
Bovine liver	14,000	?[c]	9.55	PC, PI, PE, PS, PG, SM, Chol	Crain and Zilversmit (1980a); Westerman and Wirtz (1985)
Human liver	14,000	20	?	?	Van Amerongen et al. (1987); Yamamoto et al. (1991)
Goat liver	12,000	?	8.65	PC, PI, PE, ?	Basu et al. (1988)
Mouse liver	13,000	20	?	Chol, ?	Moncecchi et al. (1991)
Rat ovary	13,200	?	9	Chol, ?	Tanaka et al. (1984)
Rat hepatoma	11,200	?	5.2	PC, PI, PE, SM	Dyatlovitskaya et al. (1978)
Chicken liver	12,000	?	?	—	Reinhart et al. (1991)
Castor bean A	9,300	yes	>10.5	PC, PI, PE, PG, PA, MGDG, DGDG	Takishima et al. (1986); Tsuboi et al. (1989)
Castor bean B	9,800	yes	?	PC, PI, PE, PG, PA, MGDG, DGDG	Takishima et al. (1988); Tsuboi et al. (1989)
Castor bean C	9,600	24	?	PC, PI, PE, PG, PA, MGDG, DGDG	Takishima et al. (1988)

	M_r	No. of residues[b]	pI	Lipids	References
Castor bean D	9,300	yes	?	PC, PI, PE, PG, PA, MGDG, DGDG	Tsuboi et al. (1989,1991)
Spinach leaves	8,800	26	9.0	PC, PE, PI, PG	Kader et al. (1984); Bouillon et al. (1987); Bernhard et al. (1990)
Maize seeds	9,000	27	8.8	PC, PE, PI	Douady et al. (1985); Tchang et al. (1988)
Barley seeds	10,000	26	?	PC, ?	Mundy and Rogers (1986); Svensson et al. (1986); Breu et al. (1989)
Finger millet seeds	9,300	?	>10	?	Campos and Richardson (1984); Bernhard and Somerville (1989)
Rice seeds	8,900	?	?	?	Yu et al. (1988)
Carrot somatic embryo	9,700	26	8.86	several phospholipids	Sterk et al. (1991)
Wheat seeds	9,600	26	?	?	Désormeaux et al. (1992); Dieryck et al. (1992)[d]
Tomato seeds	9,000	23	8.85	?	Torres-Schumann et al. (1992)
Radish seeds	9,000	?	>10.5	?	Terras et al. (1992)
Sunflower seeds	9,000	?	9.5	PC>PI>PE	Vergnolle et al. (1992)
Tobacco flower	9,000	23	?	PC,?	Masuta et al. (1992)
Tobacco shoots	9,200	23	11.3	?	Fleming et al. (1992)

[a]Abbreviations: PC, phosphatidylcholine; PI, phosphatidylinositol; PE, phosphatidylethanolamine; PS, phosphatidylserine; PG, phosphatidylglycerol; SM, sphingomyelin; Chol, cholesterol; PA, phosphatidic acid; MGDG, monogalactosyl diacylglycerol; DGDG, digalactosyl diacylglycerol.

[b]Number of amino acid residues.

[c]?, Unknown.

[d]Désormeaux et al. (1992) used *T. aestivum* and Dieryck et al. (1992) used *T. durum*, but sequences are identical.

orf and Assmann, 1991). The 1.1-kb mRNA encodes nsL-TP and the 2.4-kb mRNA encodes a 58-kDa protein related to nsL-TP (see subsqequent discussion). However, that additional mRNA species exist cannot be excluded. Comparison of mRNA levels with protein levels in the same tissue has failed to show a direct relationship. For instance, the mRNA species are as abundantly present in rat kidney as in liver whereas a great difference is seen in nsL-TP levels, most likely because of differences in the regulation of translation.

Southern blot analysis of rat genomic DNA suggested that the mRNAs originate from a single gene (Ossendorp *et al.*, 1991; Seedorf and Assmann, 1991). However, the possible existence of multiple nsL-TP genes also was reported (Moncecchi *et al.*, 1991; Mori *et al.*, 1991). A study on mouse genomic DNA has revealed the presence of a truncated processed pseudogene in the mouse genome (Seedorf *et al.*, 1993). Cross-hybridization with this pseudogene could explain some of the multiple restriction fragments observed in Southern blot analysis of mouse genomic DNA. The gene encoding human nsL-TP and its related proteins was determined to reside on chromosome 1 (He *et al.*, 1991).

2. Proteins that Contain the Complete nsL-TP Sequence

In search of cDNA clones encoding nsL-TP, several groups found additional larger clones encoding a 58.7-kDa protein (Mori *et al.*, 1991; Ossendorp *et al.*, 1990,1991; Seedorf and Assmann, 1991). This protein contains the complete sequence of nsL-TP at its C terminus. Further, a rat liver cDNA clone was described encoding a 29-kDa protein that also contains the complete nsL-TP sequence (Billheimer *et al.*, 1990). Human cDNA clones encoding the 15-kDa nsL-TP precursor and the 29- and 58-kDa proteins were identified as well (He *et al.*, 1991; Yamamoto *et al.*, 1991). The cellular function of the 29- and 58-kDa proteins is not known, nor is it known whether these proteins have lipid transfer activity.

By comparing the 58-kDa protein sequence with the sequences in the protein identification resources (PIR) protein database, researchers established that the part of the 58-kDa protein not identical to nsL-TP (i.e., about 400 amino acid residues) has a high sequence similarity to rat mitochondrial and peroxisomal 3-oxoacyl-CoA thiolases (Mori *et al.*, 1991; Ossendorp *et al.*, 1991). This similarity includes the substrate binding site of which Cys 94 is a part. In another study, investigators assessed that the N-terminal portion of the mammalian 58-kDa protein has a greater similarity to the *Escherichia coli* 3-oxoacyl-CoA thiolase than to the mammalian enzyme (Baker *et al.*, 1991). Thus, the gene encoding the 58-kDa protein was suggested to be formed by fusion of an ancestral 3-oxoacyl-CoA thiolase gene and an nsL-TP-gene. Expression of the 58-kDa protein in mouse liver was found to be regulated developmentally in a sex-specific pattern (Roff *et al.*, 1992). The amounts of hepatic 58-kDa protein increased 4-fold during male sexual development but decreased substantially during female development. Expression of hepatic mouse nsL-TP is not sex specific, and the levels of this protein did not vary significantly during male and female development (Roff *et al.*, 1992). Interestingly, in bile from rats as well as chickens, an immunoreactive 55- to 60-kDa protein was detected with anti-nsL-TP antibodies (Reinhart *et al.*, 1991).

3. Plant LTP

In the last few years, research on plant LTPs has shown very rapid development with the identification of LTP-encoding cDNA clones from maize (Tchang *et al.*, 1988), spinach (Bernhard *et al.*, 1990), castor bean (Tsuboi *et al.*, 1991), carrot (Sterk et al., 1991), barley (Mundy and Rogers, 1986), tomato (Torres-Schumann *et al.*, 1992), wheat (Dieryck *et al.*, 1992), and tobacco (Masuta *et al.*, 192). These cDNAs recognize a single mRNA band of 0.6–0.9 kb on Northern blot. In the maize varieties W64A (pure inbred line) and E41 (double hybrid line), two LTP-encoding mRNAs were characterized that differ by a 74-bp insertion at the end of the coding region (Tchang *et al.*, 1988; Arondel *et al.*, 1991). This insertion would give rise to eight additional C-terminal amino acid residues in the protein. These mRNAs were suggested to represent isoforms of LTP. For castor bean, two cDNAs encoding LTP isoforms were isolated (Tsuboi *et al.*, 1991). However, these cDNAs are much less alike than the maize cDNAs, which differ only in the 3' part of the coding region. A gene encoding the barley LTP was characterized (*Ltp*1) (Linnestad *et al.*, 1991; Skriver *et al.*, 1992). By Southern blot analysis, this barley LTP was shown to be encoded by a single gene, but the barley genome was shown to contain one or two other genes with weak homology to *Ltp*1. Subsequently, the gene product of a low temperature-induced barley gene, *blt*4, was shown to have 42% homology with the maize LTPs (Hughes *et al.*, 1992). The homology with the *Ltp*1 gene product was much lower. At least two LTP genes are present in the tobacco genome (Fleming *et al.*, 1992). Also, in the tomato genome, several LTP-related genes are present (Torres-Schumann *et al.*, 1992). Whether one LTP gene can generate multiple isoforms of LTP remains to be established.

C. Structure, Shape, and Lipid Binding Site

1. Mammalian nsL-TP

Because of a lack of crystals, structural predictions have been based on the analysis of the primary sequence. The better part of nsL-TP from rat liver (residues 35-115) was observed to have a significant sequence similarity to the variable domains of the heavy chain of immunoglobulin G (IgG) (Pastuszyn *et al.*, 1987). In analogy with the known structure of IgG, the segment consisting of residues 35–95 was predicted to be composed largely of an antiparallel β-structure core. Based on the hydrophobicity plot, this core was proposed to have sufficient hydrophobic domains to accommodate a cholesterol molecule. Similarly, a predicted amphipathic α-helix (residues 21-34) was proposed to be a suitable site of interaction with phospholipid (Pastuszyn *et al.*, 1987). The interaction of the fluorescent sterol dehydroergosterol (DHE) with nsL-TP was determined by measuring effects on light scattering, lifetime heterogeneity, and energy transfer. By this approach, researchers estimated that the sterol binds to nsL-TP with a 1:1 molar stoichiometry (Schroeder *et al.*, 1990). In this study, measurements were carried out on nsL-TP mixed with pure DHE. However, titration of vesicles consisting of egg PC, DHE (10–50 mol%), and the internal quencher N-(2,4,6-trinitrophenyl)phosphatidylethanolamine (TNP-PE; 10 mol%) with nsL-TP failed to show any fluorescence increase. This result implies that, under these conditions, DHE does not bind to nsL-TP (Gadella and Wirtz, 1991). The binding stoichiometry could not be confirmed either by injecting nsL-TP at the air–water interface under a phospholipid monolayer that contained highly radiolabeled cholesterol (10 mol%) or 25-hydroxycholesterol (5 mol%). Although nsL-TP would transfer these sterol molecules to phospholipid vesicles in the subphase, direct binding of these sterols to nsL-TP was not observed (Van Amerongen *et al.*, 1989b).

Binding of lipids to nsL-TP has been investigated further making use of phospholipids that carry a fluorescent moiety at the *sn*-2-acyl chain (for formulas, see Fig. 2). These fluorescent analogs have the property that, on excitation, the emission yield increases when the probe molecule leaves the donor vesicle and binds to the protein. The lower yield in the vesicles may be because of self-quenching, as in the case of 1-palmitoyl-2-[3(*p*-diphenylhexatrienyl)-propionyl]-*sn*-glycero-3-phosphocholine (DPHp-PC), or because of the presence of the internal quencher TNP-PE, as in the case of 1-palmitoyl-2-[pyrenylacyl]-*sn*-glycero-3-phosphocholine (Pyr-PC). In agreement with the initial observation by Nichols (1987), nsL-TP can bind PC. Addition of nsL-TP to an excess of vesicles consisting of Pyr(6)-PC (90 mol%) and TNP-PE (10 mol%) was observed to result in a large increase of pyrene fluorescence. Under these conditions, an estimated 8% of the nsL-TP molecules contained a Pyr(6)-PC molecule (Gadella and Wirtz, 1991). Interestingly, binding decreased dramatically with increasing acyl chain length to as little as 0.3% of the nsL-TP molecules when Pyr(14)-PC was used. This limited binding reflects the equilibrium of the PC molecule between the vesicles and the low-affinity binding

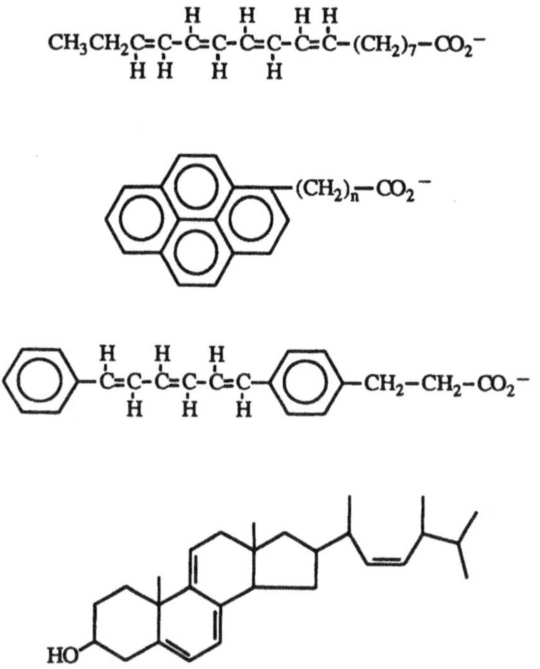

$$CH_3CH_2C{=}C{-}C{=}C{-}C{=}C{-}C{=}C{-}(CH_2)_7{-}CO_2^-$$

cis-parinaroyl

$$-(CH_2)_n{-}CO_2^-$$

Pyrenyl'acyl
(n=5,7,9,11,13)

$$-C{=}C{-}C{=}C{-}C{=}C{-}\bigcirc{-}CH_2{-}CH_2{-}CO_2^-$$

DPH-propionyl

dehydroergosterol

FIGURE 2 Fluorescent lipid structures.

site. Because of the relatively low affinity, the PC–nsL-TP complex disso-
ciates when the vesicles are removed. On the other hand, when bound to
nsL-TP, the fluorophore appears shielded from the medium since the
collisional quenchers iodide and acrylamide failed to quench the pyrene
fluorescence effectively (Gadella *et al.*, 1991). Note that the extent of
binding to nsL-TP is dictated not only by the acyl chain length, but also
by the negative charge on the lipid molecule. For instance, the binding
of Pyr(10)-PIP$_2$ (phosphatidylinositol 4,5-bis phosphate) is an order of
magnitude higher than that of Pyr(10)-PC. This increased binding most
likely reflects the net positive charge on nsL-TP.

Using time-resolved fluorescence techniques, conclusions could be
drawn about the shape of nsL-TP (Gadella *et al.*, 1991). The anisotropy
decay of the single tryptophan residue was described by a single exponential function yielding a rotational correlation time of 15 nsec. A similar
analysis of the 1,6-diphenyl-1,3,5-hexatriene (DPH) fluorescence of
DPHp-PC bound to nsL-TP yielded a correlation time of 7.4 nsec. The
difference between these two correlation times is compatible with a molecular model in which nsL-TP has an axial ratio of 2.8 with the emission

fluorescence dipole of tryptophan parallel and that of DPH perpendicular to the long symmetry axis (Fig. 3).

2. Plant LTP

Although the homology among plant LTPs is generally rather low, eight cysteine residues are always present at the same position in the sequence. As was reported for castor bean LTP, these cysteine residues form four internal disulfide bridges (Takishima *et al.*, 1988). Dimer formation has been reported as well (Douady *et al.*, 1982; Terras *et al.*, 1992). From two- and three-dimensional ¹H-NMR studies, researchers concluded that wheat LTP is organized mainly as helical fragments connected by disulfide bridges, except for the C-terminal portion (Simorre *et al.*, 1991). The conformation of wheat LTP also has been studied by Raman and Fourier transform infrared spectroscopy (Désormeaux *et al.*, 1992). Infrared results show that the wheat protein contains 41% α-helix and 19% β-sheet structures with 40% of the conformation undefined or composed of turns. Raman spectroscopy shows that three disulfide bridges adopt a *gauche–gauche–gauche* conformation whereas the other exhibits a *gauche–gauche–trans* conformation, and that the two tyrosine residues are hydrogen bonded to water molecules. The cleavage of the disulfide bridges affects the conformation of the protein significantly; the extended conformation is increased by 15% at the expense of the α-helix content. On the other hand, the interaction of the protein with lyso-PC micelles leads to an increase of 8% in the α-helix content, accompanied by a decrease of 4% in the β-sheet content (Désormeaux *et al.*, 1992). Previously, the α-helix content of bovine liver PC-TP also was found to

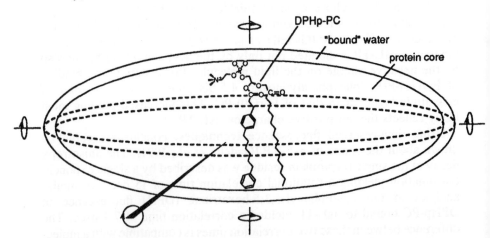

FIGURE 3 Spatial model of mammalian nsL-TP. The position of the lipid in the protein is indicated (see Section II,C,1). Reproduced from Gadella *et al.* (1991) with permission.

increase significantly in the presence of lyso-PC micelles at pH 3.6, as observed by circular dichroism (Wirtz and Moonen, 1977). LTP may bind phospholipids (Arondel and Kader, 1990) as well as fatty acids and their CoA esters (Rickers *et al.*, 1985). The lipid binding site has been argued to involve the amphipathic helix and adjacent basic residues located in the center of the sequence. In this model, adsorption to phospholipid interfaces would occur through helix and extended structures at the N and C termini of the LTPs (Désormeaux *et al.*, 1992). Crystals of wheat LTP have been obtained by adding phospholipids to the protein solution (12 mg/ml) in a molar ratio of 2:1 (Pebay-Peyroula *et al.*, 1992). Diffraction analysis of the crystals (3-Å resolution) provides evidence for the existence of two dimers in the asymmetric unit.

D. Mechanism of Action

As mentioned earlier, nsL-TP is nonspecific because it mediates the transfer of a wide range of phospholipids, cholesterol, and glycolipids between membranes. By measuring the transfer of phospholipids with different polar head groups, researchers observed that the rate of nsL-TP-mediated transfer was correlated highly with the rate of their spontaneous intermembrane transfer (Nichols and Pagano, 1983). This observation was confirmed for a series of pyrene-labeled PC species carrying acyl chains of different length at the *sn*-1 or the *sn*-2 position (Van Amerongen *et al.*, 1989b). A high correlation between nsL-TP-mediated and spontaneous transfer also was observed for a series of oxysterol derivatives using the monolayer–vesicle assay (Van Amerongen *et al.*, 1989b). This striking correlation between passive and nsL-TP-mediated transfer of lipids is thought to indicate that the actual lipid binding step is passive, that is, the natural tendency of lipids to leave the membrane as controlled by hydrophobicity and lipid charge determines the extent of binding to nsL-TP. Despite its passivity, formation of the lipid–nsL-TP complex is the key intermediate in the mode of action of nsL-TP (Fig. 4). This passive behavior explains why the activity of nsL-TP is restricted to the lipids of the outer leaflet (Crain and Zilversmit, 1980b; Wirtz *et al.*, 1986b). The diffusion barrier, as indicated in Fig. 4, is based on the observation that negatively charged membranes hardly affect the rotational mobility of nsL-TP (Gadella *et al.*, 1991), whereas this negative surface charge increases the association of nsL-TP with the membrane (T. W. J. Gadella, Jr., unpublished observation). As a result, this supposedly electrostatic interaction accelerates the equilibration of lipid monomers between membrane and nsL-TP, leading to an increased rate of lipid transfer (Billheimer and Gaylor, 1990; Butko *et al.*, 1990). Apparently, nsL-TP can interact

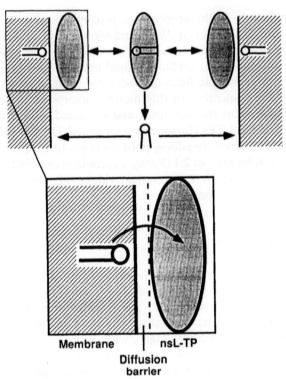

FIGURE 4   Mode of action of mammalian nsL-TP. Lipid monomers spontaneously cross a diffusion barrier before binding to the low-affinity lipid binding site of nsL-TP. nsL-TP either transfers the bound lipid to a membrane or releases it into the aqueous phase (see Section II,D).

with membranes without being hindered substantially in its rotational mobility. The proposed diffusion barrier is believed to control the passive lipid binding step, that is, the lipids must transverse this barrier spontaneously before binding to nsL-TP. Another interesting feature of the mode of action of nsL-TP is the possibility of releasing a lipid monomer into the aqueous phase (Gadella and Wirtz, 1991). This release action appears to be significant for sterols but not for PC, reflecting the marginal binding of sterols to nsL-TP (Gadella and Wirtz, 1991).

With respect to the physiological significance of this protein, note that nsL-TP is able to catalyze a net mass transfer of lipids. As observed in the monolayer–vesicle assay, nsL-TP may catalyze a net mass transfer of lipids from the vesicles to the monolayer (Van Amerongen *et al.*, 1989b). Similarly, nsL-TP was shown to mediate a net mass transfer of phospholipids from PC/PI multilamellar vesicles to intact or delipidated human high density lipoprotein, as well as from PC unilamellar vesicles to rat liver mitoplasts (Crain and Zilversmit, 1980c). Additionally, nsL-TP induced drastic changes in the lipid content and cholesterol:phospholipid ratio of synaptic plasma membranes, depending on whether these membranes were incubated with PC vesicles or PC/cholesterol vesicles (North and Fleischer, 1983). Similar observations were made when nsL-TP was incubated with human erythrocytes (Franck *et al.*, 1984). Apparently, under conditions in which donor and acceptor membrane differ in lipid composition and content, the activity of nsL-TP is directed toward eliminating this "chemical gradient" by mediating both a net mass transfer and a random exchange of lipids. On the other hand, nsL-TP is found to catalyze a strict exchange of phospholipids between intact erythrocytes and acceptor vesicles consisting of a lipid composition equivalent to that of the erythrocyte outer leaflet (Crain and Zilversmit, 1980b). In conclusion, the ability of nsL-TP to transfer lipids down a chemical gradient is inherent to its low-affinity lipid binding site; the bound lipid favors the membrane with the highest lipid uptake capacity.

Rat and human pre-nsL-TP were overexpressed in *E. coli* and purified from the bacterial lysate in high yield (Ossendorp *et al.*, 1992; Matsuura *et al.*, 1993). Interestingly, *in vitro* pre-nsL-TP was found to have the same transfer activity as the mature nsL-TP toward both PC and cholesterol. This result indicates that cleavage of the presequence is not necessary to activate the protein. Although no structural homology exists, all available evidence indicates that the mode of action of plant LTP is comparable with that of mammalian nsL-TP (Arondel and Kader, 1990). Unexpectedly, maize LTP was unable to mediate the transfer of a spin-labeled cholesterol analog (Geldwerth *et al.*, 1991).

E. Tissue Distribution and Subcellular Localization

1. Mammalian nsL-TP
Virtually all distribution and localization studies have been carried out on rat tissues using affinity-purified polyclonal antibodies against nsL-TP. Initially, levels of nsL-TP were measured only in the cytosol because its presence in this cell fraction was consistent with its presumed role as an LTP. As determined by enzyme-linked immunosorbent assays (ELISA), levels were high in liver and intestinal cytosol, but low in all other tissues tested (Teerlink *et al.*, 1984). However, subsequent immunoblot analysis of total homogenates demonstrated that nsL-TP levels were highest in liver, lung, and adrenal (Van Heusden *et al.*, 1990b). The reason for the

apparent discrepancy was that, in all tissues except for liver, the bulk of nsL-TP was membrane bound (Chanderbhan *et al.*, 1986; Van Amerongen *et al.*, 1989a; Van Heusden *et al.*, 1990b; Van Haren *et al.*, 1992). In fact, nsL-TP was found predominantly in the lysosomal/peroxisomal fractions and to a lesser extent in the microsomal and mitochondrial fractions. This membrane-bound nsL-TP had the characteristics of a peripheral membrane protein, since it could be released by resuspending the lysosomal/peroxisomal fraction in 0.1 M Na_2CO_3, pH 11.5 (Van Heusden *et al.*, 1990b).

Localization studies also were performed by immunogold labeling. In rat liver and adrenal gland, a dense labeling of peroxisomes was observed (Van der Krift *et al.*, 1985; Tsuneoka *et al.*, 1988; Keller *et al.*, 1989; Van Amerongen *et al.*, 1989a). In addition, the cytoplasm, the inside of mitochondria, and the cytoplasmic surface of the endoplasmic reticulum were labeled distinctly. At first, labeling of the peroxisomes was ascribed to the immunoreactive 58-kDa protein rather than to nsL-TP (Van der Krift *et al.*, 1985). However, later studies indicated that the peroxisomal fraction also contained substantial amounts of nsL-TP (Tsuneoka *et al.*, 1988; Keller *et al.*, 1989). On the other hand, density gradient centrifugation of rat liver, rat testes, and CHO cell homogenates confirmed that the 58-kDa protein was present in the same fractions as catalase (a peroxisomal marker enzyme) whereas nsL-TP was present in a fraction of somewhat lower density (Van Heusden *et al.*, 1990a,b). The relationship of nsL-TP with peroxisomes is discussed later in this chapter.

2. Plant LTP

The four isoforms of LTP identified in castor bean are distributed in a tissue-specific manner: cotyledons contain LTP-A and -C, the endosperm LTP-B, and the axis LTP-D (Tsuboi *et al.*, 1989). By immunogold labeling, researchers observed that in castor bean cotyledons gold particles were predominantly present in the glyoxysome matrix and the vessel wall (Tsuboi *et al.*, 1992). In maize, LTP was present in coleoptile (mainly in the outer epidermis), in mesocotyl, and in scutellum, but not in roots (Sossountov *et al.*, 1991). Interestingly, plant LTPs have a leader sequence that normally is associated with proteins destined to be incorporated into membranes or to be secreted. *In vitro* transcription and translation of the cDNA for spinach LTP in the presence of microsomes indicated that proteolytic processing of the precursor is associated with cotranslational insertion into the microsomal membranes (Bernhard *et al.*, 1990). In support of this concept, barley LTP was targeted to the lumen of the endoplasmic reticulum (Madrid, 1991). These observations are at odds with the purification of plant LTPs from the cytosol and their supposed role in the intracellular transfer of lipids, but they agree with secretion of LTP into the medium by embryogenic cell cultures of carrot (Sterk *et al.*, 1991). In addition, carrot LTP was found to be present in cell walls. In carrot plants, the gene encoding this LTP was expressed in meristematic regions of the shoot and the flower, but not in the root (Sterk *et al.*, 1991). The plant LTP genes studied to date in carrot (Sterk *et al.*, 1991), maize (Sossountov *et al.*, 1991), and tobacco (Fleming *et al.*, 1992) differ in their spatial pattern of expression. As in the case of castor bean, tissue-specific LTP isoforms may exist in these plants. LTP expression is under environmental and developmental control. The effects of developmental and environmental signals on tomato LTP expression are additive (Torres-Schumann *et al.*, 1992). The level of tomato LTP mRNA increases during seed germination and after salt stress or heat shock. Developmental control is in accordance with the accumulation of maize LTP in embryos and endosperms during seed maturation (Sossountov *et al.*, 1991). Environmental regulation also was reported for the LTP gene from barley, *blt*4 (Hughes *et al.*, 1992). The gene could be induced by a low positive temperature treatment of shoot meristems, as well as by drought stress. Expression of LTP also was induced by a single foliar application of abscisic acid (a plant growth factor).

F. Relationship with Peroxisomes and Glyoxysomes

1. Mammalian nsL-TP

The peroxisome was the last "true" subcellular organelle to be discovered in mammalian cells. This organelle plays an important role in lipid metabolism (reviewed by Lazarow and Fujiki, 1985; Van den Bosch *et al.*, 1992). Specifically, these organelles contain the enzymes involved in fatty acid β-oxidation (Lazarow and De Duve, 1976) and in the synthesis of bile acid and ether-linked phospholipids, including plasmalogens (Hajra and Bishop, 1982; Heymans *et al.*, 1983; Krisans *et al.*, 1985). Enzymes involved in the synthesis of cholesterol and dolichol also were found in peroxisomes (Thompson *et al.*, 1987; Appelkvist *et al.*, 1990). With respect to the enzymes and substrates, peroxisomal β-oxidation is distinct from that in mitochondria. The peroxisomal β-oxidation system degrades very-long-chain fatty acids (Singh *et al.*, 1984), polyunsaturated fatty acids (Hovik and Osmundsen, 1987), dicarboxylic acids (Kølvraa and Gregersen, 1986), prostaglandins (Schepers *et al.*, 1988), and xenobiotics with an acyl side chain (Yamada *et al.*, 1986). The importance of functional peroxisomes is emphasized by the fact that either a defect in the biogenesis of peroxisomes or a deficiency in one or more peroxisomal enzymes leads to very serious, and often lethal, diseases (for review, see Lazarow and Moser, 1989).

The first indication of a link between peroxisomes and nsL-TP was the observation that concomitant with a greatly reduced number of peroxisomes (Mochizuki *et al.*, 1971), Morris hepatoma 7777 cells almost completely lack nsL-TP (Poorthuis *et al.*, 1980; Trzaskos and Gaylor, 1983; Teerlink *et al.*, 1984). In support of this observation, investigators showed by immunogold labeling that, in rat liver, peroxisomes were labeled by nsL-TP-specific antibodies (Van der Krift *et al.*, 1985). Direct evidence for a link followed from studies on infants with the cerebro-hepato-renal (Zellweger) syndrome, who characteristically lack the ability to assemble peroxisomes. In the liver of these patients, nsL-TP was found to be absent (Van Amerongen *et al.*, 1987). nsL-TP was also absent from mutant CHO cells that lack functional peroxisomes (Van Heusden *et al.*, 1990a). In support of these observations, mass spectroscopic analyses indicated that the C terminus of rat nsL-TP contained the peroxisomal targeting signal Ala–Lys–Leu (Morris *et al.*, 1988). This sequence was confirmed for nsL-TP from other species by cDNA analysis. In addition to this targeting signal, nsL-TP contains a leader sequence of 20 amino acid residues (Trzeciak *et al.*, 1987; Fujiki *et al.*, 1989). Although the significance of this leader sequence is not known (see subsequent discussion), somatic cell fusion using Zellweger fibroblasts from different genetic groups clearly demonstrated that the processing of the nsL-TP precursor depends on the presence of functional peroxisomes, and that this precursor is degraded rapidly when peroxisomes are absent (Suzuki *et al.*, 1990). Also of interest is that a major peroxisomal polypeptide from *Candida tropicalis* was found to have 33% sequence identity to rat liver nsL-TP (Tan *et al.*, 1990). This polypeptide was suggested to have a role in the β-oxidation of long-chain fatty acids. However, although nsL-TP now generally is considered to be a protein present in the peroxisomal matrix, some discrepancies remain:

1. Immunogold labeling studies using anti-nsL-TP antibody showed a significant accumulation of gold particles inside the mitochondria, as well as associated with the endoplasmic reticulum and the cytosol (Tsuneoka *et al.*, 1988; Keller *et al.*, 1989).
2. In all tissues except for liver, the majority of nsL-TP is membrane bound. Membrane-bound nsL-TP in testis and in CHO cells was sensitive to trypsin treatment, suggesting that the protein is exposed to the cytosol (Van Heusden *et al.*, 1990a,b).
3. The amino acid sequences of bovine and rat liver nsL-TP, purified from liver cytosol, lack the C-terminal residue(s) that is (are) essential for peroxisomal targeting (Westerman and Wirtz, 1985; Pastuszyn *et*

al., 1987). This absence may be the result of specific posttranslational processing to direct nsL-TP to sites in the cell other than the peroxisomes.

The nature of nsL-TP raises important questions about its intracellular targeting and processing, since it has both a C-terminal peroxisomal targeting signal and an amino acid sequence near the N-terminal processing site (position -4 to -2, and + 1) that is highly identical to that of 3-oxoacyl-CoA thiolase, the other peroxisomal protein synthesized as a precursor (Mori *et al.*, 1991). The peroxisomal targeting signal of this thiolase, however, resides in the N-terminal presequence (Swinkels *et al.*, 1991); 3-oxoacyl-CoA thiolase lacks the C-terminal targeting signal. In fibroblasts from patients with the peroxisomal disease rhizomelic chondrodysplasia punctata (RCDP), import into peroxisomes and maturation of 3-oxoacyl-CoA thiolase are impaired. In contrast, nsL-TP is present in its mature form in the peroxisomes of both control and RCDP cells (Heikoop *et al.*, 1992), indicating that the protease responsible for processing pre-nsL-TP is active in the RCDP cells. If indeed pre-thiolase and pre-nsL-TP are processed by the same protease, the impaired import of 3-oxoacyl-CoA thiolase into peroxisomes of RCDP patients is most likely to be caused by a defect in the import machinery used by proteins that lack the C-terminal peroxisomal targeting signal. Moreover, provided the same protease is involved, import of pre-nsL-TP into the peroxisomes precedes processing or, at least, import and processing take place simultaneously.

The nsL-TP presequence has been suggested to be a mitochondrial targeting signal (Keller *et al.*, 1989). In general, these targeting signals lack a consensus sequence, but do represent a positively charged amphiphilic α-helix (Roise and Schatz, 1988; Horwich, 1990). The presequence of nsL-TP has no surplus of positively charged residues, but otherwise could be a mitochondrial targeting sequence. In this sense, note that processing of mitochondrial targeting signals often occurs in two steps in the mitochondrial matrix. In the first step, the bond between the residues at positions -9 and -8 is cleaved, with a typical arginine residue at position -10. In the second step, the remaining octapeptide is removed (Hendrick *et al.*, 1989; Von Heijne *et al.*, 1989). The presequence of nsL-TP resembles the mitochondrial targeting signal because it contains arginine at position -10 and serine at position -5 (Moncecchi *et al.*, 1991). However, histidine is present at position -8 where, according to the consensus sequence, a hydrophobic residue should be. Another point of contention is that nsL-TP is supposed to be present in the mitochondrial intermembrane space rather than in the matrix (Vahouny *et al.*, 1983). To clarify this issue, *in vitro* import studies should be carried out to establish whether, indeed, the nsL-TP presequence is able to target the protein to mitochondria. If so, the nsL-TP precursor must contain targeting signals for both peroxisomes and mitochondria. A similar situation has been observed for the enzyme L-alanine:glyoxylate aminotransferase I, which is present in peroxisomes in primates (including humans), in mitochondria in carnivores, and in both organelles in rodents (Takada *et al.*, 1990). The subtlety of the targeting of proteins follows from the observation that the replacement of histidine at position -17 by arginine, lysine, leucine, or valine in the presequence of peroxisomal 3-ketoacyl-CoA thiolase prevents targeting of the enzyme to the peroxisomes, but reroutes the protein to the mitochondria (Osumi *et al.*, 1992).

2. Plant LTP

Organelles in plants that are comparable to peroxisomes are glyoxysomes. Whether a relationship also exists between LTPs and glyoxysomes is not yet clear. The only evidence is from studies on castor bean (Tsuboi *et al.*, 1992). By sucrose density gradient fractionation of castor bean cotyledon homogenates, and subsequent immunoblotting, 13% of LTP isoforms A and C was shown to be present in the glyoxysomes. Further, by an *in vitro* import assay, pre-LTP-C was shown to be imported into glyoxysomes and subsequently processed to the mature LTP. In contrast to the results for spinach (Bernhard *et al.*, 1990) and barley (Madrid, 1991) LTPs, import into microsomes could not be demonstrated for castor bean pre-LTP-C (Tsuboi *et al.*, 1992). In the same study, researchers showed that glyoxysomal acyl-CoA oxidase activity was increased by LTP-A in a time- and dose-dependent manner, suggesting a relationship between LTP and β-oxidation.

G. Possible Functions in Vivo

1. Mammalian nsL-TP

Various aspects of cholesterol metabolism were reported to be stimulated by mammalian nsL-TP, for example, the microsomal conversion of lanosterol into cholesterol (Noland *et al.*, 1980; Trzaskos and Gaylor, 1983), cholesterol esterification (Gavey *et al.*, 1981; Poorthuis and Wirtz, 1982), bile acid formation (Seltman *et al.*, 1985), and steroid hormone synthesis (Chanderbhan *et al.*, 1982; Vahouny *et al.*, 1983). Dolichol biosynthesis was found to be stimulated by nsL-TP as well (Ericsson *et al.*, 1991). The stimulatory effect of nsL-TP in these processes is most likely the result of increased transfer of the lipid precursor to the site of product formation. However, all these observations were made under conditions *in vitro*. To date, attempts to correlate the cellular content of nsL-TP with rates of cholesterol metabolism have failed. In hepatoma cells, the nsL-TP levels generally are observed to be substantially decreased, both in the cytosol (Poorthuis *et al.*, 1980; Trzaskos and Gaylor, 1983) and in other subcellular fractions (Lyons *et al.*, 1991). For example, nsL-TP levels in the cytosol of Reuber H35 hepatoma cells are lowered 16-fold compared with the cytosol of hepatocytes (Van Heusden *et al.*, 1985). Despite this difference, both synthesis and esterification of cholesterol are increased 2-fold in these hepatoma cells. In agreement with this observation, cholesterol synthesis was found to be increased 2- to 3-fold in mutant CHO cells that lack nsL-TP (Van Heusden *et al.*, 1992). Further, no correlation was found between the amount of nsL-TP in rat liver cytosol and widely varying rates of bile acid synthesis (Geelen *et al.*, 1987). However, since whether the delivery of cholesterol to the site of metabolism is the rate-limiting step in cholesteryl ester and bile acid formation is not known, the level of nsL-TP may not have a direct bearing on these processes (Billheimer and Reinhart, 1990). On the other hand, in steroid hormone synthesis, the transport of cholesterol to the side-chain cleavage enzyme (cytochrome $P450_{scc}$) in the mitochondrial inner membrane is the rate-limiting step. In this respect, the finding that, concomitant with the stimulation of corticosteroid synthesis, a 3-fold increase in nsL-TP synthesis was observed in adrenocortical cells treated with adrenocorticotrophic hormone (ACTH) for 48 hr is of great interest (Trzeciak *et al.*, 1987). Injection of ACTH into rats 15 min before killing them had no effect on the level of nsL-TP in the adrenal cytosol (McNamara and Jefcoate, 1989). In view of the fact that ACTH is known to have both acute and long-term effects on steroidogenesis, the effect on nsL-TP is most likely to be long term. In support of nsL-TP involvement in steroidogenesis, fusion of adrenocortical cells with liposomael-encapsulated anti-nsL-TP antibody resulted in a 40–65% reduction in ACTH-stimulated corticosterone production (Chanderbhan *et al.*, 1986). Similarly, expression of human nsL-TP-cDNA in COS African green monkey kidney cells, in conjuction with expression vectors for cholesterol side-chain cleavage enzyme and adrenodoxin, resulted in a 2.5-fold enhancement of pregnenolone/progesterone synthesis relative to synthesis when the steroidogenic enzyme system was expressed alone (Yamamoto *et al.*, 1991). This result clearly suggests that an nsL-TP-mediated step is important in steroidogenesis.

A study on the pattern and regulation of nsL-TP gene expression in the rat ovary suggests a role for nsL-TP in cell differentiation (Rennert *et al.*, 1991). The rat ovary contains several compartments of varying steroidogenic activity. By *in situ* hybridization, nsL-TP mRNA was found primarily in the cell types that are most active in steroidogenesis. In a transformed rat granulosa cell line that displays steroidogenic activity, the cyclic AMP analog 8-Br-cAMP increased all nsL-TP mRNAs 5-fold after 24 hr of treatment and increased the nsL-TP level 4-fold (Rennert *et al.*, 1991). This result agrees with a previous observation that synthesis of nsL-TP in adrenocortical cells is induced by ACTH and that this induction is mediated by cyclic AMP (Trzeciak *et al.*, 1987). However, in a transformed rat granulosa cell line that does not synthesize steroid hormones, 8-Br-cAMP increased the levels of nsL-TP mRNA to the same extent. This observation was interpreted to indicate that nsL-TP gene expression in the ovary is correlated with the state of differentiation of granulosa cells, and that coupling to steroidogenic activity is not obligatory. These findings agree with our observation that nsL-TP is present in high amounts in Sertoli cells, a cell type that is not active in steroidogenesis (Van Haren *et al.*, 1992). Evidently, more studies must be carried out to clarify the connection between nsL-TP and the ability of cells to produce steroid hormones.

Type C Niemann–Pick disease is characterized by an accumulation of low density liporotein-derived cholesterol in lysosomes (Liscum *et al.*, 1989). This accumulation is believed to be the result of a lesion in the cholesterol transport pathway involving the lysosomes. Because of the possible role of nsL-TP in intracellular cholesterol transfer, nsL-TP levels were determined by immunoblotting in livers of normal and of mutant Niemann–Pick Type C mice (Roff *et al.*, 1992). Levels were near normal in affected immature males and subnormal in affected immature females, but declined during sexual maturation. As a result, adult Niemann–Pick Type C mice were deficient in hepatic nsL-TP. Since affected prepubertal mice already showed abnormal lipid processing, these results strongly suggest that the reduced levels of nsL-TP in Niemann–Pick Type C mice were a consequence of the disorder and not the cause of the lysosomal cholesterol accumulation.

2. Plant LTP

The function of plant LTPs as intracellular carriers of lipids was questioned when evidence suggested that several of these proteins were secreted (Bernhard *et al.*, 1990; Madrid, 1991; Sterk *et al.*, 1991). Since then, several hypotheses concerning the physiological role of these proteins have been formulated. Based on the extracellular location in carrot embryos and the expression pattern of the encoding gene, LTP was proposed to play a role in the transport of cutin monomers through the extracellular matrix to sites of cutin synthesis (Sterk *et al.*, 1991). The purification of an antifungal protein from radish was described (Terras *et al.*, 1992). The first 43 amino acids of the purified protein were determined by automated Edman degradation, and were found to be homologous to LTPs. Further, from barley leaves, four LTP isoforms were purified that all potently inhibited bacterial and fungal plant pathogens (Molina *et al.*, 1993). A maize LTP very similar to the one identified earlier (Tchang *et al.*, 1988) was found to possess the same properties (Molina *et al.*, 1993). These findings strongly indicate that plant LTPs in general play an important role in the defense against pathogens. LTP may exert this function in either of two ways: indirectly by being involved in the formation of a mechanical cutin barrier and directly by expressing an intrinsic antifungal activity (Terras *et al.*, 1992). Also, both these aspects may be involved.

In addition to having a function on the outside of the cell, cytosolic forms of LTP exist that may have various functions inside the cell. Based on the presence of castor bean LTP in the glyoxysome, LTP was proposed to be involved in the regulation of fatty acid β-oxidation through the enhancement of acyl-CoA oxidase activity (Tsuboi *et al.*, 1992). Acyl-CoA oxidase is the rate-limiting enzyme in peroxisomal β-oxidation, suggesting that LTP has a function in the degradation of storage lipids.

III. PHOSPHATIDYLINOSITOL TRANSFER PROTEIN

A. Purification and Biochemical Characterization

The first phospholipid transfer protein to be purified partially was from bovine heart cytosol (Wirtz and Zilversmit, 1970). This protein was noted for stimulating the transfer of PC and PI. Subsequently, a homogeneous preparation of phospholipid transfer protein was obtained from this source (DiCorleto *et al.*, 1979). Ultimately, this protein had all the characteristics of PI-TP which, in the meantime, was purified from bovine brain (Helmkamp *et al.*, 1974; Wirtz, 1982). To date, PI-TP has been found in every eukaryotic cell investigated, including a great variety of mammalian tissues, yeast, fungus, and *Xenopus laevis* (for reviews, see Helmkamp 1990; Wirtz, 1991; Helmkamp *et al.*, 1992). As shown in Table II, PI-TP has been purified from many of these sources. The various PI-TP have several properties in common: a molecular mass of 33–35 kDa, an acidic isoelectric point of 5.1–5.6, and the capacity to bind one molecule of either PI or PC. *In vitro*, PI-TP functions as a regular phospholipid exchange protein that is, for every phospholipid molecule transferred to the acceptor membrane, one phospholipid molecule is transferred back to the donor membrane. On the other hand, because of its dual specificity, PI-TP can catalyze a net transfer of PI to an acceptor membrane that is PI deficient in exchange for PC, which is transported in the opposite direction (Van

TABLE II
Properties of the Phosphatidylinositol Transfer Proteins from Mammalian Sources and Microorganisms

Source	Molecular weight	Isoelectric point	Substrate specificity[a]	References
Bovine brain	32,500	5.3, 5.6	PI > PC = PG	Helmkamp et al. (1974); Somerharju et al. (1983)
Bovine heart	33,500	5.3, 5.6	PI > PC >> SM	DiCorleto et al. (1979)
Rat liver	35,000	5.1, 5.3	PI > PC	Lumb et al. (1976)
Rat brain	36,000	4.9, 5.3	PI > PC	Venuti and Helmkamp (1988)
Human platelet	29,000	5.6, 5.9	PI > PC = PG	George and Helmkamp (1985)
Rabbit lung	32,000	6.5, 6.8	PI = PC	Tsao et al. (1992)
Saccharomyces cerevisiae	35,000	4.6	PI > PC	Daum and Paltauf (1984); Szolderits et al. (1989)
Mucor mucedo	24,000	5.05	PI > PC > PE	Grondin et al. (1990)
Neurospora crassa	38,000	4–5	PI > PC >> PS	Basu et al. (1992)

[a]See Table I for abbreviations.

Paridon *et al.*, 1987a). Indeed, the activity of PI-TP generally is determined by measuring the transfer of PI (either fluorescent or radioactive) unidirectionally from donor to acceptor membranes. This characteristic ability of PI-TP to give rise to a net transfer of PI to membranes that are PI deficient is believed to be a true reflection of its cellular function. However, to date definitive proof that this is indeed the case is lacking.

Detailed studies on PI-TP from brain and *Saccharomyces cerevisiae* demonstrated a remarkable similarity in molecular mass, isoelectric point, and lipid specificity (see Table II). However, when the cDNAs encoding rat brain PI-TP and PI-TP from *S. cerevisiae* were compared, no similarity was found (Dickeson *et al.*, 1989; Aitken *et al.*, 1990; Bankaitis *et al.*, 1990; Salama *et al.*, 1990). In agreement with this finding, antibodies against bovine brain PI-TP failed to cross-react with yeast PI-TP (Szolderits *et al.*, 1989). On the other hand, antibodies against bovine brain PI-TP and antibodies against synthetic peptides representing predicted epitopes in rat brain PI-TP were cross-reactive with PI-TP from many tissues and cell types (Venuti and Helmkamp, 1988; Dickeson *et al.*, 1989; Snoek *et al.*, 1992). Antibodies against PI-TP from *S. cerevisiae* also were cross-reactive with PI-TP from other yeast strains (Bankaitis *et al.*, 1989). In plants, no protein could be detected that was cross-reactive with the antibody against either yeast PI-TP or mammalian PI-TP, although PI transfer activity was present (P. Laurent, personal communication). Basu *et al.* (1992) isolated PI-TP from *Neurospora crassa* that also did not bind antibodies against bovine brain PI-TP. These observations strongly suggest that PI-TP is highly conserved among animal species but that conservation has disappeared (if it ever existed) among animals, plants, and microorganisms.

In mammalian tissues, additional proteins have been detected that are cross-reactive with antibodies against PI-TP. Thomas *et al.* (1989) reported the presence of a cross-reactive 41-kDa protein in mature rat testis that was shown to have PI/PC transfer activity. A possible precursor of PI-TP was isolated from bovine brain (Van den Akker *et al.*, 1990). This cross-reactive 38-kDa protein almost completely lacked the capacity to transfer or bind PI. However, when this protein was incubated with trypsin, it was converted into a 35-kDa protein with a strongly increased affinity for PI. The cDNA encoding rat brain PI-TP has been analyzed, showing that the deduced amino acid sequence defines a protein with a molecular mass of 35 kDa (Dickeson *et al.*, 1989). However, DNA and mRNA hybridization experiments have indicated that other related genes are present. Therefore, that one of these related genes encodes the 38-kDa protein is to be expected. In addition to bovine brain, the 38-kDa protein also was found in mouse 3T3 fibroblast cells. Whether this protein actually functions as a precursor of PI-TP in the cell still remains to be established.

In mouse 3T3 fibroblast cells, still another protein (M_r of 36 kDa) was detected that was cross-reactive with antibodies against PI-TP and against synthetic peptides representing predicted epitopes (Snoek et al., 1992). Computer analysis (PC Gene program) of the amino acid sequence of rat brain PI-TP (Dickeson et al., 1989) indicated the presence of five potential phosphorylation sites for protein kinase C. In addition to being a substrate for protein kinase C in vitro, some of the PI-TP in mouse 3T3 fibroblast cells was found to be already phosphorylated (Snoek et al., 1993a,b). This phosphorylation was demonstrated by two independent methods: (1) analysis of PI-TP immunoprecipitated from 3T3 cells after labeling the cells with [^{32}P]phosphate and (2) separation of 3T3 cellular proteins on isoelectric focusing gels followed by Western blot analysis. These analyses demonstrated that the phosphorylated form of PI-TP in 3T3 cells is identical to the cross-reactive 36-kDa protein. The phosphorylated and nonphosphorylated forms of PI-TP have not been separated to date. Therefore, the properties of PI-TP that are affected by the phosphorylation remain unknown.

B. Structural Features of the Lipid Binding Site

Isolation of PI-TP from several tissues yielded two subforms (PI-TP I and II) that were identical in molecular mass (35 kDa) but different in isoelectric point. PI-TPI (pI of 5.5) was shown to contain one molecule of PI and PI-TP II (pI of 5.7) was shown to contain one molecule of PC (Van Paridon et al., 1987a,b). Given the fact that bovine brain contains about 10 times more PC than PI, the finding that 65% of the PI-TP isolated from this tissue contained a PI molecule (PI-TP I) and 35% contained a PC molecule (PI-TP II) was of interest. This result indicates that the affinity of PI-TP for PI is clearly much higher than that for PC. In fact, from competition binding experiments, the affinity of PI-TP for PI could be estimated to be 17-fold higher than that for PC (Van Paridon et al., 1987a,b). This observation was confirmed by carrying out monolayer experiments designed to measure the binding of PI and PC to PI-TP. When this protein was injected at the air–water interface under a monolayer spread from an equimolar mixture of [^{14}C]PC/PI or PC/[^{14}C]PI PI-TP was found to bind 10 times more PI than PC (Demel et al., 1977).

The binding of phospholipid to PI-TP was analyzed extensively using fluorescent analogs of PI and PC that carry pyrenylacyl or parinaroyl chains (see Fig. 2). Steady state fluorescent measurements showed that PI-TP has specific binding sites for the sn-1 and sn-2 acyl chains; each site has a characteristic acyl chain preference (Van Paridon $et\ al.$, 1988). Further, these studies supported the model that at least the sn-2 acyl chains of PI and PC bind to the same site in the protein. This result was in agreement with previous findings that indicated only one lipid binding site in PI-TP and that binding of PI and PC is mutually exclusive (Zborowski and Demel, 1982; Van Paridon $et\ al.$, 1987b). To obtain further information on the 2-acyl binding site, time-resolved fluorescence studies were carried out with the sn-2 parinaroyl derivatives of PI and PC bound to PI-TP (Van Paridon $et\ al.$, 1987b). Measuring the fluorescence anisotropy decay clearly demonstrated that the sn-2 parinaroyl chains of both PI and PC are strongly immobilized in the lipid binding site of PI-TP. In this sense, PI-TP is remarkably similar to the PC-specific transfer protein (Berkhout $et\ al.$, 1984).

C. Tissue Distribution and Subcellular Localization

PI-TP can be detected in every mammalian tissue, although at highly varying concentrations. Brain is the richest source of PI-TP, whereas adipose tissue and skeletal muscle contain very low amounts (Wirtz $et\ al.$, 1976; Venuti and Helmkamp, 1988). In developing rat brain, PI transfer activity increases gradually from fetal day 16 to postnatal day 9. Transfer activities then decline and are maintained at or slightly above fetal levels to day 60 (Nyquist and Helmkamp, 1989). When cells are homogenized, PI-TP is found almost exclusively in the membrane-free cytosol. However, when the intracellular localization of PI-TP is assessed by indirect immunofluorescence, a substantial amount of PI-TP in mouse 3T3 fibroblast cells is associated with the Golgi structures (Fig. 5). PI-TP also is detected in the nucleus and in the cytoplasm, where it is present in a diffuse and aggregated form (Snoek $et\ al.$, 1992). Hence, disruption of the cells may lead to a release of PI-TP from the membrane structures. Cellular levels of PI-TP and localization also are found to be dependent on the growth conditions. In comparison with quiescent 3T3 cells, an enhanced labeling of PI-TP is observed in the cytosol and in the Golgi structures of exponentially growing cells. Subfractionation studies of yeast homogenates indicate that the predominant fraction of PI-TP (which is identical to SEC14p; see subsequent discussion) is found in the membrane-free cytosol, as demonstrated by transfer activity assays and Western blot analysis (Daum

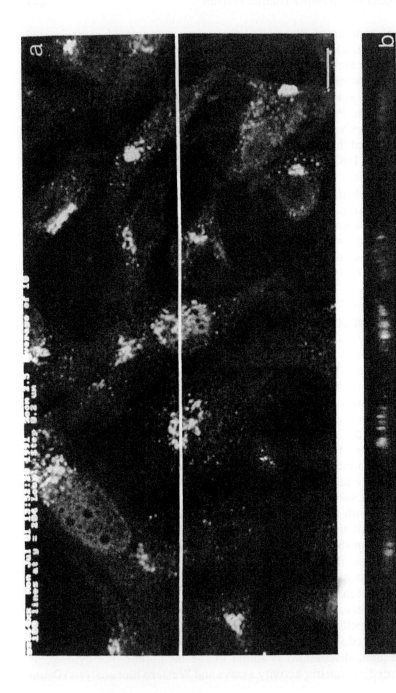

FIGURE 5 Immunofluorescent labeling of PI-TP in Swiss mouse 3T3 cells analyzed by confocal laser scanning microscopy. Exponentially growing cells were fixed and labeled with anti-PI-TP and goat-anti-rabbit fluorescein isothiocyanate (FITC) as described by Snoek *et al.* (1992). (A) Scan of PI-TP labeling of 3T3 cells. (B) Scan of PI-TP labeling in a plane perpendicular to that of panel A along the drawn line. Bar, 3 μm.

and Paltauf, 1984; Bankaitis *et al.*, 1989). However, in contrast to mammalian cells, 20–40% of PI-TP in *S. cerevisiae* remained reproducibly associated with membrane material (Bankaitis *et al.*, 1989). This material may include the Golgi complex to which, as was shown by immunofluorescence, a substantial fraction of PI-TP in yeast is associated (Cleves *et al.*, 1991a).

D. Possible Functions in Vivo

1. Mammalian PI-TP

The extensive efforts to characterize the mammalian PI-TP have produced much information about the biochemical and physical properties of the protein but have given no definite indication of its actual cellular function. With its high affinity for PI and lower affinity for PC, PI-TP is able to accomplish *in vitro* a net transfer of PI between membranes (see previous discussion). This ability has led to the hypothesis that PI-TP is ideally suited to maintaining the correct PI levels in cellular membranes (Wirtz *et al.*, 1978, Van Paridon *et al.*, 1987a,c). Thus, investigators suggested that PI-TP has a function in replenishing the PI pool of the plasma membrane during agonist-induced breakdown of PIP_2 by phospholipase C. In this process, PIP_2 is formed continuously by the subsequent phosphorylation of PI by PI kinase and of phosphatidylinositol 4-phosphate (PIP) by PIP kinase. These steps occur in the plasma membrane, as a result of which the PI pool in the plasma membrane becomes depleted. PI-TP would be instrumental in transferring PI from its site of synthesis (i.e., the endoplasmic reticulum) to the depleted membrane. Although popular as a hypothesis, the possible connection between PI turnover and PI-TP-mediated PI transfer still remains to be established (Helmkamp, 1990).

To achieve greater insight into the possible relationship between PI-TP and agonist-induced PI turnover, the localization of PI-TP was determined in quiescent and in agonist-stimulated 3T3 mouse fibroblast cells. In quiescent (serum-starved or confluent) 3T3 cells, the majority of PI-TP was found to be associated with the Golgi complex. When these cells were stimulated with bombesin, an agonist that stimulates PI turnover in the plasma membrane (Cook *et al.*, 1990), PI-TP was redistributed rapidly throughout the cell (Snoek *et al.*, 1993a). Under these circumstances, no evidence was found for an accumulation of PI-TP near or at the plasma membrane. Therefore, a direct insertion of PI molecules into the plasma membrane by PI-TP appears unlikely. These localization studies, however, do suggest that PI-TP is functionally active at the level of the Golgi system. Perhaps by exchanging PI for PC, PI-TP has a direct effect on the PI content of the secretory vesicles leaving the Golgi complex, thereby regulating the PI content of the plasma membrane. Also, investigators have observed that the stimulation of 3T3 cells by phorbol esters resulted in an increased phosphorylation of PI-TP, with a concomitant extensive intracellular redistribution of this protein (Snoek *et al.*, 1993a). Speculating that phosphorylation of PI-TP may regulate its cellular localization, its affinity for PI and PC, and/or the kinetics of the lipid transfer reaction is tempting. Confirmation of these possibilities awaits the isolation of phosphorylated PI-TP.

2. Yeast PI-TP

In terms of appreciating the function of PI-TP, much progress has been made in yeast. When the sequence of the cDNA encoding PI-TP of yeast was compared with that of the cDNA encoding SEC14p, it was clear that these two proteins were identical (Bankaitis *et al.*, 1989,1990; Aitken *et al.*, 1990). This result was surprising since SEC14p was known as a protein required for the transport of secretory proteins at the late Golgi stage (Novick *et al.*, 1980; Salama *et al.*, 1990). The role of SEC14p in the secretory process has become better understood with the availability of temperature-sensitive mutants (*sec14-1*^*ts*^). Protein secretion in *sec14-1*^*ts*^ mutants was inhibited when these cells were grown at the nonpermissive temperature (37°C). On return to the permissive temperature (25°C), protein secretion and cell growth again proceeded normally. This cellular response is most likely to be related directly to the activity of PI-TP/SEC14p. PI-TP in the cytosol of the *sec14-1*^*ts*^ yeast strain is inhibited when PI transfer activity is assayed at 37°C, but is fully active at 25°C (Bankaitis *et al.*, 1990). At this time, the connection between PI-TP transfer activity and Golgi function is not clear. On the other hand, some evidence suggests that the ability to bind or transfer phospholipids, not some as yet unknown property of PI-TP, is critical for secretory competence of the Golgi complex.

Deletion of the gene encoding PI-TP/SEC14p was shown to be lethal to *S. cerevisiae* (Aitken *et al.*, 1990). However, the normally essential SEC14p requirement can be bypassed by mutations in any one of at least six genes (Cleves *et al.*, 1991a). Specifically, mutations that block PC synthesis via the CDP-choline pathway rendered the yeast cell independent of the SEC14p requirement, whereas mutations that inactivated the alternative pathway for PC synthesis (the methylation pathway) did not bypass the cellular SEC14p requirement. In view of these observations and of the fact that SEC14p is localized at the Golgi complex, two models were proposed for the mode of action of SEC14p (Cleves *et al.*, 1991a,b).

In the first model, PC synthesis via the CDP-choline pathway is localized exclusively in the Golgi complex. Proper functioning of the Golgi would require that PC is removed by SEC14p in exchange for PI, resulting in a PI:PC ratio that is critically important for the secretory competence of this membrane system. In the case of defects in the CDP-choline pathway of PC biosynthesis, a proper PI:PC ratio in the Golgi would ensue without SEC14p. In the second model, the CDP-choline pathway is localized in the endoplasmic reticulum. Inhibition of this pathway would result in an increased PI:PC ratio which, via vesicle flow, would be imposed on the Golgi, thereby again bypassing SEC14p.

The high viability of the yeast double mutants that bypass the SEC14p requirement leads to the conclusion that SEC14p/PI-TP is not essential for intracellular phospholipid transport. This suggestion was confirmed when the transfer of PI and PC from internal membranes to the plasma membrane was determined in the temperature-sensitive mutant *S. cerevisiae sec 14* (Gnamusch *et al.*, 1992). Cultivation of the SEC14 mutant strain at 37°C did not inhibit the appearance of radiolabeled PI or PC in the plasma membrane, indicating that PI-TP/SEC14p is not involved directly in the transfer of phospholipids to the plasma membrane. However, PI-TP/SEC14p still may be able to regulate the PI traffic to the plasma membrane indirectly by manipulating the PI:PC ratio of vesicles competent for protein or lipid secretion.

IV. FUTURE DIRECTIONS

Since their discovery in 1968, a great deal of progress has been made in the study of phospholipid transfer proteins. These proteins now have been purified from a many different sources and some of the cDNAs have been cloned. An understanding of the intracellular function of these proteins has been elusive, although we can anticipate considerable progress along this line in the future. In particular, research on lipid transfer proteins in plants and yeast seems very promising because of the effective genetic tools available for unraveling the secrets of these organisms. A clear breakthrough is the observation that plant LTPs are potent pathogen inhibitors (Terras *et al.*, 1992; Molina *et al.*, 1993). Another promising breakthrough emerged from the apparent similarities of yeast and mammalian PI-TP, not so much in the biochemical characteristics as in the connection with the Golgi complex. A role for mammalian PI-TP in Golgi function that is similar to that for yeast PI-TP/SEC14p is attractive. The lack of sequence similarity, however, argues against such a similarity in function (Griff *et al.*, 1992; Rothman and Orci, 1992). In the coming years, the role of phosphorylation and dephosphorylation of PI-TP may be better understood. Phosphorylation of mammalian PI-TP presents an intriguing possibility for regulating the activity and/or the intracellular localization of the protein. Phosphorylation and dephosphorylation of PI-TP also emphasizes the occurrence of a possible, as yet poorly understood, connection between PI-TP and protein kinases in signal transduction pathways.

In the near future, we can expect that the multidisciplinary approach to determining the functional aspects of phospholipid transfer proteins will enlarge our knowledge of this fascinating class of proteins greatly.

Acknowledgments

We greatly acknowledge the many contributions by our previous collaborators to the research described in this chapter.

References

Aitken, J. F., Van Heusden, G. P. H., Temkin, M., and Dowhan, W. (1990). The gene encoding the phosphatidylinositol transfer protein is essential for cell growth. *J. Biol. Chem.* **265**, 4711–4717.

Appelkvist, E.-L., Reinhart, M., Fischer, R., Billheimer, J., and Dallner, G. (1990). Presence of individual enzymes of cholesterol biosynthesis in rat liver peroxisomes. *Arch. Biochem. Biophys.* **282**, 318–325.

Arondel, V., and Kader, J.-C. (1990). Lipid transfer in plants. *Experienta* **46**, 579–585.

Arondel, V., Tchang, F., Baillet, B., Vignols, F., Grellet, F., Delseny, M., Kader, J.-C., and Puigdomenech, P. (1991). Multiple mRNA coding for phospholipid-transfer protein from *Zea mays* arise from alternative splicing. *Gene* **99**, 133–136.

Baker, M. E., Billheimer, J. T., and Strauss, J. F., III (1991). Similarity between the amino-terminal portion of mammalian 58-kDa sterol carrier protein (SCPx) and *Escherichia coli* acetyl-CoA acyltransferase: Evidence for a gene fusion in SCPx. *DNA Cell Biol.* **10**, 695–698.

Bankaitis, V. A., Malehorn, D. E., Emr, S. D., and Greene, R. (1989). The *Saccharomyces cerevisiae* SEC14 gene encodes a cytosolic factor that is required for transport of secretory proteins from the yeast Golgi complex. *J. Cell Biol.* **108**, 1271–1281.

Bankaitis, V. A., Aitken, J. F., Cleves, A. E., and Dowhan, W. (1990). An essential role for a phospholipid transfer protein in yeast Golgi function. *Nature (London)* **347**, 561–562.

Basu, J., Kundu, M., Bhattacharya, U., Mazumder, C., and Chakrabarti, P. (1988). Purification and characterization of a non-specific lipid transfer protein from goat liver. *Biochim. Biophys. Acta* **959**, 134–142.

Basu, J., Kundu, M., and Chakrabarti, P. (1992). Purification of a phosphatidylinositol/phosphatidylcholine transfer protein from *Neurospora crassa*. *Biochim. Biophys. Acta* **1126**, 286–290.

Berkhout, T. A., Visser, A. J. W. G., and Wirtz, K. W. A. (1984). Static and time-resolved fluorescence studies of fluorescent phosphatidylcholine bound to the phosphatidylcholine transfer protein. *Biochemistry* **23**, 1505–1513.

Bernhard, W. R., and Somerville, C. R. (1989). Coidentity of putative amylase inhibitors from barley and finger millet with phospholipid transfer proteins inferred from amino acid sequence homology. *Arch. Biochem. Biophys.* **269**, 695–697.

Bernhard, W. R., Thoma, S., Botella, J., and Somerville, C. R. (1990). Isolation of a cDNA clone for spinach lipid transfer protein and evidence that the protein is synthesized by the secretory pathway. *Plant Physiol.* **95**, 164–170.

Billheimer, J. T., and Gaylor, J. L. (1990). Effect of lipid composition on the transfer of sterols mediated by non-specific lipid transfer protein (sterol carrier protein 2). *Biochim. Biophys. Acta* **1046**, 136–143.

Billheimer, J. T., and Reinhart, M. P. (1990). Intracellular trafficking of sterols. *Subcell. Biochem.* **16**, 301–331.

Billheimer, J. T., Strehl, L. L., Davis, G. L., Strauss J. F., III, and Davis, L. G. (1990). Characterization of a cDNA encoding rat sterol carrier protein 2. *DNA Cell Biol.* **9**, 159–165.

Bloj, B., and Zilversmit, D. B. (1977). Rat liver proteins capable of transferring phosphatidylethanolamine. Purification and transfer activity for other phospholipids and cholesterol. *J. Biol. Chem.* **252**, 1613–1619.

Bouillon, P., Drischel, C., Vergnolle, C., Duranton, H., and Kader, J.-C. (1987). The primary structure of spinach-leaf phospholipid-transfer protein. *Eur. J. Biochem.* **166**, 387–391.

Breu, V., Guerbette, F., Kader, J. C., Kannangara, C. G., Svensson, B., and von Wettstein-Knowles, P. (1989). A 10-kD barley basic protein transfers phosphatidylcholine from liposomes to mitochondria. *Carlsberg Res. Commun.* **54**, 81–84.

Butko, P., Hapala, I., Scallen, T. J., and Schroeder, F. (1990). Acidic phospholipids strikingly potentiate sterol carrier protein 2 mediated intermembrane sterol transfer. *Biochemistry* **29**, 4070–4077.

Campos, F. A. P., and Richardson, M. (1984). The complete amino acid sequence of the α-amylase inhibitor I-2 from seeds of ragi (Indian finger millet, *Eleusine coracana* Gaertn.). *FEBS Lett.* **167**, 221–225.

Chanderbhan, R., Noland, B. J., Scallen, T. J., and Vahouny, G. V. (1982). Sterol carrier protein$_2$. Delivery of cholesterol from adrenal lipid droplets to mitochondria for pregnenolone synthesis. *J. Biol. Chem.* **257**, 8928–8934.

Chanderbhan, R. F., Kharroubi, A. T., Noland, B. J., Scallen, T. J., and Vahouny, G. V. (1986). Sterol carrier protein$_2$: Further evidence for its role in adrenal steroidogenesis. *Endocrine Res.* **12**, 351–370.

Cleves, A. E., McGee, T. P., Whitters, E. A., Champion, K. M., Aitken, J. R., Dowhan, W., Goebl, M., and Bankaitis, V. A. (1991a). Mutations in the CDP-choline pathway for phospholipid biosynthesis bypass the requirement for an essential phospholipid transfer protein. *Cell* **64**, 789–800.

Cleves, A. E., McGee, T. P., and Bankaitis, V. A. (1991b). Phospholipid transfer proteins: A biological debut. *Trends Cell Biol.* **1**, 30–34.

Cook, S. J., Palmer, S., Plevin, R., and Wakelam, M. J. O. (1990). Mass measurement of inositol 1,4,5-triphosphate and sn-1,2-diacylglycerol in bombesin stimulated Swiss 3T3 mouse fibroblasts. *Biochem. J.* **265**, 617–620.

Crain, R. C., and Zilversmit, D. B. (1980a). Two nonspecific phospholipid exchange proteins from beef liver. 1. Purification and characterization. *Biochemistry* **19**, 1433–1439.

Crain, R. C., and Zilversmit, D. B. (1980b). Two nonspecific phospholipid exchange proteins from beef liver. 2. Use in studying the asymmetry and the transbilayer movement of phosphatidylcholine, phosphatidylethanolamine and sphingomyelin in intact rat erythrocytes. *Biochemistry* **19**, 1440–1447.

Crain, R. C., and Zilversmit, D. B. (1980c). Net transfer of phospholipid by the nonspecific phospholipid transfer proteins from bovine liver. *Biochim. Biophys. Acta* **620**, 37–48.

Daum, G., and Paltauf, F. (1984). Phospholipid transfer in yeast. Isolation and partial characterization of a phospholipid transfer protein from yeast cytosol. *Biochim. Biophys. Acta* **794**, 385–391.

Demel, R. A., Kalsbeek, R., Wirtz, K. W. A., and van Deenen, L. L. M. (1977). The protein mediated net transfer of phosphatidylinositol in model systems. *Biochim. Biophys. Acta* **466**, 10–22.

Désormeaux, A., Blochet, J.-E., Pézolet, M., and Marion, D. (1992). Amino acid sequence of a non-specific wheat phospholipid transfer protein and its conformation as revealed by infrared and Raman spectroscopy. Role of disulfide bridges and phospholipids in the stabilization of the α-helix structure. *Biochim. Biophys. Acta* **1121**, 137–152.

Dickeson, S. K., Lim, C. N., Schuyler, G. T., Dalton, T. P., Helmkamp, G. M., and Yarbrough, L. R. (1989). Isolation and sequence of cDNA clones encoding rat phosphatidylinositol transfer protein. *J. Biol. Chem.* **264**, 16557–16564.

DiCorleto, P. E., Warach, J. B., and Zilversmit, D. B. (1979). Purification and characterization of two phospholipid exchange proteins from bovine heart. *J. Biol. Chem.* **254**, 7795–7802.

Dieryck, W., Gautier, M.-F., Lullien, V., and Joudrier, P. (1992). Nucleotide sequence of a cDNA encoding a lipid transfer protein from wheat (*Triticum durum* Desf.). *Plant Mol. Biol.* **19**, 707–709.

Douady, D., Grosbois, M., Guerbette, F., and Kader, J. C. (1982). Purification of a basic phospholipid transfer protein from maize seedlings. *Biochim. Biophys. Acta* **710**, 143–153.

Douady, D., Guerbette, F., and Kader, J.-C. (1985). Purification of phospholipid transfer protein from maize seeds using a two-step chromatographic procedure. *Physiol. Veg.* **23**, 373–380.

Dyatlovitskaya, E. V., Timofeeva, N. G., and Bergelson, L. D. (1978). A universal lipid exchange protein from rat hepatoma. *Eur. J. Biochem.* **82**, 463–471.

Dyatlovitskaya, E. V., Timofeeva, N. G., Yakimenko, E. F., Barsukov, L. I., Muzya, G. I., and Bergelson, L. D. (1982). A sphingomyelin transfer protein in rat tumors and fetal liver. *Eur. J. Biochem.* **123**, 311–315.

Ericsson, J., Scallen, T. J., Chojnacki, T., and Dallner, G. (1991). Involvement of sterol carrier protein-2 in dolichol biosynthesis. *J. Biol. Chem.* **266**, 10602–10607.

Fleming, A. J., Mandel, T., Hofmann, S., Sterk, P., de Vries, S. C., and Kuhlemeier, C. (1992). Expression pattern of a tobacco lipid transfer protein gene within the shoot apex. *Plant J.* **2**, 855–862.

Franck, P. F. H., de Ree, J. M., Roelofsen, B., and Op den Kamp, J. A. F. (1984). Modification of the erythrocyte membrane by a non-specific lipid transfer protein. *Biochim. Biophys. Acta* **778**, 405–411.

Fujiki, Y., Tsuneoka, M., and Tashiro, Y. (1989). Biosynthesis of nonspecific lipid transfer protein (sterol carrier protein 2) on free polyribosomes as a large precursor in rat liver. *J. Biochem.* **106**, 1126–1131.

Gadella, T. W. J., Jr., and Wirtz, K. W. A. (1991). The low-affinity lipid binding site of the non-specific lipid transfer protein. Implications for its mode of action. *Biochim. Biophys Acta* **1070**, 237–245.

Gadella, T. W. J., Jr., Bastiaens, P. I. H., Visser, A. J. W. G., and Wirtz, K. W. A. (1991). Shape and lipid-binding site of the nonspecific lipid-transfer protein (sterol carrier protein 2): A steady-state and time-resolved fluorescence study. *Biochemistry* **30**, 5555–5564.

Gavey, K. L., Noland, B. J., and Scallen, T. J. (1981). The participation of sterol carrier protein$_2$ in the conversion of cholesterol ester by rat liver microsomes. *J. Biol. Chem.* **256**, 2993–2999.

Geelen, M. J. H., Beynen, A. C., and Wirtz, K. W. A. (1987). Cholesterol metabolism and sterol carrier protein-2 (non-specific lipid transfer protein). *Int. J. Biochem.* **19,** 619–623.

Geldwerth, D., De Kermel, A., Zachowski, A., Guerbette, F., Kader, J.-C., Henry, J. P., and Devaux, P. F. (1991). Use of spin-labeled and fluorescent lipids to study the activity of the phospholipid transfer protein from maize seedlings. *Biochim. Biophys. Acta* **1082,** 255–264.

George, P. Y., and Helmkamp, G. M., Jr. (1985). Purification and characterization of a phosphatidylinositol transfer protein from human platelets. *Biochim. Biophys. Acta* **836,** 176–184.

Gnamusch, E., Kalans, C., Hrastik, C., Paltauf, F., and Daum, G. (1992). Transport of phospholipids between cellular membranes of wild-type yeast cells and of the phosphatidylinositol transfer protein-deficient strain *Saccharomyces cerevisiae* sec14. *Biochim. Biophys. Acta* **1111,** 120–126.

Griff, I. C., Schekman, R., Rothman, J. E., and Kaiser, C. A. (1992). The yeast SEC17 gene product is functionally equivalent to mammalian α-SNAP protein. *J. Biol. Chem.* **267,** 12106–12115.

Grondin, P., Vergnolle, C., Chavant, L., and Kader, J. C. (1990). Purification and characterization of a novel phospholipid transfer protein from filamentous fungi. *Int. J. Biochem.* **22,** 93–98.

Hajra, A. K., and Bishop, J. E. (1982). Glycerolipid biosynthesis in peroxisomes via the acyl dihydroxyacetone phosphate pathway. *Ann. N.Y. Acad. Sci.* **386,** 170–182.

He, Z., Yamamoto, R., Furth, E. E., Schantz, L. J., Naylor, S. L., George, H., Billheimer, J. T., and Strauss, J. F., III. (1991). cDNAs encoding members of a family of proteins related to human sterol carrier protein 2 and assignment of the gene to human chromosome 1 p21→pter. *DNA Cell Biol.* **10,** 559–569.

Heikoop, J. C., Ossendorp, B. C., Wanders, R. J. A., Wirtz, K. W. A., and Tager, J. M. (1992). Subcellular localisation and processing of non-specific lipid transfer protein are not aberrant in Rhizomelic Chondrodysplasia Punctata fibroblasts. *FEBS Lett.* **299,** 201–204.

Helmkamp, G. M., Jr. (1986). Pospholipid transfer proteins: mechanism of action. *J. Bioenerg. Biomembr.* **18,** 71–91.

Helmkamp, G. M., Jr. (1990). Transport and metabolism of phosphatidylinositol in eukaryotic cells. *Subcell. Biochem.* **16,** 129–174.

Helmkamp, G. M., Jr., Harvey, M. S., Wirtz, K. W. A., and Van Deenen, L. L. M. (1974). Phospholipid exchange between membranes. Purification of bovine brain proteins that preferentially catalyze the transfer of phosphatidylinositol. *J. Biol. Chem.* **249,** 6382–6389.

Helmkamp, G. M., Jr., Venuti, S. E., and Dalton, T. P. (1992). Phosphatidylinositol transfer proteins from higher eukaryotes. *Meth. Enzymol.* **209,** 504–514.

Hendrick, J. P., Hodges, P. E., and Rosenberg, L. E. (1989). Survey of amino-terminal proteolytic cleavage sites in mitochondrial precursor proteins: Leader peptides cleaved by two matrix proteases share a three-amino acid motif. *Proc. Natl. Acad. Sci. U.S.A.* **86,** 4056–4060.

Heymans, H. S. A., Schutgens, R. B. H., Tan, R., Van den Bosch, H., and Borst, P. (1983). Severe plasmalogen deficiency in tissues of infants without peroxisomes (Zellweger syndrome). *Nature (London)* **306,** 69–70.

Horwich, A. (1990). Protein import into mitochondria and peroxisomes. *Curr. Opin. Cell Biol.* **2,** 625–633.

Hovik, R., and Osmundsen, H. (1987). Peroxisomal β-oxidation of long-chain fatty acids possessing different extents of unsaturation. *Biochem. J.* **247,** 531–535.

Hughes, M. A., Dunn, M. A., Pearce, R. S., White, A. J., and Zhang, L. (1992). An abscisic-acid-responsive, low temperature barley gene has homology with a maize phospholipid transfer protein. *Plant Cell Environ.* **15**, 861–865.

Kader, J. C. (1975). Proteins and the intracellular exchange of lipids. I. Stimulation of phospholipid exchange between mitochondria and microsomal fractions by proteins isolated from potato tuber. *Biochim. Biophys. Acta* **380**, 31–44.

Kader, J.-C. (1990). Intracellular transfer of phospholipids, galactolipids, and fatty acids in plant cells. *Subcell. Biochem.* **16**, 69–111.

Kader, J.-C., Douady, D., and Mazliak, P. (1982). Phospholipid transfer proteins. *In* "Phospholipids" (J. N. Hawthorne and G. B. Ansell, eds.), pp. 279–311. Elsevier, Amsterdam.

Kader, J. C., Julienne, M., and Vergnolle, C. (1984). Purification and characterization of a spinach-leaf protein capable of transferring phospholipids from liposomes to mitochondria or chloroplasts. *Eur. J. Biochem.* **139**, 411–416.

Keller, G.-A., Scallen, T. J., Clarke, D., Maher, P. A., Krisans, S. K., and Singer, S. J. (1989). Subcellular localization of sterol carrier protein-2 in rat hepatocytes: Its primary localization to peroxisomes. *J. Cell Biol.* **108**, 1353–1361.

Kølvraa, S., and Gregersen, N. (1986). *In vitro* studies on the oxidation of medium-chain dicarboxylic acids in rat liver. *Biochim. Biophys. Acta* **876**, 515–525.

Koumanov, K., Boyanov, A., Neicheva, T., Markovska, T., Momchilova, A., Gavazova, E., and Chelibonova-Lorer, H. (1982). Phospholipid composition of subcellular fractions and phospholipid-exchange activity in chicken liver and MC-29 hepatoma. *Biochim. Biophys. Acta* **713**, 23–28.

Krisans, S. K., Thompson, S. L., Pena, L. A., Kok, E., and Javitt, N. B. (1985). Bile acid synthesis in rat liver peroxisomes: metabolism of 26-hydroxycholesterol to 3β-hydroxy-5-cholenoic acid. *J. Lipid Res.* **26**, 1324–1332.

Lazarow, P. B., and De Duve, C. (1976). A fatty acyl-CoA oxidizing system in rat liver peroxisomes; enhancement by clofibrate, a hypolipidemic drug. *Proc. Natl. Acad. Sci. U.S.A.* **73**, 2043–2046.

Lazarow, P. B., and Fujiki, Y. (1985). Biogenesis of peroxisomes. *Ann. Rev. Cell. Biol.* **1**, 489–530.

Lazarow, P. B., and Moser, H. W. (1989). Disorders of peroxisome biogenesis. *In* "The Metabolic Basis of Inherited Disease" (C. R. Scriver, A. L. Beaudet, W. S. Sly, and D. Valle, eds.), pp. 1479–1509. McGraw–Hill, New York.

Linnestad, C., Lönneborg, A., Kalla, R., and Olsen, O.-A. (1991). Promoter of a lipid transfer protein gene expressed in barley aleurone cells contains similar *myb* and *myc* recognition sites as the maize *Bz-McC* allele. *Plant Physiol.* **97**, 841–843.

Liscum, L., Ruggiero, R. M., and Faust, J. R. (1989). The intracellular transport of low density lipoprotein-derived cholesterol is defective in Niemann–Pick type C fibroblasts. *J. Cell Biol.* **108**, 1625–1636.

Lumb, R. H., Kloosterman, A. D., Wirtz, K. W. A., and Van Deenen, L. L. M. (1976). Some properties of phospholipid exchange proteins from rat liver. *Eur. J. Biochem.* **69**, 15–22.

Lyons, H. T., Kharroubi, A., Wolins, N., Tenner, S., Chanderbhan, R. F., Fiskum, G., and Donaldson, R. P. (1991). Elevated cholesterol and decreased sterol carrier protein-2 in peroxisomes from AS-30D hepatoma compared to normal rat liver. *Arch. Biochem. Biophys.* **285**, 238–245.

Madrid, S. M. (1991). The barley lipid transfer protein is targeted into the lumen of the endoplasmic reticulum. *Plant Physiol. Biochem.* **29**, 695–703.

Masuta, C., Furunu, M., Tanaka, H., Yamada, M., and Koiwai, A. (1992). Molecular cloning of a cDNA clone for tobacco lipid transfer protein and expression of the functional protein in *Escherichia coli*. *FEBS Lett.* **311**, 119–123.

Matsuura, J. E., George, H. J., Ramachandran, G. N., Alavarez, J. G., Strauss, J. F., III, and Billheimer, J. T. (1993). Expression of the mature and the pro-form of human sterol carrier protein 2 in *Escherichia coli* alters bacterial lipids. *Biochemistry* **32**, 567–572.

McNamara, B. C., and Jefcoate, C. R. (1989). The role of sterol carrier protein 2 in stimulation of steroidogenesis in rat adrenal mitochondria by adrenal cytosol. *Arch. Biochem. Biophys.* **275**, 53–62.

Megli, F. M., De Lisi, A., Van Amerongen, A., Wirtz, K. W. A., and Quagliariello, E. (1986). Nonspecific lipid transfer protein (sterol carrier protein 2) is bound to rat liver mitochondria: Its role in spontaneous intermembrane phospholipid transfer. *Biochim. Biophys. Acta* **861**, 463–470.

Mochizuki, Y., Hruban, Z., Morris, H. P., Slesers, A., and Vigil, E. L. (1971). Microbodies of Morris hepatomas. *Cancer Res.* **31**, 763–773.

Molina, A., Segura, A., and García-Olmedo, F. (1993). Lipid transfer proteins (nsLTPs) from barley and maize leaves are potent inhibitors of bacterial and fungal plant pathogens. *FEBS Lett.* **316**, 119–122.

Moncecchi, D., Pastuszyn, A., and Scallen, T. J. (1991). cDNA sequence and bacterial expression of mouse liver sterol carrier protein-2. *J. Biol. Chem.* **266**, 9885–9892.

Mori, T., Tsukamoto, T., Mori, H., Tashiro, Y., and Fujiki, Y. (1991). Molecular cloning and deduced amino acid sequence of nonspecific lipid transfer protein (sterol carrier protein 2) of rat liver: A higher molecular mass (60 kDa) protein contains the primary sequence of nonspecific lipid transfer protein as its C-terminal part. *Proc. Natl. Acad. Sci. U.S.A.* **88**, 4338–4342.

Morris, H. R., Larsen, B. S., and Billheimer, J. T. (1988). A mass spectrometric study of the structure of sterol carrier protein SCP₂ from rat liver. *Biochem. Biophys. Res. Commun.* **154**, 476–482.

Mundy, J., and Rogers, J. C. (1986). Selective expression of a probable amylase/protease inhibitor in barley aleurone cells: Comparison to the barley amylase/subtilisin inhibitor. *Planta* **169**, 51–63.

Nichols, J. W. (1987). Binding of fluorescent-labeled phosphatidylcholine to rat liver nonspecific lipid transfer protein. *J. Biol. Chem.* **262**, 14172–14177.

Nichols, J. W., and Pagano, R. F. (1983). Resonance energy transfer of protein-mediated lipid transfer between vesicles. *J. Biol. Chem.* **258**, 5368–5371.

Noland, B. J., Arebalo, R. E., Hansbury, E., and Scallen, T. J. (1980). Purification and properties of sterol carrier protein₂. *J. Biol. Chem.* **255**, 4282–4289.

North, P., and Fleischer, S. (1983). Use of a non-specific lipid transfer protein to modify the cholesterol content of synaptic membranes. *Methods Enzymol.* **98**, 599–613.

Novick, P., Field, C., and Schekman, R. (1980). Identification of 23 complementation groups required for post-translational events in the yeast secretory pathway. *Cell* **21**, 205–215.

Nyquist, D. A., and Helmkamp, G. M. (1989). Developmental patterns in rat brain of phosphatidylinositol synthetic enzymes and phosphatidylinositol transfer protein. *Biochim. Biophys. Acta* **987**, 165–170.

Ossendorp, B. C., Van Heusden, G. P. H., and Wirtz, K. W. A. (1990). The amino acid sequence of rat liver non-specific lipid transfer protein (sterol carrier protein 2) is present in a high molecular weight protein: Evidence from cDNA analysis. *Biochem. Biophys. Res. Commun.* **168**, 631–636.

Ossendorp, B. C., Van Heusden, G. P. H., De Beer, A. L. J., Bos., K., Schouten, G. L., and Wirtz, K. W. A. (1991). Identification of the cDNA clone which encodes the 58-kDa protein containing the amino acid sequence of rat liver non-specific lipid-transfer protein (sterol carrier protein 2). Homology with rat peroxisomal and mitochondrial 3-oxoacyl-CoA thiolases. *Eur. J. Biochem.* **201**, 233–239.

Ossendorp, B. C., Geijtenbeek, T. B. H., and Wirtz, K. W. A. (1992). The precursor form of the rat liver non-specific lipid-transfer protein, expressed in *Escherichia coli*, has lipid transfer activity. *FEBS Lett.* **296,** 179–183.

Osumi, T., Tsukamoto, T., and Hata, S. (1992). Signal peptide for peroxisomal targeting: replacement of an essential histidine residue by certain amino acids converts the amino-terminal presequence of peroxisomal 3-ketoacyl-CoA thiolase to a mitochondrial signal peptide. *Biochem. Biophys. Res. Commun.* **186,** 811–818.

Paltauf, F., and Daum, G. (1990). Phospholipid transfer in microorganisms. *Subcell. Biochem.* **16,** 279–299.

Pastuszyn, A., Noland, B. J., Bazan, J. F., Fletterick, R. J., and Scallen, T. J. (1987). Primary sequence and structural analysis of sterol carrier protein 2 from rat liver: Homology with immunoglobulins. *J. Biol. Chem.* **262,** 13219–13227.

Paulussen, J. A., and Veerkamp, J. H. (1990). Intracellular fatty-acid-binding proteins. Characteristics and function. *Subcell. Biochem.* **16,** 175–226.

Pebay-Peyroula, E., Cohen-Addad, C., Lehmann, M. S., and Marion, D. (1992). Crystallographic data for the 9000 Dalton wheat non-specific phospholipid transfer protein. *J. Mol. Biol.* **226,** 563–564.

Poorthuis, B. J. H. M., and Wirtz, K. W. A. (1982). Increased cholesterol esterification in rat liver microsomes by purified non-specific phospholipid transfer protein. *Biochim. Biophys. Acta* **710,** 99–105.

Poorthuis, B. J. H. M., Van der Krift, T. P., Teerlink, T., Akeroyd, R., Hostetler, K. Y., and Wirtz, K. W. A. (1980). Phospholipid transfer activities in Morris hepatomas and the specific contribution of the phosphatidylcholine exchange protein. *Biochim. Biophys. Acta* **600,** 376–386.

Poorthuis, B. J. H. M., Glatz, J. F. C., Akeroyd, R., and Wirtz, K. W. A. (1981). A new high-yield procedure for the purification of the non-specific phospholipid transfer protein from rat liver. *Biochim. Biophys. Acta* **665,** 256–261.

Reinhart, M. P., Avart, S. J., and Foglia, T. (1991). Purification, characterization and comparison with mammalian SCP_2 of a chicken SCP_2-like protein. *Comp. Biochem. Physiol.* **100B,** 243–248.

Rennert, H., Amsterdam, A., Billheimer, J. T., and Strauss, J. F., III (1991). Regulated expression of sterol carrier protein 2 in the ovary: a key role for cyclic AMP. *Biochemistry* **30,** 11280–11285.

Rickers, J., Spener, F., and Kader, J.-C. (1985). A phospholipid transfer protein that binds long-chain fatty acids. *FEBS Lett.* **180,** 29–32.

Roff, C. F., Pastuszyn, A., Strauss, J. F., III, Billheimer, J. T., Vanier, M. T., Brady, R. O., Scallen, T. J., and Pentchev, P. G. (1992). Deficiencies in sex-regulated expression and levels of two hepatic sterol carrier proteins in a murine model of Niemann-Pick type C disease. *J. Biol. Chem.* **267,** 15902–15908.

Roise, D., and Schatz, G. (1988). Mitochondrial presequences. Minireview. *J. Biol. Chem.* **263,** 4509–4511.

Rothman, J. E., and Orci, L. (1992). Molecular dissection of the secretory pathway. *Nature (London)* **355,** 409–415.

Salama, S. R., Cleves, A. E., Malehorn, D. E., Whitters, E. A., and Bankaitis, V. A. (1990). Cloning and characterization of *Kluyveromyces lactis* SEC14, a gene whose product stimulates Golgi secretory function in *Saccharomyces cerevisiae*. *J. Bacteriol.* **172,** 4510–4521.

Scallen, T. J., Srikantaiah, M. V., Seetharam, B., Hansbury, E., and Gavey, K. L. (1974). Sterol carrier protein hypothesis. *Fed. Proc.* **33,** 1733–1746.

Scallen, T. J., Pastuszyn, A., Noland, B. J., Chanderbhan, R., Kharroubi, A., and Vahouny, G. V. (1985). Sterol carrier and lipid transfer proteins. *Chem. Phys. Lipids* **38,** 239–261.

Schepers, L., Casteels, M., Vamecq, J., Parmentier, G., Van Veldhoven, P. P., and Mannaerts, G. P. (1988). β-Oxidation of the carboxylic side chain of prostaglandin E2 in rat liver peroxisomes and mitochondria. *J. Biol. Chem.* **263**, 2724–2731.

Schroeder, F., Butko, P., Nemecz, G., and Scallen, T. J. (1990). Interaction of fluorescent 5,7,9(11),22-ergostatetraen-3β-ol with sterol carrier protein-2. *J. Biol. Chem.* **265**, 151–157.

Seedorf, U., and Assmann, G. (1991). Cloning, expression and nucleotide sequence of rat liver sterol carrier protein 2 cDNAs. *J. Biol. Chem.* **266**, 630–636.

Seedorf, U., Raabe, M., and Assmann, G. (1993). Cloning, expression and sequences of mouse sterol-carrier protein-x-encoding cDNAs and a related pseudogene. *Gene* **123**, 165–172.

Seltman, H., Diven, W., Rizk, M., Noland, B. J., Chanderbhan, R., Scallen, T. J., Vahouny, G., and Sanghvi, A. (1985). Regulation of bile-acid synthesis. Role of sterol carrier protein₂ in the biosynthesis of 7α-hydroxycholesterol. *Biochem. J.* **230**, 19–24.

Simorre, J.-P., Caille, A., Marion, D., Marion, D., and Ptak, M. (1991). Two- and three-dimensional ¹H NMR studies of a wheat phospholipid transfer protein: Sequential resonance assignments and secondary structure. *Biochemistry* **30**, 11600–11608.

Singh, I., Moser, A. B., Goldfischer, S., and Moser, H. W. (1984). Lignoceric acid is oxidized in the peroxisome: Implications for the Zellweger cerebro-hepato-renal syndrome and adrenoleukodystrophy. *Proc. Natl. Acad. Sci. U.S.A.* **81**, 4203–4207.

Skriver, K., Leah, R., Müller-Uri, F., Olsen, F.-L., and Mundy, J. (1992). Structure and expression of the barley lipid transfer protein gene *Ltp1*. *Plant Mol. Biol.* **18**, 585–589.

Snoek, G. T., De Wit, I. S. C., Van Mourik, J. H. G., and Wirtz, K. W. A. (1992). The phosphatidylinositol transfer protein in 3T3 mouse fibroblast cells is associated with the Golgi system. *J. Cell. Biochem.* **49**, 339–348.

Snoek, G. T., Westerman, J., Wouters, F. S. and Wirtz, K. W. A. (1993a). Phosphorylation and redistribution of the phosphatidylinositol-transfer protein in phorbol 12-myristate 13-acetate- and bombesin-stimulated Swiss mouse 3T3 fibroblasts. *Biochem. J.* **291**, 649–656.

Snoek, G. T., De Wit, I. S. C. and Wirtz, K. W. A. (1993b). Properties and intracellular localization of phosphatidylinositol transfer protein in Swiss mouse 3T3 cells. *Biochem. Soc. Trans.* **21**, 244–247.

Somerharju, P. J., Van Paridon, P., and Wirtz, K. W. A. (1983). Phosphatidylinositol transfer protein from bovine brain. Substrate specificity and membrane binding properties. *Biochim. Biophys. Acta* **731**, 186–195.

Sossountov, L., Ruiz-Avila, L., Vignois, F., Jolliot, A., Arondel, V., Tchang, F., Grosbois, M., Guerbette, F., Miginiac, E., Delseny, M., Puigdomenèch, P., and Kader, J.-C. (1991). Spatial and temporal expression of a maize lipid transfer protein gene. *Plant Cell* **3**, 923–933.

Steinschneider, A., McLean, M. P., Billheimer, J. T., Azhar, S., and Gibori, G. (1989). Protein kinase C catalyzed phosphorylation of sterol carrier protein 2. *Endocrinology* **125**, 569–571.

Sterk, P., Booij, H., Schellekens, G. A., Van Kammen, A., and de Vries, S. C. (1991). Cell-specific expression of the carrot EP2 lipid transfer protein gene. *Plant Cell* **3**, 907–921.

Suzuki, Y., Yamaguchi, S., Orii, T., Tsuneoka, M., and Tashiro, T. (1990). Nonspecific lipid transfer protein (sterol carrier protein-2) defective in patients with deficient peroxisomes. *Cell Struct. Funct.* **15**, 301–308.

Svensson, B., Asanao, K., Jonassen, I., Poulsen, F. M., Mundy, J., and Svendsen, I. (1986). A 10-kD barley seed protein homologous with an α-amylase inhibitor from indian finger millet. *Carlsberg Res. Commun.* **51**, 493–500.

Swinkels, B. W., Gould, S. J., Bodnar, A. G., Rachubinski, R. A., and Subramani, S. (1991). A novel, cleavable peroxisomal targeting signal at the amino-terminus of the rat 3-ketoacyl-CoA thiolase. *EMBO J.* **10**, 3255–3262.

Szolderits, G., Hermetter, A., Paltauf, F., and Daum, G. (1989). Membrane properties modulate the activity of a phosphatidylinositol transfer protein from the yeast *Saccharomyces cerevisiae*. *Biochim. Biophys. Acta* **986**, 301–309.

Takada, Y., Kaneko, N., Esumi, H., Purdue, P. E., and Danpure, C. J. (1990). Human peroxisomal L-alanine:glyoxylate aminotransferase. Evolutionary loss of a mitochondrial targeting signal by point mutation of the initiation codon. *Biochem. J.* **268**, 517–520.

Takishima, K., Watanabe, S., Yamada, M., and Mamiya, G. (1986). The amino acid sequence of the nonspecific lipid transfer protein from germinated castor bean endosperms. *Biochim. Biophys. Acta* **870**, 248–255.

Takishima, K., Watanabe, S., Yamada, M., Suga, T., and Mamiya, G. (1988). Amino acid sequences of two nonspecific lipid-transfer proteins from germinated castor bean. *Eur. J. Biochem.* **177**, 241–249.

Tan, H., Okazaki, K., Kubota, I., Kamiryo, T., and Utiyama, H. (1990). A novel peroxisomal nonspecific lipid-transfer protein from *Candida tropicalis*. Gene structure, purification and possible role in β-oxidation. *Eur. J. Biochem.* **190**, 107–112.

Tanaka, T., Billheimer, J. T., and Strauss, J. F., III (1984). Luteinized rat ovaries contain a sterol carrier protein. *Endocrinology* **114**, 533–540.

Tchang, F., This, P., Stiefel, V., Arondel, V., Morch, M.-D., Pages, M., Puigdomenech, P., Grellet, F., Delseny, M., Bouillon, P., Huet, J.-C., Guerbette, F., Beauvais-Cante, F., Duranton, H., Pernollet, J.-C., and Kader, J.-C. (1988). Phospholipid transfer protein: full-length cDNA and amino acid sequence in maize. Amino acid sequence homologies between plant phospholipid transfer proteins. *J. Biol. Chem.* **263**, 16849–16855.

Teerlink, T., Van der Krift, T. P., Van Heusden, G. P. H., and Wirtz, K. W. A. (1984). Determination of nonspecific lipid transfer protein in rat tissues and Morris hepatomas by enzyme immunoassay. *Biochim. Biophys. Acta* **793**, 251–259.

Terras, F. R. G., Goderis, I. J., Van Leuven, F., Vanderleyden, J., Cammue, B. P. A., and Broekaert, W. F. (1992). In vitro antifungal activity of a radish (*Raphanus sativus* L.) seed protein homologous to nonspecific lipid transfer proteins. *Plant Physiol.* **100**, 1055–1058.

Thomas, P. J., Wendelburg, B. E., Venuti, S. E., and Helmkamp, G. M., Jr. (1989). Mature rat testis contains a high molecular weight species of phsohpatidylinositol transfer protein. *Biochim. Biophys. Acta* **982**, 24–30.

Thompson, S. L., Burrows, R., Laub, R. J., and Krisans, S. K. (1987). Cholesterol synthesis in rat liver peroxisomes. *J. Biol. Chem.* **262**, 17420–17425.

Torres-Schumann, S., Godoy, J. A., and Pintor-Toro, J. A. (1992). A probable lipid transfer protein is induced by NaCl in stems of tomato plants. *Plant Mol. Biol.* **18**, 749–757.

Traszkos, J. M., and Gaylor, J. L. (1983). Cytosolic modulators of activities of microsomal enzymes of cholesterol biosynthesis. Purification and characterization of a non-specific lipid-transfer protein. *Biochim. Biophys. Acta* **751**, 52–65.

Trzeciak, W. H., Simpson, E. R., Scallen, T. J., Vahouny, G. V., and Waterman, M. R. (1987). Studies on the synthesis of sterol carrier protein-2 in rat adrenocortical cells in monolayer culture. Regulation by ACTH and dibutyryl cyclic 3′,5′-AMP. *J. Biol. Chem.* **262**, 3713–3717.

Tsao, F. H. C., Tian, Q., and Strickland, M. S. (1992). Purification, characterization and substrate specificity of rabbit lung phospholipid transfer proteins. *Biochim. Biophys. Acta* **1125**, 321–329.

Tsuboi, S., Watanabe, S.-I., Ozeki, Y., and Yamada, M. (1989). Biosynthesis of nonspecific lipid transfer proteins in germinating castor bean seeds. *Plant Physiol.* **90**, 841–845.

Tsuboi, S., Suga, T., Takishima, K., Mamiya, G., Matsui, K., Ozeki, Y., and Yamada, M. (1991). Organ-specific occurrence and expression of the isoforms of nonspecific lipid transfer protein in castor bean seedlings, and molecular cloning of a full-length cDNA for a cotyledon-specific isoform. *J. Biochem.* **110**, 823–831.

Tsuboi, S., Osafune, T., Tsugeki, R., Nishimura, M., and Yamada, M. (1992). Nonspecific lipid transfer protein in castor bean cotyledon cells: Subcellular localization and a possible role in lipid metabolism. *J. Biochem.* **111**, 500–508.

Tsuneoka, M., Yamamoto, A., Fujiki, Y., and Tashiro, Y. (1988). Nonspecific lipid transfer protein (sterol carrier protein-2) is located in rat liver peroxisomes. *J. Biochem.* **104**, 560–564.

Vahouny, G. V., Chanderbhan, R., Noland, B. J., Irwin, D., Dennis, P., Lambeth, J. D., and Scallen, T. J. (1983). Sterol carrier protein$_2$. Identification of adrenal sterol carrier protein$_2$ and site of action for mitochondrial cholesterol utilization. *J. Biol. Chem.* **258**, 11731–11737.

Van Amerongen, A., Teerlink, T., Van Heusden, G. P. H., and Wirtz, K. W. A. (1985). The non-specific lipid transfer protein (sterol carrier protein 2) from rat and bovine liver. *Chem. Phys. Lipids* **38**, 195–204.

Van Amerongen, A., Helms, J. B., Van der Krift, T. P., Schutgens, R. B. H., and Wirtz, K. W. A. (1987). Purification of nonspecific lipid transfer protein (sterol carrier protein 2) from human liver and its deficiency in livers from patients with cerebro-hepato-renal (Zellweger) syndrome. *Biochim. Biophys. Acta* **919**, 149–155.

Van Amerongen, A., Van Noort, M., Van Beckhoven, J. R. C. M., Rommerts, F. F. G., Orly, J., and Wirtz, K. W. A. (1989a). The subcellular distribution of the nonspecific lipid transfer protein (sterol carrier protein 2) in rat liver and adrenal gland. *Biochim. Biophys. Acta* **1001**, 243–248.

Van Amerongen, A., Demel, R. A., Westerman, J., and Wirtz, K. W. A. (1989b). Transfer of cholesterol and oxysterol derivatives by the nonspecific lipid transfer protein (sterol carrier protein 2): A study on its mode of action. *Biochim. Biophys. Acta* **1004**, 36–43.

Van den Akker, W. M. R., Westerman, J., Gadella, T. W. J., Wirtz, K. W. A., and Snoek, G. T. (1990). Proteolytic activation of a bovine brain protein with phosphatidylinositol transfer activity. *FEBS Lett.* **276**, 123–126.

Van den Bosch, H., Schutgens, R. B. H., Wanders, R. J. A., and Tager, J. M. (1992). Biochemistry of peroxisomes. *Annu. Rev. Biochem.* **61**, 157–197.

Van der Krift, T. P., Leunissen, J., Teerlink, T., Van Heusden, G. P. H., Verkleij, A. J., and Wirtz, K. W. A. (1985). Ultrastructural localization of a peroxisomal protein in rat liver using the specific antibody against the non-specific lipid transfer protein (sterol carrier protein 2). *Biochim. Biophys. Acta* **812**, 387–392.

Van Haren, L., Teerds, K. J., Ossendorp, B. C., Van Heusden, G. P. H., Orly, J., Stocco, D. M., Wirtz, K. W. A., and Rommerts, F. F. G. (1992). Sterol carrier protein 2 (non-specific lipid-transfer protein) is localized in membranous fractions of Leydig cells and Serboli cells but not in germ cells. *Biochim. Biophys. Acta* **1124**, 288–296.

Van Heusden, G. P. H., Souren, J., Geelen, M. J. H., and Wirtz, K. W. A. (1985). The synthesis and esterification of cholesterol by hepatocytes and H35 hepatoma cells are independent of the level of nonspecific lipid transfer protein. *Biochim. Biophys. Acta* **846**, 21–25.

Van Heusden, G. P. H., Bos, K., Raetz, C. R. H., and Wirtz, K. W. A. (1990a). Chinese hamster ovary cells deficient in peroxisomes lack the nonspecific lipid transfer protein (sterol carrier protein 2). *J. Biol. Chem.* **265**, 4105–4110.

Van Heusden, G. P. H., Bos, K., and Wirtz, K. W. A. (1990b). The occurrence of soluble and membrane-bound non-specific lipid transfer protein (sterol carrier protein 2) in rat tissues. *Biochim. Biophys. Acta* **1046**, 315–321.

Van Heusden, G. P. H., Van Beckhoven, J. R. C. M., Thieringer, R., Raetz, C. R. H., and Wirtz, K. W. A. (1992). Increased cholesterol synthesis in Chinese hamster ovary cells deficient in peroxisomes. *Biochim. Biophys. Acta* **1126**, 81–87.

Van Paridon, P. A., Gadella, T. W. J., Somerharju, P. J. and Wirtz, K. W. A. (1987a). On the relationship between the dual specificity of the bovine brain phosphatidylinositol transfer protein and membrane phosphatidylinositol levels. *Biochim. Biophys. Acta* **903**, 68–77.

Van Paridon, P. A., Visser, A. J. W. G., and Wirtz, K. W. A. (1987b). Binding of phospholipids to the phosphatidylinositol transfer protein from bovine brain as studied by steady-state and time-resolved fluorescence spectroscopy. *Biochim. Biophys. Acta* **898**, 172–180.

Van Paridon, P., Somerharju, P., and Wirtz, K. W. A. (1987c). Phosphatidylinositol transfer protein and cellular phosphatidylinositol metabolism. *Biochem. Soc. Trans.* **15**, 321–323.

Van Paridon, P. A., Gadella, T. W. J., Jr., Somerharju, P. J., and Wirtz, K. W. A. (1988). Properties of the binding sites for the *sn*-1 and *sn*-2 acyl chains on the phosphatidylinositol transfer protein from bovine brain. *Biochemistry* **27**, 6208–6214.

Venuti, S. E., and Helmkamp, G. M. (1988). Tissue distribution, purification and characterization of rat phosphatidylinositol transfer protein. *Biochim. Biophys. Acta* **946**, 119–128.

Vergnolle, C., Arondel, V., Jolliot, A., and Kader, J.-C. (1992). Phospholipid transfer proteins from higher plants. *Methods Enzymol.* **209**, 522–530.

Von Heijne, G., Steppuhn, J., and Hermann, R. G. (1989). Domain structure of mitochondrial and chloroplast targeting peptides. *Eur. J. Biochem.* **180**, 535–545.

Watanabe, S., and Yamada, M. (1986). Purification and characterization of a non-specific lipid transfer protein from germinated castor bean endosperms which transfers phospholipids and galactolipids. *Biochim. Biophys. Acta* **876**, 116–123.

Westerman, J., and Wirtz, K. W. A. (1985). The primary structure of the nonspecific lipid transfer protein (sterol carrier protein 2) from bovine liver. *Biochem. Biophys. Res. Commun.* **127**, 333–338.

Wirtz, K. W. A. (1982). Phospholipid transfer proteins. *Lipid–Protein Interact.* **1**, 151–231.

Wirtz, K. W. A. (1991). Phospholipid transfer proteins. *Annu. Rev. Biochem.* **60**, 73–99.

Wirtz, K. W. A., and Moonen, P. (1977). Interaction of the phosphatidylcholine exchange protein with phospholipids. A fluorescence and circular dichroism study. *Eur. J. Biochem.* **77**, 437–443.

Wirtz, K. W. A., and Zilversmit, D. B. (1968). Exchange of phospholipids between liver mitochondria and microsomes in vitro. *J. Biol. Chem.* **243**, 3596–3602.

Wirtz, K. W. A., and Zilversmit, D. B. (1970). Partial purification of phospholipid exchange protein from beef heart. *FEBS Lett.* **7**, 44–46.

Wirtz, K. W. A., Kamp, H. H., and Van Deenen, L. L. M. (1972). Isolation of a protein from beef liver which specifically stimulates the exchange of phosphatidylcholine. *Biochim. Biophys. Acta* **274**, 606–617.

Wirtz, K. W. A., Jolles, J., Westerman, J., and Neys, F. (1976). Phospholipid exchange proteins in synapsome and myelin fraction from rat brain. *Nature* (*London*) **260**, 354–355.

Wirtz, K. W. A., Helmkamp, G. M., Jr., and Demel, R. A. (1978). The phosphatidylinositol exchange protein from bovine brain. *In* "Protides of the Biological Fluids" (H. Peters, ed.), pp. 25–31. Pergamon Press, Oxford.

Wirtz, K. W. A., Op den Kamp, J. A. F., and Roelofsen, B. (1986a). Phosphatidylcholine transfer protein: properties and applications in membrane research. *Progr. Protein–Lipid Interact.* **2**, 221–265.

Wirtz, K. W. A., Westerman, J., Van Amerongen, A., and Van der Krift, T. P. (1986b). Structure and function of the nonspecific lipid transfer protein (sterol carrier protein

2). *In* "Enzymes of Lipid Metabolism II" (L. Freysz, H. Dreyfus, R. Massarelli, and S. Gatt, eds.), pp. 429–435. Plenum Press, New York.

Yamada, J., Itoh, S., Horie, S., Watanabe, T., and Suga, T. (1986). Chain-shortening of a xenobiotic acyl compound by the peroxisomal β-oxidation system in rat liver. *Biochem. Pharmacol.* **35**, 4363–4368.

Yamamoto, R., Kallen, C. B., Babalola, G. O., Rennert, H., Billheimer, J. T., and Strauss, J. F. III (1991). Cloning and expression of a cDNA encoding human sterol carrier protein 2. *Proc. Natl. Acad. Sci. U.S.A.* **88**, 463–467.

Yu, Y. G., Chung, C. H., Fowler, A., and Suh, S. W. (1988). Amino acid sequence of a probable amylase/protease inhibitor from rice seeds. *Arch. Biochem. Biophys.* **265**, 466–475.

Zborowski, J., and Demel, R. A. (1982). Transfer properties of the bovine brain phospholipid transfer protein. Effect of charged phospholipids and of phosphatidylcholine fatty acid composition. *Biochim. Biophys. Acta* **688**, 381–387.

CHAPTER 10

Fatty Acid Binding Proteins

Torsten Börchers and Friedrich Spener
Department of Biochemistry, University of Münster, D-48149 Münster, Germany

I. INTRODUCTION

In 1972, fatty acid binding proteins (FABPs) were described for the first time (Ockner *et al.*, 1972).These intracellular cytosolic proteins with relatively low molecular masses (14–15 kDa) were discovered in a variety of tissues because of their ability to bind radioactively labeled fatty acids. Since then, a wealth of information has been collected on the level of primary as well as tertiary structure that permits the classification of the different fatty acid binding proteins as a protein family comprising at least five members or types (Table I). In conjunction with proteins that bind retinoids and a bile acid binding protein, the FABPs form the family of small intracellular hydrophobic ligand binding proteins. This family is clearly distinguishable from the lipocalins, a family of extracellular lipid-binding proteins containing, among others, serum retinol binding protein, microglobulin, β-lactoglobulin, and bilin binding protein.

From the ability of FABPs to bind fatty acids with high affinity and from their abundance in tissues with "active lipid metabolism" such as

TABLE I

Family of Intracellular Hydrophobic Ligand-Binding Proteins

Protein[a]	Ligand	Homology[b] (%)	Tissue occurrence
H-FABP	Fatty acid	100	Heart, skeletal muscle, mammary gland, kidney, stomach, aorta, lung
A-FABP	Fatty acid	63	Adipose tissue
M-FABP	Fatty acid	61	Peripheral nerve myelin
CRABP	Retinoic acid	43	Widespread
CRABP II	Retinoic acid	43	Widespread
CRBP	Retinol	34	Liver, lung, kidney, testis
CRBP II	Retinol, retinal	40	Intestine
I-FABP	Fatty acid	30	Intestine
ILBP	Bile acids	27	Intestine
L-FABP	Fatty acids, others	27	Liver, intestine, kidney

[a] Abbreviations: FABP, fatty acid binding protein; H, heart; A, adipocyte; M, myelin; I, intestinal; L, liver; CRABP, cellular retinoic acid binding protein; CRBP, cellular retinol binding protein; ILBP, ileal lipid binding protein.

[b] The homology was calculated relative to H-FABP for human proteins, except for ILBP, which was from pig.

heart, liver, intestine, and adipose tissue, in which they constitute up to 5% of cytosolic proteins, a role in lipid metabolism was inferred for these proteins. However, no conclusive evidence *in vivo* for this function has been presented to date, and the concepts of how FABPs may modulate the activity of the enzymes of lipid metabolism are still speculative. For reviews on these aspects, see Spener *et al.* (1989) and Veerkamp *et al.* (1991). The regulation of FABP expression on the gene level and the so-called functions of the "second generation"-like signal transduction and involvement in differentiation processes have attracted much attention. This chapter reviews the structural peculiarities of the various types of FABPs, their regulation, and the relevance of these findings to the different functional concepts.

II. FAMILY OF FATTY ACID BINDING PROTEINS

Originally, FABPs were named according to the tissue of their first isolation. Immunological studies then revealed the presence of distinct types of FABPs, namely, intestinal type (I-FABP), liver type (L-FABP), and heart type (H-FABP). The specific reaction of the respective antibodies reflects homologies of ~30% among the different FABP types. On the other hand, immunological cross-reactivity among orthologous FABPs (i.e., FABPs of the same type but from different mammalian species) correlates with homologies of ~90%. Later, binding studies and tertiary structures (see Section IV) suggested that two other proteins are also members of the FABP family, namely, the adipocyte type (A-FABP) that historically was designated p422, aP2, p15, or adipocyte lipid-binding protein (ALBP) and the myelin type (M-FABP) that originally was termed mP2. The latter types are more related to H-FABP than to the others (Table I).

FABPs from other tissues such as muscle, kidney, mammary gland, and brain also have been regarded as distinct FABP types. However, careful sequence analysis in these cases unravelled identity to one of the already mentioned types. FABP from human muscle has been cloned (Peeters et al., 1991) and was found to be identical to H-FABP from human heart (Börchers et al., 1990). The presumed renal type FABP actually was an N-terminally processed α_{2u}-globulin, a male-specific protein of the lipocalin family, secreted from liver and endocytosed by the proximal tubules of the kidney (Kimura et al., 1991). With immunological and polymerase chain reaction (PCR) techniques, H-FABP and L-FABP have been identified as the fatty acid binding proteins present in kidney (Maatman et al., 1992). Work in our laboratory revealed that the mammary gland FABP (P. D. Jones et al., 1988), also called mammary-derived growth inhibitor (see Section IV), actually is H-FABP (Nielsen et al., 1993). Northern blot analysis indicated the presence of mRNA specific for H-FABP in brain. The existence of a brain type FABP that is highly homologous (95%) to bovine H-FABP (Billich et al., 1988), as postulated by Schoentgen et al. (1989), is not very likely, particularly since the protein used for sequencing apparently was heterogeneous in isoelectric focusing. However, in this regard, the description of a heretofore unknown brain-specific FABP with ~70% identity to H-FABP is of great interest (T. Müller, personal communication). This protein was discovered during screening of brain mRNA for H-FABP-related proteins by PCR. The protein is expressed specifically in the olfactory bulb and, thus, tentatively is termed O-FABP. Heterologous expression of this protein will facilitate analysis of ligand binding and generation of antibodies for immunohistochemical investigations.

An unsettled issue at present is the newly discovered FABP from skin epidermis. This protein first was identified in two-dimensional gel electrophoresis of cultured human psoriatic keratinocytes as one of the proteins that is highly up-regulated under these pathological conditions. This protein was sequenced partially directly from the blot of a two-dimensional gel and cloned. The complete sequence of the cDNA revealed 48% identity to H-FABP (Madsen *et al.*, 1992). Note that the skin FABP contains 6 cysteines (compared with a maximum of 2 in the other established FABP types; see Fig. 1) and seems to be secreted by keratinocytes. Independently, (the same?) epidermal FABP was identified by autoradioblotting after incubation of cytosolic proteins from normal epidermis, cultured keratinocytes, and psoriatic epidermis and subsequent gel electrophoresis under nondenaturing conditions (Siegenthaler *et al.*, 1993). In this study, the highest FABP concentration was found under psoriatic conditions as well. Our understanding of this interesting new FABP type, linked to a pathological situation, should increase by additional investigations including quantitative analysis of ligand binding, structure, tissue localization, and, eventually, regulation of expression at the gene level (see Section IV).

In addition to the five established FABP types and the two newly described putative olfactory type and epidermal type FABPs, proteins capable of retinoid binding—the cellular retinol binding proteins (CRBPs) and the cellular retinoic acid binding proteins (CRABPs)—form a subfamily of the intracellular hydrophobic ligand binding proteins (Table I). In addition to their binding specificity, cellular retinoid binding proteins differ from FABPs by their considerably lower abundance. The concentration of CRBP II, the most abundant cellular retinoid binding protein, in the intestinal mucosa is still almost an order of magnitude lower than that of I-FABP and L-FABP. A comparative survey of the different functional concepts of FABPs and cellular retinoid binding proteins was done by Bass (1993).

The last member of the family of hydrophobic ligand binding proteins, the ileal lipid binding protein (ILBP), originally was discovered as a porcine ileal polypeptide (PIP) that was expressed specifically in the villus-associated enterocytes. The hypothesis that this protein was an intestinal phase hormone that stimulated gastric acid secretion led to the name gastrotropin. However, cloning and sequence analysis revealed a striking 37% homology to L-FABP (Gantz *et al.*, 1989). The protein was shown not to be secreted and to have no influence on acid secretion of isolated gastric parietal cells. Despite the similarity to L-FABP, ILBP has only low affinity for fatty acids; instead, this protein binds bile salts with high affinity and is thought to be involved in their absorption from the intestine (Miller and Cistola, 1993).

The different members of the family of intracellular hydrophobic ligand binding proteins, their predominant ligand, and their tissue occurrence (see Section III) are summarized in Table I. The proteins are listed in order

```
                  10        20        30        40        50        60        70
                   .         .         .         .         .         .         .

E-FABP   ATVQQLEGRWRLVDSKGFDEYMKELGVGIALRKMGAMAKPDCIITCDGKNLTIKTESTLKTTQFSCTLGE
A-FABP   --CDAFVGTWKLVSSENFDDYMKEVGVGFATRKVAGMAKPNMIISVNGDVITIKSESTFKNTEISFILGQ
M-FABP   --SNKFLGTWKLVSSENFDDYMKALGVGLATRKLGNLAKPTVIISKKGDIITIRTESTFKNTEISFKLGQ
H-FABP   --VDAFLGTWKLVDSKNFDDYMKSLGVGFATRQVASMTKPTTIIEKNGDILTLKTHSTFKNTEISFKLGV
O-FABP   --VDAFCATWKLTDSQNFDEYMKALGVGFATRQVGNVTKPTVIISQEGGKVVIRTQCTFKNTEINFQLGE
I-FABP   ------AFDSTWKVDRSENYDKFMEKMGVNIVKRKLAAHDNLKLTITQEGNKFTVKESSAFRNIEVVFELGV
L-FABP   ---MSFSGKYQLQSQENFEAFMKAIGLPEELIQKGKDIKGVSEIVQNGKHFKFTITAGSKVIQNEFTVGE


E-FABP   KFEETTADGRKTQTVCNFTDGALVQHQEW--DGKESTITRKLKDGKLVVECVMNNVTCTRIYEKVE
A-FABP   EFDEVTADDRKVKSTITLDGGVLVHVQKW--DGKSTTIKRKREDDKLVVECVMKGVTSTRVYER-A
M-FABP   EFEETTADNRKTKSIVTLQRYSLNQVQRW--NGKETTIKRKLVDGKMVAECKMKGVVCTRIYEK-V
H-FABP   EFDETTADDRKVKSIVTLDGGKLVHLQKW--DGQETTLVRELIDGKLILTLTHGTAVCTRTYQKEA
O-FABP   EFDETSIDDRNCKSVVRLDGDKLIHVQKW--DGKETNCTREIKDGKMVVTLTFGDIVAVRCYEK-A
I-FABP   TFNYNLADGTELRGTWSLEGNKLIGKFKRTDNGNELNTVREIIGDELVQTYVYEGVEAKRIFKKD-
L-FABP   ECELETMTGEKVKTVVQLEGDN-----KLVTTFKNIKSVTELNGDIITNTMTLGDIVFKRI-SKRI
```

FIGURE 1 Alignment of fatty acid binding proteins (FABPs). All FABPs are of human origin, except O-FABP, which is from mouse. Only identities are marked.

of their homology to H-FABP, giving a rough estimate of the evolutionary relationship among the proteins, since proteins that are listed next to each other in this table are more related. For detailed analysis, a phylogenetic tree was constructed (Fig. 2). This analysis shows that cellular retinoid binding proteins and the subfamily comprising A-, M-, H-, O-, and E-FABP have a common progenitor that diverged early from I-FABP and from the L-FABP/ILBP subfamily. Thus, the specialized function of retinoid binding has evolved from fatty acid binding. This process can be reversed in the laboratory by site-directed mutagenesis, as demonstrated in Section IV. Note that the L-FABP/ILBP subfamily has the longest evolutionary distance from the other members, as reflected by their binding characteristics. A common ancestral origin of the hydrophobic ligand

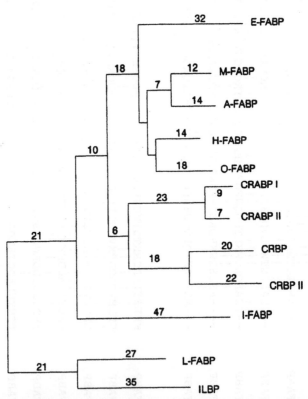

FIGURE 2 Phylogenetic tree of the family of fatty acid binding proteins (FABPs). The program package TREE, based on algorithms described by Feng and Doolittle (1987), was used. First an approximate phylogenetic order is determined by a series of pairwise alignments with the Needleman–Wunsch method. Any combinations of subclusters are prealigned before the final alignment in the form of a dendrogram is plotted.

binding protein family also is supported by the conserved intron–exon structure. All genes of this family analyzed to date contain 4 exons and 3 similarly positioned introns of variable size. A calculation by Matarese *et al.* (1989) indicated that the different subfamilies diverged from the common ancestral genes about 700 million years ago, that is, before the invertebrate–vertebrate divergence.

III. OCCURRENCE AND PHYSIOLOGICAL ASPECTS

Table I compiles the occurrence of particular FABPs in different tissues. Proteins have been identified using FABP-specific antibodies in enzyme-linked immunosorbent assay (ELISA), Western blotting, or immunohisto-chemistry; corresponding mRNA was detected by hybridization with respective cDNAs or oligonucleotides in Northern blotting or *in situ* hybridization. Most FABPs are rather tissue specific, whereas H-FABP has a widespread distribution. The presence of more than one FABP type in a particular tissue—as observed for the intestine, kidney, stomach, and probably brain (see previous discussion)—generally is regarded as a functional diversification of FABPs.

H-FABP is found in heart, different muscle types, mammary gland, kidney, aorta, stomach, and placenta. One hypothesis links the H-FABP content with the β-oxidative capacity of the respective tissue. This idea is supported by a comparison of the H-FABP concentration in myocardium, slow-twitch red muscle fibers, and fast-twitch white muscle fibers. The decrease of oxidative capacity and the concomitant increase in capacity for glycogenesis in these tissues is paralleled by a decrease in H-FABP. Concentrations of H-FABP in, for example, soleus and red gastrocnemius were ~30% of those in heart, whereas only minute amounts of H-FABP were measured in extensor digitorum longis and white gastrocnemius (Crisman *et al.*, 1987). In agreement with these findings, the H-FABP level of the tibialis anterior muscle, which contains intermediate amounts of H-FABP, can be raised almost to those of the cardiac muscle by chronic low frequency stimulation (Kaufmann *et al.*, 1989). Changes of the H-FABP content in the developing heart also reflect the shift in metabolism that occurs at birth. The fetal rat heart, in which H-FABP is first detectable at the end of gestation (day 17) and does not exceed 10–20% of the maximum value found in adult heart, relies on a predominantly anaerobic glycolytic metabolism. In the adult heart, fatty acids are the preferred substrate and H-FABP mRNA consistently rises 3-fold from day 19 until the day after parturition (Heuckeroth *et al.*, 1987) and increases gradually to the maximum level. This concept that H-FABP funnels the large flux of fatty acids from the plasma membrane to mitochondrial β-oxidation is challenged by the finding of H-FABP in mammary gland, in which the mass flow of fatty acids is directed toward triglyceride synthesis, and in brain, which does not oxidize fatty acids for energy production.

In the intestine, four members of the family of intracellular hydrophobic ligand binding proteins are expressed in a distinct cell-, region-, and developmentally specific manner: I-FABP, L-FABP, ILBP, and CRBP II. This expression pattern reflects the role of the intestine in entry of nutritional lipids to cellular metabolism. L-FABP and I-FABP form gradients along the two axes of the small intestine. L-FABP and I-FABP concentrations decrease from the proximal to distal parts of the intestine, with highest concentrations measured in the jejunum. Virtually no L-FABP or I-FABP is expressed in the colon. With respect to the vertical axis, both FABPs are absent from the crypts and are abundant in the differentiated enterocytes of the villus tip. Of the various cell types present in the intestine, only the absorptive enterocytes and a small subpopulation of enteroendocrine cells contain L-FABP, whereas I-FABP is confined to enterocytes and to exocrine goblet cells. Since the villus tips of the jejunum are the predominant region of fatty acid uptake, the observed horizontal and vertical gradients are consistent with a role for FABPs in fatty acid metabolism. By the same token ILBP, which according to *in vitro* studies (discussed in Section IV) binds chenodeoxycholic acid with higher affinity than fatty acids, is most abundant in the ileal epithelium (Gantz *et al.,* 1989). This finding coincides with the fact that the more distal part of the small intestine is involved in the absorption of bile salts. Since ILBP is not confined to exocrine cells but is also present in absorptive epithelial cells, a gastrotropin function, that is, the induction of acid secretion in gastric parietal cells, is unlikely. How the complex pattern of FABP expression within the rapidly and continuously regenerating intestine is maintained is discussed in Section V.

A cell- and region-specific expression also was observed in liver, the other L-FABP-containing tissue. Only hepatocytes—not nonparenchymal cells such as bile ductular epithelium, sinusoidal endothelial cells, or Kupffer cells—express L-FABP. A declining gradient from the portal to the central zones of the liver acinus is observed (Bass *et al.,* 1989). The concentration of L-FABP in portal hepatocytes of male rats is 1.6 times higher than that in central hepatocytes. This ratio is less pronounced in female rats and is not affected by the hypolipidemic drug clofibrate (see Section V), which increases the L-FABP concentration proportionally in both zones. Since no correlation with the capacity for utilization or synthesis of fatty acids was observed, this acinar gradient may reflect differences in the flux of fatty acids or fatty acid metabolites.

In human kidney, H-FABP and L-FABP are similarly abundant (0.5% and 1% of cytosolic proteins, respectively). For unknown reasons, the concentration of L-FABP in male as well as in female rat kidney is one order of magnitude lower than that in human kidney or that of H-FABP in rat kidney (Maatman *et al.*, 1992). Immunoperoxidase staining of rat renal tissue revealed L-FABP in proximal and distal tubules, whereas H-FABP was found only in distal tubules.

In contrast to the tissues already discussed, different FABP types, that is, H-FABP and I-FABP, are present in the different cells of adult rat stomach. Immunocytochemistry revealed a predominant localization of H-FABP in parietal cells and of I-FABP in surface mucous cells (Iseki *et al.*, 1991). Additionally, during the last days of embryogenesis and during the suckling period, L-FABP is present in a variety of stomach cells as a third FABP type.

Developmental changes have been reported for most FABPs. In the rat intestine, mRNA for L-FABP as well as for I-FABP is first detectable during late fetal life, that is, day 17 to 19 of gestation (Gordon *et al.*, 1985). The concentration of these mRNAs sharply increases several times immediately after birth and stays relatively constant during the suckling and weanling period (up to 35 days postpartum). The abrupt induction of FABPs may be related to the high fat diet during the suckling period. In adult rat, the intestinal concentration is increased further by a factor 2. Similar results have been observed for L-FABP mRNA in liver. Note that, already in late fetal life, a proximal to distal gradient of L-FABP is established in the intestine. In contrast, at that time this protein is distributed uniformly in the liver; no acinar gradient could be observed.

Although primarily cytosolic, a portion of FABPs is also organelle bound. By immunogold staining of electron microscopy samples and by other immunochemical techniques, H-FABP was detected on myofibrils and within the mitochondrial matrix (Fournier and Rahim, 1985; Börchers et al., 1989), whereas L-FABP was bound preferentially to membranes of mitochondria and endoplasmic reticulum (Bordewick *et al.*, 1989; Spener *et al.*, 1990). This type-specific subcellular localization of FABPs may be related to functional differences, for example, β-oxidation vs triglyceride synthesis. In light of a possible involvement of FABPs in signal transduction (see Section VI), note that H-FABP as well as L-FABP also has been found in the nucleus (Börchers *et al.*, 1989, Bordewick et al., 1989; Fahimi *et al.*, 1990).

IV. STRUCTURE AND LIGAND BINDING

The sequence homologies among the different types of FABPs constitute an important classification principle within this protein family, as was discussed in Section II. Figure 1 summarizes the sequences of the five well-known FABP types and the two new types from epidermis and olfactory bulb. Some of the most conserved amino acids are the glycines, asparagines, and aspartates that are necessary for the formation of narrow β-turns. Also highly conserved is an arginine that is involved in fatty acid binding (Arg 126 of H-FABP).

In addition to the occurrence of different types, another level of diversity has been observed since for some FABPs isoforms have been described. The term "isoform" pertains to a relatively small change in the primary structure (usually only one amino acid) that often results in a difference in the isoelectric point. Other heterogeneities are caused by the status of lipidation, posttranslational modifications, or changes in conformation. An explanation for the latter relates to the presence of partially unfolded states of the protein; for example, for A-FABP, several imunochemically identical forms have been observed after ion exchange chromatography. In an analysis of these forms by denaturing isoelectric focusing, A-FABP exhibited different isoelectric points depending on the urea concentration. With progressive unfolding, the pI of the protein shifted from an acidic pH to pH 8–9, consistent with predictions from the amino acid composition (Matarese et al., 1989). The authors conclude that this behavior is not a result of delipidation but of splitting of intramolecular salt bridges and the emergence of previously buried amino acids. Similar shifts were observed for H-FABP as well (Nielsen et al., 1990).

In the bovine heart, two isoforms of H-FABP have been described: pI 4.9 H-FABP and pI 5.1 H-FABP. These isoforms are not interconvertible; their pI is not altered by lipidation and both forms are detectable in denaturing isoelectric focusing gels. The isoforms have been obtained in homogeneous form after separation by anion exchange chromatography and the heterogeneity has been traced back to the level of primary structure. The only difference between the isoforms is an asparagine–aspartate exchange at position 98 (Unterberg et al., 1990). Interestingly, the previously cloned cDNA for bovine H-FABP encoded for Asn 98 (Billich et al., 1988). We also were able to elucidate the combination of events that cause the heterogeneic pattern of bovine L-FABP in isoelectric focusing. As reported earlier (Haunerland et al., 1984), apparently two isoforms of apo L-FABP exist (pI 6.0 L-FABP and pI 7.0 L-FABP) that, on lipidation, bind one and two molecules of fatty acid, respectively. The pI of the lipidated isoforms shifts to 5.0 in either case. A possible explanation for this tremendous shift is the formation of a salt bridge between an exposed positive charge on the protein with the carboxyl group of the fatty acid. That the negative charge of the fatty acid per se is responsible for the pI shift seems rather unlikely, since this behavior is a unique property of L-FABP. In the other FABP types, the fatty acid and, hence, the Arg 106 that interacts with the carboxyl group (see subsequent text) is buried deeply within the hydrophobic pocket of the protein. The separation of the two apo-isoforms of L-FABP by preparative isoelectric focusing in stretched pH gradients revealed another microheterogeneity of the pI 7.0 isoform. By careful sequencing of the protein and mass spectrometric analysis of peptides, the difference between the isoforms could be attributed to an asparagine–aspartate exchange at position 105, similar to the one observed for bovine H-FABP, and to a covalent modification of Cys 69 of the pI 7.0 isoform by either cysteine or glutathione (Dörmann et al., 1993). These phenomena are not specific to bovine tissues, since heterogeneity also was reported for rat and human L-FABP and for rat H-FABP (Ockner et al., 1982; P. D. Jones, et al., 1988).

At present, neither the origin of these asparagine–aspartate exchanges that lead to H-FABP and L-FABP isoforms nor their functional relevance is clear; at least affinity of fatty acid binding is not affected by this phenomenon. However, the exchange seems to alter some properties of the binding site of L-FABP, as documented by the decreased binding stoichiometry. On the one hand, spontaneous deamidation of asparagine may reflect an aging process of proteins, particularly since asparagine precedes glycine in both cases. On the other hand, the existence of isoforms may be the result of distinct mRNAs and may represent a further adaptation of FABP to interaction with certain membranes, enzymes, or receptors. In the case of bovine H-FABP, the latter opinion is supported by several lines of evidence.

1. Analysis by denaturing isoelectric focusing of immunoprecipitated [^{35}S]methionine-labeled products translated in vitro from total mRNA as well as from positive hybrid-selected bovine H-FABP mRNA yielded two proteins that comigrated with the authentic pI 5.1 and pI 4.9 H-FABP isoforms. In control experiments, in vitro transcription/translation of recombinant pI 5.1 H-FABP cDNA produced only the pI 5.1 isoform (Bartetzko et al., 1993).
2. The existence of different mRNAs apparently also applies to M-FABP, since in this case asparagine was found at position 98 of the protein of human origin whereas aspartate was encoded by the cDNA (Hayasaka et al., 1991). Spontaneous amidation of aspartyl residues in proteins has not been described in the literature.
3. In both FABP types, the exchange position is located prominently within the β-turn between β-strands G and H and protrudes considerably from the surface of the protein, emphasizing its possible function as a recognition signal.
4. Interestingly, only the pI 4.9 isoform of H-FABP is found in the mitochondrial matrix (Unterberg et al., 1990), suggesting an involvement of Asp 98 in a specific import mechanism.

Despite the only moderate similarity among the members of the family of hydrophobic ligand binding proteins on the level of primary structure, a striking coincidence exists among the tertiary structures. To date, tertiary structures for members of all established FABP types except L-FABP have been reported. The structures of bovine M-FABP (T. A. Jones *et al.*, 1988), rat I-FABP (Scapin *et al.*, 1992), bovine H-FABP (Müller-Fahrnow *et al.*, 1991), human H-FABP (Zanotti *et al.*, 1992), and murine A-FABP (Xu *et al.*, 1992) are assembled from 10 antiparallel β-strands, arranged as two orthogonal β-sheets, and from 2 short α-helices located between β-strand A and B. Since the two β-sheets have the overall shape of a clam shell, this structural motif is also called β-clam. The similarity among these structures is remarkable; the C_α backbones are superimposable without major deviations. The mechanism of binding of the fatty acids has been deciphered progressively using high resolution structures of rat I-FABP and human H-FABP, which have been crystallized as holo proteins with bound ligands (see Fig. 3). Comparison with a refined structure of rat apo I-FABP at a resolution of 0.12 nm was possible. Surprisingly, pronounced differences with respect to the conformation of the bound ligand exist. In I-FABP, the fatty acid spans almost the whole protein in a slightly bent conformation and interacts with Arg 106. As a key hydrophobic interaction, Tyr 70 and Tyr 117 sandwich the fatty acid, thus stabilizing its conformation. Phe 57 interacts with the ω-methylene group of the fatty acid and serves as a kind of lid for the binding pocket (Sacchettini *et al.*, 1992). In contrast, in H-FABP, the fatty acid has an "u-shaped" conformation and is wound around Phe 16 of α-helix I. Hence, the fatty acid has more contact with the upper part of the cavity, mostly with hydrophobic amino acids of the α-helices that represent the upper boundary of the binding site. The lower part of the cavity is filled completely by side chains of hydrophobic and aromatic amino acids. Tyr 70 and Tyr 117, which are important in I-FABP, are not conserved in H-FABP. Moreover, this type also has a different hydrogen bonding network with the carboxyl group of the fatty acid, as displayed in Fig. 3A and B. The fatty acid interacts with Arg 126 and Tyr 128 and, via an ordered water molecule, with Arg 106 of H-FABP (Zanotti *et al.*, 1992).

Although X-ray crystallography gives a principally static picture, some clues may be obtained about the binding process. The following scenario has been proposed for I-FABP (Gordon *et al.*, 1991). The fatty acid, headed by its carboxyl group, enters the cavity through a portal area bordered by α_{II}, the turn between β_E and β_F, and Phe 55 of the turn between β_C and β_D. Most of the ordered solvent molecules that can be seen in the apo I-FABP structure are expelled through a gap between β_D and β_E (see Fig. 3). The carboxylate moves through the binding cavity until

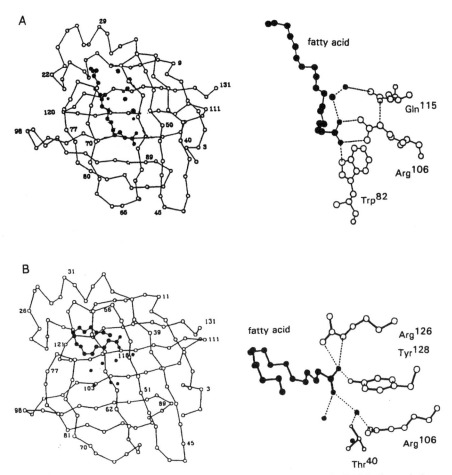

FIGURE 3 Localization of fatty acid in intestinal (A) and heart (B) FABP. Open circles represent the C_α chain, filled circles the fatty acid and internal solvent. Reproduced from Sacchettini *et al.* (1992) and Zanotti *et al.* (1992) with permission.

it reaches Arg 106, where it displaces the water molecule that maintains the hydrogen bonding network. The driving force for the binding of the fatty acid may be the entropic effect connected with the displacement of ordered water molecules. The structure of H-FABP indicates an opening near Phe 57, which is equivalent to Phe 55 in I-FABP, and reveals several ordered water molecules, suggesting that the fatty acid may enter H-FABP in a manner comparable to its entering I-FABP. Analysis of fatty acid binding to some FABPs by titration calorimetry gave access to the enthalpic and entropic components of the binding energy (Miller and Cistola, 1993). At least for I-FABP, the entropic term TΔS is small compared with the binding enthalpy ΔH. Thus, these thermodynamic data do not support the model just described.

From an evolutionary perspective, speculating that the relatively symmetric organization of the β-barrel originated from a doubling of a common ancestral gene is tempting. This view is supported by the intriguing finding that, in all FABP genes, intron 2 is positioned exactly in the middle of β-strand F, commonly shared by the two orthogonal β-sheets of the protein. According to another hypothesis, put forward by Petrou et al. (1993), this β-clam structure also may be present as a distinct domain in the N-methyl-D-aspartate (NMDA) receptor and may supply the known modulation of this ion channel by fatty acids. A comparison of residues 263–393 of the NMDA receptor with bovine H-FABP revealed a 33% identity. Amino acids that are important for formation of β-turns (Gly, Asp) as well as those involved in fatty acid binding (Arg 126, Tyr 128, Phe 57; see previous discussion) are conserved in both proteins.

An interesting phenomenon is the evolution of retinoid binding from fatty acid binding (cf. Section II). The binding affinity of retinoid binding proteins is about two orders of magnitude greater than that of FABPs for their ligands. This high affinity binding seems to have evolved to insure effective sequestration of even small amounts of the very potent retinoids by their respective binding proteins. The major difference in the primary structures of the CRBPs is the presence of glutamines in positions 109 and 129 rather than the arginines in equivalent positions of FABPs and CRABPs. The crystal structure of CRBP II was solved (Winter et al., 1993) and indicated that the protein, as expected, has the same β-clam topology described for the fatty acid binding proteins and that the hydroxyl group of the retinol is anchored at Gln 109 via a hydrogen bond. Whether Gln 109 is the key determinant for retinol binding was tested by reciprocal mutations of CRBP II (Q109R) and I-FABP (R106Q). Indeed, fatty acid binding properties of CRBP II (Q109R) were identical to those of wild-type I-FABP (K_d 0.2 μM; one fatty acid per protein molecule) whereas retinol binding was abolished. The I-FABP (R106Q) mutant still bound one fatty acid but with 20-fold lower affinity (K_d 4 μM); instead, it now bound retinol with high affinity (Jakoby et al., 1993). Interestingly, substitution of neither Gln 109 nor Gln 129 alone with arginine generated a retinoic acid binding protein (Cheng et al., 1991). Presumably, both arginines that are present in CRABP at positions 111 and 131 are necessary to form a more extended hydrogen bonding network similar to that in H-FABP (Zhang et al., 1992). Mutants in which these arginines are replaced by glutamine bind retinoic acid with low affinity.

As already discussed, (see Fig. 2), L-FABP is related most distantly to the other FABP types. Note that both L-FABP and ILBP do not contain

Arg 106. L-FABP has a threonine in this position and ILBP an alanine. In both proteins Arg 122, which is equivalent to Arg 126 of H-FABP (see Fig. 1), is conserved. This special status is emphasized by the different ligand binding characteristics of these proteins. Unfortunately, no crystal structure of L-FABP or ILBP has been solved to date, and information about ligand binding to these proteins has been gained only by indirect means. Some properties of the binding site can be disclosed by the ligands that are bound. ILBP binds bile acids such as chenodeoxycholic acid with high affinity but has negligible affinity for fatty acids (Miller and Cistola, 1993). L-FABP, on the other hand, differs from the other FABP types inasmuch as it binds two molecules of fatty acid (Haunerland *et al.*, 1984; Miller and Cistola, 1993) and various other bulky ligands. Thus, the binding site of L-FABP is larger than that of the other FABP types and is able to accommodate fatty acids with fluorescent side chains (Haunerland *et al.*, 1984), heme (Vincent and Muller-Eberhard, 1985; Börchers and Spener, 1993), lysophosphatidic acid (Vancura and Haldar, 1992), and some eicosanoids (Raza *et al.*, 1989). A common characteristic of these ligands is the negatively charged head group, pin-pointing an involvement of a positively charged amino acid in L-FABP. This hypothesis is supported by the fact that, in contrast to lysophosphatidic acid, monoglycerides do not bind to this protein (A. Meyjohann and F. Spener, unpublished observations), nor do fatty acid methyl esters (Haunerland *et al.*, 1984). Further, the binding affinity of L-FABP for fatty acids and heme decreased by more than 50% on modification of arginine by phenylglyoxal. Modification experiments in the presence of fatty acid revealed the protection of one of the two arginines of L-FABP. By peptide mapping and Edman degradation, Arg 122 was identified as the counterpart of the carboxyl group of the fatty acid (Börchers and Spener, 1993). Thus, anchoring of the ligand appears to occur by the same mechanism in L-FABP as in the other FABP types. The binding of two other hydrophobic ligands, acyl CoA and cholesterol, often is referred to in the literature. However, compelling evidence now suggests that acyl CoA is bound by a specialized transport protein, the acyl-CoA binding protein, that is structurally unrelated to FABP, and that L-FABP cannot compete with this protein for acyl CoAs (Rasmussen *et al.*, 1990). Controversial data exist regarding the binding of cholesterol. Binding of cholesterol to recombinant L-FABP was reported (Nemecz and Schroeder, 1991), but other investigators could not demonstrate affinity of L-FABP for cholesterol by a variety of methods (Scallen *et al.*, 1985; B. Rolf and F. Spener, unpublished results). Reliable binding experiments are hampered by the very low aqueous solubility of this hydrophobic molecule.

To increase our knowledge about the interactions of the hydrophobic tail of the fatty acid, two approaches were undertaken. First, fluorescent probes were attached at different positions of the fatty acid backbone to explore hydrophobicity and rigidity of the binding site of L-FABP (Storch *et al.*, 1989). Not surprisingly, large fluorescence quantum yields and long lifetimes for the excited state indicated a highly hydrophobic binding site. The emission maxima suggested that the fatty acid is constrained within the binding site, which is inaccessible to the aqueous phase as judged by quenching of fluorescence by acrylamide. When the fluorescent reporter is attached near the carboxyl group of the fatty acid, the relatively low quantum yield is consistent with localization of this part of the fatty acid near the aqueous surface of L-FABP, as was deduced also by Cistola *et al.* (1989) using ^{13}C-NMR spectroscopy. Second, in our laboratory we aimed to identify amino acids that are in close contact with the hydrophobic tail of the fatty acid. For this purpose, photoaffinity labeling with two different photoreactive radioactive fatty acids was employed. The photolabile group resided in a position equivalent to C16 of fatty acids (T. Börchers, M. Burow, and F. Spener, unpublished observations). In both cases, the photoreactive fatty acid attached to the same region of L-FABP, namely, amino acids 60 to 68, corresponding to the end of β_D and the following turn (if L-FABP is assumed to have the same overall structure as the other FABPs). With one of the photoreactive fatty acids, 4-azido-2,4-dinitrophenyl aminoundecanoic acid, three alternating amino acids (Phe 63, Leu 65, and Asp 67) were labeled specifically, indicative of a β-strand. These results, in conjunction with the interaction of Arg 122 with the carboxyl group, implicate that the first of the two fatty acids that bind to L-FABP covers the "lower part" of the binding site and has a different orientation as the fatty acid bound, for example, to I-FABP. Additional clarification of this issue awaits X-ray or NMR structural analysis of L-FABP.

V. REGULATION OF FATTY ACID BINDING PROTEIN EXPRESSION

The availability of FABP genes, particularly of large promoter regions, stimulated research that was dedicated to the disclosure of those *cis*-acting elements—and possible *trans*-acting factors—that govern cell-specific as well as time- and differentiation-dependent expression. In contrast to the large number of protein and cDNA sequences that have been elaborated in the last several years, genomic sequences of members of this family, that is, for L-FABP (Sweetser *et al.*, 1986), I-FABP (Sweetser *et al.*, 1987), A-FABP (Hunt *et al.*, 1986), and CRBP II (Demmer *et al.*, 1987) are reported more scarcely. Only recently did Narayanan *et al.* (1991) report the structure of the mouse myelin type FABP gene and did Müller and collaborators (personal communication) succeed in cloning the mouse H-FABP gene, facilitating analysis of the regulation of a widespread member of the FABP family.

Some information may be discerned by comparative analysis of the 5'-nontranscribed nucleotides of different genes. Comparison of the upstream regions of the human and rat I-FABP gene reveals several common 14-bp sequences that consist of two direct 7 nucleotide repeats (consensus: 5'-*TGAACTT TGAACTT*-3'). Interestingly, this sequence also is found in other genes that are expressed in the intestine, for example, CRBP II and apolipoprotein AII, whereas it is not present in the promoter of the L-FABP and A-FABP genes. Similarly, the A-FABP gene promoter contains a 13-bp consensus sequence termed "fat specific element 1" (FSE 1; 5'-GGCT/ACTGGTCAG/TG-3') that also is found in the 5'-nontranscribed regions of glycerolphosphate dehydrogenase and adipsin. A second sequence, FSE 2 (5'-GACTCAGAGGAAAC-3'), located at position −118, also is common to the A-FABP and glycerolphosphate dehydrogenase genes (Hunt et al., 1986). A putative glucocorticoid responsive element (GRE) is located at −393 and a CCAAT enhancer binding sequence is found at position −140.

However, finding conclusive evidence that these sequences indeed confer cell-specific expression or response to some factors or reagents necessitates transfection experiments with either cultured cell lines or transgenic mice. In these studies, the promoter of the FABP analyzed was fused with a reporter gene: chloramphenicol acetyltransferase (CAT) in the case of cell cultures and human growth hormone (hGH) in the case of most transgenes. In the CAT assay, enzymatic conversion of chloramphenicol to radioactively labeled acetylated forms is measured. The presence of reporter in transgenes is assessed by Northern analysis and immunocytochemical methods.

A rather detailed mapping of the A-FABP promoter was achieved by transfecting adipocytes with a series of deletion mutations around putative responsive elements. This approach was aided greatly by the availability of cell culture models such as 3T3-L1 adipocytes or related cell lines, which reproduce the differentiation process of fat cell development. Fibroblast-like preadipocytes can be stimulated by various agents to differentiate into mature adipocytes, accompanied by accumulation of lipid droplets and the induction of a number of adipose specific genes such as those for the serine protease adipsin, glycerolphosphate dehydrogenase, and A-FABP (Spiegelman et al., 1983). The differentiation program of the 3T3 cells can be initiated by the phosphodiesterase inhibitor 3-butyl-1-methylxanthine, indicating involvement of cAMP, the glucocorticoid dexamethasone, and insulin and insulin-like growth factor. During differentiation, A-FABP is activated transcriptionally, leading to ~100-fold increased accumulation of its mRNA. The cognate responsive elements of the various inducers of differentiation have been identified by mapping the A-FABP promoter linked to CAT. After transfection of preadipocytes or adipocytes, even a proximal promoter starting from nucleotide −168 could direct CAT expression in a differentiation-dependent manner, that is, in differentiated adipocytes but not in preadipocytes. Progressive 5′-deletion of the A-FABP promoter up to −141 diminished CAT activity in adipocytes; further deletion up to −120 then induced CAT expression in preadipocytes since the FSE 2 was removed partially or completely (A-FABP −93 to +21/CAT). Further deletion (A-FABP −73 to +21/CAT) totally abolished reporter expression in preadipocytes and in adipocytes (Distel *et al.*, 1987). These results are explained as follows: first, a positive regulatory element is removed—the CCAAT enhancer that binds the *trans*-activator C/EBP, which is known to activate several genes during differentiation (Herrera *et al.*, 1989). Then the adjacent FSE 2—a negative regulatory element that binds the AP-1 complex, a dimer comprising c-*fos* and c-*jun* products or related transcription factors—is eliminated (Fig. 4). This inhibitory effect could be relieved by competition with synthetic FSE 2 oligomer. In line with these findings, C/EBP mRNA is elevated on adipocyte differentiation and c-*fos* mRNA is known to be induced by insulin in adipocytes. The cAMP-mediated activation of A-FABP is observed only in confluent preadipocytes and probably is conferred by relieving the inhibition by the negative regulatory element. In accordance with the putative GRE at position −393, only a construct containing the first 858 nucleotides of the A-FABP promoter but not one starting at −248 could be stimulated by dexamethasone (Cook *et al.*, 1988). Agents such as retinoic acid (Stone and Bernlohr, 1990) that attenuate adipocyte differentiation also inhibit the expression of A-FABP and stearoyl-CoA desaturase. Concomitantly, a prolonged expression of c-*jun* mRNA but a decrease in c-*fos* and *jun*-B is observed, indicating the importance of the relative composition of the transcription factor AP-1.

Surprisingly, studies conducted in transgenic mice demonstrated that up to −1.7 kb of the A-FABP promoter was not sufficient to direct reporter expression to the adipose tissue (Ross *et al.*, 1990). Instead, a 518-bp enhancer was identified at −5.4 kb containing several responsive elements that could induce high CAT activity in adipose tissue of transgenic mice when linked to a fragment of the A-FABP promoter starting at −63 bp (Fig. 4). As expected from cell culture experiments, this upstream truncated promoter alone was not effective, although it contained the proximal FSE 1. Apparently, the promoter binding sites for AP-1 and C/EBP are neither

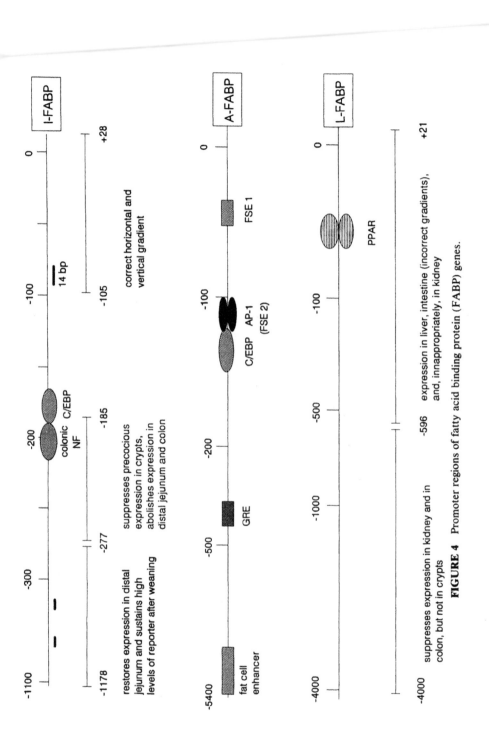

FIGURE 4 Promoter regions of fatty acid binding protein (FABP) genes.

sufficient nor necessary for adipose-specific reporter expression *in vivo*, but may be involved in the modulation of promoter activity. The fat cell enhancer recognizes various nuclear factors, as established by DNA mobility shift binding assays. One of the responsive elements present in this enhancer, ARE 6, accounted for the cell-type specificity of the enhancer since it stimulated differentiation-dependent CAT expression. In this experiment, multiple copies of ARE 6 activated the reporter from an enhancerless SV40 early promoter only in adipocytes, not in preadipocytes. The cognate nuclear factor ARF 6 is only present in nuclear extracts of adipocytes and may be a key regulator of adipogenic gene regulation (Graves *et al.*, 1992).

The gut epithelial cell lineage consists of four different cell types that originate from a single multipotential stem cell anchored in the crypts of intestinal villi. During upward migration of stem cell descendants to the villus tip, differentiation into absorptive enterocytes, mucus-producing goblet cells, and enteroendocrine cells occurs. The lysozyme-producing Paneth cells differentiate during downward migration to the crypt base (Gordon *et al.*, 1991). As already mentioned, FABP expression in the intestine is confined to enterocytes as well as to some enteroendocrine cells (L-FABP) and goblet cells (I-FABP), and is organized spatially along two axes (crypt to villus and proximal to distal). Given the complex differentiation program that leads to this cell-specific expression, which cannot be reproduced by cell cultures, transgenic mice were used to analyze the cell-specific, regional, and temporal expression of FABP genes. Various regions of the I-FABP promoter were linked to the reporter hGH. In a first study, Sweetser *et al.* (1988) showed that *cis*-acting elements present in the first 277 nucleotides of the rat I-FABP promoter—which are relatively conserved in the orthologous rat, mouse, and human I-FABP genes—were sufficient to limit hGH expression to the small intestine. The (I-FABP -277 to $+28$/hGH) construct induced reporter expression only after the epithelial cells left the crypt, and hGH was not detected in Paneth cells or enteroendocrine cells. However, hGH mRNA inappropriately was almost absent in the distal jejunum and in the ileum. With the longer promoter construct (I-FABP -1178 to $+28$/hGH), the proximal to distal gradient of I-FABP mRNA expression was mimicked correctly since reporter gene transcription was restored in the distal parts of the small intestine. In another study, the short promoter was dissected further (Cohn *et al.*, 1992). A promoter containing merely nucleotides -103 to $+28$ directed proper cell- and region-specific reporter expression. Note that the 14-bp consensus sequence closest to the transcription start is located within this minimal promoter. A (I-FABP -185 to $+28$/hGH) chimera still established a correct horizontal gradient but, in contrast to the (I-FABP $-277/+28$/hGH) construct, epithelial cells of the crypt now expressed hGH. With respect to this finding, the fact that removal of nucleotides -277 to -185 eliminated an element that binds a nuclear factor present in the colonic but not the small intestinal epithelial cells, as well as a part of the CCAAT enhancer binding protein sequence, may be relevant. The effects of the various portions of the I-FABP promoter are summarized in Fig. 4.

Although only two chimeras of the L-FABP promoter, of which a considerably larger part up to -4000 bp is available, and the hGH reporter have been analyzed to date (reviewed by Gordon *et al.*, 1991), pronounced differences from the I-FABP promoter exist. Nucleotides -596 to $+21$ of the L-FABP promoter could initiate expression of hGH in liver and small intestine but failed to suppress expression in the crypt cells. Further, the colon and renal epithelial cells expressed the reporter protein. Addition of nucleotides -4000 to -597 in either polarity abolished this inappropriate regional expression, but still could not establish a correct vertical gradient. The identification of reasonable amounts of L-FABP in human kidney (in contrast to relatively low amounts in rat; see previous discussion) indicates that regulatory elements may indeed exist that can direct expression of L-FABP in renal epithelial cells and that, for unknown reasons, suppress this expression in rodents. Considering the high copy number of L-FABP promoter–hGH constructs incorporated into the genome of transgenic mice (up to 600), a dose effect may be responsible for the precocious expression of hGH in the crypt cells. Alternatively, *cis*-acting suppressor elements may exist outside the 4000 nucleotides of the L-FABP promoter. With both promoters, hGH was expressed in a variety of enteroendocrine cells from which L-FABP is normally absent.

The developmental regulation of L-FABP expression in gut has been studied as well. The endogenous FABP gene is activated at day 16–18 of gestation, when the gut epithelium differentiates into a monolayer with nascent villi. At that time, the short and long promoter chimeras are induced as well. Interestingly, in fetal transgenic mice, the reporter was absent from the colon. Not until 6 weeks after parturition did colonic crypts incorrectly begin production of hGH, indicating that different regulatory mechanisms are effective in fetal and adult mice. A progressive wave of extinction of transgene expression proceeds (in particular with the long promoter) from the distal to the proximal parts of the intestine, abolishing hGH expression in the colon of 10-month-old transgenic mice (Cohn *et al.*, 1991). To exclude an influence of the reporter gene itself, a study with a different reporter that is not secreted but is presented on the cell surface was conducted (Hansbrough *et al.*, 1991). The decay accelerating factor was fused with the transmembrane and cytosolic domains of the human

HLA-B44 antigen. Essentially, the same results were obtained with this reporter as with hGH.

Other aspects of L-FABP regulation include sex-specific expression and dietary regulation; for example, L-FABP concentration in liver is 60% higher in females than in males and is increased 25% by a high fat diet (reviewed by Bass, 1988). Much interest was attracted by the regulation of L-FABP by peroxisome proliferators. Early studies by Kawashima *et al.* (1983) indicated that L-FABP expression was induced 2- to 3-fold by the hypolipidemic drug clofibrate. The mechanism by which fibrates or other peroxisome proliferators act was unknown. However, researchers have shown (Brandes *et al.*, 1990) that L-FABP is activated transcription-ally in response to bezafibrate and that the induction of L-FABP mRNA is preceded by that of peroxisomal β-oxidation. The eventual goal of this and other studies was to establish a connection between the regulation of L-FABP expression and the concomitantly observed induction of per-oxisomal β-oxidation. Investigators speculated that L-FABP somehow mediates the effects of the low molecular weight hypolipidemic drugs to the level of the gene activation. The cloning of a peroxisome proliferator activated receptor (PPAR) gives the first insight into this missing link (Issemann and Green, 1990). PPAR belongs to the family of ligand acti-vated transcription factors that, among others, includes the nuclear steroid hormone receptors and the retinoic acid receptors (RAR). Transactivation assays, analysis of methylation interference, and mobility shift assays revealed that PPAR, on induction by peroxisome proliferators, recognizes a response element (PPRE) in the promoter of acyl-CoA oxidase, the key enzyme for peroxisomal β-oxidation (Tugwood *et al.*, 1992). Interestingly, nearly the same PPRE is present in the L-FABP promoter (see Fig. 4), located between nucleotides −68 and −56, and contains the imperfect direct repeat 5'-*TGACCTATGGCCT*-3' (Issemann *et al.*, 1992). The en-dogenous ligand for PPAR is still unknown, and binding of peroxisome proliferators to PPAR could not be demonstrated (Issemann and Green, 1990). The structural similarity of peroxisome proliferators, often hy-drophobic molecules carrying a carboxylic group, and fatty acids suggests that these xenobiotics act as fatty acid analogs. Issemann *et al.* (1992) reported that fatty acids, particularly a nonmetabolizable thio-substituted fatty acid analog, activate the acyl-CoA oxidase reporter gene (CAT) in cells that were transfected with the expression vector for PPAR. In the model proposed by the authors, a peroxisome proliferator displaces fatty acids from L-FABP which then bind to the PPAR and, thus, *trans*-activates genes that carry the cognate response elements. This hypothesis is in accordance with the finding that peroxisome proliferators such as bezafibrate indeed bind to L-FABP, albeit with lower affinity (Brandes *et al.*, 1990). A somewhat different hypothesis was put forward by Kaikaus *et al.* (1993), who reported that in hepatocytes the induction of cytochrome P450 4A1, a microsomal enzyme responsible for ω-hydroxylation of fatty acids, precedes that of L-FABP and acyl-CoA oxidase. Since peroxisome proliferators also inhibit the mitochondrial β-oxidation, the ω-oxidation pathway leads to the eventual accumulation of dicarboxylic acids that, in this model, would be the primary inducers of L-FABP and peroxisomal β-oxidation. Such accumulation could occur during substrate overload of the mitochondrial β-oxidation, for example, in response to a high fat diet. Indeed, dicarboxylic acids were able to induce L-FABP and acyl-CoA oxidase mRNA transcription. Fatty acids were effective under conditions in which they were not metabolized, for example, in the presence of tetradecylglycidic acid, an inhibitor of carnitine acyltransferases. The suggestion that long-chain dicarboxylic acids act via the PPAR, however, is in conflict with the work of Göttlicher *et al.* (1992), who only found activation of the PPAR by peroxisome proliferators and by fatty acids but not by dicarboxylic acids. Note that this system resembles the regulation of differentiation by retinoic acid and the respective receptor, which belongs to the same family of nuclear receptors (see Bass, 1993). In that case, CRABP probably attenuates the effects of retinoic acid by sequestering this very potent ligand in the cytosol.

The other example of regulation of gene expression by fatty acids is the induction of A-FABP. During conversion of preadipocytes to adipocytes, an increased uptake of fatty acids is observed that precedes the induction of fatty acid synthase and A-FABP mRNA expression. Under conditions in which no endogenous fatty acid synthesis occurs (absence of glucose and biotin in the medium), exogenous fatty acids are necessary to sustain the A-FABP mRNA level of differentiated adipocytes (Abumrad *et al.*, 1991). Similar results were obtained by Amri *et al.* (1991) for preadipocytes (1 day postconfluence) that are committed to differentiation and have very low fatty acid synthesis. In these cells, A-FABP mRNA accumulation is triggered by exposure to α-linolenate but not to short-chain fatty acids. Expression of other markers of adipocyte differentiation, such as adipsin and glycerolphosphate dehydrogenase, is not affected by fatty acids. Moreover, the transcription factor C/EBP, which is known to transactivate A-FABP (see preceding discussion), is not expressed in preadipocytes exposed to fatty acids and, hence, does not participate in this regulatory pathway. On the other hand, regulation of A-FABP by fatty acid requires protein biosynthesis since it is prevented by cycloheximide, suggesting the involvement of a *trans*-acting factor that presumably mediates the response to fatty acids via the fat cell enhancer (see Fig. 4; Amri *et al.*, 1991). In another study, Grimaldi *et al.* (1992) addressed the question of whether the fatty acids themselves or their metabolites exert these effects. Using the α-bromo derivative of palmitate, which is not incorporated into cell lipids and is not metabolized in preadipocytes, an even stronger induction of A-FABP was observed than with palmitate. The activation is reversible since, after removal of the bromopalmitate, the amount of A-FABP mRNA decreased with a half-life of 12 hr, in agreement with the turnover of A-FABP mRNA. Thus, activation of A-FABP is performed by unprocessed fatty acid.

VI. ROLE OF FATTY ACID BINDING PROTEINS IN FATTY ACID METABOLISM, DIFFERENTIATION, AND SIGNAL TRANSDUCTION

Proposed functions for FABPs refer mostly to an involvement of these proteins in intracellular fatty acid transport and metabolism, and can be summarized as follows:

1. to serve as a pool for solubilized fatty acids
2. to promote the cellular uptake of fatty acids and their utilization
3. to protect enzymes against detergent effects of fatty acids
4. to modulate enzyme activities
5. to facilitate targetted transport of fatty acids to specific metabolic pathways

The relevance of these functions is supported by the correlation of the cellular FABP content and the necessity of lipid handling, either in different tissues (see Section II) or in one tissue depending on differentiation, as in adipocytes or mammary gland (see subsequent discussion). Differences in primary structure and in the binding specificity suggest functional differences for FABP types, as exemplified by the intestinal enterocyte. Distinct roles are self-evident for ILBP and CRBP II, but are still elusive for I-FABP and L-FABP. Whereas modulation of a variety of enzymes by FABP and fatty acids has been reported (reviewed by Bass, 1988), no unequivocal conclusion can be drawn from these *in vitro* experiments. Many expectations thus have been directed to cell culture models, in particular in connection with transfection experiments. In adipocytes and in the hepatocyte-derived Hep G2 cell line that contains L-FABP, the *in vivo* binding of fatty acid to A-FABP and to L-FABP, respectively, was proven for the first time with the aid of photoaffinity labeling using a radioactive iodinated fatty acid analog (Waggoner and Bernlohr, 1990; Waggoner *et al.*, 1991). The regulation of A-FABP in the preadipocyte/ adipocyte cell model by exogenous fatty acid application was discussed in Section V.

In the small intestinal epithelium, morphological polarization of the absorptive enterocytes is accompanied by a functional polarization. Whereas luminal fatty acids are predominantly esterified to triglycerides, the plasma-derived fatty acids preferentially undergo β-oxidation (in particular during starvation) or are utilized for phopholipid synthesis. A good model for studying the intracellular compartmentalization of intestinal lipid metabolism is the colon-derived enterocyte-like cell line Caco-2. This cell line is able to form a morphologically polarized monolayer in culture. However, Caco-2 cells only mimic some aspects of intestinal lipid metabolism (e.g., triglyceride synthesis, lipoprotein secretion) but do not channel basolateral fatty acids to β-oxidation or phospholipid synthesis. Regardless of the mode of oleic acid presentation—whether apical (corresponding to luminal) or basolateral (corresponding to plasma)—triglyceride synthesis prevailed (Levin et al., 1992). The authors hypothesize a correlation with the observed deficiency of I-FABP in Caco-2 cells, which express only L-FABP. Transfection with an expression vector for I-FABP in this case would induce the ability of Caco-2 cells to biosynthesize phospholipid.

In other studies, cell lines that initially did not express this protein were transfected with FABP to obtain clues to possible functions. However, no profound changes with respect to lipid metabolism have been found to date. Transfection experiments with L-cell fibroblasts suggested that FABPs may modulate lipid homeostasis not only by effects on the fatty acid pool or by direct interaction of fatty acids with enzymes, but also by altering the membrane structure (Jefferson et al., 1990). Thus, L-cells with increased L-FABP concentrations exhibit small but possibly significant changes in membrane fluidity responsible for a decreased Na^+/K^+-ATPase activity (Incerpi et al., 1992). However, the increased fluidity of the outer leaflet of the plasma membrane could not be connected simply with the reduced cholesterol content observed for the inner leaflet. The latter may be caused by binding of cholesterol to the transfected L-FABP, as presumed by the authors, but also may be caused by activation of other cholesterol-consuming processes (cf. the controversial matter of cholesterol binding to L-FABP). In another study, a rat hepatoma cell line not endogenously expressing L-FABP also was transfected with L-FABP. In this case, L-FABP in concert with linoleic acid stimulated growth of hepatoma cells almost 3-fold over L-FABP$^-$ clones. No significant elevation of DNA synthesis was observed with other fatty acids and growth stimulation was abolished by lipoxygenase inhibitors, indicating that the effects are mediated by oxygenated metabolites of linoleic acid such as eicosanoids (Keler et al., 1992). In the context of these experiments, the identification of L-FABP as a target protein for a carcinogen that shows markedly elevated expression during mitosis and the reported binding of C_{20}-eicosanoids to L-FABP (Raza et al., 1989) are noteworthy. As a general drawback of all these studies, the cell lines that are transfected do not necessarily contain the enzymatic machinery for those metabolic processes that might be affected by FABP. Additional experiments are needed in which the influence of FABP content on defined pathways of lipid metabolism (e.g., β-oxidation, phospholipid and triglyceride biosynthesis) is studied in a cell culture model marked by "physiological concentrations" of the endogenous FABP. Alterations in the FABP content may be obtained by transfection with either sense or antisense constructs.

Interestingly, another FABP type as well is connected with cell growth. From bovine lactating mammary gland, a factor was isolated that inhibited growth of various normal and transformed mammary epithelial cells after they became committed to differentiation (Böhmer et al., 1987). The concentration of the so-called mammary-derived growth inhibitor (MDGI) is increased dramatically in terminally differentiated mammary epithelial cells compared with proliferating cells of virgin and early pregnant mammary gland. The published sequence of this MDGI deviates in only 7 positions from that of bovine H-FABP (Böhmer et al., 1987; Billich et al., 1988; Spener and Börchers, 1992). However, several lines of evidence now suggest that MDGI actually is identical to H-FABP.

1. Efforts to clone a cDNA specific for bovine MDGI were not successful, since only a cDNA encoding the pI 5.1 isoform of bovine H-FABP was obtained (Kurtz et al., 1990; C. Meinken and A. G. Lezius, unpublished results).

2. The FABP from fully differentiated rat mammary gland, which is 80-fold increased (up to 59 ± 19 $\mu g/mg$) over the FABP content of mammary gland of virgin rats, is identical to H-FABP from rat heart (Nielsen et al., 1993).

3. Growth inhibitory activity on different mammary epithelial cell lines, Ehrlich ascites tumor cells, and organ cultures of mammary gland also was demonstrated for recombinant bovine H-FABP (T. Müller and E. Spitzer, personal communication).

A role for H-FABP in differentiation does not necessarily preclude an involvement of this protein in lipid metabolism. During lactation, large amounts of triglycerides are secreted by the mammary epithelial cell, requiring an enhanced intracellular flux of fatty acids. Accordingly, the mammary gland is the tissue with the largest amount of H-FABP.

An interesting issue is the tissue-dependent regulation of H-FABP in mammary gland on the one hand and in cardiac or skeletal muscle on the other hand. Analysis of the 5'-nontranscribed region of the H-FABP gene indicates the presence of responsive elements for MyoD1-like proteins (M. Treuner and T. Müller, unpublished observation). This transcription factor serves as a switch for several muscle-specific genes and probably is responsible for the constitutive expression of H-FABP in heart. Additionally, a consensus sequence for factors that regulate several milk proteins was identified in the mouse H-FABP gene, which could confer the hormone- and differentiation-dependent expression of H-FABP in mammary gland. In the organ culture system (mammary gland explants), lactogenic hormones such as prolactin, insulin, and cortisol are necessary for H-FABP expression. Respective mRNA levels can be suppressed strongly by growth factors such as EGF, FGF, or PDGF in a reversible manner concomitant with increased DNA synthesis (Binas *et al.*, 1992; E. Spitzer, personal communication). The repression of H-FABP synthesis by growth factors probably is mediated by two putative AP-1 binding sites within the H-FABP promoter, since these growth factors induce production of the *fos* and *jun* encoded components of the AP-1 complex. The presence of AP-1 sites in the promoters of three FABP genes, namely the adipocyte, myelin, and heart type, is remarkable.

Currently we are establishing a cell model that allows the examination of another example for the differentiation-dependent expression of H-FABP. Mouse C2C12 myoblasts can be stimulated by certain nutritional and hormonal conditions to fuse into multinuclear myotubes. This differentiation is accompanied by a dramatic 40-fold increase of H-FABP, as detected by ELISA. Inspection by immunofluorescence confined H-FABP to the differentiated myotubes. This increase may reflect the different metabolic needs of mature muscle cells (R. Rump and F. Spener, unpublished observations). With C2C12 cells as a suitable cell model, we now can map the promoter of H-FABP and further define those *cis*-acting elements that govern its expression in muscle development.

The involvement of FABP in signal transduction pathways was inferred from the finding that A-FABP is the target of the insulin receptor tyrosine kinase in 3T3-L1 adipocytes (Bernier *et al.*, 1987). Insulin-dependent phosphorylation occurred at Tyr19, which is part of the well-known consensus sequence Asn–Phe–Asp–Asp–Tyr for tyrosine kinases. The sequence motif is also present in H-FABP and M-FABP; we demonstrated that, in a primary culture of myocytes from adult rat heart, a fraction of H-FABP becomes phosphorylated after stimulation with insulin (Nielsen and Spener, 1993). Also in this case, Tyr19 became phosphorylated. In adipocytes and myocytes, tyrosine phosphatase inhibitors such as vanadate and phenylarsineoxide are necessary for the accumulation of reasonable amounts of phosphorylated FABP. Even under these conditions, the phosphorylated species that migrate toward a slightly more acidic pH than the unmodified FABPs in two-dimensional gel electrophoresis account for ~0.2% of the total FABP. In agreement with the finding that mammary epithelial cells also contain a large number of insulin receptor molecules, on stimulation a phosphorylated H-FABP can be identified by two-dimensional gel electrophoresis and autoradiography of immunopurified protein in these cells as well (Nielsen *et al.*, 1993). An interesting phenomenon that may provide clues for a possible function is the finding that *in vitro* phosphorylation of A-FABP virtually abolishes fatty acid binding (Buelt *et al.*, 1991). On the other hand, A-FABP with bound fatty acid is a much better substrate than the apo protein for the insulin receptor kinase (Hresko *et al.*, 1990). Binding of fatty acids led to at least a 10-fold reduction of K_m for phosphorylaton of A-FABP by the tyrosine kinase. The authors speculate that this activation is the result of enhanced accessibility of the previously buried Tyr19 of α-helix I on fatty acid binding, and also is reflected by an increased susceptibility to anti-phosphotyrosine antibody and to the adipocyte phosphatase that is responsible for the turnover of the phosphate group. These observations led to the hypothesis (Buelt *et al.*, 1991) that phosphorylation of A-FABP regulates fatty acid uptake and trafficking in adipocytes. Unphosphorylated A-FABP could become loaded with fatty acid at the plasma membrane. By phosphorylation, the exchange of ligand is inhibited as a consequence of the very low binding affinity of phospho-A-FABP (assuming that the fatty acid enters and leaves the cavity of FABP through the same port and that this process is affected). The phosphorylated holo-A-FABP then may be targeted to microsomal membranes, where dephosphorylation by a tyrosine phosphatase allows dissociation of the fatty acid and its subsequent esterification. Although this theory links lipid binding capacity and phosphorylation properties, it is hampered by the observed low stoichiometry of phosphorylation. Catalytic fatty acid trafficking by phospho-FABP in the presence of 500-fold excess unmodified binding protein is hard to imagine. In our working hypothesis (Nielsen and Spener, 1993), gross flow of fatty acids is maintained by unmodified FABP. Depending on the fatty acid status of the cell, the signal transduction chain is modulated at the intermediate step "phosphorylation of FABP." For the propagation of the insulin signal, minute amounts of phosphorylated FABP are sufficient. In this model, FABP senses the intracellular fatty acid content and passes this information on to more distal parts of the signal transduction chain. Considering the small but significant amount of H-FABP that is also found in the nucleus (Börchers *et al.*, 1989; see Section II), and that Tyr19 is part of a helix–turn–helix motif also present in prokaryotic and eukaryotic transcription factors, phosphorylated FABP may act in *trans* on responsive elements of certain genes for lipolytic enzymes that are regulated by insulin.

References

Abumrad, N. A., Forest, C. C., Regen, D. M., and Sanders, S. (1991). Increase in membrane uptake of long chain fatty acids early during preadipocyte differentiation. *Proc. Natl. Acad. Sci. U.S.A.* **88**, 6008–6012.

Amri, E. Z., Ailhaud, G., and Grimaldi, P. (1991). Regulation of adipose cell differentiation. II. Kinetics of induction of the aP2 gene by fatty acids and modulation by dexamethasone. *J. Lipid Res.* **32**, 1457–1462.

Bartetzko, N., Lezius, A. G., and Spener, F. (1993). Isoforms of fatty-acid-binding protein in bovine heart are coded by distinct mRMAs. *Eur. J. Biochem.* **251**, 555–559.

Bass, N. M. (1988). The cellular fatty acid-binding proteins. Aspects of structure, regulation, and function. *Int. Rev. Cytol.* **111**, 143–184.

Bass, N. M. (1993). Cellular binding proteins for fatty acids and retinoids. Similar or specialized functions? *Mol. Cell. Biochem.* **123**, 191–202.

Bass, N. M., Barker, M. E., Manning, J. A., Jones, A. L., and Ockner, R. K. (1989). Acinar heterogeneity of fatty acid binding protein expression in the livers of male, female, and clofibrate-treated rats. *Hepatology* **9**, 12–21.

Bernier, M., Laird, D. M., and Lane, M. D. (1987). Insulin-activated tyrosine phosphorylation of a 15-kilodalton protein in intact 3T3-L1 adipocytes. *Proc. Natl. Acad. Sci. U.S.A.* **84**, 1844–1848.

Billich, S., Wissel, T., Kratzin, H., Hahn, U., Hagenhoff, B., Lezius, A. G., and Spener, F. (1988). Cloning of a full length complementary DNA for fatty acid binding protein from bovine heart. *Eur. J. Biochem.* **175**, 549–556.

Binas, B., Spitzer, E., Zschiesche, W., Erdmann, B., Kurtz, A., Müller, T., Niemann, C., Blenau, W., and Grosse, R. (1992). Hormonal induction of functional differentiation and mammary-derived growth inhibitor expression in cultured mouse mammary gland explants. *In Vitro Cell. Dev. Biol.* **28**, 625–634.

Böhmer, F. D., Kraft, R., Otto, A., Wernstedt, C., Hellman, U., Kurtz, A., Müller, T., Rohde, K., Etzold, G., Lehmann, W., Langen, P., Heldin, C. H., and Grosse, R. (1987). Identification of a polypeptide growth inhibitor from bovine mammary gland. Sequence homology to fatty acid and retinoid binding proteins. *J. Biol. Chem.* **262**, 15137–15143.

Börchers, T., and Spener, F. (1993). Involvement of arginine in the binding of heme and fatty acids to fatty acid-binding protein from bovine liver. *Mol. Cell. Biochem.* **123**, 23–27.

Börchers, T., Unterberg, C., Rüdel, H., Robenek, H., and Spener, F. (1989). Subcellular distribution of cardiac fatty acid binding protein in bovine heart muscle and quantitation with an enzyme linked immunosorbent assay. *Biochim. Biophys. Acta* **1002**, 54–61.

Börchers, T., Højrup, P., Nielsen, S. U., Roepstorff, P., Spener, F., and Knudsen, J. (1990). Revision of the amino acid sequence of human heart fatty acid binding protein. *Mol. Cell. Biochem.* **98**, 127–33.

Bordewick, U., Heese, M., Börchers, T., Robenek, H., and Spener, F. (1989). Compartmentation of hepatic fatty-acid-binding protein in liver cells and its effect on microsomal phosphatidic acid biosynthesis. *Biol. Chem. Hoppe-Seyler* **379**, 229–238.

Brandes, R., Kaikaus, R. M., Lysenko, N., Ockner, R. K., and Bass, N. M. (1990). Induction of fatty acid binding protein by peroxisome proliferators in primary hepatocyte cultures and its relationship to the induction of peroxisomal beta-oxidation. *Biochim. Biophys. Acta* **1034**, 53–61.

Buelt, M. K., Shekels, L. L., Jarvis, B. W., and Bernlohr, D. A. (1991). In vitro phosphoryla-
     tion of the adipocyte lipid binding protein (p15) by the insulin receptor. Effects of fatty
     acid on receptor kinase and substrate phosphorylation. *J. Biol. Chem.* **266,** 12266–12271.
Cheng, L., Qian, S., Rothschild, C., d'Avignon, A., Lefkowith, J. B., Gordon, J. I., and
     Li, E. (1991). Alteration of the binding specificity of cellular retinol-binding protein II
     by site-directed mutagenesis. *J. Biol. Chem.* **266,** 24404–24412.
Cistola, D. P., Sacchettini, J. C., Banaszak, L. J., Walsh, M. T., and Gordon, J. I. (1989).
     Fatty acid interactions with rat intestinal and liver fatty acid-binding proteins expressed
     in *Escherichia coli.* A comparative ^{13}C NMR study. *J. Biol. Chem.* **264,** 2700–2710.
Cohn, S. M., Roth, K. A., Birkenmeier, E. H., and Gordon, J. I. (1991). Temporal and
     spatial patterns of transgene expression in aging adult mice provide insights about the
     origins, organization, and differentiation of the intestinal epithelium. *Proc. Natl. Acad.
     Sci. U.S.A.* **88,** 1034–1038.
Cohn, S. M., Simon, T. C., Roth, K. A., Birkenmeier, E. H., and Gordon, J. I. (1992).
     Use of transgenic mice to map cis-acting elements in the intestinal fatty acid binding
     protein gene (fabpi) that control its cell lineage-specific and regional patterns of expres-
     sion along the duodenal-colonic and cryt-villus axes of the gut epthelium. *J. Cell Biol.*
     119, 27–44.
Cook, J. S., Lucas, J. J., Sibley, E., Bolanowski, M. A., Christy, R. J., Kelly, T. J., and
     Lane, M. D. (1988). Expression of the differentiation-induced gene for fatty acid-binding
     protein is activated by glucocorticoid and cAMP. *Proc. Natl. Acad. Sci. U.S.A.* **85,**
     2949–2953.
Crisman, T. S., Claffey, K. P., Saouaf, R., Hanspal, J., and Brecher, P. (1987). Measurement
     of rat heart fatty acid binding protein by ELISA. Tissue distribution, developmental
     changes and subcellular distribution. *J. Mol. Cell. Cardiol.* **19,** 423–431.
Demmer, L. A., Birkenmeier, E. H., Sweetser, D. A., Levin, M. S., Zollman, S., Sparkes,
     R., Mohandas, T., Lusis, A. J., and Gordon, J. I. (1987). The cellular retinol binding
     protein II gene. *J. Biol. Chem.* **262,** 2458–2467.
Distel, R. J., Ro, H. S., Rosen, B. S., Groves, D. L., and Spiegelman, B. M. (1987).
     Nucleoprotein complexes that regulate gene expression in adipocyte differentiation.
     Direct participation of c*fos. Cell* **49,** 835–844.
Dörmann, P., Börchers, T., Korf, U., Højrup, P., Roepstorff, P., and Spener, F. (1993).
     Amino acid exchange and covalent modification by cysteine and glutathione explain
     isoforms of fatty acid-binding proteins occurring in bovine liver. *J. Biol. Chem.* **268,**
     16286–16292.
Fahimi, H. D., Voelkl, A., Vincent, S. H., and Muller-Eberhard, U. (1990). Localization
     of the heme-binding protein in the cytoplasm and of a heme-binding protein-like immuno-
     reactive protein in the nucleus of rat liver parechymal cells. *Hepatology* **11,** 859–
     865.
Feng, D. F., and Doolittle, R. F. (1987). Progressive sequence alignment as a prerequisite
     to correct phylogenetic trees. *J. Mol. Evol.* **25,** 351–360.
Fournier, N. C., and Rahim, M. (1985). Control of energy production in the heart. A new
     function for fatty acid binding protein. *Biochemistry* **24,** 2387–2396.
Gantz, I., Nothwehr, S. F., Lucey, M., Sacchettini, J. C., Del Valle, J., Banaszak, L. J.,
     Naud, M., Gordon, J. I., and Yamada, T. (1989). Gastrotropin: not an enterooxyntin
     but a member of a family of cytoplasmic hydrophobic ligand binding proteins. *J. Biol.
     Chem.* **264,** 20248–20254.
Göttlicher, M., Widmark, E., Li, Q., and Gustafsson, J. A. (1992). Fatty acids activate a
     chimera of the clofibric acid-activated receptor and the glucocorticoid receptor. *Proc.
     Natl. Acad. Sci. U.S.A.* **89,** 4653–4657.
Gordon, J. I., Elshourbagy, N., Lowe, J. B., Liao, W. S., Alpers, D. H., and Taylor,

J. M. (1985). Tissue specific expression and developmental regulation of two genes coding for rat fatty acid binding proteins. *J. Biol. Chem.* **260**, 1995–1998.

Gordon, J. I., Sacchettini, J. C., Ropson, I. J., Frieden, C., Li, E., Rubin, D. C., Roth, K. A., and Cistola, D. (1991). Intracellular fatty-acid-binding proteins and their genes. Useful models for diverse biological questions. *Curr. Opin. Lipidol.* **2**, 125–137.

Graves, R. A., Tontonoz, P., and Spiegelman, B. M. (1992). Analysis of a tissue-specific enhancer. ARF6 regulates adipogenic gene expression. *Mol. Cell. Biol.* **12**, 1202–1208.

Grimaldi, P. A., Knobel, S. M., Whitesell, R. R., and Abumrad, N. D. (1992). Induction of aP2 gene expression by nonmetabolized long-chain fatty acids. *Proc. Natl. Acad. Sci. U.S.A.* **89**, 10930–10934.

Hansbrough, J.R., Lublin, D. M., Roth, K. A., Birkenmeier, E. A., and Gordon, J. I. (1991). Expression of a liver fatty acid binding protein/human decay accelerating factor/HLA B44 chimeric gene in transgenic mice. *Am. J. Physiol.* **260**, G929–G939.

Haunerland, N., Jagschies, G., Schulenberg, H., and Spener, F. (1984). Fatty-acid-binding proteins. Occurrence of two fatty-acid-binding proteins in bovine liver cytosol and their binding of fatty acids, cholesterol, and other lipophilic ligands. *Hoppe-Seyler's Z. Physiol. Chem.* **365**, 365–376.

Hayasaka, K., Nanao, K., Tahara, M., Sato, W., Takada, G., Miura, M., and Uyemura, K. (1991). Isolation and sequence determination of cDNA encoding P2 protein of human peripheral myelin. *Biochem. Biophys. Res. Commun.* **181**, 204–207.

Herrera, R., Ro, H. S., Robinson, G. S., Xanthopoulos, K. G., and Spiegelman, B. M. (1989). A direct role for C/EBP and the AP-1 binding site in gene expression linked to adipocyte differentiation. *Mol. Cell. Biol.* **9**, 5331–5339.

Heuckeroth, R. O., Birkenmeier, E. H., Levin, M. S., and Gordon, J. I. (1987). Analysis of the tissue specific expression, developmental regulation, and linkage relationships of a rodent gene encoding heart fatty acid binding protein. *J. Biol. Chem.* **262**, 9709–9717.

Hresko, R. C., Hoffman, R. D., Flores-Riveros, J. R., and Lane M. D. (1990). Insulin receptor tyrosine kinase-catalyzed phosphorylation of 422(aP2) protein. Substrate activation by long-chain fatty acid. *J. Biol. Chem.* **265**, 21075–21085.

Hunt, C., Ro, J. H. S., Dobson, D. E., Min, H. Y., and Spiegelman, B. M. (1986). Adipocyte P2 gene. Developmental expression and homology of 5′-flanking sequences among fat cell-specific genes. *Proc. Natl. Acad. Sci U.S.A.* **83**, 3786–3790.

Incerpi, S., Jefferson, J. R., Wood, W. G., Ball, W. J., and Schroeder, F. (1992). Na pump and plasma membrane structure in L-cell fibroblasts expressing rat liver fatty acid binding proteinn. *Arch. Biochem. Biophys.* **298**, 35–42.

Iseki, S., Kanda, T., Hitomi, M., and Ono, T. (1991). Ontogenic appearance of three fatty acid binding proteins in the rat stomach. *Anat. Rec.* **229**, 51–60.

Issemann, I., and Green, S. (1990). Activation of a member of the steroid hormone receptor super-family by peroxisome proliferators. *Nature (London)* **347**, 645–650.

Issemann, I., Prince, R., Tugwood, J., and Green, S. (1992). A role for fatty acids and liver fatty acid binding protein in peroxisome proliferation? *Biochem. Soc. Trans.* **20**, 824–827.

Jakoby, M. G., Miller, K. R., Toner, J. J., Bauman, A., Cheng, L., Li, E., and Cistola, D. P. (1993). Ligand-protein electrostatic interactions govern the specificity of retinol- and fatty acid-binding proteins. *Biochemistry* **32**, 872–878.

Jefferson, J. R., Powell, D. M., Rymaszewski, Z., Kukowska-Latallo, J., Lowe, J. B., and Schroeder, F. (1990). Altered membrane structure in transfected mouse L cell fibroblasts expressing rat liver fatty acid binding protein. *J. Biol. Chem.* **265**, 11062–11068.

Jones, P. D., Carne, A., Bass, N. M., and Grigor, M. R. (1988). Isolation and characterization of fatty acid binding proteins from mammary tissue of lactating rats. *Biochem. J.* **251**, 919–925.

Jones, T. A., Bergfors, T., Sedzik, J., and Unge, T. (1988). The three-dimensional structure of the P₂ myelin protein. *EMBO J.* **7**, 1597–1604.

Kaikaus, R. M., Chan, W. K., Ortiz de Montellano, P. R., and Bass, N. M. (1993). Mechanisms of regulation of liver fatty-acid-binding protein. *Mol. Cell. Biochem.* **123**, 93–100.

Kaufmann, M., Simoneau, J. A., Veerkamp, J. H., and Pette, D. (1989). Electrostimulation induced increases in fatty acid binding protein and myoglobin in rat fast twitch muscle and comparison with tissue levels in heart. *FEBS Lett.* **245**, 181–184.

Kawashima, Y., Nakagawa, S., Tachibana, Y., and Kozuka, H. (1983). Effects of peroxisome proliferators on fatty acid-binding proteins in rat liver. *Biochim. Biophys. Acta* **754**, 21–27.

Keler, T., Baker, C. S., and Sorof, S. (1992). Specific growth stimulation by linoleic acid in hepatoma cell lines transfected with the target protein of a liver carcinogen. *Proc. Natl. Acad. Sci. U.S.A.* **89**, 4830–4834.

Kimura, H., Odani, S., Nishi, S., Sato, H., Arakawa, M., and Ono, T. (1991). Primary structure and cellular distribution of two fatty acid binding proteins in adult rat kidneys. *J. Biol. Chem.* **266**, 5963–5972.

Kurtz, A., Vogel, F., Funa, K., Heldin, C. H., and Grosse, R. (1990). Developmental regulation of mammary-derived growth inhibitor expression in bovine mammary tissue. *J. Cell Biol.* **110**, 1779–1789.

Levin, M. S., Venugopal, D. T., Gordon, J. I., and Stenson, W. F. (1992). Trafficking of exogenous fatty acids within Caco-2 cells. *J. Lipid Res.* **33**, 9–19.

Maatman, R. G. H. J., van de Westerlo, E. M. A., van Kuppevelt, T. H. M. S. M., and Veerkamp, J. H. (1992). Molecular identification of the liver- and the heart-type fatty acid-binding proteins in human and rat kidney. *Biochem. J.* **288**, 285–290.

Madsen, P., Rasmussen, H. H., Leffers, H. Honore, B., and Celis J. E. (1992). Molecular cloning and expression of a novel keratinocyte protein (psoriasis-associated fatty acid-binding protein [PA-FABP]) that is highly upregulated in psoriatic skin and that shares similarity to fatty acid-binding proteins. *J. Invest. Dermatol.* **99**, 299–305.

Matarese, V., Stone, R. L., Waggoner, D. W., and Bernlohr, D. A. (1989). Intracellular fatty acid trafficking and the role of cytosolic lipid binding proteins. *Prog. Lipid Res.* **28**, 245–272.

Miller, K. R., and Cistola, D. P. (1993). Titration calorimetry as a binding assay for lipid-binding proteins. *Mol. Cell. Biochem.* **123**, 29–37.

Müller-Fahrnow, A., Egner, U., Jones, T. A., Rüdel, H., Spener, F., and Saenger, W. (1991). Three dimensional structure of fatty acid binding protein from bovine heart. *Eur. J. Biochem.* **199**, 271–276.

Narayanan, V., Kaestner, K. H., and Tennekoon, G. I. (1991). Structure of the mouse myelin P2 protein gene. *J. Neurochem.* **57**, 75–80.

Nemecz, G., and Schroeder, F. (1991). Selective binding of cholesterol by recombinant fatty acid binding proteins. *J. Biol. Chem.* **266**, 17180–17186.

Nielsen, S. U., and Spener, F. (1993). The fatty acid-binding protein from rat heart is phosphorylated on Tyr¹⁹ in response to insulin stimulation. *J. Lipid Res.* **34**, 1355–1366.

Nielsen, S. U., Vorum, H., Spener, F., and Brodersen, R. (1990). Two-dimensional electrophoresis of the fatty acid binding protein from human heart. Evidence for a thiol group which can form an intermolecular disulfide bond. *Electrophoresis* **11**, 870–877.

Nielsen, S. U., Rump, R., Højrup, P., Roepstorff, P., and Spener, F. (1993). Differentiational regulation and phosphorylation of the fatty acid-binding protein from rat mammary epithelial cells. *Biochim. Biophys. Acta* (*in press*).

Ockner, R. K., Manning, J. A., Poppenhausen, R. B., and Ho, W. K. L. (1972). A binding protein for fatty acids in cytosol of intestinal mucosa, liver, myocardium, and other tissues. *Science* **177**, 56–58.

Ockner, R. K., Manning, J. A., and Kane, J. P. (1982). Fatty acid binding proteins. Isolation from rat liver, characterization and immunochemical quantification. *J. Biol. Chem.* **257**, 7872–7878.

Peeters, R. A., Veerkamp, J. H., van Kessel A. G., Kanda, T., and Ono, T. (1991). Cloning of the cDNA encoding human skeletal-muscle fatty acid-binding protein, its peptide sequence and chromosomal localization. *Biochem. J.* **276**, 203–207.

Petrou, S., Ordway, R. W., Singer, J. J., and Walsh, J. V. (1993). A putative fatty acid-binding domain of the NMDA Receptor. *Trends Biochem. Sci.* **18**, 41–42.

Rasmussen, J. T., Börchers, T., and Knudsen, J. (1990). Comparison of the binding affinities of acyl CoA binding protein and fatty acid binding protein for long chain acyl CoA esters. *Biochem. J.* **265**, 849–855.

Raza, H., Pongubala, J. R., and Sorof, S. (1989). Specific high affinity binding of lipoxygenase metabolites of arachidonic acid by liver fatty acid binding protein. *Biochem. Biophys. Res. Commun.* **161**, 448–455.

Ross, S. R., Graves, R. A., Greenstein, A., Platt, K. A., Shyu, H. L., Mellovitz, B., and Spiegelman, B. M. (1990). A fat specific enhancer is the primary determinant of gene expression for adipocyte P2 in vivo. *Proc. Natl. Acad. Sci. U.S.A.* **87**, 9590–9594.

Sacchettini, J. C., Scapin, G., Gopaul, D., and Gordon, J. I. (1992). Refinement of the structure of Escherichia coli-derived rat intestinal fatty acid binding protein with bound oleate to 1.75-Å resolution. Correlation with the structures of the apoprotein and the protein with bound palmitate. *J. Biol. Chem.* **267**, 23534–23545.

Scallen, T. J., Noland, B. J., Gavey, K. L., Bass, N. M., Ockner, R. K., Chanderbhan, R., and Vahouny, G. V. (1985). Sterol carrier protein 2 and fatty acid-binding protein. Separate and distinct physiological functions. *J. Biol. Chem.* **260**, 4733–4739.

Scapin, G., Gordon, J. I., and Sacchettini, J. C. (1992). Refinement of the structure of recombinant rat intestinal fatty acid-binding apoprotein at 1.2-Å resolution. *J. Biol. Chem.* **267**, 4253–4269.

Schoentgen, F., Pignède, G., Bonanno, L. M., and Jollès, P. (1989). Fatty acid binding protein from bovine brain. Amino acid sequence and some properties. *Eur. J. Biochem.* **185**, 35–40.

Siegenthaler, G., Hotz, R., Chatellard-Gruaz, D., Jaconi, S., and Saurat J. H. (1993). Characterization and expression of a novel human fatty acid-binding protein: The epidermal type (E-FABP). *Biochem. Biophys. Res. Commun.* **190**, 482–487.

Spener, F., and Börchers, T. (1992). Structural and multifunctional properties of cardiac fatty acid-binding protein. From fatty acid binding to cell growth inhibition. *Biochem. Soc. Trans.* **20**, 806–811.

Spener, F., Börchers, T., and Mukherjea, M. (1989). On the role of fatty acid binding proteins in fatty acid transport and metabolism. *FEBS Lett.* **244**, 1–5.

Spener, F., Unterberg, C., Börchers, T., and Grosse, R. (1990). Characteristics of fatty acid-binding proteins and their relation to mammary-derived growth inhibitor. *Mol. Cell. Biochem.* **98**, 52–68.

Spiegelman, B. M., Franck, M., and Green, H. (1983). Molecular cloning of mRNA from 3T3 adipocytes. Regulation of mRNA content for glycerolphosphate dehydrogenase and other differentiation-dependent proteins during adipocyte development. *J. Biol. Chem.* **258**, 10083–10089.

Stone, R. L., and Bernlohr, D. A. (1990). The molecular basis for inhibition of adipose conversion of murine 3T3 L1 cells by retinoic acid. *Differentiation* **45**, 119–127.

Storch, J., Bass, N. M., and Kleinfeld, A. M. (1989). Studies of the fatty acid-binding site of rat liver fatty acid-binding protein using fluorescent fatty acids. *J. Biol. Chem.* **264**, 8708–8713.

Sweetser, D., Lowe, J. B., and Gordon, J. I. (1986). The nucleotide sequence of the rat liver fatty acid binding protein gene. *J. Biol. Chem.* **261**, 5553–5561.

Sweetser, D. S., Birkenmeier, E. H., Klisak, I. J., Zollman, S., Sparkes, R. S., Mohandas, T., Lusis, A. J., and Gordon, J. I. (1987). The human and rodent intestinal fatty acid binding protein genes. A comparative analysis of their structure, expression, and linkage relationships. *J. Biol. Chem.* **262**, 16060–16071.

Sweetser, D. A., Hauft, S. M., Hoppe, P. C., Birkenmeier, E. H., and Gordon, J. I. (1988). Transgenic mice containing intestinal fatty acid binding protein human growth hormone fusion genes exhibit correct regional and cell specific expression of the reporter gene in their small intestine. *Proc. Natl. Acad. Sci. U.S.A.* **85**, 9611–9615.

Tugwood, J. D., Issemann, I., Anderson, R. G., Bundell, K. R., McPheat, W. L., and Green, S. (1992). The mouse peroxisome proliferator acivated receptor recognizes a response element in the 5′ flanking sequence of the rat acyl CoA oxidase gene. *EMBO J.* **11**, 433–439.

Unterberg, C., Börchers, T., Højrup, P., Roepstorff, P., Knudsen, J., and Spener, F. (1990). Cardiac fatty acid binding proteins. Isolation and characterization of the mitochondrial fatty acid binding protein and its structural relationship with the cytosolic isoforms. *J. Biol. Chem.* **265**, 16255–16261.

Vancura, A., and Haldar D. (1992). Regulation of mitochondrial and microsomal phospholipid synthesis by liver fatty acid-binding protein. *J. Biol. Chem.* **267**, 14353–14359.

Veerkamp, J. H., Peeters, R. A., and Maatman, R. G. (1991). Structural and functional features of different types of cytoplasmic fatty acid binding proteins. *Biochim. Biophys. Acta* **1081**, 1–24.

Vincent, S. H., and Muller-Eberhard, U. (1985). A protein of the Z class of liver cytosolic proteins in the rat that preferentially binds heme. *J. Biol. Chem.* **260**, 14521–14528.

Waggoner, D. W., and Bernlohr, D. A. (1990). In situ labeling of the adipocyte lipid binding protein with 3-[^{125}I]iodo-4-azido-N-hexadecylsalicylamide. Evidence for a role of fatty acid binding proteins in lipid uptake. *J. Biol. Chem.* **265**, 11417–11420.

Waggoner, D. W., Manning, J. A., Bass, N. M., and Bernlohr, D. A. (1991). In situ binding of fatty acids to the liver fatty acid binding protein. Analysis using 3-[^{125}I]iodo-4-azido-N-hexadecylsalicylamide. *Biochem. Biophys. Res. Commun.* **180**, 407–415.

Winter, N. S., Bratt, J., and Banaszak, L. J. (1993). Crystal structures of holo and apocellular retinol-binding protein II. *J. Mol. Biol.* **230**, 1247–1259.

Xu, Z., Bernlohr, D. A., and Banaszak, L. J. (1992). Crystal structure of recombinant murine adipocyte lipid-binding protein. *Biochemistry* **31**, 3484–3492.

Zanotti, G., Scapin, G., Spadon, P., Veerkamp, J. H., and Sacchettini, J. C. (1992). Three-dimensional structure of recombinant human muscle fatty acid-binding protein. *J. Biol. Chem.* **267**, 18541–18550.

Zhang, J., Liu, Z. P., Jones, T. A., Gierasch, L. M., and Sambrook, J. F. (1992). Mutating the charged residues in the binding pocket of cellular retinoic acid-binding protein simultaneously reduces its binding affinity to retinoic acid and increases its thermostability. *Proteins* **13**, 87–99.

CHAPTER 11

Lipid-Tagged Proteins

Chiara Zurzolo* and Enrique Rodriguez-Boulan
Department of Cell Biology and Anatomy, Cornell University Medical College, New York, New York 10021

I. INTRODUCTION

The existence of lipid modification of proteins has been known for about 30 years (Schlesinger, 1981; Schultz *et al.*, 1988), but only in the last few years has the structural and functional diversity of these modifications been studied in detail. Three modes of fatty acid linking to proteins have been described in eukaryotic cells (Towler and Gordon, 1988): (1) the covalent attachment of long-chain acyl groups (C_{14} fatty acid myristilate or C_{16} palmitate), (2) modification by long-chain prenoid groups, and (3) C-terminal addition of a phosphatidylinositol-containing glycan moiety. Such modifications also have been found in unicellular eukaryotes such

---

* Present address: Dipartimento di Biologia e Patologia Cellulare e Molecolare, CEOS, II Policlinico, 80131, Napoli, Italy.

as yeast. Their evolutionary conservation presumably testifies to their biological importance (Deschenes *et al.*, 1990; Glomset *et al.*, 1990; Magee, 1990).

The covalent attachment of fatty acid chains to nascent or mature polypeptides dramatically alters their hydrophobic properties and can facilitate their interaction with membranes. The different localizations of these lipid-tagged proteins also supports a role for them in targeting. These alternative methods of membrane association could be more flexible than amino acid transmembrane motifs within the polypeptide structure itself, since lipid tags are potentially reversible. This feature also may play a key role in controlling cellular signaling events as well as in controling growth and differentiation.

II. PROTEIN ACYLATION

The first demonstration of protein acylation was by Hantke and Braun (1973), who showed that the N-terminus of the outer membrane mucein lipoprotein of *Escherichia coli* was bound covalently to palmitic acid chains. However, at that time, fatty acylation of proteins was thought to be a rare phenomenon. The subsequent finding that the acylation and esterification of amino acids with palmitic acid in the glycoprotein of Sindbis virus (Bracha *et al.*, 1977) suggested that the enzyme catalyzing this lipid modification was provided by the host cell, indicating the existence of a more widespread phenomenon (Schlesinger *et al.*, 1980). Also, myristic acid has been found bound to the N-terminal amino acid of several retroviral and cellular proteins (Aitken *et al.*, 1982; Carr *et al.*, 1982; Henderson *et al.*, 1983). These and subsequent studies have confirmed that protein acylation is a widespread modification mechanism in prokaryotic and eukaryotic organisms (James and Olson, 1990; McIlhinney, 1990; Gordon *et al.*, 1991).

The most common fatty acids covalently attached to proteins are myristic acid (14:0), palmitic acid (16:0), and stearic acid (18:0). Using radiolabeled myristate and palmitate, different studies have shown that labeling of several proteins occurs in eukaryotic cells and that different pathways are involved (Magee and Courtneidge, 1985a; Olson *et al.*, 1985; McIlhinney *et al.*, 1987). The attachment of myristic acid to an N-terminal glycine by amide linkage occurs cotranslationally. In contrast, palmitoylation is a posttranslational mechanism that occurs usually via an ester or thioester bond (Sefton and Buss, 1987). Moreover, all palmitoylated proteins appear to be membrane bound, although they are localized differently (e.g., on the nuclear envelope, Golgi, and plasma membrane). Myristate-containing proteins, on the other hand, also are found in the cytosol and nucleoplasm (Magee and Courtneidge, 1985a; Olson *et al.*, 1985; McIlhinney *et al.*, 1987). The structural and subcellular differences among myristoylated and palmitoylated proteins are summarized in Table I. Immediately evident is that different functions may be ascribed to these two types of protein acylation (see subsequent discussion for further details).

III. MYRISTOYLATION: BIOCHEMISTRY AND BIOLOGICAL FUNCTION

Myristoylation has been found to occur on penultimate N-terminal glycine residues and requires the prior removal of the initial methionine residue. This myristoylation is an early event in acyl protein biosynthesis and can be blocked immediately by inhibiting protein biosynthesis (Olson and Spizz, 1986). Myristate is found attached to nascent polypeptides of less than 100 amino acids (Wilcox *et al.*, 1987).

Point mutation studies of the polypeptide $p60^{v\text{-}src}$, the myristoylated transforming protein of the Rous sarcoma virus (RSV), have shown that the signal for myristoylation is contained within the first 7–10 amino acids (Pellman *et al.*, 1985). Glycine at position 1 is essential for myristoylation but, since not all proteins with an N-terminal glycine are myristoylated, some additional determinant(s) for protein myristoylation must exist. A weak consensus sequence for protein myristoylation has been proposed (Towler and Gordon, 1988) that is formed by an exopeptide core that is presumed to contain much of the information necessary for recognition by *N*-myristoyl transferase (NMT), the enzyme that catalyzes the attachment of the myristoyl moiety to the protein.

NMT has been purified from yeast and mammals and its properties have been studied in detail (Towler and Gordon, 1988). This soluble enzyme catalyzes the cotranslational myristoylation of appropriate protein substrates. Using myristoyl CoA as cosubstrate, NMT selects its acyl CoA

TABLE I

Characteristics of Acylated Proteins

Myristoylation	Palmitoylation
Amide linked to N-terminal glycine	Ester linked mainly to cysteine
Slow turnover	Fast turnover
Cotranslational addition	Posttranslational addition
Cytosolic or plasma membrane proteins	Membrane associated proteins

substrates by chain length rather than by hydrophobicity (Heuckeroth *et al.*, 1988). Acyl CoA species of the incorrect chain length have been found to alter the K_m of the enzyme for its peptide substrates dramatically, resulting in the reduction of acyl CoA-peptide production. On the other hand, analogs with reduced hydrophobicity (by substitution of the methylene group with an oxygen or a sulfur atom) can be incorporated into myristoylated proteins *in vivo*, resulting in their partial redistribution from the membrane-bound fraction to the soluble fraction.

This type of interference in the membrane targeting of some myristoylated proteins could be of pharmacological use for the treatment of human diseases. For example, these analogs have been shown to disrupt targeting of transforming pp60[src] proteins. They also interfere with the replication of human immunodeficiency virus (HIV) by preventing membrane localization and processing of the viral structural protein gag (Bryant *et al.*, 1989). The high specificity of the enzyme NMT, in conjunction with the fact that myristate is a relatively rare fatty acid, suggests a more specialized function for myristoylation than providing an alternative mechanism for membrane association. Further, the existence of myristoylated proteins in the cytosol indicates that myristoylation alone is not sufficient for membrane localization. Therefore, it cannot be the only function of protein myristoylation. Table II lists some examples of myristoylated proteins, as well as their subcellular localization.

One of the best characterized myristoylated proteins is pp60[src] of RSV. This protein is found primarily in the plasma membrane, particularly in association with adhesion plaques. Myristoylation of pp60[src] is required

TABLE II

Examples of Myristoylated Proteins and
Their Subcellular Localization

Protein	Cell localization
cAMP-dependent protein kinase	Cytoplasm
Calcineurin B (protein phosphatase)	Cytoplasm
Cytocrome b_5 reductase	ER,[a] mitochondria
G proteins (α subunits of G_i and G_o)	Plasma membrane
Murine lymphoma p56	Plasma membrane
ADP-ribosylation factors (ARFs)	ER, golgi, endosomes

[a]ER, Endoplasmic reticulum.

for stable association of the protein with cell membranes and for the consequent transformation of the cell; nonmyristoylated variants of pp60^{src} (made by mutation) remain in the cytoplasm and the virus is unable to transform cells (Kamps *et al.*, 1986).

The specific localization of pp60^{src} in the plasma membrane cannot, however, be explained by the hydrophobic nature of this modification; rather, this localization must involve a specific receptor that is localized in the plasma membrane. In this case, myristic acid could provide a signal that promotes transport of the protein to specific domains within the cell. A 32-kDa plasma membrane protein has been identified that binds the myristoylated N terminus of pp60^{src}, this protein is, perhaps, likely to be a myristoylated-src receptor (Resh and Ling, 1990).

These findings are consistent with the hypothesis that myristate regulates protein–protein interactions. This hypothesis is consistent with the fact that many myristoylated proteins are subunits of large protein complexes (e.g., protein kinase A, α subunits of G_i and G_o proteins).

As previously suggested, myristoylation is also important for the assembly of viral particles. Experiments in which the N-terminal attachment site for myristic acid was mutated have been performed with RSV, poliovirus, and HIV (Gottlinger *et al.*, 1989; Marc *et al.*, 1989), resulting in the prevention of viral replication and virion budding. This phenomenon clearly offers enormous potential for the development of antiviral drugs.

The presence of myristoylated protein receptors might be a mechanism for targeting different proteins to different cellular compartments. Further, because many myristoylated proteins are cytosolic and are involved in growth control and signal transduction, these functions could be explained if myristoylation were to play a major role in promoting protein–protein interactions.

A fascinating role for myristoylation has been proposed for the ADP-ribosylating factors (ARF), a group of more than a dozen ras-related proteins that are involved in membrane trafficking (Kahn *et al.*, 1991; Tsuchiya *et al.*, 1991). All the ARF proteins tested to date have been found to be myristoylated; this modification is involved in membrane binding (Kahn *et al.*, 1992). Although myristoylation cannot be responsible for the specificity of targeting of the ARF proteins to different subcellular locations (Balch *et al.*, 1992; Lenhard *et al.*, 1992; Taylor *et al.*, 1992), the fatty acid group has been shown to be able to interact with an adjacent specific protein sequence (Tsuchiya *et al.*, 1991; Kahn *et al.*, 1992) that forms a distinct structure that is likely to be involved in the recognition of a "receptor protein" within the target membrane (reviewed by Pfeffer, 1992).

IV. PALMITOYLATION: BIOCHEMISTRY AND BIOLOGICAL FUNCTION

Palmitoylation is the other type of protein fatty acid acylation. The esterification of palmitate can occur either on serine or on threonine residues to form an oxyester or, more frequently, on cysteine residues to form the corresponding thioester (reviewed by Schultz *et al.*, 1988; Towler and Gordon, 1988; James and Olson, 1990). In contrast with myristoylation, palmitoylation is a posttranslational event, as originally shown for the vesicular-stomatitis virus (VSV) G and the Semliki Forest virus (SFV) E1 glycoproteins (Schmid and Schlesinger, 1980). In these studies, palmitoylation was found to occur shortly before the acquisition of endo H resistance, suggesting that the process occurs in the late endoplasmic reticulum (ER) or *cis*-Golgi. Further, in many cases this link is biologically labile and the turnover of the fatty acid is faster than that of the protein. Hence, palmitoylation also can occur at a later stage in the life of a protein, as shown for rhodopsin and the transferrin receptor (Omary and Trowbridge, 1981; Jing and Trowbridge, 1987).

Table III presents some of the palmitoylated proteins discovered to date. The majority of them are membrane anchored, such as transmembrane proteins (e.g., HLA-DR, insulin receptor, transferrin receptor), but others have been also found, particularly proteins that otherwise would be hydrophilic. The result of palmitoylation is that these proteins associate with the cytoplasmic face of cellular membranes (e.g., $p21^{ras}$, GAP 48, ankyrin, vinculin) (Magee and Courtneidge, 1985b; Staufenbiel, 1987; Skene and Virag, 1989). For all these proteins, the turnover of the fatty acid is very fast; this event is likely to be highly regulated, because both acylation and deacylation appear to be enzymatic processes.

Although little is known about the protein–fatty acyl esterase that removes palmitate from proteins, much work has focused on the characterization of the protein acyltransferase that is responsible for palmitoylation. Kasinathan *et al.* (1990) have described a protein acyltransferase from the "rough microsomal fraction" of rat gastric mucosa. The active enzyme has a molecular mass of 234 kDa and is composed of two subunits of 65 and 67 kDa that are associated tightly with the membrane of the rough ER and are exposed topologically to the cytoplasm. Palmitoyl CoA is the best acyl donor, although in fact little evidence exists for any specific requirements for the enzyme substrate. Palmitoylation apparently can occur at any position within the primary structure of the protein, although the modification is found most often close to the transmembrane region of membrane spanning proteins, particularly on the cytoplasmic side.

Most work that has addressed the site of palmitoylation has been done with the protein encoded by the *ras* oncogene, in which palmitoylation

TABLE III

Examples of Palmitoylated Proteins and
Their Subcellular Localization

Protein	Cell localization
Transferrin receptor	Plasma membrane
Insulin and IGF-1 receptor	Plasma membrane
Interleukin 1 receptor	Plasma membrane
β Adrenergic receptor	Plasma membrane
Nicotinic acetylcholine receptor	Plasma membrane
p21 (and *ras* superfamily proteins)	Plasma membrane
Rhodopsin	Disc membranes
Ankyrin	Cytoskeleton
Vinculin	Cytoskeleton
Galactosyl transferase	Golgi
Mannosidase II	Golgi
Ca^{2+}–ATPase	Sarcoplasmic reticulum
GAP 43	Growth cones

provides a mechanism for membrane association. Mutational analysis originally suggested that palmitoylation of H-*ras* occurred on a cysteine residue close to the C terminus. This cysteine residue is part of the sequence CAAX (where A is alipathic and X is any amino acid), which was proposed to be a consensus sequence for palmitoylation (Willumsen *et al.*, 1984). Subsequent studies have shown that the palmitoylation process is, in fact, more complex and that the "CAAX box" serves as a signal for protein isoprenylation (see subsequent discussion) rather than for palmitoylation (Hancock *et al.*, 1989). Palmitoylation occurs on a second cysteine residue that is close to the CAAX box, but this process appears to be dependent on prior isoprenylation of the CAAX box cysteine residue (Schafer *et al.*, 1989). These results thus explain why a mutation in the CAAX box can block palmitoylation of the H-*ras* gene product (Willumsen *et al.*, 1984).

The intracellular function of palmitoylation is not fully understood. Site-specific mutagenesis of the VSV G protein (Balch *et al.*, 1984) and the transferrin receptor (Jing and Trowbridge, 1987) indicate that palmitoylation is not essential for the transport or compartmentalization of these proteins in the membrane, or for their function; palmitoylation therefore has functions other anchorage.

Other clues again have come from studies of *ras*-encoded proteins. In these experiments, mutation of the palmitoylated cysteine residue was found to have no effect on the biological activity of the protein, although the mutated protein attached to the membrane with lower avidity than did the native protein. Therefore, palmitoylation may regulate the affinity of membrane association. Considering that the rate of turnover is faster for the palmitate moiety than for the protein itself, activity may be regulated by a modulation of the degree of acylation. Palmitoylated Ras proteins may cycle between the Golgi complex and the plasma membrane, thus adding another layer to the hierarchy of regulation of the signal transduction mechanism.

Reversible acylation and deacylation of proteins required for membrane fusion can be involved in the regulation of vesicle trafficking within the cell. Using a vesicle fusion system (Balch *et al.*, 1984), Glick and Rothman (1987) showed that fusion is stimulated dramatically in the presence of palmitoyl CoA.

Some evidence that palmitoylation is involved in membrane targeting has been obtained from experiments performed by Zuber *et al.* (1989) on GAP 43, a protein localized exclusively within growth cones. GAP 43 is known to be palmitoylated at its N terminus. A fusion protein containing the truncated N-terminal domain of GAP 43 fused to chloramphenicol acetyl transferase (CAT) still is targeted to the growth cones. Hence, palmitoylation of GAP 43 appears to serve as a transport signal as well as to promote membrane anchorage.

In some cases, palmitoylation has been found to be necessary for protein function rather than for protein localization, as is the case for the β-adrenergic receptor. A mutagenized receptor that can no longer be palmitoylated still can be targeted to the plasma membrane, where it can bind ligand normally, although it is no longer competent for signal transduction (O'Down *et al.*, 1989). In this case, the role of palmitate could be to hold the receptor in a conformation that permits interaction between the receptor and the G protein. In this novel mechanism, acylation–deacylation cycling would regulate receptor function. Unfortunately, the turnover of the palmitate attached to the β-adrenergic receptor has not yet been measured.

V. ISOPRENYLATION: BIOCHEMISTRY AND BIOLOGICAL FUNCTION

The first isoprenylated peptide identified was the 11-amino-acid yeast α mating factor that contains a farnesyl group (15:0) attached to the C-terminal cysteine residue (Kamiya *et al.*, 1978). This modification was found to be common to other fungal mating factors and, further, to be essential for their activity (Youji *et al.*, 1981). The sequence of the α mating factor of *Saccharomyces cerevisiae* contains a C-terminal sequence CAAX (as previously described) that was found to be necessary for isoprenylation (Hancock *et al.*, 1989).

The sequence of events that occurs during isoprenylation is rather complicated and involves, first, the addition of a 15-carbon isoprenol lipid farnesyl to the C-terminal cysteine residue. This addition is followed by the removal of the AAX amino acids and by the carboxymethylation of the cysteine.

In 1989, Hancock *et al.* showed that these are the same modifications that occur during the processing of ras-related proteins. In fact, all processing steps appeared to be conserved between the mammalian and yeast proteins. To date, more than 40 intracellular proteins have been shown to be prenylated, including the nuclear lamin A and lamin B proteins (Farnsworth *et al.*, 1989).

Table IV lists some example of prenylated proteins, as well as the consensus sequence at the C-terminal domain. Like the ras proteins and the yeast mating factors, these proteins also possess a CAAX motif. Mutation of this sequence prevents farnesylation and blocks nuclear assembly (Holtz *et al.*, 1989). The lamin proteins also are known to contain signal sequences for nuclear targeting; the combination of these and the CAAX box farnesylation site is responsible for correct association of the proteins with the membranous nuclear envelope (Holtz *et al.*, 1989).

TABLE IV

Examples of Prenylated Proteins

Protein	Terminal sequence
Lamin A	QAPQNCSIM
Lamin B	SGNKNCAIM
G protein, γ subunit	EKKFFCAIL
H-ras	CMSCKCVLS
α Factor	FWAPACVIA
Transducin, γ subunit	ELKGGCVIS
rap 1 A	PKKKSCLLL
ypt 1	NTGGGCC
rab 2	QAGGGCC
sec 4	SSKSNCC

All the proteins just described contain a farnesyl moiety as the prenylated modification. This group is derived from the cholesterol precursor mevalonate (for review, see Sinensky and Lutz, 1992). However, another isoprenyl group, the 20-carbon geranyl-geranyl isoprenoid has been identified (Farnsworth *et al.*, 1990; Rilling *et al.*, 1990) that is, in fact, more common than the farnesyl group. Although the consensus sequence for this other type of modification is not fully understood, recent work has shown that the CAAX box is not the only C-terminal sequence that marks proteins for prenylation. Further, no absolute requirement exists for the two alipathic amino acids (A), although the last amino acid (X) plays an important role in designating which isoprenyl donor (farnesyl or geranyl) becomes attached (Reiss *et al.*, 1991).

Two reports have shown that the γ subunits of the mammalian heterotrimeric G proteins are geranyl-geranylated (Mumby *et al.*, 1990; Yamane *et al.*, 1990). Interestingly, these proteins contain the sequence CAIL at the C terminus. Further, another subset of ras-related proteins, termed rab proteins, have been shown to be modified by a C_{20} isoprenoid (Khosravi-Far *et al.*, 1991). This phenomenon is of particular interest since these are small GTP-binding proteins that are involved in membrane trafficking. This lipid moiety has been proposed to play a role in regulating the targeting of these proteins (for review, see Magee and Newman, 1992; Pfeffer, 1992; subsequent discussion).

Sequences that regulate geranyl-geranylation have been found, but rules have not yet been assigned. The rab proteins usually contain C-terminal motifs of either CC or CXC (Table IV). Rab3A and Ypt 5 contain the CXC motif and geranyl groups have been found on both cysteines. Further, the C-terminal cysteine is carboxymethylated (Newman *et al.*, 1992). For proteins ending with the CC motif, whether both cysteines are prenylated is not clear, although they have been shown to lack carboxymethylation (Newman *et al.*, 1992; Wei *et al.*, 1992), which would suggest that the C-terminal cysteine is not prenylated.

Studies of enzymes involved in the prenylation of proteins are also in progress. A protein prenyltransferase that can transfer farnesyl groups onto ras proteins has been purified and characterized (Reiss *et al.*, 1990). The observation that only CAAX (and not CAIL) peptides of various lengths could inhibit the farnesylation of ras proteins competitively *in vitro* suggested the existence of two different enzymes that catalyze farnesylation and geranylation reactions (Reiss *et al.*, 1991). Subsequent studies have demonstrated that the enzymes exist as heterodimers; the α subunit is common to both the farnesyl- and geranyltransferase (Seabra *et al.*, 1992), whereas the β subunits are specific to each enzyme and presumably control the specificity of the reactions that they carry out.

Concerning the biological role of farnesylation, clearly this C-terminal modification has major consequences, altering the hydrophobic character of the protein and promoting membrane binding as well as protein–protein interactions. For example, in the case of the rab proteins, the addition of one or two geranyl groups, plus or minus a farnesyl group, might change the avidity of the protein for the membrane, which may be a determinant in specifying different subcellular localizations. However, the exchange of the lipid modification sites of Rab5 and Rab7 has been shown to have no effect on their subcellular location (Chavrier *et al.*, 1991). Further, the major determinants of rab protein targeting are located in sequences that are 30–40 amino acids upstream of the C-terminus (Chavrier *et al.*, 1991). Conversely, the lipid modification site at the C terminus of ras is not a primary determinant for its subcellular localization (Hancock *et al.*, 1989). Hence, the role of lipid modification of proteins would appear to be more subtle than originally proposed.

VI. GLYCOSYL-PHOSPHATIDYLINOSITOL-ANCHORED PROTEINS

In glycosyl-phosphatidylinositol (GPI)-anchored proteins, the C-terminal amino acid is linked by an amide bond to ethanolamine, which in turn is linked by a phosphodiester bond to a mannosyl glycosaminyl core glycan anchored to the membrane by phosphatidylinositol (Ferguson and Williams, 1988; Low and Saltiel, 1988; Cross, 1990). GPI-anchored proteins usually are identified by their sensitivity to cleavage by PI-specific phospholipase (PLC); this assay, however, is not perfect since acylation of the inositol ring (generally by palmitate) renders GPI insensitive to the enzyme (Roberts *et al.*, 1988; discussed by Low and Saltiel, 1988). Considerable variability of the GPI structure can be generated by substitutions of the glycan with ethanolamine phosphate or by sugars (aGal, aMan, or bGalNAc) (Deeg *et al.*, 1992a; Ferguson and Williams, 1988).

The addition of GPI to the protein is directed by a C-terminal signal that consists of a hydrophobic segment followed N-terminally by specific sequences (Berger *et al.*, 1988; Caras *et al.*, 1989; Gerber *et al.*, 1992). A transamidation event takes place in the lumen of the rough endoplasmic reticulum (ER) after cleavage of the signal. A major advance has been the reconstitution of this process *in vitro*, which should allow the identification and characterization of the enzymes involved (Roberts *et al.*, 1988; Menon *et al.*, 1990; Mayor *et al.*, 1991; Kodukula *et al.*, 1992).

More than 50 GPI-anchored proteins have been identified in the last 5 years. A few functions have been assigned to the GPI anchor that fall into the categories of sorting, receptor-mediated endocytosis, and signal transduction.

VII. APICAL SORTING IN EPITHELIA

An examination of the list of GPI-anchored proteins identifies several that are apically localized in eptihelial cells but no basolateral ones (Table V).

The correlation between apical localization and membrane anchorage via GPI led Lisanti *et al.* (1988,1990) to study the localization of all endogenous GPI-anchored proteins in several epithelial cell lines. Two kidney epithelial cell lines, Madin–Darby canine kidney (MDCK) and LLC-PK1, displayed 6 and 9 GPI-anchored proteins, all in the apical surface. Since the polarity assay utilized depended on the use of PI-PLC, the possibility exists that basolateral GPI proteins exist but are not sensitive to PI-PLC. This option is unlikely since, when antibodies against known GPI-

TABLE V

Apical Localization of Glycosyl-Phosphatidylinositol-Anchored
Proteins in Epithelial Cells

GPI-anchored protein	Localization	Cell type	Reference
5′ Nucleotidase	Apical	Intestine, kidney	Hooper and Turner (1988)
Trehalase	Apical	Intestine, kidney	Hooper and Turner (1988)
Alkaline phosphatase	Apical	Intestine, kidney, MDCK[a]	Hooper and Turner (1988)
Renal dipeptidase	Apical	Kidney	Hooper and Turner (1988)
N-CAM[b] (GPI-anchored)	Apical	MDCK[a]	Powell *et al.* (1991)
Decay accelerating factor	Apical	MDCK, [a]Caco-2, SKCO-15	Lisanti *et al.* (1990)
Carcinoembryonic antigen	Apical	Intestine, Caco-2, SKCO15	Lisanti *et al.* (1990)
Thy-1	Apical	MDCK[a]	Powell *et al.* (1991)
gD1-DAF	Apical	MDCK[a]	Lisanti *et al.* (1989)
VSV G-PLAP	Apical	MDCK[a]	Brown *et al.* (1989)
PLAP	Apical	MDCK[a]	Brown and Rose (1992)

[a]Transfected GPI-anchored proteins.

anchored proteins were used, these proteins were found localized to the apical surface of MDCK cells and the intestinal cell lines Caco-2 and SK-CO-15. Several GPI-anchored proteins also have been found to be enriched in bile canaliculi, the apical pole of hepatocytes (Ali and Evans, 1990).

Additional support for the apical sorting role of GPI was provided by the transfection of cDNAs encoding transmembrane and GPI-anchored isoforms of the neural cell adhesion molecule N-CAM (Powell *et al.*, 1991). Only the GPI-anchored isoform was targeted to the apical surface. Final proof was provided by the construction of hybrid cDNAs encoding fusion proteins that consisted of the ectodomain of a basolateral protein and the C-terminal domain of GPI-anchored proteins (directing GPI addition). The ectodomains of two viral envelope glycoproteins that are basolaterally localized in infected cells, herpes simplex gD1 and VSV G protein, were targeted basolaterally after fusion with decay accelerating factor (DAF) and Placental Alkaline Phosphatase (PLAP) GPI-anchoring signals (Brown *et al.*, 1989; Lisanti *et al.*, 1989).

How does GPI anchoring lead to apical localization? Work from several laboratories over the past several years has demonstrated that a major site of sorting of apical and basolateral glycoproteins is a distal compartment of the Golgi apparatus, the *trans*-Golgi network (TGN). At this level, the proteins are incorporated into distinct vesicles destined to fuse with opposite poles of the cell. Since certain glycosphingolipids (GSLs) are sorted apically in MDCK cells and tend to form tight clusters when they reach a high molar fraction of the bilayer lipids, van Meer and Simons (1988) have proposed that apical proteins associate with GSL clusters prior to incorporation into apical vesicles (see Chapter 21). Lisanti and Rodriguez-Boulan (1990) proposed that GPI may have affinity for GSL clusters, leading to apical localization.

Support for the "cluster hypothesis" for apical targeting has been provided by other experiments. For some time, GPI-anchored proteins have been known to show resistance to dissociation by certain mild nonionic detergents (TX-100, TX-114) at low temperature (Hoessli and Runngger-Brandle, 1985; Hooper and Turner, 1988). Brown and Rose (1992) followed the maturation of a GPI-anchored protein, PLAP, transfected into MDCK cells and made the striking observation that this protein became insoluble in TX-100 at 4°C as it left the ER and entered the Golgi apparatus. The insoluble aggregates did not interact with the cytoskeleton but could be purified by flotation in sucrose density gradients due to a high content of sphingomyelin (SM) and GSLs. Since these lipids are produced in the Golgi apparatus, these experiments suggest that clusters of GPI-anchored proteins and GSLs are formed in the Golgi apparatus. However, clustering as detected by detergent insolubility is not exclusive of epithelial cells and, therefore, cannot be the only mechanism responsible for apical sorting (Mescher *et al.*, 1981; Stefanova and Horejsi, 1991; Stefanova, *et al.*, 1991; Cinek and Horejsi, 1992).

An experiment that adds a new dimension to this problem was reported by Hannan *et al.* (1993). These authors used two biophysical techniques to measure the mobility and clustering of a GPI-anchored protein, gD1-DAF, transfected into MDCK cells. Fluorescence recovery after photobleaching (FRAP) was used to measure the diffusion of gD1-DAF in the apical surface and fluorescence energy transfer (FET) was used to detect the degree of proximity of the protein molecules (clustering). Two populations of molecules were examined, those that had resided in the cell surface for a long time (stable population) and those that recently had reached the cell surface after release from an intracellular transport block (new population). The experiments were carried out in wild-type MDCK cells and in a concanavalin A-resistant (Con A^r) mutant MDCK line (Meiss *et al.*, 1982) with a deficient sorting of GPI-anchored proteins (Lisanti *et al.*, 1990). The results, shown in Table VI, indicate two major findings: (1) both stable and new populations of gD1-DAF are clustered as measured by FET and (2) molecules recently delivered to the cell surface are immobile (R~40%) in wild-type cells but are fully mobile (R~90%) in MDCK-ConA^r cells. After long periods of residence at the cell surface, the new molecule population acquires characteristics of the stable population, that is, the R value reaches ~90%. These experiments suggest that clusters of GPI-anchored proteins being delivered to the cell surface are linked to the sorting machinery in wild-type cells but are free in MDCK-ConA^r cells. A putative transmembrane "sorting receptor" may be involved in linking the GSL–GPI protein aggregates (present in the luminal leaflet of the bilayer) to the vesicle-forming machinery in the cytoplasmic side.

These results indicate that clustering is not sufficient to determine the apical targeting of GPI-anchored proteins. Is it necessary? The answer to

TABLE VI

Immobilized gD1-DAF Clusters Reach the Apical Surface of MDCK Cells[a]

Cell type	Localization	Stable population		New surface population	
		FET	FRAP	FET	FRAP
MDCK	Apical	Cluster	Mobile	Cluster	Immobile
MDCK-ConA^r	Unsorted	Cluster	Mobile	Cluster	Mobile

[a]Data from Hannan *et al.* (1993).

this question may be provided by a different cell system, the Fischer rat thyroid (FRT) cell line. These cells have been shown to display similar polarity to MDCK cells regarding the distribution of different apical and basolateral transmembrane proteins (Zurzolo *et al.*, 1992b) and some viral glycoproteins (Zurzolo *et al.*, 1992a). However, GPI-anchored proteins are mainly basolateral (Zurzolo *et al.*, 1993a) in contrast to MDCK cells. In particular, exogenous gD1-DAF is targeted basolaterally in these cells whereas exogenous DAF is unpolarized. New results (Zurzolo *et al.*, 1993b) indicate that GPI-anchored proteins are never sedimentable, at least under the low *g* force of the Brown and Rose (1992) assay, after treatment with Triton X-100 at low temperature. However, both FRT and MDCK cells assemble GSLs into Tx 100 insoluble complexes with identical isopycnic densities, but, differently from MDCK cells, GPI-DAF does not cluster with them in FRT cells. The clustering defect correlates with the absence of VIP21/caveolin in FRT cells (Zurzolo *et al.*, 1993b), suggesting that this protein may have a role in the formation of GPI-anchored protein/GSL clusters.

However, other factors might be involved in the differential targeting of GPI-proteins in MDCK and FRT cells, e.g., different GPI structure, different GSL composition, or alteration in the luminal pH of FRT cells.

VIII. RECEPTOR-MEDIATED ENDOCYTOSIS IN CAVEOLAE

GPI-anchored proteins are concentrated in caveolae, small invaginated pits or vesicles with a diameter of ~50 nm. The cytoplasmic surface of these vesicles/pits is covered with a characteristic striated coat composed of delicate filaments; a major protein component of these filaments is a 21-kDa protein called caveolin (Rothberg *et al.*, 1992). High affinity folate receptors populate the caveolae of certain cultured cells, such as the kidney epithelial line MA104, and play a role in the uptake of folate. Only half the receptors bind 5-methyltetrahydrofolate at 0°C; the other half slowly exchanges with the surface receptors every hour (Rothberg *et al.*, 1990a). The caveolar pathway does not communicate with the clathrin-coated endocytic pathway (Anderson *et al.*, 1992). Caveolar structure is disrupted by cholesterol-binding drugs such as filipin as well as in cholesterol-depleted cells; thus, cholesterol seems to be essential for the clustering of GPI proteins in caveolae (Rothberg *et al.*, 1990b; Chang *et al.*, 1991). The trapping of GPI proteins in caveolae may act as a mechanism to concentrate the specific substrate (many of the apical GPI proteins are enzymes), which would favor transport across the membrane by a specific system. This mode of internalization of small molecules has been termed potocytosis and constitutes an alternative to traffic through the endomembrane system provided by the classic clathrin-mediated pathway (Anderson *et al.*, 1992). Interestingly, a protein identical to caveolin, VP21 (vesicle protein of 21 kDa), has been isolated in association with both apical and basolateral vesicles that bud from the TGN of MDCK cells (Kurzchalia *et al.*, 1992).

IX. ROLE OF GLYCOSYL-PHOSPHATIDYLINOSITOL IN SIGNALING: HYDROLYSIS AND THE PRODUCTION OF MESSENGERS FOR GROWTH FACTORS

Saltiel and collaborators initially presented evidence for a role for free GPI as an intracellular messenger for insulin (Saltiel and Cuatrecasas, 1986,1988, Saltiel *et al.*, 1987). According to this hypothesis, binding of insulin to its receptor leads to activation of a PLC that cleaves free GPI at the plasma membrane, generating two intracellular signals: diacylglycerol and inositol phosphoglycans (IPGs). Later work also identified a similar signaling mechanism for nerve growth factor (NGF) receptor and interleukin 2 (Chan *et al.*, 1989). In apparent support of this proposal, extracts containing IPGs simulate the action of insulin in several cell types (Romero, 1991). The insulin-sensitive glycolipids are cleavable by PI-PLC and nitrous acid, and can be labeled with tritiated anchor precursors such as [^{3}H]glucosamine, [^{3}H]galactose, [^{3}H]myoinositol, or [^{3}H]myristic acid but not with [^{3}H]mannose or [^{3}H]ethanolamine (Mato *et al.*, 1987; Gaulton *et al.*, 1988; Suzuki *et al.*, 1991). However, careful analysis of free GPI isolated from H35 hepatoma cells demonstrated that it can be labeled with [^{3}H]mannose and [^{3}H]ethanolamine, but cannot be labeled with [^{3}H]glucosamine (Deeg *et al.*, 1992b). Further, free GPI cannot be cleaved with PI-PLC unless it is pretreated with base to remove acyl groups from inositol. The lipid labeled with [^{3}H]glucosamine may be distinct from GPI anchors in proteins and may correspond to a novel class of free GPIs with a core glycan structure considerably different, that is, with a galactose-containing core (Turco *et al.*, 1989; Deeg *et al.*, 1992a,b). Additional studies are needed to determine precisely the role of GPI in signaling, the nature of the lipase(s) and glycolipids involved, and the orientation of these lipids in the bilayer. Note that, if this putative glycolipid is external, mechanisms would be needed to prevent diffusive loss of the released soluble phosphoglycans and to facilitate their transport across the membrane. A simple solution would be that these events occur after endocytosis in caveolae, as proposed for folate transport (Anderson *et al.*, 1990). However, no phospholipase that might participate in such a process has been identified in caveolae (or at the plasma membrane) to date.

X. ACTIVATION OF T LYMPHOCYTES AND OTHER SIGNALING EVENTS IN HEMATOPOIETIC CELLS

Antibodies against surface components of T lymphocytes lead to activation of these cells, that is, to the production of lymphokines and to the expression of other effector functions that regulate the immune response (Robinson, 1991). Although the central player in this process is the multicomponent T-cell receptor (TCR), cross-linking of other surface components (e.g., by antibodies) such as Thy-1 may lead to similar effects. T-cell activation by either pathway is characterized by rapid activation of the protein tyrosine kinase pathway and tyrosine phosphorylation of multiple substrates (June et al., 1990; Klausner and Samelson, 1991; Robinson, 1991). Later events include phospholipid hydrolysis, production of inositol 1,4,5-triphosphate (IP$_3$), increase in intracellular calcium, and activation of serine/threonine kinases. A surprising finding is that, in addition to Thy-1, seven other GPI-anchored proteins have been shown to cause activation of T cells when cross-linked by specific antibodies, including DAF and Qa-2 (MacDonald et al., 1985; Davis et al., 1988; Hahn and Soloski, 1989; Robinson et al., 1989). Linkage to GPI is essential for activation, since activation is abolished after treatment with PI-PLC (Thomas and Samelson, 1992) and is observed after transgenic expression of GPI-anchored Qa-2 but not transmembrane-anchored Qa-2 in mouse T cells (Robinson et al., 1989). Various experiments indicate that clustering of GPI-anchored proteins is required for activation. How could cross-linking of GPI proteins, which are located entirely on the external leaflet of the membrane, lead to activation of tyrosine phosphorylation on the cytosolic side? Several experiments have shown that members of the src family of tyrosine kinases, such as lck and fyn, are found in association with GPI-anchored proteins; this association can be observed by immunoprecipitation with antibodies against GPI-anchored proteins. Interestingly, this association is abolished by octyl-glycoside but not by TX-100, thus showing detergent specificity similar to that of the complexes isolated by Brown and Rose from MDCK cells (discussed by Brown, 1992). The size of these complexes is similar to that of the smallest insoluble vesicles in MDCK cells, ~100 nm (Cinek and Horejsi, 1992).

As described earlier for GPI-anchored proteins in epithelial cells, a clustering event in the exoplasmic side leads to an event in the cytosolic side. Whereas sorting in the TGN involves interaction of the GPI protein–GSL patch with the sorting machinery, interaction of the GPI proteins of the T cell with the signaling machinery in the cytoplasm leads to a cascade of second messengers and activation of a differentiated program. More recent work has demonstrated an important role for trimeric G proteins (Bomsel and Mostov, 1992; Pfeffer, 1992) in apical and basolateral protein delivery from the TGN to the cell surface (Stow *et al.*, 1991; Pimplikar and Simons, 1993). Thus, mechanisms previously thought to be restricted to the plasma membrane are found to play an important role in intracellular traffic. Finding an important role for tyrosine kinases in the regulation of intracellular events should not be very surprising.

Acknowledgments

We thank Anant Menon (Rockefeller University) for critical reading of the manuscript and Grace Papaseraphim for secretarial work. This work was supported by grants from the NIH (GM-34107 and GM-41771) to E. Rodriguez-Boulan. C. Zurzolo is also supported as Ricercatore at Naples University, Dipartamiento di Biologia Cellulare e Molecolare, II Policlinico, by the Italian Government.

References

Aitken, A., Cohen, P., Santikarn, S., Williams, D. H., Calder, A. G., Smith, A., and Klee, C. B. (1982). Identification of the NH$_2$-terminal blocking group of calcineurin B as myristic acid. *FEBS Lett.* **150**, 314–318.
Ali, N., and Evans, W. H. (1990). Priority targeting of GPI-anchored proteins to the bile-canalicular (apical) plasma membrane of hepatocytes. *Biochem. J.* **271**, 193–199.
Anderson, D., Guerin, C., Matsumoto, B., and Pfeffer, B. (1990). Identification and localization of a beta 1 receptor from the integrin family in mammalian RPE cells. *Invest. Ophthalmol. Vis. Sci.* **31**, 81–93.
Anderson, R. G. W., Kamen, B. A., Rothberg, K. G., and Lacey, S. W. (1992). Potocytosis: Sequestration and transport of small molecules by Caveolae. *Science* **255**, 410–411.
Balch, W. E., Dunphy, W. G., Braell, W. A., and Rothman, J. E. (1984). Reconstitution of the transport of protein between successive compartments of the Golgi measured by the coupled incorporation of *N*-acetylglucosamine. *Cell* **39**, 405–416.
Balch, W. E., Kahn, R. A., and Schwaninger, R. (1992). ADP-ribosylation factor is required for vesicular trafficking between the endoplasmic reticulum and the cis-Golgi compartment. *J. Biol. Chem.* **267**, 13053–13061.
Berger, J., Howard, A. D., Brink, L., Gerber, L., Haubert, J., Cullen, B. R., and Udenfriend, S. (1988). COOH-terminal requirements for the correct processing of a phosphatidyl-inositol-glycan anchored membrane protein. *J. Biol. Chem.* **263**, 10016–10021.
Bomsel, M., and Mostov, K. (1992). Role of heterotrimeric G proteins in membrane traffic. *Mol. Biol. Cell* **3**, 1317–1328.
Bracha, M., Sagher, D., and Schlesinger, M. J. (1977). Reaction of the protease inhibitor *p*-nitrophenyl-*p*-guanidinobenzoate with Sindbis virus. *Virology* **83**, 246–253.
Brown, D. A. (1992). Interactions between GPI-anchored proteins and membrane lipids. *Trends Cell Biol.* **2**, 338–343.
Brown, D. A., and Rose, J. K. (1992). Sorting of GPI-anchored proteins to glycolipid enriched membrane subdomains during transport to the apical cell surface. *Cell* **68**, 533–544.
Brown, D. A., Crise, B., and Rose, J. K. (1989). Mechanism of membrane anchoring affects polarized expression of two proteins in MDCK cells. *Science* **245**, 1499–1501.
Bryant, M. L., Heuckeroth, R. O., Kimata, J. T., Ratner, L., and Gordon, J. I. (1989). Replication of human immunodeficiency virus 1 and Moloney murine leukemia virus is inhibited by different heteroatom-containing analogs of myristic acid. *Proc. Natl. Acad. Sci. U.S.A.* **86**, 8655–8659.

Caras, I. W., Weddell, G. N., and Williams, S. R. (1989). Analysis of the signal for attachment of a glycosylphospholipid membrane anchor. *J. Cell Biol.* **108**, 1387–1396.

Carr, S. A., Biemann, K., Shoji, S., Parmele, D. C., and Titani, K. (1982). *n*-Tetradecanoyl is the NH$_2$-terminal blocking group of the catalytic subunit of cyclic AMP-dependent protein kinase from bovine cardiac muscle. *Proc. Natl. Acad. Sci. U.S.A.* **79**, 6128–6131.

Chan, B. L., Chao, M. V., and Saltiel, A. R. (1989). NGF stimulates the hydrolysis of glycosyl phosphatidylinositol in PC12 cells: A novel mechanism of protein kinase C regulation. *Proc. Natl. Acad. Sci. U.S.A.* **86**, 1756–1760.

Chang, W.-J., Rothberg, K. G., Kamen, B. A., and Anderson, R. G. W. (1991). Lowering the cholesterol content of MA104 cells inhibits receptor-mediated transport of folate. *J. Cell Biol.* **118**, 63–69.

Chavrier, P., Gorvel, J.-P., Stelzer, E., Simons, K., Gruenberg, J., and Zerial, M. (1991). Hypervariable C-terminal domain of rab proteins acts as a targeting signal. *Nature* (*London*) **353**, 769–772.

Cinek, T., and Horejsi, V. (1992). The nature of large noncovalent complexes containing glycosyl-phosphatidylinositol-anchored membrane glycoproteins and protein tyrosine kinases. *J. Immunol.* **149**, 2262–2270.

Cross, G. A. M. (1990). Glycolipid anchoring of plasma membrane proteins. *Ann. Rev. Cell Biol.* **6**, 1–39.

Davis, L. S., Patel, S. S., Atkinson, J. P., and Lipsky, P. E. (1988). Decay-accelerating factor functions as a signal transducing molecule for human T cells. *J. Immunol.* **141**, 2246–2252.

Deeg, M. A., Humphrey, D. R., Yang, S. H., Ferguson, T. R., Reinhold, V. N., and Rosenberry, T. L. (1992a). Glycan components in the glycoinositol phospholipid anchor of human erythrocyte acetylcholinesterase. *J. Biol. Chem.* **267**, 18573–18580.

Deeg, M. A., Murray, N. R., and Rosenberry, T. L. (1992b). Identification of glycoinositol phospholipids in rat liver by reductive radiomethylation of amines but not in H4IIE hepatoma cells or isolated hepatocytes by biosynthetic labeling with glucosamine. *J. Biol. Chem.* **267**, 18581–18588.

Deschenes, R. J., Resh, M. D., and Broach, J. R. (1990). Acylation and prenylation of proteins. *Curr. Opin. Cell Biol.* **2**, 1108–1113.

Farnsworth, C. C., Wolda, S. L., Gelb, M. H., and Glomset, J. A. (1989). Human lamin B contains a farnesylated cysteine residue. *J. Biol. Chem.* **264**, 20422–20429.

Farnsworth, C. C., Gelb, M. H., and Glomset, J. A. (1990). Identification of geranylgeranyl-modified proteins in HeLa cells. *Science* **247**, 320–322.

Ferguson, M. A. J., and Williams, A. F. (1988). Cell-surface anchoring of proteins via glycosyl-phosphatidylinositol structures. *Annu. Rev. Biochem.* **57**, 285–320.

Gaulton, G. N., Kelly, K. L., Pawlowski, J., Mato, J. M., and Jarett, L. (1988). Regulation and function of an insulin-sensitive glycosyl-phosphatidylinositol during T lymphocyte activation. *Cell* **53**, 963–970.

Gerber, L. D., Kodukula, K., and Udenfriend, S. (1992). Phosphatidylinositol glycan (PI-G) anchored membrane proteins. *J. Biol. Chem.* **267**, 12168–12173.

Glick, B. S., and Rothman, J. E. (1987). Possible role for fatty acyl-coenzyme A in intracellular protein transport. *Nature* (*London*) **326**, 309–312.

Glomset, J. A., Gelb, M. H., and Farnsworth, C. C. (1990). Prenyl proteins in eukaryotic cells: a new type of membrane anchor. *Trends Biochem. Sci.* **15**, 139–142.

Gordon, J. I., Duronio, R. J., Rudnick, D. A., Adams, S. P., and Gokel, G. W. (1991). Protein-*N*-myristoylation. *J. Biol. Chem.* **266**, 8647–8650.

Gottlinger, H. G., Sodroski, J. G., and Haseltine, W. A. (1989). Role of capsid precursor processing and myristoylation in morphogenesis and infectivity of human immunodeficiency virus type 1. *Proc. Natl. Acad. Sci. U.S.A.* **86,** 5781–5785.

Hahn, A. B., and Soloski, M. J. (1989). Anti-Qa-2 induced T cell activation: The parameters of activation, the definition of mitogenic and nonmitogenic antibodies and the differential effects on CD4+ vs CD8+ T cells. *Am. Assoc. Immunol.* **143,** 407–413.

Hancock, J. F., Magee, A. I., Childs, J. E., and Marshall, C. J. (1989). All ras proteins are polyisoprenylated but only some are palmitoylated. *Cell* **57,** 1167–1177.

Hannan, L. A., Lisanti, M. P., Rodriguez-Boulan, E., and Edidin, M. (1993). Correctly sorted molecules of a GPI-anchored protein are clustered and immobile when they arrive at the apical surface of MDCK cells. *J. Cell Biol.* **120,** 353–358.

Hantke, K., and Braun, V. (1973). Covalent binding of lipid to protein. Diglyceride and amide-linked fatty acid at the N-terminal end of the murein-lipoprotein of the Escherichia coli outer membrane. *Eur. J. Biochem.* **34,** 284–296.

Henderson, L. E., Krutzsch, H. C., and Oroszlan, S. (1983). Myristyl amino-terminal acylation of murine retrovirus proteins: An unusual post-translational protein modification. *Proc. Natl. Acad. Sci. U.S.A.* **80,** 339–343.

Heuckeroth, R. O., Glaser, L., and Gordon, J. I. (1988). Heteroatom-substituted fatty acid analogs as substrates for *N*-myristoyltransferase: An approach for studying both the enzymology and function of protein acylation. *Proc. Natl. Acad. Sci. U.S.A.* **85,** 8795–8799.

Hoessli, D., and Rungger-Brandle, E. (1985). Association of specific cell-surface glycoproteins with a Triton X-100 resistant complex of plasma membrane proteins isolated from T-lymphoma cells (P1798). *Exp. Cell Res.* **156,** 239–250.

Holtz, D., Tanaka, R. A., Hartwig, J., and McKeon, F. (1989). The CAAX motif or lamin A functions in conjunction with the nuclear localization signal to target assembly to the nuclear envelope. *Cell* **59,** 969–977.

Hooper, N. M., and Turner, A. J. (1988). Ectoenzymes of the kidney microvillar membrane. *Biochem. J.* **250,** 865–869.

James, G., and Olson, E. N. (1990). Fatty acylated proteins as components of intracellular signaling pathways. *Biochemistry* **29,** 2623–2634.

Jing, S., and Trowbridge, I. S. (1987). Identification of the intermolecular disulfide bonds of the human transferrin receptor and its lipid-attachment site. *EMBO J.* **6,** 327–331.

June, C. H., Fletcher, M. C., Ledbetter, J. A., and Samelson, L. E. (1990). Increases in tyrosine phosphorylation are detectable before phospholipase C activation after T cell receptor stimulation. *J. Immunol.* **144,** 1591–1599.

Kahn, R. A., Kern, F. G., Clark, J., Gelmann, E. P., and Rulka, C. (1991). Human ADP-ribosylation factors. *J. Biol. Chem.* **266,** 2606–2614.

Kahn, R. A., Randazzo, P., Serafini, T., Weiss, O., Rulka, C., Clark, J., Amherdt, M., Roller, P., Orci, L., and Rothman, J. E. (1992). The amino terminus of ADP-ribosylation factor (ARF) is a critical determinant of ARF activities and is a potent and specific inhibitor of protein transport. *J. Biol. Chem.* **267,** 13039–13046.

Kamiya, Y., Sakurai, A., Tamura, S., and Takahashi, N. (1978). Structure of rhodotorucine A, a novel lipopeptide, inducing mating tube formation in *Rhodosporidium toruloides*. *Biochem. Biophys. Res. Commun.* **83,** 1077–1083.

Kamps, M. P., Buss, J. E., and Sefton, B. M. (1986). Rous sarcoma virus transforming protein lacking myristic acid phosphorylates known polypeptide substrates without inducing transformation. *Cell* **45,** 105–112.

Kasinathan, C., Grzelinska, E., Okazaki, K., Slomiany, B. L., and Slomiany, A. (1990). Purification of protein fatty acyltransferase and determination of its distribution and topology. *J. Biol. Chem.* **265,** 5139–5144.

Khosravi-Far, R., Lutz, R. J., Cox, A. D., Conroy, L., Bourne, J. R., Sinenski, M., Balch, W. E., Buss, J. E., and Der, C. J. (1991). Isoprenoid modification of *rab* proteins terminating in CC or CXC motifs. *Proc. Natl. Acad. Sci. U.S.A.* **88**, 6264–6268.

Klausner, R. D., and Samelson, L. E. (1991). T cell antigen receptor activation pathways: The tyrosine kinase connection. *Cell* **64**, 875–878.

Kodukula, K., Amthauer, R., Cines, D., Yeh, E. T. H., Brink, L., Thomas, L. J., and Udenfriend, S. (1992). Biosynthesis of phosphatidylinositol-glycan (PI-G) anchored membrane proteins in cell-free systems: PI-G is an obligatory cosubstrate for COOH-terminal processing of nascent proteins. *Proc. Natl. Acad. Sci. U.S.A.* **89**, 4982–4985.

Kurzchalia, T. V., Dupree, P., Parton, R. G., Kellner, R., Virta, H., Lehnert, M., and Simons, K. (1992). VIP21, a 21-kD membrane protein is an integral component of Trans-Golgi-network-derived transport vesicles. *J. Cell Biol.* **118**, 1003–1014.

Lenhard, J. M., Kahn, R. A., and Stahl, P. D. (1992). Evidence for ADP-ribosylation factor (ARF) as a regulator of in vitro endosome-endosome fusion. *J. Biol. Chem.* **267**, 13047–13052.

Lisanti, M., and Rodriguez-Boulan, E. (1990). Glycophospholipid membrane anchoring provides clues to the mechanism of protein sorting in polarized epithelial cells. *Trends Biochem. Sci.* **15**, 113–118.

Lisanti, M., Sargiacomo, M., Graeve, L., Saltiel, A., and Rodriguez-Boulan, E. (1988). Polarized apical distribution of glycosyl phosphatidylinositol anchored proteins in a renal epithelial line. *Proc. Natl. Acad. Sci. U.S.A.* **85**, 9557–9561.

Lisanti, M., Caras, I. P., Davitz, M. A., and Rodriguez-Boulan, E. (1989). A glycophospholipid membrane anchor acts as an apical targeting signal in polarized epithalial cells. *J. Cell Biol.* **109**, 2145–2156.

Lisanti, M. P., Le Bivic, A., Saltiel, A., and Rodriguez-Boulan, E. (1990). Preferred apical distribution of glycosyl-phosphatidylinositol (GPI) anchored proteins: a highly conserved feature of the polarized epithelial cell phenotype. *J. Memb. Biol.* **113**, 155–167.

Low, M. G., and Saltiel, A. R. (1988). Structural and functional roles of glycosyl-phosphatidylinositol in membranes. *Science* **239**, 268–275.

MacDonald, H. R., Bron, C., Rousseaux, M., Horvath, C., and Cerottini, J. C. (1985). Production and characterization of monoclonal anti-Thy-1 antibodies that stimulate lymphokine production by cytolytic T cell clones. *Eur. J. Immunol.* **15**, 495–501.

Magee, A. I. (1990). Lipid modification of proteins and its relevance to protein targeting. *J. Cell Sci.* **97**, 581–584.

Magee, A. I., and Courtneidge, S. A. (1985a). Two classes of fatty acid acylated proteins exist in eukaryotic cells. *EMBO J.* **4**, 1137–1144.

Magee, A. I., and Courtneidge, S. A. (1985b). Two classes of fatty acid acylated proteins exist in eukaryotic cells. *EMBO J.* **4**, 1137–1144.

Magee, T., and Newman, C. (1992). The role of lipid anchors for small G proteins in membrane trafficking. *Trends Cell Biol.* **2**, 318–323.

Marc, D., Drugeon, G., Haenni, A.-L., Girard, M., and van der Werf, S. (1989). Role of myristoylation of poliovirus capsid protein VP4 as determined by site-directed mutagenesis of its N-terminal sequence. *EMBO J.* **8**, 2661–2668.

Mato, J. M., Kelly, K. L., Abler, A., and Jarett, L. (1987). Identification of a novel insulin-sensitive glycophospholipid from H35 hepatoma cells. *J. Biol. Chem.* **15**, 2131–2137.

Mayor, S., Menon, A. K., and Cross, G. A. M. (1991). Transfer of glycosyl-phosphatidylinositol membrane anchors to polypeptide acceptors in a cell-free system. *J. Cell Biol.* **114**, 61–71.

McIlhinney, R. A. J. (1990). The fats of life: The importance and function of protein acylation. *Trends Biochem. Sci.* **15**, 387–391.

McIlhinney, R. A. J., Chadwick, J. K., and Pelly, S. J. (1987). Studies on the cellular location, physical properties and endogenously attached lipids of acylated proteins in human squamous-carcinoma cell lines. *Biochem. J.* **244**, 109–115.

Meiss, H. K., Green, R. F., and Rodriguez-Boulan, E. (1982). Lectin-resistant mutants of polarized epithelial cells. *Mol. Cell. Biol.* **2**, 1287–1294.

Menon, A. K., Schwarz, R. T., Mayor, S., and Cross, G. A. M. (1990). Cell-free synthesis of glycosyl-phosphatidylinositol precursors for the glycolipid membrane anchor of *Trypanosoma brucei* variant surface glycoproteins. *J. Biol. Chem.* **265**, 9033–9042.

Mescher, M. F., Jose, M. J. L., and Balk, S. P. (1981). Actin-containing matrix associated with the plasma membrane of murine tumour and lymphoid cells. *Nature (London)* **289**, 139–144.

Mumby, S. M., Casey, P. J., Gilman, A. G., Gutowski, S., and Sternweis, P. C. (1990). G protein γ subunits contain a 20-carbon isoprenoid. *Proc. Natl. Acad. Sci. U.S.A.* **87**, 5873–5877.

Newman, C. M. H., Giannakouros, T., Hancock, J. F. Fawell, E. H., Armstrong, J., and Magee, A. I. (1992). Post-translational processing of *Schizosaccharomyces pombe* YPT proteins. *J. Biol. Chem.* **267**, 11329–11336.

O'Down, B. F., Hnatowich, M., Caron, M. G., Lefkowitz, R. J., and Bouvier, M. (1989). Palmitoylation of the human B2-adrenergic receptor. *J. Biol. Chem.* **264**, 7564–7569.

Olson, E. N., and Spizz, G. (1986). Fatty acylation of cellular proteins. *J. Biol. Chem.* **261**, 2458–2466.

Olson, E. N., Towler, D. A., and Glaser, L. (1985). Specificity of fatty acid acylation of cellular proteins. *J. Biol. Chem.* **260**, 3784–3790.

Omary, M. B., and Trowbridge, I. S. (1981). Biosynthesis of the human transferrin receptor in cultured cells. *J. Biol. Chem.* **256**, 12888–12892.

Pellman, D., Garber, E. A., Cross, F. R., and Hanafusa, H. (1985). Fine structural mapping of a critical NHx-terminal region of p60^{src}. *Proc. Natl. Acad. Sci. U.S.A.* **82**, 1623–1627.

Pfeffer, S. R. (1992). GTP-binding proteins in intracellular transport. *Trends Cell Biol.* **2**, 41–46.

Pimplikar, S. W., and Simons, K. (1993). Apical transport in epithelial cells is regulated by a Gs class of heterotrimeric G protein. *Nature (London)* **362**, 456–458.

Powell, S. K., Cunningham, B. A., Edelman, G. M., and Rodriguez-Boulan, E. (1991). Targeting of transmembrane and GPI-anchored forms of N-Cam to opposite domains of a polarized epithelial cell. *Nature (London)* **353**, 76–77.

Reiss, Y., Goldstein, J. L., Seabra, M. C., Casey, P. J., and Brown, M. S. (1990). Inhibition of purified p21^{ras} farnesyl : protein transferase by Cys-AAX tetrapeptides. *Cell* **62**, 81–88.

Reiss, Y., Stradley, S. J., Gierasch, L. M., Brown, M. S., and Goldstein, J. L. (1991). Sequence requirement for peptide recognition by rat brain p21ras protein farnesyltransferase. *Biochemistry* **88**, 732–736.

Resh, M. D., and Ling, H.-P. (1990). Identification of a 32K plasma membrane protein that binds to the myristylated amino-terminal sequence of p60^{v-src}. *Nature (London)* **346**, 84–86.

Rilling, H. C., Breunger, E., Epstein, W. W., and Crain, P. F. (1990). Prenylated proteins: The structure of the isoprenoid group. *Science* **247**, 318–320.

Roberts, W. L., Santikarn, S., Reinhold, V. N., and Rosenberry, T. L. (1988). Structural characterization of the glycoinositol phospholipid membrane anchor of human erythrocyte acetylcholinesterase by fast atom bombardment mass spectrometry. *J. Biol. Chem.* **263**, 18776–18784.

Robinson, P. J. (1991). Phosphatidylinositol membrane anchors and T-cell activation. *Immunol. Today* **12**, 35–41.

Robinson, P. J., Millrain, M., Antoniou, J., Simpson, E., and Mellor, A. L. (1989). A glycophospholipid anchor is required for Qa-2 mediated T cell activation. *Nature (London)* **342**, 85–87.

Romero, G. (1991). Inositolglycans and cellular signalling. *Cell Biol. Int. Rep.* **15**, 827–852.

Rothberg, K. G., Ying, Y., Kolhouse, J. F., Kamen, B. A., and Anderson, R. G. W. (1990a). The glycophospholipid-linked folate receptor internalizes folate without entering the clathrin-coated pit endocytic pathway. *J. Cell Biol.* **110**, 637–648.

Rothberg, K. G., Ying, Y.-S., Kamen, B. A., and Anderson, R. G. W. (1990b). Cholesterol controls the clustering of the glycophospholipid-anchored membrane receptor for 5-methyltetrahydrofolate. *J. Cell Biol.* **111**, 2931–2938.

Rothberg, K. G., Heuser, J. E., Donzell, W. C., Ying, Y.-S., Glenney, J. R., and Anderson, R. G. W. (1992). Caveolin, a protein component of caveolae membrane coats. *Cell* **68**, 673–682.

Saltiel, A. R., and Cuatrecasas, P. (1986). Insulin stimulates the generation from hepatic plasma membranes of modulators derived from an inositol glycolipid. *Proc. Natl. Acad. Sci. U.S.A.* **83**, 5793–5797.

Saltiel, A. R., and Cuatrecasas, P. (1988). In search of a second messenger for insulin. *Am. J. Physiol.* **255**, C1–C11.

Saltiel, A. R., Sherline, P., and Fox, J. A. (1987). Insulin-stimulated diacylglycerol production results from the hydrolysis of a novel phosphatidylinositol glycan. *J. Biol. Chem.* **262**, 1116–1121.

Schafer, W. R., Kim, R., Sterne, R., Thorner, J., Kim, S.-H., and Rine, J. (1989). Genetic and pharmacological suppression of oncogenic mutations in RAS genes of yeast and humans. *Science* **247**, 379–385.

Schlesinger, M. J. (1981). Proteolipids. *Ann. Rev. Biochem.* **50**, 193–206.

Schlesinger, M. J., Magee, A. I., and Schmidt, M. F. G. (1980). Fatty acid acylation of proteins in cultured cells. *J. Biol. Chem.* **255**, 10021–10024.

Schmid, M. F. G., and Schlesinger, M. J. (1980). Relation of fatty acid attachment to the translation and maturation of vesicular stomatitits and sindbis virus membrane glycoproteins. *J. Biol. Chem.* **255**, 3334–3339.

Schultz, A. M., Henderson, L. E., and Oroszlan, S. (1988). Fatty acylation of proteins. *Ann. Rev. Cell Biol.* **4**, 611–647.

Seabra, M. C., Goldstein, J. L., Sudhof, T. C., and Brown, M. S. (1992). Rab geranylgeranyl transferase. *J. Biol. Chem.* **267**, 14497–14503.

Sefton, B. M., and Buss, J. E. (1987). The covalent modification of eukaryotic proteins with lipid. *J. Cell Biol.* **104**, 1449–1453.

Sinensky, M., and Lutz, R. J. (1992). The prenylation of proteins. *BioEssays* **14**, 25–31.

Skene, J. H. P., and Virag, I. (1989). Posttranslational membrane attachment and dynamic fatty acylation of a neuronal growth cone protein. GAP-43. *J. Cell Biol.* **108**, 613–624.

Staufenbiel, M. (1987). Ankyrin-ound fatty acid turns over rapidly at the erythrocyte plasma membrane. *Mol. Cell. Biol.* **7**, 2981–2984.

Stefanova, I., and Horejsi, V. (1991). Association of the CD59 and CD55 cell surface glycoproteins with other membrane molecules. *J. Immunol.* **147**, 1587–1592.

Stefanova, I., Horejsi, V., Ansotegui, I. J., Knapp, W., and Stockinger, H. (1991). GPI-anchored cell-surface molecules complexed to protein tyrosine kinases. *Science* **254**, 1016–1019.

Stow, J. L., DeAlmeida, J. B., Narula, N., Holtzman, E. J., Ercolani, L., and Ausiello, D. A. (1991). A heterotrimeric G protein, G alpha i-3, on Golgi membranes regulates the secretion of a heparan sulfate proteoglycan in LLC-PK1 epithelial cells. *J. Cell Biol.* **114**, 1113–1124.

Suzuki, S., Sugawara, K., Satoh, Y., and Toyota, T. (1991). Insulin stimulates the generation of two putative insulin mediators, inositol-glycan and diacylglycerol in BC3H-1 myocytes. *J. Biol. Chem.* **266,** 8115–8121.

Taylor, T. C., Kahn, R. A., and Melancon, P. (1992). Two distinct members of the ADP-ribosylation factor family of GTP-binding proteins regulate cell-free intra-golgi transport. *Cell* **70,** 69–79.

Thomas, P. M., and Samelson, L. E. (1992). The glycophosphatidylinositol-anchored Thy-1 molecule interacts with the p60^{fyn} protein tyrosine kinase in T cells. *J. Biol. Chem.* **267,** 12317–12322.

Towler, D. A., and Gordon, J. I. (1988). The biology and enzymology of eukaryotic protein acylation. *Ann. Rev. Biochem.* **57,** 69–99.

Tsuchiya, M., Price, S. R., Tsai, S.-C., Moss, J., and Vaughan, M. (1991). Molecular identification of ADP-ribosylation factor mRNAs and their expression in mammalian cells. *J. Biol. Chem.* **266,** 2772–2777.

Turco, S. J., Orlandi, P. A., Homans, S. W., Ferguson, M. A. J., Dwek, R. A., and Rademacher, T. W. (1989). Structure of the phosphosaccharide-inositol core of the *Leishmania donovani* lipophosphoglycan. *J. Biol. Chem.* **264,** 6711–6715.

van Meer, G., and Simons, K. (1988). Lipid polarity and sorting in epithelial cells. *J. Cell Biochem.* **36,** 51–58.

Wei, C., Lutz, R., Sinensky, M., and Macara, I. A. (1992). p23^{rab2}, a *ras*-like GTPase with a –GGGCC C terminus, is isoprenylated but not detectably carboxymethylated in NIH 3T3 cells. *Oncogene* **7,** 467–473.

Wilcox, C., Hu, J.-S., and Olson, E. N. (1987). Acylation of proteins with myristic acid occurs cotranslationally. *Science* **238,** 1275–1278.

Willumsen, B. M., Norris, K., Papageorge, A. G., Hubbert, N. L., and Lowy, D. R. (1984). Harvey murine sarcoma virus p21 ras protein: Biological and biochemical significance of the cysteine nearest the carboxy terminus. *EMBO J.* **3,** 2581–2585.

Yamane, H. K., Farnsworth, C. C., Xie, H., Howald, W., Fung, B. K. K., Clarke, S., Gelb, M. H., and Glomset, J. A. (1990). Brain g protein γ subunits contain an all-trans-geranylgeranyl-cysteine methyl ester at their carboxyl termini. *Proc. Natl. Acad. Sci. U.S.A.* **87,** 5868–5872.

Youji, S., Mitsuru, Y., Akira, I., and Akinori, S. (1981). Peptidal sex hormones inducing conjugation tube formation in compatible mating-type cells of *Tremella mesenterica*. *Science* **212,** 1525–1527.

Zuber, M. X., Strittmatter, S. M., and Fishman, M. C. (1989). A membrane-targeting signal in the amino terminus of the neuronal protein GAP-43. *Nature (London)* **341,** 345–348.

Zurzolo, C., Polistina, C., Saini, M., Gentile, R., Aloj, L., Migliaccio, G., Bonatti, S., and Nitsch, L. (1992a). Opposite polarity of virus budding and of viral envelope glycoprotein distribution in epithelial cells derived from different tissues. *J. Cell Biol.* **117,** 551–564.

Zurzolo, C., Le Bivic, A., Quaroni, A., Nitsch, L., and Rodriguez-Boulan, E. (1992b). Modulation of transcytotic and direct targeting pathways in a polarized thyroid cell line. *EMBO J.* **11,** 2337–2344.

Zurzolo, C., Lisanti, M. P., Caras, I. W., Nitsch, L., and Rodriguez-Boulan, E. (1993a). GPI anchor targets proteins to the basolateral surface in Fischer rat thyroid (FRT) epithelial cells. **121,** 1031–1039.

Zurzolo, C., Van't Hof, W., Van Meer, G., and Rodriguez-Boulan, E. (1993b). VIP21/Caveolin, glycosphingolipid clusters and the sorting of glycosyl phosphatidylinositol-anchored proteins in epithelial cells. *EMBO J.,* **13.**

CHAPTER 12

Modulation of Protein Function by Lipids

Alain Bienvenüe and Josette Sainte Marie
Laboratoire de Biologie Physico-chimique, CNRS URA 530 "interactions membranaires", Université Montpellier II, 34095 Montpellier, France

I. AIMS AND SCOPE

Enzymatic activities and ligand binding as well as protein, nucleic acid, or mixed assemblies depend very strongly on the composition and physico-chemical properties of all intra- or extracellular aqueous media (cytoplasm, organelles, nucleus, plasma, and so on). For example, viscosity, pH, ionic strength, and more precisely ionic composition (mainly calcium, magnesium, and potassium . . .) directly act on metabolism, cell shape, cell response to many signals, and cell–cell interactions. Rapid and precise regulation of aqueous biological media is of primary importance for correct cell functioning.

A good indication of the importance of lipids in cell functioning is that cells need large amounts of different lipids with strictly regulated locations and compositions. Cell membranes fulfill five essential functions:

1. acting as a semipermeable frontier with respect to ions, metabolites, peptides, proteins, and larger assemblies
2. participating in cell organization and movement by bending, fusion, and budding processes
3. solvating hydrophobic parts of integral proteins, stabilizing them in native conformation and correct lateral organization
4. interacting with extrinsic proteins to regulate their enzymatic activity or their binding and aggregation properties
5. stocking metabolites (fatty acids such as arachidonic acid, alkyl phospholipids, diacyl glycerol, inositol phosphates)

This chapter is devoted to the functional role of lipid–protein interactions, which has been shown to influence all these properties strongly in such a way that membrane composition and structure must be regulated as precisely as the aqueous medium. Since this field is large and rapidly moving, this chapter deals only with some recent advances in understanding how lipids govern important biological processes. In the first part, the major data concerning lipid interactions with proteins are summarized. The subsequent three parts focus on lipid effects on molecular, supramolecular, and cellular functions, respectively.

Only the major lipid components of biological membranes (phospholipids, cholesterol, and their derivatives or metabolites) will be considered. Other lipids and lipid-like compounds such as peptidolipids, glycolipids, and acylated proteins are excluded from this chapter since they are treated extensively in other chapters of this book. One general question is asked. Are the observed functional effects related to specific interactions, with known stoichiometry and lifetime? As shown in the following discussion, this question is much more complicated in relation to lipid–protein interactions than for any other compounds (ions, substrate, effector, agonist, and so on) for which the notion of specific interaction sites is theoretically and technically convenient. For a clear answer, lipid–protein interactions should be considered from two complementary points of view: lipid effects on protein properties such as catalytic activity, binding, structure, movement, and aggregation state and protein effects on lipid properties including effects on structure, lipid movement, and, eventually, phase separation. In fact, very few papers (if any) have discussed both these topics.

II. LIPID–PROTEIN INTERACTIONS AS SEEN BY LIPIDS

Only some of the major results concerning interactions between lipids and integral or peripheral proteins are discussed in this section, since they are addressed in considerable detail in other chapters of this book.

A. Integral Membrane Protein Modulation of Lipid Structure

The major lipid feature in model membranes is the bilayer arrangement, accompanied by fast lateral and slow transverse mobility, fatty acid and head group fast movement, and disorder, especially when distance from the glycerophosphoryl moiety increases. The main issue is whether these properties differ significantly when native membranes are compared with extracted lipids. Broadly speaking, the answer is no (Gennis, 1989), except in very few cases such as the bilayer stabilizing effect of integral proteins on membranes enriched in phosphatidylethanolamine (PE) (Taraschi *et al.*, 1982).

B. Lipids Surrounding Integral Membrane Proteins

Taking into account the lipid : protein weight ratio (about 1 : 1), up to 50% of the lipids would be adjacent to an integral protein molecule, on both sides of the membrane. The critical questions are four. (1) Do intrinsic proteins bind lipids and what is the lifetime of the complex? (2) How many lipids enter into the complex? (3) Do intrinsic proteins interact specifically with given phospholipids? (4) What is the molecular structure of the fatty acid chain and polar head group of the lipids surrounding the proteins? Only a few proteins have been examined, generally in reconstituted systems containing purified proteins included in well-defined vesicle membranes (Devaux, 1983; Lee, 1987). One report (Horvath *et al.*, 1990) shows once more that even so-called immobilized lipids have a very rapid exchange rate between boundary and free positions (about 10^7 sec^{-1}), similar to the events between two positions in bulk lipids. Only acyl chain carbon atoms near the polar head group are restricted motionally; their order parameter decreases by interaction with the rough protein surface. Generally, anionic phospholipids (phosphatidylserine, PS; phosphatidic acid,

PA) have slightly higher affinities than neutral lipids; their stoichiometry allows hydrophobic spanning peptides to be surrounded completely. Such specificity could be related to the asymmetric distribution of charged residues on both sides of the bacterial photosynthetic center polypeptides (Michel *et al.*, 1986). In the same vein, basic residues are found statistically more often in aqueous loops on the periplasmic side of the bacterial inner membrane proteins (Von Heijne, 1986); the stop transfer signal preferentially contains positively charged amino acids (Dalbey, 1990). Note that very few different integral proteins have been studied to date, so the results summarized here should be considered cautiously. According to different theoretical models (Jähnig, 1981; Mouritsen and Bloom, 1984; Sperotto and Mouritsen, 1988) and in agreement with some data (see subsequent discussion), the proposed major lipid–protein interaction could be governed by matching the lipid thickness with the protein hydrophobic span.

C. Phospholipid Binding to Purified Integral Proteins

A specific lipid–membrane protein interaction often is considered proven by the fact that some special phospholipid(s) remain(s) firmly associated with detergent-solubilized integral proteins during their purification. This conclusion seems doubtful, however, since the physical state of the proteins (aggregation) and lipids (phase segregation, for example) is generally ill defined. Although apparently homogeneous at the macroscopic level, the mixture contains micelles of very heterogeneous composition, often giving enzymatic or binding properties that are strongly dependent on the history of the assay sample (nature of the detergent, mixture procedures, pH, salt composition, temperature, etc.) (see Gennis and Jonas, 1977).

D. Binding of Peripheral Proteins on Membrane Lipids

Peripheral protein binding to a membrane is assayed easily by simple separation procedures. As discussed later, this characteristic has many important functional implications. From a structural perspective, complementary phospholipid binding to peripheral protein is very difficult to prove clearly. Polar head group movements are modified barely by protein binding, as shown by ^{2}H-NMR studies (Roux *et al.*, 1989; Bloom *et al.*, 1991). However, phase exclusion seems to occur when phospholipids cluster on interaction with polycationic molecules (Hartmann and Galla,

1978; Carrier and Pezolet, 1986; Ikeda *et al.*, 1990; Kim *et al.*, 1991; Mosior and McLaughlin, 1991), cytochrome *c* (Birrel and Griffith, 1976), myelin basic protein (Boggs *et al.*, 1977), and vesicular stomatitis virus (VSV) protein M (Wiener *et al.*, 1985). The same results are obtained with annexins (see subsequent discussion).

III. MODULATION OF CATALYTIC ACTIVITY AND BINDING PROPERTIES OF PROTEINS BY LIPIDS

In addition to their function in binding and adhesion, integral plasma membrane proteins essentially act as receptors of external signals or as enzymes (with hydrophilic endo- or ectodomain substrates or with hydrophobic substrates). Since these proteins are embedded in the phospholipid bilayer, the question of the modulation of their activity by the surrounding lipids naturally is raised. Some results are given in Table I.

A. Receptors Implied in Cell Signaling

Peptide and hormone receptors transmit a message through the plasma membrane, generally by modifying their interaction with cytoplasmic trimeric G proteins after binding their natural ligands. The effects of their environmental composition (lipid components or plasma membrane fluidity) have been studied extensively (Spector and Yorek, 1985; Yeagle, 1985). Some new results confirm the diversity of lipid effects. According to Spector and Yorek (1985), some alteration in the quaternary structure of the insulin receptor (IR) would be caused by a change in the unsaturation of phospholipids, in agreement with results of Berlin *et al.* (1989). Reciprocally, the ligand-induced conformational change of the receptor seems to be accompanied by a change in membrane lipid packing (Bergelson, 1992). Interestingly, oxidatively fragmented derivatives of phosphatidylcholines (PCs) have been found to activate neutrophils by interacting with platelet-activating factor receptor (Smiley *et al.*, 1991).

B. Enzymes

The modulation of integral enzymes is the current subject of active research, in spite (or because) of the long controversy over the lipid annulus around calcium ATPase (see the discussion in section I). Cholesterol seems to be the lipid that has the greatest influence on ion pump

TABLE I

Lipid Modulation of Protein Functions in Purified and Reconstituted Membranes

Receptor or enzyme	Cell	Effector concentration[a]	Effect[a]	Reference
Insulin-R	Friend erythroleukemia Ehrlich ascites	Unsaturated fatty acids (+)	Number (+); K_d (−)	Spector and Yorek (1985)
	Erythrocyte (human) (ALC disease)	Cholesterol (+)	Number (+)	Maehara (1991)
	Erythrocyte (human)	Unsaturated fatty acids (+)	Binding (+); number (+)	Berlin et al. (1989)
Na^+/K^+–ATPase	Membrane vesicles	Cholesterol (+)	Biphasic modulation	Yeagle (1988,1991)
	Proteoliposomes	Cholesterol + phospholipid	Activity dependent on cholesterol and phospholipid	Vemuri and Philipson (1989)
Na^+/Ca^{2+} exchange	Proteoliposomes	Cholesterol + phospholipid	Activity dependent on cholesterol and phospholipid	Vemuri and Philipson (1989)
Ca^{2+}–ATPase	Sarcoplasmic reticulum	Cholesterol (+)	None	Vemuri and Philipson (1989)
	Sarcolemmal reticulum	Cholesterol (+)	Activity (+)	Vemuri and Philipson (1989)
UDP-glucuronosyl transferase	Proteoliposomes	Cholesterol (+)	Activity (−)	Rotenberg and Zakim (1991) Brenner (1990)
	Rat liver microsomes	Cholesterol (+)	K_m (−); V_{max} (+)	
NADPH-cytochrome c reductase/aniline hydroxylase	Liver microsomes	FA : 20 : 3 (+); 18 : 2, 20 : 4 (−)	Activity (−)	Léger et al. (1989)
GABA transporter	Proteoliposomes	Cholesterol (+)	GABA transport rate (+)	Shouffani and Kanner, (1990)
Acetylcholine receptor	Reconstituted membrane	Cholesterol + phospholipid	Ion gating activity (+)	Fong and Mc Namee, (1986,1987)

[a] +, Increase; −, decrease.

activities such as Na^+/K^+; Na^+/Ca^{2+}-, and Ca^{2+}-ATPases in purified or reconstituted membranes. A large increase in the molefraction of cholesterol inhibits Na^+/K^+-ATPase activity (Broderick *et al.*, 1989), whereas a smaller increase can activate the same pump (Yeagle, 1988,1991). The cholesterol effect actually seems to depend on the membrane phospholipid composition. For Ca^{2+}-ATPase, the effect also depends on the tissue used as the source of transporter (Vemuri and Philipson, 1989).

On the basis of the correspondence between discontinuities in energies of activation of integral membrane enzymes and the main phase transition of the lipid part of the membrane, researchers currently consider that lipid fluidity regulates enzymatic catalysis. However, the *discontinuous* large changes in lipid organization at the phase transition have been shown not to be able to account for the alterations of enzyme activities since the latter generally change *continuously* (Zakim *et al.*, 1992).

In summary, modulation of many protein properties has been demonstrated in several different models. However, no clear rule exists that rationalizes the effect of lipid composition on ligand binding or enzyme activity. For example cholesterol, which has been widely studied, may increase or decrease very similar activities (Table I). The same conclusions were suggested several years ago by Melchior and colleagues (For a review, see Carruthers and Melchior, 1986), as derived from a large series of experiments on reconstituted membranes containing the passive sugar transporter of human erythrocytes. Head group structure, fatty acid length, unsaturation, lipid fluidity and lateral pressure, and lipid–water interface polarity simultaneously influence integral membrane protein activities in this model. Considering each protein as a different model interacting with its own lipid environment, a theoretical breakthrough is necessary to explain how lipids regulate membrane protein activities.

IV. LIPIDS IN SUPRAMOLECULAR ASSEMBLIES

Unlike the other major cell components (proteins, nucleic acids, glycans), which usually can be separated into simple pure molecules without difficulty, phospholipids generally are assembled in complex organized structures. Thus, as discussed in the previous section, most protein–lipid interactions do not have a precise stoichiometry and the "interaction site" definition, which is essential in classical molecular biology, is probably not relevant in many cases. On the contrary, the notion of the existence of a complex lipid–water or lipid–protein interface is much more operational, as shown in the next sections.

A. Integral Protein Organization within Membranes

Protein–lipid interactions are responsible for preserving the functional integrity of integral proteins. These are polar interactions between phospholipid head groups and hydrophilic portions of proteins (leading to some specificity of the phospholipids surrounding some proteins). Hydrophobic matching of lipid and protein thicknesses is probably of major importance (see Section II). When lipids are not suitable for protein accommodation (Sperotto and Mouritsen, 1988; Mouritsen and Briltonen, 1992), aggregation and conformational changes may occur, giving rise to the modulation of the function (inhibition or stimulation, depending on the protein functioning mechanism). Bovine rhodopsin and bacteriorhodopsin (Lewis and Engelman, 1983; Sperotto and Mouritsen, 1991a,b) are, by far, the best studied models for protein aggregation within the membrane plane, since they are easy to purify in large amounts and can be reconstituted readily in synthetic membranes of many different compositions. Rhodopsin freely diffuses in bilayers of dimyristoyl phosphatidylcholine (DMPC) above the gel–liquid crystalline transition temperature. On the contrary, rhodopsin assembles in large aggregates at a lower temperature, as seen by electron microscopy and electron spin resonance (ESR) spectroscopy (Devaux, 1983). The state of aggregation of rhodopsin in membranes containing phosphatidylcholine molecules of different chain lengths has been studied by standard and saturation transfer ESR spectroscopy (Kusumi and Hyde, 1982; Ryba and Marsh, 1992). These authors agree on a minimum of rhodopsin aggregation in membrane containing C_{15} lipid chain lengths, but the protein maintains its photochemical activity in all recombinants. Since retinal rod outer segment disk membranes mostly contain longer polyunsaturated phospholipids, a very low fraction of the rhodopsin will be in an aggregated state. This conclusion seems to apply to all other membrane proteins, as long as they are not forced to aggregate by interaction with cytoskeleton proteins, since sufficiently different phospholipids exist in all biological membranes to fit the hydrophobic portion of integral proteins. From this perspective, integral proteins could interact preferentially with microdomains composed of thickness-matched lipids, while probably exchanging lipids with other membrane parts on a short time scale.

B. Extrinsic Protein–Membrane Lipid Assembly

Among many extrinsic proteins interacting with phospholipids, only five classes (phospholipases, protein kinase C, annexins, coagulation cascade, and cytoskeleton-associated proteins) are discussed in this section since they are highly representative of the results obtained to date.

1. Phospholipase A_2

Phospholipases and phospholipids interact as enzymes do. However, before attaining the catalytic site, the interface properties determine the enzyme–substrate complex to be formed (Pernas et al., 1992). Phospholipase A_2 (PLA$_2$) found in snake and bee venom nonselectively hydrolyzes phospholipids independent of their polar head group (Schalkwijk et al., 1990). However, full catalytic activity depends on the phospholipid organization (in liposomes, micelles, or mono- and bilayer structures), since the limiting step in the overall kinetics is enzyme binding to an ordered phospholipid layer. The head group charges considerably modify the apparent phospholipase surface binding affinity: 0.1 pM and 1 μM for anionic lipids and uncharged lipids, respectively (Ramirez and Jain, 1991). This binding is responsible for the so-called interface activation of phospholipase; the protein structure changes very little, if at all, since it is maintained by 6–7 disulfide bridges. One important point is the lifetime of the phospholipid–phospholipase complex, which seems to be very long at the interface (Berg et al., 1991), the enzyme covering ~35 phospholipid molecules (Ramirez and Jain, 1991). However, the enzyme diffuses quickly over the entire surface, probably interacting for a very short time with given phospholipids. The second step is phospholipid entry through a hydrophobic channel into the catalytic site, provided a calcium ion is bound to phospholipase as well to position the substrate correctly and to stabilize one of the intermediary compounds (Holland et al., 1990). Since the calcium affinity constant is about 0.1 mM, only external or stimulated cell internal media can provide this high calcium concentration. Another crucial factor for catalytic activity is any structural defect in the bilayer caused by alcohols, detergents, diacylglycerol, lysoderivatives, and fatty acids or by proximity of the gel–liquid crystalline transition temperature (see Zidovetzki et al., 1992, and references cited therein).

The best characterized intracellular PLA$_2$ is very specific for arachidonic acid-containing phospholipids. This enzyme also needs calcium to translocate to membranes, probably by its C-terminal sequence which is homologous to those of other translocatable proteins (Coussens et al., 1986). Other phospholipases with different localizations and specificities are found from time to time (Aarsman et al., 1989; Cassama-Diagne et al., 1989; Hazen and Gross, 1991). Interestingly enough, one cytosolic PLA$_2$ was characterized that contained a calcium-dependent domain analogous to the protein kinase C (PKC) anionic lipid binding domain (Clark et al., 1991).

2. Protein Kinase C

One of the main signal transduction systems is based on protein phosphorylation and dephosphorylation, acting as an amplification and distribution mechanism through a set of protein kinases and phosphatases. Numerous reviews on PKC have emphasized the role of this enzyme family in the regulation of many major cell responses (for example, see Housley, 1991; Azzi et al., 1992). Activation of PKC by increased calcium concentration involves enzyme translocation from cytosol in resting cells to plasma membrane after stimulation. Phospholipids, notably PS, function as essential cofactors whereas diacylglycerol (another membrane compound) acts as a PKC activator. A single diacylglycerol molecule interacts with PKC (at the same site as phorbol esters) and stimulates kinase activity, perhaps in conjunction with phosphatidylinositol 4,5-bisphosphate (PIP_2) binding to another site (Lee and Bell, 1991) and with cis-unsaturated fatty acids (Shinomura et al., 1991). On the other hand, many PS molecules are needed cooperatively to insure full kinase activity (Newton and Koshland, 1989). However, the exact nature of the PS–PKC interaction is not known, probably because the experimental results (either catalytic activity or membrane binding) have been determined under so many different conditions (substrate, interface composition and nature, activator, calcium), thus making a comparison of apparently contradictory results rather meaningless. This case is analogous to observations in the 1960s of two-substrate enzymatic kinetics that were misunderstood until a very simple theory was developed. However, lipid–PKC interactions are much more difficult to delineate because many more parameters must be controlled, and because no conceptual breakthroughs have been made to clarify the discussion.

Bazzi and Nelsestuen partially clarified the understanding of PKC–lipid interactions. PKC forms two different associated states: the first is calcium dependent and reversible on calcium chelation; the second is irreversible in the presence of high calcium concentrations, as well as in the presence of phorbol esters (Bazzi and Nelsestuen, 1988a). Activity of this membrane-associated PKC is no longer dependent on calcium or phorbol ester (Bazzi and Nelsestuen, 1988b). Calcium binding depends on the presence of phospholipids (Bazzi and Nelsestuen, 1990). At least 8 calcium ions bind cooperatively per PKC molecule, so the simultaneous exchange of all calcium ions in the complex is a very unlikely event, despite the rapid exchange rate of individual calcium ions from each binding site (Bazzi and Nelsestuen, 1991b). Moreover, PKC binds to phospholipids with high affinity in a sequence of essentially irreversible steps, leading to clustering of more than 10 acid phospholipids per PKC molecule, as shown by fluorescence quenching of NBD-labeled PA molecules (Bazzi and Nelsestuen, 1991a). Finally, an important finding demonstrated that neutral phospholipids such as PC and PE do not play the same role in binding (Bazzi and Nelsestuen, 1992a): membranes containing 20% PS/60% PE provided optimum conditions for binding and were as effective as membranes composed of 100% PS. Surprisingly, PE/PC and even PC membranes are able to bind PKC in the presence of a much higher calcium concentration than PS-containing membranes (by more than 1000-fold). In conclusion, the functional role of phospholipids in PKC activity within cells is probably more complicated than expected: PKC binding to the membrane depends on at least three factors—PKC and calcium concentrations, and membrane composition—operating in a highly cooperative manner. A fourth factor could be the binding site of phorbol esters (Kazaniev *et al.*, 1992). Thus, establishing the exact *in vivo* enzyme concentration on the membrane is difficult, since other highly concentrated intracellular proteins (annexins, for example; see subsequent discussion) can compete for the same acidic phospholipids. A direct relationship between calcium- and membrane-binding and PKC phosphorylation activity can be expected, but substrate phosphorylation also depends on the substrate–membrane interaction and on the colocalization of enzymes and substrates in the same membrane. In the Myristoylated Alanine Rich C Kinase Substrate (MARCKS) family, presumably the most abundant endogenous PKC substrates, membrane anchoring is insured by the hydrophobic myristoyl moiety, but MARCKS targeting to the cytoplasmic side of the plasma membrane seems to be the result of association with a specific receptor (Veillette *et al.*, 1988, Resh, 1989, Resh and Ling, 1990). Much research is required to determine whether PE and PS on the cytoplasmic side of organelle and plasma membranes could be involved in putative targeting of PKC.

3. Annexins

Annexins are members of a protein family consisting of at least 8 members with high sequence homology and well-characterized physicochemical properties (calcium binding, interaction with cytoskeleton as well as phospholipids and membranes). However, the functions of these proteins seem to be broad and still relatively ill defined (Crumpton and Dedman, 1990; Römisch and Pâques, 1991). Intracellular annexins could play a role in membrane fusion during exocytosis (Creutz, 1992), in the regulation of inflammatory processes (Russo-Marie, 1992), in membrane–cytoskeleton attachment (Sobue *et al.*, 1989), in cell growth, signal transduction, and differentiation processes (Römisch and Pâques, 1991), or in some calcium ion channel activity (Pollard *et al.*, 1992). Annexin III and V also have been found in extracellular spaces, for example, in placenta and plasma, especially after myocardial infarction, although they have no N-terminal leader sequence. In this case, these proteins seem to participate in blood coagulation regulation (Funakoshi et al., 1987). Annexin–lipid interactions are probably of the same kind as those described for PKC, if results of studies on annexin VI can be applied generally to the whole family. For example, association of annexin VI with membranes was found to induce extensive clustering of acidic phospholipids in a calcium-dependent manner (Bazzi and Nelsestuen, 1991a). Contrary to what happens with PC, membranes containing PE show a much slower response to acidic phospholipid cluster formation and dissipation for annexin VI binding and dissociation, respectively. This slow lateral diffusion of phospholipids specific to PE-containing membranes assumes phase separation in some microdomains (Bazzi and Nelsuesren, 1992b), often proposed in biological membranes but yet to be proven definitively (Tocane et al., 1989; see also Chapters 5 and 6). Other important annexin properties (except for annexin V) involve their ability to induce membrane aggregation (Meers et al., 1992). The aggregates are fairly stable, but intermembrane fusion can occur quickly in the presence of a small amount of cis-unsaturated fatty acid (Creutz, 1981). The specificity of annexin binding for different organelle membranes (Creutz et al., 1992) reinforces their role in exocytosis (Creutz, 1992) as well as in membrane fusion that occurs during all membrane traffic. Annexin binding to the membrane is so strong that the entire surface of the cytosolic face of the cell membrane can be occupied by this protein as triskelion assemblies (Brisson et al., 1991; Mosser et al., 1991); the intracellular annexin concentration then is $\sim 100 \ \mu M$. Other proteins (e.g., PLA_2, PKC, cytoskeleton-associated proteins) could lose their access to the membrane surface, thus explaining the role of annexin in inflammation and cell shape maintenance. Finally, the calcium channel function of some annexins is poorly understood; much work is required to determine the process that allows nonintegral proteins to induce ion channels through bilayers (Huber et al., 1992; Pollard et al., 1992).

4. Coagulation Cascade Proteins

The plasma membrane of stimulated platelets provides the primary support for the coagulation cascade that quickly occurs after vessel injury (Toti et al., 1992). Activation of factors XII, XI, C, and X and of prothrombin is catalyzed by bringing proteolytic enzymes and their substrates closer on the platelet surface (Davie et al., 1991). As a consequence of this local enrichment, the K_m of the prothrombinase complex decreases 1400-fold (Krishnawamy et al., 1987). Many proteins involved in the cascade contain γ-carboxyglutamic acid-rich N-terminal sequence. The other proteins also have anion-rich regions. All these proteins probably form a calcium bridge with acidic phospholipid head groups (Davie et al., 1991), as discussed earlier.

However, in contrast to PKC and annexins, the prothrombinase complex (Jones and Lentz, 1986), factor X, and protein Z (Bazzi and Nelsestuen, 1991a) do not require extensive regions of phase-separated acidic phospholipids despite the high affinity of the final protein for such phospholipids. Indeed, PS is much more efficient in binding these proteins than are other anionic phospholipids (Rosing et al., 1988), since the surface potential is not the primary factor governing the strength of the interaction. A PS mole fraction of 5% seems to be best for binding ~1 prothrombinase complex per 50 phospholipids. A direct correlation exists between the PS composition of stimulated platelet plasma membrane and activation of prothrombinase activity (Comfurius et al., 1985). No research has been done to date on the possible effects of other phospholipids, such as PE, which are known to appear on the external platelet surface under the same conditions (Bassé et al., 1993).

5. Phospholipid–Cytoskeleton Interactions

Cytoskeleton structure and dynamics are very important in many respects, since they control cell shape and movement, membrane stability, and domain organization (Luna and Hitt, 1992). Cytoskeleton proteins such as spectrin, actin, tubulin, and other associated proteins essentially interact with integral membrane proteins through a set of intermediary proteins possessing multidomain structures that are able to provide specific sites for binding. However, some of these cytoskeletal proteins directly bind pure phospholipid membranes, which generally contain anionic phospholipids such as PIP_2 and PS molecules. For example, regulation of the F-actin gelating activity of α-actinin requires the presence of PIP_2 (Fukami et al., 1992). This phospholipid also can potentiate both the release of assembly-competent actin monomers from a complex with profilin and the uncovering of barbed filament ends blocked by gelsolin or other filament-capping proteins (Goldschmidt-Clermont and Jeanmey, 1991; Hartwig and Kwiatkowski, 1991). Interestingly, a neutral lipid such as diacylglycerol, which is included in plasma membranes of *Dictyostelium discoideum*, also is able to induce actin polymerization, probably mediated by a peripheral protein other than PKC (Shariff and Luna, 1992). Clathrin, vinculin, and talin have been reported to bind to acidic phospholipids, such as PS, directly (Isenberg, 1991; Goldman et al., 1992; Seppen et al., 1992).

V. LIPID MODULATION OF CELL FUNCTIONS

Addressing the modulation of cell functions by lipids, the following questions must be considered. How can lipid distribution within and between cell membranes be achieved? What is the influence of this distribution on cell functions such as cell shape; intermembrane fusion; blebbing or budding processes during endocytosis, exocytosis, and virus invasion; and interactions of cells with external medium (hemostasis, adhesion)?

A. Lipid Transverse Distribution and Lipid Assembly in Cell Membranes

A rapid count of the total lipid composition in cells of lower and higher organisms reveals that 100–1000 distinct phospholipid molecules are contained within each cell, distributed differently in organelle and plasma membranes (Raetz, 1982). Moreover, lipids seem to be distributed asymmetrically in each membrane. The plasma membrane of eukaryotic cells essentially contains choline head group phospholipids in the external face, whereas aminophospholipids are located mainly on the cytoplasmic leaflet (Devaux, 1991; see also Chapter 1). Some of the organelle membranes (chromaffin granules, synaptic vesicles, reticulocyte exosomes) also seem to have an asymmetric distribution of aminophospholipids, which are located preferentially in the cytoplasmic leaflet (Vidal *et al.*, 1989; Zachowski *et al.*, 1989; Zachowski and Morrot Gaudry-Talarmain, 1990). The external (or luminal) to cytoplasmic side movement of aminophospholipids (Seigneuret and Devaux, 1984) depends on the ATP concentration in the cytoplasm, and can be inhibited by calcium, vanadate anion, and sulfhydryl reactants such as N-ethylmaleimide (Zachowski *et al.*, 1986, see also Chapter 2). Endoplasmic reticulum (ER) exhibits very fast phospholipid flip-flop of PC molecules, catalyzed by a PC transporter in a non-energy-requiring manner (Bishop and Bell, 1985), whereas the fate of other phospholipids (PE, PS, and PI) is unknown. Assuming that differences in molecular structure most likely correspond to different functions, concluding that this composition is essential for cell life is tempting. This point of view is strengthened by the vital effect of genetic regulation of lipid transport, as reported in yeast by Aitken *et al.* (1990). The mechanism of this process must be very efficient to sort rapidly diffusing and mixing lipids (in the membrane plane), and to address them to the proper target membrane from their synthesis site or from their point of entry into the cell.

Different means for lipids to move from a donor to an acceptor membrane have been considered (Part VI), including monomer transfer through the cytoplasm (facilitated or not by lipid transfer protein), vesicle transport, and membrane fusion processes. Direct monomer transfer through cytosol obviously is reserved for hydrophilic diacylglycerol, fatty acid, some glycolipids, and steroid molecules. More hydrophobic molecules such as endogenous phospholipids are known to interact in a 1 : 1 ratio with specific or nonspecific phospholipid-exchange proteins (Wirtz, 1990; see also Chapter 9). When lipid transfer occurs via vesicle shuttle, a sorting mechanism is necessary to determine the lipid composition of different membranes. In this case, a long-lasting lipid–protein interaction or specialized lipid domain is required for sorting to occur (Voelker, 1991; Van Meer, 1989; see also Chapter 21). In conclusion, lipid sorting and targeting into the cell seems to be related strongly to lipid interactions, either with soluble transfer proteins or with intrinsic vesicle proteins. Lipid distribution in the different membranes has a physiological role in the life of the whole cell, by different mechanisms discussed in the following sections.

B. Lipids and Cell Shape

Overall cell shape obviously is maintained by the cytoskeleton, which is composed of many different proteins, organized in quite a stable array, although they are able to rearrange to participate in all cell movements such as membrane traffic, exocytosis, endocytosis, mitosis, or overall displacement on solid substrates. However, in some cases, the lipid distribution within membranes directly influences cell shape. Several authors have shown that changes in the plasma membrane bilayer balance can influence erythrocyte shapes (for example, see Seigneuret and Devaux, 1984; Daleke and Huestis, 1989), in agreement with the bilayer couple hypothesis proposed by Sheetz and Singer (1974). According to this model, an excess of lipid on one side of a membrane is counterbalanced by a convex membrane curvature on the same side (Farge and Devaux, 1992). Incorporation of amphiphile molecules in the external (internal) leaflet has been shown to induce the echinocyte (stomatocyte) shape of red blood cells. Indeed, the steady state imposed by the aminophospholipid translocase in the normal physiological state of erythrocytes could generate a small excess of lipids in the internal leaflet (Devaux, 1991). Conversely, the spicules appearing in echinocytes were found to allow the membrane to accommodate a small excess of lipids in the external face (Daleke and Huestis, 1985). A larger excess induces some vesicle shedding in erythrocytes (Frenkel et al., 1986; Bütikofer et al., 1987), as in many other cells (Beaudoin and Grondin, 1991). Similar results were found in platelets (Bevers et al., 1983; Kobayashi et al., 1984; Ferrell et al., 1988; Suné and Bienvenüe, 1988), but with an important morphological difference. Instead of the spicules observed in erythrocytes, externally localized amphiphiles were found to induce much longer filipodia in platelets containing actin filaments. When aminophospholipid molecules (spin-labeled PS or PE) were added to platelets, the initial filipodia disappeared quickly with the same rate as phospholipid internalization (Suné and Bienvenüe, 1988). An actin rearrangement thus seems to accompany the appearance or disappearance of filipodia, but with no net change in the long actin filament proportion (Brunauer *et al.*, 1989).

In conclusion, a cytoskeleton–membrane interaction seems to occur, with some differences according to the cell type. In erythrocytes, spicule blebbing could appear only within the meshes of the cytoskeleton net; in platelets, the cytoskeleton would be in a more unstable equilibrium, and a small defect in membrane curvature provokes a cytoskeletal reorganization; in lymphocytes, in which the cytoskeleton seems to be much more stable, no shape change accompanies a transverse lipid asymmetry in the plasma membrane (Suné *et al.*, 1988).

C. Endocytosis and Cell Stimulation

Endocytosis is characterized by specific binding of ligands to a plasma membrane receptor, allowing the ligand to be internalized by cells. This is the only mechanism by which cells obtain iron (through transferrin, Tf) and the major part of the cholesterol (through low density lipoprotein, LDL) needed for their development and multiplication. Receptors are gathered together in clathrin-coated pits, which invaginate to form intracellular vesicles called coated vesicles. After uncoating, the vesicles mature into organelles called endosomes, in which the ligands are processed directly (transferrin) or are directed to lysozomes (LDL), while the receptors are recycled to the plasma membrane (Dautry-Varsat *et al.*, 1983; Goldstein *et al.*, 1985). Down-regulation of the insulin receptor in the presence of a large excess of the substrate follows a similar pathway. The binding of 5-methyltetrahydrofolate to a membrane receptor bound to glycosyl-phosphatidylinositol (GPI) also is followed by internalization of the ligand, by means of plasma membrane invaginations (called caveolae) that differ from clathrin-coated pits (Rothberg *et al.*, 1990; Hooper, 1992). Using these four ligands (Tf, LDL, insulin, and 5-methyltetrahydrofolate), many results have shown that a modification of the lipid composition of cells leads to a major change in the number of binding sites on their surface, the affinity of receptors for their natural ligand, and/or the rate of endocytosis (Table II). Other data on the lipid modulation of some metabolic pathways occurring in whole cells also are reported in this section.

1. Receptors

In chronic lymphocytic leukemic lymphocytes and in Friend MuLV cells (clone Z27), a relationship was found (Krueger et al., 1987; Daefler and Krueger, 1989) between membrane fluidity (as measured by fluorescence depolarization) and the number of transferrin receptor (TfR) sites. In the same way, dibucaine and benzyl alcohol (two local anesthetics) were shown to perturb TfR endocytosis strongly, without any change in the number of sites or their affinity for Tf (Hagiwara and Azawa, 1990; Sainte Marie et al., 1990). Interestingly, inhibition by benzyl alcohol was shown to be rapid and fully reversible by simple washing (Sainte Marie et al., 1990).

In many cells, the number of LDL receptors (LDLR) (not their affinity for LDL) is modulated according to needs in cholesterol and the amount of this lipid in the intra- and extracellular pools (Schneider, 1989). However, many more subtle changes occur when the lipid composition is modified. LDLRs of macrophages grown in delipidated culture medium (Knight and Soutar, 1986) increase their affinity for LDL 5-fold. A delipidated medium decreases the intracellular cholesterol concentration in NS1 mouse myeloma cells without enhancing the number of LDLRs. In contrast, LDL binding decreases by 70–90%; cholesterol complementation of the medium restores the "normal" state (Chen and Li, 1987). LDL binding to its receptor also was observed to decrease in human CEM lymphoblastic cells after their cholesterol content had been decreased (Sainte Marie et al., 1989). Lipids other than cholesterol are able to modify LDLR properties: recycling time decreases when unsaturated fatty acids are introduced into fibroblast phospholipids (Gavignan and Knight, 1981). In mononuclear cells (Loscalzo et al., 1987), monocytes U937 (Kuo et al., 1990a), and hepatocytes Hep62 (Kuo et al., 1990b), an increase in the number of LDLRs and in their affinity for LDL occurs when more unsaturated fatty acids are integrated into membrane phospholipids, leading to greater LDL degradation in these types of cells. The reverse was shown when cells were enriched with stearate and palmitate. An increase in the cell cholesterol content was shown to modify the internalization rate (Santini et al., 1992), the K_d (Léger et al., 1989), or the number (Maehara, 1991) of insulin receptors.

The 5-methyltetrahydrofolate receptor (MTHFR) linked to GPI seems to cluster in certain domains on the plasma membrane (see also Chapter 11). Unknown is how this clustering occurs, since only the phospholipid part of the receptor is included in the membrane bilayer. In cells treated with sterol-binding agents, under which conditions the free cholesterol content decreases, the MTHFR was shown to uncluster, leading to a

TABLE II

Lipid Modulation of Cell Functions: Receptor-Mediated Endocytosis and Metabolic Pathways

Receptor or metabolic pathway	Cell	Effector concentration[a]	Effect[a]	Reference
Fluid-phase endocytosis	U937 macrophage-like	Cholesterol (−)	Fluid-phase endocytosis (−)	Esfahani et al. (1986)
5-Methyl tetrahydrofolate receptor	MA 104 cell (epithelial)	Cholesterol (−)	Clustering (−)	Rothberg et al. (1990)
LDLR[b]	Macrophage	Lipoprotein-deficient serum	Affinity (+)	Knight and Soutar (1986)
	NS$_1$ (mouse myeloma auxotroph)	Cholesterol (−)	Binding (−)	Chen and Li (1987)
	Lymphoblasts C.E.M.	Cholesterol (−)	Number (−); K_d, turnover (−)	Sainte-Marie et al. (1989)
	Fibroblasts	Unsaturated fatty acid (+)	Uptake and degradation of LDL (+)	Gavignan and Knight (1981)
	Mononuclear cells	Unsaturated fatty acid (+)	Uptake and degradation of LDL (+)	Loscalzo et al. (1987)
	U937 macrophage-like	Unsaturated fatty acid (+)	Uptake and degradation of LDL (+)	Kuo et al. (1990b)
	Hepatocyte Hep92	Unsaturated fatty acid (+)	Uptake and degradation of LDL (+)	Kuo et al. (1990a)
TfR[c]	Hepatocyte Hep92 Z27	Unsaturated fatty acid (+)	None	Kuo et al. (1990a)
	Lymphocytes (CLL disease)	Membrane fluidity (−)	Number (−)	Krueger et al. (1987)
		Membrane fluidity (+)	Number (+)	Daefler and Krueger (1989)
LDLR and TfR	Lymphoblasts L$_2$C	Benzyl alcohol (+)	Receptor internalization (−)	Sainte-Marie et al. (1990)
TfR	Rat muscle L6	Dibucain (+)	Receptor internalization (−)	Hagiware and Ozawa (1990)

Rhodopsin/cGMP cascade	Vertebrate retinal rod outer segment disc	Cholesterol (−)	Activation of cGMP cascade	Boesze-Battaglia and Albert (1990)
β Adrenergic/adenylate cyclase system	Membrane of ventricular tissue	Cholesterol (+)	Affinity (+) adenyl cyclase activity (+)	McMurchie and Patten (1988)
β Adrenergic/adenylate cyclase system	Rabbit arterial smooth muscle	Cholesterol (+)	Sensitization to norepinephrine	Broderick et al. (1989)
	Myocyte ventricularis	ConA modified fluidity	K_d (+); down regulation (−); cAMP generation (+)	Rocha-Singh et al. (1991)
Na$^+$/K$^+$–ATPase	Rabbit arterial smooth muscle	Cholesterol (+)	Activity (−)	Broderick et al. (1989)
Ca^{2+} Transport	Arterial segment	Cholesterol (+)	Ca^{2+} uptake (+)	Bialecki and Tulenko (1989)
Phosphatase	Rat intestinal microvillus	Cholesterol (+)	Activity (−)	Brasitus et al. (1988)
Amino phospholipid translocase	Erythrocytes	Benzyl alcohol (+)	Activity (+)	Bassé et al. (1992a)
D-Glucose transport	Kidney brush border membrane	Benzyl alcohol (+)	Transport (+)	Carrière and Le Grimellec (1986)
Adenylate cyclase system	Bovine adrenal cortex	Benzyl alcohol (+)	Forskolin stimulated activity (−)	De Foresta et al. (1987); Friedlander et al. (1990)
	MDCK	Benzyl alcohol (+)	Biphasic modulation	Friedlander et al. (1987)
Transport systems	MDCK and LLC-PK	Benzyl alcohol (+)	Pi uptake (+); hexose uptake (−)	Friedlander et al. (1988)
Insulin receptor	Erythrocyte	Cholesterol (+)	Down regulation (−)	Santini et al. (1992)

[a] +, Increases; −, decreases.
[b] LDLR, Low density lipoprotein receptor.
[c] TfR, Transferrin receptor.

reduced number of receptors per caveola (Rothberg et al., 1990). A lipid–cholesterol interaction could be the cause of such receptor clustering. Other proteins bound to GPI, such as differentiation antigen Thy-1, scrapie prion proteins, and alkaline phosphatase, also have been shown to appear in caveolar structures identical to those containing MTHFR (Hooper, 1992). Interestingly, a stabilizing effect of cholesterol was demonstrated previously in some endocytotic processes (Barrière et al., 1988; Bal and Bird, 1991) and in so-called fluid-phase endocytosis (Esfahani et al., 1986).

2. Lipid Modulation of Cell Responses to Stimulation

Cholesterol inhibits light-induced activation of the cGMP cascade in retina cells (Boesze-Battaglia and Albert, 1990). Inversely, in heart ventricular tissue (McMurchie and Patten, 1988), higher amounts of cholesterol enhance β-adrenergic receptor affinity and some features of adenylate cyclase activity (Broderick et al., 1989). In conclusion, the entry of many important cell metabolites and overall cell functioning depend to some extent on the lipid content of cell membranes. However, delineating the mechanisms responsible for these effects is difficult since they are very complex processes in which many different proteins and membranes are involved.

D. Membrane Fusion

Fusion of two bilayers to form one membrane, with complete intermixing of lipids and proteins, generally is considered an unlikely event as far as the lipids are concerned. The main prerequisite for lipid mixing is that water must move away from the lipid interface, so the hydrophobic parts of the membrane will be very close to each other (see also Chapter 8). The aqueous spacing between two PC leaflets is 3 nm, whereas the spacing is only 1.5 nm between PE or PS bilayers (Rand and Parsegian, 1989). Fusion presumably would be easier in these latter cases, although pure lipid bilayers never fuse except under nonphysiological conditions (such as high concentrations of polyethylene glycol, which removes water from the lipid–water interface). Indeed, fusion of biological membranes is essential to membrane trafficking and to enveloped virus invasion. Viral protein-induced intermembrane fusion is the best known mechanism at the molecular level (Hoekstra, 1990), essentially by the model of hemagglutinin (HA) proteins of the influenza virus membrane. Almost all viral fusion proteins (Sechoy et al., 1987; Stegmann et al., 1989), are oligomers containing a so-called N-terminal hydrophobic fusion peptide. The main conclusions about the mechanism of viral-induced fusion can be summarized as follows:

1. Fusion peptides play an important role in the fusion mechanism, although the fusion part of the fusion protein can be located far from the binding site, present in the same protein, and mediating viral attachment to the target membrane.
2. After pH decreases in endosomes or some other unknown mechanism in non-pH-sensitive fusion proteins, the fusion peptides, initially buried in the protein stem, often are exposed while globular heads of the proteins dissociate from one another.
3. Provided that the viral and target membranes are apposed, this shape change increases protein hydrophobicity, initiating immediate protein attachment to the target membrane, while it continues to be included in the viral envelope by its transmembrane hydrophobic peptide.
4. After a lag phase—probably caused by additional protein conformational changes, aggregation, and formation of a fusion pore—the lipid mixing rapidly occurs. Whether the pore contains lipids, possibly in a nonbilayer structure, is currently under debate (see Chapter 8).
5. In some viruses, other proteins may participate in the overall mechanism for apposing the two membranes, but the so-called fusion proteins are sufficient to induce the intermembrane fusion. Cell injection of the internal content of liposomes reconstituted with only Sendaï virus F proteins in their membranes (provided that the two membranes are apposed by antibodies, for example, rather than by the natural binding protein HN) may be considered evidence for the idea (Sechoy et al., 1989; Compagnon et al., 1992) that the fusion protein alone suffices. Indeed, a monoclonal antifusion antibody fully inhibits the biological effects of the fusion process, namely the entry of poly(rI)–poly(rC) molecules stimulating the antiviral effect of interferon. Some experiments (Asano and Asano, 1988; Braunwald et al., 1991; Phalen and Kielian, 1991) provided evidence that specific target membrane lipids are involved in the virus-induced fusion mechanism. This point is discussed in Chapter 8).

Intracellular sorting and targeting of vesicles is a matter that concerns proteins (many reviews have been devoted to this topic; see, for example, Shekman, 1992), the more important being a set of related small GTP-binding proteins. Vesicle fusion is governed by another set of proteins, including cytoplasmic SNAP, N-ethylmaleimide-sensitive factor (NSF), and Soluble NSF Attachment Protein (SNAP) receptors in the target membrane (Rothman and Orci, 1992; Whiteheart et al., 1992), all involved in a so-called fusion machine (Malhotra *et al.*, 1988). In comparison to viral fusion proteins, they probably assemble in a very large multimeric complex (White, 1992). Whether or not such a complex is responsible for forming the so-called early fusion pores (Spruce *et al.*, 1990) detected in mast cells by electrophysiological methods during exocytosis (Oberhauser *et al.*, 1992) is not known. Conversely, the hypothesis of a purely lipid pore has been proposed (Nanavati *et al.*, 1992).

In both viral fusion and fusion occurring in exocytosis, bilayer destabilization is obviously a crucial step. The fact that large protein assemblies participate in this event has not clarified this point greatly. In any case, protein–lipid interactions are essential. Some experimental findings have shown clearly that very simple structures can be as efficient in fusing membranes into sophisticated aggregates. N-Terminal peptides of viral fusion proteins are able to destabilize pure phospholipid membranes considerably (Düzgünes and Shavnin, 1992; see also Chapter 8). A pH decrease in endosomes is sufficient to destabilize liposomes containing oleic acid and phosphatidylethanolamine, allowing the endosome and liposome bilayer to fuse and the internal liposome contents to enter the cell (Milhaud *et al.*, 1992). The endosome membrane also is destabilized efficiently at low pH by complexes containing polylysine, transferrin, and DNA as well as the N-terminal hydrophobic virus peptides known to be fusion peptides, as discussed earlier (Wagner *et al.*, 1992). Finally, membrane fusion also occurs between negatively charged membranes (containing PS essentially) under control of calcium ion. This process is so simple and efficient that Devaux (1991) proposed that the main characteristic of fusion-competent membranes could be the presence of PS on the cytoplasmic face, whereas other factors are involved in targeting or apposing membranes.

E. Lipids and Hemostasis

1. Lipids and Coagulation Cascade

Hemostasis is the control of blood circulation in vessels by a very complex and fast cascade of proteolytic reactions. Indeed, blood seems to be in a dynamic hemostatic equilibrium, with each main activation factor (factors IX, X, and II) at a very low but definite activation level (Davie *et al.*, 1991; Toti *et al.*, 1991,1992). However, the coagulation cascade does not occur under normal conditions because it is under the control of some inhibitory factors. In the case of vascular lesions, the tissue factor (TF) is expressed by adipocytes, subendothelial, and smooth muscle cells (Drake *et al.*, 1989) and a major activator complex is formed between TF and factor VII. At the same time, platelets adhering to subendothelial components secrete their granule contents and activate other platelets in a chain reaction. These platelets aggregate in turn. Vitamin K-dependent factors and PS interact by a calcium bridge, whereas Va and VIIIa cofactors probably stick to membranes mainly by hydrophobic interactions. Finally, the procoagulant power of PS-containing membranes depends on their ability to assemble tenase and prothrombinase complexes and to protect activated factors against endogenous anticoagulating factors (Mann *et al.*, 1990).

During stimulation, some cytoskeleton proteins are proteolyzed and many vesicles (up to 20% of the phospholipid content of the cell) are shed from platelets; the surface becomes anionic by the appearance of PS. This latter point is of major importance since the plasma anionic surface is the site at which different complexes, required for the entire hemostatic cascade to proceed, assemble and greatly increase their catalytic rate. In this respect, the transverse distribution of PS in platelets is one of the key requirements for blood clotting.

2. PS Inside–Outside Translocation during Platelet Stimulation

In the resting state, PS is localized completely in the cytoplasmic leaflet of the plasma membrane, in platelets as in all other types of cells studied to date (Bevers *et al.*, 1983; Devaux, 1991; see also Chapter 1). The distribution of phospholipids is determined much more easily with labeled phospholipids than with endogenous phospholipids, but similar conclusions have been drawn concerning phospholipid transverse mobility and distribution when paramagnetic (Suné and Bienvenüe, 1988; Morrot *et al.*, 1989; Devaux, 1991), fluorescent (Tilly *et al.*, 1986), radiolabeled (Daleke and Huestis, 1989), and endogenous phospholipids were used. The aminophospholipid distribution is controlled by a translocase using the energy of ATP hydrolysis (Seigneuret and Devaux, 1984; Suné *et al.*, 1987; Zachowski and Devaux, 1990; see also Chapter 2). Using paramagnetic analogs of different phospholipids, the distribution and rate of transfer between the two leaflets of the plasma membrane of platelets can be studied, either at rest or after activation *in vitro* by two agonists: thrombin in the absence of calcium or A23187 in the presence of 1 mM calcium (Bassé *et al.*, 1992b,1993). In thrombin-activated platelets, translocase activity is stimulated slightly, PC passive diffusion is enhanced slightly, but no cytoskeleton proteolysis occurs and no vesicles are shed into the external medium. In calcium ionophore-stimulated cells, a very sharp rise in the internal calcium concentration causes cytoskeleton proteolysis, vesicle shedding, and very strong translocase inhibition (Bassé *et al.*, 1992b). Moreover, phospholipid transverse movements also were measured experimentally. Previously internalized aminophospholipids and externally located PC can be used as probes to follow the transverse movement of phospholipids from the internal and external leaflets, respectively, *during* platelet stimulation by the same agonists (Bassé *et al.*, 1993). No outward (even transient) movement of internally located spin-labeled aminophospholipids was observed during thrombin-induced activation, whereas the influx of externally located probes increased slightly and slowly. During A23187-mediated activation, a similar slightly increased influx was observed, whereas 40–50% of the initially internally located spin-labeled aminophospholipids (PS and PE) appeared on the outer leaflet. The sudden exposure on the outer face was dependent on an increase in intracellular calcium and was achieved in less than 2 min at 37°C. When lipid probes were incorporated into the outer face of the plasma membrane in resting platelets, they still were fully accessible from the extracellular medium after ionophore-induced activation. Moreover, these probes were distributed between the vesicles and remnant cells in proportion to their phospholipid content, suggesting that no scrambling of plasma membrane leaflets occurred during vesicle blebbing, contrary to a suggestion by Sims *et al.* (1989) and other experimental evidence obtained by more indirect methods (Comfurius *et al.*, 1990; Bevers *et al.*, 1992). More important is the finding that the spin-labeled aminophospholipid exposure rate and amplitude were unchanged when vesicle formation was inhibited strongly by calpeptin, a calpain-inhibitor. All these results indicate that loss of asymmetry (thus inducing generation of a catalytic surface of the external face of the platelet plasma membrane) is not the consequence of vesicle formation. Conversely, we propose that vesicle shedding is an effect of phospholipid transverse redistribution and calpain-mediated cytoskeleton proteolysis during activation. According to this view, the calcium–A23187-induced platelet stimulation very rapidly causes a phospholipid excess in the external leaflet of the plasma membrane from PS and PE initially located on the internal leaflet. This large extra surface can be counterbalanced only by very small bending radius shapes (see Section IV,B,5) such as filipodia (Ferrell *et al.*, 1988; Suné and Bienvenüe, 1988). These structures are very taut so the loss of membrane vesicles is a very likely event, provided that the cytoskeleton is not able to maintain cell integrity as, for example, after calpain-catalyzed proteolysis of the cytoskeleton.

VI. CONCLUSION

Lipid–protein interactions (based on matching of hydrophobic regions of lipids and proteins and on polar head group effects) allow integral proteins to maintain their native conformations with no uncontrolled aggregation. Lipids are also important in forming convenient large anionic surfaces on which to assemble active enzymes with their substrate, a two-dimensional diffusion increasing the catalytic rate, and with which to modulate cell structure by interacting with cytoskeleton-associated proteins. The functional role of the lipid chemical structure and location is thus of major importance in the regulation of many metabolic pathways, of signal transduction mechanisms, of cell shape, and, thus, of harmonizing of cell function. New properties can be provided to cells (such as platelets in the hemostasis reaction) by lipid redistribution. The unravelled paradox is that such large effects are not caused by strictly defined interactions in terms of sites, stoichiometry, and lifetime other than in very few cases. The notion of interface between lipid media and other media is probably more relevant, but many more experiments are needed to understand the molecular mechanism of protein modulation by lipids in terms of lifetime of the lipid–protein complexes and of protein conformational changes according to different environments.

Changes in lipid composition or location often accompany a cell dysfunction. However, cell deregulation as occurs in cancer, for example, merely could induce lipid changes without any requirement for the development of pathological cells. In this way, lipid changes should, at best, be a signature of cell disease and should be used in diagnosing cancer only if lipid assays are found to be easier and more sensitive than existing immunological or genetic methods.

Certainly that some major pathologies are accompanied by fundamental lipid metabolism modifications (see Chapter 1). Candidosis infections are increasingly prevalent because immunosuppressive drugs are widely used and AIDS is spreading (Mishra *et al.*, 1992). Parasite development also strongly depends on fast phospholipid biosynthesis. Our laboratory currently is pursuing a very promising approach concerning antiparasite therapy, particularly malarial cures (Vial and Ancelin, 1992). The intraerythrocyte development of a malarial agent (*Plasmodium*) is associated with an efficient conversation between red blood cells and parasites, producing many changes in the erythrocyte membrane to insure nutrition, maturation, and survival of parasites despite the host immune system. Growth of the parasite requires a large increase in total membrane surface. During parasite maturation, complex membranous material accumulates in the future merozoites, suggesting that a large amount of lipids capable of assembling into membranes is formed and stored in a short time (24 to 48 hr, according to the parasite) (Holz, 1977; Sherman, 1979,1984). That many differences exist between the lipid composition of malaria parasites and that of the original host erythrocyte has been clearly established.

With respect to the metabolic pathway of phospholipids, noninfected cells have a very low turnover of their membrane lipids, whereas the biosynthetic activity of *Plasmodium* is intense, at the expense of precursors originating from the plasma, mainly fatty acids and choline. Incorporation of various precursors into each phospholipid represents a large part of parasite requirements and does not require a large net transfer of external phospholipids, although the transverse diffusion rate for all phospholipids is found to be faster in infected than in noninfected erythrocytes (Beaumelle *et al.*, 1988). In summary, phospholipid synthesis by *Plasmodium* places the parasite midway between prokaryotic and eukaryotic organisms, allowing use of this metabolism as a target for new malaria chemotherapy. Three types of putative drugs have been investigated: fatty acids (Beaumelle and Vial, 1988), PC or PE head group analogs (Vial *et al.*, 1984), and larger molecules inhibiting choline transport through the erythrocyte plasma membrane (Ancelin and Vial, 1986). This final family of drugs was revealed when observed to cause a specific decrease in the PC biosynthesis, in very close correlation with the inhibition of *Plasmodium falciparum* growth. This is currently the focus of many pharmacological studies with a potentially promising future in therapeutics.

References

Aarsman, A. J., DeJong, J. G. N., Amoldressen, E., Neys, F. W., van Wassorman, P., and van den Bosch, H. (1989). Immunoaffinity purification, partial sequence, and subcellular localization of rat liver phospholipase A2. *J. Biol. Chem.* **264,** 10008–10014.

Aitken, J., Van Heusden, G. P. H., Temkin, M., and Dowhan, W. (1990). The gene coding the phosphatidyl inositol transfer protein is essential for cell growth. *J. Biol. Chem.* **265,** 4711–4717.

Ancelin, M. L., and Vial, H. J. (1986). Quaternary ammonium compounds efficiently inhibit *Plasmodium falciparum* growth in vitro by impairment of choline transport. *Antimicrob. Agents Chemother.* **29,** 814–820.

Asano, K., and Asano, A. (1988). Binding of cholesterol and inhibitory peptide derivatives with the fusogenic hydrophobic sequence of F-glycoprotein of HVJ (Sendai virus): Possible implication in the fusion reaction. *Biochemistry* **27,** 1321–1329.

Azzi, A., Boscoboinik, D., and Hensey, C. (1992). The protein kinase C family. *Eur. J. Biochem.* **208,** 547–557.

Bal, A., and Bird, M. M. (1991). Changes in filipin-sterol binding in the rat cingulate cortex after the administration of antidepressant drugs. A freeze-fracture study. *Brain Res.* **550,** 147–151.

Barrière, H., Chailley, B., Chambard, M., Selzner, J. P., Mauchamp, J., and Gabrion, J. (1988). Cholesterol-rich microdomains in rat and porcine thyroid membranes involved in TSH-induced endocytotic processes. *Acta Anat.* **132,** 205–215.

Bassé, F., Gaffet, P., Rendu, F., and Bienvenüe, A. (1992a). Phospholipid transverse mobility modifications in plasma membranes of activated platelets: An ESR study. *Biochem. Biophys. Res. Commun.* **189,** 465–471.

Bassé, F., Sainte-Marie, J., Maurin, L., and Bienvenüe, A. (1992b). Effect of benzyl alcohol on phospholipid transverse mobility in human erythrocyte membrane. *Eur. J. Biochem.* **205,** 155–162.

Bassé, F., Gaffet, P., Rendu, F., and Bienvenüe, A. (1993). Translocation of spin labeled phospholipids through plasma membrane during thrombin- and ionophore A23187-induced platelet activation. *Biochemistry* **32**, 2337–2344.

Bazzi, M. D., and Nelsestuen, G. L. (1988a). Constitutive activity of membrane-inserted protein kinase C. *Biochem. Biophys. Res. Commun.* **152**, 336–343.

Bazzi, M. D., and Nelsestuen, G. L. (1988b). Properties of membrane-inserted protein kinase C. *Biochemistry* **27**, 7589–7593.

Bazzi, M. D., and Nelsestuen, G. L. (1990). Protein kinase C interaction with calcium: A phospholipid-dependent process. *Biochemistry* **29**, 7624–7630.

Bazzi, M. D., and Nelsestuen, G. L. (1991a). Extensive segragation of acidic phospholipids in membranes induced by protein kinase C and related proteins. *Biochemistry* **30**, 7961–7969.

Bazzi, M. D., and Nelsestuen, G. L. (1991b). Highly sequential binding or protein kinase C and related proteins to membranes. *Biochemistry* **30**, 7970–7977.

Bazzi, M. D., and Nelsestuen, G. L. (1992a). Importance of phosphatidylethanolamine for association of protein kinase C and other cytoplasmic proteins with membranes. *Biochemistry* **31**, 1125–1134.

Bazzi, M., and Nelsuesten, G. L. (1992b). Interaction of annexin VI with membranes: highly restricted dissipation of clustered phospholipids in membranes containing phosphatidylethanolamine. *Biochemistry* **31**, 10406–10413.

Beaudoin, A. R., and Grondin, G. (1991). Shedding of vesicular material from the cell surface of eukaryotic cells: Different cellular phenomena. *Biochim. Biophys. Acta* **1071**, 203–219.

Beaumelle, B. D., and Vial, H. J. (1988). Correlation of the efficiency of fatty acids derivatives in suppressing *Plasmodium falciparum* growth in culture with their inhibitory effect on acyl CoA synthetase activity. *Mol. Biochem. Parasitol.* **28**, 39–42.

Beaumelle, B. D., Vial, H. J., and Bienvenüe, A. (1988). Enhanced transbilayer mobility of phospholipids in malaria-infected monkey erythrocytes. A spin label study. *J. Cell Physiol.* **135**, 94–100.

Berg, O. G., Hu, B. Z., Rogers, J., and Jain, M. K. (1991). Interfacial catalysis by phospholipas A2: Determination of the interfacial kinetic rate constants. *Biochemistry* **30**, 7283–7297.

Bergelson, L. D. (1992). Lipid domain reorganization and receptor events. *FEBS Lett.* **297**, 212–215.

Berlin, E., Bhathena, S. J., Judd, J. T., and Taylor, P. R. (1989). Dietary lipid influence on erythrocyte membrane composition, fluidity and insulin receptor binding. *Biomembr. Nutr. Colloq. INSERM* **195**, 187–196.

Bevers, E. M., Comfurius, P., and Zwaal, R. F. A. (1983). Changes in membrane phospholipid distribution during platelet activation. *Biochim. Biophys. Acta* **736**, 57–66.

Bevers, E. M., Wiedmer, T., Comfurius, P., Shattil, S. J., Weiss, H. J., Zwaal, R. F. A., and Sims, P. J. (1992). Defective Ca^{2+}-induced microvesiculation and deficient expression of procoagulant activity in erythrocytes from a patient with a bleeding disorder: A study of the red blood cell of Scott syndrome. *Blood* **79**, 380–388.

Bialecki, R. A., and Tulenko, T. N. (1989). Excess membrane cholesterol alters calcium channels in arterial smooth muscle. *Am. J. Physiol.* **257**, C306–C314.

Birrel, G. B., and Griffith, O. H. (1976). Cytochrome c induced lateral phase separation in a phosphaticylglycerol-steroid spin label model membrane. *Biochemistry* **15**, 2925–2929.

Bishop, W. R., and Bell, R. M. (1985). Assembly of the endoplasmic reticulum phospholipid bilayer: The phosphatidylcholine transporter. *Cell* **42**, 57–60.

Bloom, M., Evans, E., and Mouritsen, O. G. (1991). Physical properties of the fluid lipid-bilayer component of cell membranes: A perspective. *Q. Rev. Biophys.* **24**, 293–397.

Boesze-Battaglia, K., and Albert, A. D. (1990). Cholesterol modulation of photoreceptor function in bovine retinal rod outer segments. *J. Biol. Chem.* **265**, 20727–20730.

Boggs, J. M., Moscarello, M. A., and Papahadjopoulos, D. (1977). Phase separation of acidic and neutral phospholipids induced by human myelin basic protein. *Biochemistry* **16**, 5420–5426.

Brasitus, T. A., Dahiya, R., Dudeja, P. K., and Bissonnette, B. M. (1988). Cholesterol modulates alkaline phosphatase activity of rat intestinal microvillus membranes. *J. Biol. Chem.* **263**, 8592–8597.

Braunwald, J., Nonnenmacher, H., Pereira, C. A., and Kirn, A. (1991). Increased susceptibility to mouse hepatitis virus type 3 (MHV3) infection induced by a hypercholesterolaemic diet with increased adsorption of MHV3 to primary hepatocyte cultures. *Res. Virol.* **142**, 5–15.

Brenner, R. R. (1990). Role of cholesterol in the microsomal membrane. *Lipids* **25**, 581–585.

Brisson, A., Mosser, G., and Huber, R. (1991). Structure of soluble and membrane-bound human Annexin V. *J. Mol. Biol.* **220**, 199–203.

Broderick, R., Bialecki, R., and Tulenko, T. N. (1989). Cholesterol-induced changes in rabbit arterial smooth muscle sensitivity to adrenergic stimulation. *Am. J. Physiol.* **257**, H170–H178.

Brunauer, L. S., Waters, S. I., Stevenson, M. J., and Husetis, W. H. (1989). Amphipath-mediated platelet shape changes occur without direct involvement of the cellular cytoskeleton. *Biophys. J.* **55**, 139a.

Bütikofer, P., Brodbeck, V., and Ott, P. (1987). Modulation of erythrocyte vesiculation by amphiphilic drugs. *Biochim. Biophys. Acta* **901**, 291–295.

Carrier, D., and Pezolet, M. (1986). Investigation of polylysine-dipalmitoylphosphatidylglycerol interactions in model membranes. *Biochemistry* **25**, 4167–4174.

Carrière, B., and Le Grimellec, C. (1986). Effects of benzyl alcohol on enzyme activities and D-glucose transport in kidney brush-border membranes. *Biochim. Biophys. Acta* **857**, 131–138.

Carruthers, A., and Melchior, D. L. (1986). How bilayer lipids affect membrane protein activity. *Trends Biochem. Sci.* **11**, 331–335.

Cassama-Diagne, A., Fauvel, J., and Chap, H. (1989). Purification of a new, calcium-independent, high molecular weight phospholipase A2/lysophospholipase (phospholipase B) from guinea pig intestinal brush-border membrane. *J. Biol. Chem.* **264**, 9470–9475.

Chen, J. K., and Li, L. (1987). Sterol depletion reduces receptor-mediated low-density lipoprotein binding in NS-1 mouse myeloma cells. *Exp. Cell Res.* **171**, 76–85.

Clark, J. D., Lin, L. L., Kriz, R. W., Ramesha, C. S., Sultzman, L. A., Lin, A. Y., Milona, N., and Knopf, J. K. (1991). A novel arachidonic acid-selective cytosolic PLA2 contains a calcium dependent translocation domain with homology to PKC and GAP. *Cell* **65**, 1043–1051.

Comfurius, P., Bevers, E. M., and Zwaal, R. F. A. (1985). The involvement of cytoskeleton in the regulation of transbilayer movement of phospholipids in human blood platelets. *Biochim. Biophys. Acta* **815**, 143–148.

Comfurius, P., Senden, J. M. G., Tilly, R. H. J., Schroit, A. J., Bevers, E. M., and Zwaal, R. F. A. (1990). Loss of membrane phospholipid asymmetry in platelets and red cells may be associated with calcium-induced shedding of plasma membrane and inhibition of aminophospholipid translocase. *Biochim. Biophys. Acta* **1026**, 153–160.

Compagnon, B., Milhaud, P., Bienvenüe, A., and Philippot, J. (1992). Targeting of poly(rI)-poly(rC) by fusogenic (F protein) immunoliposomes. *Exp. Cell Res.* **200**, 333–338.

Coussens, L., Parker, P. J., Rhee, L., Yang-Fang, T. L., Chen, E., Waterfield, M. D.,

Franke, U., and Ullrich, A. (1986). Multiple, distinct forms of bovine and human protein kinase C suggest diversity in cellular signaling pathways. *Science* **233**, 859–866.

Creutz, C. E. (1981). *cis*-Unsaturated fatty acids induce the fusion of chromaffin granules aggregated by synexin. *J. Cell Biol.* **91**, 247–256.

Creutz, C. E. (1992). The annexins and exocytosis. *Science* **258**, 924–931.

Creutz, C. E., Moss, S., Edwardson, J. M., Hide, I., and Gomperts, B. (1992). Differential recognition of secretory vesicules by annexins. *Biochem. Biophys. Res. Commun.* **184**, 347–352.

Crumpton, M. J., and Dedman, J. R. (1990). Protein terminology tangle. *Nature (London)* **345**, 212.

Daefler, S., and Krueger, G. R. F. (1989). Expression of proliferation and differentiation antigens in response to modulation of membrane fluidity in chronic lymphocytic leukemia lymphocytes. *Anticancer Res.* **9**, 501–506.

Dalbey, R. E. (1990). Positively charged residues are important determinants of membrane topology. *Trends Biochem. Sci.* **15**, 253–257.

Daleke, D. L., and Huestis, W. H. (1985). Incorporation and translocation of aminophospholipids in human erythrocytes. *Biochemistry* **24**, 5406–5416.

Daleke, D. L., and Huestis, W. H. (1989). Erythrocyte morphology reflects the transbilayer distribution of incorporated phospholipids. *J. Cell Biol.* **108**, 1375–1385.

Dautry-Varsat, A., Ciechanover, A., and Lodish, H. F. (1983). pH and the recycling of transferrin during receptor mediated endocytosis. *Proc. Natl. Acad. Sci. U.S.A.* **80**, 2258–2262.

Davie, E. W., Fujikawa, K., and Kisiel, W. (1991). The coagulation cascade: initiation, maintenance and regulation. *Biochemistry* **30**, 10363–10370.

De Foresta, B., Rogard, M., Le Maire, M., and Gallay, J. (1987). Effects of temperature and benzyl alcohol on the structure and adenylate cyclase activity of plasma membranes from bovine adrenal cortex. *Biochim. Biophys. Acta* **905**, 240–256.

Devaux, P. F. (1983). ESR and NMR studies of lipid protein interactions in membranes. *In* "Biological Magnetic Resonance" (L. J. Berliner, and J. Ruebens, eds.), pp. 183–229. Plenum Press, New York.

Devaux, P. F. (1991). Static and dynamic lipid asymetry in cell membranes. *Biochemistry* **30**, 1163–1173.

Drake, T. A., Morrissey, J. H., and Edgington, T. S. (1989). Selective cellular expression of tissue factor in human tissues. Implications for disorder of haemostasis and thrombosis. *Am. J. Pathol.* **134**, 1087–1097.

Düzgünes, N., and Shavnin, S. A. (1992). Membrane destabilization by N-terminal peptides of viral envelopope proteins. *J. Membr. Biol.* **128**, 71–80.

Esfahani, M., Scerbo, L., Lund-Katz, S., DePace, D. M., Maniglia, R., Alexander, J. K., and Phillips, M. C. (1986). Effects of cholesterol and lipoproteins on endocytosis by a monocyte-like cell line. *Biochim. Biophys. Acta* **889**, 287–300.

Farge, E., and Devaux, P. F. (1992). Shape changes of giant liposomes induced by an asymetric transmembrane distribution of phospholipids. *Biophys. J.* **61**, 347–357.

Ferrell, J. E., Mitchell, K. T., and Huestis, W. H. (1988) Membrane bilayer balance and platelet shape: Morphological and biochemical responses to amphipatic compounds. *Biochim. Biophys. Acta* **939**, 223–237.

Fong, T. M., and Mc Namee, M. G. (1986). Correlation between acetylcholine receptor function and structural properties of membranes. *Biochemistry* **25**, 830–840.

Fong, T. M., and Mc Namee, M. G. (1987). Stabilization of acetylcholine receptor secondary structure by cholesterol and negatively charged phospholipids in membranes. *Biochemistry* **26**, 3871–3880.

Frenkel, E. J., Kuyper, F. H., Op den Kamp, J. A. F., Roelofsen, B., and Ott, P. (1986). Effect of membrane cholesterol on dimyristoyl phosphatidylcholine induced vesiculation of human red blood cells. *Biochim. Biophys. Acta* **855**, 293–301.

Friedlander, G., Le Grimellec, C., Giocondi, M. C., and Amiel, C. (1987). Benzyl alcohol increases membrane fluidity and modulates cyclic AMP synthesis in intact renal epithelial cells. *Biochim. Biophys. Acta* **903**, 341–348.

Friedlander, G., Shahedi, M., Le Grimellec, C., and Amiel, C. (1988). Increase in membrane fluidity and opening of tight junctions have similar effects on sodium-coupled uptakes in renal epithelial cells. *J. Biol. Chem.* **263**, 11183–11188.

Friedlander, G., Le Grimellec, C., and Amiel, C. (1990). Increase in membrane fluidity modulates sodium-coupled uptakes and cyclic AMP synthesis by renal proximal tubular cells in primary culture. *Biochim. Biophys. Acta* **1022**, 1–7.

Fukami, K., Furuhashi, K., Inagaki, M., Endo, T., Hatano, S., and Takenawa, T. (1992). Requirement of phosphatidylinositol 4,5-bisphosphate for α-actinin function. *Nature (London)* **359**, 150–152.

Funakoshi, T., Heinmark, R. L., and Hendrikson, L. E. (1987). Human placental anticoagulant protein: Isolation and characterisation. *Biochemistry* **26**, 5572–5578.

Gavignan, S. J. P., and Knight, B. L. (1981). Catabolism of low-density lipoprotein by fibroblasts cultured in medium supplemented with saturated or unsaturated free fatty acids. *Biochim. Biophys. Acta* **665**, 632–635.

Gennis, R. B., ed. (1989). "Biomembranes, Molecular Structure and Function." Springer Verlag, Heidelberg.

Gennis, R. B., and Jonas, A. (1977). Protein lipid interactions. *Ann. Rev. Biophys. Bioeng.* **6**, 195–238.

Goldmann, W. H., Niggli, V., Kaufmann, S., and Isenberg, G. (1992). Probing actin and liposome interaction of talin and talin-vinculin complexes: A kinetic, thermodynamic and lipid labeling study. *Biochemistry* **31**, 7665–7671.

Goldschmidt-Clermont, P. J., and Jeanmey, P. A. (1991). Profilin, a weak CAP for actin and RAS. *Cell* **66**, 419–421.

Goldstein, J. L., Brown, M. S., Anderson, R. G. W., Russell, D. W., and Schneider, W. J. (1985). Receptor-mediated endocytosis: Concepts emerging from the LDL receptor system. *Ann. Rev. Cell. Biol.* **1**, 1–39.

Hagiwara, Y., and Ozawa, E. (1990). Suppression of transferrin internalization in myogenic L6 cells by dibucaine. *Biochim. Biophys. Acta* **1051**, 237–241.

Hartmann, W., and Galla, H. J. (1978). Binding of polylysine to charged bilayer membranes. Molecular organization of a peptide-lipid complex. *Biochim. Biophys. Acta* **509**, 474–490.

Hartwig, J. H., and Kwiatkowski, D. J. (1991). Actin-binding proteins. *Curr. Opin. Cell Biol.* **3**, 87–97.

Hazen, S. L., and Gross, R. W. (1991). ATP-dependent regulation of rabbit myocardial cytosolic calcium independent phospholipase A2. *J. Biol. Chem.* **266**, 14526–14534.

Hoekstra, D. (1990). Membrane fusion of enveloped viruses: especially a matter of proteins. *J. Bioenerg. Biomembr.* **22**, 121–125.

Holland, D. R., Clancy, L. L., Muchmore, S. W., Ryde, T. J., Einspahr, H. M., Finzel, B. C., Heinrikson, R. L., and Watenpaugh, K. D. (1990). The crystal structure of a lysine 49 phospholipase A2 from the venom of the cottonmouth snake at 2.0 Å resolution *J. Biol. Chem.* **265**, 17649–17656.

Holz, G. G. (1977). Lipids and the malarial parasite. *Bull. WHO* **55**, 237–248.

Hooper, N. M. (1992). More than just a membrane anchor. Potocytosis, the uptake of small molecules by cells, and intracellular targetting are just two of the functions proposed for glycosyl-phosphatidyl inositol anchors. *Cur. Biol.* **2**, 617–619.

Horvath, L. I., Drees, M., Beyer, K., Klingenberg, M., and Marsh, D. (1990). Lipid-protein interactions in ADP–ATP carrier/egg phosphatidylcholine reecombinants studied by spin label ESR spectroscopt. *Biochemistry* **29**, 10664–10669.

Housley, M. D. (1991). "Crosstalk" a pivotal role for protein kinase C in modulating relationships between signal transduction pathways. *Eur. J. Biochem.* **195**, 9–27.

Huber, R., Berendes, R., Burger, A., Schneider, M., Karshikov, A., and Luecke, H. (1992). Crystal and molecular structure of human annexin V after refinement. *J. Mol. Biol.* **223**, 683–704.

Ikeda, T., Yamaguchi, H., and Tazuke, S. (1990). Phase separation in phospholipid bilayers induced by biologically active polycations. *Biochim. Biophys. Acta* **1026**, 105–112.

Isenberg, G. J. (1991). Acting binding protein-lipid interactions. *J. Muscle Res. Cell Motil.* **12**, 136–144.

Jähnig, F. (1981). Critical effects from lipid-protein interaction in membranes. *Biophys. J.* **36**, 329–345.

Jones, M. E., and Lentz, B. R. (1986). Phospholipid lateral organization in synthetic membranes as monitored by pyrene-labeled phospholipids: effects of temperature and prothrombin fragment 1 binding. *Biochemistry* **25**, 567–584.

Kazaniev, M. G., Krausz, K. W., and Blumberg, P. M. (1992). Differential irreversible insertion of protein kinase C into phospholipid vesicles by phorbol esters and related activators. *J. Biol. Chem.* **267**, 20878–20886.

Kim, J., Mosior, M., Chung, L. A., Wu, H., and McLaughlin, S. (1991). Binding of peptides with basic residues to membranes containing acidic phospholipids. *Biophys. J.* **60**, 135–148.

Knight, B. L., and Soutar, A. K. (1986). Low apparent affinity for low-density lipoprotein of receptors expressed by human macrophages maintained with whole serum. *Eur. J. Biochem.* **156**, 205–210.

Kobayashi, T., Okamoto, H., Yamada, J. I., Setaka, M., and Kwan, T. (1984). Vesiculation of platelet plasma membranes. *Biochim. Biophys. Acta* **778**, 210–218.

Krishnawamy, S., Church, W. R., Nesheim, M. E., and Mann, K. G. (1987). Activation of human thrombin by human prothrombinase. *J. Biol. Chem.* **262**, 3291–3299.

Krueger, G. R. F., Stolzenburg, T., and Muller, C. (1987). Cell membrane lipid fluidity and receptor expression in Moloney- and Friend virus-transformed cells. *In Vivo* **1**, 343–346.

Kuo, P., Weinfeld, M., Rudd, M. A., Amarante, P., and Loscalzo, J. (1990a). Plasma membrane enrichment with cis-unsaturated fatty acids enhances LDL metabolism in U937 monocytes. *Arteriosclerosis* **10**, 111–118.

Kuo, P., Weinfeld, M., and Loscalzo, J. (1990b). Effect of membrane fatty acyl composition on LDL metabolism in Hep G2 hepatocytes. *Biochemistry* **29**, 6626–6632.

Kusumi, A., and Hyde, J. S. (1982). Spin-label saturation-transfer electron spin resonance detection of transient association of rhodopsin in reconstituted membranes. *Biochemistry* **21**, 5978–5983.

Lee, A. G. (1987). Interactions of lipids and proteins: some general principles. *J. Bioenerg. Biomembr.* **19**, 581–603.

Lee, M. H., and Bell, R. M. (1991). Mechanism of protein kinase C activation by phosphatidylinositol 4,5-bisphosphate. *Biochemistry* **30**, 1041–1049.

Léger, C. L., Christon, R., Viret, J., Daveloose, D., Mitjavila, S., and Even, V. (1989). Nutrition and biomembranes: Additional information concerning the incidence of dietary polyunsaturated fatty acids on membrane organization and biological activity. *Biochimie* **71**, 159–165.

Lewis, B. A., and Engelman, D. M. (1983). Bacteriorhodopsin remains dispersed in fluid phospholipid bilayers over a wide range of bilayer thicknesses. *J. Mol. Biol.* **166**, 203–210.

Loscalzo, J., Freedman, J., Rudd, M. A., Barsky-Vasserman, I., and Vaughan, D. E. (1987). Unsaturated fatty acids enhance low density lipoprotein uptake and degradation by peripheral blood mononuclear cells. *Arteriosclerosis* **7**, 450–455.

Luna, E. J., and Hitt, A. L. (1992). Cytoskeleton–plasma membrane interactions. *Science* **258**, 955–964.

Maehara, K. (1991). Membrane cholesterol and insulin receptor in erythrocytes. *Fukuoka Acta Medica* **82**, 586–602.

Malhotra, V., Orci, B., Glick, B. J., Block, M. R., and Rothman, J. E. (1988). Role of an *N*-ethylmaleimide-sensitive transport component in promoting fusion of transport vesicules with cisternae of the Golgi stack. *Cell* **54**, 221–227.

Mann, K. G., Nesheim, M. E., Church, W. R., Haley, P., and Krishnawamy, S. (1990). Surface dependent reactions of the vitamin K-dependent enzyme complexes. *Blood* **76**, 1–16.

McMurchie, E. J., and Patten, G. S. (1988). Dietary cholesterol influences cardiac β-adrenergic receptor adenylate cyclase activity in the marmoset monkey by changes in membrane cholesterol status. *Biochim. Biophys. Acta* **942**, 324–332.

Meers, P., Mealy, T., Pavlotsky, N., and Tauber, A. (1992). Annexin I-mediated vesicular agregation: Mechanism and role in human neutrophils. *Biochemistry* **31**, 6372–6382.

Michel, H., Weyer, K. A., Gruenberg, H., Dunger, I., Oesterhelt, D., and Lottspeich, F. (1986). The light and medium subunits of the photosynthetic center from *Rhodopseudomonas viridis:* Isolation of genes, nucleotide and amino acid sequence. *EMBO J.* **5**, 1149–1158.

Milhaud, P. G., Compagnon, B., Bienvenüe, A., and Philippot, J. R. (1992). Interferon production of L929 and HeLa cells enhanced by polyriboinosinic–polyribocytidylic acid pH-sensitive liposomes. *Bioconj. Chem.* **3**, 402–407.

Mishra, P., Bolard, J., and Prasad, R. (1992). Emerging role of lipids of *Candida albicans,* a pathogenic dimorphic yeast. *Biochim. Biophys. Acta* **1127**, 1–14.

Morrot, G., Hervé, P., Zachowski, A., Fellmann, P., and Devaux, P. F. (1989). Aminophospholipid translocase of human erythrocytes: Phospholipid specificity and effect of cholesterol. *Biochemistry* **28**, 3456–3462.

Mosior, M., and McLaughlin, S. (1991). Peptides that mimic the pseudosubstrate region of protein kinase C bind to acidic lipids in membranes. *Biophys. J.* **60**, 149–159.

Mosser, G., Ravanat, C., Freyssinet, J. M., and Brisson, A. (1991). Sub-domain structure of lipid-bound annexin V resolved by electron image analysis. *J. Mol. Biol.* **217**, 241–245.

Mouritsen, O. G., and Bloom, M. (1984). Mattress model of lipid-protein interaction in membranes. *Biophys. J.* **46**, 141–153.

Mouritsen, O. G., and Briltonen, R. L. (1992). Protein-lipid interactions and membrane heterogeneity. *In* "New Comprehensive Biochemistry" (A. Watts, ed.), pp. 34–49. Protein-lipid interactions. Elsevier, Amsterdam.

Nanavati, C., Markin, V. S., Oberhauser, A. F., and Fernandez, J. M. (1992). The exocytic fusion pore modeled as a lipidic pore. *Biophys. J.* **63**, 1118–1132.

Newton, A. C., and Koshland, D. E. (1989). High cooperativity specificity and multiplicity in the protein kinase C-lipid interaction. *J. Biol. Chem.* **264**, 14909–14915.

Oberhauser, A. F., Monck, J. F., and Fernandez, J. M. (1992). Events leading to the opening and closing of the exocytotic fusion pore have markedly different temperature dependencies. *Biophys. J.* **61**, 800–809.

Pernas, P., Olivier, J. L., Masliah, J., Etienne, J., and Bereziat, G. (1992). Les phospholipases cellulaires. *Reg. Biochimie* **2**, 41–50.

Phalen, T., and Kielian, M. (1991). Cholesterol is required for infection by Semliki forest virus. *J. Cell Biol.* **112**, 615–623.

Pollard, H. P., Guy, H. R., Arispe, N., de la Fuente, M., Lee, G., Rojas, E. M., Pollard, J. M., Srivastara, M., Zhang-Keck, Y., Merezhinskaya, C., Lee-Burns, A., and Rojas, E. (1992). Calcium channel and membrane fusion activity of synexin and other members of the annexin gene family *Biophys. J.* **62**, 15–18.

Raetz, C. R. H. (1982). Genetic control of phospholipid bilayer assembly. *In* "Phospholipids" (J. N. Hawthorne and G. B. Ansell, eds.), Vol. 4, pp. 435–477. Elsevier, Amsterdam.

Ramirez, F., and Jain, M. K. (1991). Phospholipase A2 at the bilayer interface. *Proteins* **9**, 229–239.

Rand, R. P., and Parsegian, V. A. (1989). Hydration forces between phospholipid bilayers. *Biochim. Biophys. Acta* **988**, 351–376.

Resh, M. D. (1989). Specific and saturable binding of pp60v^{-src} to plasma membranes: evidence of a myristyl-src receptor. *Cell* **58**, 281–286.

Resh, M. D., and Ling, H. (1990). Identification of a 32K plasma membrane protein that binds to the myristoylated amino-terminal sequence of p60^{-src}. *Nature (London)* **346**, 84–86.

Rocha-Singh, K. J., Hines, D. K., Honbo, N. Y., and Karliner, J. S. (1991). Concanavalin A amplifies both b-adrenergic and muscarinic cholinergic receptor–adenylate cyclase-linked pathways in cardiac myocytes. *J. Clin. Invest.* **88**, 760–766.

Römisch, J., and Pâques, E-P. (1991). Annexins: Calcium binding proteins of multifunctional importance? *Med. Microbiol. Immunol.* **180**, 109–126.

Rosing, J., Speijer, H., and Zwaal, R. F. A. (1988). Prothrombin activation on phospholipid membranes with positive electrostatic potential. *Biochemistry* **27**, 8–11.

Rotenberg, M., and Zakim, D. (1991). Effects of cholesterol on the function and thermotropic properties of pure UDP-glucuronosyltransferase. *J. Biol. Chem.* **266**, 4159–4161.

Rothberg, K. G., Ying, Y., Kamen, B. A., and Anderson, R. G. W. (1990). Cholesterol controls the clustering of the glycosphospholipid-anchored membrane receptor for 5-methyltetrahydrofolate. *J. Cell Biol.* **111**, 2931–2938.

Rothman, J. E., and Orci, L. (1992). Molecular dissection of the secretory pathway. *Nature (London)* **355**, 409–415.

Roux, M., Neumann, J. M., Hodges, R. S., Devaux, P. F., and Bloom, M. (1989). Conformational changes of phospholipid headgroups induced by a cationic integral membrane peptide as seen by deuterium magnetic resonance. *Biochemistry* **28**, 2313–2321.

Russo-Marie, F. (1992). Annexins, phospholipase A2 and the glucocorticoids. *In* "The Annexins" (S. B. Moss, ed.), pp. 158–165. Portland Press, London.

Ryba, N. J. P., and Marsh, D. (1992). Protein rotational diffusion and lipid/protein interactions in recombinants of bovine rhodopsin with saturated diacylphosphatidylcholines of different chain length by conventional and saturation-transfer electron spin resonance. *Biochemistry* **31**, 7511–7518.

Sainte-Marie, J., Vidal, M., Suné, A., Ravel, S., Philippot, J. R., and Bienvenüe, A. (1989). Modifications of LDL-receptor-mediated endocytosis rates in CEM lymphoblastic cells grown in lipoprotein-depleted fetal calf serum. *Biochim. Biophys. Acta* **982**, 265–270.

Sainte-Marie, J., Vignes, M., Vidal, M., Philippot, J. R., and Bienvenüe, A. (1990). Effects of benzyl alcohol on transferrin and low density lipoprotein receptor mediated endocytosis in leukemic guinea pig B lymphocytes. *FEBS Lett.* **262**, 13–16.

Santini, M. T., Masella, R., Cantafora, A., and Peterson, S. W. (1992). Changes in erythrocyte membrane lipid composition affect the transient decrease in membrane order which accompanies insulin receptor down-regulation. *Experientia* **48**, 36–39.

Schalkwijk, C. G., Märki, F., and Van den Bosch, H. (1990). Studies on the acyl chain selectivity of cellular phospholipase A2. *Biochim. Biophys. Acta* **1044**, 139–146.

Schneider, W. J. (1989). The low density lipoprotein receptor. *Biochim. Biophys. Acta* **988**, 303–317.

Sechoy, O., Philippot, J., and Bienvenüe, A. (1987). F-protein-F-protein interaction within the Sendai virus identified by native binding or chemical cross-linking. *J. Biol. Chem.* **262**, 11519–11523.

Sechoy, O., Philippot, J. R., and Bienvenüe, A. (1989). Targeting of loaded fusogenic vesicles bearing specific antibodies toward leukemic T cells. *Exp. Cell Res.* **185**, 122–131.

Seigneuret, M., and Devaux, P. F. (1984). ATP-dependent asymetric distribution of spin-labeled phospholipids in the erythrocyte membrane: Relation to shape changes. *Proc. Natl. Acad. Sci. U.S.A.* **81**, 3751–3755.

Seppen, J., Ramalho-Santos, J., de Carvalho, A. P., ter Beest, M., Kok, J. W., de Lima, M. C., and Hoekstra, D. (1992). Interaction of clathrin with large unilamellar phospholipid vesicles at neutral pH. Lipid dependence and protein penetration. *Biochim. Biophys. Acta* **1106**, 209–215.

Shariff, A., and Luna, E. J. (1992). Diacylglycerol-stimulated formation of actin nucleation sites at plasma membranes. *Science* **256**, 245–247.

Sheetz, M. P., and Singer, S. J. (1974). Biological membranes as bilayer couples. A molecular mechanism of drug erythrocyte interactions. *Proc. Natl. Acad. Sci. U.S.A.* **71**, 4457–4461.

Shekman, R. (1992). Genetic and biochemical analysis of vesicular traffic in yeast. *Curr. Opin. Cell Biol.* **4**, 587–592.

Sherman, I. W. (1979). Biochemistry of *Plasmodium* (malarial parasite). *Microbiol. Rev.* **43**, 453–495.

Sherman, I. W. (1984). Metabolism. *In* "Metabolism in Antimalarial Drugs" (W. Petrs and W. H. Richards, eds.), Vol. I, pp. 31–71. Springer-Verlag, Berlin.

Shinomura, T., Asaoka, Y., Oka, M., Yoshida, K., and Nishizuka, Y. (1991). Synergestic action of diacylglycerol and unsaturated fatty acid for protein kinase C activation: its possible implications. *Proc. Natl. Acad. Sci. U.S.A.* **88**, 5149–5153.

Shouffani, A., and Kanner, B. I. (1990). Cholesterol is required for the reconstitution of the sodium- and chloride-coupled, γ-aminobutyric acid transporter from rat brain. *J. Biol. Chem.* **265**, 6002–6008.

Sims, P. J., Wiedmer, T., Esmon, C. T., Weiss, H. J., and Shattil, S. J. (1989). Assembly of the platelet prothrombinase complex is linked to vesiculation of the platelet plasma membrane. *J. Biol. Chem.* **264**, 17049–17057.

Smiley, P. L., Skemler, K. E., Prescott, S. M., Zimmerman, G. A., and Mc Intyre, T. M. (1991). Oxidatively fragmented phosphatidylcholines activate human neutrophils through the receptor for platelet-activating factor. *J. Biol. Chem.* **266**, 11104–11110.

Sobue, K., Kanda, K., Miyamoto, I., Ilda, K., Yahara, I., Hirai, R., and Hiragun, A. (1989). Comparison of the regional distribution of calspectin (nonerythroid spectrin and fodrin), alpha-actinin, vinculin, nonerythroid protein 4.1 and calpactin in normal and avian sarcoma virus- or rous sarcoma virus-induced transformed cells. *Exp. Cell Res.* **181**, 256–292.

Spector, A. A., and Yorek, M. A. (1985). Membrane lipid composition and cellular function. *J. Lipid Res.* **26**, 1015–1035.

Sperotto, M. M., and Mouritsen, O. G. (1988). Dependence of lipid membrane phase transition temperature on the mismatch of proteins and lipid hydrophobic thickness. *Eur. J. Biophys.* **16**, 1–10.

Sperotto, M. M., and Mouritsen, O. G. (1991a). Monte-Carlo simulation studies of lipid order parameter profiles near integral membrane proteins. *Biophys. J.* **59**, 261–270.

Sperotto, M. M., and Mouritsen, O. G. (1991b). Mean field and Monte-Carlo simulation studies of the lateral distribution of proteins in membranes. *Eur. J. Biophys.* **19**, 157–168.

Spruce, A. E., Breckenridge, L. J., Lee, A. K., and Almers, W. (1990). Properties of the fusion pore that forms during exocytosis of a mast exocytosis secretory vesicle. *Neuron* **4**, 643–654.

Stegmann, T., Doms, R. W., and Helenius, A. (1989). Protein-mediated membrane fusion. *Ann. Rev. Biophys. Biophys. Chem.* **18**, 187–211.

Suné, A., and Bienvenüe, A. (1988). Relationship between the transverse distribution of phospholipids in plasma membrane and shape change of human platelets. *Biochemistry* **27**, 6794–6800.

Suné, A., Bette-Bobillo, P., Bienvenüe, A., Fellmann, P., and Devaux, P. F. (1987). Selective outside–inside translocation of aminophospholipids in human platelets. *Biochemistry* **26**, 2972–2978.

Suné, A., Vidal, M., Morin, P., Sainte Marie, J., and Bienvenüe, A. (1988). Evidence for bidirectional transverse diffusion of spin labeled phospholipids in the plasma membrane of guinea pig blood cells. *Biochim. Biophys. Acta* **946**, 315–327.

Taraschi, T. F., de Kruijff, B., Verkleij, A., and van Echteld, C. J. A. (1982). Effect of glycophorin on lipid polymorphism: A ^{31}P NMR study. *Biochim. Biophys. Acta* **685**, 153–161.

Tilly, L., Cribier, S., Roelofsen, B., Op den Kamp, J. A. F, and VanDeenen, L. L. M. (1986). ATP-dependent translocation of amino phospholipids across the human erythrocyte membrane. *FEBS Lett.* **194**, 21–27.

Tocane, J. F., Dupou-Cézanne, L., Lopez, A., and Tournier, J. F. (1989). Lipid lateral diffusion and membrane organization. *FEBS Lett.* **257**, 10–16.

Toti, F., Eschwège, V., Davie, E. W., Fujikawa, K., and Kisiel, W. (1991). The coagulation cascade: Initiation, maintenance and regulation. *Biochemistry* **30**, 10363–10370.

Toti, F., Eschwège, V., and Freyssinet, J. M. (1992). Thrombine et facteur Xa, cibles principales de l'anticoagulation. Données actuelles et perspectives. *Sang. Thromb. Vaisseaux* **4**, 483–494.

Van Meer, G. (1989). Lipid traffic in animal cells. *Ann. Rev. Cell Biol.* **5**, 247–275.

Veillette, A., Bookman, M. A. N., Horak, E. M., and Bolen, J. B. (1988). The CD4 and CD8 T surface antigens are associated with the internal membrane tyrosine-protein kinase p56[lck]. *Cell* **55**, 301–308.

Vemuri, R., and Philipson, K. D. (1989). Influence of sterols and phospholipids on sarcolemmal and sarcoplasmic reticular cation transporters. *J. Biol. Chem.* **264**, 8680–8685.

Vial, H. J., and Ancelin, M. L. (1992). Malarial lipids: an overview. *Subcell. Biochem.* **18**, 254–306.

Vial, H. J., Thuet, M. J., Ancelin, M. L., Philippot, J. R., and Chavis, C. (1984). Phospholipid metabolism as a new target for malaria chemotherapy. Mechanism for aciton of D-2-amino 1-butanol. *Biochem. Pharmacol.* **33**, 2761–2770.

Vidal, M., Sainte Marie, J., Philippot, J. R., and Bienvenüe, A. (1989). Asymetric distribution of phospholipids in the membrane of vesicle released during in vitro maturation of guinea-pig reticulocytes. *J. Cell Physiol.* **140**, 455–462.

Voelker, D. R. (1991). Lipid assembly into cell membranes. *In* "Biochemistry of Lipids, Lipoproteins and Membranes" (D. E. Vance and J. Vance, eds.), pp. 110–128. Elsevier, Amsterdam.

Von Heijne, G. (1986). The distribution of positively charged residues in bacterial inner membranes proteins correlate with the transmembrane topology. *Nucl. Acids Res.* **14**, 4683–4690.

Wagner, E., Plank, C., Zatloukal, K., Cotten, M., and Birnstiel, M. L. (1992). Influenza virus hemagglutinin HA-2 N-terminal fusogenic peptides augment gene transfer by transferrin-polylysine-DNA complexes: Toward a synthetic virus-like gene transfer vehicle. *Proc. Natl. Acad. Sci. U.S.A.* **89**, 7934–7938.

White, J. M. (1992). Membrane fusion. *Science* **258**, 917–924.

Whiteheart, S. W., Brunner, M., Wilson, D. W., Wiedmann, M., and Rothman, J. E. (1992). Soluble *N*-ethylmaleimide-sensitive fusion attachment proteins (SNAPs) bind to a multi-SNAP receptor complex in Golgi membranes. *J. Biol. Chem.* **267**, 12239–12243.

Wiener, J. R., Pal, R., Barenholz, Y., and Wagner, R. R. (1985). Effect of the vesicular stomatitis virus matrix protein on the lateral organization of lipid bilayers containing phosphatidylglycerol: Use of fluorescent phospholipid analogues. *Biochemistry* **24**, 7651–7658.

Wirtz, K. W. A. (1990). Phospholipid transfer proteins. *Ann. Rev. Biochem.* **60**, 73–100.

Yeagle, P. L. (1985). Cholesterol and the cell membrane. *Biochim. Biophys. Acta* **822**, 267–287.

Yeagle, P. L. (1988). Effects of cholesterol on (Na^+/K^+)-ATPase ATP hydrolyzing activity in bovine kidney. *Biochemistry* **27**, 6449–6452.

Yeagle, P. L. (1991). Modulation of membrane function by cholesterol. *Biochimie* **73**, 1303–1310.

Zachowski, A., and Devaux, P. F. (1990). Transmembrane movements of lipids. *Experientia* **46**, 644–656.

Zachowski, A., and Morrot Gaudry-Talarmain, Y. (1990). Phospholipid transverse diffusion in synaptosomes: Evidence for the involvement of the aminophospholipid translocase. *J. Neurochem.* **55**, 1352–1356.

Zachowski, A., Favre, E., Cribier, S., Hervé, P., and Devaux, P. F. (1986). Outside-inside translocation of aminophospholipids in the human erythrocyte membrane is mediated by a specific enzyme. *Biochemistry* **25**, 2585–2590.

Zachowski, A., Henry, J. P., and Devaux, P. F. (1989). Control of transmembrane asymmetry in chromaffin granules by an ATP-dependent portein. *Nature (London)* **340**, 75–76.

Zakim, D., Kavecansky, J., and Scarlata, S. (1992). Are membrane enzymes regulated by the viscosity of the membrane environment? *Biochemistry* **31**, 11589–11594.

Zidovetzki, R., Laptalo, L., and Crawford, J. (1992). Effect of diacylglycerol on the activity of cobra venom, bee venom and pig pancreatic phospholipases A2. *Biochemistry* **31**, 7683–7691.

PART III

Lipids as Modulators of Cell Functioning

Lipids as Modulators of Cell Functioning

CHAPTER 13

Lipids and the Modulation of Cell Function

Charles C. Sweeley
Department of Biochemistry, Michigan State University, East Lansing, Michigan 48824

Future chronicles of lipid research will record that discoveries made in the latter part of the twentieth century established the existence of lipid-mediated transmembrane signaling mechanisms. Prior to about 1970, lipids were thought to be important elements of structure, membrane fluidity, and permeability. Few investigators dared to imagine that certain lipids would be directly or indirectly responsible for the transfer of information from extracellular effectors to intracellular targets. However, enzymatic hydrolysis products of phosphatidylcholine, sphingomyelin, and phospho-inositides now clearly can be classified as precursors of "lipid second messengers" and their conversion to these vector molecules occurs by activation of phospholipases imbedded in the plasma membrane. Products such as diacylglycerol (DG), inositol 4,5-bisphosphate, phosphatidic acid, ceramide (Cer), sphingosine, and others have been evaluated as modulators of specific tyrosine or serine–threonine protein kinases and protein phosphatases. In this part of this book, the pathways involved in the production of lipid second messengers, their biological activities and further metabolism, and control of the pathways are discussed in detail. The field is still in its infancy, and promises to be exciting and important.

Glycosphingolipids, which are believed to be localized in the outer leaflet of the plasma bilayer membrane, appear to be equally important participants in transmembrane signaling events. In contrast to the phospholipids, however, which must be converted to bioactive vectors, the glycosphingolipids appear to interact directly via their carbohydrate chains with adjacent proteins in the same cell or with carbohydrates or proteins of other cells via cell–cell contacts. How many different glycosphingolipids are involved in signaling events is still unknown, but the occurrence of more than 75 gangliosides and more than 100 neutral glycosphingolipids suggests that structural differences in the oligosaccharide chains could accommodate a very large number of specific membrane-mediated events. The best-characterized example to date is the inhibition of growth factor receptor tyrosine phosphorylation by membrane-bound gangliosides, which appears to be an important mechanism for the regulation of growth factor stimulation of cell growth. The relatively specific behaviors of G_{M3} ganglioside toward epidermal growth factor receptor, G_{M1} ganglioside toward platelet-derived growth factor, and sialylparagloboside toward insulin receptor suggest the possibility that other receptors may be regulated similarly. The role of gangliosides in the regulation of transmembrane signaling of growth factors also is reviewed in this part of this volume.

Studies of the biological effects of lipids commonly involve the addition of exogenous lipid to cultured cells and measurement of a specific event such as protein kinase activity. When the first reports appeared on the stimulatory or inhibitory effects of such exogenously added lipids, much concern was raised about whether the results were of a physiological or a pharmacological nature. These concerns emphasize the importance of using putative lipid bioeffectors at their normal cellular concentrations. Studies in which the modulation of a target system occurs only when levels of the experimental lipid are two or more orders of magnitude higher in concentration than ever would be expected to occur in cells must be repeated at normal intracellular concentrations; in some cases in which this was done, the biological effect was not observed or was quite different. More conservative approaches—such as measurements of the turnover of radiolabeled membrane phospholipid precursors after stimulation of cells with extracellular signals, effects of deliberate modulation of the membrane hydrolases, and cell-free systems in which the interaction of lipid vector and target are studied directly—circumvent the criticism about use of exogenous lipids and add substantial support to currently accepted models.

Other concerns exist about the addition of lipids to cells to study transmembrane signaling events. To mention only a few, these lipids are certain to have a perturbing effect on the fluidity and structural organization of the membrane, perhaps including asymmetric distribution, interruption of protein–protein interactions in the membrane, and changes in the permeability of the membrane. Such effects could be the dominant cause of activation or inhibition of a specific intracellular system, that is, receptorless mechanisms of transmembrane signaling may be invoked by the addition of lipids to the plasma membrane (R. Hollingsworth, personal communication). Certainly many studies must be done to clarify the precise mechanism by which intracellular effects are brought about by exogenous lipids.

Studies of the three-dimensional structures of the membrane-bound precursor phospholipids and their hydrolysis products will lead to new insights into the molecular interactions of the lipid second messengers with their target molecules and probably to the eventual development of powerful new treatment strategies in cancer and other diseases. Models of the palmitoyl derivatives of phosphatidylcholine and sphingomyelin are shown in Fig. 1. As known for some time (Kolesnick, 1991), the hydrophobic region of the Cer and DG portions are virtually identical, as are the polar head groups. At the interface of each lipid with the membrane surface, differences in the two structures are in the number and location of hydrogen bond donors and acceptors (NH and allylic OH groups) in the Cer portion of sphingomyelin, which could affect the physical properties of sphingomyelin and Cer relative to those of phosphatidylcholine and DG, and are assumed to be the structural determinants of the differential specificities of the two phospholipids and of the lipid second messengers derived from them.

A striking parallel also exists in the metabolic pathways of phosphoinositide, phosphatidylcholine, and sphingomyelin metabolism. As pointed out by Kan and Kolesnick (1993), all three lipids are hydrolyzed by membrane-bound phospholipases, leading to lipid second messengers that modulate specific protein kinases. Thus, as shown below, phosphoinositide and phosphatidylcholine are converted by specific phospholipase C to 1,2-diacylglycerol (DG), which is a physiological activator of protein kinase C:

phosphoinositide ⟶ IP_3 + DG

phospholipase C

phosphatidylcholine ⟶ P-choline + DG

sphingomyelin ⟶ P-choline + Cer

sphingomyelinase

The level of DG in the cell can be attenuated by phosphorylation, catalyzed by a regulated DG kinase, giving phosphatidic acid:

DG ⟶ phosphatidic acid

DG-specific kinase

Cer ⟶ ceramide 1-phosphate

Cer kinase

This pathway and current knowledge about its role in transmembrane signaling is reviewed in depth in Chapters 16 and 17.

A parallel pathway of sphingomyelin metabolism (shown earlier) is initiated by a membrane-bound neutral sphingomyelinase that appears to be activated by interaction with the complex of tumor necrosis factor (TNFα) and its receptor (Kan and Kolesnick, 1993). The product, Cer, is structurally similar to DG but the two effectors are produced as a result of different extracellular signals, and are metabolized further by separate kinases. Alternatively to the pathway to Cer 1-phosphate shown, Cer can be degraded by ceramidase to free sphingosine, which is well known to be a potent inhibitor of protein kinase C (see Chapters 14 and 15).

In summary, membrane-bound phospholipids and glycosphingolipids can be classified into several important families of signal-induction systems in which exquisite control of intracellular events results from extracellular hormones and other biological effectors. The research described in the following chapters serves to bring the reader to the current state of knowledge about this field, and to highlight areas in which future developments will be made.

References

Kan, C.-C., and Kolesnick, R. N. (1993). Signal transduction via the sphingomyelin pathway. *Trends Glycosci. Glycotechnol.* **5,** 99–106.

Kolesnick, R. N. (1991). Sphingomyelin and derivatives as cellular signals. *Prog. Lipid Res.* **30,** 1–38.

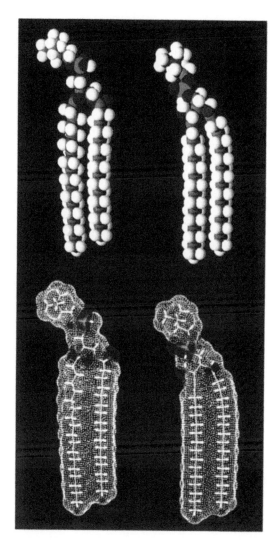

FIGURE 1. Molecular models of phosphatidylcholine (*left*) and sphingomyelin (*right*), shown in space-filling (*top*) and electron cloud (*bottom*) forms that reveal the three-dimensional similarity in these molecules. The –NH and allylic –OH groups of the sphingomyelin are shown clearly just below the yellow phosphorus atom.

FIGURE 1. Molecular models of proteins with important roles in allosteric regulation. Upper row is of spectra-filtered native protein. Lower is artificially denatured conformation. The same colors are used throughout the images. The left two images represent one configuration and the two at the right represent the complement.

CHAPTER 14

Sphingosine and Other Long-Chain Bases That Alter Cell Behavior

Alfred H. Merrill, Jr.
Department of Biochemistry, Emory University School of Medicine, Atlanta, Georgia 30322

I. INTRODUCTION

Sphingolipids comprise the free long-chain (sphingoid) bases, their simple derivatives, ceramides, phosphosphingolipids, neutral glycolipids, gangliosides, sulfatides, and many other compounds, including membrane anchors for proteins (for reviews, see Bell *et al.*, 1993). Considering the hundreds of possible head groups, at least 60 long-chain bases, and dozens of fatty acids found in sphingolipids, the number of different molecular species is likely to be in the tens of thousands, making sphingolipids the most structurally diverse class of lipids.

Sphingolipids help define the structural properties of membranes and lipoproteins; have important roles in cell–cell and cell–substratum interaction and in other forms of cell recognition, and help regulate cell growth and differentiation (Barenholz and Thompson, 1980; Hakomori, 1981,1986; Thompson and Tillack, 1985; Merrill, 1991; Zeller and Marchase, 1992; Bell *et al.*, 1993). The biochemical mechanisms by which information from sphingolipids, which are largely on the cell surface, is transmitted to intracellular targets have been difficult to elucidate. At one level, sphingolipids apparently interact with receptors that transmit signals via "conventional" signal transduction pathways (Hakomori, 1990; Zeller and Marchase, 1992). At another, sphingolipids and fragments of sphingolipids (such as sphingosine and ceramide) are the mediators (Hannun and Bell, 1989; Merrill, 1991), in analogy to the lipid second messengers that are formed from glycerolipids (diacylglycerols, unsaturated fatty acids, and their bioactive derivatives such as prostaglandins) (Dennis *et al.*, 1991). This chapter focuses on knowledge about bioactive long-chain bases and derivatives, and about recently discovered toxins and inhibitors that affect key steps of long-chain base metabolism.

II. BIOACTIVE LONG-CHAIN (SPHINGOID) BASES

A. Sphingosine, Sphinganine, and Related Long-Chain Bases

Since the discovery of sphingosine by Thudichum (1884), long-chain bases have been found to be a large group of structurally related compounds (Karlsson, 1970). Sphingosine (*trans*-4-sphingenine), the prevalent long-chain base of many mammalian sphingolipids, has 18 carbon atoms; the structure is shown in Fig. 1. Variations include longer or shorter alkyl chain length(s) (sometimes with branching), other types of unsaturation, and a hydroxyl group at position 4 (Fig. 1; Karlsson, 1970). The major derivatives are the 1-phosphate, *N*-methyl, and *N*-acyl derivatives (including ceramides and more complex sphingolipids).

Interest in long-chain bases was rekindled by the discovery that sphingosine inhibits protein kinase C *in vitro* and in intact cells (Hannun *et al.*, 1986; Merrill *et al.*, 1986; Wilson *et al.*, 1986), raising the possibility that cells might utilize hydrolysis products of sphingolipids as another type of "lipid second messenger." Since these initial discoveries, over 100 different cellular systems have been found to be affected by these compounds. This review highlights the systems that are likely to be regulated by sphin-

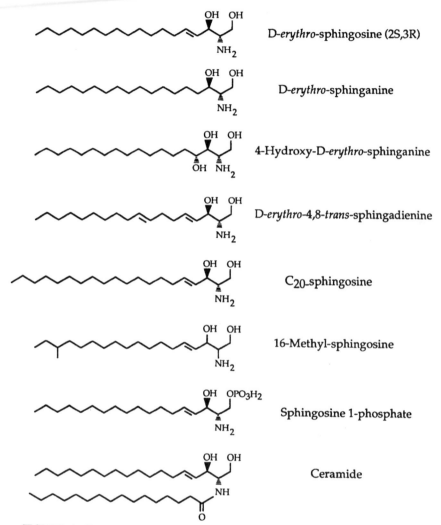

FIGURE 1 Examples of long-chain (sphingoid) bases found in animal tissues. The stereochemistry is shown of sphingosine (*trans*-4-sphingenine) and the other common long-chain bases—sphinganine (dihydrosphingosine) and 4-D-hydroxysphinganine (phytosphingosine) and their derivatives (sphingosine 1-phosphate and ceramide)—as well as of examples of less frequently encountered chain length homologs and of compounds with additional double bonds.

gosine *in vivo* (based on sensitivity to inhibition by exogenous sphingosine) as well as a few that exhibit particularly interesting features.

Sphingosine inhibits protein kinase C competitively with diacylglycerol, phorbol dibutyrate, and Ca^{2+} and blocks activation by unsaturated fatty acids and other lipids (Hannun *et al.*, 1986; Wilson *et al.*, 1986; Oishi *et al.*, 1988; El Touny *et al.*, 1990). The inhibition depends on the mole percent of sphingosine with respect to the other lipids or detergents. The exact mechanism of inhibition is not known. The positive charge of sphingosine is likely to be involved (Hannun *et al.*, 1986; Hannun and Bell, 1989; Bottega *et al.*, 1989; Merrill *et al.*, 1989); therefore, long-chain bases may interact with the acidic lipids that are required by protein kinase C (Hannun *et al.*, 1986; Bazzi and Nelsestuen, 1987; Merrill *et al.*, 1989). Note that sphingosine has a lower pK_a than do simple alkylamines; thus, sphingosine exists in both the neutral and the ionized form at physiological pH (Merrill *et al.*, 1989).

Structure–function studies (Merrill *et al.*, 1989) revealed that the most potent inhibitors have an alkyl chain length of 18 carbon atoms; subtle differences exist among the four stereoisomers of sphingosine, *N*-methyl derivatives, and simpler alkylamines (e.g., stearylamine). Because inhibition of protein kinase C is not very stereospecific, protein kinase C may not have a specific binding site for long-chain bases; however, drawing this conclusion is premature because all four stereoisomers provide the same head group conformation and vary only in the position of the alkyl chain (Merrill *et al.*, 1989).

Sphingosine and other alkylamines are potent inhibitors of phosphatidic acid phosphohydrolase (Jamal *et al.*, 1991; Mullmann *et al.*, 1991; Aridor-Piterman *et al.*, 1992), which forms diacylglycerols (and alkylacylglycerols) in glycerolipid biosynthesis and turnover via the phospholipase D pathway. Studies with human neutrophils revealed that sphingosine is about 10-fold more potent as an inhibitor of phosphatidic acid phosphohydrolase than as an inhibitor of protein kinase C (Perry *et al.*, 1992). Long-chain bases appear to be promising as inhibitors of neutrophil activation, blocking superoxide formation (Wilson *et al.*, 1986) but not oxygen-independent antimicrobial action (Stinavage and Spitznagel, 1989). Sphingosine affects other aspects of glycerolipid metabolism—it activates a phosphatidylethanolamine-specific phospholipase D (Kiss *et al.*, 1991) and activates (whereas sphingomyelin inhibits) phospholipase C delta (Pawelczyk and Lowenstein, 1992).

Long-chain bases also have been reported to affect several ion transport systems. Of particular interest is the finding that sphingosine and sphingosylphosphorylcholine induce calcium release from intracellular stores in smooth muscle cells permeabilized with saponin (Ghosh *et al.*, 1990).

Sphingosine and sphingosine 1-phosphate additionally stimulate increases in intracellular calcium in Swiss 3T3 cells (Zhang *et al.*, 1991); these observations have led to the hypothesis that a sphingosine metabolite—possibly sphingosine 1-phosphate—might act as a regulator of calcium mobilization (see subsequent discussion). In contrast, sphingosine is an inhibitor of sarcoplasmic reticulum calcium release in response to caffeine, doxorubicin, and other agents (Sabbadini *et al.*, 1992). Sphingosine inhibits Na^+K^+-ATPase with approximately the same dose response as protein kinase C inhibition (Oishi *et al.*, 1990). The effects of long-chain bases on ion homeostasis are clearly complex.

Sphingosine also has been shown to *activate* some protein kinases. Davis and co-workers (Faucher *et al.*, 1988) demonstrated that sphingosine inhibits protein kinase C, but additionally increases the phosphorylation at Thr 669 (the MAP kinase site) of the epidermal growth factor (EGF) receptor through a pathway that appears to be independent of protein kinase C. The phosphorylation of Thr 669 may be the result of the formation of ceramide and activation of a ceramide-activated kinase (Goldkorn *et al.*, 1991); nonetheless, sphingosine and *N*-methylated long-chain bases also directly induce phosphorylation on Thr 669. Sphingosine increases the activity of the cytoplasmic tyrosine kinase domain of the EGF receptor to equal or greater than that of the ligand-activated holo-EGF receptor (Wedegaertner and Gill, 1989).

Pushkareva *et al.* (1992) have reported that D-*erythro*-sphingosine induces the phosphorylation of a number of cytosolic proteins in Jurkat T cells. Several sphingosine-activated protein kinases appear to exist because the concentration dependence on sphingosine varied among the endogenous substrates and differences were seen in utilization of ATP and GTP. The *erythro*-sphinganines were less active than sphingosine, and *threo*-sphinganine was not active.

Sphingosine induces dephosphorylation of the retinoblastoma gene product (Chao *et al.*, 1992), a nuclear phosphoprotein that is thought to function as a tumor suppressor. As noted by the authors, the phosphorylated forms of retinoblastoma protein are observed in S and G_2/M phases of the cell cycle, whereas the less phosphorylated forms are found in G_0/G_1. Therefore, sphingosine may modulate phosphorylation and dephosphorylation of retinoblastoma protein and progression through the cell cycle. For most of the cells that have been studied to date, sphingosine is growth inhibitory and cytotoxic. For Chinese hamster ovary (CHO) cells, inhibition of protein kinase C appears to be involved based on concentration dependence, structural specificity, protein phosphorylation patterns, and apparent *lack* of perturbation of cellular acidic compartments—the most likely site of an artifactual effect of these types of compounds (Merrill *et al.*, 1989; Stevens *et al.*, 1990a). Interestingly, cells have a mechanism for expulsion of organic cations (which apparently is involved in multidrug resistance) (Dellinger *et al.*, 1992); perhaps a physiological function of this protein is preventing long-chain bases from being toxic to cells. As discussed in Section II,C, sphingosine is mitogenic for some cells (Zhang *et al.*, 1990); it also reverses growth inhibition caused by protein kinase C activation in vascular smooth muscle cells (Weiss *et al.*, 1991).

Virus replication, another type of "growth," also is affected by sphingosine, which inhibits the induction of DNA polymerase and DNase activities in Epstein–Barr virus-infected cells treated with phorbol esters and *n*-butyrate (Nutter *et al.*, 1987). Phorbol esters alter the sphingosine levels in Epstein–Barr virus-transformed B lymphocytes (Miccheli *et al.*, 1991). Interestingly, Van Veldhoven *et al.*, (1992) have found that ceramide, but not sphingosine, is increased in HIV-infected cells.

Considering the large number of cellular systems that can be affected by exogenously added sphingosine, one would predict that this compound is utilized as a regulator of cell functions. In contrast to ceramide, however, no studies have linked agonist-induced changes in the cellular amounts of sphingosine conclusively with a biological response. A provacative suggestion has been made that, because exogenous long-chain bases inhibit insulin-stimulated hexose transport and glucose oxidation (Robertson *et al.*, 1989), dexamethasone may inhibit glucose uptake by cells by increasing sphingosine levels (Nelson and Murray, 1989).

All cells examined to date contain on the order of 1–10 nmol free sphingosine per gm tissue, wet weight (10–100 pmol/10^6 cells) (for examples, see Merrill *et al.*, 1986, 1988; Kobayashi *et al.*, 1988b; Wilson *et al.*, 1988; Van Veldhoven *et al.*, 1989; Wertz and Downing *et al.*, 1989). These concentrations are about one-tenth the level that inhibits protein kinase C when added exogenously. Local concentrations of endogenous sphingosine might be higher because most sphingosine appears to be located in plasma membranes (Slife *et al.*, 1989). Studies with intact cells (Wilson *et al.*,1988) and isolated rat liver plasma membranes (Slife *et al.*, 1989) found that endogenous sphingolipids are hydrolyzed to free sphingosine, in the latter case via turnover of sphingomyelin by a Mg^{2+}-dependent sphingomyelinase and ceramidase. What factors determine the extent to which ceramides are cleaved to sphingosine is unclear.

In addition to these systems, sphingosine and related analogs are inhibitory for the *de novo* biosynthesis pathway for sphingolipids (Van Echten *et al.*, 1990), which may influence the usefulness of sphingosine analogs as pharmaceuticals.

B. Lysosphingolipids

Hannun and Bell (1987) reported that lysosphingolipids are inhibitory for protein kinase C. Because this group includes compounds (such as psychosine or galactosylsphingosine) that accumulate in tissues of patients with some sphingolipidoses (such as Krabbe's disease) (Kobayashi *et al.*, 1988a), accumulation of lysosphingolipids might contribute to the pathogenesis of these disorders. This hypothesis remains appealing, especially considering that additional targets have been found to be affected by long-chain bases. The list of known lysosphingolipids continues to expand, with the discovery of a new type of "psychosine"-like sphingolipid (with a galactosyl head group with 3,4- and 4,6-linked cyclic acetal linkages) (Nudelman, 1992) in extracts of human brain. Vartanian *et al.* (1989) have reported that galactosylsphingosine inhibits the phosphorylation of myelin basic protein and induces morphological changes in oligodendrocytes by mechanisms that are both protein kinase C-dependent and -independent. Additional work to test these hypotheses should be feasible because galactosylsphingosine has been observed to accumulate in one of the animal models for a metabolic neurological disorder, the Twitcher mouse (Shinoda *et al.*, 1987).

C. Sphingosine 1-Phosphate

Spiegel and co-workers have shown that sphingosine and sphingosine 1-phosphate stimulate proliferation of quiescent Swiss 3T3 fibroblasts (Zhang *et al.*, 1990), and have suggested that this behavior may be mediated through release of intracellular calcium (Zhang *et al.*, 1991) and increases in phosphatidic acid levels (Desai *et al.*, 1992). Studies in permeabilized cells also have implicated sphingosine 1-phosphate in release of calcium from intracellular stores (Ghosh *et al.*, 1990). Other work has found sphingosine 1-phosphate to inhibit cell motility and phagokinesis of B16 melanoma cells (Sadahira *et al.*, 1992). These findings underscore the likely role of sphingosine 1-phosphate as another long-chain base "second messenger." An early indication that the 1-phosphate derivative might have bioactivity was that administration of 2–10 μmol 1-phosphonate (which cannot be cleaved by phosphatases) to rats was lethal within 1 min (Stoffel and Grol, 1974).

Sphingolipids are turned over by hydrolysis to ceramides, then to the long-chain bases (Fig. 2), which are phosphorylated (Hirschberg *et al.*, 1970; Stoffel *et al.*, 1970; Stoffel and Bister, 1973; Louie *et al.*, 1976;

Buehrer and Bell, 1992) and cleaved in an aldolase-like reaction to ethanol-amine phosphate and *trans*-2-hexadecenal, as catalyzed by sphingosine 1-phosphate lyase (Stoffel *et al.*, 1969; Van Veldhoven and Mannaerts, 1991). Although initial reports suggested that the kinase phosphorylates *erythro* and *threo* long-chain bases, Buehrer and Bell (1992) have found it to be active only with *erythro* long-chain bases and to be inhibited by the *threo* stereoisomers. Thus, *threo* long-chain bases can be used to block phosphorylation of sphingosine in platelets (Buehrer and Bell, 1992).

D. N-Methyl Sphingosines

N-methyl derivatives inhibit protein kinase C (Merrill *et al.*, 1989). One report claimed commercial sphingosine is inhibitory only because it is contaminated with a small amount of the *N,N*-dimethyl derivative (Igarashi *et al.*, 1989). However, this possibility is clearly not the case because chemically synthesized D-*erythro*-sphingosine inhibits protein kinase C (Merrill *et al.*, 1989). *N,N,N*-Trimethyl-sphingosine also inhibits protein kinase C *in vitro*, but is less effective when added to intact cells (Merrill *et al.*, 1989), probably because the positive charge slows movement across the bilayer. *N*-Methylated sphingosines induce phosphorylation of the EGF receptor (Igarashi *et al.*, 1990; Goldkorn *et al.*, 1991), down-regulation of GMP-140 expression in platelets (Hanada *et al.*, 1991), and the oxidative burst and migration of human neutrophils (Kimura *et al.*, 1992). *N*-Methyl-sphingosines inhibit *in vitro* and *in vivo* growth of tumor cells in nude mice (Endo *et al.*, 1991) and inhibit metastasis (Okoshi *et al.*, 1991).

Some of the enhanced activity of the *N*-methyl-sphingosines *in vivo* is likely to be the result of their resistance to metabolism by *N*-acylation and, perhaps, phosphorylation. Relatively little is known about the factors that influence the cellular methylation of sphingosine. Felding-Habermann *et al.* (1990) have found that diversion of ceramides from their normal metabolism using a glycosyltransferase inhibitor (PDMP) enhances *N,N*,-dimethylsphingosine synthesis.

E. Ceramides

Investigations by Hannun and co-workers have discovered a sphingomyelin–ceramide cycle that meets the criteria of a lipid second messenger pathway (Fig. 2). The initial observations (Okazaki *et al.*, 1989,1990) were that treatment of HL-60 cells with 1-α, 25-dihydroxyvitamin D_3 to induce

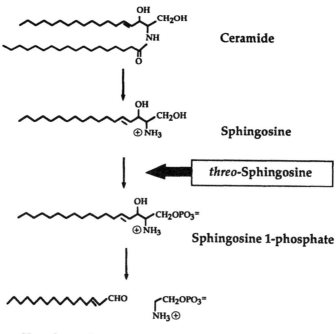

FIGURE 2 Ceramide turnover. After hydrolysis of complex sphingolipids to ceramide, the fatty acid is removed by ceramidases to release sphingosine. Sphingosine is phosphorylated to sphingosine 1-phosphate by sphingosine kinase (which is inhibited by *threo*-sphingosines), followed by cleavage to *trans*-2-hexadecenal and ethanolamine phosphate.

differentiation resulted in hydrolysis of a significant portion of the cellular sphingomyelin to ceramide and phosphorylcholine. This response, which was maximal after approximately 2 hr, was attributed to activation of a neutral sphingomyelinase. By 4 hr, the amounts of sphingomyelin and ceramide had returned to basal levels. Treatment of the cells with a bacterial sphingomyelinase, or addition of a cell-permeant ceramide (*N*-acetylsphingosine), resulted in similar changes in cell phenotype (Okazaki *et al.*, 1989). This sphingomyelin cycle appears to be involved in the action of at least two other factors that induce differentiation of HL-60 cells: tumor necrosis factor (TNF) and interferon γ (Kim *et al.*, 1991). Other agents that induce differentiation (such as retinoic acid), but to granulocytes rather than to monocytes, did not induce sphingomyelin turnover to ceramide. However, in another study, retinoic acid increased ceramide levels in GH4C1 cells (Kalen *et al.*, 1992).

Ceramide has been found to activate a serine–threonine protein phosphatase (Dobrowsky and Hannun, 1992). The characteristics of the ceramide-activated protein phosphatase (which has been abbreviated CAPP) are similar to those of the subgroup 2A protein phosphatases (i.e., cation independent and sensitive to inhibition by okadaic acid). CAPP is activated by a variety of ceramides with different hydrophobic moieties; however, studies with N-acyl phenylaminoalcohol analogs indicate that CAPP is stereospecific (Bielawska et al., 1992). Fluorescent ceramides exhibit selectivity in the intracellular movement of these compounds (Lipsky and Pagano, 1983; Pagano and Martin, 1988), a feature that may be important not only to the biosynthesis and turnover of cellular membranes but also to the second messenger properties of these compounds.

Kolesnick and co-workers have shown that ceramide also activates protein kinase(s). These studies arose from investigation of the mechanism of sphingosine activation of the EGF receptor (Goldkorn et al., 1991), and demonstrated that ceramide was able to induce phosphorylation of the receptor. Therefore, at the concentrations of sphingosine that caused activation of the receptor, these researchers hypothesized that this metabolite of sphingosine actually is active. Subsequent studies (Mathias et al., 1991) utilizing a polypeptide substrate (representing amino acids 663–681 of the EGF receptor) found a Mg^{2+}-dependent ceramide-activated kinase activity in the membrane fraction from A431 cells. Activation of this kinase activity has been demonstrated in a cell-free system (Dressler et al., 1992) in which postnuclear supernatants from HL-60 cells were treated with TNF to induce sphingomyelin turnover to ceramide, and could be mimicked by adding exogenous sphingomyelinase rather than TNF to the preparation.

These studies appear to show that multiple sphingomyelinases may be involved in this signaling pathway (Kolesnick, 1991). A neutral sphingomyelinase activity has been known to exist for some time (Rao and Spence, 1976), but its function has been unknown. This activity may be involved, but the activity that Hannun's group finds to be activated by TNF appears to be cytosolic (Y. A. Hannun, personal communication). Schutze et al. (1992) have suggested that TNF activates NF-κB by a phosphatidylcholine-specific phospholipase C-induced acidic sphingomyelinase, which may be similar to the activity that is increased by diacylglycerol but not phorbol ester (Kolesnick and Clegg, 1988).

In addition to these targets, ceramide may affect signal transduction by inhibition of diacylglycerol kinase (Younes et al., 1992), a result that is not too surprising because the bacterial kinase is able to phosphorylate both diacylglycerols and ceramides. Mammalian cells have a separate ceramide kinase (Bajjalieh et al., 1989; Kolesnick and Hemer, 1990). Little is known about the biological activity (if any) of ceramide phosphates; however, some bacteria and the brown recluse spider produce a sphingomyelinase D (which forms ceramide phosphate). Injection of this enzyme causes a prolonged inflammatory response (Rees *et al.*, 1984).

F. Ceramide-Like Compounds

Radin's group has prepared a series of ceramide-like compounds based on the structure shown in Fig. 3, with the goal of producing inhibitors of glucosylceramide synthesis (Vunnam and Radin, 1980; Inokuchi and Radin, 1987). The most active compound was found to be D-*threo*-1-phenyl-2-decanoylamino-3-morpholine-1-propanol (PDMP), although subsequent work has found that this compound also blocks sphingomyelin synthesis when added at high concentrations (Rosenwald *et al.*, 1992).

PDMP has a wide range of interesting effects on cells, including growth inhibition (Shayman *et al.*, 1991), inhibition of adherence during differentiation of HL-60 cells (Kan and Kolesnick, 1992), and antitumor activity (Inokuchi *et al.*, 1987,1989,1990). PDMP is relatively nontoxic *in vivo*, although administration to mice causes poor growth of kidneys and liver, with the kidney being the more sensitive organ (Shukla *et al.*, 1991). Studies with Madin–Darby canine kidney cells (Shayman *et al.*, 1991) found that PDMP causes ceramide accumulation and increases in free sphingosine. *In vivo*, PDMP is converted to a number of more polar compounds, apparently via cytochrome P450-type reactions (Shukla and Radin, 1991).

III. INHIBITORS OF LONG-CHAIN BASE BIOSYNTHESIS PRODUCED BY MOLDS AND OTHER ORGANISMS

Several novel sphingosine derivatives have been isolated from the sponge *Penares* sp. (Kobayashi *et al.*, 1991; Fig. 4). These molecules are

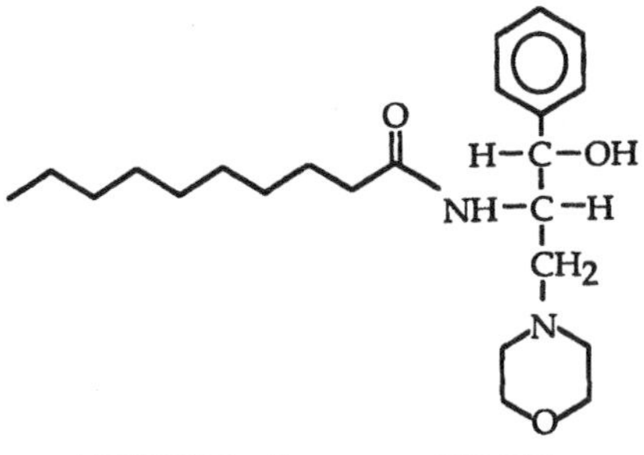

FIGURE 3   Structure of PDMP.

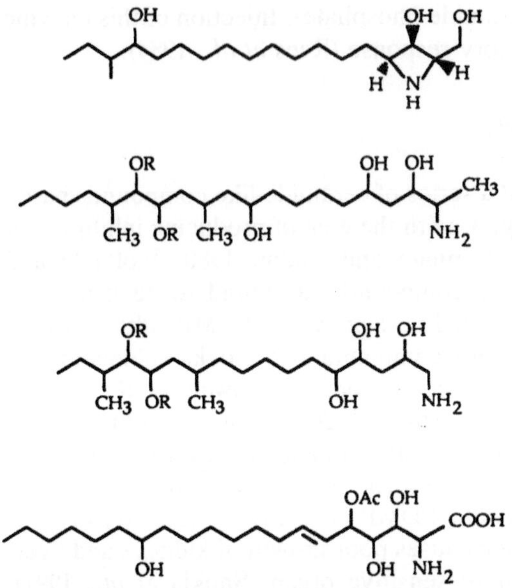

Penaresidin A

Fumonisin B_1
(R = $COCH_2CH(COOH)CH_2COOH$)

Alternaria toxin
(R = $COCH_2CH(COOH)CH_2COOH$)

Sphingofungin C

FIGURE 4 Sphingosine-like compounds produced by sponges (penaresidin A) and molds (fumonisins, *Alternaria* toxin, and sphingofungins).

interesting because they are potent activators of actomyosin ATPase and inhibitors of protein kinase C (Alvi, 1992). Further, molds produce a variety of structurally related compounds that disrupt sphingolipid metabolism, for example, the fumonisins (Wang *et al.*, 1991), *Alternaria* toxins (Merrill, 1993), and sphingofungins (Zweerink *et al.*, 1992; Fig. 4). These compounds, with PDMP and the synthetic stereoisomers, allow disruption of long-chain base biosynthesis (Fig. 5) and catabolism (Fig. 2) and evaluation of the consequences.

A. Fumonisins

Fusarium moniliforme (Sheldon) is one of the most prevalent molds on corn, sorghum, and other grains throughout the world. Consumption of *F. moniliforme*-contaminated corn causes several diseases of agricultural concern (e.g., equine leukoencephalomalacia and porcine pulmonary edema) (Kriek *et al.*, 1981; Marasas *et al.*, 1988; Ross *et al.*, 1990), has been correlated with human esophageal cancer (Lin *et al.*, 1980; Yang, 1980; Marasas, 1982), and results in hepatotoxicity and liver cancer in rats (Marasas *et al.*, 1984; Gelderblom *et al.*, 1988,1991). The agents thought to be responsible for these diseases constitute a group of mycotoxins termed fumonisins (Bezuidenhout et al., 1988). (The most prevalent compound, fumonisin B1, is shown in Fig. 4).

Because the fumonisins bear a remarkable structural similarity to sphinganine, we hypothesized that they might act by disrupting sphingolipid metabolism. Fumonisins were found (Wang et al., 1991) to inhibit the incorporation of [^{14}C]serine into [^{14}C]sphingosine by hepatocytes, with an IC_{50} of approximately 0.1 μM; the effects were selective because no reduction occurred in the radiolabeling or mass of fatty acids, phosphatidylserine, phosphatidylethanolamine, or phosphatidylcholine. The primary target of the fumonisins is ceramide synthase, as depicted in Fig. 2. The structural basis for this inhibition is unknown. However, one can speculate that similarities between the fumonisins and long-chain (sphingoid) bases allow them to be recognized as substrate (or transition state or product) analogs by ceramide synthase.

As predicted for inhibition at this site, fumonisins cause sphinganine to accumulate; indeed, the mass of sphinganine increased 110-fold when hepatocytes were treated with fumonisin B_1 for 4 days (Wang et al., 1991). This result is provocative because this cellular level of sphinganine has been shown to inhibit protein kinase C (Wilson et al., 1988). In fact, disruption of this pathway could account for the toxicity and carcinogenicity of fumonisins by a number of mechanisms, including the following:

1. The accumulation of sphinganine might be the major factor because long-chain bases are both toxic (Stevens et al., 1990a) and mitogenic at low concentrations (Zhang et al., 1990).
2. The accumulation of sphinganine may lead to formation of the 1-phosphate, which is another bioactive molecule.
3. The inhibition of the formation of ceramides and more complex sphingolipids may be major factors because these molecules are involved in the regulation of growth, differentiation, and cell function—both as determinants of membrane structure and as ligands for extracellular matrix proteins, as regulators of growth factor receptors, and so on.

Additional studies using LLC-PK1 cells have related the accumulation of sphinganine to the inhibition of cell growth and, ultimately, to cell death (Norred et al., 1992; Yoo et al., 1992). To determine whether these in vitro observations are relevant in vivo, the amounts of sphingosine, sphinganine, and total sphingolipids were measured using serum from ponies that had been given fumonisin-contaminated feed. A large increase in sphinganine and sphingosine occurred, as did a reduction in the amounts of complex sphingolipids in serum (Wang et al., 1992). Similar findings have been obtained subsequently in feeding studies with pigs, rats, chickens, and turkeys.

B. Alternaria Toxin

The presence of a sphinganine-like backbone is shared by a number of other mycotoxins, such as the host-specific phytotoxins produced by another fungus *Alternaria alternata* f. sp. *lycopersici,* which produces the macroscopic symptoms of stem canker disease in tomatoes (Bottini *et al.,* 1981). These compounds are similar to fumonisins, but lack the methyl group at position 1, and have been shown to be cytotoxic by *in vitro* assays of various mammalian cells in culture, with IC_{50} ranging from 2- to 7-fold higher than that of fumonisin B_1 (Shier *et al.,* 1991). We find that *Alternaria* toxin is also an inhibitor of sphingolipid biosynthesis with an IC_{50} of approximately 1 μM, which is about 10-fold higher than the IC_{50} for fumonisin B_1. How common this structural motif is among other fungi, plants, and organisms is not known. However, considering the fact that it is encountered in organisms as different as the *Fusarium* and *Alternaria* molds, assuming that additional naturally occurring inhibitors of this pathway are produced by other organisms is reasonable.

C. Sphingofungins

The initial precursors of *de novo* sphingolipid biosynthesis are palmitoyl CoA and serine, which are combined to form 3-ketosphinganine by serine palmitoyltransferase (Fig. 5). Zweerink *et al.* (1992) have discovered a class of compounds produced by molds (sphingofungins) (Fig. 4) that inhibit serine palmitoyltransferase. This enzyme has been shown to be critical for cell survival by work with yeast cells with a defective enzyme (selected by auxotrophy for exogenously provided long-chain bases) (Dickson *et al.,* 1990) and by studies with a temperature-sensitive enzyme in CHO cells (Hanada *et al.,* 1990).

D. Cycloserine and Mechanistically Related Compounds

As a pyridoxal 5'-phosphate-dependent enzyme, serine palmitoyltransferase is sensitive to many of the compounds that react with this cofactor as active-site directed (suicide) inhibitors, including natural products (cycloserine; Sundaram and Lev, 1984; Holleran *et al.,* 1990) and synthetic compounds (the beta-haloalanines; Medlock and Merrill, 1988; Holleran

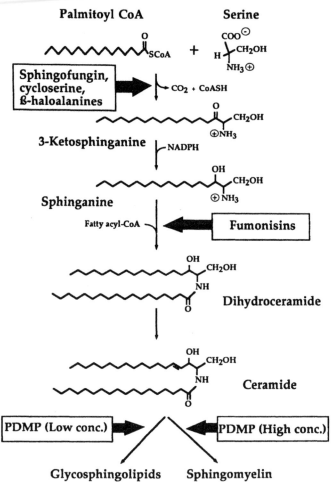

FIGURE 5 Sphingolipid biosynthesis and sites of action of inhibitors. *De novo* sphingo-lipid biosynthesis begins in the endoplasmic reticulum with the condensation of serine and palmitoyl CoA by serine palmitoyltransferase (inhibited by cycloserine, beta-haloalanines, and sphingofungins). The 3-ketosphinganine is reduced to sphinganine, which is acylated to dihydroceramides by ceramide synthase (inhibited by fumonisins). The 4-*trans* double bond of sphingosine is added at some later step. Glycosylation of ceramide is inhibited by the synthetic compound PDMP at low concentrations; at higher concentrations, sphingomyelin synthesis is blocked also.

et al., 1990). Long-term administration of L-cycloserine to mice had little effect on the levels of sphingomyelin, sulfatides, or gangliosides but re-duced cerebroside levels in brain (Sundaram and Lev, 1989). Other ways of modulating the activity of this enzyme have been devised by genetic means (Dickson *et al.*, 1990; Hanada *et al.*, 1990,1992).

IV. LONG-CHAIN BASES AND SIGNAL TRANSDUCTION

Considering the large number of bioactive long-chain bases and the multiple pathways for their formation (Figs. 2,5,6), these molecules are logical candidates for another class of lipid second messengers. Ceramides could be released from a large number of more complex sphingolipids, providing considerable potential for specificity among different cell types and even within domains of a single membrane. Some of the ceramide can be hydrolyzed to sphingosines and other long-chain base derivatives, which affect very diverse cell behaviors.

The bioactive long-chain bases apparently have "specific" effects on signal transduction through direct interaction with binding sites on proteins. These molecules may have equally important effects on cell regulation by modulating the charge of the membrane surface. These field effects can be very powerful in regulating the properties of proteins and other biological processes, but would not be defined as "specific" according to the current definition (i.e., the effects could be mimicked by other stereoisomers and compounds with very different structures provided they could interact with the membranes in a similar way). In this context, note that long-chain (sphingoid) bases are the only cationic lipids in mammalian cell membranes; therefore, many of the systems that are affected by these compounds would not appear to need to have developed a high degree of structural specificity.

To date, only one case appears to require sphingolipids to meet all the criteria of a second messenger pathway—the sphingomyelin–ceramide cycle. However, relatively few studies of ceramide and long-chain base metabolism have been conducted. The biological processes that are controlled by sphingolipids (such as growth and differentiation) are relatively difficult to study.

The findings with fumonisins and *Alternaria* toxins have shown that these mycotoxins exert their toxicity, and perhaps their carcinogenicity, through disruption of the conversion of long-chain (sphingoid) bases to ceramides. This observation establishes the importance of normal long-chain base metabolism for survival. Other examples of defective acylation of sphingosine that are associated with disease include the reports by Boiron-Sargueil *et al.* (1992) that Trembler mice have very low levels of ceramide synthase and by Goldin *et al.* (1992) that fibroblasts from patients with Niemann–Pick Type C disease have unusually high amounts of sphinganine (which usually only appears in significant amounts as an intermediate of sphingolipid synthesis). Further, Kendler and Dawson (1992) also have found that hypoxic injury to rat oligodendrocytes causes ceramide to accumulate in the endoplasmic reticulum. The findings about the roles of long-chain bases and their derivatives in pathophysiology probably

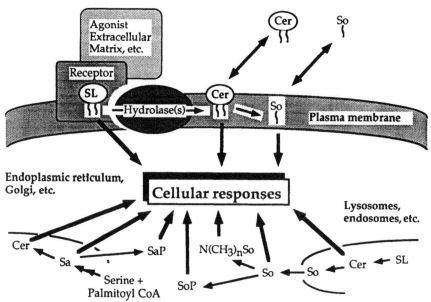

FIGURE 6 A scheme outlining steps through which cells encounter bioactive long-chain bases. The pathways of Figs. 2 and 3 have been redrawn to focus on the intracellular location of the biosynthetic and catabolic reactions that yield bioactive long-chain bases and derivatives such as sphingosine and sphinganine, the 1-phosphates, N-methyl-sphingosines, and (dihydro)ceramides. (Additional sites may exist for some of these reactions.) Sphingosine and ceramide also can be taken up by cells, although ceramides are difficult to deliver to cells unless the alkyl chain(s) have been modified to increase water solubility. As described in the text, many targets exist for these compounds.

will provide hints about the functions of these molecules in normal cell regulation.

Another aspect that warrants consideration is whether certain cell behaviors are affected *in vivo* by exposure to exogenous sphingolipids from dietary sources, cosmetics, or pharmaceutical formulations. As an inhibitor of protein kinase C, sphingosine and other long-chain bases might be antitumor agents (Hannun *et al.*, 1986). Topical application of sphingosine blocked the induction of ornithine decarboxylase by phorbol esters in mouse skin (Gupta *et al.*, 1988; Enkvetchakul *et al.*, 1989), one biochemical marker of tumor promotion. However, sphingosine did not reduce the number of tumors in a longer term study (Enkvetchakul *et al.*, 1992), although skin is already rich in sphingosine (Wertz and Downing, 1989) which might complicate such studies.

In a cell culture model of transformation (mouse C3H 10T1/2 cells) (Borek *et al.*, 1991), sphingosine and sphinganine reduced cell transformation in response to gamma irradiation and phorbol 12-myristate

13-acetate (PMA). A carcinogenesis study with mice treated with *N,N*-dimethylhydrazine (DMH) to induce colon tumors (Dillehay *et al.*, 1994) showed that including milk sphingomyelin in the diet reduced the number of aberrant colonic crypts in short-term studies and decreased tumor incidence by approximately one-third. Long-chain bases also have been found to induce differentiation and, when used in combination with other agents that induce differentiation, to result in cells with a more fully differentiated phenotype (Stevens *et al.*, 1989,1990b; Yung *et al.*, 1992). Sphingosines, *N*-methylated sphingosines, and sphingosine 1-phosphate reduce tumor growth and metastasis (Sadahira *et al.*, 1992). Collectively, these findings suggest that long-chain bases still offer promise in the prevention and/or treatment of cancer. Based on the diverse signal transduction pathways affected by sphingolipids, one can envision that these compounds—or, more likely, new compounds based on the major pharmacophors of the bioactive sphingolipids—will prove useful in the prevention and treatment of a wide range of diseases.

Acknowledgments

The author gratefully acknowledges the contributions of his collaborators and the authors of other works cited in this review to the ideas that have been presented. Work by the author's laboratory that was cited in this review was supported by NIH Grants GM33369 and GM46368, USDA Grant 92022509, and funds from the National Dairy Council.

References

Alvi, K. A., Palmer, W., and Crews, P. (1992). Protein kinase C inhibitory alkaloids from the marine sponge *Penares sollasi*. Abstracts of the 33rd Annual Meeting of the American Society of Pharmacognosy, July 26–31, Williamsburg, Virginia.

Aridor-Piterman, O., Lavie, Y., and Liscovitch, M. (1992). Bimodal distribution of phosphatidic acid phosphohydrolase in NG108-15 cells. Modulation by the amphiphilic lipids oleic acid and sphingosine. *Eur. J. Biochem.* **204**, 561–568.

Bajjalieh, S. M., Martin, T. F. J., and Floor, E. (1989). Synaptic vesicle ceramide kinase. A calcium-stimulated lipid kinase that co-purifies with brain synaptic vesicles. *J. Biol. Chem.* **264**, 14354–14360.

Barenholz, Y., and Thompson, T. E. (1980). Sphingomyelins in bilayers and biological membranes. *Biochim. Biophys. Acta* **604**, 129–158.

Bazzi, M. D., and Nelsestuen, G. L. (1987). Mechanism of protein kinase C inhibition by sphingosine. *Biochem. Biophys. Res. Commun.* **146**, 203–207.

Bell, R. M., Hannun, Y. A., and Merrill, A. H., Jr., eds. (1993). "Advances in Lipid Research: Sphingolipids and Their Metabolites," Vol. 25–26. Academic Press, Orlando, Florida.

Bezuidenhout, C. S., Gelderblom, W. C. A., Gorstallman, C. P., Horak, R. M., Marasas, W. F. O., Spiteller, G., and Vleggaar, R. (1988). Structure elucidation of the fumonisins, mycotoxins from *Fusarium moniliforme*. *J. Chem. Soc. Commun.* 743–745.

Bielawska, A., Linardic, C. M., and Hannun, Y. A. (1992). Ceramide-mediated biology: Determination of structural and stereospecific requirements through the use of *N*-acylphenylaminoalcohol analogs. *J. Biol. Chem.* **267**, 18493–18497.

Boiron-Sargueil, F., Heape, A., and Cassagne, C. (1992). Lack of in vitro ceramide formation in the PNS of Trembler mice. *J. Neurochem.* **59,** 652–656.

Borek, C., Ong, A., Stevens, V. L., Wang, E., and Merrill, A. H., Jr. (1991). Long-chain (sphingoid) bases inhibit multistage carcinogenesis in mouse C3H/10T1/2 cells treated with radiation and phorbol 12-myristate 13-acetate. *Proc. Natl. Acad. Sci. U.S.A.* **88,** 1953–1957.

Bottega, R., Epand, R. M., and Ball, E. H. (1989). Inhibition of protein kinase C by sphingosine correlates with the presence of positive charge. *Biochem. Biophys. Res. Commun.* **164,** 102–107.

Bottini, A. T., Bowen, J. R., and Gilchrist, D. G. (1981). Phytotoxins II. Characterization of a phytotoxic fraction from *Alternaria alternata* f. sp. *lycopersici. Tetr. Lett.* **22,** 2723–2726.

Buehrer, B. M., and Bell, R. M. (1992). Inhibition of sphingosine kinase in vitro and in platelets. Implications for signal transduction pathways. *J. Biol. Chem.* **267,** 3154–3159.

Chao, C., Khan, W., and Hannun, Y. A. (1992). Retinoblastoma protein dephosphorylation induced by D-*erythro*-sphingosine. *J. Biol. Chem.* **267,** 23459–23462.

Dellinger, M., Pressman, B. C., Calderon-Higginson, C., Savaraj, N., Tapiero, H., Kolonias, D., and Lampidis, T. J. (1992). Structural requirements of simple organic cations for recognition by multidrug-resistant cells. *Cancer Res.* **52,** 6385–6389.

Dennis, E. A., Rhee, S. G., Billah, M. M., and Hannun, Y. A. (1991). Role of phospholipases in generating lipid second messengers in signal transduction. *FASEB J.* **5,** 2068–2077.

Desai, N. N., Zhang, H., Olivera, A., Mattie, M. E., and Spiegel, S. (1992). Sphingosine-1-phosphate, a metabolite of sphingosine, increases phosphatidic acid levels by phospholipase D activation. *J. Biol. Chem.* **267,** 23122–23128.

Dickson, R. C., Wells, G. B., Schmidt, A., and Lester, R. L. (1990). Isolation of mutant *Saccharomyces cerevisiae* strains that survive without sphingolipids. *Mol. Cell. Biol.* **10,** 2176–2181.

Dillehay, D. L., Crall, K. J., Webb, S. K., Schmelz, E., and Merrill, A. H., Jr. (1994). Dietary sphingomyelin inhibits 1,2-dimethylhydrazine-induced colon cancer in CF1 mice. *J. Nutr.* (*in press*).

Dobrowsky, R. T., and Hannun, Y. A. (1992). Ceramide stimulates a cytosolic protein phosphatase. *J. Biol. Chem.* **267,** 5048–5051.

Dressler, K. A., Mathias, S., and Kolesnick, R. N. (1992). Tumor necrosis factor-alpha activates the sphingomyelin signal transduction pathway in a cell-free system. *Science* **255,** 1715–1718.

El Touny, S., Khan, W., and Hannun, Y. A. (1990). Regulation of platelet protein kinase C by oleic acid. Kinetic analysis of allosteric regulation and effects on autophosphorylation, phorbol ester binding, and susceptibility to inhibition. *J. Biol. Chem.* **265,** 16437–16443.

Endo, K., Igarashi, Y., Nisar, M., Zhou, Q. H., and Hakomori, S.-I. (1991). Cell membrane signaling as target in cancer therapy: Inhibitory effect of *N,N*-dimethyl and *N,N,N*-trimethyl sphingosine derivatives on in vitro and in vivo growth of human tumor cells in nude mice. *Cancer Res.* **51,** 1613–1618.

Enkvetchakul, B., Merrill, A. H., Jr., and Birt, D. F. (1989). Inhibition of the induction of ornithine decarboxylase activity by 12-*O*-tetredecanoylphorbol-13-acetate in mouse skin by sphingosine sulfate. *Carcinogenesis* **10,** 379–381.

Enkvetchakul, B., Barnett, T., Liotta, D. C., Geisler, V., Menaldino, D., Merrill, A. H., Jr., and Birt, D. F. (1992). Influences of sphingosine on two-stage skin tumorigenesis in Sencar mice. *Cancer Lett.* **62,** 35–42.

Faucher, M., Girones, N., Hannun, Y. A., Bell, R. M., and Davis, R. A. (1988). Regulation of the epidermal growth factor receptor phosphorylation state by sphingosine in A431 human epidermoid carconoma cells. *J. Biol. Chem.* **263**, 5319–5327.

Felding-Habermann, B., Igarashi, Y., Fenderson, B. A., Park, L. S., Radin, N. S., Inokuchi, J.-I., Strassmann, G., Hanada, K., and Hakomori, S.-I. (1990). A ceramide analogue inhibits T cell proliferative response through inhibition of glycosphingolipid synthesis and enhancement of *N,N*-dimethylsphingosine synthesis. *Biochemistry* **29**, 6314–6322.

Gelderblom, W. C. A., Jaskiewicz, K., Marasas, W. F. O., Thiel, P. G., Dorak, R. M., Vleggaar, R., and Kriek, N. P. J. (1988). Cancer promoting potential of different strains of *Fusarium moniliforme* in a short-term cancer initiation/promotion assay. *Carcinogenesis* **9**, 1405–1409.

Gelderblom, W. C. A., Kriek, N. P. J., Marasas, W. F. O., and Thiel, P. G. (1991). Toxicity and carcinogenicity of the *Fusarium moniliforme* metabolite, fumonisin B1, in rats. *Carcinogenesis* **12**, 1247–1251.

Ghosh, T. K., Bian, J., and Gill, D. L. (1990). Intracellular calcium release mediated by sphingosine derivatives generated in cells. *Science* **248**, 1653–1656.

Goldin, E., Roff, C. F., Miller, S. P., Rodriguez-Lafrasse, C., Vanier, M. T., Brady, R. O., and Pentchev, P. G. (1992). Type C Niemann–Pick disease: A murine model of the lysosomal cholesterol lipidosis accumulates sphingosine and sphinganine in liver. *Biochim. Biophys. Acta* **1127**, 303–311.

Goldkorn, T., Dressler, K. A., Muindi, J., Radin, N. S., Mendelsohn, J., Menaldino, D., Liotta, D., and Kolesnick, R. N. (1991). Ceramide stimulates epidermal growth factor receptor phosphorylation in A431 human epidermoid carcinoma cells: Evidence that ceramide may mediate sphingosine action. *J. Biol. Chem.* **266**, 16092–16097.

Gupta, A. K., Fischer, G. J., Elder, J. T., Nickoloff, B. J., and Voorhees, J. J. (1988). Sphingosine inhibits phorbol ester-induced inflammation, ornithine decarboxylase activity, and activation of protein kinase C in mouse skin. *J. Invest. Dermatol.* **91**, 486–491.

Hakomori, S.-I. (1981). Glycosphingolipids in cellular interaction, differentiation, and oncogenesis. *Ann. Rev. Biochem.* **50**, 733–764.

Hakomori, S.-I. (1986). Glycosphingolipids. *Sci. Am.* **254**, 44–53.

Hakomori, S.-I. (1990). Bifunctional roles of glycosphingolipids. Modulators for transmembrane signalling and mediators for cellular interactions. *J. Biol. Chem.* **265**, 18713–18716.

Hanada, K., Nishijima, M., and Akamatsu, Y. (1990). A temperature-sensitive mammalian cell mutant with thermolabile serine palmitoyltransferase for the sphingolipid biosynthesis. *J. Biol. Chem.* **265**, 22137–22142.

Hanada, K., Igarashi, Y., Nisar, M., and Hakomori, S.-I. (1991). Downregulation of GMP-140 (CD62 or PADGEM) expression on platelets by *N,N*-dimethyl and *N,N,N*-trimethyl derivatives of sphingosine. *Biochemistry* **30**, 11682–11686.

Hanada, K., Nishijima, M., Kiso, H., Hasegawa, A., Fujita, S., Ogawa, T., and Akamatsu, Y. (1992). Sphingolipids are essential for the growth of Chinese hamster ovary cells. Restoration of the growth of a mutant defective in sphingoid base biosynthesis with exogenous sphingolipids. *J. Biol. Chem.* **267**, 23527–23533.

Hannun, Y. A., and Bell, R. M. (1987). Lysosphingolipids inhibit protein kinase C: Implications for sphingolipidoses. *Science* **235**, 670–674.

Hannun, Y. A., and Bell, R. M. (1989). Functions of sphingolipids and sphingolipid breakdown products in cellular regulation. *Science* **243**, 500–507.

Hannun, Y., Loomis, C. R., Merrill, A., and Bell, R. M. (1986). Sphingosine inhibition of protein kinase C activity and of phorbol bibutyrate binding in vitro and in human platelets. *J. Biol. Chem.* **261**, 12604–12609.

Hirschberg, C. B., Kisic, A., and Schroepfer, G. J. (1970). Enzymatic formation of dihydrosphingosine-1-phosphate. *J. Biol. Chem.* **245**, 3084–3090.

Holleran, W. M., Williams, M. L., Gao, W. N., and Elias, P. M. (1990). Serine palmitoyltransferase activity in cultured human keratinocytes. *J. Lipid Res.* **31**, 1655–
1661.

Igarashi, Y., Hakomori, S.-I., Toyokuni, T., Dean, B., Fujita, S., Sugimoto, M., Ogawa,
T., El-Ghendy, K., and Racker, E. (1989). Effect of chemically well-defined sphingosine
and its N-methyl derivatives on protein kinase C and src kinase activities. *Biochemistry*
28, 6796–6800.

Igarashi, Y., Kitamura, K., Toyokuni, T., Dean, B., Fenderson, B., Ogawa, T., and Hakomori, S.-I. (1990). A specific enhancing effect of N,N-dimethylsphingosine on epidermal
growth factor receptor autophosphorylation. *J. Biol. Chem.* **265**, 5385–5389.

Inokuchi, J., and Radin, N. S. (1987). Preparation of the active isomer of 1-phenyl-2-
decanoylamino-3-morpholino-1-propanol, inhibitor of murine glucocerebroside synthetase. *J. Lipid Res.* **28**, 565–571.

Inokuchi, J., Mason, I., and Radin, N. S. (1987). Antitumor activity in mice of an inhibitor
of glucocerebroside biosynthesis. *Cancer Lett.* **38**, 23–30.

Inokuchi, J., Momosaki, K., Shimeno, H., Nagamatsu, A., and Radin, N. S. (1989). Effects
of D-*threo*-PDMP, an inhibitor of glucocerebroside biosynthesis, on expression of cell
surface glycolipid antigen and binding of adhesive proteins by B16 melanoma cells.
J. Cell Physiol. **141**, 573–583.

Inokuchi, J., Jumbo, M., Momosaki, K., Shimeno, H., Nagamatsu, A., and Radin, N. S.
(1990). Inhibition of experimental metastasis of murine Lewis lung carcinoma by an
inhibitor of glucocerebroside biosynthesis and its possible mechanism of action. *Cancer
Res.* **50**, 6731–6737.

Jamal, H., Martin, A., Gomez-Munoz, A., and Brindley, D. N. (1991). Plasma membrane
fractions from rat liver contain a phosphatidate phosphohydrolase distinct from that in
the endoplasmic reticulum and cytosol. *J. Biol. Chem.* **266**, 2988–2996.

Kalen, A., Borchardt, R. A., and Bell, R. M. (1992). Elevated ceramide levels in GH_4C_1
cells treated with retinoic acid. *Biochim. Biophys. Acta* **1125**, 90–96.

Kan, C.-C., and Kolesnick, R. N. (1992). A synthetic ceramide analog, D-*threo*- 1-phenyl-
2-decanoylamino-3-morpholino-1-propanol, selectively inhibits adherence during macrophage differentiation of human leukemia cells. *J. Biol. Chem.* **267**, 9663–9667.

Karlsson, K.-A. (1970). Sphingolipid long chain bases. *Lipids* **5**, 878–891.

Kendler, A., and Dawson, G. (1992). Hypoxic injury to oligodendrocytes: reversible inhibition of ATP-dependent transport of ceramide from the endoplasmic reticulum to the
Golgi. *J. Neurosci. Res.* **31**, 205–211.

Kim, M.-Y., Linardic, C., Obeid, L., and Hannun, Y. A. (1991). Identification of sphingomyelin turnover as an effector mechanism for the action of tumor necrosis factor alpha
and gamma-interferon. *J. Biol. Chem.* **266**, 484–489.

Kimura, S., Kawa, S., Ruan, F., Nisar, M., Sadahira, Y., Hakomori, S.-I., and Igarashi,
Y. (1992). Effect of sphingosine and its N-methyl derivatives on oxidative burst, phagoki-
netic activity, and trans-endothelial migration of human neutrophils, *Biochem. Pharmacol.* **44**, 1585–1595.

Kiss, Z., Crilly K., and Chattopadhyay, J. (1991). Ethanol potentiates the stimulatory effects
of phorbol ester, sphingosine and 4-hydroxynonenal on the hydrolysis of phosphatidylethanolamine in NIH 3T3 cells. *Eur. J. Biochem.* **197**, 785–790.

Kobayashi, J., Cheng, J.-F., Ishibashi, M., Walchli, M. R., Yamamura, S., and Ohizumi,
Y. (1991). Penaresidin A and B, two novel azetidine alkaloids with potent actomyosin
ATPase-activating activity from the Okinawan marine sponge Penares sp. *J. Chem.
Soc. Perkin Trans.* **1**, 1135–1137.

Kobayashi, T., Goto, I., Yamanaka, T., Suzuki, Y., Nakano, T., and Suzuki, K. (1988a).

Infantile and fetal globoid cell leukodystrophy: Analysis of galactosylceramide and galactosylsphingosine. *Ann. Neurol.* **24,** 517–522.

Kobayashi, T., Mitsuo, K., and Goto, I. (1988b). Free sphingoid bases in normal murine tissues. *Eur. J. Biochem.* **171,** 747–752.

Kolesnick, R. N. (1987). 1,2-Diacylglycerols but not phorbol ester stimulate sphingomyelin hydrolysis in GH3 pituitary cells. *J. Biol. Chem.* **262,** 16759–16762.

Kolesnick, R. N. (1991). Sphingomyelin and derivatives as cellular signals. *Prog. Lipid Res.* **30,** 1–38.

Kolesnick, R. N., and Clegg, S. (1988). 1,2-Diacylglycerols but not phorbol esters activate a potlential inhibitory pathway for protein kinase C in GH3 pituitary cells. Evidence for involvement of a sphingomyelinase. *J. Biol. Chem.* **263,** 6534–6537.

Kolesnick, R. N., and Hemer, M. R. (1990). Characterization of a ceramide kinase activity from human leukemia (HL-60) cells. Separation from diacylglycerol kinase activity. *J. Biol. Chem.* **265,** 18803–18808.

Kriek, N. P. J., Kellerman, T. S., and Marasas, W. F. O. (1981). A comparative study of the toxicity of *Fusarium verticillioides* (= F. moniliforme) to horses, primates, pigs, sheep, and rats. *Onderstepoort J. Vet. Res.* **48,** 129–131.

Lin, M., Lu, S., Ji, C., Wang, Y., Wang, M., Cheng, S., and Tian, G. (1980). Experimental studies on the carcinogenicity of fungus-contaminated food from Linxian County. *In* "Genetic and Environmental Factors in Experimental and Human Cancer" (H. V. Gelboin, ed.), pp. 139–148. Japan Science Society Press, Tokyo.

Lipsky, N. G., and Pagano, R. E. (1983). Sphingolipid metabolism in cultured fibroblasts: Microscopic and biochemical studies employing a fluorescent ceramide analogue. *Proc. Natl. Acad. Sci. U.S.A.* **80,** 2608–2612.

Louie, D. D., Kisic, A., and Schroepfer, G. J. (1976). Sphingolipid base metabolism. Partial purification and properties of sphinganine kinase of brain. *J. Biol. Chem.* **251,** 4557–4564.

Marasas, W. F. O. (1982). Mycotoxicological investigations on corn produced in oesophageal cancer areas in Transkei. *In* "Cancer of the Oesophagus" (C. J. Pfeiffer, ed.), Vol. 1, pp. 29–40. CRC Press, Boca Raton, Florida.

Marasas, W. F. O., Kriek, N. P. J., Fincham, J. E., and van Rensburg, S. J. (1984). Primary liver cancer and oesophageal basal cell hyperplasia in rats caused by *Fusarium moniliforme. Int. J. Cancer* **34,** 383–387.

Marasas, W. F. O., Kellerman, T. S., Gelderblom, W. C. A., Coetzer, J. A. W., Thiel, P. G., and van der Lugt, J. J. (1988). Leukoencephalomalacia in a horse induced by fumonsin B₁ isolated from *Fusarium moniliforme. Onderstepoort J. Vet. Res.* **55,** 197–203.

Mathias, S., Dressler, K. A., and Kolesnick, R. N. (1991). Characterization of a ceramide-activated protein kinase: Stimulation by tumor necrosis factor alpha. *Proc. Natl. Acad. Sci. U.S.A.* **88,** 10009–10013.

Medlock, K. A., and Merrill, A. H., Jr. (1988). Inhibition of serine palmitoyltransferase in vitro and long-chain base biosynthesis in intact Chinese hamster ovary cells by beta-chloroalanine. *Biochemistry* **27,** 7079–7084.

Merrill, A. H., Jr. (1991). Cell regulation by sphingosine and more complex sphingolipids. *J. Bioeng. Biomembr.* **23,** 83–104.

Merrill, A. H., Jr. (1993). Fumonisins and other inhibitors of de novo sphingolipid biosynthesis. *In* "Advances in Lipid Research: Sphingolipids and Their Metabolites" (R. M. Bell, Y. A. Hannun, and A. H. Merrill, Jr., eds.), Vol. 26, pp. 215–234. Academic Press, San Diego, California.

Merrill, A. H., Jr., Sereni, A. M., Stevens, V. L., Hannun, Y. A., Bell, R. M., and Kinkade, J. M., Jr. (1986). Inhibition of phorbol ester-dependent differentiation of human promyelocytic leukemic (HL-60) cells by sphinganine and other long-chain bases. *J. Biol. Chem.* **261**, 12610–12615.

Merrill, A. H., Jr., Wang, E., Mullins, R. E., Jamison, W. C. L., Nimkar, S., and Liotta, D. C. (1988). Quantitation of free sphingosine in liver by high-performance liquid chromatography. *Anal. Biochem.* **171**, 373–381.

Merrill, A. H., Jr., Nimkar, S., Menaldino, D., Hannun, Y. A., Loomis, C., Bell, R. M., Tyagi, S. R., Lambeth, J. D., Stevens, V. L., Hunter, R., and Liotta, D. C. (1989). Structural requirements for long-chain (sphingoid) base inhibition of protein kinase C *in vitro* and for the cellular effects of these compounds, Biochemistry **28**, 3138–3145.

Miccheli, A., Ricciolini, R., Lagana, A., Piccolella, E., and Conti, F. (1991). Modulation of the free sphingosine levels in Epstein–Barr virus transformed human B lymphocytes by phorbol dibutyrate. *Biochim. Biophys. Acta* **1095**, 90–92.

Mullmann, T. J., Siegel, M. I., Egan, R. W., and Billah, M. M. (1991). Sphingosine inhibits phosphatidate phosphohydrolase in human neutrophils by a protein kinase C-independent mechanism. *J. Biol. Chem.* **266**, 2013–2016.

Nelson, D. H., and Murray, D. K. (1989). Dexamethasone and sphingolipids inhibit concanavalin A stimulated glucose uptake in 3T3-L1 fibroblasts. *Endocrine Res.* **14**, 305–318.

Norred, W. P., Wang, E., Yoo, H., Riley, R. T., and Merrill, A. H., Jr. (1992). In vitro toxicology of fumonisins and the mechanistic implications. *Mycopathologia* **117**, 73–78.

Nudelman, E. D., Levery, S. B., Igarashi, Y., and Hakomori, S.-I. (1992). Plasmalopsychosine, a novel plasmal (fatty aldehyde) conjugate of psychosine with cyclic acetal linkage. Isolation and characterization from human brain white matter. *J. Biol. Chem.* **267**, 11007–11016.

Nutter, L. M., Grill, S. P., Li, J. S., Tan, R. S., and Cheng, Y. C. (1987). Induction of virus enzymes by phorbol esters and n-butyrate in Epstein–Barr virus genome-carrying Raji cells. *Cancer Res.* **47**, 4407–4412.

Oishi, K., Raynor, R. L., Charp, P. A., and Kuo, J. F. (1988). Regulation of protein kinase C by lysophospholipids. Potential role in signal transduction. *J. Biol. Chem.* **263**, 6865–6871.

Oishi, K., Zheng, B., and Kuo, J. F. (1990). Inhibition of Na,K-ATPase and sodium pump by protein kinase C regulators sphingosine, lysophosphatidylcholine, and oleic acid. *J. Biol. Chem.* **265**, 70–75.

Okazaki, T., Bell, R. M., and Hannun, Y. A. (1989). Sphingomyelin turnover induced by 1-alpha, 25-dihydroxyvitamine D_3 in HL-60 cells. Role in cell differentiation. *J. Biol. Chem.* **264**, 10976–19080.

Okazaki, T., Bielawska, A., Bell, R. M., and Hannun, Y. A. (1990). Role of ceramide as a lipid mediator of 1α-25-dihydroxyvitamin D_3-induced HL-60 cell differentiation. *J. Biol. Chem.* **265**, 15823–15831.

Okoshi, H., Hakomori, S.-I., Nisar, M., Zhou, Q. H., Kimura, S., Tashiro, K., and Igarshi, Y. (1991). Cell membrane signaling as target in cancer therapy. II. Inhibitory effect of N,N,N-trimethylsphingosine on metastatic potential of murine B16 melanoma cell line through blocking of tumor cell-dependent platelet aggregation. *Cancer Res.* **51**, 6019–6024.

Pagano, R. E., and Martin, O. C. (1988). A series of fluorescent N-acylsphingosines: Synthesis, physical properties, and studies in cultured cells. *Biochemistry* **27**, 4439–4445.

Pawelczyk, T., and Lowenstein, J. M. (1992). Regulation of phospholipase C delta activity by sphingomyelin and sphingosine. *Arch. Biochem. Biophys.* **297**, 328–333.

Perry, D. K., Hand, W. L., Edomndson, D. E., and Lambeth, J. D. (1992). Role of phospholipase D-derived diradylglycerol in the activation of the human neutrophil respiratory burst oxidase. Inhibition by phosphatidic acid phosphohydrolase inhibitors. *J. Immunol.* **149,** 2749–2758.

Pushkareva, M. Y., Khan, W. A., Alessenko, A. V., Sahyoun, N., and Hannun, Y. A. (1992). Sphingosine activation of protein kinases in Jurkat T cells. In vitro phosphorylation of endogenous protein substrates and specificity of action. *J. Biol. Chem.* **267,** 15246–15251.

Rao, B. G., and Spence, M. W. (1976). Sphingomyelinase activity at pH 7.4 in human brain and a comparison to activity at pH 5.0. *J. Lipid Res.* **17,** 506–515.

Rees, R. S., Nanney, L. B., Yates, R. A., and King, L. J. (1984). Interaction of brown recluse spider venom on cell membranes: The inciting mechanism? *J. Invest. Dermatol.* **83,** 270–275.

Robertson, D. G., DiGirolama, M., Merrill, A. H., Jr., and Lambeth, J. D. (1989). Insulin-stimulated hexose transport and glucose oxidation in rat adipocytes is inhibited by sphingosine at a step after insulin binding. *J. Biol. Chem.* **264,** 6773–6779.

Rosenwald, A. G., Machamer, C. E., and Pagano, R. E. (1992). Effects of a sphingolipid synthesis inhibitor on membrane transport through the secretory pathway. *Biochemistry* **31,** 3581–3590.

Ross, F. F., Nelson, P. E., Richard, J. L., Osweiler, G. D., Rice, L. G., Plattner, R. D., and Wilson, T. M. (1990). Production of fumonisins by *Fusarium moniliforme* and *Fusarium proliferatum* isolates associated with equine leukoencephalomalacia and a pulmonary edema syndrome in swine. *Appl. Environ. Microbiol.* **56,** 3225–3226.

Sabbadini, R. A., Betto, R., Teresi, A., Fachechi-Cassano, G., and Salviati, G. (1992). The effects of sphingosine on sarcoplasmic reticulum membrane calcium release. *J. Biol. Chem.* **267,** 15475–15484.

Sadahira, Y., Ruan, F., Hakomori, S., and Igarashi, Y. (1992). Sphingosine 1-phosphate, a specific endogenous signaling molecule controlling cell motility and tumor cell invasiveness. *Proc. Natl. Acad. Sci. U.S.A.* **89,** 9686–9690.

Schutze, S., Potthoff, K., Machleidt, T., Berkovic, D., Wiegmann, K., and Kronke, M. (1992). TNF activates NF-kB by phosphatidylcholine-specific phospholipase C-induced "acidic" sphingomyelin breakdown. *Cell* **71,** 765–776.

Shayman, J. A., Deshmukh, G. D., Mahdiyoun, S., Thomas, T. P., Wu, D., Barcelon, F. S., and Radin, N. S. (1991). Modulation of renal epithelial cell growth by glucosylceramide. Association with protein kinase C, sphingosine, and diacylglycerol. *J. Biol. Chem.* **266,** 22968–22974.

Shier, W. T., Abbas, H. K., and Mirocha, C. J. (1991). Toxicity of the mycotoxins fumonisins B1 and B2 and *Alternaria alternata* f. sp. *lycopersici* toxin (AAL) in cultured mammalian cells. *Mycopathologia* **116,** 97–104.

Shinoda, H., Kobayashi, T., Katayama, M., Goto, I., and Nagara, H. (1987). Accumulation of galactosylsphingosine (psychosine) in the Twitcher mouse: Determination by HPLC. *J. Neurochem.* **49,** 92–99.

Shukla, A., and Radin, N. S. (1991). Metabolism of D-[^{3}H]*threo*-1-phenyl-2-decanoylamino-3-morpholino-1-propanol, an inhibitor of glucosylceramide synthesis, and the synergistic action of an inhibitor of microsomal monooxygenase. *J. Lipid Res.* **32,** 713–722.

Shukla, G., Shukla, A., Inokuchi, J.-I., and Radin, N. S. (1991). Rapid kidney changes resulting from glucosphingolipid depletion by treatment with a glucosyltransferase inhibitor. *Biochim. Biophys. Acta* **1083,** 101–108.

Slife, C. W., Wang, E., Hunter, R., Wang, S., Burgess, C., Liotta, D. C., and Merrill, A. H., Jr. (1989). Free sphingosine formation from endogenous substrates by a liver plasma membrane system with a divalent cation dependence and a neutral pH optimum. *J. Biol. Chem.* **264,** 10371–10377.

Stevens, V. L., Winton, E. F., Smith, E. E., Owens, N. E., Kinkade, J. M., Jr., and Merrill, A. H., Jr. (1989). Differential effects of long-chain (sphingoid) bases on the monocytic differentiation of human leukemia (HL-60) cells induced by phorbol esters, 1alpha, 25-dihydroxyvitamin D_3, or ganglioside G_{M3}. *Cancer Res.* **49**, 3229–3234.

Stevens, V. L., Nimkar, S., Jamison, W. C., Liotta, D. C., and Merrill, A. H., Jr. (1990a). Characteristics of the growth inhibition and cytotoxicity of long-chain (sphingoid) bases for Chinese hamster ovary cells: Evidence for an involvement of protein kinase C. *Biochim. Biophys. Acta* **1051**, 37–45.

Stevens, V. L., Owens, N. E., Winton, E. F., Kinkade, J. M., Jr., and Merrill, A. H., Jr. (1990b). Modulation of retinoic acid-induced differentiation of human leukemia (HL-60) cells by serum factors and sphinganine. *Cancer Res.* **50**, 222–226.

Stinavage, P., and Spitznagel, J. K. (1989). Oxygen-independent and antimicrobial action in sphingosine-treated neutrophils. *J. Immunol. Meth.* **124**, 267–275.

Stoffel, W., and Bister, K. (1973). Stereospecificities in the metabolic reactions of the four isomeric sphinganines (dihydrosphingosines) in rat liver, *Hoppe Seyler's Z. Physiol. Chem.* **354**, 169–181.

Stoffel, W., and Grol, M. (1974). Chemistry and biochemistry of 1-desoxysphinganine 1-phosphonate (dihydrosphingosine 1-phosphonate). *Chem. Phys. Lipids* **13**, 372–388.

Stoffel, W., LeKim, D., and Sticht, G. (1969). Metabolism of sphingosine bases. XI. Distribution and properties of dihydrosphingosine-1-phosphate aldolase (sphinganine-1-phosphate alkanal-lyase). *Hoppe-Seyler's Z. Physiol. Chem.* **350**, 1233–1241.

Stoffel, W., Assmann, G., and Binczek, E. (1970). Metabolism of sphingosine bases. XIII. Enzymatic synthesis of 1-phosphate esters of 4t-sphingenine (sphingosine), sphinganine (dihydrosphingosine), 4-hydroxysphinganine (phytosphingosine), and 3-dehydrosphingosine by erythrocytes. *Hoppe Seyler's Z. Physiol. Chem.* **351**, 635–642.

Sundaram, K. S., and Lev, M. (1984). Inhibition of sphingolipid synthesis by cycloserine in vitro and in vivo. *J. Neurochem.* **42**, 577–581.

Sundaram, K. S., and Lev, M. (1989). The long-term administration of L-cycloserine to mice: Specific reduction of cerebroside level. *Neurochem. Res.* **14**, 245–248.

Thompson, T. E., and Tillack, T. W. (1985). Organization of glycosphingolipids in bilayers and plasma membranes of mammalian cells. *Ann. Rev. Biophys. Biophys. Chem.* **14**, 361–386.

Thudichum, J. L. W. (1884). "A Treatise on the Chemical Constitution of Brain." Bailliere, Tindall, Cox, London.

Van Echten, G., Birk, R., Brenner-Weiss, G., Schmidt, R. R., and Sandhoff, K. (1990). Modulation of sphingolipids biosynthesis in primary cultured neurons by long chain bases. *J. Biol. Chem.* **265**, 9333–9339.

Van Veldhoven, P. P., and Mannaerts, G. P. (1991). Subcellular localization and membrane topology of sphingosine-1-phosphate lyase in rat liver. *J. Biol. Chem.* **266**, 12502–12507.

Van Veldhoven, P. P., Bishop, W. R., and Bell, R. M. (1989). Enzymatic quantification of sphingosine in the picomole range in cultured cells. *Anal. Biochem.* **183**, 177–189.

Van Veldhoven, P. P., Matthews, T. J., Bolognesi, D. P., and Bell, R. M. (1992). Changes in bioactive lipids, alkylacylglycerol and ceramide, occur in HIV-infected cells, *Biochem. Biophys. Res. Commun.* **187**, 209–216.

Vartanian, T., Dawson, G., Soliven, B., Nelson, D. J., and Szuchet, S. (1989). Phosphorylation of myelin basic protein in intact oligodendrocytes: Inhibition by galactosylsphingosine and cyclic AMP. *Glia* **2**, 370–379.

Vunnam, R. R., and Radin, N. S. (1980). Analogs of ceramide that inhibit glucocerebroside synthetase in mouse brain. *Chem. Phys. Lipids* **26**, 265–278.

Wang, E., Norred, W. P., Bacon, C. W., Riley, R. T., and Merrill, A. H., Jr. (1991). Inhibition of sphingolipid biosynthesis by fumonisins. Implications for diseases associated with *Fusarium moniliforme. J. Biol. Chem.* **266**, 14486–14490.

Wang, E., Ross, P. F., Wilson, T. M., Riley, R. T., and Merrill, A. H., Jr. (1992). Increases in serum sphingosine and sphinganine and decreases in complex sphingolipids in ponies given feed containing fumonisins, mycotoxins produced by *Fusarium moniliforme*. *J. Nutr.* **122**, 1706–1716.

Wedegaertner, P. B., and Gill, G. N. (1989). Activation of the purified protein tyrosine kinase domain of the epidermal growth factor receptor. *J. Biol. Chem.* **264**, 11346–11353.

Weiss, R. H., Huang, C.-H., and Ives, H. E. (1991). Sphingosine reverses growth inhibition caused by activation of protein kinase C in vascular smooth muscle cells. *J. Cell. Physiol.* **149**, 307–312.

Wertz, P. W., and Downing, D. T. (1989). Free sphingosines in porcine epidermis. *Biochim. Biophys. Acta* **1002**, 213–217.

Wilson, E., Olcott, M. C., Bell, R. M., Merrill, A. H., Jr., and Lambeth, J. D. (1986). Inhibition of the oxidative burst in human neutrophils by sphingoid long-chain bases. *J. Biol. Chem.* **261**, 12616–12623.

Wilson, E., Wang, E., Mullins, R. E., Liotta, D. C., Lambeth, J. D., and Merrill, A. H., Jr. (1988). Modulation of the free sphingosine levels in human neutrophils by phorbol esters and other factors. *J. Biol. Chem.* **263**, 9304–9309.

Yang, C. S. (1980). Research on esophageal cancer in China: A review. *Cancer Res.* **40**, 2633–2644.

Yoo, H., Norred, W. P., Wang, E., Merrill, A. H., Jr., and Riley, R. T. (1992). Sphingosine inhibition of *de novo* sphingolipid biosynthesis and cytotoxicity are correlated in LLC-PK₁ cells. *Toxicol. Appl. Pharmacol.* **114**, 9–15.

Younes, A., Kahn, D. W., Besterman, J. M., Bittman, R., Byun, H. S., and Kolesnick, R. N. (1992). Ceramide is a competitive inhibitor of diacylglycerol kinase *in vitro* and in intact human leukemia (HL-60) cells. *J. Biol. Chem.* **267**, 842–847.

Yung, B. Y., Luo, K. J., and Hui, E. K. (1992). Interaction of antileukemia agents adriamycin and daunomycin with sphinganine on the differentiation of human leukemia cell line HL-60. *Cancer Res.* **52**, 3593–3597.

Zeller, C. B., and Marchase, R. B. (1992). Gangliosides as modulators of cell function. *Am. J. Physiol.* **262**, (*Cell Physiol.* **31**), C1341–1355.

Zhang, H., Buckley, N. E., Gibson, K., and Spiegel, S. (1990). Sphingosine stimulates cellular proliferation via a protein kinase C-independent pathway. *J. Biol. Chem.* **265**, 76–81.

Zhang, H., Desai, N. N., Olivera, A., Seki, T., Booker, G., and Spiegel, S. (1991). Sphingosine-1-phosphate, a novel lipid, involved in cellular proliferation. *J. Cell Biol.* **114**, 155–167.

Zweerink, M. M., Edison, A. M., Wells, G. B., Pinto, W., and Lester, R. L. (1992). Characterization of a novel, potent, and specific inhibitor of serine palmitoyltransferase. *J. Biol. Chem.* **267**, 25032–25038.

CHAPTER 15

Glycosphingolipids as Effectors of Growth and Differentiation

Eric G. Bremer*

Department of Immunology/Microbiology, Rush University, Rush-Presbyterian-St. Luke's Medical Center, Chicago, Illinois 60612

I. INTRODUCTION

Sphingolipids are found in virtually all cells of vertebrate organisms. These substances are composed of a lipophilic ceramide moiety, which is embedded in the external leaflet of the plasma membrane, and a hydrophilic head group that extends out into the extracellular space. Sphingomyelin (ceramide phosphorylcholine) is a common constituent of plasma membrane and is an example of a phosphosphingolipid. Sphingomyelin now is believed to serve as a cellular precursor of two bioactive metabolites: ceramide and sphingosine (Merrill, 1991; Hannun and Bell, 1989). In

* Present Address: Chicago Institute for Neurosurgery and Neuroresearch, Chicago, Illinois 60614

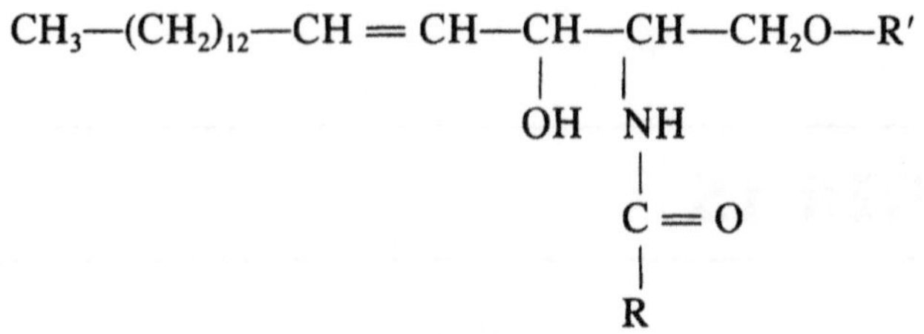

$$CH_3\text{---}(CH_2)_{12}\text{---}CH = CH\text{---}CH\text{---}CH\text{---}CH_2O\text{---}R'$$

$$
\begin{array}{cc}
| & | \\
OH & NH \\
 & | \\
 & C = O \\
 & | \\
 & R
\end{array}
$$

SCHEME 1

contrast, the diverse structures of the extracellular glycans (carbohydrate chains) of glycosphingolipids provide biological specificity for numerous putative cellular functions such as cell–cell recognition, cell growth regulation, development and differentiation, and oncogenic transformation (Hakomori, 1981). Considering sphingolipids only as structural components of the plasma membrane is no longer adequate. Instead, they should be viewed as bioactive molecules involved in transmembrane signaling events by direct binding or by turnover to products that can serve as intracellular second messengers. This chapter focuses on the current status of research on the role of cell surface glycosphingolipids as effectors of growth and differentiation.

The lipid structure of glycosphingolipids is based on the long-chain sphingosine (Scheme 1) rather than on the glycerol backbone of phospholipids.

In the parent compound sphingosine, both the nitrogen and the R' are primary amine hydrogens. The addition of a fatty acyl chain to the primary amine creates ceramide (N-acyl sphingosine). The hydrocarbon tail of the acyl chain (R) varies from C_{16} to C_{24} and may be saturated or unsaturated. The acyl chain also may contain an alpha-hydroxy group. Substitution at R' with phosphorylcholine yields sphingomyelin. Glycosphingolipids can be formed by the addition of O-glycan chains at R'.

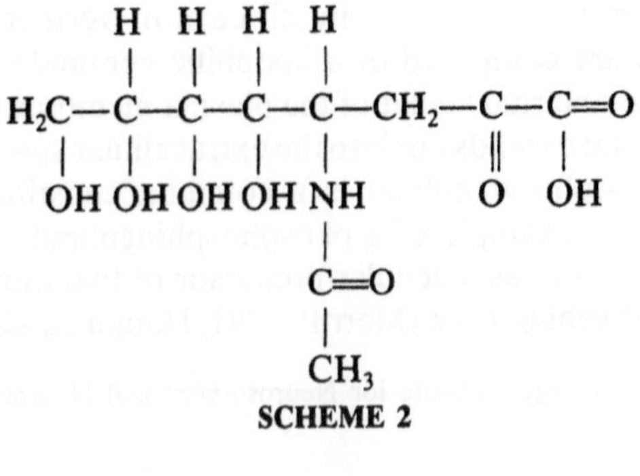

$$
\begin{array}{c}
\quad H\ \ H\ \ H\ \ H \\
\quad |\ \ \ |\ \ \ |\ \ \ | \\
H_2C\text{---}C\text{---}C\text{---}C\text{---}C\text{---}CH_2\text{---}C\text{---}C{=}O \\
\ \ |\ \ \ |\ \ \ |\ \ \ |\ \ \ |\ \qquad \|\ \ \ | \\
OH\ OH\ OH\ OH\ NH\qquad O\ \ OH \\
\qquad\qquad\qquad | \\
\qquad\qquad\qquad C{=}O \\
\qquad\qquad\qquad | \\
\qquad\qquad\qquad CH_3
\end{array}
$$

SCHEME 2

TABLE I

Common Core Carbohydrate Structures of Glycosphingolipids

Core series name	Structure	Abbreviation[a]
Ganglio-	Galβ1 $\rightarrow$ 3GalNAcβ1 $\rightarrow$ 4Galβ1 $\rightarrow$ 4GlcCer	G$_9$Ose$_4$Cer
Globo-	GalNAcβ1 $\rightarrow$ 3Galα1 $\rightarrow$ 4Galβ1 $\rightarrow$ 4GlcCer	GbOse$_4$Cer
Lacto-	Galβ1 $\rightarrow$ 3GlcNAcβ1 $\rightarrow$ 3Galβ1 $\rightarrow$ 4GlcCer	LcOse$_4$Cer
Neolacto-	Galβ1 $\rightarrow$ 4GlcNAcβ1 $\rightarrow$ 3Galβ1 $\rightarrow$ 4GlcCer	nLcOse$_4$Cer

[a] Abbreviated symbols are from the IUPAC Commission on Biochemical Nomenclature Recommendations.

The glycosphingolipids can be divided conveniently into two classes on the basis of the presence or absence of sialic acid (Scheme 2) as a constituent of the carbohydrate chain. Sialic acid confers acidity on a glycosphingolipid; these acidic glycosphingolipids commonly are referred to as gangliosides.

N-Acetyl neuraminic acid (Scheme 2) is the member of the sialic acid family most commonly found in gangliosides. Those without sialic acid are neutral glycosphingolipids. The glycosphingolipids also can be distinguished by the type of "core" carbohydrate structure (Table I).

G$_{M1}$ (II^3 NeuAcGg$_4$Cer) (Scheme 3) is an example of a ganglioside consisting of a ganglio core structure and one sialic acid residue on the internal galactose.[1]

II. GROWTH REGULATION THROUGH GLYCOSPHINGOLIPIDS

Since the first clear demonstration of glycosphingolipid (GSL) changes associated with oncogenic transformation in 1968 (Hakomori and Murakami, 1968), a great deal of work has been done to understand the biological

$$\text{Gal}\beta1 \rightarrow 3\text{GalNAc}\beta1 \rightarrow 4\text{Gal}\beta1 \rightarrow 4\text{GlcCer}$$
$$3$$
$$\uparrow$$
$$\text{NeuAc}\alpha2$$

SCHEME 3

[1] Ganglioside nomenclature used in this chapter is based on that described by Svennerholm (1963, 1964).

significance of this phenomenon. Three general categories of GSL change are associated with oncogenic transformation: (1) incomplete synthesis and accumulation of precursor GSL, (2) neosynthesis of previously unexpressed GSLs, and (3) organizational rearrangement or exposure of glycolipids at the cell surface. These changes have been described in a variety of transformed cells induced by oncogenic DNA viruses, RNA viruses, and chemical carcinogens, as well as in cells taken from spontaneous tumors, including a variety of human cancers (see Hakomori, 1981; Hakomori and Kannagi, 1983, for reviews).

Changes in glycolipid metabolism and expression associated with oncogenic transformation have suggested the possibility that GSLs may be associated with cell growth regulation. This possibility was suggested also by "GSL contact response" closely related to "contact inhibition" of cell growth. Synthesis of Gb_3Cer in BHK (Hakomori, 1970) and Gb_5 Cer and G_{M3} in NIL (Sakiyama et al., 1972; Critchley and Macpherson, 1973) was enhanced greatly prior to density-dependent growth inhibition. The enzymatic basis of this response (Kijimoto and Hakomori, 1971; Chandrabose et al., 1976) and the loss of this response have been well correlated with the loss of "contact inhibition" in the oncogenic transformants of these cells (Hakomori, 1981). Changes in the exposure of GSLs at the cell surface (Gahmberg and Hakomori, 1975) also have been correlated with the cell cycle. Collectively, these data suggest that GSLs may have an important role in the control of cell growth.

A. Cell Growth Control

To test the possibility that cell surface GSLs may be involved in the control of cell proliferation experimentally, four different approaches have been undertaken (for reviews, see Hakomori, 1981,1990): (1) exogenous addition of GSL to cells in culture, (2) application of antibodies directed against GSL, (3) measurement of GSL changes caused by differentiation inducers, and (4) addition of inhibitors of GSL synthesis.

1. Addition of Glycosphingolipids to Cell Cultures

Addition of exogenous GSLs to cultured cells has been the most important approach used to study the relationships between GSLs and cellular growth behavior. Hamster NIL cells cultured in media containing Gb_4Cer showed more than a twofold enhancement of Gb_4Cer at the plasma membrane. Consequently, cell adhesiveness increased, morphology of polyoma virus-transformed NIL cells resembled normal NIL cells, and the prereplicative period, particularly G_1 phase, became twice as long as before treatment (Laine and Hakomori, 1973). Similarly, the addition of several gangliosides to culture media reduced both the growth rate and the saturation density of SV40-transformed and nontransformed 3T3 cells (Keenan *et al.*, 1975). In the 3T3 cells, G_{M1} was the most effective GSL, followed by G_{D1b} and G_{T1b}; however, ceramide and G_{D1a} were ineffective.

In the cases just described, the exogenously added glycolipid accumulated at the cell surface (Keenan *et al.*, 1975). Although some of the glycolipid could be released by trypsin treatment (Callies *et al.*, 1977; Bremer *et al.*, 1984), considerable evidence suggests that the ceramide moiety of the added glycolipids is inserted in the lipid layer (Kanda *et al.*, 1982; Schwarzmann *et al.*, 1983).

2. Added Ganglioside Alters Cell Response to Growth Factors

To proliferate, cells require stimulation by specific growth factors (Barnes and Saito, 1980). The effect of exogenous ganglioside (sialic acid-containing GSLs) on the mitogenic response to individual growth factors has been determined to be a means of explaining the growth inhibition caused by added GSLs. The effects of individual gangliosides on growth factor receptors are summarized in Table II and described briefly in this section.

Baby hamster kidney fibroblasts (BHK C-21) can be grown in a serum-free medium containing insulin, transferrin, hydrocortisone, and fibroblast growth factor (FGF) (Barnes and Saito, 1980). BHK cell growth in the

TABLE II

Modulation Growth Factor Response by Glycosphingolipids

Growth factor receptor[a]	Glycosphingolipid[b]	Reference
FGF	$G_{M3}( )$	Bremer and Hakomori (1982)
PDGF	G_{M3}, G_{M1} (−)	Bremer *et al.* (1984)
	Ganglio series (−)	Yates *et al.* (1993)
EGF	G_{M3} (−)	Bremer *et al.* (1986)
	Lyso-G_{M3} (−)	Hanai *et al.* (1988)
Insulin	SPG[c] (−)	Nojiri *et al.* (1991)
EGF	de-*N*-Acetyl GM3 (+)	Hanai *et al.* (1988)
	DMS[d] (+)	Igarashi *et al.* (1990)

[a] Abbreviations: FGF, fibroblast growth factor; PDGF, platelet-derived growth factor; EGF, epidermal growth factor.
[b] +, stimulatory; −, inhibitory.
[c] SPG, Sialosyl paragloboside.
[d] DMS, *N,N*-Dimethyl-D-*erythro*-sphingenine.

presence of FGF was inhibited specifically by culturing the cells in the presence of G_{M3}, but not in the presence of G_{M1}, G_{D1a}, and other gangliosides. G_{M3} inhibited BHK cells, which became refractory to FGF growth stimulation and showed much higher FGF binding at the cell surface, suggesting that G_{M3} growth inhibition may be related to alteration of FGF binding (Bremer and Hakomori, 1982).

Swiss 3T3 cells show dependence on platelet-derived growth factor (PDGF) and epidermal growth factor (EGF) but not FGF. The effect of gangliosides on the responses of cells to individual growth factors has been examined more extensively in 3T3 cells (Bremer et al., 1984), with the following results.

1. Cell growth (cell number increase) in serum-free medium was inhibited specifically by the presence of G_{M1} and, to a lesser extent, by G_{M3}, but not by IV^3 $NeuAcnLc_4Cer$, although the gangliosides were incorporated equally well into cell membranes. G_{M3} inhibited both PDGF- and EFG-stimulated mitogenesis, as determined by thymidine incorporation. On the other hand, G_{M1} could inhibit only PDGF-stimulated mitogenesis. IV^3 $NeuAcnLc_4Cer$ had no effect on mitogen-stimulated thymidine incorporation.
2. The concentration-dependent binding of ^{125}I-labeled PDGF to cells indicated that cells whose growth was inhibited by G_{M1} or G_{M3} showed an increased affinity for PDGF over cells grown without addition of ganglioside, whereas the total number of receptors stayed the same. Addition of ganglioside did not affect the binding of ^{125}I-labeled EGF.
3. No direct interaction was observed between gangliosides and growth factors, as evidenced by the lack of competition by ganglioside-containing liposomes for cellular binding of ^{125}I-labeled growth factors.
4. G_{M1} and G_{M3}, but neither IV^3 $NeuAcnLc_4Cer$ nor Gb_4Cer, reduced the PDGF-stimulated tyrosine phosphorylation of a 170,000 molecular weight protein that is probably the PDGF receptor.

Yates et al. (1993) performed a more extensive survey of 3T3 growth inhibition by gangliosides and found that, in addition to G_{M1} most of the brain ganglio-series gangliosides could inhibit the PDGF response and PDGF receptor autophosphorylation.

A similar inhibition of tyrosine phosphorylation of the EGF receptor by G_{M3} alone and G_{M3}-induced cell growth inhibition in A431 (human epidermoid carcinoma) cells have been observed (Bremer et al., 1986). The major EGF-dependent tyrosine-phosphorylated band of A431 cell membranes has a molecular weight of 170,000 and corresponds to the EGF receptor (for reviews on the EGF receptor, see Carpenter, 1987; Yarden and Ullrich, 1988). The intensity of this tyrosine-phosphorylated band was diminished greatly (about 50% of control) when the membrane fraction was incubated in the presence of 10 nmol G_{M3}, but only slightly or not at all with 40 nmol G_{M1}, IV^3 NeuAcnLc$_4$Cer, or Gb$_4$Cer. Inhibition of EGF receptor autophosphorylation by G_{M3} also could be observed in intact cells with anti-phosphotyrosine antibodies (Fig. 1). Immunoprecipi-

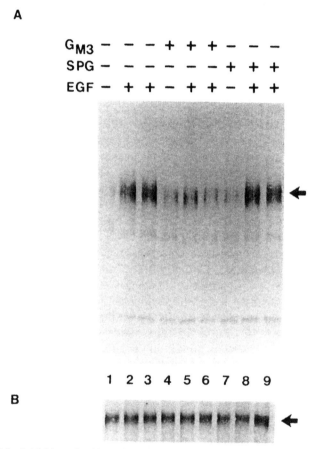

FIGURE 1 Inhibition of epidermal growth factor (EGF) receptor autophosphorylation by G_{M3} occurs in human foreskin fibroblast (HFF) cell cultures. HFF cells were cultured in the presence or absence of 50 μM ganglioside, similar to the conditions described elsewhere for A431 and KB cells (Bremer *et al.*, 1986). The cells were stimulated with 10 ng/ml EGF for 10 min. The cell membranes then were solubilized with 1% Triton X-100. Soluble proteins were separated by SDS–PAGE (100 μg protein loaded per lane) and then transferred to nitrocellulose. Identical nitrocellulose blots were probed with either anti-phosphotyrosine MAb (A) or anti-EGF receptor MAb (B). The results indicate that, in intact cells, G_{M3} but not sialosyl paragloboside (SPG) is able to inhibit EGF-stimulated receptor autophosphorylation (A) and that the reduced phosphotyrosine content cannot be explained by changes in EGF receptor levels.

tation of the EGF receptor, followed by a phosphorylation assay, suggested that G_{M3} might affect the EGF receptor directly. The data described
earlier strongly indicated that gangliosides, especially ganglioside G_{M3},
can be a potent pharmacological regulator of EGF receptor function as
well as of other growth factor receptors. This pharmacological regulation
of growth factor receptors could explain the growth inhibition observed
when GSLs are added to cell cultures.

A significant question remains, however, of whether G_{M3} is a physiological regulator of the EGF receptor, because the addition of exogenous
gangliosides to tissue culture media may increase incorporation of gangliosides into the plasma membrane beyond their physiological range or may
have effects on the normal turnover and biosynthesis of gangliosides. To
test the hypothesis that expression of gangliosides at physiological levels
may modulate signal transduction of the EGF receptor, Davis and colleagues (Weis and Davis, 1990) employed a strategy using a mutant Chinese hamster ovary (CHO) cell line that possesses a reversible defect in
the biosynthesis of complex carbohydrates. This mutant CHO cell line
(ldlD) lacks the 4-epimerase enzyme that converts glucose to galactose
(Kingsley et al., 1986). This mutation renders the cells incapable of the
synthesis of UDP-galactose and UDP-GalNAc when maintained in tissue
culture media containing only glucose. As a result, these cells have no
glycolipid biosynthesis and altered protein glycosylation when maintained
in a galactose-free medium. On addition of galactose and GalNAc to the
tissue culture media, the synthesis of gangliosides proceeds normally. The
ldlD cell line was transfected with the human EGF receptor gene by Weis
and Davis (1990) to examine the effects of endogenously synthesized G_{M3}
on the EGF receptor. Similar to the studies in which ganglioside was
added to cell culture media, changes in ganglioside expression in the ldlD
mutants were not observed to cause any significant alterations in the
affinity or number of EGF receptors detected at the cell surface. Decreases
in ganglioside expression, however, were associated with increased EGF
receptor autophosphorylation and increased the ability of EGF to stimulate
cellular proliferation. This inverse correlation between the level of ganglioside expression and signal transduction by the EGF receptor is consistent
with the hypothesis that the function of the EGF receptor can be regulated
physiologically by gangliosides.

3. Inhibitors of Glycosphingolipid Synthesis

Contrary to enrichment of GSLs by addition to cell culture, reduction
of cellular GSLs by inhibitors of GSL biosynthesis has supported a growth
regulatory role for GSLs. The ceramide analog D-threo-1-phenyl-2-
decanoylamino-3-morpholine-1-propanol (PDMP), which inhibits UDP-

Glc : ceramide glucosyltransferase (Inokuchi and Radin, 1987), greatly reduces the synthesis of all GSLs derived from GlcCer (Inokuchi and Radin, 1987; Felding-Habermann et al., 1990). The addition of PDMP to interleukin 2-dependent CTLL cells (Felding-Habermann et al., 1990) and EGF-dependent A431 cells (Igarashi et al., 1990) alters the proliferation of these cells. These data are complementary to observations that G_{M3} addition to A431 cells results in growth inhibition (Bremer et al., 1986). In addition to the inhibition of GSL synthesis, PDMP also enhances the synthesis of other sphingosine-containing products, particularly N,N-dimethyl-D-erythro-sphingenine (DMS). This compound has been shown to modulate the activity of protein kinase C (PKC) in A431 cells (Igarashi et al., 1989). Therefore, whether the growth modulatory effects of PDMP are caused by decreased levels of G_{M3}, by increased levels of DMS, or both remains unclear.

B. Induction of Differentiation

The expression of gangliosides is developmentally regulated. For example, hematopoietic cell lines have been found to have cell line-specific ganglioside patterns that can be very complex, containing more than 100 different components (Rosenfelder et al., 1982). During development of the avian or mammalian brain, distinct ganglioside patterns are associated with the neural tube stage, proliferation of glial and neural progenitor cells, neuritogenesis and synaptogenesis, and finally during myelination (for review, see Schengrund, 1990; Zeller and Marchase, 1992). These changes in developmental pattern have led to the idea that gangliosides may be able to influence the differentiation pattern of different cell types.

The promyelocitic leukemic cell line HL-60 has been shown to differentiate along a monocytic pathway by exogenous addition of ganglioside G_{M3} (Nojiri et al., 1986), suggesting the idea that gangliosides themselves can influence differentiation of cells. This differentiation is accompanied by an increase in G_{M3} synthesis and a decrease in PKC activity (Kreutter et al., 1987). The HL-60 promyelocitic cell line is a pleuripotent cell line. In addition to differentiation along the monocytic pathway, it can differentiate through a granulocytic pathway. This pathway has been shown to be induced by the addition of neolactogangliosides (Nojiri et al., 1988). Granulocytic differentiation in response to GSLs has been shown to occur by a mechanism independent of retinoid-induced granulocytic differentiation of these cells (Nakamura et al., 1991). The monocytic differentiation of HL-60 cells has been suggested to be due to inhibition of insulin receptor signal transduction (Nojiri et al., 1991). These data suggest that GSL-induced differentiation is mechanistically similar to GSL inhibition of EGF- or PDGF-dependent cell growth.

The induction of neurite outgrowth by gangliosides in neuronal cell culture has been used extensively for the study of GSL-induced differentiation (Schengrund, 1990; Zeller and Marchase, 1992). Gangliosides and anti-ganglioside antibodies have been shown to stimulate neurite outgrowth in cell culture (Cannella *et al.*, 1988; Barletta *et al.*, 1991; Doherty *et al.*, 1992; Chatterjee, *et al.*, 1992). Gangliosides G_{M3}, G_{M1}, and G_{Q1b} all have been shown to have this property (Cannella *et al.*, 1988; Doherty *et al.*, 1992), as have anti-G_{M1} antibody or the beta subunit of cholera toxin (Barletta *et al.*, 1991). In the pheochromocytoma cell line PC12, the neural cell adhesion molecule (N-CAM), and *N*-cadherin-dependent neurite outgrowth, is enhanced by ganglioside G_{M1} (Doherty *et al.*, 1992), presumably by promoting cell adhesion molecule-induced calcium influx into the neurons.

Contrary to the data suggesting that gangliosides act as differentiation agents, ganglioside G_{M3} can inhibit keratinocyte proliferation in a reversible manner without inducing keratinocyte differentiation (Paller *et al.*, 1992,1993). In this system, differentiation is assessed by cornified envelope production and expression of desmoplakin or involucrin. Although growth arrest of keratinocytes often occurs as a consequence of keratinocyte differentiation, decreased proliferation does not necessarily result in the onset of differentiation. Isoleucine starvation, for example, results in a reversible inhibition of proliferation in G_1 phase of the cell cycle without associated increase of keratinocyte differentiation (Pittelkow *et al.*, 1986; Wilke *et al.*, 1988). The inhibition of keratinocyte cell growth by gangliosides suggests that the inhibition of cell growth and the induction of differentiation by gangliosides may proceed by very different mechanisms.

III. MECHANISMS OF GROWTH MODULATION BY GLYCOSPHINGOLIPIDS

The experiments described in the preceding sections show that GSLs may influence cell growth and differentiation by modulation of cell surface receptors. Mechanisms for how GSLs may effect receptor activities are now beginning to emerge.

A. Effect on Growth Factor Receptors

The EGF receptor is among the most widely studied growth factor receptors (Carpenter, 1987; Yarden and Ullrich, 1988). The tyrosine kinase activity of this receptor is inhibited by G_{M3} (Bremer *et al.*, 1986). The mechanism for this inhibition is beginning to be understood in some detail.

1. Carbohydrate Specificity

The specificity of G_{M3} inhibition of the EGF receptor appears to reside in the carbohydrate moiety of G_{M3}. As can be seen in Table III, inhibition of immunoprecipitated EGF receptor kinase activity is restricted to ganglioside G_{M3}. Removal of sialic acid from G_{M3} to form lactosyl ceramide results in a complete loss of inhibitory activity. Ganglio-series gangliosides are not effective as inhibitors. Sialosyl paragloboside (SPG; IV^3 NeuAcnLc$_4$Cer) also was noninhibitory. This result is significant since SPG has a terminal trisaccharide structure very similar to that of G_{M3}. Further, the carboxyl group of G_{M3} sialic acid appears to be essential for inhibition of the EGF receptor (K. S. Kochhar and E. G. Bremer,

TABLE III

Inhibition of EGF Receptor Autophosphorylation by Different Gangliosides

Gangliosides	Structure	Minimum amount required for inhibition
G_{M3}	Gal$\beta \rightarrow$ 4GlcCer 3 $\uparrow$ α2 NANA	$<50 \ \mu M$
LacCer	Galβ1 $\rightarrow$ 4GlcCer	NI[a]
IV^3 NeuAcnLcCer	Galβ1 $\rightarrow$ 4GlcNAcβ1 $\rightarrow$ 3Ga1β1 $\rightarrow$ 4GlcCer 3 $\uparrow$ α2 NANA	$>450 \ \mu M$
G_{DIA}	Galβ1 $\rightarrow$ 3GalNAcβ1 $\rightarrow$ 4Galβ1 $\rightarrow$ 4GlcCer 3 3 $\uparrow$ $\uparrow$ α2 α2 NANA NANA	$>330 \ \mu M$
G_{MI}	Galβ1 $\rightarrow$ 3GalNAcβ1 $\rightarrow$ 4Galβ1 $\rightarrow$ 4GlcCer 3 $\uparrow$ α2 NANA	$>400 \ \mu M$

[a] NI, No inhibition

unpublished observation). When G_{M3} is converted to the nonulosamine derivative (the sialic acid carboxyl group in structure 2 is reduced to a primary alcohol), the molecule is no longer able to inhibit the EGF receptor (Fig. 2). The lactone, or internal ester, form of G_{M3} also was tested for its ability to inhibit the EGF receptor kinase. In contrast to the nonulosamine derivative, the lactone form of G_{M3} was found to be a more effective inhibitor than G_{M3} (Fig. 2). These data suggest that a specific conformation, in addition to the change on sialic acid, may be required for inhibition of the EGF receptor kinase. Removal of the acetate from the N-acetyl group on sialic acid (Scheme 2) also resulted in some loss of inhibition (K. S. Kochhar, and E. G. Bremer, unpublished observation). These results are especially interesting since Hakomori and colleagues (Hanai *et al.*, 1988) previously described the presence of this same modification (de-N-acetyl-G_{M3}) in A431 cells and suggested that this molecule stimulates the EGF receptor kinase. Results from our laboratory, however, suggest that de-N-acetyl-G_{M3} does not stimulate. Experiments from the laboratory of Rintoul also suggested that de-N-acetyl-G_{M3} may work by different mechanisms (Song *et al.*, 1991).

2. Effect on Growth Factor Receptor Dimerization

Dimerization of the EGF receptor has been described as a mechanism for activation of the kinase activity and signal transduction through the EGF receptor (Yarden and Schlessinger, 1987; Schlessinger, 1988). This proposed mechanism for activation is somewhat controversial, however, since the EGF receptor kinase can be activated without dimerization (Davis *et al.*, 1988). As shown in Fig. 3, G_{M3} is able to inhibit formation of EGF receptor dimers after stimulation of cells with EGF. These data suggest that G_{M3} might be able to act on EGF receptors by preventing activation of the EGF receptor. This particular mechanism also may be a general mechanism for ganglioside inhibition of growth factor receptors. The PDGF receptor also is thought to be activated by dimerization (Heldin *et al.*, 1989). Van Brocklyn and colleagues (1993) have reported that ganglio-series gangliosides can inhibit dimerization of the PDGF receptor as well as inhibit kinase activity of this receptor in both 3T3 cells and NG108 neuroblastoma cells. Hakomori (1990) also suggested that ganglio-sides may act by inhibiting dimer formation of growth factor receptors.

B. Effect on Cell Adhesion

The effect of GSLs on the regulation of adhesion processes has been suggested in several experimental systems. Substratum adhesion sites for

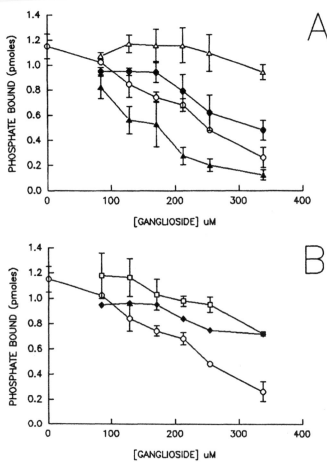

FIGURE 2   Effect of increasing concentration of G_{M3} and G_{M3} carbohydrate derivatives on *in vitro* phosphorylation of EGF receptor. Solubilized membrane proteins (25 μg in 2% Triton from A431 cells were incubated in phosphorylation buffer containing increasing concentration of G_{M3} and its derivatives, and 0.16 μM EGF for 10 min at 25°C. The reaction was started by the addition of 0.36 μM [γ-^{32}P] ATP for 4 min at 0°C. The reaction was terminated by the addition of 50 μl Laemmli's sample buffer. Aliquots were subjected to 8% SDS–PAGE. The gel was dried and visualized by autoradiography. The region containing the EGF receptor (170 kDa) was excised and ^{32}P activity was determined by a liquid scintillation counter. (A) O, G_{M3}; ●, methyl-ester G_{M3}; △, reduced-methyl-ester G_{M3}; ▲, inner-ester G_{M3}; (B) ◆, de-*N*-acetyl G_{M3}; □, de-*N*-acetyl lyso-G_{M3}; O, G_{M3}.

fibroblast (Mugnai *et al.*, 1984), hepatoma (Barletta *et al.*, 1989), and melanoma (Okada *et al.*, 1984) cells contain unique ganglioside distributions. As described in the previous section on induction of differentiation, neurite outgrowth has been shown to be associated with adhesion. These

F4 mAb   − +   + +   + +   + +

BS^3 + −   + +   + +   + +

EGF   + +   + −   + −   + −

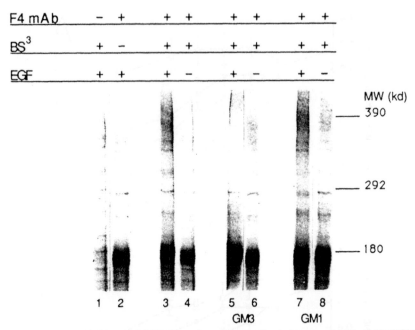

MW (kd)
— 390

— 292

— 180

1   2   3   4   5   6   7   8

GM3   GM1

FIGURE 3 Inhibition of EGFR dimer formation by G_{M3} in intact cells. The A1S (EGF-responsive) clone of A431 cells were cultured in the presence or absence of gangliosides for 3 days. The cultures then were labeled metabolically with [^{35}S]methionine. After stimulation with EGF, dimers were trapped on the surface of the cells by cross-linking with Bis(sulfo-succinimidyl)suberate (BS^3). The cells then were solubilized, immunoadsorbed with anti-EGFR MAb, and separated by SDS–PAGE. G_{M3} but not G_{M1} was able to prevent EGF receptor dimerization.

experiments indicate that gangliosides may play a role in the adhesion process, perhaps by binding directly to extracellular matrix proteins such as fibronectin (Thompson *et al.*, 1986) or by serving as modulators of the interaction of cell surface receptors with fibronectin. Cell attachment to fibronectin matrix occurs via the cell binding domain of fibronectin and the integrin receptor on the cell surface. The interaction of fibronectin and integrin receptors on the cell surface can be inhibited by small peptides containing the sequence Arg–Gly–Asp (RGD). The majority of the data is in favor of a modulating role for gangliosides in fibronectin-mediated adhesion and neurite outgrowth. Disialogangliosides enhance interactions between integrin receptors and vitronectin (Cheresh *et al.*, 1986,1987; Burns *et al.*, 1988). Addition of anti-G_{M1} ganglioside antibodies can shift neuritogenesis from an RGD peptide-sensitive process to a peptide-insensitive one in dorsal root ganglion neuron hybrid clones (Barletta *et al.*, 1991). These data suggest that ganglioside G_{M1} may modulate integrin receptor activity. Ganglioside G_{D2} can be coimmunoprecipitated with the vitronectin receptor. The presence of G_{D2} in this complex appears to enhance the interaction of the vitronectin receptor with vitronectin (Cheresh et al., 1987). Since vitronectin receptor and fibronectin receptor are closely related integrin heterodimers, both of which bind RGD sequences in their respective ligands, a similar ganglioside receptor complex seems likely to exist in the case of the fibronectin receptor. Zheng et al. (1992,1993) found that ganglioside G_{M3} can modulate the fibronectin receptor function ($\alpha 5, \beta 1$ integrin). This conclusion was based on mouse mammary carcinoma mutant cell lines characterized by the presence or absence of ganglioside G_{M3}. Both cell lines expressed approximately the same quantity of integrin receptors, but the cell line that contained a high content of G_{M3} showed a much stronger adhesion to fibronectin-coated plates. Further, liposomes containing phosphatidylcholine, cholesterol, $\alpha 5, \beta 1$ integrin, and ganglioside G_{M3} were able to adhere to fibronectin-coated plates with much greater efficiency than liposomes without G_{M3}.

In addition to the ability of GSLs to modulate the activity of integrin receptors, GSLs can participate directly in adhesion processes. The selectin family of adhesion molecules has been implicated in the interactions between leukocytes and the vascular endothelium, leading to lymphocyte homing, platelet binding, and neutrophil extravasation. The three known selectins are L selectin, E selectin, and P selectin (for review, see Springer, 1990). All share structural features that include a calcium-dependent lectin binding domain. Through the calcium-dependent lectin domain, these receptors are able to interact directly with the Lewis blood group-related carbohydrate epitopes, including the sialyl Lewis X (SLe^x) antigen (Phillips et al., 1990; Foxall et al., 1992) and sialyl Lewis A (SLe^a) antigen (Berg et al., 1991).

A third mechanism by which GSLs may participate in cell adhesion is through GSL–GSL interactions. Kojima and Hakomori (1989,1991) have described that cell lines expressing various levels of ganglioside G_{M3} at the cell surface show different degrees of adhesion and spreading on solid phases coated with GSLs. The degree of cell adhesion in spreading was greatest on Gg$_3$. The spreading and motility of G_{M3}-expressing cells on Gg$_3$-coated solid phases were inhibited by treatment of the cells with anti-G_{M3} antibodies or sialidase. Further, Kojima et al. (1992) have suggested that, for cell adhesion in a dynamic flow system (used as a model for blood flow through a vessel), GSL–GSL interactions may predominate over both lectin- and integrin-based mechanisms of adhesion of B16 melanoma cells to nonactivated endothelial cells. The nonactivated endothelial cells express integrin receptors but do not express the selectin family of receptors. Again, the adhesion could be inhibited by liposomes containing G_{M3} or by pretreatment of the cells with anti-G_{M3}.

C. Signal Transduction by Glycosphingolipid Metabolites

GSLs have been shown to influence a variety of signal transduction events such as ion transport and kinase activation or inhibition (for reviews, see Hakomori, 1990; Zeller and Marchase, 1992). This influence may occur through modulation, as described earlier for growth factor or adhesion receptors, or through the generation of second messengers. Products that could arise from the metabolism from glycosphingolipids such as ceramide, sphingosine, or dimethylsphingosine have been shown to be modulators of specific tyrosine or serine–theonine protein kinases (Hakomori, 1990). In addition, these molecules can act as second messengers in other systems. For example, the binding of tumor necrosis factor (TNFα) to its receptor results in the production of ceramide from sphingomyelin. The released ceramide then has been shown to act on a ceramide-dependent kinase. The metabolism and signaling properties of these ceramides is described in Chapter 14 and will not be described in detail here. Some of the effects of added glycosphingolipids to cell cultures may be due to the effect of ceramide or its metabolites rather than to the glycosphingolipid itself. When glycosphingolipids are added to cell cultures, typically about 50 μM added ganglioside is required because of the rather inefficient uptake of glycosphingolipids by the cells (Bremer et al., 1984; Paller et al., 1993; Yates et al., 1993). On the other hand, the effects on signal transduction seen by ceramide require 1 μM or less, suggesting that only a small percentage of contamination by ceramide or sphingosine in the glycosphingolipid preparation could result in an effect of proliferation or cell function caused by ceramide or sphingosine and not the GSL. This possibility emphasizes the point brought up earlier in this review that highly purified glycosphingolipids are required to study the growth behavior of cells in response to GSLs.

IV. REGULATION OF GLYCOSPHINGOLIPID EXPRESSION

If GSLs are involved in cell growth and differentiation, as the data described earlier would suggest, the regulation of their expression on the cell surface is probably a very tightly controlled event. The specific changes in GSLs seen during differentiation and oncogenic transformation are consistent with precise regulation of GSL expression in the cell. The expression of certain GSLs such as G_{M3} can be controlled by the expression of the required glycosyltransferase required for synthesis of a particular GSL. The expression of hydrolytic enzymes such as cell surface sialidase for G_{M3} also must be considered in the regulation of GSL expression.

Finally, posttranslational modifications of metabolic enzymes, such as phosphorylation or other modifications, can affect enzyme activity.

A. Glycosyltransferases

Over the past few years, several glycosyltransferase enzymes have been cloned (for review, see Paulson and Colley, 1989). Although most of the enzymes cloned are enzymes used in glycoprotein biosynthesis, an understanding of these enzymes should reveal significant information about the enzymes used in GSL biosynthesis. For example, the cloning and sequence analysis of $\alpha 2,3$ sialyltransferase enzymes have revealed evidence for a conserved protein motif in the sialyltransferase gene family (Wen *et al.*, 1992). Previous data had indicated a lack of sequence homology among the glycosyltransferases cloned (Paulson and Colley, 1989). The data on sialyltransferases suggest that these enzymes may be grouped into families, thus enhancing the search for other sialyltransferases.

The availability of cloned transferase enzymes should facilitate the understanding of the regulation of expression of these enzymes as they relate to cell growth and differentiation. The $\alpha 2,6$ sialyltransferase, for example, is tissue restricted with the highest levels found in liver (Svensson *et al.*, 1990). Its expression is regulated transcriptionally by multiple promoters. The expression of the $\alpha 2,6$ sialyltransferase can be regulated by the liver-enriched transcription factors HNF1, DBP, and LAP (Svensson *et al.*, 1992). These results indicate that tissue-specific glycosylation can be regulated at the level of transcription by the same factors involved in the expression of a number of other tissue-specific genes (Svensson *et al.*, 1990, 1992).

B. Cell Surface Sialidases

The presence of hydrolytic enzymes on the cell surface also can regulate the expression of GSLs. A plasma membrane-bound sialidase has been suggested to be involved in cell growth and transformation (Vaheri *et al.*, 1972; Yogeeswaran and Hakomori, 1975). The occurrence of sialidase activity in the culture medium of human skin fibroblast has been reported (Usuki *et al.*, 1988a,b). The sialidase activity found in cell culture supernatants was bimodal with respect to pH, with optima for ganglioside cleavage. An acidic form with an optimum at pH 4.5 and a neutral form with an optimum at pH 6.5 have been described. The sialidase activity at pH 6.5 was found to correlate with the growth activity of the cultured cells, and was present during logarithmic growth of the cells and virtually absent in the medium of contact-inhibited cells. Usuki *et al.* (1988b) have suggested that this change in sialidase activity might account for the positive and negative modulation of the EGF receptor tyrosine kinase activity by G_{M3}. According to their model, the inhibitory effect of G_{M3} can be abolished by sialidase-catalyzed conversion of cell-surface G_{M3} to lactosyl ceramide (LacCer). Support for this model was suggested by the finding that added sialidase from *Clostridium perfringens* was able to stimulate the growth of human skin fibroblast (Ogura and Sweeley, 1992). When fibroblast cells were incubated for 24–48 hr in the presence of the sialidase, DNA synthesis increased approximately 2-fold and the cell density was stimulated 1.5-fold over controls. These studies are consistent with the model proposed by Usuki *et al.* (1988b) and consistent with the idea that regulation of GSLs at the cell surface may be very important in cell growth regulation.

C. Regulation of Enzyme Activity

In addition to transcriptional regulation of transferase activity or hydrolytic enzyme activity, posttranslational modifications of GSL metabolic enzymes or inhibitors of GSL synthesis may be involved in the regulation of GSL expression on the cell surface. The G_{M3} ganglioside-forming enzyme CMP-sialic acid : lactosyl ceramide $\alpha 2,3$ sialyltransferase (SAT-1) has been purified to apparent homogeneity from rat liver Golgi apparatus (Melkerson-Watson and Sweeley, 1991). This enzyme can be detected with an anti-phosphotyrosine monoclonal antibody on Western Blot (L. J. Melkerson-Watson and C. C. Sweeley, personal communication), suggesting the possibility that the tyrosine phosphorylation state of the cell may alter the activity of this sialyltransferase enzyme. Consistent with this idea, the synthesis of G_{M3} has been shown to be enhanced by the treatment of granulosa cells with okadaic acid, a phosphatase inhibitor (Hattori and Horiuchi, 1992). Modulation of SAT-1 activity by tyrosine phosphorylation is a very attractive hypothesis, since such an event would provide a quick way to regulate the expression of G_{M3} on the cell surface in response to EGF and suggests the possibility of a feedback mechanism: the stimulation of tyrosine kinase activity by EGF stimulation results in phosphorylation of SAT-1, which can increase the synthesis of G_{M3}, thus inhibiting the tyrosine kinase activity of the EGF receptor.

Glycosyltransferase enzymes also may be regulated by inhibitors of their activity. Koul *et al.* (1990) have isolated the UDP-galactose : globoside galactosyltransferase from mouse kidneys. This enzyme is among those responsible for generating the stage-specific embryonic antigens 1 and 3 (SSEA3 and SSEA1). These researchers found an almost 2-fold greater enzyme activity in male than in female DBA/2 mice. In the presence of the detergent sodium cholate, however, the activity was similar in both male and female preparations. These data suggest the presence of an enzyme modulator in the female DBA/2 mice. These data also suggest the possibility that the difference in male and female SSEA3 and SSEA1 expression is not the result of differential expression of the galactosyltransferase enzyme but of the expression of an enzyme inhibitor in the female mice. Inhibitors of glycosyltransferase activity also have been described for other glycosyltransferases (Quiroga et al., 1985; Quiroga and Caputto, 1988; Martin et al., 1990).

The localization of the glycosyltransferase in the Golgi apparatus also may play an important role in directing terminal glycosylation of glycolipids or glycoproteins. The $\alpha2,6$ sialyltransferase enzyme has been described as an acute phase reactant (Lammers and Jamieson, 1988), meaning after an injury the level of $\alpha2,6$ sialyltransferase increases in the serum at the expense of the trans-Golgi population of the $\alpha2,6$ sialyltransferase. This observation also suggests that glycoconjugates assembled after an injury or an inflammatory event are more likely to have a different terminal glycosylation than those assembled prior to the injury.

V. SUMMARY

Over the past few years, glycosphingolipids have been demonstrated to be more than structural components of the plasma membrane. Glycosphingolipids have been described as playing an important role in cell–cell recognition (Hakomori, 1981), modulation of cell growth (Hakomori, 1990), and possibly as second messenger signal transduction (Merrill, 1991). In terms of cell growth and differentiation, glycosphingolipids may function primarily as modulators of cell surface receptors. The data described in this chapter, the ability of gangliosides to modulate the activity of integrin receptors, and the ability of gangliosides to modulate growth factor receptor signal transduction have suggested this emerging theme for a role glycosphingolipids may play on the cell surface. The ability of membrane lipids to modulate the activity of membrane bound enzymes is not necessarily a new idea (for review, see Yeagle, 1987). The diversity of carbohydrate structures found on glycosphingolipids suggests that the carbohydrate may provide specificity for lipid modulation of membrane proteins. That the carbohydrate moiety of glycosphingolipids is responsible for creating the proper lipid microenvironment for regulating growth factor function or for concentrating sphingolipid second messengers so they can be metabilized on stimulation is an intriguing area for further research.

References

Barletta, E., Mugnai, G., and Ruggieri, S. (1989). Morphological characteristics and ganglioside composition of substratum adhesion sites in a hepatoma cell line (CMH5123) during different phases of growth. *Exp. Cell. Res.* **182**, 394–402.

Barletta, E., Bremer, E. G., and Culp, L. A. (1991). Neurite outgrowth in dorsal root neuronal hybrid clones modulated by ganglioside GM1 and disintegrins. *Exp. Cell Res.* **193**, 101–111.

Barnes, D., and Saito, G. (1980). Methods for growth of cultured cells in serum-free medium. *Anal. Biochem.* **102**, 255–270.

Berg, E. L., Robinson, M. K., Mansson, O., Butcher, E. C., and Magnani, J. L. (1991). A carbohydrate domain common to both sialyl Le^a and sialyl Le^x is recognized by the endothelial cell leukocyte adhesion molecule ELAM-1. *J. Biol. Chem.* **266**, 14869–14872.

Bremer, E. G., and Hakomori, S. (1982). GM3 ganglioside induces hamster fibroblast growth inhibition in chemically-defined medium: Ganglioside may regulate growth factor receptor function. *Biochem. Biophys. Res. Commun.* **106**, 711–728.

Bremer, E. G.,. Hakomori, S., Bowen-Pope, D. F., Raines, E., and Ross, R. (1984). Ganglioside-mediated modulation of cell growth, growth factor binding and receptor phosphorylation. *J. Biol. Chem.* **259**, 6818–6825.

Bremer, E. G., Schlessinger, J., and Hakomori, S. (1986). Ganglioside-mediated modulation of cell growth: Specific effects of GM3 on tyrosine phosphorylation of the epidermal growth factor receptor. *J. Biol. Chem.* **261**, 2434–2440.

Burns, G. F., Lucas, C. M., Krissansen, G. W., Werkmeister, J. A., Scanlon, D. B., Simpson, D. B., and Vadas, M. A. (1988). Synergism between membrane gangliosides and Arg-Gly-Asp-directed glycoprotein receptors in attachment to matrix proteins by melanoma cells. *J. Cell Biol.* **107**, 1225–1230.

Callies, R., Schwarzmann, G., Radsak, K., Siegert, R., and Wiegandt, H. (1977). Characterization of the cellular binding of exogenous gangliosides. *Eur. J. Biochem.* **80**, 425–432.

Cannella, M. S., Roisen, F. J., Ogawa, T., Sugimoto, M., Ledeen, R. W. (1988). Comparison of epi-GM3, with GM3 and GM1 as stimulators of neurite outgrowth. *Brain Res.* **467**, 137–143.

Carpenter, G. (1987). Receptors of epidermal growth factor and other polypeptide mitogens. *Annu. Rev. Biochem.* **56**, 881–914.

Chandrabose, K. A., Graham, J. M., and Macpherson, I. A. (1976). Glycolipid glycosyl transferases of a hamster cell line in culture. II. Subcellular distribution and the effect of culture age and density. *Biochim. Biophys. Acta.* **429**, 112–122.

Chatterjee, D., Chakraborty, M., and Anderson, G. M. (1992). Differentiation of Neuro-2a neuroblastoma cells by an antibody to GM3 ganglioside. *Brain Res.* **583(1–2)**, 31–44.

Cheresh, D. A., Pierschbacher, M. D., Herzig, M. A., and Mujoo. (1986). Disialogangliosides GD2 and GD3 are involved in the attachment of human melanoma and neuroblastoma cells to extracellular matrix proteins. *J. Cell Biol.* **102**, 688–694.

Cheresh, D. A., Pytela, R., Pierschbacher, M. D., Klier, F. G., Ruoslahti, E., and Reisfeld, R. A. (1987). An Arg-Gly-Asp-directed receptor on the surface of human melanoma cells exists in a divalent cation-dependent functional complex with the disialoganglioside GD2. *J. Cell. Biol.* **105**, 1163–1173.

Critchley, D. R., and Macpherson, I. A. (1973). Cell density-dependent glycolipids in NIL2 hamster cells derived from an alignment and transformed cell lines. *Biochim. Biophys. Acta* **296**, 145–159.

Davis, R. J., Girones, N., and Fancher, M. F. (1988). Two alternative mechanisms control the interconversion of functional states of the epidermal growth factor receptor. *J. Biol. Chem.* **263**, 5373–5379.

Doherty, P., Ashton, S. V., Skaper, S. D., Leon, A., and Walsh, F. S. (1992). Ganglioside modulation of neural cell adhesion molecule and N-cadherin-dependent neurite outgrowth. *J. Cell. Biol.* **117**, 1093–1099.

Felding-Habermann, B., Igarashi, Y., Fenderson, B. A., Park, L. S., Radin, N. S., Inokuchi, J., Strassmann, G., Handa, K., and Hakomori, S. (1990). A ceramide analogue inhibits T cell proliferative response through inhibition of glycosphingolipid synthesis and enhancement of N_1,N-dimethylsphingosine synthesis. *Biochemistry* **29**, 6314–6322.

Foxall, C., Watson, S. R., Dowbenko, D., Fennie, C., Lasky, L. A., Kiso, M., Hasegawa, A., Asa, D., and Brandley, B. K. (1992). The three members of the selectin receptor family recognize a common carbohydrate epitope, the sialyl Lewis(x) oligosaccharide. *J. Cell. Biol.* **117**, 895–902.

Gahmberg, C. G., and Hakomori, S. (1975). Surface carbohydrates of hamster fibroblasts. I. Chemical characterization of surface-labeled glycosphingolipids and a special ceramide tetrasaccharide for transformants. *J. Biol. Chem.* **250**, 2438–2446.

Hakomori, S. (1970). Cell density-dependent changes of glycolipid concentrations in fibroblasts and loss of this response in virus transformed cells. *Proc. Natl. Acad. Sci. U.S.A.* **67**, 1741–1747.

Hakomori, S. (1981). Glycosphingolipids in cellular interaction, differentiation and oncogenesis. *Annu. Rev. Biochem.* **50**, 733–764.

Hakomori, S. (1990). Bifunctional roles of glycosphingolipids. Modulators for transmembrane signalling and mediators for cellular interactions. *J. Biol. Chem.* **265**, 18713–18716.

Hakomori, S., and Kannagi, R. (1983). Glycosphingolipids as tumor-associated and differentiation markers. *J. Natl. Cancer Inst.* **71**, 231–251.

Hakomori, S., and Murakami, W. T. (1968). Glycolipids of hamster fibroblasts and derived malignant-transformed cell lines. *Proc. Natl. Acad. Sci. U.S.A.* **59**, 254–261.

Hanai, N., Dohi, T., Nores, G. A., and Hakomori, S. (1988). A novel ganglioside, de-N-acetyl-GM3 (II^{3}NeuNH$_2$LacCer), acting as a strong promoter for epidermal growth factor receptor kinase and as a stimulator for cell growth. *J. Biol. Chem.* **263**, 6296–6301.

Hannun, Y. A., and Bell, R. M. (1989). Functions of sphingolipids and sphingolipid breakdown products in cellular regulation. *Science* **243**, 500–507.

Hattori, M., and Horiuchi, R. (1992). Enhancement of ganglioside GM3 synthesis in okadaic-acid-treated granulosa cells. *Biochim. Biophys. Acta* **1137(1)**, 101–106.

Heldin, C.-H., Ernlund, A., Rorsman, C., and Ronnstrand, L. (1989). Dimerization of B-type platelet-derived growth factor receptors occurs after ligand binding and is closely associated with receptor kinase activation. *J. Biol. Chem.* **264**, 8905–8912.

Igarashi, Y., Hakomori, S., Toyokuni, T., Dean, B., Fujita, S., Sugimoto, M., Ogawa, T., El-Ghendy, K., and Racker. E. (1989). Effect of chemically well-defined sphingosine and its N-methyl derivatives on protein kinase C and src kinase activities. *Biochemistry* **28**, 6796–6800.

Igarashi, Y., Kitamura, K., Toyokuni, T., Dean, B., Fenderson, B. A., Ogawa, T., and Hakomori, S. (1990). A specific enhancing effect of N,N-dimethylsphingosine on epidermal growth factor receptor autophosphorylation: Demonstration of its endogenous occurrence (and the virtual absence of unsubstituted sphingosine) in human epidermoid carcinoma A431 cells. *J. Biol. Chem.* **265**, 5385–5389.

Inokuchi, J., and Radin, N. S. (1987). Preparation of the active isomer of 1-phenyl-2-decanoylamino-3-morpholino-1-propanol inhibitor of murine glucocerebroside synthetase. *J. Lipid Res.* **28,** 565–571.

Kanda, S., Inoue, K., Nojima, S., Utsumi, H., and Weigandt, H. (1982). Incorporation of spin-labeled ganglioside analogues into cell and liposomal membranes. *J. Biochem. (Tokyo)* **91,** 1707–1718.

Keenan, T. W.,. Schmid, E., Franke, W. W., and Wiegandt, H. (1975). Exogenous ganglioside suppresses growth rate of transformed and untransformed 3T3 mouse cells. *Exp. Cell Res.* **92,** 259–270.

Kijimoto, S., and Hakomori, S. (1971). Enhanced glycolipid: Alpha-galactosyltransferase activity in contact-inhibited hamster cells, and loss of this response in polyoma transformants. *Biochem. Biophys. Res. Commun.* **44,** 557–563.

Kingsley, D. M., Kozarsky, K. F., Hobbie, L., and Kreiger, M. (1986). Reversible defects in O-linked glycosylation and LDL receptor expression in a UDP-Gal/UDP-GalNAc-4-epimerase mutant. *Cell* **44,** 749–759.

Kojima, N., and Hakomori, S. (1989). Specific interaction between gangliotriaosylceramide (Gg3) and sialosyllactosylceramide (GM3) as a basis for specific cellular recognition between lymphoma and melanoma cells. *J. Biol. Chem.* **264(34),** 20159–20162.

Kojima, N., and Hakomori, S. (1991). Cell adhesion, spreading, and motility of GM3-expressing cells based on glycolipid-glycolipid interaction. *J. Biol. Chem.* **266(26),** 17552–17558.

Kojima, N., Shiota, M., Sadahira, Y., Handa, K., and Hakomori, S. (1992). Cell adhesion in a dynamic flow system as compared to static system. Glycosphingolipid-glycosphingolipid interaction in the dynamic system predominates over lectin- or integrin-based mechanisms in adhesion of B16 melanoma cells to non-activated endothelial cells. *J. Biol. Chem.* **267(24),** 17264–17270.

Koul, O., Prada-Maluf, M., and McCluer, R. H. (1990). UDP-galactose:globoside galactosyltransferase in murine kidney. *J. Lipid Res.* **31(12),** 2227–2234.

Kreutter, D., Kim, J. Y. H., Goldenring, J. R., Rasmussen, H., Ukomadu, C., DeLorenzo, R. J., and Yu, R. K. (1987). Regulation of protein kinase C activity by gangliosides. *J. Biol. Chem.* **262,** 1633–1637.

Laine, R. A., and Hakomori, S. (1973). Incorporation of exogenous glycosphingolipids in plasma membranes of cultured hamster cells and concurrent change of growth behavior. *Biochem. Biophys. Res. Commun.* **54,** 1039–1045.

Lammers, G., and Jamieson, J. C. (1988). The role of a cathepsin D-like activity in the release of Galβ1 → 4 GlcNAc α2 → 6-sialyltransferase from rat liver Golgi membranes during the acute-phase response. *Biochem. J.* **256,** 623–631.

Martin, A., Ruggiero-Lopez, D., Biol, M. C., and Louisot, P. (1990). Evidence for the presence of an endogenous cytosolic protein inhibitor of intestinal fucosyltransferase activities. *Biochem. Biophys. Res. Commun.* **166,** 1024–1034.

Melkerson-Watson, L. J., and Sweeley, C. C. (1991). Purification to apparent homogeneity by immunoaffinity chromatography and partial characterization of the GM3 ganglioside-forming enzyme, CMP-sialic acid:lactosylceramide alpha 2,3-sialyltransferase (SAT-1), from rat liver Golgi. *J. Biol. Chem.* **266(7),** 4448–4457.

Merrill, A. H. (1991). Cell regulation by sphingosine and more complex sphingolipids. *J. Bioenerg. Biomembr.* **23,** 83–104.

Mugnai, G., Tombaccini, D., and Ruggieri, S. (1984). Ganglioside composition of substrate-adhesion sites of normal and virally-transformed BALB/c 3T3 cells. *Biochem. Biophys. Res. Commun.* **125,** 142–148.

Nakamura, M., Kirito, K., Yamanoi, J., Wainai, T., Noijiri, H., and Saito, M. (1991). Ganglioside GM3 can induce megakaryocytoid differentiation of human leukemia cell line K562 cells. *Cancer Res.* **51**, 1940–1945.

Nojiri, H., Takaku, F., Terui, Y., Miura, Y., and Saito, M. (1986). Ganglioside GM3: an acidic membrane component that increases during macrophage-like cell differentiation can induce monocytic differentiation of human myeloid and monocytoid leukemic cell lines HL-60 and U937. *Proc. Natl. Acad. Sci. U.S.A.* **83**, 782–786.

Nojiri, H., Kitagawa, S., Nakamura, M., Kirito, K., Enomoto, Y., and Saito, M. (1988). Neolacto-series gangliosides induce granulocytic differentiation of human promyelocytic leukemia cell line HL-60. *J. Biol. Chem.* **263**, 7443–7446.

Nojiri, H., Stroud, M., and Hakomori, S. (1991). A specific type of ganglioside as a modulator of insulin-dependent cell growth and insulin receptor tyrosine kinase activity: Possible association of ganglioside-induced inhibition of insulin receptor function and monocytic differentiation induction of HL-60 cells. *J. Biol. Chem.* **266**, 4531–4537.

Ogura, K., and Sweeley, C. C. (1992). Mitogenic effects of bacterial neuroaminidase and lactosylceramide on human cultured fibroblasts. *Exp. Cell Res.* **199(1)**, 169–73.

Okada, Y., Mugnai, G., Bremer, E. G., and Hakomori, S. (1984). Glycosphingolipids in detergent-insoluble substrate attachment matrix (DISAM) prepared from substrate attachment material (SAM). *Exp. Cell. Res.* **155**, 448–456.

Paller, A. S., Arnsmeier, S. K., Robinson, J. K., and Bremer, E. G. (1992). Alteration in keratinocyte ganglioside content in basal cell carcinomas. *J. Invest. Dermatol.* **998**, 226–232.

Paller, A. S., Arnsmeier, S. K., Alvarez-Franco, M., and Bremer, E. G. (1993). Ganglioside GM3 inhibits the proliferation of cultured keratinocytes. *J. Invest. Derm.* **100**, 841–845.

Paulson, J. C., and Colley, K. J. (1989). Glycosyltransferases. Structure, localization, and control of cell type-specific glycosylation. *J. Biol. Chem.* **264**, 17615–17618.

Phillips, M. L., Nudelman, E., Gaeta, F. C., Perez, M., Singhal, A. K., Hakomori, S., and Paulson, J. C. (1990). ELAM-1 mediates cell adhesion by recognition of a carbohydrate ligand, sialyl-Lex. *Science* **250**, 1130–1132.

Pittelkow, M. R., Wille, J. J., and Scott, R. E. (1986). Two functionally distinct classes of growth arrest states in human prokeratinocytes that regulate clonogenic potential. *J. Invest. Dermatol.* **86**, 410–417.

Quiroga, S., and Caputto, R. (1988). An inhibitor of the UDP-N-acetyl galactosaminyltransferase: Purification and properties, and preparation of an antibody to this inhibitor. *J. Neurochem.* **50**, 1695–1700.

Quiroga, S., Caputto, B., and Caputto, R. (1985). Inhibition of the chicken retinal UDP-GalNAc: GM3, N-acetylgalactosaminyltransferase by blood serum and by pineal gland extracts. *J. Neurosci. Res.* **12**, 269–276.

Rosenfelder, G., Ziegler, A., Wernet, P., and Braun, D. G. (1982). Ganglioside patterns: New biochemical markers for human hematopoietic cell lines. *J. Natl. Cancer Inst.* **68**, 203–209.

Sakiyama, H., Gross, S. K., and Robbins, P. W. (1972). Glycolipid synthesis in normal and virus-transformed hamster cell lines. *Proc. Natl. Acad. Sci. U.S.A.* **69**, 872–876.

Schengrund, C. L. (1990). The role(s) of gangliosides in neural differentiation and repair: A perspective. *Brain Res. Bull.* **24**, 131–141.

Schlessinger, J. (1988). The epidermal growth factor receptor as a multifunctional allosteric protein. *Biochemistry* **27**, 3119–3123.

Schwarzmann, G., Hoffmann-Bleihauer, P., Schubert, J., Sandhoff, K., and Marsh, D. (1983). Incorporation of ganglioside analogues into fibroblast cell membranes: A spin-label study. *Biochemistry* **22**, 5041–5048.

Song, W. X., Vacca, M. F., Welti, R., and Rintoul, D. A. (1991). Effects of gangliosides GM3 and De-*N*-acetyl GM3 on epidermal growth factor receptor kinase activity and cell growth. *J. Biol. Chem.* **266(16),** 10174–10181.

Springer, T. A. (1990). Adhesion receptors of the immune system. *Nature (London)* **346,** 425–434.

Svennerholm, L. (1963). Chromatographic separation of human brain gangliosides. *J. Neurochem.* **10,** 613–623.

Svennerholm, L. (1964). The gangliosides. *J. Lipid Res.* **5,** 134–155.

Svensson, E. C., Soreghan, B., and Paulson, J. C. (1990). Organization of the beta-galactoside alpha 2,6-sialyltransferase gene. Evidence for the transcriptional regulation of terminal glycosylation. *J. Biol. Chem.* **265,** 20863–20868.

Svensson, E. C., Conley, P. B., and Paulson, J. C. (1992). Regulated expression of α2,6-sialyltransferase by the liver-enriched transcription factors HNF-1, DBP, and LAP. *J. Biol. Chem.* **267,** 3466–3472.

Thompson, L. K., Horowitz, P. M., Bentley, K. L., Thomas, D. D., Alderete, J. F., and Klebe, R. J. (1986). Localization of the ganglioside-binding site of fibronectin. *J. Biol. Chem.* **261.** 5209–5214.

Usuki, S., Lyu, S-C., and Sweeley, C. C. (1988a). Sialidase activities of cultured human fibroblasts and the metabolism of GM3 ganglioside. *J. Biol. Chem.* **263,** 6847–6853.

Usuki, S., Hoops, P., and Sweeley, C. C. (1988b). Growth control of human foreskin fibroblasts and inhibition of extracellular sialidase activity by 2-deoxy-2,3-dehydro-*N*-acetylneuraminic acid. *J. Biol. Chem.* **263,** 10595–10599.

Vaheri, A., Rouslahti, E., and Nordling, S. (1972). Neuraminidase stimulates division and sugar uptake in density-inhibited cell cultures. *Nature New Biology* **238,** 211–212.

Van Brocklyn, J., Bremer, E. G., and Yates, A. J. (1993). Gangliosides inhibit PDGF-stimulated receptor dimerization in human glioma U-1242MG and Swiss 5T3 cells. *J. Neurochem.* **61,** 371–374.

Weis, F. M. B., and Davis, R. J. (1990). Regulation of epidermal growth factor receptor signal transduction: Role of gangliosides. *J. Biol. Chem.* **265,** 12059–12066.

Wen, D. X., Livingson, B. D., Medzihradszky, K. F., Kelm, S., Burlingame, A. L., and Paulson, J. C. (1992). Primary structure of Galβ1 → 3,(4)G1cNAcα2 → 3-sialyltransferase determined by mass spectrometry sequence analysis and molecular cloning. Evidence for a protein motif in the sialyltransferase gene family. *J. Biol. Chem.* **267,** 21011–21019.

Wilke, M. S., Hsu, B. M., Wille, J. J., Pittelkow, M. R., and Scott, R. E. (1988). Biologic mechanisms for the regulation of normal human keratinocyte proliferation and differentiation. *Am. J. Pathol.* **131,** 171–181.

Yarden, Y., and Schlessinger, J. (1987). Self-phosphorylation of epidermal growth factor receptor: Evidence for a model of intermolecular allosteric activation. *Biochemistry* **26,** 1434–1442.

Yarden, Y., and Ullrich, A. (1988). Growth factor receptor tyrosine kinases. *Annu. Rev. Biochem.* **57,** 443–478.

Yates, A. J., Van Brocklyn, J., Saqr, H. E., Guan, Z., Stokes, B. T., and O'Dorisio, M. S. (1993). Mechanisms through which gangliosides inhibit PDGF-stimulated mitogenesis in intact Swiss 3T3 cells: Receptor tyrosine phosphorylation, intracellular calcium, and receptor binding. *Exp. Cell Res.* **204,** 38–45.

Yeagle, P. (1987). "The Membranes of Cells." Academic Press, Orlando, Florida.

Yogeeswaran, G., and Hakomori, S. (1975). Cell contact-dependent ganglioside changes in mouse 3T3 fibroblasts and suppressed sialidase activity on cell contact. *Biochemistry* **14,** 2151–2156.

Zeller, C. B., and Marchase, R. B. (1992). Gangliosides as modulators of cell function. *Am. J. Physiol.* **262**, 1341–1355.

Zheng, M., Tsuruoka, T., Tsuji, T., and Hakomori, S. (1992). Regulatory role of GM3 ganglioside in integrin function, as evidenced by its effect on function of $\alpha 5,\beta 1$-liposomes: A preliminary note. *Biochem. Biophys. Res. Commun.* **186**, 1397–1402.

Zheng, M., Fang, H., Tsuruoka, T., Tsuji, T., Sasaki, T., and Hakomori, S. (1993). Regulatory role of GM3 ganglioside in $\alpha 5\beta 1$ integrin receptor for fibronectin-mediated adhesion of FUA169 cells. *J. Biol. Chem.* **268**, 2217–2222.

CHAPTER 16

Generation and Attenuation of Lipid Second Messengers in Intracellular Signaling

Wim J. van Blitterswijk, Dick Schaap, and Rob van der Bend
Division of Cellular Biochemistry, The Netherlands Cancer Institute, 1066 CX
Amsterdam, The Netherlands

I. INTRODUCTION

Many phospholipid-derived products are implicated as mediators and second messengers in signal transduction, and more are likely to be discovered (Ferguson and Hanley, 1991). This chapter focuses on diacylglycerol (DG) as an established second messenger, and phosphatidic acid (PA) as a putative one. DG, the activator of protein kinase C (PKC), is generated by receptor-mediated hydrolysis of phosphatidylinositol 4,5-bisphosphate (PIP$_2$) and, to a larger extent, of other phospholipids, mainly phosphatidylcholine (PC) (Exton, 1990). PA is produced in most cell systems by two enzymatic pathways—DG kinase and phospholipase D (PLD). An exception is interleukin 1-stimulated mesangial cells, in which PA is produced rapidly by a lyso-PA acyltransferase in the plasma membrane (Bursten *et al.*, 1991). Note that the formation, role, and topology of PA and DG in signal transduction are distinct from these compounds as intermediates in *de novo* lipid biosynthesis.

Why is PA thought to be a second messenger? The rapid formation of high concentrations of this lipid during stimulation with agonists strongly suggests that PA has signaling functions. In neutrophils, the time course of PA, rather than DG, formation coincides with that of the respiratory burst and secretion stimulated by fMet–Leu–Phe (fMLP). Abrogation of PA formation by certain inhibitors (Section II,C) or by adding ethanol to the cells inhibits these physiological responses (Cockcroft, 1992; Gélas *et al.*, 1992). Further, PA may stimulate the activity of small G proteins (ras and rho) by inhibition of their associated GTPase-activating proteins (GAPs; Tsai *et al.*, 1990; see, however, Section V). PA is likely to be a direct activator of a protein kinase (Bocckino *et al.*, 1991), possibly PKCζ (Nakanishi and Exton, 1992; C. Limatola and W. van Blitterswijk, unpublished data), the only PKC isotype known to date that is not activated by DG or phorbol ester. Exogenous PA as well as lyso-PA is a potent inducer of DNA synthesis (Moolenaar *et al.*, 1986; van Corven *et al.*, 1989; see Chapter 17). However, this effect of PA/lyso-PA is initiated at the external side of the plasma membrane, possibly via a specific receptor (van der Bend *et al.*, 1992a), and should be distinguished from receptor-stimulated endogenous PA formation inside the cell.

The enzymes directly involved in receptor-mediated DG formation are phospholipases C specific for phosphoinositides (PLC$_i$) and PC (PLC$_c$). In addition, DG is generated by the sequential activation of PLD and PA phosphohydrolase. To date, only cDNAs of PLC$_i$ have been cloned (Rhee, 1991). The different subfamilies are activated by different receptors. For example, PLCγ1 is activated directly by tyrosine kinases, such as the epidermal growth factor (EGF) and platelet-derived growth factor (PDGF) receptors (Section II,D); PLCβ1 is activated through G$_q$ and G$_{11}$, G proteins that couple to heptahelical receptors (Taylor *et al.*, 1991; Wu *et al.*, 1992). A major metabolic route by which second messenger DG levels are attenuated is that of DG kinase(s) in the phosphatidylinositol (PI) cycle. One mammalian DG kinase isotype has been cloned (Sakane *et al.*, 1990; Schaap *et al.*, 1990). Section IV deals with structural, regulatory, and substrate selectivity aspects of this enzyme. DG kinase and PA phosphohydrolase seem to be counteracting enzymes, generating PA and DG and vice versa, respectively. However, evidence suggests that this is not the case. These reactions are likely to be topologically restricted or "channeled" in the cell (Sections III; IV,C). For the same reason, PA generated by DG kinase may not have the same (putative) second messenger function as PA generated by PLD.

In the following sections, we discuss various aspects of generation and attenuation of DG and PA after stimulation of various types of cell surface receptors, including G protein-coupled (heptahelical) receptors, cytokine receptors (Section II,B), and receptors with intrinsic tyrosine kinase activity (Section II,D). Rather than presenting an exhaustive list of these receptors and the pertaining effector mechanisms, we address some selected aspects that have been covered only briefly in other reviews (Billah and Anthes, 1990; Exton, 1990; Cook and Wakelam, 1991; Dennis *et al.*, 1991; Cockcroft, 1992; Liscovitch, 1992).

II. SIGNALING THROUGH PHOSPHOLIPASES C AND D

A. Biphasic Diacylglycerol Generated from Different Phospholipids

Receptor-activated formation of DG is often biphasic. Initial PIP_2 hydrolysis, within seconds, is transient and generally is followed (after 1–2 min) by a second longer-lasting phase of DG formation from a different phospholipid, PC (Billah and Anthes, 1990; van Blitterswijk *et al.*, 1991a; van der Bend *et al.*, 1992b). Evidence for the existence of these two sources of DG is derived from their different fatty acid composition (Pessin and Raben, 1989). [^{3}H]Arachidonate selectively incorporates into PI (PIP_2) and [^{14}C]palmitate or [^{14}C]myristate into PC. In such double-labeled cells, whether DG originates from PI or PC (denoted DG_i and DG_c, respectively) can be determined by measuring $^3H : {}^{14}C$ ratios (Martinson *et al.*, 1989; van Blitterswijk *et al.*, 1991a). Further, agonist-induced release of phosphocholine or choline from PC can be demonstrated, indicative of PLC_c or PLD activity, respectively.

In contrast to the rapidly and transiently formed DG_i, sustained DG_c does not activate PKC in IIC9 fibroblasts stimulated with α-thrombin (Leach *et al.*, 1991), nor in GH3 cells (Martin *et al.*, 1990) and GH4C1 cells (Kiley *et al.*, 1991) stimulated with thyrotropin-releasing hormone. However, in the latter case, $PKC\varepsilon$ is down-regulated, suggesting that it first was activated for a prolonged time. In complement receptor-stimulated neutrophils (Fällmann *et al.*, 1992) and in IgE-stimulated mast cells (Lin *et al.*, 1992), sustained DG_c does activate PKC.

Receptor-mediated phospholipid breakdown is controlled by PKC, which may act as a switch. Activation of PKC with phorbol ester inhibits PIP_2 hydrolysis but stimulates PLD. Conversely, down-regulation of PKC by prolonged treatment of cells with phorbol ester may lead to enhanced agonist-induced PLC activity and consequent DG formation, but blocks PLD activity (van Blitterswijk *et al.*, 1991a,b). Further, Rat 6 fibroblasts that overexpress PKCβ1 show up-regulation of α-thrombin-induced PLD and down-regulation of phosphoinositide hydrolysis (Pachter *et al.*, 1992). *et al.*, 1992).

PLC_c activation is overt for certain cytokine-activated cell systems, activated receptor protein tyrosine kinases (e.g., for EGF and PDGF; Section II,D), and muscarinic receptors (Diaz-Meco *et al.*, 1989; Pacini *et al.*, 1993) but negligible in α-thrombin-stimulated platelets (Huang *et al.*, 1991). In bradykinin-stimulated fibroblasts, PLC_c activity is masked. Here, intracellular release of phosphocholine cannot be detected easily against the high background, but a release of phosphocholine into the extracellular medium is readily detectable (van Blitterswijk *et al.*, 1991b), suggesting that PC hydrolysis occurs in the outer leaflet of the plasma membrane. Similar observations were made in other types of cells (Table I).

TABLE I

PLC_c Activation in Cells Resulting in Rapid Release of Phosphocholine into the Extracellular Medium

		Time		
Agonist[a]	Cell type	Intracellular	Extracellular	Reference
Muscarinic receptors				
ACho/CCho	Swiss 3T3	No	1–2 min	Diaz-Meco *et al.* (1989)
Carbachol	1321N1[b]	nd[c]	2 min	Martinson *et al.* (1989)
CCK/CCho	Pancreatic acini	nd	1–60 min	Matozaki and Williams (1989)
Carbachol	SK-N-BE(2)[d]	No	15–60 sec	Pacini *et al.* (1993)
Other receptors				
IL-1	Jurkat T cells	3–5 min	1 min	Rosoff *et al.* (1988)
Epinephrine	MDCK-D1[e]	nd	0.5–10 min	Slivka *et al.* (1988)
Bradykinin	Fibroblasts	No	2 min	van Blitterswijk *et al.* (1991b)
AngII	VSMC[f]	2 min	2–20 min	Lassègue *et al.* (1991)
IgE	Mast cells	Decrease	0.3–3 hr	Dinh and Kennerly (1991)

[a] Abbreviations: ACho, acetylcholine; CCho, carbamoylcholine; CCK, cholecystokinin; AngII, angiotensin II.
[b] Astrocytoma cells.
[c] nd, Not determined.
[d] Neuroblastoma cells.
[e] Canine kidney cells.
[f] VSMC, Vascular smooth muscle cells.

B. Phosphatidylcholine Hydrolysis in the Absence of Phosphatidylinositol-Bisphosphate Breakdown

Agonist-stimulated cell systems exist (listed in Table II) in which PIP_2 hydrolysis, measured by DG_i formation and Ca^{2+} mobilization, is not detectable, but which still exhibit PC breakdown, mostly by PLC_c. This event is found particularly in white blood cell lines stimulated with cytokines [interleukins 1 and 3 (IL-1 and -3), tumor necrosis factor (TNFα), interferon α (IFNα)]. The resulting DG_c, generated within a few minutes, is generally capable of activating PKC (Table II), presumably the novel PKC isotypes that do not need Ca^{2+} for activation (Parker et al., 1989). In addition to PKC activation, this DG_c may serve another important function in the cell, that is, the (PKC-independent) activation of a sphingomyelinase, generating a different novel second messenger, ceramide (Kolesnick, 1992). This event would stimulate a ceramide-specific protein kinase (Mathias et al., 1993) that can trigger rapid induction of a nuclear transcription factor (Schütze et al., 1992).

C. Mechanisms of Phospholipase D Activation

PLD is activated rapidly in many cells by many agonists, generally in association with PIP_2 hydrolysis (Billah and Anthes, 1990). In that case, PLD activation may be downstream from PKC activation and is blocked

TABLE II

Agonist-Induced Phosphatidylcholine Breakdown in the Absence of PIP_2 Hydrolysis

Agonist[a]	Cell type	PLC_c[b]	PLD[b]	PKC[b]	Reference
TNFα	U937	+	−	+	Schütze et al. (1991,1992)
IFNα	HeLa, Daudi	+	−	+	Pfeffer et al. (1990,1991)
IL-1	Jurkat	+	nd	nd	Rosoff et al. (1988)
IL-3	FDCP-Mix 1	nd	nd	+	Whetton et al. (1988)
IL-3	R6-XE.4	+?	nd	nd	Duronio et al. (1989)
α_2-C10[c]	Rat-1 transfected	−	+	+	McNulty et al. (1992)
CSF-1	Monocytes	+	nd	+	Imamura et al. (1990)
Insulin	Myocytes; adipocytes	+	nd[d]	nd	Hoffman et al. (1991)

[a] Abbreviations: TNF, tumor necrosis factor; IFN, interferon; IL, interleukin; CSF, colony stimulating factor.

[b] +, Activation; −, no activation; nd, not determined.

[c] α_2-C10 adrenergic receptor transfected in Rat-1 cells, stimulated with agonist UK14304.

[d] DG generated from PI glycans.

when PKC is down-regulated in the cell (van Blitterswijk *et al.*, 1991b; Kester *et al.*, 1992; van der Bend *et al.*, 1992b). Conricode *et al.* (1992) confirmed, in a fibroblast membrane system, that PKC can activate PLD but, unexpectedly, this activation did not require ATP-dependent phosphorylation. In contrast, PLD activation in permeabilized neutrophils or HL-60 cells did depend on ATP (Olson *et al.*, 1991; Geny and Cockcroft, 1992).

PLD also can be activated independently of PKC, by a G protein that is thought to act on PLD directly (Geny and Cockcroft, 1992; Uings *et al.*, 1992), and by elevated cytosolic Ca^{2+} (Huang *et al.*, 1991). In platelet membranes and in permeabilized HL-60 or NG108-15 cells, a marked synergism exists between the effect of phorbol ester treatment and GTPγS to activate PLD (van der Meulen and Haslam, 1990; Geny and Cockcroft, 1992; Liscovitch, 1992). In these cell types, full activation of PLD may require the interplay of PKC, increased Ca^{2+}, and a G protein. Thus, two receptor-stimulated pathways, dependent and independent of PKC, may converge in PLD activation. The PKC-independent pathway can be inhibited selectively by the fungal metabolite wortmannin (Thompson *et al.*, 1991; Gélas *et al.*, 1992; Yatomi *et al.*, 1992) and by a certain protease inhibitor (Kessels *et al.*, 1991). These inhibitors do not act on PLD itself, but act at a site between the activated receptor and the G protein that couples to PLD. Remarkably, in rabbit peritoneal neutrophils, the PKC-independent pathway was stimulated by high concentrations of staurosporin in the absence of a natural agonist (Kanaho *et al.*, 1992). This PKC-independent PLD activation was mediated by a pertussis toxin-sensitive G protein. In human neutrophils, the activation of PLD by the chemotactic peptide fMLP is also sensitive to pertussis toxin (Bourgoin *et al.*, 1992). The PKC-independent pathway of activating PLD also requires tyrosine phosphorylation, at least in neutrophils and HL-60 cells, since inhibitors of tyrosine kinases inhibited PLD whereas a phosphotyrosine phosphatase inhibitor, pervanadate, was found to stimulate PLD (Bourgoin and Grinstein, 1992; Bourgoin *et al.*, 1992; Uings *et al.*, 1992). Moreover, the addition of granulocyte–macrophage colony-stimulating factor (GM-CSF) to neutrophils primes PLD to subsequent stimulation by fMLP or phorbol 12-myristate 13-acetate (PMA). This GM-CSF pretreatment accelerates the tyrosine phosphorylation response to fMLP. In contrast, cyclic AMP-elevating agents that block the fMLP-induced respiratory burst also inhibit receptor-mediated PKC-independent PLD activation at a site proximal to PLD (Tyagi *et al.*, 1991), again indicating that phosphorylations (in this case, via A kinase) at such a site may regulate PLD activity.

PLD activity has been found to be associated with (plasma) membranes and cytosol (Wang *et al.*, 1991). In HL-60 and neutrophil membranes,

PLD activity requires unknown cytosolic protein factors (Anthes *et al.*, 1991; Olson *et al.*, 1991). Permeabilized HL-60 cells gradually lose GTSγS-stimulated PLD activity because of leakage of cytosolic components (Geny and Cockcroft, 1992) that could include cytosolic PLD, a tyrosine kinase (described earlier), or some other regulatory protein. Permeabilization in the presence of GTPγS partially retains this activity, suggesting that these factors can be recruited in a G protein-dependent manner; these factors then play a role in obtaining the full PLD response (Cockcroft, 1992).

The relative impact of PLD- over DG kinase-mediated PA formation may vary depending on the cell system. In fMLP-stimulated neutrophils (Cockcroft, 1992) and IgE-stimulated mast cells (Dinh and Kennerly, 1991), rapid PA formation (within 30 sec) occurs mainly via PLD. The amount of PA generated is calculated to be as much as 1.5 and 2.8%, respectively, of total cellular PC. In contrast, Huang *et al.* (1991) concluded that, in platelets stimulated with α-thrombin for 5 min, 87% of the PA arose via PLC$_i$/DG kinase and the remaining 13% via PLD.

D. Activation by Receptor Protein Tyrosine Kinases

Receptor protein tyrosine kinases such as those for PDGF, CSF-1 (c-*fms* protein), EGF, insulin, and fibroblast growth factor (FGF) are related structurally (Ullrich and Schlessinger, 1990; Cantley *et al.*, 1991; Jaye *et al.*, 1992). All have an extracellular ligand binding domain, a single transmembrane domain, and an intracellular kinase domain that may or may not have an insert. Stimulation of these receptors enhances their intrinsic tyrosine kinase activity, resulting in autophosphorylation, which creates binding sites for recruitment of enzymes such as ras GTPase-activating protein (GAP), PI 3-kinase, PLCγ, and others. These enzymes then transduce signals to the cell interior that cause a variety of responses needed for cell division. The recruitment mechanism utilizes association between certain phosphotyrosines of the receptor and so-called src homology 2 (SH-2) regions of the particular enzymes (Mayer and Baltimore, 1993). Not every receptor type associates with each of the enzymes mentioned. PI 3-kinase and GAP do not seem to interact with the activated FGF receptor, whereas PLCγ is not associated with or activated by the CSF-1 and insulin receptors. PLCγ activation by stimulated PDGF or FGF receptors is not essential for cell proliferation (Hill *et al.*, 1990; Mohammadi *et al.*, 1992; Peters *et al.*, 1992). In these cases, Ca^{2+} mobilization seems not to be essential for cell proliferation although PKC still can be activated, if necessary, by DG$_c$ generated from PC (see subsequent discussion) instead of PIP$_2$ (Nanberg *et al.*, 1990).

Stimulation of receptor protein tyrosine kinases induces PC hydrolysis. The early study by Besterman *et al.*, (1986) already indicated that PDGF stimulation of 3T3-Li cells resulted in rapid (within 30 sec) PC breakdown by PLC_c. This process seems partially independent of PKC, as concluded from experiments executed in PKC-down-regulated cells. In later studies, PDGF-stimulated DG formation in other types of fibroblasts was found to be biphasic, consistent with a rapid and transient DG level generated by PIP_2 hydrolysis, followed by a second sustained phase of DG_c, from about 2 min to at least 30 min (Pessin *et al.*, 1990; Plevin *et al.*, 1991; Fukami and Takenawa, 1992). This DG_c gradient was generated mainly by the PLD/PA phosphohydrolase pathway and was dependent on PKC.

Responses to EGF were different from those to PDGF. In EGF-stimulated IIC9 fibroblasts, Wright *et al.* (1988,1992) found a rapid and transient (up to 1 min) PLD activation and a sustained monophasic DG_c that did not result from the initial PLD activation but from PLC_c activity. In HeLa and A431 cells, EGF stimulated both PIP_2 hydrolysis and PLD, resulting in monophasic DG formation (Kaszkin *et al.*, 1992). PA was generated by DG kinase and PLD and peaked 5 min after stimulation. Activation of PLD, as measured by the formation of phosphatidylbutanol, was not affected by long-term pretreatment of cells with phorbol ester, indicating that PKC was not involved.

Stimulation of the FGF receptor in transfected L6 myoblasts induces PI turnover and activates PLD (M. van Dijk and W. van Blitterswijk, unpublished observations). Both activities were abrogated in cells transfected with an FGF receptor mutant (Y766F) that is unable to bind to and activate $PLC\gamma$ (Mohammadi *et al.*, 1992), suggesting that PLD in this system is downstream of $PLC\gamma$ or at least tightly connected with $PLC\gamma$ activation.

CSF-1-stimulated monocytes (Imamura *et al.*, 1990) and bone marrow-derived macrophages (Veis and Hamilton, 1991) also show PC hydrolysis, probably by PLC_c (in the absence of PI turnover; Table II). The resulting DG_c is capable of activating PKC (Imamura *et al.*, 1990). Unlike PDGF-stimulated cells, this CSF-1-stimulated PC hydrolysis as well as the concomitant DNA synthesis was sensitive to pertussis toxin, indicative of the involvement of a G protein. How this G protein forms the apparently unconventional functional link between a receptor protein tyrosine kinase and a phospholipase is unknown.

E. Role of Phosphatidylcholine-Phospholipase C in Cell Proliferation

Several observations indicate that receptor-stimulated sustained increase in the level of DG_c is connected tightly with, if not essential to, cell proliferation. Larrodera *et al.* (1990) stimulated Swiss 3T3 cells with PDGF and found a delayed significant elevation of intracellular phosphocholine and DG levels (from 4 to over 14 hr) prior to DNA synthesis (starting after 12 hr). The simple addition of PLC$_c$ from *Bacillus cereus* was sufficient to elicit a potent mitogenic response in these cells, to the same level induced by PDGF. Optimal concentrations of the bacterial PLC$_c$ and PDGF were nonadditive, whereas the bacterial PLC$_c$ and insulin were synergistic in the mitogenic response. These results indicate the importance of PC breakdown in the mitogenic signaling cascade activated by PDGF. EGF-stimulated fibroblasts also generate sustained DG$_c$ prior to DNA synthesis (Wright *et al.*, 1990). In both the PDGF- and the EGF-stimulated cells, the concentration dependence of ligand-stimulated DG$_c$ production and DNA synthesis were similar (Larrodera *et al.*, 1990; Wright *et al.*, 1990). In agreement with these data, thyroid-stimulating hormone and insulin-like growth factor 1 synergistically stimulate DNA synthesis in thyroid cells and also synergize to elevate DG levels in these cells (Brenner-Gati *et al.*, 1988). In bone marrow-derived macrophages, CSF-1 stimulates DNA synthesis and induces DG$_c$ formation without PIP$_2$ hydrolysis. In contrast, in resident peritoneal macrophages that show a poor proliferative response to CSF-1, phospholipid hydrolysis is not induced (Veis and Hamilton, 1991), again suggesting that generation of DG$_c$ may be necessary for, in this case, macrophage proliferation. Data presented by Rangan *et al.* (1991) also support a role for sustained DG in the mitogenic response to α-thrombin. Suzuki *et al.* (1989) found that microinjection of 1,2-dioleoylglycerol into BALB 3T3 cells induces DNA synthesis. Finally, in *Xenopus* oocytes, insulin-induced meiotic cell division was preceded by an essential increase in the levels of DG (Garcia de Herreros *et al.*, 1991; Stith *et al.*, 1991).

An important question is whether PKC activation is necessary for cell proliferation induced by these receptors. Researchers customarily downregulate PKC by prolonged treatment with phorbol ester to determine whether a cellular effect is "PKC independent." In doing so, PDGF- or serum-induced PC hydrolysis is attenuated only partially (Besterman *et al.*, 1986), and the mitogenic response activated by bacterial PLC$_c$ is not affected at all (Larrodera *et al.*, 1990). However, these so-called PKC-independent effects still may be mediated by PKC isoforms that are not down-regulated by phorbol esters. Such a PKC isoform also may reside in a compartment of the cell in which it is not readily accessible to or recruited by phorbol ester at the cell surface, or degraded by proteases, but in which it can be reached readily by DG$_c$. The phorbol ester-insensitive PKCζ appeared to be involved in maturation of *Xenopus* oocytes by "channeling" the mitogenic signal induced by insulin, ras protein (see Section II,F), and PLC$_c$ (Dominquez *et al.*, 1992). This signal could be inhibited specifically by microinjection of a peptide corresponding to the pseudo-substrate region of PKCζ or of an antisense RNA for this kinase. However, since the PKCζ was found not to be activated by DG, how PLC$_c$-mediated DG$_c$ formation in these oocytes should be linked to PKCζ activation is not clear.

F. Phosphatidylcholine-Phospholipase C Connection to ras

ras acts as a signal transducer in many cell types (Satoh *et al.*, 1992). Tyrosine kinases, intrinsic to or associated with receptors, play an important role in the accumulation of ras–GTP in response to cell stimulation. A GDP–GTP exchange regulator and a GTPase activating protein (GAP) are thought to transduce signals from these kinases. Somehow the activated ras protein, or perhaps the GAP–ras protein complex that associates with an activated receptor protein tyrosine kinase, stimulates the PLC$_c$-mediated PC breakdown that precedes DNA synthesis in PDGF- or EGF-stimulated fibroblasts (Cai *et al.*, 1992) or meiosis in insulin-stimulated *Xenopus* oocytes (Dominguez *et al.*, 1991; Garcia de Herreros *et al.*, 1991). ras-mediated PLC$_c$ activation appears to be a more general phenomenon, as indicated by several findings. Cells that are transformed by oncogenic ras display elevated levels of DG$_c$ and phosphocholine (Lacal *et al.*, 1987; Wolfman and Macara, 1987; Matyas and Fishman, 1989; Price *et al.*, 1989). This elevation may depend on PKC activation (Price *et al.*, 1989) and also leads to permanent translocation of PKC to the plasma membrane (Huang *et al.*, 1988; Diaz-Laviada *et al.*, 1990). This continuous activation of PKC by oncogenic ras does not lead or only partially leads, to down-regulation of PKC as determined by *in vitro* activity assays (Huang *et al.*, 1988), *in vivo* phosphorylation of MARCKS (80-kDa) protein (Wolfman and Macara, 1987), immunofluorescence, and immunoblotting (Diaz-Laviada *et al.*, 1990). These results suggest that DG formation is an important step in the regulation of PKC by *ras* oncogene products. Scrape-loading or microinjection of cells with normal or oncogenic ras protein also leads to PC breakdown (DG$_c$ formation) and activation of PKC in the absence of phosphoinositide hydrolysis (Morris *et al.*, 1989; Price *et al.*, 1989; Lacal, 1990). Scrape-loaded Val 12-p21^{ras}, in conjunction with insulin stimulation, resulted in DNA synthesis that was blocked when PKC was down-regulated (Morris *et al.*, 1989), indicating that PKC cooperates with ras in stimulating cell growth. Lopez-Barahona *et al.* (1990) studied the kinetics of ras-induced PC breakdown using a temperature-sensitive mutant of Ki-*ras*. On shift to the permissive temperature, products of the activated PLC$_c$ were detected by 30 min and reached maximal levels by 1–2 hr. The fact that at least 4 hr are required for PDGF or serum to activate this PLC_c suggests that the *ras* oncogene product might be involved in the late steps of the mitogenic signaling cascade (Larrodera *et al.*, 1990; Lopez-Barahona *et al.*, 1990).

Not only mitogens such as PDGF and insulin but also growth inhibitors such as transforming growth factor β (TGFβ) transmit their signals via ras and PLC_c, in this case in a negative way. TGFβ inhibits the coupling of ras to PC breakdown, both in EGF-induced proliferation of keratinocytes and in insulin-induced maturation of *Xenopus* oocytes (Diaz-Meco *et al.*, 1992). TGFβ inhibits late events in mitogenic signaling. The inhibition could be overcome by treatment or microinjection of cells with bacterial PLC_c.

Finally, oncogenic ras also directs the route of PC hydrolysis, as induced by G protein-coupled receptors (Fu *et al.*, 1992). In normal NIH 3T3 cells, bradykinin-induced DG formation is biphasic, the second phase being caused mainly by the initial PLD activity, in agreement with the results of van Blitterswijk *et al.* (1991a,b). In K-*ras*-transformed cells, bradykinin-induced PC hydrolysis is shifted from PLD to PLC_c (Fu *et al.*, 1992).

III. ATTENUATION OF SECOND MESSENGER DIACYLGLYCEROL LEVELS

Receptor-mediated PIP_2 hydrolysis rapidly generates DG_i. This lipid second messenger is known to be converted by DG kinase and DG lipase, generating PA and monoacylglycerol, respectively (Fig. 1). The activity of DG lipase may vary among cell types, being substantial in stimulated platelets (Bishop and Bell, 1986; Majerus *et al.*, 1986) but low in other cell types such as fibroblasts stimulated with bradykinin (van Blitterswijk *et al.*, 1991a). DG kinase, being active in the PI cycle, is generally a more important attenuator than DG lipase of rapidly formed DG_i. Neither enzyme converts the second-phase DG_c originating from PC, at least in bradykinin-stimulated fibroblasts (van Blitterswijk *et al.*, 1991a). How this second burst of DG is quenched is not known. By virtue of its ability to stimulate cytidylyltransferase, the rate-limiting enzyme in PC biosynthesis (Pelech and Vance, 1989), sustained DG is likely to stimulate its own conversion to PC (Florin-Christensen *et al.*, 1992,1993) with the eventual termination of the signal (Fig. 1).

van Blitterswijk and Hilkmann (1993) have discovered a novel mechanism of attenuation of rapidly formed DG. The principle of this mechanism is transphosphatidylation by PLD, hitherto regarded as a nonphysiological reaction that depends on the presence of a primary alcohol, generating

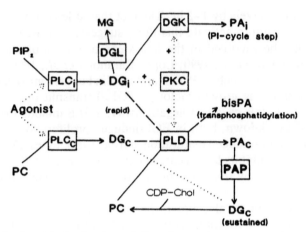

FIGURE 1 Agonist-stimulated metabolic pathways that generate and attenuate diacylglycerols (DG_i and DG_c) from phosphoinositides and phosphatidylcholine (PC), respectively. Rapidly generated levels of second messenger DG are attenuated by DG kinase (DGK), DG lipase (DGL), and phospholipase D (PLD)-mediated bisPA formation (drawn with broken line). Sustained DG_c levels obtained directly by PLC_c and/or indirectly by sequential activation of PLD and phosphatidic acid phosphohydrolase (PAP) eventually may decrease by reconversion into PC. Enzymes are drawn in boxes. Stimulatory effects are indicated with dotted arrows.

a phosphatidyl alcohol (van Blitterswijk *et al.*, 1991b). In bradykinin-stimulated human fibroblasts, PLD was found to mediate transphosphatidylation from PC (donor) to the endogenous "alcohol" DG (acceptor), yielding bis(1,2-diacylglycero)-3-*sn*-phosphate (bisphosphatidic acid, bisPA) (Fig. 2). This uncommon phospholipid is a condensation product of the phospholipase C (PLC) and PLD signaling pathways, in which PLC produces DG and PLD couples this DG to a phosphatidyl moiety. Thus, bisPA formation is found not only to attenuate DG formation substantially but also to prevent the generation of the putative second messenger PA. Long-term phorbol ester treatment, resulting in down-regulation of PKC, blocks bradykinin-induced activation of PLD and consequent bisPA formation, thereby unveiling rapid formation of DG. BisPA formation occurs rapidly (15 sec) and transiently (peaks at 2–10 min) and also is induced by other stimuli capable of raising DG and activating PLD simultaneously, for example, endothelin, lysophosphatidic acid, fetal bovine serum, phorbol ester, dioctanoylglycerol, or bacterial PLC. This novel metabolic route thus counteracts rapid accumulation of DG and PA, and assigns a physiological role to the transphosphatidylation activity of PLD, that is, signal attenuation (Fig. 1).

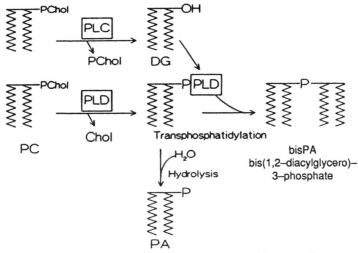

FIGURE 2 bisPA formation from phosphatidylcholine (PC) by phospholipase D-mediated transphosphatidylation. Agonist-induced diacylglycerol (DG), generated by phospholipase C, serves as the nucleophilic primary alcohol reacting with the phosphatidyl-PLD intermediate. bisPA is the condensation product of two signaling pathways, that is, those mediated by the enzymes phospholipase C (PLC) and D (PLD).

IV. ROLE OF DIACYLGLYCEROL KINASE IN CELLULAR SIGNALING

In receptor-stimulated cells, the enzyme DG kinase attenuates the levels of second messenger DG by conversion to PA (Kanoh *et al.*, 1990). Thus, the enzyme may regulate the action of PKC by controlling cellular levels of DG. Further, PA, the product of DG kinase, may function as a second messenger in its own right (Section I). PA also has been found to be an activator of PI-4-P kinase (Moritz *et al.*, 1992).

Several DG kinases have been purified from various tissues. The cDNAs of a procine and a human 86-kDa DG kinase have been cloned and appear to be 93% homologous (Sakane *et al.*, 1990; Schaap *et al.*, 1990). This isozyme is expressed highly in thymus and moderately in brain and spleen.

A. Structure–Activity Relationship

The 86-kDa DG kinase contains conserved structures including a double EF-hand (binds Ca^{2+}), a double cysteine repeat ("Zn^{2+} fingers"), and two putative ATP binding motifs. The consensus sequence of ATP binding motifs in protein kinases is $GXGXXGX_{14}K$ (X being any amino acid)

(Hanks *et al.*, 1988); the lysine residue is essential for activity. Mutation of DG kinase at these sites, however, does not abrogate enzyme activity (D. Schaap, unpublished observations). DG kinase can bind Ca^{2+} *in vitro* and thereby is translocated to membranes and activated (Sakane *et al.*, 1991a). Site-directed mutagenesis also shows that deletion of the double EF-hand makes the enzyme Ca^{2+} independent and leads to constitutive membrane association *in vitro* (Sakane *et al.* 1991b). This result suggests that, in the wild-type DG kinase, a phospholipid binding site is masked in the absence of Ca^{2+} and is unveiled when intracellular Ca^{2+} is raised by cell stimulation, as was suggested for classical PKCs (Parker *et al.*, 1989). The double cysteine repeats of DG kinase are highly homologous to those in PKC. The latter are known to bind phorbol esters and DG (Ono *et al.*, 1989; Burns and Bell, 1991). The double cysteine repeats of DG kinase, however, do not bind phorbol ester (D. Schaap, unpublished observations).

B. Regulation

DG kinase is activated via stimulation of G protein-coupled receptors (e.g., Rider and Baquet, 1988; Lee *et al.*, 1991; van Blitterswijk *et al.*, 1991a; van der Bend *et al.*, 1992b), the T cell receptor/CD3 complex in lymphocytes (R. L. van der Bend and W. van Blitterswijk, unpublished observations), and receptor protein tyrosine kinases for EGF (Kato *et al.*, 1985; Pike and Eakes, 1987; Payrastre *et al.*, 1991), PDGF (McDonald *et al.*, 1988a), and FGF (M. van Dijk and W. van Blitterswijk, unpublished observations).

Phorbol ester induces the translocation of DG kinase in Swiss 3T3 cells and neutrophils, presumably via PKC activation. DG kinase activity thereby is increased in the membrane fraction and decreased in the cytosol but overall activity remains unchanged (Ishitoya *et al.*, 1987; Maroney and Macara, 1989). The proto-oncogene product $pp60^{src}$ may activate DG kinase, as indicated by studies in fibroblasts infected with temperature-sensitive mutants (Sugimoto *et al.*, 1984). ras transformation of cells generally increases DG levels and activates PKC (Section II,F) by a decrease in membrane-bound DG kinase (Huang *et al.*, 1988; Kato *et al.*, 1988).

We have transfected human DG kinase cDNA into Rat-1 fibroblasts, resulting in a stable 10- to 20-fold overexpression of DG kinase in these cells. However, stimulation of these transfected cells with endothelin, a potent activator of DG kinase in these cells, did not result in higher PA formation than in stimulated control cells (D. Schaap and W. van Blitterswijk, unpublished observations). This result indicates that cellular

DG kinase, irrespective of its quantity, is "dormant" and requires a regulatory mechanism (see subsequent discussion) to be activated. Apparently, overexpressed DG kinase is not recruited in a signaling cascade.

DG kinase activity in the cell may be regulated by phosphorylation. *In vitro* studies with purified porcine 86-kDa DG kinase revealed that both PKA and PKC are capable of phosphorylating the enzyme. This event did not increase activity but did increase association of the protein with phosphatidylserine membranes and stabilized the enzyme as well (Kanoh *et al.*, 1989). Thus, phosphorylation and a rise in Ca^{2+} (Section IV,A) may contribute to enzyme activation. Schaap *et al.* (1993) found that the 86-kDa DG kinase, transiently overexpressed in COS-7 cells, is phosphorylated on serine residues by stimulation of the cells by IBMX/forskolin, which activates cAMP-dependent protein kinase, or by phorbol ester, which activates PKC. These two kinases phosphorylate the enzyme within the same tryptic phosphopeptides, suggesting that they exert similar control over DG kinase. Cotransfections with PKCα or PKCε revealed that both protein kinases, when stimulated, were able to phosphorylate DG kinase *in vivo*. Schaap *et al.* (1993) also found that DG kinase transfected into these cells was phosphorylated on tyrosine on stimulation of the EGF receptor. Since this receptor has an intrinsic tyrosine kinase activity, this finding implies that DG kinase may be a direct substrate of the activated EGF receptor. This suggestion has been substantiated further by the finding that the EGF receptor and DG kinase can be co-immunoprecipitated when overexpressed in COS cells (D. Schaap, unpublished observation). These studies could not correlate increased DG kinase phosphorylation with activity because of limitations of the transfection system used. Collectively, these results suggest that multiple signaling routes may converge in regulating DG kinase.

C. Substrate Selectivity

DG kinase converts 1,2-*sn*-diacylglycerols to PA. *In vitro*, the purified 86-kDa DG kinase shows a preference for ester-linked DGs that contain an unsaturated fatty acid at the *sn*-2 position, but no preference for either DG_i or DG_c (Schaap *et al.*, 1990). *In vivo*, the activated membrane-bound enzyme prefers the endogenous substrate 1-stearoyl, 2-arachidonoyl-*sn*-glycerol generated from phosphoinositides (McDonald *et al.*, 1988a,b; Lemaitre *et al.*, 1990). Exogenous short-chain DGs such as dioctanoylglycerol are taken up rapidly by cells and are converted, to a large extent, to PA (Bishop and Bell, 1986). The DG kinase responsible for this conversion of short-chain DGs is not membrane bound but is located in the cytosol

(McDonald *et al.*, 1988b). Endogenous DG_c derived from PC, at a somewhat later stage (several minutes) of cell stimulation, is not a likely substrate for DG kinase (Lee *et al.*, 1991; van Blitterswijk *et al.*, 1991a).

We have studied the substrate selectivity of DG kinase in [^{3}H]arachidonate-labeled cells in which the level of endogenous DG was raised artificially by treatment of the cells with bacterial PI-specific PLC (R. L. van der Bend and W. van Blitterswijk, unpublished observation). This treatment enhanced the levels of radiolabeled DG 5- to 10-fold, but did not result in PA formation. DG generated in this way was not used as substrate by DG kinase activated by stimulation of cell surface receptors with endothelin or bradykinin (in fibroblasts) and anti-CD3 (in Jurkat T cells). However, in deoxycholate-solubilized cells, DG kinase was fully capable of converting DG generated by prior bacterial PLC treatment. These results indicate that DG kinase converts DG in a topologically restricted fashion; it does not convert DG that has been generated randomly in the plasma membrane by exogenous PLC, but converts only the DG that directly results from receptor stimulation and consequent endogenous PLC activation. This observation supports the notion that DG kinase is regulated tightly in a signaling complex in association with the receptor and PLCγ, in which second messenger DG is "channeled."

V. CONCLUSIONS AND FUTURE PROSPECTS

Various types of cell surface receptor in many cell types can activate various metabolic pathways that generate and subsequently attenuate the lipid messengers DG and PA (summarized in Fig. 1). However, as long as the PC-specific phosphodiesterases PLC_c and PLD have not been cloned, elucidating the complex interactions between these pathways and precisely how they are coupled to an activated receptor seems impossible. However, several novel features emerge from the present overview.

Agonist-stimulated PLC_c does not necessarily require initial hydrolysis of PIP_2, particularly for certain cytokine receptors. PLC_c may hydrolyze PC in the outer leaflet of the plasma membrane, which is a quite provocative thought. Rapidly produced DG_c not only may activate PKC but also may activate a sphingomyelinase, oriented externally in the plasma membrane, where its substrate sphingomyelin is located also (Mathias *et al.*, 1993). The product, ceramide, appears to be a novel second messenger, the impact of which undoubtedly will be explored in the near future.

Although rapidly generated DGs from PIP_2 and PC can activate PKC, DG_c produced at a later stage of cell stimulation may not do so in some cases. To reach a consensus on this issue, we must investigate this behavior in more depth in more cell systems. Various studies have revealed that PIP$_2$ hydrolysis is not essential for cell proliferation. In contrast, PLC$_c$ activation (DG$_c$ formation) in connection with ras activation has been proposed to play a dominant role in late steps of the mitogenic process. This concept has been put forward strongly by Moscat's group, but needs confirmation from other studies.

PLD is activated rapidly in many types of receptor-stimulated cells. Maximal activation may require an interplay of a G protein, PKC, Ca^{2+}, and a tyrosine kinase or other accessory cytosolic protein(s) in different proportions, depending on the cell system. Apart from generating the putative second messenger PA, PLD also can attenuate signals. By virtue of its unique property of transphosphatidylation, PLD can generate bisPA rapidly, at the cost of DG and PA. An intriguing question is whether this bisPA functions only as a "sink" of lipid messengers or whether it might be a second messenger in its own right.

A general feature that is emerging in an increasing number of cell signaling systems is that a "signal," such as DG or PA, is "channeled" via a physical complex of enzymes, in which the "signal" is propelled from one enzyme to the next in a cascade of metabolic reactions. The occurrence of such "signal transfer particles" already has been proposed for activated receptor protein tyrosine kinase systems (Ullrich and Schlessinger, 1990). Now DG kinase also seems to be connected in this way to these receptors as well as to other (G protein-coupled) receptors.

A major challenge for future studies is to establish firmly or to disprove the role of PA as a second messenger. One approach is to investigate, preferably *in vivo*, whether PA specifically activates known protein kinases such as PKCζ, a likely candidate for activation by PA. Overexpression of PKCζ by cDNA transfection in COS cells, for instance, has revealed PA specificity for this kinase over other PKC isotypes (C. Limatola and W. van Blitterswijk, unpublished observations). PA-specific phosphorylation(s) could reveal known or novel PKC substrates. A different and potentially powerful approach would be photoaffinity labeling (cf. van der Bend *et al.*, 1992) with [^{32}P]-labeled diazirine-PA to detect specific PA binding proteins in the cell. Confirming the proposed regulatory function of PA for the GTPase activity of small G proteins is also important (Tsai *et al.*, 1990). *In vivo* experiments should evaluate whether GAP inhibition by PA is physiologically relevant or an *in vitro* artifact caused by binding of GAP to micellar structures (Serth *et al.*, 1991). Obviously, much work remains to be done to define the physiological significance of agonist-stimulated PA formation and lipid turnover in general.

Acknowledgments

We thank our colleagues Marc van Dijk, Henk Hilkmann, Cristina Limatola, José van der Wal, and John de Widt, who contributed to our own experimental work reviewed in this chapter. This work was supported by the Dutch Cancer Society.

References

Anthes, J. C., Wang, P., Siegel, M. I., Egan, R. W., and Billah, M. M. (1991). Granulocyte phospholipase D is activated by a guanine nucleotide dependent protein factor. *Biochem. Biophys. Res. Commun.* **175**, 236–243.

Besterman, J. M., Duronio, V., and Cuatrecasas, P. (1986). Rapid formation of diacylglycerol from phosphatidylcholine: A pathway for generation of a second messenger. *Proc. Natl. Acad. Sci. U.S.A.* **83**, 6785–6789.

Billah, M. M., and Anthes, J. C. (1990). The regulation and cellular functions of phosphatidylcholine hydrolysis. *Biochem. J.* **269**, 281–291.

Bishop, W. R., and Bell, R. M. (1986). Attenuation of *sn*-1,2-diacylglycerol second messengers. *J. Biol. Chem.* **261**, 12513–12519.

Bocckino, S. B., Wilson, P. B., and Exton, J. H. (1991). Phosphatidate-dependent protein phosphorylation. *Proc. Natl. Acad. Sci. U.S.A.* **88**, 6210–6213.

Bourgoin, S., and Grinstein, S. (1992). Peroxides of vanadate induce activation of phospholipase D in HL-60 cells. Role of tyrosine phosphorylation. *J. Biol. Chem.* **267**, 11908–11916.

Bourgoin, S., Poubelle, P. E., Liao, N. W., Umezawa, K., Borgeat, P., and Naccache, P. H. (1992). Granulocyte-macrophage colony-stimulating factor primes phospholipase D activity in human neutrophils in vitro: Role of calcium, G-proteins and tyrosine kinases. *Cell. Signalling* **4**, 487–500.

Brenner-Gati, L., Berg, K. A., and Gershengorn, M. C. (1988). Thyroid-stimulating hormone and insulin-like growth factor-1 synergize to elevate 1,2-diacylglycerol in rat thyroid cells. *J. Clin. Invest.* **82**, 1144–1148.

Burns, D. J., and Bell, R. M. (1991). Protein kinase C contains two phorbol ester binding domains. *J. Biol. Chem.* **266**, 18330–18338.

Bursten, S. L., Harris, W. E., Bomsztyk, K., and Lovett, D. (1991). Interleukin-1 rapidly stimulates lysophosphatidate acyltransferase and phosphatidate phosphohydrolase activities in human mesangial cells. *J. Biol. Chem.* **266**, 20732–20743.

Cai, H., Erhardt, P., Szeberényi, J., Diaz-Meco, M. T., Johansen, T., Moscat, J., and Cooper, G. M. (1992). Hydrolysis of phosphatidylcholine is stimulated by ras proteins during mitogenic signal transduction. *Mol. Cell. Biol.* **12**, 5329–5335.

Cantley, L. C., Auger, K. R., Carpenter, C., Duckworth, B., Graziani, A., Kapeller, R., and Soltoff, S. (1991). Oncogenes and signal transduction. *Cell* **64**, 281–302.

Cockcroft, S. (1992). G-protein-regulated phospholipases C, D and A_2-mediated signalling in neutrophils. *Biochim. Biophys. Acta* **1113**, 135–160.

Conricode, K. M., Brewer, K. A., and Exton, J. H. (1992). Activation of phospholipase D by protein kinase C. *J. Biol. Chem.* **267**, 7199–7202.

Cook, S. J., and Wakelam, M. J. O. (1991). Stimulated phosphatidylcholine hydrolysis as a signal transduction pathway in mitogenesis. *Cell. Signalling* **3**, 273–282.

Dennis, E. A., Rhee, S. G., Billah, M. M., and Hannun, Y. A. (1991). Role of phospholipases in generating lipid second messengers in signal transduction. *FASEB J.* **5**, 2068–2077.

Diaz-Laviada, I., Larrodera, P., Diaz-Meco, M. T., Cornet, M. E., Guddal, P. H., Johansen, T., and Moscat, J. (1990). Evidence for a role of phosphatidylcholine-hydrolysing phospholipase C in the regulation of protein kinase C by ras and src oncogenes. *EMBO J.* **9**, 3907–3912.

Diaz-Meco, M. T., Larrodera, P., Lopez-Barahona, M., Cornet, M. E., Barreno, P. G., and Moscat, J. (1989). Phospholipase C-mediated hydrolysis of phosphatidylcholine is activated by muscarinic agonists. *Biochem. J.* **263**, 115–120.

Diaz-Meco, M. T., Dominguez, I., Sanz, L., Municio, M. M., Berra, E., Cornet, M. E., Garcia de Herreros, A., Johansen, T., and Moscat, J. (1992). Phospholipase C-mediated hydrolysis of phosphatidylcholine is a target of transforming growth factor β1 inhibitory signals. *Mol. Cell. Biol.* **12**, 302–308.

Dinh, T. T., and Kennerly, D. A. (1991). Assessment of receptor-dependent activation of phosphatidylcholine hydrolysis by both phospholipase D and phospholipase C. *Cell Regul.* **2**, 299–309.

Dominguez, I., Marshall, M. S., Gibbs, J. B., Garcia de Herreros, A., Cornet, M. E., Graziani, G., Diaz-Meco, M. T., Johansen, T., McCormick, F., and Moscat, J. (1991). Role of GTPase activating protein in mitogenic signalling through phosphatidylcholine-hydrolysing phospholipase C. *EMBO J.* **10**, 3215–3220.

Dominguez, I., Diaz-Meco, M. T., Municio, M. M., Berra, E., Garcia de Herreros, A., Cornet, M. E., Sanz, L., and Moscat, J. (1992). Evidence for a role of protein kinase C ζ subspecies in maturation of Xenopus laevis oocytes. *Mol. Cell. Biol.* **12**, 3776–3783.

Duriono, V., Nip, L., and Pelech, S. L. (1989). Interleukin 3 stimulates phosphatidylcholine turnover in a mast/megakaryocyte cell line. *Biochem. Biophys. Res. Commun.* **164**, 804–808.

Exton, J. H. (1990). Signaling through phosphatidylcholine breakdown. *J. Biol. Chem.* **265**, 1–4.

Fällman, M., Gullberg, M., Hellberg, C., and Andersson, T. (1992). Complement receptor-mediated phagocytosis is associated with accumulation of phosphatidylcholine-derived diglyceride in human neutrophils. *J. Biol. Chem.* **267**, 2656–2663.

Ferguson, J. E., and Hanley, M. R. (1991). The role of phospholipases and phospholipid-derived signals in cell activation. *Curr. Biol.* **3**, 206–212.

Florin-Christensen, J., Florin-Christensen, M., Delfino, J. M., Stegmann, T., and Rasmussen, H. (1992). Metabolic fate of plasma membrane diacylglycerols in NIH 3T3 fibroblasts. *J. Biol. Chem.* **267**, 14783–14789.

Florin-Christensen, J., Florin-Christensen, M., Delfino, J. M., and Rasmussen, H. (1993). New patterns of diacylglycerol metabolism in intact cells. *Biochem. J.* **289**, 783–788.

Fu, T., Okano, Y., and Nozawa, Y. (1992). Differential pathways (phospholipase C and phospholipase D) of bradykinin-induced biphasic 1,2-diacylglycerol formation in non-transformed and K-*ras*-transformed NIH-3T3 fibroblasts. *Biochem. J.* **283**, 347–354.

Fukami, K., and Takenawa, T. (1992). Phosphatidic acid that accumulates in platelet-derived growth factor-stimulated *BALB*/c 3T3 cells is a potential mitogenic signal. *J. Biol. Chem.* **267**, 10988–10993.

Garcia de Herreros, A., Dominguez, I., Diaz-Meco, M. T., Graziani, G., Cornet, M. E., Guddal, P. H., Johansen, T., and Moscat, J. (1991). Requirement of phospholipase C-catalyzed hydrolysis of phosphatidylcholine for maturation of *Xenopus laevis* oocytes in response to insulin and ras p21. *J. Biol. Chem.* **266**, 6825–6829.

Gélas, P., Von Tscharner, V., Record, M., Baggiolini, M., and Chap, H. (1992). Human neutrophil phospholipase D activation by *N*-formylmethionyl-leucylphenylalanine reveals a two-step process for the control of phosphatidylcholine breakdown and oxidative burst. *Biochem. J.* **287**, 67–72.

Geny, B., and Cockcroft, S. (1992). Synergistic activation of phospholipase D by protein kinase C- and G-protein-mediated pathways in streptolysin O-permeabilized HL60 cells. *Biochem. J.* **284**, 531–538.

Hanks, S. K., Quinn, A. M., and Hunter, T. (1988). The protein kinase family: Conserved features and deduced phylogeny of the catalytic domains. *Science* **241**, 42–52.

Hill, T. D., Dean, N. M., Mordan, L. J., Lau, A. F., Kanemitsu, M. Y., and Boyton, A. L. (1990). PDGF-induced activation of phospholipase C is not required for induction of DNA synthesis. *Science* **248**, 1660–1663.

Hoffman, J. M., Standaert, M. L., Nair, G. P., and Farese, R. V. (1991). Differential effects of pertussis toxin on insulin-stimulated phosphatidylcholine hydrolysis and glycerolipid synthesis de novo. Studies in BC3H-1 myocytes and rat adipocytes. *Biochemistry* **30**, 3315–3322.

Huang, M., Chida, K., Kamata, N., Nose, K., Kato, M., Homma, Y., Takenawa, T., and Kuroki, T. (1988). Enhancement of inositol phospholipid metabolism and activation of protein kinase C in *ras*-transformed rate fibroblasts. *J. Biol. Chem.* **263**, 17975–17980.

Huang, R., Kucera, G. L., and Rittenhouse, S. E. (1991). Elevated cytosolic Ca^{2+} activates phospholipase D in human platelets. *J. Biol. Chem.* **266**, 1652–1655.

Imamura, K., Dianoux, A., Nakamura, T., and Kufe, D. (1990). Colony-stimulating factor 1 activates protein kinase C in human monocytes. *EMBO J.* **9**, 2423–2429.

Ishitoya, J., Yamakawa, A., and Takenawa, T. (1987). Translocation of diacylglycerol kinase in response to chemotactic peptide and phorbol ester in neutrophils. *Biochem. Biophys. Res. Commun.* **144**, 1025–1030.

Jaye, M., Schlessinger, J., and Dionne, C. A. (1992). Fibroblast growth factor receptor tyrosine kinases: Molecular analysis and signal transduction. *Biochim. Biophys. Acta* **1135**, 185–199.

Kanaho, Y., Takahashi, K., Tomita, U., Iiri, T., Katada, T., Ui, M., and Nozawa, Y. (1992). A protein kinase C inhibitor, staurosporine, activates phospholipase D via pertussis toxin-sensitive GTP-binding protein in rabbit peritoneal neutrophils. *J. Biol. Chem.* **267**, 23554–23559.

Kanoh, H., Yamada, K., Sakane, F., and Imaizumi, T. (1989). Phosphorylation of diacylglycerol kinase *in vitro* by protein kinase C. *Biochem. J.* **258**, 455–462.

Kanoh, H., Yamada, K., and Sakane F. (1990). Diacylglycerol kinase: A key modulator of signal transduction? *Trends Biochem. Sci.* **15**, 47–50.

Kaszkin, M., Seidler, L., Kast, R., and Kinzel, V. (1992). Epidermal-growth-factor-induced production of phosphatidylalcohols by HeLa cells and A431 cells through activation of phospholipase D. *Biochem. J.* **287**, 51–57.

Kato, M., Homma, Y., Nagai, Y., and Takenawa, T. (1985). Epidermal growth factor stimulates diacylglycerol kinase in isolated plasma membrane vesicles from A431 cells. *Biochem. Biophys. Res. Commun.* **129**, 375–380.

Kato, H., Kawai, S., and Takenawa, T. (1988). Disappearance of diacylglycerol kinase translocation in *ras*-transformed cells. *Biochem. Biophys. Res. Commun.* **154**, 959–966.

Kessels, G. C. R., Gervaix, A., Lew, P. D., and Verhoeven, A. J. (1991). The chymotrypsin inhibitor carbobenzyloxy-leucine-tyrosine-chloromethylketone interferes with phospholipase D activation induced by formyl-methionyl-leucyl-phenylalanine in human neutrophils. *J. Biol. Chem.* **266**, 15870–15875.

Kester, M., Simonson, M. S., McDermott, R. G., Baldi, E., and Dunn, M. J. (1992). Endothelin stimulates phosphatidic acid formation in cultured rat mesangial cells: Role of a protein kinase C-regulated phospholipase D. *J. Cell. Physiol.* **150**, 578–585.

Kiley, S. C., Parker, P. J., Fabbro, D., and Jaken, S. (1991). Differential regulation of protein kinase C isozymes by thyrotropin-releasing hormone in GH_4C_1 cells. *J. Biol. Chem.* **266**, 23761–23768.

Kolesnick, R. (1992). Ceramide: a novel second messenger. *Trends Cell Biol.* **2**, 232–236.

Lacal, J. C. (1990). Diacylglycerol production in Xenopus laevis oocytes after microinjection of $p21^{ras}$ proteins is a consequence of activation of phosphatidylcholine metabolism. *Mol. Cell. Biol.* **10**, 333–340.

Lacal, J. C., Moscat, J., and Aaronson, S. A. (1987). Novel source of 1,2-diacylglycerol elevated in cells transformed by Ha-*ras* oncogene. *Nature* (*London*) **330**, 269–272.

Larrodera, P., Cornet, M. E., Diaz-Meco, M. T., Lopez- Barahona, M., Diaz-Laviada, I., Guddal, P. H., Johansen, T., and Moscat, J. (1990). Phospholipase C-mediated hydrolysis of phosphatidylcholine is an important step in PDGF-stimulated DNA synthesis. *Cell* **61**, 1113–1120.

Lassègue, B., Alexander, R. W., Clark, M., and Griendling, K. K. (1991). Angiotensin II-induced phosphatidylcholine hydrolysis in cultured vascular smooth-muscle cells. *Biochem. J.* **276**, 19–25.

Leach, K. L., Ruff, V. A., Wright, T. M., Pessin, M. S., and Raben, D. M. (1991). Dissociation of protein kinase C activation and sn-1,2-diacylglycerol formation. *J. Biol. Chem.* **266**, 3215–3221.

Lee, C., Fisher, S. K., Agranoff, B. W., and Hajra, A. K. (1991). Quantitative analysis of molecular species of diacylglycerol and phosphatidate formed upon muscarinic receptor activation of human SK-N-SH neuroblastoma cells. *J. Biol. Chem.* **266**, 22837–22846.

Lemaitre, R. N., King, W. C., MacDonald, M. L., and Glomset, J. A. (1990). Distribution of distinct arachidonoyl-specific and nonspecific isoenzymes of diacylglycerol kinase in baboon (*Papio cynocephalus*) tissues. *Biochem. J.* **266**, 291–299.

Lin, P., Fung, W.-J. C., and Gilfillan, A. M. (1992). Phosphatidylcholine-specific phospholipase D-derived 1,2-diacylglycerol does not initiate protein kinase C activation in the RBL 2H3 mast-cell line. *Biochem. J.* **287**, 325–331.

Lisovitch, M. (1992). Crosstalk among multiple signal-activated phospholipases. *Trends Biochem. Sci.* **17**, 393–399.

Lopez-Barahona, M., Kaplan, P. L., Cornet, M. E., Diaz-Meco, M. T., Larrodera, P., Diaz-Laviada, I., Municio, A. M., and Moscat, J. (1990). Kinetic evidence of a rapid activation of phosphatidylcholine hydrolysis by Ki-*ras* oncogene. Possible involvement in late steps of the mitogenic cascade. *J. Biol. Chem.* **265**, 9022–9026.

MacDonald, M. L., Mack, K. F., Richardson, C. N., and Glomset, J. A. (1988a). Regulation of diacylglycerol kinase reaction in Swiss 3T3 cells. *J. Biol. Chem.* **263**, 1575–1583.

MacDonald, M. L., Mack, K. F., Williams, B. W., King, W. C., and Glomset, J. A. (1988b). A membrane-bound diacylglycerol kinase that selectively phosphorylates arachidonoyl-diacylglycerol. *J. Biol. Chem.* **263**, 1584–1592.

MacNulty, E. E., McClue, S. J., Carr, I. C., Jess, T., Wakelam, M. J. O., and Milligan, G. (1992). α_2-C10 adrenergic receptors expressed in Rat 1 fibroblasts can regulate both adenylylcyclase and phospholipase D-mediated hydrolysis of phosphatidylcholine by interacting with pertussis toxin-sensitive guanine nucleotide-binding proteins. *J. Biol. Chem.* **267**, 2149–2156.

Majerus, P. W., Connally, T. M., Deckmyn, H., Ross, T. S., Bross, T. E., Ishii, H., Bansal, V. S., and Wilson, D. (1986). The metabolism of phosphoinositide-derived messenger molecules. *Science* **234**, 1519–1526.

Maroney, A. C., and Macara, I. G. (1989). Phorbol ester-induced translocation of diacylglycerol kinase from the cytosol to the membrane in Swiss 3T3 fibroblasts. *J. Biol. Chem.* **264**, 2537–2544.

Martin, T. F. J., Hsieh, K.-P., and Porter, B. W. (1990). The sustained second phase of hormone-stimulated diacylglycerol accumulation does not activate protein kinase C in GH₃ cells. *J. Biol. Chem.* **265**, 7623–7631.

Martinson, E. A., Goldstein, D., and Brown, J. H. (1989). Muscarinic receptor activation of phosphatidylcholine hydrolysis. *J. Biol. Chem.* **264**, 14748–14754.

Mathias, S., Younes, A., Kan, C.-C., Orlow, I., Joseph, C., and Kolesnick, R. N. (1993). Activation of the sphingomyelin signaling pathway in intact EL4 cells and in a cell-free system by IL-1β. *Science* **259**, 519–522.

434   Wim J. van Blitterswijk *et al.*

Matozaki, T., and Williams, J. A. (1989). Multiple sources of 1,2-diacylglycerol in isolated
    rat pancreatic acini stimulated by cholecystokinin. *J. Biol. Chem.* **264**, 14729–14734.
Matyas, G. R., and Fishman, P. H. (1989). Lipid signalling pathways in normal and ras-
    transfected NIH/3T3 cells. *Cell. Signalling* **1**, 395–404.
Mayer, B. J., and Baltimore, D. (1993). Signalling through SH2 and SH3 domains. *Trends
    Cell Biol.* **3**, 8–13.
Mohammadi, M., Dionne, C. A., Li, W., Li, N., Spivak, T., Honegger, A. M., Jaye, M., and
    Schlessinger, J. (1992). Point mutation in FGF receptor eliminates phosphatidylinositol
    hydrolysis without affecting mitogenesis. *Nature (London)* **358**, 681–684.
Moolenaar, W. H., Kruijer, W., Tilly, B. C., Verlaan, I., Bierman, A. J., and de Laat,
    S. W. (1986). Growth factor-like action of phosphatidic acid. *Nature (London)* **323**,
    171–173.
Moritz, A., De Graan, P. N. E., Gispen, W. H., and Wirtz, K. W. A. (1992). Phosphatidic
    acid is a specific activator of phosphatidylinositol-4-phosphate kinase. *J. Biol. Chem.*
    276, 7207–7210.
Morris, J. D. H., Price, B., Lloyd, A. C., Self, A. J., Marshall, C. J., and Hall, A. (1989).
    Scrape-loading of Swiss 3T3 cells with ras protein rapidly activates protein kinase C
    in the absence of phosphoinositide hydrolysis. *Oncogene* **4**, 27–31.
Nakanishi, H., and Exton, J. H. (1992). Purification and characterization of the ζ isoform
    of protein kinase C from bovine kidney. *J. Biol. Chem.* **267**, 16347–16354.
Nanberg, E., Morris, C., Higgins, T., Vara, F., and Rozengurt, E. (1990). Fibroblast growth
    factor stimulates protein kinase C in quiescent 3T3 cells without Ca^{2+} mobilization or
    inositol phosphate accumulation. *J. Cell. Physiol.* **143**, 232–242.
Olson, S. C., Bowman, E. P., and Lambeth, J. D. (1991). Phospholipase D activation in a
    cell-free system from human neutrophils by phorbol 12-myristate 13-acetate and guano-
    sine 5'-*O*-(3-thiotriphosphate). *J. Biol. Chem.* **266**, 17236–17242.
Ono, Y., Fujii, T., Igarashi, K., Kuno, T., Tanaka, C., Kikkawa, U., and Nishizuka, Y.
    (1989). Phorbol ester binding to protein kinase C requires a cysteine-rich zinc-finger-
    like sequence. *Proc. Natl. Acad. Sci. U.S.A.* **86**, 4868–4871.
Pachter, J. A., Pai, J.-K., Mayer-Ezell, R., Petrin, J. M., Dobek, E., and Bishop, W. R.
    (1992). Differential regulation of phosphoinositide and phosphatidylcholine hydrolysis
    by protein kinase C-β1 overexpression. *Biochem. J.* **267**, 9826–9830.
Pacini, L., Limatola, C., Frati, L., Luly, P., and Spinedi, A. (1993). Muscarinic stimulation
    of SK-N-BE9(2) human neuroblastoma cells elicits phosphoinositide and phosphatidyl-
    choline hydrolysis: Relationship to diacylglycerol and phosphatidic acid accumulation.
    Biochem. J. **289**, 269–275.
Parker, P. J., Kour, G., Marais, R. M., Mitchell, F., Pears, C., Schaap, D., Stabel, S., and
    Webster, C. (1989). Protein kinase C- a family affair. *Mol. Cell. Endocrinol.* **65**, 1–11.
Payrastre, B., van Bergen en Henegouwen, P. M. P., Breton, M., den Hartigh, J. C.,
    Plantavid, M., Verkleij, A. J., and Boonstra, J. (1991). Phosphoinositide kinase, diacyl-
    glycerol kinase, and phospholipase C activities associated to the cytosceleton: Effect
    of epidermal growth factor. *J. Cell Biol.* **115**, 121–128.
Pelech, S. L., and Vance, D. E. (1989). Signal transduction via phosphatidylcholine cycles.
    Trends Biochem. Sci. **14**, 28–30.
Pessin, M. S., and Raben, D. M. (1989). Molecular species analysis of 1,2-diglycerides
    stimulated by α-thrombin in cultured fibroblasts. *J. Biol. Chem.* **264**, 8729–8738.
Pessin, M. S., Baldassare, J. J., and Raben, D. M. (1990). Molecular species analysis of
    mitogen-stimulated 1,2-diglycerides in fibroblasts. Comparison of α-thrombin, EGF and
    PDGF. *J. Biol. Chem.* **265**, 7959–7966.
Peters, K. G., Marie, J., Wilson, E., Ives, H. E., Escobedo, J., Del Rosario, M., Mirda,

D., and Williams, L. T. (1992). Point mutation of an FGF receptor abolishes phosphatidylinositol turnover and Ca^{2+} flux but not mitogenesis. *Nature (London)* **358**, 678–684.

Pfeffer, L. M., Strulovici, B., and Saltiel, A. R. (1990). Interferon-α selectively activates the β isoform of protein kinase C through phosphatidylcholine hydrolysis. *Proc. Natl. Acad. Sci. U.S.A.* **87**, 6537–6541.

Pfeffer, L. M., Eisenkraft, B. L., Reich, N. C., Improta, T., Baxter, G., Daniel-Issakani, S., and Strulovici, B. (1991). Transmembrane signaling by interferon α involves diacylglycerol production and activation of the ε isoform of protein kinase C in Daudi cells. *Proc. Natl. Acad. Sci. U.S.A.* **88**, 7988–7992.

Pike, L. J., and Eakes, A. T. (1987). Epidermal growth factor stimulates the production of phosphatidylinositol monophosphate and the breakdown of polyphosphoinositides in A431 cells. *J. Biol. Chem.* **262**, 1644–1651.

Plevin, R., Cook, S. J., Palmer, S., and Wakelam, M. J. O. (1991). Multiple sources of sn-1,2-diacylglycerol in platelet-derived-growth-factor-stimulated Swiss 3T3 fibroblasts. *Biochem. J.* **279**, 559–565.

Price, B. D., Morris, J. D. H., Marshall, C. J., and Hall, A. (1989). Stimulation of phosphatidylcholine hydrolysis, diacylglycerol release, and arachidonic acid production by oncogenic ras is a consequence of protein kinase C activation. *J. Biol. Chem.* **264**, 16638–16643.

Rangan, L. A., Wright, T. M., and Raben, D. M. (1991). Differential dependence of early and late increases in 1,2-diacylglycerol on the presence of catalytically active α-thrombin: Evidence for regulation at the level of 1,2-diacylglycerol generation. *Cell Regul.* **2**, 311–316.

Rhee, S. G. (1991). Inositol phospholipid-specific phospholipase C: Interaction of the γ_1 isoform with tyrosine kinase. *Trends Biochem. Sci.* **16**, 297–301.

Rider, M. H., and Baquet, A. (1988). Activation of rat liver plasma-membrane diacylglycerol kinase by vasopressin and phenylephrine. *Biochem. J.* **255**, 923–928.

Rosoff, P. M., Savage, M., and Dinarello, C. A. (1988). Interleukin-1 stimulates diacylglycerol production in T lymphocytes by a novel mechanism. *Cell* **54**, 73–81.

Sakane, F., Yamada, K., Kanoh, H., Yokoyama, C., and Tanabe, T. (1990). Porcine diacylglycerol kinase sequence has zinc finger and E-F hand motifs. *Nature (London)* **344**, 345–348.

Sakane, F., Yamada, K., Imai, S., and Kanoh, H. (1991a). Porcine 80-kDa diacylglycerol kinase is a calcium-binding and calcium/phospholipid-dependent enzyme and undergoes calcium-dependent translocation. *J. Biol. Chem.* **266**, 7096–7100.

Sakane, F., Imai, S., Yamada, K., and Kanoh, H. (1991b). The regulatory role of EF-hand motifs of pig 80K diacylglycerol kinase as assessed using truncation and deletion mutants. *Biochem. Biophys. Res. Commun.* **181**, 1015–1021.

Satoh, T., Nakafuku, M., and Kaziro, Y. (1992). Function of Ras as a molecular switch in signal transduction. *J. Biol. Chem.* **267**, 24149–24152.

Schaap, D., de Widt, J., van der Wal, J., Vandekerckhove, J., van Damme, J., Gussow, D., Ploegh, H. L., van Blitterswijk, W. J., and van der Bend, R. L. (1990). Purification, cDNA-cloning and expression of human diacylglycerol kinase. *FEBS Lett.* **275**, 151–158.

Schaap, D., van der Wal, J., van Blitterswijk, W. J., van der Bend, R. L., and Ploegh, H. L. (1993). Diacylglycerol kinase is phosphorylated in vivo upon stimulation of the epidermal growth factor receptor and serine/threonine kinases, including protein kinase C-ε. *Biochem. J.* **289**, 875–881.

Schütze, S., Berkovic, D., Tomsing, O., Unger, C., and Krönke, M. (1991). Tumor necrosis factor induces rapid production of 1,2-diacylglycerol by a phosphatidylcholine-specific phospholipase C. *J. Exp. Med.* **174**, 975–988.

Schütze, S., Potthoff, K., Machleidt, T., Berkovic, D., Wiegmann, K., and Kröncke, M. (1992). TNF activates NF-κB by phosphatidylcholine-specific phospholipase C-induced "acidic" sphingomyelin breakdown. *Cell* **71,** 765–776.

Serth, J., Lautwein, A., Frech, M., Wittinghofer, A., and Pingoud, A. (1991). The inhibition of the GTPase activating protein–Ha-*ras* interaction by acidic lipids is due to physical association of the terminal domain of the GTPase activating protein with micellar structures. *EMBO J.* **10,** 1325–1330.

Slivka, S. R., Meier, K. E., and Insel, P. A. (1988). α_1-Adrenergic receptors promote PC hydrolysis in MDCK-D1 cells. A mechanism for rapid activation of protein kinase C. *J. Biol. Chem.* **263,** 12242–12246.

Stith, B. J., Kirkwood, A. J., and Wohnlich, E. (1991). Insulin-like growth factor 1, insulin, and progesterone induce early and late increases in *Xenopus* oocyte *sn*-1,2-diacylglycerol levels before meiotic cell division. *J. Cell. Physiol.* **149,** 252–259.

Sugimoto, Y., Whitman, M., Cantley, L. C., and Erikson, R. L. (1984). Evidence that the Rous sarcoma virus transforming gene product phosphorylates phosphatidylinositol and diacylglycerol. *Proc. Natl. Acad. Sci. U.S.A.* **81,** 2117–2121.

Suzuki-Sekimori, R., Matuoka, K., Nagai, Y., and Takenawa, T. (1989). Diacylglycerol, but not inositol 1,4,5-trisphosphate, accounts for platelet-derived growth factor-stimulated proliferation of BALB 3T3 cells. *J. Cell. Physiol.* **140,** 432–438.

Taylor, S. J., Chae, H. Z., Rhee, S. G., and Exton, J. H. (1991). Activation of the β1 isozyme of phospholipase C by α subunits of the G_q class of G proteins. *Nature (London)* **350,** 516–518.

Thompson, N. T., Bonser, R. W., and Garland, L. G. (1991). Receptor-coupled phospholipase D and its inhibition. *Trends Pharmacol. Sci.* **12,** 404–407.

Tsai, M.-H., Yu, C.-L., and Stacey, D. W. (1990). A cytoplasmic protein inhibits the GTPase activity of H-*ras* in a phospholipid-dependent manner. *Science* **250,** 982–985.

Tyagi, S. R., Olson, S. C., Burnham, D. N., and Lambeth, J. D. (1991). Cyclic AMP-elevating agents block chemoattractant activation of diradylglycerol generation by inhibiting phospholipase D activation. *J. Biol. Chem.* **266,** 3498–3504.

Uings, I. J., Thompson, N. T., Randall, R. W., Spacey, G. D., Bonser, R. W., Hudson, A. T., and Garland, O. G. (1992). Tyrosine phosphorylation is involved in receptor coupling to phospholipase D but not phospholipase C in the human neutrophil. *Biochem. J.* **281,** 597–600.

Ullrich, A., and Schlessinger, J. (1990). Signal transduction by receptors with tyrosine kinase activity. *Cell* **61,** 203–212.

Van Blitterswijk, W. J., and Hilkmann, H. (1993). Rapid attenuation of receptor-induced diacylglycerol and phosphatidic acid by phospholipase D-mediated transphosphatidylation: formation of bisphosphatidic acid. *EMBO J.* **12,** 2655–2662.

Van Blitterswijk, W. J., Hilkmann, H., de Widt, J., and van der Bend, R. L. (1991a). Phospholipid metabolism in bradykinin-stimulated human fibroblasts. I. Biphasic formation of diacylglycerol from phosphatidylinositol and phosphatidylcholine, controlled by protein kinase C. *J. Biol. Chem.* **266,** 10337–10343.

Van Blitterswijk, W. J., Hilkmann, H., de Widt, J., and van der Bend, R. L. (1991b). Phospholipid metabolism in bradykinin-stimulated human fibroblasts. II. Phosphatidylcholine breakdown by phospholipases C and D; involvement of protein kinase C. *J. Biol. Chem.* **266,** 10344–10350.

Van Corven, E., Groenink, A., Jalink, K., Eichholtz, T., and Moolenaar, W. H. (1989). Lysophosphatidate-induced cell proliferation: Identification and dissection of signalling pathways mediated by G proteins. *Cell* **59,** 45–54.

Van der Bend, R. L., Brunner, J., Jalink, K., van Corven, E. J., Moolenaar, W. H., and van Blitterswijk, W. J. (1992a). Identification of a putative membrane receptor for the bioactive phospholipid, lysophosphatidic acid. *EMBO J.* 11, 2495–2501.

Van der Bend, R. L., de Widt, J., van Corven, E. J., Moolenaar, W. H., and van Blitterswijk, W. J. (1992b). The biologically active phospholipid, lysophosphatidic acid, induces phosphatidylcholine breakdown in fibroblasts via activation of phospholipase D. Comparison with the response to endothelin. *Biochem. J.* 285, 235–240.

Van der Meulen, J., and Haslam, R. J. (1990). Phorbol ester treatment of intact rabbit platelets greatly enhances both the basal and guanosine 5'-[γ-thio]triphosphate-stimulated phospholipase D activities of isolated platelet membranes. *Biochem. J.* 271, 693–700.

Veis, N., and Hamilton, J. A. (1991). Colony stimulating factor-1 stimulates diacylglycerol generation in murine bone marrow-derived macrophages, but not in resident peritoneal macrophages. *J. Cell. Physiol.* 147, 298–305.

Wang, P., Anthes, J. C., Siegel, M. I., Egan, R. W., and Billah, M. M. (1991). Existence of cytosolic phospholipase D. *J. Biol. Chem.* 266, 14877–14880.

Whetton, A. D., Monk, P. N., Consalvey, S. D., Huang, S. J., Dexter, T. M., and Downes, C. P. (1988). Interleukin 3 stimulates proliferation via protein kinase C activation without increasing inositol lipid turnover. *Proc. Natl. Acad. Sci. U.S.A.* 85, 3284–3288.

Wolfman, A., and Macara, I. G. (1987). Elevated levels of diacylglycerol and decreased phorbol ester sensitivity in ras-transformed fibroblasts. *Nature (London)* 325, 359–361.

Wright, T. M., Rangan, L. A., Shin, H. S., and Raben, D. M. (1988). Kinetic analysis of 1,2-diacylglycerol mass levels in cultured fibroblasts. Comparison of stimulation by α-thrombin and epidermal growth factor. *J. Biol. Chem.* 263, 9374–9380.

Wright, T. M., Shin, H. S., and Raben, D. M. (1990). Sustained increase in 1,2-diacylglycerol precedes DNA synthesis in epidermal-growth-factor stimulated fibroblasts. *Biochem. J.* 267, 501–507.

Wright, T. M., Willenberger, S., and Raben, D. M. (1992). Activation of phospholipase D by α-thrombin or epidermal growth factor contributes to the formation of phosphatidic acid, but not to observed increases in 1,2-diacylglycerol. *Biochem. J.* 285, 395–400.

Wu, D., Lee, C. H., Rhee, S. G., and Simon, M. I. (1992). Activation of phospholipase C by the α subunits of the G_q and G_{11} proteins in transfected COS-7 cells. *J. Biol. Chem.* 267, 1811–1817.

Yatomi, Y., Hazeki, O., Kume, S., and Ui, M. (1992). Suppression by wortmannin of platelet responses to stimuli due to inhibition of pleckstrin phosphorylation. *Biochem. J.* 285, 745–751.

CHAPTER 17

Lysophosphatidic Acid as a Novel Lipid Mediator

Wouter H. Moolenaar, Kees Jalink, Thomas Eichholtz, Peter L. Hordijk, Rob van der Bend, Wim J. van Blitterswijk, and Emile van Corven
Division of Cellular Biochemistry, The Netherlands Cancer Institute, 1066 CX Amsterdam, The Netherlands

I. INTRODUCTION

Phospholipids and their bioactive metabolites continue to attract much interest in studies on receptor-mediated signal transduction. Breakdown products of some phospholipids can function not only as intracellular second messengers, but also as extracellular agonists that modulate cell function. Well-known examples of phospholipid-derived signaling molecules include diacylglycerol, inositol triphosphate, and arachidonate and its active metabolites, all of which are produced rapidly in activated cells. Relatively little attention has been paid to lysophospholipids as potential signaling molecules, perhaps not too surprisingly in view of the detergent-like and lytic properties of at least some of the lysolipids, particularly lysophosphatidylcholine. However, the simplest naturally occurring phospholipid, lysophosphatidic acid (LPA), shows striking biological activities when added to appropriate target cells at doses far below its critical micelle concentration. Although LPA has been known as a critical precursor in *de novo* lipid biosynthesis for more than three decades (for review, see Bishop and Bell, 1988), the possibility that this simple phospholipid may function as a hormone or local mediator has been appreciated only for a few years. The intention of this chapter is to summarize briefly the current state of knowledge and understanding of the multiple biochemical and biological activities of LPA, with emphasis on its formation, site of action, and activation of G protein-mediated signal transduction cascades.

II. LYSOPHOSPHATIDIC ACID: FORMATION AND RELEASE

When cells are stimulated by agonists, the level of phosphatidic acid (PA) in the plasma membrane often rises significantly. PA can be generated via phosphorylation of diacylglycerol (DG) by DG kinase or, more directly, through the action of phospholipase D (PLD) on phosphatidylcholine (PC) and phospahtidylethanolamine (PE) (Exton, 1990). In the latter case, the level of PA often rises higher and more rapidly than that of DG. Newly produced PA can function as (1) a source of DG (the second messenger for protein kinase C), (2) a potential second messenger in its own right, (3) a precursor for cytidine diphospho-diacylglycerol (CDP-DG), or (4) a precursor for LPA. The final pathway has not been examined thoroughly to date, but increasing evidence indicates that LPA is generated rapidly in at least some cell types on receptor stimulation. Thus, LPA is produced rapidly in thrombin-activated platelets and in growth factor-stimulated fibroblasts (Gerrard and Robinson, 1989; Fukami and Takenawa, 1992; Eichholtz *et al.*, 1993). In thrombin-activated platelets, LPA formation is secondary to the formation of PA through the action of a specific phospholipase A$_2$ (PLA$_2$) (Billah *et al.*, 1981), although alternative metabolic routes leading to LPA formation cannot be excluded at present. A schematic outline of PA and LPA formation in activated cells is illustrated in Fig. 1.

Further, LPA appears to be a normal constituent of serum (Jalink *et al.*, 1993) but is not detectable in platelet-poor plasma, suggesting that LPA is secreted into the extracellular environment on platelet activation and blood clotting. Our findings provide direct support for that view. Eichholtz *et al.* (1993) have shown that ~90% of the newly formed LPA in thrombin-activated platelets is released into the medium. Whether newly produced LPA is secreted via a distinct transport mechanism or whether it simply diffuses out of the platelets during their aggregation remains to be investigated. In any case, serum-borne LPA has fatty acid composition

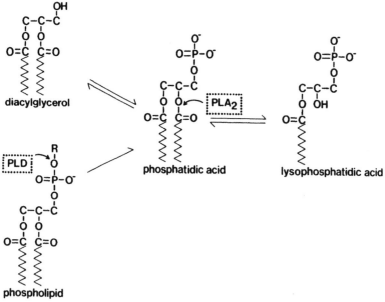

FIGURE 1 Proposed scheme of generation of lysophosphatidic acid (LPA) from newly formed phosphatidic acid during platelet activation. PLA$_2$, Phospholipase A$_2$; PLD, phospholipase D; R, phospholipid head group. The contribution of other pathways to LPA generation remains to be explored.

similar to platelet-derived LPA and occurs in an albumin-bound form (estimated concentration in serum, ~1 μM (Eichholtz *et al.*, 1993). Preliminary results suggest that LPA release is not restricted to activated platelets but also may occur in mitogenically stimulated fibroblasts (Fukami and Takenawa, 1992; C. Limatola and W. H. Moolenaar, unpublished observations). Collectively, current evidence strongly suggests that LPA is released from activated cells into the extracellular space where it may act as a multifunctional lipid mediator.

III. BIOLOGICAL AND BIOCHEMICAL ACTIVITIES

Exogenous LPA, at submicromolar doses, evokes a remarkably wide range of hormone- and growth factor-like effects in many different cell types, as summarized in Table I. Although exogenous LPA is metabolized partially to monoacylglycerol (van der Bend *et al.*, 1992a), the latter metabolite has no biological activity. Cells as diverse as mammalian fibro-

TABLE I

Rapid Cellular Responses to Lysophosphatidic Acid[a]

Cell type	Response
Many cell types	PIP^2 hydrolysis/Ca^{2+} signaling
MDCK epithelial cells	Cl^- secretion[b]
Smooth muscle	Contraction
Neuronal cells	Neurite retraction
Fibroblasts	Stress fiber formation[c]
Platelets	Aggregation
Xenopus oocytes	Cl^- inward current
Dictyostelium	Chemotaxis

[a] Other references can be found in the text.
[b] H. de Jonge, personal communication.
[c] Ridley and Hall (1992); K. Jalink, unpublished observations.

blasts, neuronal cells, platelets, and *Xenopus laevis* oocytes share a common rapid response to LPA, that is, stimulation of phosphatidylinositol 4,5-bisphosphate (PIP_2) hydrolysis and Ca^{2+} mobilization in a G- protein-dependent manner (van Corven *et al.*, 1989; Jalink *et al.*, 1990; Fernhout *et al.*, 1992). However, responsiveness to LPA is not a universal trait of all cell types; for example, neutrophils, monocytes, and mast cells fail to show a detectable Ca^{2+} signal in response to LPA, even when challenged with a high concentration of the lipid (Jalink *et al.*, 1990). Further, note that amebas of the slime mold *Dictyostelium* respond chemotactically to exogenous LPA (Jalink *et al.*, 1993a). Although the physiological significance of this phenomenon remains to be explored, this finding implies that LPA responsiveness is not restricted to vertebrate cells.

A. Growth Factor-Like Action

When added to quiescent fibroblasts, LPA stimulates long-term DNA synthesis and cell division (van Corven *et al.*, 1989,1992; Moolenaar, 1991). Naturally occurring 1-oleoyl PLA stimulates thymidine incorporation into fibroblasts with a half-maximal effect observed at about 10 μM, a dose that is significantly higher than that required for evoking the early events listed in Table I, which are half-maximal at 10–50 nM. The discrepancy between these dose–response relationships remains unexplained at present. The mitogenic response to LPA appears to require its long-term presence in the medium; when LPA is removed a few hours after its addition to the cells, entry into S phase does not occur. Note that the growth-promoting activity of LPA does not require the presence of synergizing peptide growth factors. In this respect, LPA action differs from the mitogenic activity of certain neuropeptides such as serotonin or vasopressin, which fail to sustain cell proliferation unless insulin or epidermal growth factor (EGF) is present (Moolenaar, 1991).

A somewhat puzzling finding is that PA can mimic the mitogenic activity of LPA (Moolenaar *et al.*, 1986; van Corven *et al.*, 1992), whereas PA fails to evoke LPA-like early responses as listed in Table I (provided that PA does not contain contaminating traces of LPA, which is often the case; Jalink *et al.*, 1990). The observed dose–response relationships for PA and LPA mitogenicity largely overlap, implying that the mitogenic action of PA cannot be explained simply by LPA impurities in the PA used. Natural lipids other than PA and LPA are incapable of stimulating DNA synthesis in quiescent fibroblasts. The mitogenic potency of LPA, as well as of PA, is dependent on the fatty acid chain length. Thus, the oleoyl and palmitoyl species show the highest activity, whereas the potency decreases with decreasing chain length; lauroyl (L)PA shows hardly any mitogenic activity (van Corven *et al.*, 1992).

B. Signal Transduction and Receptor Identification

Several "classic" signal tansduction pathways in the action of LPA have been identified (van Corven *et al.*, 1989; Jalink *et al.*, 1990). As summarized in Fig. 2, these include (1) GTP-dependent activation of phospholipase C, (2) release of arachidonic acid, presumably as a result of PLA_2 activation, (3) activation of phospholipase D (van der Bend *et al.*, 1992b), and (4) pertussis toxin-sensitive inhibition of adenylate cyclase (van Corven *et al.*, 1989). In addition, studies on neuronal cells have revealed that LPA causes rapid changes in cytoskeletal organization (Jalink *et al.*, 1993b), a process that is not attributable to the second messenger cascades, as is discussed later. Collectively, current evidence strongly suggests that LPA activates one or more specific G protein-coupled receptors at the cell surface.

In another study, van der Bend *et al.*, (1992c) succeeded in identifying a putative high-affinity LPA receptor. A ^{32}P-labeled LPA analog containing a photoreactive fatty acid, diazirine LPA, labels a membrane protein of apparent molecular mass 38–40 kDa in various LPA-responsive cell types; labeling was most prominent in neuronal cells and brain homogenates. Labeling is specific in that only unlabeled LPA but not other phsopholipids

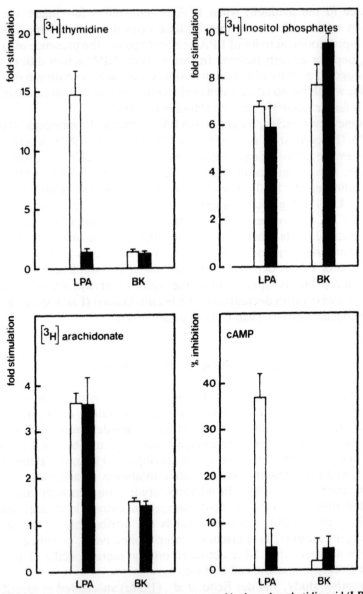

FIGURE 2 Second messenger pathways generated by lysophosphatidic acid (LPA) and bradykinin (BK) in human fibroblasts in the presence (filled bars) or absence (open bars) of 100 ng/ml pertussis toxin. For measurement of adenylate cyclase inhibition, cells first were stimulated with prostaglandin E_1 (1 μM) for 10 min to raise intracellular cAMP levels several-fold above control values. For details, see van Corven *et al.* (1989).

inhibit incorporation of diazirine LPA into the 38- to 40-kDa protein band; half-maximal inhibition is observed at approximately 10–20 nM unlabeled LPA. The 38- to 40-kDa protein presumably represents a specific LPA cell surface receptor mediating at least some of the multiple cellular responses to LPA (for further details, see van der Bend *et al.*, 1992c). The putative LPA receptor is not detectable in human neutrophils, which are biologically unresponsive to LPA (Jalink *et al.*, 1990). The polyanionic drug suramin was found to inhibit not only LPA receptor binding but also all biological responses to LPA (van Corven *et al.*, 1992; van der Bend *et al.*, 1992c; Jalink *et al.*, 1993b).

1. Changes in Neuronal Cell Shape: A Novel Signaling Mechanism

In several experiments, we used the neuronal cell lines N1E-115 and NG108-15 as model systems in which to explore the neurobiological actions of LPA. These cells constitute a convenient model system in which to investigate early signaling events in the proliferative response. After serum starvation, these cells stop growing and subsequently begin to acquire various differentiated properties of mature neurons, including the formation of long neurites. Addition of 1-oleoyl LPA (1 μM) to growth factor-starved N1E-115 or NG108-15 cells causes rapid and dramatic changes in cell shape reminiscent of those observed during mitosis (Jalink *et al.*, 1993b). Virtually every flattened cell begins to round up as early as 5–10 sec after LPA addition; rounding is complete within ~1 min. Almost simultaneously, growth cones begin to collapse and developing neurites retract. The effects of LPA on neuronal cell shape are dose dependent, with some rounding detectable at doses as low as 10 nM and maximal responses at 0.5–1 μM. In the continuous presence of LPA, cells maintain their rounded shape for 5–10 min. At 10–20 min after the addition of LPA, however, most cells gradually resume a flattened morphology, but neurite outgrowth remains suppressed for at least a few hours. Remarkably, a second application of LPA to such respread cells leaves the shape unaltered. LPA-induced cell rounding thus is subject to homologous desensitization, as would be expected for a receptor-mediated process. Additional evidence for the receptor-mediated nature of the morphological response includes the finding that (1) the drug suramin, which blocks LPA–receptor interaction, inhibits LPA-induced shape changes; (2) other related lipids have no effect; (3) microinjected LPA fails to mimic the action of extracellularly applied LPA; and (4) the morphological response to LPA is mimicked fully by a synthetic peptide agonist of the cloned G protein-coupled thrombin receptor, whereas LPA and thrombin clearly act through separate receptors (Jalink and Moolenaar, 1992; Jalink *et al.*, 1993b).

LPA- and thrombin-induced changes in neuronal cell shape appear to be mediated by "contraction" of the cortical actin cytoskeleton with no direct involvement of microtubules. As for the signaling mechanism responsible for this novel action of LPA and thrombin, in N1E-115 cells as in many other cells, LPA stimulates phosphoinositide hydrolysis, leading to calcium mobilization and activation of protein kinase C. However, the following findings indicate that the PLC–Ca^{2+}–PKC pathway is not responsible for the observed changes in cell shape: (1) phosphoinositide-hydrolyzing neurotransmitters, Ca^{2+} ionophores, and PKC-activating phorbol ester all fail to mimic LPA in inducing cell rounding; (2) LPA-induced shape changes are not prevented by down-regulating PKC by long-term treatment of the cells with phorbol ester; and (3) cells depleted of internal Ca^{2+} (by addition of ionophore in Ca^{2+}-free medium) show a normal morphological response to LPA, despite the absence of a Ca^{2+} transient. Similarly, bacterial toxin-sensitive G proteins, adenylate cyclase, and cyclic nucleotides have no apparent role in mediating LPA-induced shape changes, since LPA action is neither inhibited nor mimicked by treatment of the cells with pertussis or cholera toxin, cAMP and cGMP analogs, or forskolin (10 μM).

Collectively, these results suggest that LPA- and thrombin-induced cell rounding is not mediated by known G protein-linked second messenger cascades. By analogy with the cell rounding induced by active protein tyrosine kinases (Jove and Hanafusa, 1987), specific phosphorylations and dephosphorylations of certain actin-binding proteins (Pollard and Cooper, 1986; Stossel, 1989) may be responsible for the morphological effects of LPA. In support of this hypothesis, we observed that LPA-induced cell rounding is blocked by both microinjected vanadate and exogenously added pervanadate, a membrane-permeant form of vanadate and a widely used inhibitor of protein tyrosine phosphatases. Further, brief preincubation with such (nonspecific) protein kinase inhibitors as genistein (50 μM), quercetin (50 μM), and staurosporine (1 μM) similarly inhibits cell rounding (Jalink *et al.*, 1993b).

We have begun to analyze the possible role of the src protein tyrosine kinase in the rapid neuronal shape changes. LPA and thrombin were found to induce a small but statistically significant increase in src kinase activity in N1E-115 cells, as measured in an immunocomplex kinase assay (Jalink *et al.*, 1993b). Although we cannot draw conclusions yet about cause–effect relationships, the results obtained are intriguing and deserve further study.

2. Activation of p21^{ras}

The product of the *ras* proto-oncogene (p21^{ras}) is a focal point of signal transduction by growth factor receptors, notably those with intrinsic tyrosine kinase activity. Given the mitogenic potency of LPA, we have examined the activation state of the ras protein in quiescent fibroblasts after treatment with LPA. LPA causes the rapid accumulation of the GTP-bound active form of ras within 1 min (van Corven *et al.*, 1993). The kinetics of activation are transient; the response subsides after about 10 min. This early response to LPA is similar to that elicited by EGF, although the extent of p21^{ras} activation by LPA is somewhat weaker than that observed with EGF (van Corven *et al.*, 1993). Like the other early events in LPA action, p21–GTP accumulation conforms to all the criteria of a receptor-mediated process, including its dose dependency (half-maximal effects close to 10 nM) and complete inhibition by suramin. Importantly, LPA-induced p21–GTP accumulation is blocked by prior treatment of the cells with pertussis toxin, whereas EGF-induced p21–GTP accumulation remains unaffected (Fig. 3). Thus, a heterotrimeric G protein of the G$_i$ subclass regulates p21^{ras} through an as yet unknown effector pathway that appears to be independent of known G protein-mediated second messenger pathways (van Corven *et al.*, 1993).

Additional experiments have revealed that activation of p21^{ras} is not observed uniquely with the putative LPA receptor, but is a more general response to stimulation of G$_i$ protein-coupled receptors. Activation of the endogenous thrombin receptor in hamster lung fibroblasts and agonist stimulation of the transfected α_2 adrenergic receptor in Rat-1 cells similarly leads to pertussis toxin-sensitive activation of p21^{ras}. Also, in these cases,

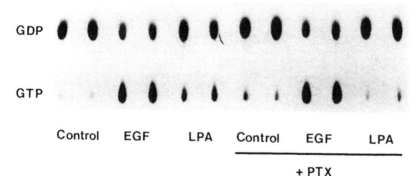

FIGURE 3 Activation of p21^{ras} by lysophosphatidic acid (LPA) and epidermal growth factor (EGF) in Rat-1 cells and inhibition by pertussis toxin (PTX). Phosphate-labeled cells were treated with EGF (50 ng/ml) or LPA (10 μM); p21^{ras} was immunoprecipitated using monoclonal 259; radiolabeled GDP and GTP were separated by thin layer chromatography (van Corven *et al.*, 1993). PTX (100 ng/ml) was added 3 hr prior to EGF and LPA. Results shown are from repeat experiments.

ras activation correlates well with the ultimate stimulation of DNA synthesis (van Corven *et al.*, 1993).

Although the downstream targets of ras are still unknown, we found that LPA-induced p21^{ras} activation correlates well with increased phosphorylation of cytosolic MAP kinases (Hordijk *et al.*, 1994), both with respect to kinetics of activation and phosphorylation and to pertussis toxin sensitivity. These results are consistent with MAP kinase activation being secondary to p21^{ras} activation.

3. Modulation of Membrane Potential

Mitogenic activation of quiescent cells generally is accompanied by rapid changes in ionic conductance resulting in alterations in electrical membrane potential. Our results indicate that LPA evokes a rapid long-lasting membrane depolarization in fibroblasts and neuroblastoma cells that is strikingly similar to that evoked by whole serum. LPA-induced membrane depolarization is likely to be caused by receptor-mediated opening of nonselective ion channels, but the prescise underlying signaling event and physiological significance remain to be explored (Moolenaar and Jalink, 1992).

A schematic outline of the various signaling pathways in the action of LPA is illustrated in Fig. 4.

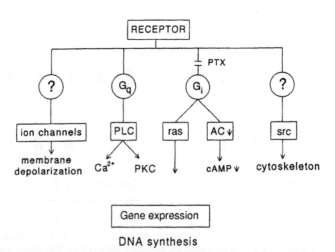

FIGURE 4 Scheme of receptor-mediated signal transduction pathways in the action of lysophosphatidic acid (LPA). PTX, Pertussis toxin; G$_q$, G protein activating phospholipase C (PLC); G$_i$, G protein(s) inhibiting adenylate cyclase and stimulating p21^{ras} (ras) activation; PKC, protein kinase C; AC, adenylate cyclase; src, p60^{src} protein tyrosine kinase.

IV. CONCLUDING REMARKS

The discovery that LPA exerts profound effects on many cell types suggests that LPA may have a previously unrecognized role as a lipid mediator. LPA appears to be released by cells (notably activated platelets), perhaps in a manner similar to the secretion of platelet-activating factor, prostaglandins, and other lipid agonists by activated cells. The released LPA then activates target cells in a paracrine or autocrine fashion. At present, speculating further on the normal physiological (or perhaps pathophysiological) roles of LPA is premature. A major challenge for future studies is to establish the amino acid sequence of the putative LPA cell surface receptor, either through protein purification or, perhaps more straight forwardly, by expression cloning in COS cells. Another challenge is to delineate the unique signal transduction events, involving activation of both the *src* and *ras* proto-oncogene products, that are responsible for neurite retraction and reversal of the differentiated phenotype in neuroblastoma cells, and for the initiation of DNA synthesis and cell division in fibroblasts.

Acknowledgments

We thank José Overwater for preparation of this manuscript. Research related to this chapter was supported by the Netherlands Cancer Foundation and by the Netherlands Organization for Scientific Research (NWO).

References

Billah, M. M., Lapetina, E. G., and Cuatsecasas, P. (1981). Phospholipase A_2 activity specific for phosphatidic acid. *J. Biol. Chem.* **256**, 5399–5403.

Bishop, W. R., and Bell, R. M. (1988). Assembly of phospholipids into cellular membranes: Biosynthesis, transmembrane movement and intracellular translocation. *Annu. Rev. Cell Biol.* **4**, 579–610.

Eichholtz, T., Jalink, K., Fahrenfort, I., and Moolenaar, W. H. (1993). The bioactive phospholipid lysophosphatidate is released from activated platelets. *Biochem. J.* **291**, 677–680.

Exton, J. H. (1990). Signaling through phosphatidylcholine hydrolysis. *J. Biol. Chem.* **265**, 1–4.

Fernhout, B. J., Dijcks, F. A., Moolenaar, W. H., and Ruigt, G. S. F. (1992). Cytophosphatidic acid induces inward currents in *Xenopus laevis* oocytes. *Eur. J. Pharmacol.* **213**, 313–315.

Fukami, K., and Takenawa, T. (1992). Formation of phosphatidic acid in PDGF-stimulated 3T3 cells is a potential mitogenic signal. *J. Biol. Chem.* **267**, 10988–10993.

Gerrard, J. M., and Robinson, P. (1989). Identification of the molecular species of lysophosphatidic acid produced when platelets are stimulated by thrombin. *Biochim. Biophys. Acta* **1001**, 282–285.

Hordijk, P. L., Verlaan, I., van Corven, E. J., and Moolenaar, W. H. (1994). Protein tyrosine phosphorylation induced by lysophosphatidic acid in Rat-1 fibroblasts. Evidence that phosphorylation of MAP kinase is mediated by the G_i-p21^{ras} pathway. *J. Biol. Chem.* **269** (*in press*).

Jalink, K., and Moolenaar, W. H. (1992). Thrombin receptor activation causes rapid neural cell rounding and neurite retraction independent of classic second messengers. *J. Cell Biol.* **118**, 411–419.

Jalink, K., van Corven, E. J., and Moolenaar, W. H. (1990). Lysophosphatidic acid, but not phosphatidic acid, is a potent Ca^{2+}-mobilizing stimulus for fibroblasts. *J. Biol. Chem.* **265**, 12232–12239.

Jalink, K., Moolenaar, W. H., and van Duijn, B. (1993a). Lysophosphatidic acid is a chemoattractant for *Dictyostelium amoebae*. *Proc. Natl. Acad. Sci. U.S.A.* **90**, 1857–1861.

Jalink, K., Eichholtz, T., Postma, F. R., van Corven, E. J., and Moolenaar, W. H. (1993b). Lysophosphatidic acid indures neuronal shape changes via a novel, receptor-mediated signaling pathway. *Cell Growth Differ.* **4**, 247–255.

Jove, R., and Hanafusa, H. (1987). Cell transformation by the viral *src* oncogene. *Annu. Rev. Cell Biol.* **3**, 31–56.

Moolenaar, W. H. (1991). G protein-coupled receptors, phosphoinositide hydrolysis and cell proliferation. *Cell Growth Differ.* **2**, 359–364.

Moolenaar, W. H., and Jalink, K. (1992). Membrane potential changes in the action of growth factors. *Cell. Physiol. Biochem.* **2**, 189–195.

Moolenaar, W. H., Kruijer, W., Tilly, B. C., Verlaan, I., Bierman, A. J., and de Laat, S. W. (1986). Growth factor-like action of phosphatidic acid. *Nature (London)* **323**, 171–173.

Pollard, T. D., and Cooper, J. A. (1986). Actin and actin-binding proteins. *Annu. Rev. Biochem.* **55**, 987–1035.

Ridley, A. J., and Hall, A. (1992). The small GTP-binding protien rho regulates the assembly of focal adhesions and actin stress fibers in response to growth factors. *Cell* **70**, 384–394.

Stossel, T. P. (1989). From signal to pseudopod. How cells control cytoplasmic actin assembly. *J. Biol. Chem.* **264**, 18261–18264.

van Corven, E. J., Groenink, A., Jalink, K., Eichholtz, T., and Moolenaar, W. H. (1989). Lysophosphatidate-induced cell proliferation: Identification and dissection of signaling pathways mediated by G proteins. *Cell* **59**, 45–54.

van Corven, E. J., van Rijswijk, A., Jalink, K., van der Bend, R., van Blitterswijk, W. J., and Moolenaar, W. H. (1992). Mitogenic action of lysophosphatidic acid and phosphatidic acid on fibroblasts. *Biochem. J.* **281**, 163–169.

van Corven, E. J., Hordijk, P. L., Medema, R. H., Bos, J. L., and Moolenaar, W. H. (1993). Pertussis toxin-sensitive activation of p21^{ras} by G protein-coupled receptor agonists in fibroblasts. *Proc. Natl. Acad. Sci. U.S.A.* **90**, 1257–1261.

van der Bend, R. L., de Widt, J., van Corven, E. J., Moolenaar, W. H., and van Blitterswijk, W. J. (1992a). Metabolic conversion of the biologically active phospholipid, lysophosphatidic acid, in fibroblasts. *Biochim. Biophsy. Acta* **1125**, 110–112.

van der Bend, R. L., de Widt, J., van Corven, E. J., Moolenaar, W. H., and van Blitterswijk, W. J. (1992b). The biologically active phospholipid, lysophosphatidic acid, induces phosphatidylcholine breakdown via activation of phospholipase D. *Biochem J.* **285**, 235–240.

van der Bend, R. L., Brunner, J., Jalink, K., van Corven, E. J., Moolenaar, W. H., and van Blitterswijk, W. J. (1992c). Identification of a putative membrane receptor for the bioactive phospholipid, lysophosphatidic acid. *EMBO J.* **11**, 2495–2501.

PART IV

Intracellular Transport and
Traffic of Lipids

CHAPTER 18

Mechanisms of Intracellular Membrane Lipid Transport

Jonathan C. McIntyre and Richard G. Sleight
Department of Molecular Genetics, Biochemistry, and Microbiology, University of
Cincinnati College of Medicine, Cincinnati, Ohio 45267

I. Introduction
II. Relationship between Lipid and Protein Intracellular Transport
III. Vesicular Transport of Lipids
   A. Endoplasmic Reticulum to Golgi Apparatus
   B. Golgi Apparatus to Endoplasmic Reticulum
   C. Transport between Golgi Elements
   D. Golgi Apparatus to Plasma Membrane
   E. Endoplasmic Reticulum to Plasma Membrane
   F. Transport from the Plasma Membrane
IV. Monomer Transport
   A. Endoplasmic Reticulum to Plasma Membrane
   B. Transport between Endoplasmic Reticulum and Mitochondria
   C. Transport from Lysosomes
   D. Transport from the Plasma Membrane
V. Lateral Diffusion and Contact-Mediated Transport
VI. Transmembrane Movement
VII. Conclusions
   References

I. INTRODUCTION

Much recent enthusiasm in membrane cell biology is centered on lipids and lipid metabolites acting as modulators of biological regulatory processes (Pagano and Sleight, 1985; Whatley *et al.*, 1990; Dennis *et al.*, 1991; Kolesnick, 1991; Merrill, 1992; see also Part V). This interest places new focus on the role of individual lipid species in membrane function

Current Topics in Membranes, Volume 40

and, therefore, generates renewed interest in intracellular lipid transport and metabolism as it relates to organelle membrane formation and homeostasis. This chapter reviews experimental evidence establishing the mechanisms and pathways by which membrane-forming lipids are transported throughout the cell. Several previous reviews of this topic are available and should be consulted for a historical perspective (Sleight, 1987; Pagano, 1990; Voelker, 1990b,1991).

Biological membranes are composed of three major classes of lipids; glycerophospholipids, sphingolipids, and sterols (McMurray, 1973; Raetz, 1982). Structural diversity exists within each class, resulting in the appearance of over 1000 chemically distinct species in eukaryotic cells. These species are not distributed uniformly throughout the cell (Colbeau et al., 1970; McMurray, 1973). In eukaryotes, the concentrations of the different lipid species vary according to organelle membrane. For example, the content of phosphatidylcholine in rat liver nuclear membrane is approximately 60%, whereas sphingomyelin constitutes 4% of total lipid (McMurray, 1973). A much different lipid composition exists at the plasma membrane, in which phosphatidylcholine constitutes 35% of the total lipid and sphingomyelin constitutes 18% (McMurray, 1973).

The compositional diversity of individual membranes is complicated further by the asymmetric distribution of lipids across membrane bilayers (Op den Kamp, 1979). This asymmetry is demonstrated most clearly in the plasma membrane, where the majority of phosphatidylcholine and sphingomyelin is localized to the outer leaflet of the membrane whereas phosphatidylserine and phosphatidylethanolamine are concentrated in the inner leaflet (Op den Kamp, 1979; see also Chapters 1 and 2). This asymmetric distribution of lipids is not confined solely to classification by head group. The inner leaflet of the plasma membrane possesses phospholipids containing more unsaturated fatty acid chains than those of the outer leaflet (Sandra and Pagano, 1978; Cogan and Schacter, 1981; Seigneuret et al., 1984).

The enzymes responsible for the synthesis of lipids are not distributed evenly throughout the cell. Although some lipid synthetic enzymes are found in mitochondria, peroxisomes, and the Golgi apparatus (Dennis and Kennedy, 1972; Reinhart et al., 1987; Bishop and Bell, 1988), the vast majority of lipid synthesis occurs at the cytosolic face of the endoplasmic reticulum (Bell et al., 1981; Bishop and Bell, 1988). In this respect, the synthesis and distribution of lipids throughout the cell presents problems analogous to the trafficking of newly synthesized membrane proteins. Proteins are sorted and trafficked in cells by a combination of chemical modification steps, recognition by receptors, and trapping based on conformational properties (Pfeffer and Rothman, 1987). These mechanisms have not been demonstrated for the intracellular sorting and transport of lipids. The hydrophobic nature of lipids renders them poorly soluble in the aqueous cytosol. Therefore, most lipids cannot simply diffuse through the cell. To distribute different classes of lipids among cellular membranes, cells use multiple mechanisms of transport.

Several mechanisms for transporting lipids between cellular membranes have been identified. Vesicular transport is the budding of vesicles from a donor membrane and subsequent movement to an acceptor membrane, followed by fusion of the two membranes. Monomer transport is the movement of individual lipid molecules between donor and acceptor membranes through an aqueous phase. This method of transport can be either spontaneous or protein mediated. Lateral diffusion is a process in which lipids move in the plane of the membrane bilayer from donor to acceptor membranes via specialized membrane bridges. The transfer of lipids during collision of donor and acceptor membranes is called contact-mediated transport. Transmembrane movement is the transport of individual lipids from one leaflet of a bilayer to the other. Like monomer transport, this process can be spontaneous or protein mediated.

The compositional diversity of biological membranes may not be the result of lipid transport processes alone. The varied lipid compositions of organelles may arise from degradation or interconversion of specific lipid species. Presently, little experimental information is available regarding this aspect of membrane lipid content regulation. Some evidence for the remodeling of phospholipid fatty acid constituents exists (Schmid et al., 1991); however, the extent and location of these activities are poorly understood.

II. RELATIONSHIP BETWEEN LIPID AND PROTEIN INTRACELLULAR TRANSPORT

The routes of intracellular protein transport through biosynthetic and endocytic pathways are well characterized (Steinman et al., 1983; Pfeffer and Rothman, 1987). A simplified view of the intracellular transport of membrane and secretory proteins is presented in Fig. 1. Secretory, glycosylated, and transmembrane proteins move throughout the cell in membrane vesicles that bud from pre-existing membranes, are transported through the cytosol, and subsequently fuse with other membranes. These processes concomitantly transport membrane lipids throughout the cell. The extent to which this mechanism contributes to the general movement of membrane lipids, or whether compositional enrichment of specific lipid classes in organelle membranes is derived in part by these processes, is

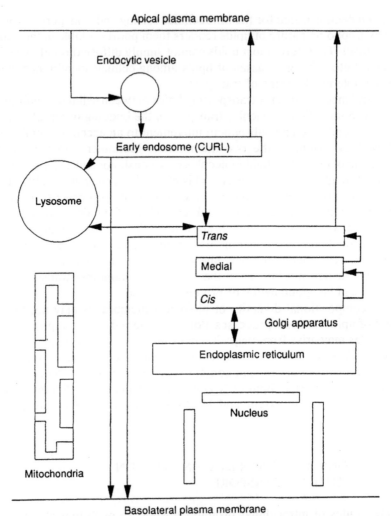

FIGURE 1 Pathways of vesicular protein transport.

not known. Early evidence for the vesicular transport of lipids between membranes was derived solely by inference from protein transport studies.

III. VESICULAR TRANSPORT OF LIPIDS

The known intracellular transport pathways of lipids by vesicular mechanisms are diagrammed in Fig. 2. Although similar to the pathways for protein transport shown in Fig. 1, several differences exist. These differ-

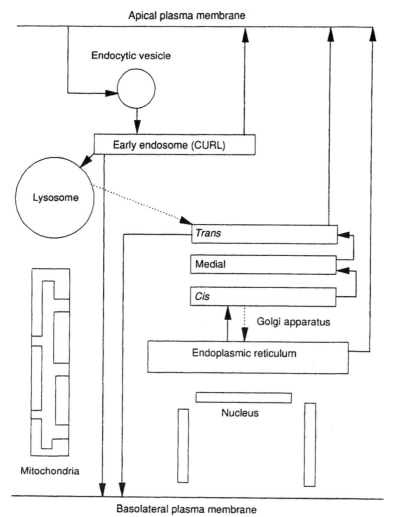

FIGURE 2   Pathways of vesicular lipid transport.

ences are likely to be resolved as our knowledge of lipid and protein trafficking increases. Evidence for each lipid transport pathway is presented in the following sections.

A. Endoplasmic Reticulum to Golgi Apparatus

Until recently, our knowledge of vesicular transport between the endoplasmic reticulum and the Golgi apparatus was confined to the transport of proteins. From these studies, a requirement for ATP, CA^{2+}, and cyto-solic factors was demonstrated, as was sensitivity to brefeldin A, N-ethylmaleimide (NEM), and GTPγS (Balch, 1990). Only recently was lipid transport between these membranes demonstrated.

In an elegant set of experiments, Moreau, Morré, and colleagues recon-stituted vesicular transport of lipids between the endoplasmic reticulum and cis-Golgi elements (Moreau and Morré, 1991). Isolated endoplasmic reticulum, enriched with respect to transition elements and labeled with radioactive lipid precursors, was incubated with Golgi apparatus immobi-lized on nitrocellulose. Radiolabeled lipids transferred from the endoplas-mic reticulum to the Golgi apparatus by two kinetically distinct mecha-nisms. One mechanism was stimulated by cytosol and ATP. Lipid transfer was reduced greatly below 25°C and was inhibited by NEM and GTPγS. These characteristics are indicative of a vesicular mechanism of transport similar to that reported for proteins (Pfeffer and Rothman, 1987; Balch, 1990; Rodman $et\ al.$, 1990). The second mechanism required cytosol, but was not ATP dependent and did not exhibit a sharp temperature-dependent transition in activity. The characteristics of the second mechanism suggest that it may represent the activity of phospholipid exchange proteins in the cytosol.

Endoplasmic reticulum transition vesicles 50–70 nm in size, isolated by free flow electrophoresis, are 5–6 times more efficient than transitional endoplasmic reticulum in transferring lipid to the Golgi apparatus (Moreau and Morré, 1991). The 50 to 70 nm transition vesicles are enriched in phosphatidylcholine and cholesterol and are depleted in triglyceride rela-tive to endoplasmic reticulum, suggesting that some form of lipid sorting or processing occurs during transition vesicle formation.

Vesicular lipid transport from the endoplasmic reticulum occurs most effectively with pure cis-Golgi as acceptor membranes (Moreau and Morré, 1991). Medial and $trans$-Golgi elements are comparatively poor acceptors. Phosphatidylcholine transferred to cis-Golgi elements is con-verted to lysophosphatidylcholine by the action of a phospholipase A (PLA) (Moreau $et\ al.$, 1991). Since the phospholipase activity is located in the lumen of the Golgi apparatus, membrane fusion has been suggested to be required to allow hydrolysis of the transferred phosphatidylcholine.

When semi-intact ("perforated") cells are incubated with an ATP-generating system and 30 to 80 nm unilamellar liposomes composed of phosphatidylcholine, fusion of the vesicles with the Golgi apparatus occurs (Kobayashi and Pagano, 1988). This process is inhibited by NEM and does not occur with vesicles enriched in phosphatidylserine or sphingomyelin. Liposome fusion to the Golgi apparatus may occur by the same process as endoplasmic reticulum transition vesicle fusion. If so, lipid composition of the endoplasmic reticulum transition vesicles rather than specific proteins may target the vesicles to the *cis*-Golgi.

B. Golgi Apparatus to Endoplasmic Reticulum

Considerable indirect evidence for vesicular recycling of membrane components between the Golgi apparatus and endoplasmic reticulum has been obtained. Wieland and colleagues (1987) estimated that the rate of lipid synthesis at the endoplasmic reticulum is insufficient to maintain the enormous flow of lipid from the endoplasmic reticulum to the Golgi apparatus as part of the protein biosynthetic machinery. These investigators postulated that recycling of lipid from the Golgi apparatus to the endoplasmic reticulum must exist. Evidence in favor of this pathway is derived from protein transport studies in the presence of brefeldin A (Fujiwara *et al.*, 1988; Doms *et al.*, 1989; Lippincott-Schwartz *et al.*, 1989,1990), a known inhibitor of protein transport from the endoplasmic reticulum to the Golgi apparatus. Brefeldin A causes the Golgi apparatus to disassemble and fuse with the endoplasmic reticulum, as evidenced by the redistribution of Golgi marker enzymes and the accumulation of secretory proteins in the endoplasmic reticulum (Fujiwara *et al.*, 1988; Doms *et al.*, 1989; Lippincott-Schwartz *et al.*, 1989). This Golgi apparatus-to-endoplasmic reticulum redistribution is inhibited by energy poisons and by reduced temperature (Lippincott-Schwartz *et al.*, 1990). Nocodazole, which depolymerizes microtubules and inhibits vesicular transport, also eliminates redistribution of the Golgi apparatus (Lippincott-Schwartz *et al.*, 1990).

Morphological studies have identified an endoplasmic reticulum–Golgi intermediate structure called an "exosome" (Lodish *et al.*, 1984; Saraste *et al.*, 1987). Biochemical markers and antibodies directed against a 53-kDa antigen (Schweizer *et al.*, 1988) and a 58-kDa antigen (Saraste *et al.*, 1987) have identified a membrane fraction distinguishable from the Golgi apparatus and the endoplasmic reticulum. Together, these data imply a vesicular route of transport from the Golgi apparatus to the endoplasmic reticulum.

Little direct evidence for the transport of lipids from the Golgi apparatus to the endoplasmic reticulum is available. Matyas and Morré (1987) observed the transport of radiolabeled gangliosides from the Golgi apparatus to the endoplasmic reticulum. Radiolabeled galactose was incorporated first into glycosphingolipids in the Golgi apparatus of rat liver. Radiolabeled glycosphingolipids were found in isolated endoplasmic reticulum fractions 30 min after appearance in the Golgi apparatus. The mechanism of this transfer was not investigated, but may involve routing through endosomes.

C. Transport between Golgi Elements

Using Chinese hamster ovary (CHO) Lec2 and Lec8 mutants as sources of donor and acceptor membranes, respectively, Wattenberg (1990) showed equivalent kinetics for glycoprotein and glycolipid transport in a cell-free system. Transport requires ATP, cytosolic factors, and elevated temperatures. Both protein and lipid transport are inhibited by NEM and GTPγS, suggesting a vesicular mechanism of transport. To date, intermediate transport vesicles derived from Golgi fractions have not been identified.

D. Golgi Apparatus to Plasma Membrane

The conversion of ceramide to sphingomyelin occurs at the luminal leaflet of the *cis* and medial elements of the Golgi apparatus (Voelker and Kennedy, 1982; Futerman *et al.*, 1990; Futerman and Pagano, 1991). The final steps of glycosphingolipid synthesis also are confined to the Golgi apparatus (Schwarzmann and Sandhoff, 1990; Futerman and Pagano, 1991). However, these lipids are found primarily in the plasma membrane (McMurray, 1973; Schwarzmann and Sandhoff, 1990), and therefore must be transported from the Golgi apparatus to the plasma membrane.

Miller-Podroza and Fishman (1982) determined a half-time of 20 min for the transport of newly synthesized gangliosides from the Golgi apparatus to the outer leaflet of the plasma membrane. This transport was monitored by the biosynthetic incorporation of radiolabeled galactose into gangliosides at the Golgi apparatus. Identification of radioactive outer leaflet plasma membrane glycosphingolipids was made by reaction with the impermeable reagents $NaIO_4$ and dinitrophenylhydrazine, followed by thin-layer chromatography (TLC). The long transport time is interpreted as evidence for a vesicular transport mechanism. Karrenbauer *et al.* (1990) measured the rate of sphingomyelin transport from the Golgi apparatus to the plasma membrane. A short-chain ceramide analog was given to cells and subsequently converted to sphingomyelin in the Golgi apparatus. The transport of this water-soluble sphingomyelin analog from the Golgi apparatus to the plasma membrane has a $t_{1/2}$ of 10–14 min.

The studies just described, and many other studies outlined in this chapter, make an assumption relating the time course of lipid movement to the mechanism of delivery, that is, when *in vivo* lipid transport is measured and demonstrated to have a time lag of 10–30 min or more, transport is assumed to have occurred by a vesicle-mediated process. Other mechanisms of intracellular lipid transport—such as phospholipid exchange, protein-mediated movement, or lateral diffusion—are believed to occur more rapidly. Although this assumption is sometimes reasonable, conclusions based on this assumption alone should be made with caution. Insensitive measurement techniques will produce results with an artifactual time lag because of the requirement that the transported lipid must accumulate to a level sufficient for detection.

Pagano and co-workers pioneered the use of a fluorescent C_6-NBD-ceramide analog for the study of the intracellular transport of sphingomyelin and glycosphingolipids (Lipsky and Pagano, 1983,1985; Pagano et al., 1989). When this analog is transferred from liposomes to cells at 2°C, it accumulates in mitochondria (Lipsky and Pagano, 1983,1985; Pagano et al., 1989). After incubating cells for 30 min at 37°C, the fluorescent analog is found primarily in the Golgi apparatus, where it is converted to fluorescent sphingomyelin and glucosylceramide. At time intervals longer than 30 min, fluorescence also is observed at the plasma membrane. The half-time for Golgi to plasma membrane transport is 20–30 min. The mechanism of transport is thought to be vesicular because it is eliminated during mitosis and inhibited by monensin (Lipsky and Pagano, 1983; Kobayashi and Pagano, 1989). Rosenwald et al. (1992) showed that PDMP [1-phenyl-2-(decanoylamino)-3-morpholino-1-propanol], an inhibitor of sphingolipid synthesis, slows the rate of protein transport through the Golgi elements. The inhibitor affects neither cell morphology nor microtubule formation. These data suggest that protein transport may be coupled to sphingolipid synthesis and transport.

Apical and basolateral membranes of polarized cells have different sphingolipid compositions. The apical membrane is enriched in glycosphingolipids, whereas sphingomyelin is found equally in both membranes (Kawai et al., 1974; Simons and Wandinger-Ness, 1990). Using C_6-NBD-labeled ceramide analogs, Simons and colleagues showed that fluorescent sphingomyelin and glucosylceramide synthesized in the Golgi apparatus are sorted during transport to the plasma membrane (van Meer et al., 1987). When Madin–Darby canine kidney (MDCK) cells, grown on polycarbonate filters, are incubated with C_6-NBD-ceramide, the fluorescent ceramide is converted to sphingomyelin and glucosylceramide in the Golgi apparatus. After warming the cells to 37°C for 1 hr, fluorescent lipid appears in both apical and basolateral plasma membrane domains. The apical surface is 2- to 4-fold enriched in fluorescent C_6-NBD-glucosylceramide, whereas C_6-NBD-sphingomyelin is distributed equally between apical and basolateral membranes. This observation is consistent with lipid sorting in the Golgi apparatus and differential transport pathways from the Golgi apparatus to the plasma membrane.

Simons and van Meer (1988) suggested that lipid and protein sorting in the *trans*-Golgi is mediated by microdomains formed by lateral migration and segregation of glycosphingolipids. This theory is supported by a study by Brown and Rose (1992). Membrane protein extraction experiments show that glycerol-phophatidylinositol-anchored proteins targeted for the apical surface are associated with detergent-resistant glycosphingolipid-rich membrane microdomains in the *trans*-Golgi. These microdomains are depleted in basolateral membrane proteins, and may represent a form of lipid and protein sorting in the *trans*-Golgi.

Specific vesicles for delivering membrane proteins to apical and basolateral surfaces have been identified by Simons and van Meer in MDCK cells (Wandinger-Ness *et al.*, 1990). Virally infected cells expressing specific apical and basolateral membrane markers were used to isolate targeted *trans*-Golgi-derived vesicles. Golgi-derived vesicles targeted for the apical surface require an NEM-sensitive cytosolic factor for microtubule association (Van der Sluijs *et al.*, 1990), as do protein biosynthetic transport vesicles (Balch, 1990). The transport of fluorescent sphingomyelin and glucosylceramide from the Golgi apparatus to the apical plasma membrane was shown to be dependent on ATP, temperature, and cytosolic factors (Kobayashi *et al.*, 1992). These findings suggest that the transport of sphingolipids from the Golgi apparatus to the plasma membrane is vesicular, and that lipid sorting in the *trans*-Golgi plays a role in segregating transport vesicles destined for apical and basolateral membranes.

E. Endoplasmic Reticulum to Plasma Membrane

Transport of phospholipids and cholesterol from their site of synthesis in the endoplasmic reticulum to the plasma membrane has been examined by several groups. Vesicles mediate transport of phospholipids from endoplasmic reticulum, to the Golgi apparatus, to collapsed vesicles, and finally to the plasma membrane in *Acanthamoeba palestinensis* (Chlapowski and Band, 1971). In *Dictyostelium discoidium*, rapid synthesis and transport of phospholipids to the plasma membrane is induced by chemotaxis (De-Silva and Siu, 1980,1981). In the later studies, isolated transport vesicles had surprisingly high lipid to protein ratios. Whether transport from the endoplasmic reticulum to the plasma membrane of *D. discoidium* is direct or via the Golgi apparatus is unknown. In *Acanthamoeba castellanii*, a lag time of 30 min exists between synthesis of phospholipids and cholesterol and appearance of these molecules at the plasma membrane (Mills *et al.*, 1984). As described earlier, this lag time is assumed to be a characteristic of vesicular transport.

Simoni and colleagues have characterized the transport of cholesterol from the endoplasmic reticulum to the plasma membrane. Cholesterol transport is energy dependent and is not inhibited by colchicine, monensin, cycloheximide, cytochalasin B, or NH_4Cl (Kaplan and Simoni, 1985b). Transport is dependent on temperature with a sharp loss of transport below 15°C. Immediately after synthesis, cholesterol is localized to the endoplasmic reticulum and to a vesicle population of low density. Cholesterol transport, unlike the transport of integral plasma membrane proteins, is not inhibited by brefeldin A, suggesting that the newly synthesized cholesterol transport vesicles bypass the Golgi apparatus and are routed directly to the plasma membrane (Urbani and Simoni, 1990). These data suggest that *de novo* synthesized cholesterol is transported to the plasma membrane by a novel veiscular pathway. Current information about this pathway suggests that it differs significantly from recognized pathways that deliver newly synthesized proteins to the plasma membrane.

F. Transport from the Plasma Membrane

During endocytosis, cells internalize extracellular ligands and solutes through the process of vesicle formation at the plasma membrane (Steinman *et al.*, 1983; Rodman *et al.*, 1990). Considerable effort has been made to identify the intracellular routing of the internalized vesicles. By tracking fluid phase markers, labeled ligands, and labeled receptors, a complex map of pathways has been defined. Initially, the endocytic vesicles are transported to and fuse with early endosomes. From the early endosomes, contents of the vesicles can be transported to one of four destinations: prelysosomes, the *trans*-Golgi, the plasma membrane (recycling), or, in polarized cells, the opposite plasma membrane surface (transcytosis). Most lipid internalized during endocytosis has been postulated to be recycled to the plasma membrane (Dunn *et al.*, 1989; Koval and Pagano, 1991).

Sleight and Pagano (1984) characterized the internalization of phospholipids from the plasma membrane of V-79 cells. A C_6-NBD-labeled analog of phosphatidylcholine was inserted into the outer leaflet of the plasma membrane at 2°C by spontaneous transfer from liposomes. The analog remains at the plasma membrane until the cells are warmed to 37°C. On warming, the fluorescent lipid is internalized in small punctate vesicles that accumulate near the Golgi apparatus. ATP-depleted cells are unable to internalize the fluorescent lipid. When cells labeled with fluorescent lipid at 2°C are warmed to 16°C, fluorescent lipid is observed in endocytic vesicles randomly dispersed throughout the cell.

A variety of cell lines, analyzed in the same manner, produced four patterns of intracellular labeling (Sleight and Abanto, 1989). The fluorescent phosphatidylcholine remains at the plasma membrane of cells with very low endocytosis rates. In several cell lines, such as CHO-K1, the labeling pattern is identical to that described for V-79 cells. BHK, WI-38, and other cell lines have extensive labeling in endocytic vesicles with little or no labeling near the Golgi apparatus. SV40-transformed WI-38 cells (VA-2) accumulate fluorescent phosphatidylcholine in mitochondria, nuclear envelope, and endocytic vesicles. This labeling pattern suggests that the lipid is internalized both by endocytosis and by transmembrane movement, followed by intracellular monomer transport. This second type of transport is discussed later in the text. These data indicate that different cell types may have somewhat different intracellular lipid trafficking routes or that, at equilibrium, the pool sizes of transport intermediates may differ in various cells.

Kok *et al.* (1990) monitored fluorescent lipid internalization from the plasma membrane of BHK cells using a C_6-NBD-labeled analog of phosphatidylcholine and a rhodamine-labeled derivative of phosphatidylethanolamine with the fluorophore attached to the head group amine (N-Rh-PE). These lipids internalize from the plasma membrane at 37°C by vesicular transport at different rates and travel to different intracellular compartments. Colocalization and gradient analysis show that the phosphatidylcholine analog is confined to endocytic vesicles, as described previously (Sleight and Pagano, 1984), whereas the phosphatidylethanolamine analog is transported to lysosomes. The different rates of internalization of these analogs from the plasma membrane suggest that some form of lipid sorting occurs at the plasma membrane. This difference may be explained by the finding that the two probes exist in the membrane in different physical forms. Whereas the C_6-NBD-labeled analog is present as free monomers, the rhodamine derivative is present as aggregates. The rhodamine derivative may be transported actively to lysosomes by a scavenger mechanism that somehow recognizes inappropriate or damaged materials at the plasma membrane. However, this possibility does not explain why the NBD-labeled analog, believed to be internalized along the pathway of receptor-mediated endocytosis, never appears in lysosomes.

Recently, we measured the rate of C_6-NBD-labeled phosphatidylcholine recycling through the endocytic pathway of CHO-K1 cells. A halftime of 6–10 min was determined and is comparable to that of receptor recycling. These data, obtained using a fluorescent probe unable to undergo transmembrane movement, support our hypothesis that the bulk internalization of lipids from the outer leaflet of the plasma membrane occurs via a vesicle-mediated recycling pathway.

The transport of sphingolipids from the plasma membrane has been assessed using C_6-NBD-labeled, spin-labeled, and biotinylated analogs (Spiegel *et al.*, 1984,1985; Sanderfeld *et al.*, 1985; Koval and Pagano, 1989,1990; Schwarzmann and Sandhoff, 1990). C_6-NBD-sphingomyelin is internalized from the plasma membrane in endocytic vesicles that accumulate in the perinuclear and Golgi regions (Koval and Pagano, 1989). The analog recycles to the plasma membrane with a half-time of 40 min (Koval and Pagano, 1989). Internalization and recycling are not inhibited by monensin treatment, but the fluorescent sphingomyelin is confined to endocytic vesicles and does not appear in the region of the Golgi apparatus. Transport of this analog in normal and Niemann–Pick Type A human skin fibroblasts shows similar kinetics for internalization and recycling (Koval and Pagano, 1990). However, the fluorescent sphingomyelin analog accumulates in lysosomes of Niemann–Pick Type A cells and not in normal cells. Niemann–Pick Type A cells do not contain acidic sphingomyelinase, which suggests that sphingomyelin hydrolysis may be required for escape from the degradative pathway to the recycling pathway. The rate of sphingomyelin delivery to lysosomes is 18–19 times slower than the recycling rate. Rapid hydrolysis of the fluorescent sphingomyelin in lysosomes of normal, but not Niemann–Pick Type A, cells is responsible for the different fluorescent labeling patterns of the cells. The comparison of normal and Niemann–Pick Type A fibroblasts represents an extreme example of how lipid metabolism, not just lipid trafficking, plays a role in determining the lipid composition of organelles.

Glycosphingolipid internalization from the plasma membrane of BHK cells was examined with a C_6-NBD-labeled analog of glucosylceramide (Kok *et al.*, 1989). Initially, the fluorescent analog is internalized rapidly from the plasma membrane at 37°C and becomes associated with early endosomes. The analog returns to the plasma membrane via the recycling pathway after an additional 30-min incubation. In other studies, evidence is presented suggesting that a small amount of ganglioside, internalized from the plasma membrane, is routed to the *trans*-Golgi network (Fishman *et al.*, 1983; Sanderfeld *et al.*, 1985). This movement of gangliosides may be the result of a flow of membrane between lysosomes and *trans*-Golgi that is produced by the recycling of the mannose 6-phosphate receptor (Dahms *et al.*, 1989).

The transcytosis of sphingolipids occurs in Caco-2 cells with a half-time of 1 hr. Transcytosis is not required for the establishment of the different lipid compositions of apical and basolateral surfaces with respect to sphingolipids in these cells, and is minimal compared with lipid endocytosis (van't Hof and van Meer, 1990). The extent and significance of lipid transcytosis have not been investigated for other membrane-forming lipids.

IV. MONOMER TRANSPORT

The spontaneous transport of lipid monomers between membranes has been studied in model systems (Duckwitz-Peterlein et al., 1977; Kremer et al., 1977; Decuyper et al., 1980; Roseman and Thompson, 1980; McLean and Phillips, 1981,1982; Barenholz et al., 1983). Multiple techniques for monitoring transport show that the rate of spontaneous transfer of phospholipids between vesicles is very slow, with half-times ranging from 2 to 48 hr. The rate-limiting step in this process is the desorption of monomers from the donor membrane population (Roseman and Thompson, 1980). The slow spontaneous transfer of neutral and phospholipid monomers suggests that this mechanism is unlikely to contribute significantly to intracellular transport. More polar lysophospholipids move more rapidly between membranes because of their increased solubility. Although the spontaneous movement of lysophospholipids has been postulated to play a role in intracellular transport (Haldar and Lipfert, 1990), this idea has not been demonstrated unambiguously.

The transfer of phospholipid monomers between membranes is stimulated by a class of small cytosolic proteins called phospholipid exchange or transfer proteins (Wirtz and Zilversmit, 1968; Wirtz, 1982; see also Chapter 9). Several of these proteins with different lipid transfer specificities have been isolated and characterized. A gene encoding a phosphatidyl-inositol-specific transfer protein was shown to be required for yeast cell growth (Aitken et al., 1990; Bankaitis et al., 1990). The protein is required for the transport of secretory proteins through the biosynthetic pathway (Novick et al., 1980; Bankaitis et al., 1989). Although this finding represents the first demonstration of a biological role for a phospholipid transfer protein, the intracellular action of this protein remains unknown.

We have observed phospholipid transfer protein-mediated lipid transport microscopically in vivo (Sleight and Hopper, 1991). Liposomes containing a self-quenching concentration of head group-labeled N-Rh-PE were microinjected into CHO-K1 cells. The fluorescent analog was observed 10 min later in nuclear envelope and mitochondria. Contents mixing experiments showed that the injected liposomes had not fused with intracellular membranes. Since the rhodamine-labeled analog is unable to move spontaneously, but can transfer between membranes in the presence of nonspecific phospholipid transfer protein, we concluded that the movement between microinjected liposomes and organelles is most likely caused by the action of cytosolic transfer proteins.

Evidence for a monomer mechanism of lipid transport often is based solely on rapid transport kinetics. Rapid transport kinetics also are expected for lipids moved by lateral diffusion. Therefore, kinetic analysis without additional supporting data can be misleading. The proposed routes of lipid transport via spontaneous and protein-mediated monomer transport are diagrammed in Fig. 3.

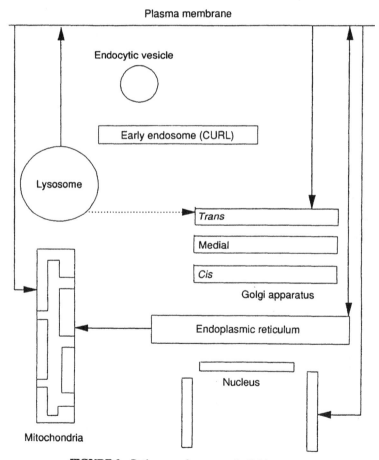

FIGURE 3 Pathways of monomeric lipid transport.

A. Endoplasmic Reticulum to Plasma Membrane

A direct route of transport from the endoplasmic reticulum to the plasma membrane exists for newly synthesized phosphatidylcholine and phosphatidylethanolamine (Sleight and Pagano, 1983; Kaplan and Simoni, 1985a). Newly synthesized phosphatidylethanolamine is transported from the endoplasmic reticulum to the outer leaflet of the plasma membrane without a significant lag time (Sleight and Pagano, 1983). This transport is not inhibited by depletion of ATP, inhibitors of protein secretion, or brefeldin A (Sleight and Pagano, 1983; Vance *et al.*, 1991) but is eliminated at 2°C. Whereas vesicular transport is suspended during mitosis, phosphatidylethanolamine transport to the plasma membrane is not inhibited in mitotic cells (Kobayashi and Pagano, 1989).

In CHO cells, the transport of newly synthesized phosphatidylcholine to the plasma membrane has a half-time of 1–2 min (Kaplan and Simoni, 1985a). Phosphatidylcholine transport is not inhibited by energy poisons or protein secretion inhibitors, but is eliminated at 2°C. The rapid kinetics of phosphatidylcholine and phosphatidylethanolamine transport from endoplasmic reticulum to the plasma membrane, combined with the lack of inhibition by agents known to block vesicular transport, suggest that the movement of these lipids is by monomer transfer.

B. Transport between Endoplasmic Reticulum and Mitochondria

Yaffe and Kennedy (1983) measured the rate of phosphatidylcholine, phosphatidyl-N-propyl-N,N-dimethylethanolamine (PDME), and phosphatidylethanolamine transport from endoplasmic reticulum to mitochondria in BHK cells and in a reconstituted system. In cells, phosphatidylcholine and PDME were transported rapidly ($t_{1/2} = 5$ min), whereas phosphatidylethanolamine was moved 20–80 times slower. Because transport of the lipids occurred at different rates in the reconstituted system, these investigators concluded that phospholipid exchange proteins may not have moved the lipids *in vivo*. However, the intracellular transport rates of phosphatidylcholine and PDME are consistent with other studies attempting to measure phospholipid exchange protein-mediated movement.

Paltauf and co-workers have measured the kinetics of phosphatidylcholine and phosphatidylethanolamine transport between the endoplasmic reticulum and mitochondria in yeast (Daum *et al.*, 1986). Phosphatidylethanolamine is transported from mitochondria to the endoplasmic reticulum by an energy-dependent process, whereas energy-dependent and energy-independent transport of phosphatidylcholine from the endoplasmic reticulum to mitochondria occurs. Phospholipid exchange protein activities, specific for phosphatidylcholine and phosphatidylinositol but not phosphatidylethanolamine, have been identified in yeast (Daum and Paltauf, 1984). Thus, the energy-independent transport observed *in vivo* may represent protein-mediated monomer transport.

C. Transport from Lysosomes

Monomer transport of phospholipids from lysosomes to the plasma membrane has not been reported. Hydrolysis of cholesterol esters in lysosomes is followed by free cholesterol translocation through the cell, with a primary destination of the plasma membrane (Johnson *et al.*, 1990; Brasaemle and Attie, 1990). Some free cholesterol has been observed to move to the Golgi apparatus, suggesting that transport of cholesterol from lysosomes to Golgi apparatus may occur (Blanchette-Mackie *et al.*, 1988). Free cholesterol transport from lysosomes is rapid, energy independent, and not inhibitable by brefeldin A, NH_4Cl, leupeptin, or agents that disrupt cytoskeletal architecture (Liscum, 1990; Butler *et al.*, 1991). Several investigators have hypothesized that transport of free cholesterol from lysosomes to the plasma membrane is mediated by sterol carrier proteins. We can expect to see an increasing number of investigations into this area based on two observations. First, Niemann–Pick Type C cells are unable to transport cholesterol from lysosomes to plasma membrane (Liscum *et al.*, 1989; Pentchev *et al.*, 1986,1987); second, transport in normal cells is blocked by progesterone and U18666A (Butler *et al.*, 1991; Liscum, 1990; Liscum and Faust, 1989).

D. Transport from the Plasma Membrane

NBD-labeled diacylglycerol rapidly translocates from the inner leaflet of the plasma membrane to the endoplasmic reticulum, mitochondria, and nuclear membrane (Pagano *et al.*, 1981,1983; Pagano and Longmuir, 1985; Ting and Pagano, 1990). Unlike C_6-NBD-labeled phospholipids, C_6-NBD-labeled diacylglycerol cannot transfer between membranes spontaneously (Nichols, 1983). Therefore, the rapid transport of NBD-diacylglycerol from the plasma membrane to intracellular membranes is most likely the result of monomer transport by a carrier protein or lateral diffusion across membrane bridging sites. The latter possibility is discussed in another section of the text.

Analogs of phosphatidylethanolamine and phosphatidylserine with a C_6-NBD fatty acid in the *sn*-2 position move rapidly between artificial vesicles as monomers in solution (Nichols and Pagano, 1982). These analogs also move quickly from the inner leaflet of the plasma membrane to intracellular organelles by a nonvesicular mechanism (Sleight and Pagano, 1985; Martin and Pagano, 1987). Although spontaneous movement from the inner leaflet of the plasma membrane might be expected to result in labeling of all organelles, C_6-NBD-labeled lipids appear to accumulate almost exclusively in mitochondrial and nuclear membranes. This accumulation is not caused by metabolic conversion to other products. As stated earlier, when liposomes containing a head group-labeled N-Rh-PE are microinjected into cells, the labeled phosphatidylethanolamine moves to mitochondrial and nuclear membranes (Sleight and Hopper, 1991), although the rhodamine-labeled phosphatidylethanolamine does not move spontaneously between membranes. These data indicate that some type of directional transport of the lipid analogs occurs to maintain specific organelle labeling. Since all these fluorescent lipid analogs are substrates for phospholipid exchange proteins *in vitro,* exchange proteins may mediate their transport *in vivo* (Nichols, 1983).

Most studies using lipids labeled with NBD have been performed with analogs containing the NBD moiety linked to a 6-carbon amino fatty acid (C_6-NBD). Kobayashi and Arakawa (1991) have reported on the transport of NBD-labeled analogs with the label linked to the end of a 12-carbon amino fatty acid (C_{12}-NBD). Although these analogs are able to transfer spontaneously between membranes (Nichols, 1983; Kobayashi and Arakawa, 1991), they move 150–200 times more slowly than C_6-NBD analogs. Insertion and transport of C_{12}-NBD-phosphatidylserine in CHO-K1 cells and human fibroblasts differ in three ways from that of C_6-NBD-phosphatidylserine (Martin and Pagano, 1987): (1) C_{12}-NBD-phosphatidylserine does not insert at low temperature; (2) its insertion is inhibited at 37°C by ATP depletion, NEM, or gluteraldehyde; and (3) internalization of the probe occurs by an ATP-dependent process and results in accumulation of the probe exclusively in the Golgi apparatus. C_{12}-NBD-phosphatidylserine is postulated to undergo an ATP-dependent translocation from outer to inner leaflet of the plasma membrane and then to move to the Golgi apparatus by the action of phospholipid exchange proteins. Similar analogs of phosphatidate, phosphatidylethanolamine, and phosphatidylcholine do not behave in the same manner. The results of Kobayashi and Arakawa (1991) clearly demonstrate the need for future studies examining the effect of lipid (and lipid probe) structure on transport.

V. LATERAL DIFFUSION AND CONTACT-MEDIATED TRANSPORT

Lipids laterally diffuse within the plane of membrane bilayers at approximately 10^{-8} cm^2/sec (Houslay and Stanley, 1982). In theory, lipids could be transported rapidly throughout the cell by diffusion along the bilayer at specific membrane bridging connections. Pathways of lipid transport by lateral diffusion and contact-mediated transfer are shown in Fig. 4. A large body of microscopic evidence suggests that this mechanism of transport is responsible for the movement of free fatty acids from the plasma membrane to the endoplasmic reticulum in parenchymal cells and hepatocytes (Eggens *et al.*, 1979; Blanchette-Mackie and Scow, 1981; Scow *et al.*, 1983; Scow and Blanchette-Mackie, 1991). Most of this evidence was

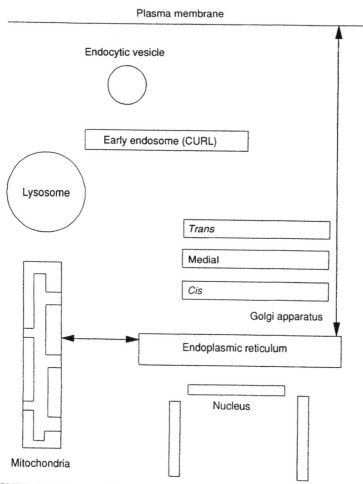

FIGURE 4 Pathways of lipid transport by diffusion at membrane contact sites and contact-mediated transfer.

obtained using cells fixed with tannic acid, lanthanum, or glutaraldehyde (Blanchette-Mackie and Scow, 1982), processes that may introduce cytoskeletal structural artifacts. Sites of continuity between membranes have been observed between sarcoplasmic reticulum and mitochondria in heart (Bowman, 1967), endoplasmic reticulum and mitochondria in rat liver (Frink and Kartenbach, 1971; Eggens *et al.*, 1979), and lipid droplets and mitochondria in heart (Wetzel and Scow, 1984). Very little biochemical evidence for lipid transport across these proposed membrane contact sites has been reported.

Adjacent *trans*-Golgi elements connected by tubulovesicular processes that can form stable interconnections have been observed in astrocytes (Cooper *et al.*, 1990). These interconnections provide direct pathways for the diffusion of lipid probes between joined *trans*-Golgi element membranes. For these connections to allow lateral diffusion without completely randomizing all membrane lipids, some type of membrane filter must retard certain lipids selectively. No such filter has been identified.

Although the rate of spontaneous phospholipid monomer transfer normally is considered to be low, the movement of phospholipids between liposomes can be accelerated greatly at high vesicle concentration (Jones and Thompson, 1989,1990). This accelerated transfer occurs during collisions between donor and acceptor liposomes (Jones and Thompson, 1990). Therefore, *in vivo*, membrane collisions or close contacts between membranes are expected to play a role in lipid transport. Although these phenomena have not been demonstrated unequivocally in living cells, several model systems are under study.

One system follows the movement of phosphatidylserine and phosphatidylethanolamine between the outer and inner membranes of mitochondria. Transport appears to occur at specific contact points between the two membranes (Simbeni *et al.*, 1990; Ardial *et al.*, 1991). In a second system, elegant methods have been devised to examine the transport of phosphatidylserine and phosphatidylethanolamine between the endoplasmic reticulum and the outer membrane of mitochondria (Voelker, 1985,1989a,b,1990a). One of the mechanisms of transport between these organelles appears to involve a two-step process. The first step is energy dependent whereas the second is not. The first step involves the production of specialized transport vesicles called "fraction X" by Vance (1990). The second step is thought to involve a collisional event between these vesicles and mitochondria. These transport studies are complicated by the fact that multiple routes of transport probably occur at the same time. This difficulty is exemplified by reports indicating that newly synthesized phospholipids leave the endoplasmic reticulum faster than other phospholipids in the membrane (Vance, 1991).

VI. TRANSMEMBRANE MOVEMENT

Many biological membranes have an asymmetric distribution of lipids across the bilayer (Op den Kamp, 1979). Most lipids are unable to move spontaneously across the membrane bilayer at rapid rates because of their polar head groups (Homan and Pownall, 1988). The half-times for spontaneous transmembrane movement of phospholipids are on the order days (Rousselet *et al.*, 1976; van Meer and Op den Kamp, 1982). Exceptions to this rule include diacylglycerol and ceramide, which do not possess large or charged head groups (Pagano *et al.*, 1981; Ganong and Bell, 1984). Two well-documented examples exist of protein-mediated transmembrane movement of phospholipids (see also Chapters 1 and 2). Several researchers have observed rapid ATP-independent transmembrane movement of lipids across microsomal membranes (Zilversmit and Hughes, 1977; van den Besselaar *et al.*, 1978; Bishop and Bell, 1985; Kawashima and Bell, 1987; Herrman *et al.*, 1990). This activity is thought to be protein mediated because it can be inhibited by structural analogs and by protein modifying reagents (Bishop and Bell, 1985). Backer and Dawidowicz (1987) reported the reconstitution of this activity into artificial vesicles. To date, the proteins responsible for transmembrane movement in microsomes have not been identified. An ATP-dependent transmembrane movement of aminophospholipids has been reported in the plasma membrane of several cell types (Seigneuret and Devaux, 1984; Sleight and Pagano, 1985; Zachowski *et al.*, 1986, 1987; Kobayashi and Arakawa, 1991). This activity has been well characterized without the benefit of reconstitution and isolation. Lipid asymmetry and the transmembrane movement of lipids are discussed in detail in other chapters in this volume.

VII. CONCLUSIONS

The intracellular transport of membrane-forming lipids is a complex process that cannot be ascribed to a single mechanism. Although multiple mechanisms and pathways of lipid transport have been identified, we still have a poor understanding of the movement of lipids throughout the cell relative to our understanding of protein transport. Additional studies are needed to identify all the mechanisms and pathways of lipid transport, as well as to understand the significance of each process in the establishment and regulation of lipid composition in individual membranes. To understand the regulation of membrane lipid composition, the role of site-specific lipid synthesis and degradation also must be determined. Areas of interest in the future are likely to be genetic studies of phospholipid exchange proteins, identification of phospholipid translocating enzymes, and studies of the role of lipid segregation into microdomains as a method of sorting. The significance of vesicular transport pathways for lipid movement will become apparent as more information about protein trafficking and mechanisms of vesicle budding and fusion are acquired. Interest in lipid transport will be broadened when defects in trafficking pathways are linked to disease states.

References

Aitken, J. F., van Heusden, G. P. H., Temkin, M., and Dowhan, W. (1990). The gene encoding the phosphatidylinositol transfer protein is essential for cell growth. *J. Biol. Chem.* **265**, 4711–4717.

Ardial, D., Lerme, F., and Louisot, P. (1991). Involvement of contact sites in phosphatidylserine import into liver mitochondria. *J. Biol. Chem.* **266**, 7978–7981.

Backer, J. M., and Dawidowicz, E. A. (1987). Reconstitution of a phospholipid flippase from rat liver microsomes. *Nature (London)* **327**, 341–343.

Balch, W. E. (1990). Molecular dissection of early stages of the eukaryotic secretory pathway. *Curr. Opin. Cell Biol.* **2**, 634–641.

Bankaitis, V. A., Malehorn, D. E., Emr, S. D., and Greene, R. (1989). The *Saccharomyces cerevisiae* SEC14 gene encodes a cytosolic factor that is required for transport of secretory proteins from the yeast Golgi complex. *J. Cell Biol.* **109**, 1271–1281.

Bankaitis, V. A., Aitken, J. R., Cleves, A. E., and Dowhan, W. (1990). An essential role for a phospholipid transfer protein in yeast Golgi complex. *Nature (London)* **347**, 561–562.

Barenholz, Y., Lichtenberg, D., and Thompson, T. E. (1983). Spontaneous transfer of sphingomyelin between phospholipid bilayers. *Biochemistry* **22**, 5647–5651.

Bell, R. M., Ballas, L. M., and Coleman, R. A. (1981). Lipid topogenesis. *J. Lipid Res.* **22**, 391–403.

Bishop, W. R., and Bell, R. M. (1985). Assembly of the endoplasmic reticulum phospholipid bilayer: The phosphatidylcholine transporter. *Cell* **42**, 51–60.

Bishop, W. R., and Bell, R. M. (1988). Assembly of phospholipids into cellular membranes: Biosynthesis transmembrane movement and intracellular translocation. *Annu. Rev. Cell Biol.* **4**, 579–610.

Blanchette-Mackie, E. J., and Scow, R. O. (1981). Membrane continuities within cells and intracellular contacts in white adipose tissue of young rats. *J. Ultrastruct. Res.* **77**, 277–294.

Blanchette-Mackie, E. J., and Scow, R. O. (1982). Continuity of intracellular channels with extracellular space in adipose tissue and liver: Demonstrated with tannic acid and lanthanum. *Anat. Rec.* **203**, 205–219.

Blanchette-Mackie, E. J., Dwyer, N. K., Amende, L. M., Kruth, H. S., Butler, J. D., Sokol, J., Comly, M. E., Vanier, M. T., August, J. T., Brady, R. O., and Pentchev, P. G.(1988). Type C Niemann–Pick disease: Low density lipoprotein uptake is associated with premature cholesterol accumulation in the Golgi complex and excessive cholesterol storage in lysosomes. *Proc. Natl. Acad. Sci. U.S.A.* **85**, 8022–8026.

Bowman, R. W. (1967). Miochondrial connections in canine myocardium. *Texas Rep. Biol. Med.* **25**, 5988–5991.

Brasaemle, D. L., and Attie, A. D. (1990). Rapid intracellular transport of LDL-derived cholesterol to the plasma membrane in cultured fibroblasts. *J. Lipid Res.* **31**, 103–110.

Brown, D. A., and Rose, J. K. (1992). Sorting of GPI-anchored proteins to glycolipid-enriched membrane subdomains during transport to the apical cell surface. *Cell* **68**, 533–544.

Butler, J. D., Blanchette-Mackie, J., Goldin, E., O'Neill, R. R., Carstea, G., Roff, C. F., Patterson, M. C., Patel, S., Comly, M. E., Cooney, A., Vanier, M. T., Brady, R. O., and Pentchev, P. G. (1991). Progesterone blocks cholesterol translocation from lysosomes. *J. Biol. Chem.* **267**, 23797–23805.

Chlapowski, F. J., and Band, R. N. (1971). Assembly of lipids into membranes in *Acanthamoeba palestinensis. J. Cell Biol.* **50**, 634–651.

Cogan, U., and Schacter, D. (1981). Asymmetry of lipid domains in human erythrocyte membranes studied with impermeant fluorophores. *Biochemistry* **20**, 6396–6403.

Colbeau, A., Machbaur, J., and Vignais, P. M. (1970). Enzymic characterization and lipid composition of rat liver subcellular membranes. *Biochim. Biophys. Acta* **249**, 462–492.

Cooper, M. S., Cornell-Bell, A. H., Chernjavsky, A., Dani, J. W., and Smith, S. J. (1990). Tubulovesicular processes emerge from trans-Golgi cisternea, extend along microtubules, and interlink adjacent trans-Golgi elements into a reticulum. *Cell* **61**, 135–145.

Dahms, N. M., Lobel, P., and Kornfeld, S. (1989). Mannose 6-phosphate receptors and lysosomal enzyme targeting. *J. Biol. Chem.* **264**, 12115–12118.

Daum, G., and Paltauf, F. (1984). Phospholipid transfer in yeast: Isolation and partial characterization of a phospholipid transfer protein from yeast cytosol. *Biochim. Biophys. Acta* **794**, 385–391.

Daum, G., Heidorn, E., and Paltauf, F. (1986). Intracellular transfer of phospholipids in the yeast, *Sacchromyces cerevisiae. Biochim. Biophys. Acta* **878**, 93–101.

Decuyper, M., Joniau, M., and Gangreau, H. (1980). Spontaneous phospholipid transfer between artificial vesicles followed by free-flow electrophoresis. *Biochem. Biophys. Res. Commun.* **95**, 1221–1230.

Dennis, E. A., and Kennedy, E. P. (1972). Intracellular sites of lipid synthesis and the biogenesis of mitochondria. *J. Lipid Res.* **13**, 263–267.

Dennis, E. A., Rhee, S. G., Billah, M. M., and Hannun, Y. A. (1991). Role of phospholipase in generating lipid second messengers in signal transduction. *Fed. Am. Soc. Exp. Biol. J.* **5**, 2068–2077.

DeSilva, N. S., and Siu, C. H. (1980). Preferential incorporation of phospholipids into plasma membranes during cell aggregation of *Dictyostelium discoideum. J. Biol. Chem.* **255**, 8489–8496.

DeSilva, N. S., and Siu, C. H. (1981). Vesicle-mediated transfer of phospholipids to plasma membrane during cell aggregation of *Dictyostelium discoideum. J. Biol. Chem.* **256**, 5845–5850.

Doms, R. W., Russ, G., and Yewdell, J. W. (1989). Brefeldin A redistributes resident and itinerant Golgi proteins to the endoplasmic reticulum. *J. Cell Biol.* **109**, 61–72.

Duckwitz-Peterlein, G., Eilenberger, G., and Overath, P. (1977). Phospholipid exchange between bilayer membranes. *Biochim. Biophys. Acta* **469**, 311–325.

Dunn, K. W., McGraw, T. E., and Maxfield, F. R. (1989). Iterative fractionation of recycling receptors from lysosomally destined ligands in an early sorting endosome. *J. Cell Biol.* **109**, 3303–3314.

Eggens, I., Valterson, C., Dallner, G., and Ernster, L. (1979). Transfer of phospholipids between the endoplasmic reticulum and mitochondria in rat hepatocytes *in vivo. Biochem. Biophys. Res. Commun.* **91**, 709–714.

Fishman, P. H., Bradley, P. H., Hom, B. E., and Moss, J. (1983). Uptake and metabolism of exogenous gangliosides by cultured cells: Effect of chloragen on the turnover of GM1. *J. Lipid Res.* **24**, 1002–1011.

Frink, W. W., and Kartenbach, J. (1971). Outer mitochondrial membrane continuous with endoplasmic reticulum. *Protoplasma* **73**, 35–41.

Fujiwara, T., Oda, K., Yokota, S., Takatsu, A., and Ikehara, Y. (1988). Brefeldin A causes disassembly of the Golgi complex and accumulation of secretory proteins in the endoplasmic reticulum. *J. Biol. Chem.* **263**, 18545–18552.

Futerman, A. H., and Pagano, R. E. (1991). Determination of the intracellular sites and topology of glucosylceramide synthesis in rat liver. *Biochem. J.* **280**, 295–302.

Futerman, A. H., Steiger, B., Hubbard, A. L., and Pagano, R. E. (1990). Sphingomyelin synthesis in rat liver occurs predominately at the cis and medial cisternae of the Golgi apparatus. *J. Biol. Chem.* **265**, 8650–8657.

Ganong, B. R., and Bell, R. M. (1984). Transmembrane movement of phosphatidylglycerol and diacylglycerol sulfhydryl analogs. *Biochemistry* **23**, 4977–4983.

Haldar, D., and Lipfert, L. (1990). Export of mitochondrially synthesized lysophosphatidic acid. *J. Biol. Chem.* **265**, 11014–11016.

Herrman, A., Zachowski, A., and Devaux, P. F. (1990). Protein mediated phospholipid translocation in the endoplasmic reticulum with a low lipid specificity. *Biochemistry* **29**, 2023–2027.

Homan, R., and Pownall, H. J. (1988). Transbilayer movement of phospholipids: Dependence on headgroup structure and acyl chain content. *Biochim. Biophys. Acta* **938**, 155–166.

Houslay, M. D., and Stanley, K. K. (1982). Mobility of the lipid and protein components of biological membranes. In "Dynamics of Biological Membranes: Influence on Synthesis" pp. 39–91. Wiley, New York.

Johnson, W. J., Chacko, G. K., Phillips, M. C., and Rothblat, G. H. (1990). The efflux of lysosomal cholesterol from cells. *J. Biol. Chem.* **265**, 5546–5553.

Jones, J. D., and Thompson, T. E. (1989). Spontaneous phosphatidylcholine transfer by collision between vesicles at high lipid concentration. *Biochemistry* **28**, 129–134.

Jones, J. D., and Thompson, T. E. (1990). Mechanism of spontaneous concentration-dependent phospholipid transfer between bilayers. *Biochemistry* **29**, 1593–1600.

Kaplan, M. R., and Simoni, R. D. (1985a). Intracellular transport of phosphatidylcholine to the plasma membrane. *J. Cell Biol.* **101**, 441–445.

Kaplan, M. R., and Simoni, R. D. (1985b). Transport of cholesterol from the endoplasmic reticulum to the plasma membrane. *J. Cell Biol.* **101**, 446–453.

Karrenbauer, A., Jeckel, D., Just, W., Birk, R., Schmidt, R. R., Rothman, J. E., and Wieland, F. T. (1990). The rate of bulk flow from the Golgi to the plasma membrane. *Cell* **63**, 259–267.

Kawai, K., Michiya, F., and Nakao, M. (1974). Lipid components of two different regions of an intestinal epithelial cell membrane of mouse. *Biochim. Biophys. Acta* **369**, 222–233.

Kawashima, Y., and Bell, R. M. (1987). Assembly of the endoplasmic reticulum bilayer. Transporter for phosphatidylcholine and metabolites. *J. Biol. Chem.* **262**, 16495–16502.

Kobayashi, T., and Arakawa, Y. (1991). Transport of exogenous fluorescent phosphatidylserine analogue to the Golgi apparatus in cultured fibroblasts. *J. Cell Biol.* **113**, 235–244.

Kobayashi, T., and Pagano, R. E. (1988). ATP-dependent fusion of liposomes with the Golgi apparatus of perforated cells. *Cell* **55**, 797–805.

Kobayashi, T., and Pagano, R. E. (1989). Lipid transport during mitosis. *J. Biol. Chem.* **264**, 5966–5973.

Kobayashi, T., Pimplikar, S. W., Parton, R. G., Bhakdi, S., and Simons, K. (1992). Sphingolipid transport from the trans-Golgi network to the apical surface in permeabilized MDCK cells. *Fed. Eur. Biochem. Soc.* **300**, 227–231.

Kok, J. W., Eskelinen, S., Hoekstra, K., and Hoekstra, D. (1989). Salvage of glucosylceramide by recycling after internalization along the pathway of receptor-mediated endocytosis. *Proc. Natl. Acad. Sci. U.S.A.* **86**, 9896–9900.

Kok, J. W., ter Beest, M., Scherpof, G., and Hoekstra, D. (1990). A non-exchangeable fluorescent phospholipid analog as a membrane traffic marker of the endocytic pathway. *Eur. J. Cell Biol.* **53,** 173–184.

Kolesnick, R. N. (1991). Sphingomyelin and derivatives as cellular signal. *Progr. Lipid Res.* **30,** 1–38.

Koval, M., and Pagano, R. E. (1989). Lipid recycling between the plasma membrane and intracellular components: Transport and metabolism of fluorescent sphingomyelin analogues in cultured fibroblasts. *J. Cell Biol.* **108,** 2169–2181.

Koval, M., and Pagano, R. E. (1990). Sorting of an internalized plasma membrane lipid between recycling and degradative pathways in normal and Niemann–Pick, Type A fibroblasts. *J. Cell Biol.* **111,** 429–442.

Koval, M., and Pagano, R. E. (1991). Intracellular transport and metabolism of sphingomyelin. *Biochim. Biophys. Acta* **1082,** 113–125.

Kremer, J. M. H., Kops-Werkoven, M. M., Pathomamnoharan, C., Gijzeman, O. L. J., and Wiersema, P. H. (1977). Phase diagrams and the kinetics of phospholipid exchange for vesicles of different composition and radius. *Biochim. Biophys. Acta* **471,** 177–188.

Lippincott-Schwartz, J., Yuan, L. C., Bonfacino, J. S., and Klausner, R. D. (1989). Rapid redistribution of Golgi proteins into the endoplasmic reticulum in cells treated with Brefeldin A: Evidence for membrane recycling from Golgi to ER. *Cell* **54,** 209–220.

Lippincott-Schwartz, J., Donaldson, J. G., Schweizer, A., Berger, E. G., Hauri, H-P., Yaun, L. C., and Klausner, R. D. (1990). Microtubule-dependent retrograde transport of proteins in the ER in the presence of Brefeldin A suggests an ER recycling pathway. *Cell* **60,** 821–836.

Lipsky, N. G., and Pagano, R. E. (1983). Sphingolipid metabolism in cultured fibroblasts: Microscopic and biochemical studies employing a fluorescent ceramide analogue. *Proc. Natl. Acad. Sci. U.S.A.* **80,** 2608–2612.

Lipsky, N. G., and Pagano, R. E. (1983). A vital stain for the Golgi apparatus. *Science* **228,** 745–747.

Liscum, L. (1990). Pharmacological inhibition of the intracellular transport of low-density lipoprotein-derived cholesterol in Chinese hamster ovary cells. *Biochim. Biophys. Acta* **1045,** 40–48.

Liscum, L., and Faust, J. R. (1989). The intracellular transport of low density lipoprotein-derived cholesterol is inhibited in Chinese hamster ovary cells cultured with 3-beta-[2-(di-ethylamino)ethoxy]androst-5-en-17-one. *J. Biol. Chem.* **264,** 11796–11806.

Liscum, L., Ruggiero, R. M., and Faust, J. R. (1989). The intracellular transport of low density lipoprotein-derived cholesterol is defective in Niemann–Pick Type C fibroblasts. *J. Cell Biol.* **108,** 1625–1636.

Lodish, H. F., Kong, N., Hirani, S., and Rasmussen, J. (1984). A vesicular intermediate in the transport of hepatoma secreting proteins from the rough endoplasmic reticulum to the Golgi complex. *J. Cell Biol.* **104,** 221–230.

Martin, O. C., and Pagano, R. E. (1987). Transbilayer movement of fluorescent analogs of phosphatidylserine and phosphatidylethanolamine at the plasma membrane of cultured cells. *J. Biol. Chem.* **262,** 5890–5898.

Matyas, G. R., and Morré, D. J. (1987). Subcellular distribution and biosynthesis of rat liver gangliosides. *Biochim. Biophys. Acta* **921,** 599–614.

McLean, L. R., and Phillips, M. C. (1981). Mechanism of cholesterol and phosphatidylcholine exchange or transfer between unilamellar vesicles. *Biochemistry* **20,** 2893–2900.

McLean, L. R., and Phillips, M. C. (1982). Cholesterol desorption from clusters of phosphatidylcholine and cholesterol unilamellar vesicles during lipid transfer or exchange. *Biochemistry* **21,** 4053–4059.

McMurray, W. C. (1973). "Form and Function of Phospholipids." Elsevier, Amsterdam.

Merrill, A. H. (1992). Ceramide: A new "second messenger." *Nutr. Rev.* **50**, 78–80.

Miller-Podraza, H., and Fishman, P. H. (1982). Translocation of newly synthesized gangliosides to the cell surface. *Biochemistry* **21**, 3265–3270.

Mills, J. T., Furlong, S. T., and Dawidowicz, E. A. (1984). Plasma membrane biogenesis in eukaryotic cells: Translocation of newly synthesized lipid. *Proc. Natl. Acad. Sci. U.S.A.* **81**, 1385–1388.

Moreau, P., and Morré, D. J. (1991). Cell-free transfer of membrane lipids. *J. Biol. Chem.* **266**, 4329–4333.

Moreau, P., Rodriguez, M., Cassagne, C., Morré, D. M., and Morré, D. J. (1991). Trafficking of lipids from the endoplasmic reticulum to the Golgi apparatus in a cell-free system from rat liver. *J. Biol. Chem.* **266**, 4322–4328.

Nichols, J. W. (1983). Resonance energy transfer assay of protein-mediated lipid transfer. *J. Biol. Chem.* **258**, 5368–5371.

Nichols, J. W., and Pagano, R. E. (1982). Use of resonance energy transfer to study the kinetics of amphiphile transfer between vesicles. *Biochemistry* **21**, 1720–1726.

Novick, P., Field, C., and Scheckman, R. (1980). Identification of 23 complementation groups required for post-translational events in the yeast secretory pathway. *Cell* **21**, 205–215.

Op den Kamp, J. A. F. (1979). Lipid asymmetry in membranes. *Annu. Rev. Biochem.* **48**, 47–71.

Pagano, R. E. (1990). Lipid traffic in eukaryotic cells: Mechanisms for intracellular transport and organelle-specific enrichment of lipids. *Curr. Opin. Cell Biol.* **2**, 652–663.

Pagano, R. E., and Longmuir, K. J. (1985). Phosphorylation, transbilayer movement, and facilitated intracellular transport of diacylglycerol are involved in the uptake of a fluorescent analog of phosphatidic acid by cultured fibroblasts. *J. Biol. Chem.* **260**, 1909–1916.

Pagano, R. E., and Sleight, R. G. (1985). Emerging problems in the cell biology of lipids. *Trends Biochem. Sci.* **10**, 421–425.

Pagano, R. E., Longmuir, K. J., Martin, O. C., and Struck, D. K. (1981). Metabolism and intracellular localization of a fluorescently labeled intermediate in lipid biosynthesis within cultured fibroblasts. *J. Cell Biol.* **91**, 872–877.

Pagano, R. E., Longmuir, K. J., and Martin, O. C. (1983). Intracellular translocation and metabolism of a fluorescent phosphatidic acid analogue in cultured fibroblasts. *J. Biol. Chem.* **258**, 2034–2040.

Pagano, R. E., Sepanski, M. A., and Martin, O. C. (1969). Molecular trapping of a fluorescent ceramide analogue at the Golgi apparatus of fixed cells: Interaction with endogenous lipids provides a trans-Golgi marker for both light and electron microscopy. *J. Cell Biol.* **109**, 2067–2079.

Pentchev, P. G., Comly, M. E., Kruth, H. S., Butler, M. T., Vanier, D. A., and Patel, S. (1986). Type C Niemann–Pick disease. A parallel loss of regulatory response in both the uptake and esterification of low density lipoprotein-derived cholesterol in cultured fibroblasts. *J. Biol. Chem.* **261**, 16775–16780.

Pentchev, P. G., Comly, M. E., Kruth, H. S., Tokoro, T., Butler, J. Sokol, J., Filling-Katz, M., Quirk, J. M., Marshall, D. C., Patel, S., Vanier, M. T., and Brady, R. O. (1987). Group C Niemann–Pick disease: Faulty regulation of low density lipoprotein uptake and cholesterol storage in cultured fibroblasts. *Fed. Am. Soc. Exp. Biol.* **1**, 40–45.

Pfeffer, S. R., and Rothman, J. E. (1987). Biosynthetic protein transport and sorting by the endoplasmic reticulum and Golgi. *Annu. Rev. Biochem.* **56**, 829–852.

Raetz, C. R. H. (1982). Genetic control of phospholipid bilayer assembly. *In* "Phospholipids" (J. N. Hawthorne and G. B. Ansell, eds.), pp. 435–477. Elsevier Biomedical, New York.

Reinhart, M. P., Billheimer, J. T., Faust, J. R., and Gaylor, J. L. (1987). Subcellular localization of the enzymes of cholesterol biosynthesis and metabolism in rat liver. *J. Biol. Chem.* **262**, 9649–9655.

Rodman, J. S., Mercer, R. W., and Stahl, P. D. (1990). Endocytosis and transcytosis. *Curr. Opin. Cell Biol.* **2**, 664–672.

Roseman, M. A., and Thompson, T. E. (1980). Mechanism of the spontaneous transfer of phospholipids between bilayers. *Biochemistry* **19**, 439–444.

Rosenwald, A. G., Machamer, C. E., and Pagano, R. E. (1992). Effects of a sphingolipid synthesis inhibitor on membrane transport through the secretory pathway. *Biochemistry* **31**, 3581–3590.

Rousselet, A., Guthman, A., Matricon, J., Bienvenue, A., and Devaux, P. F. (1976). Study of the transverse diffusion of spin labeled phospholipids in biological membranes. *Biochim. Biophys. Acta* **426**, 357–371.

Sanderfeld, S., Conzelmann, E., Schwarzmann, G., Burg, J., and Sandhoff, K. (1985). Incorporation and metabolism of ganglioside GM2 in skin fibroblasts from normal and GM2 gangliosidosis subjects. *Eur. J. Biochem.* **149**, 247–255.

Sandra, A., and Pagano, R. E. (1978). Phospholipid asymmetry in LM cell plasma membrane derivatives: Polar head group and acyl chain distributions. *Biochemistry* **17**, 332–338.

Saraste, J., Palade, G. E., and Farquhar, M. G. (1987). Antibodies to rat pancreas Golgi subfractions: Identification of a 58 kD cis-Golgi protein. *J. Cell Biol.* **105**, 2021–2029.

Schmid, P. C., Johnson, S. B., and Schmid, H. H. O. (1991). Remodeling of rat hepatocyte phospholipids by selective acyl turnover. *J. Biol. Chem.* **266**, 13690–13697.

Schwarzmann, G., and Sandhoff, K. (1990). Metabolism and intracellular transport of glyco-sphingolipids. *Biochemistry* **29**, 10865–10871.

Schweizer, A., Fransen, J. A. M., Bachi, T., Ginsel, L., and Hauri, H. P. (1988). Identification, by a monoclonal antibody of a 53-kD protein associated with a tubolo-vesicular compartment of the cis Golgi apparatus. *J. Cell Biol.* **107**, 1643–1653.

Scow, R. O., and Blanchette-Mackie, E. J. (1991). Transport of fatty acids and monoacylglyc-erols in white and brown adipose tissues. *Brain Res. Bull.* **27**, 487–491.

Scow, R. O., Blanchette-Mackie, E. J., Wetzel, M. G., and Reinela, A. (1983). Lipid transport in tissue by lateral movement in cell membrane. *In* "Adipocyte and Obesity" (A. Angel, C. H. Hollenberg, and D. A. K. Roncari, eds.), pp. 165–169. Raven Press, New York.

Seigneuret, M., and Devaux, P. F. (1984). ATP-dependent asymmetric distribution of spin-labeled phospholipids in the erythrocyte membrane: Relation to shape changes. *Proc. Natl. Acad. Sci. U.S.A.* **81**, 3751–3755.

Seigneuret, M., Zachowski, A., Herrmann, A., and Devaux, P. F. (1984). Asymmetric lipid fluidity in human erythrocyte membrane: New spin-label evidence. *Biochemistry* **23**, 4271–4275.

Simbeni, R., Paltauf, F., and Daum, G. (1990). Intramitochondrial transfers of phospholipids in the yeast, *Saccharomyces cerevisiae*. *J. Biol. Chem.* **265**, 281–285.

Simons, K., and Van Meer, G. (1988). Lipid sorting in epithelial cells. *Biochemistry*, **27**, 6197–6202.

Simons, K., and Wandinger-Ness, A. (1990). Polarized sorting in epithelia. *Cell* **62**, 207–210.

Sleight, R. G. (1987). Intracellular lipid transport in eukaryotes. *Annu. Rev. Physiol.* **49**, 193–208.

Sleight, R. G., and Abanto, M. N. (1989). Differences in intracellular transport of a fluorescent phosphatidylcholine analog in established cell lines. *J. Cell Sci.* **93**, 363–374.

Sleight, R. G., and Hopper, K. (1991). Evidence for the activity of a phospholipid exchange protein in vivo. *Biochim. Biophys. Acta* **1067**, 259–263.

Sleight, R. G., and Pagano, R. E. (1983). Rapid appearance of newly synthesized phosphatidylethanolamine at the plasma membrane. *J. Biol. Chem.* **258**, 9050–9058.

Sleight, R. G., and Pagano, R. E. (1984). Transport of a fluorescent phosphatidylcholine analog from the plasma membrane to the Golgi apparatus. *J. Cell Biol.* **99**, 742–751.

Sleight, R. G., and Pagano, R. E. (1985). Transbilayer movement of a fluorescent phosphatidylethanolamine analogue across the plasma membrane of cultured mammalian cells. *J. Biol. Chem.* **260**, 1146–1154.

Spiegel, S., Kassis, S., Wilchek, M., and Fishman, P. H. (1984). Direct visualization of redistribution and capping of fluorescent gangliosides on lymphocytes. *J. Cell Biol.* **99**, 1575–1581.

Spiegel, S., Yamada, K. M., Hom, B. E., Moss, J., and Fishman, P. H. (1985). Fluorescent gangliosides as probes for the retention and organization of fibronectin by ganglioside-deficient mouse cells. *J. Cell Biol.* **100**, 721–726.

Steinman, R. M., Mellman, I. S., Muller, W. A., and Cohn, Z. A. (1983). Endocytosis and the recycling of plasma membrane. *J. Cell Biol.* **96**, 1–27.

Ting, A. E., and Pagano, R. E. (1990). Detection of a phosphatidylinositol-specific phospholipase C at the surface of Swiss 3T3 cells and its potential role in the regulation of cell growth. *J. Biol. Chem.* **265**, 5337–5340.

Urbani, L., and Simoni, R. D. (1990). Cholesterol and vesicular stomatitis virus G protein take separate routes from the endoplasmic reticulum to the plasma membrane. *J. Biol. Chem.* **265**, 1919–1923.

Vance, J. E. (1990). Phospholipid synthesis in a membrane fraction associated with mitochondria. *J. Biol. Chem.* **265**, 7248–7256.

Vance, J. E. (1991). Newly synthesized phosphatidylserine and phosphatidylethanolamine are preferentially translocated between rat liver mitochondria and endoplasmic reticulum. *J. Biol. Chem.* **266**, 89–97.

Vance, J. E., Aasman, E. J., and Szarka, R. (1991). Brefeldin A does not inhibit the movement of phosphatidylethanolamine from its sites of synthesis to the cell surface. *J. Biol. Chem.* **266**, 8241–8247.

Van den Besselaar, A. M. H. P., DeKruijff, B., Van den Bosch, H., and Van Deenen, L. L. M. (1978). Phosphatidylcholine mobility in liver microsomal membranes. *Biochim. Biophys. Acta* **510**, 242–256.

Van der Sluijs, P., Bennett, M. K., Anthony, C., Simons, K., and Kries, T. E. (1990). Binding of exocytic vesicles from MDCK cells to microtubules in vitro. *J. Cell Sci.* **95**, 545–553.

Van Meer, G., and Op den Kamp, J. A. F. (1982). Transbilayer movement of various phosphatidylcholine species in intact human erythrocytes. *J. Cell. Biochem.* **19**, 193–204.

Van Meer, G., Stelzer, E. H. K., Wijnaendts-van-Resandt, R. W., and Simons, K. (1987). Sorting of sphingolipids in epithelial (Madin-Darby canine kidney) cells. *J. Cell Biol.* **105**, 1623–1635.

van't Hof, W., and Van Meer, G. (1990). Generation of lipid polarity in intestinal epithelial (Caco-2) cells: Sphingolipid synthesis in the Golgi complex and sorting before vesicular traffic to the plasma membrane. *J. Cell Biol.* **111**, 977–986.

Voelker, D. R. (1985). Disruption of phosphatidylserine translocation to the mitochondria in baby hamster kidney cells. *J. Biol. Chem.* **260**, 14671–14676.

Voelker, D. R. (1989a). Reconstitution of phosphatidylserine import into rat liver mitochondria. *J. Biol. Chem.* **264**, 1018–1025.

Voelker, D. R. (1989b). Phosphatidylserine translocation to the mitochondrion is an ATP-dependent process in permeabilized animal cells. *Proc. Natl. Acad. Sci. U.S.A.* **86**, 9921–9925.

Voelker, D. R. (1990a). Characterization of phosphatidylserine synthesis and translocation in permeabilized animal cells. *J. Biol. Chem.* **265,** 14340–14346.

Voelker, D. R. (1990b). Lipid transport pathways in mammalian cells. *Experientia* **46,** 569–577.

Voelker, D. R. (1991). Organelle biogenesis and intracellular lipid transport in eukaryotes. *Microbiol. Rev.* **55,** 543–560.

Voelker, D. R., and Kennedy, E. P. (1982). Cellular and enzymic synthesis of sphingomyelin. *Biochemistry* **21,** 2753–2759.

Wandinger-Ness, A., Bennett, M. K., Anthony, C., and Simons, K. (1990). Distinct transport vesicles mediate the delivery of plasma membrane proteins to the apical and basolateral domains of MDCK cells. *J. Cell Biol.* **111,** 987–1000.

Wattenberg, B. W. (1990). Glycolipid and glycoprotein transport through the Golgi complex are similar biochemically and kinetically. Reconstitution of glycolipid transport in a cell free system. *J. Cell Biol.* **111,** 421–428.

Wetzel, M. G., and Scow, R. O. (1984). Lipolysis and fatty acid transport in rat heart: Electron microscopic study. *Am. J. Physiol.* **246,** C467–C485.

Whately, R. E., Zimmerman, G. A., McIntyre, T. M., and Prescott, S. M. (1990). Lipid metabolism and signal transduction in endothelial cells. *Progr. Lipid Res.* **29,** 45–63.

Wieland, F. T., Gleason, M. L., Serafani, T. A., and Rothman, J. E. (1987). The rate of bulk flow from the endoplasmic reticulum to the plasma membrane. *Cell* **50,** 289–300.

Wirtz, K. W. A. (1982). Phospholipid transfer proteins. *Lipid–Protein Interact.* Vol. 1, pp. 151–231.

Wirtz, K. W. A., and Zilversmit, D. B. (1968). Exchange of phospholipids between liver mitochondria and microsomes in vitro. *J. Biol. Chem.* **243,** 3596–3602.

Yaffe, M. P., and Kennedy, E. P. (1983). Intracellular phospholipid movement and the role of phospholipid transfer proteins in animal cells. *Biochemistry* **22,** 1497–1507.

Zachowski, A., Favre, E., Cribier, S., Herve, P., and Devaux, P. F. (1986). Outside–inside translocation of aminophospholipids in the human erythrocyte membrane is mediated by a specific enzyme. *Biochemistry* **25,** 2585–2590.

Zachowski, A., Herrman, A., Paraf, A., and Devaux, P. F. (1987). Phospholipid outside-inside translocation in lymphocyte plasma membranes is a protein mediated phenomenon. *Biochim. Biophys. Acta* **897,** 197–200.

Zilversmit, D. B., and Hughes, M. E. (1977). Extensive exchange of rat liver microsomal phospholipids. *Biochim. Biophys. Acta* **469,** 99–110.

CHAPTER 19

Flow and Distribution of Cholesterol—Effects of Phospholipids

J. Peter Slotte, M. Isabella Pörn, and Ann-Sofi Härmälä
Department of Biochemistry and Pharmacy, Åbo Akademi University, 20520 Turku, Finland

I. INTRODUCTION

Cholesterol is an essential structural and regulatory component of cells. Because of the hydrophobic structure of cholesterol, this molecule is found predominantly in membranous structures (in cell membranes and circulating lipoprotein particles). Cholesterol affects many important properties of the membrane structure, including solute permeability (Papahadjopoulos et al., 1971; Demel et al., 1972b; Szabo, 1974; Yeagle et al., 1977), phospholipid acyl chain mobility (Gally et al., 1976; Stockton and Smith, 1976), lateral phase separations (Smutzer and Yeagle, 1985), lipid packing properties (Chapman et al., 1969; Demel et al., 1972a; Lund-Katz *et al.*, 1988), and the effective free volume in a membrane (Straume and Litman, 1987). These membrane effects of cholesterol, in turn, indirectly affect the behavior and function of membrane-embedded enzymes and other proteins (Yeagle, 1989, and references therein). In addition to the bulk effects of cholesterol on membranes, indirect evidence suggests that cholesterol also is likely to bind directly to and modulate the activity of various membrane-bound enzymes, thus regulating their metabolic function more specifically (Yeagle, 1991, and references therein; see also Chapter 12). For a detailed discussion of the effects of cholesterol on membrane lipids and proteins, the reader is referred to reviews on these topics (Yeagle, 1985, 1989, 1991; Sweet and Schroeder, 1988; Finean, 1990; Schroeder *et al.*, 1991).

The purpose of this chapter is to address questions related to the utilization, intracellular transport, and mass distribution of cholesterol in eukaryotic cells, and to discuss a possible interconnection between the distribution of specific phospholipids and cholesterol.

II. DISTRIBUTION AND FLOW OF CELL CHOLESTEROL

A. Cellular Cholesterol Distribution

The distribution of cholesterol among the membranes of a eukaryotic cell is highly disproportionate. Cell fractionation studies on rat liver suggest that the plasma membrane has the highest sterol level, whereas the outer mitochondrial membrane contains very little sterol; cholesterol is almost undetectable in the inner mitochondrial membrane (Colbeau *et al.*, 1971). This study also reveals very low sterol levels in the rough endoplasmic reticulum, known to contain the enzymes needed for cholesterol synthesis. Similar results have been reported for cholesterol distribution in baby hamster kidney cells (BHK-21; Renkonen *et al.*, 1972). De-Grella and Simoni (1982) determined the cholesterol level in Chinese hamster ovary (CHO) cells and found the cholesterol to phospholipid ratio in the plasma membrane to be 2- to 3-fold higher than in other membrane fractions.

A technique that has been used often to study cholesterol distribution in cells utilizes cholesterol oxidase (Gottlieb, 1977; Lange and Ramos, 1983). Cholesterol oxidase converts cholesterol (cholest-5-en-3β-ol) to cholestenone (cholest-4-en-3-one) at the cell surface (Lange, 1992). Since cholesterol in native membranes is a poor substrate for cholesterol oxidase (Grönberg and Slotte, 1990), different manipulations of membrane lipid packing have been developed to make plasma membrane cholesterol susceptible to cholesterol oxidase (Patzer *et al.*, 1978; Lange *et al.*, 1984;

Slotte *et al.*, 1989). The limits and benefits of this technique have been reviewed elsewhere (Lange, 1992).

Using cholesterol oxidase as a probe for plasma membrane cholesterol, Lange and Ramos (1983) demonstrated that as much as 94% of cell cholesterol is located in the plasma membrane compartment of human fibroblasts. In CHO cells and rat liver hepatocytes, 92% and 80%, respectively, of the plasma membrane cholesterol was susceptible to oxidation (Lange and Matthies, 1984; Lange and Steck, 1985). The suggestion of the almost quantitative distribution of cellular unesterified cholesterol to the plasma membrane compartment has, however, been contradicted by van Meer (1987). A high cholesterol distribution (e.g., 90%) to the plasma membrane would require 30–40% of the cellular phospholipids to be present in the membrane (Dawidowicz, 1987). The finding by Kaplan and Simoni (1985) that at least 30% of the total cellular phospholipids are present in the plasma membranes of CHO cells thus would be consistent with the suggestion that plasma membranes contain a very high proportion of cell cholesterol.

According to Lange (1991), the major fraction of the non-plasma-membrane-associated intracellular cholesterol in fibroblasts (~10%) would be located in endocytic membranes. Studies using filipin staining and electron microscopy also support the notion that cholesterol is found in organelles of the endocytic pathway (Orci *et al.*, 1981; McGookey *et al.*, 1983).

B. Routes of Cholesterol Flow

Since most lipids of the plasma membrane are synthesized intracellularly or are derived from exogenous lipoproteins (Goldstein and Brown, 1977), a continuously functioning lipid transport system is of vital importance for all living cells (see Fig. 1). Intracellular lipid transport can occur by several different mechanisms. Vesicles containing lipids and proteins can bud off from one membrane and travel through the cytosol to another membrane, where they subsequently are incorporated. Spontaneous or protein–mediated transport of lipid monomers also occurs between membranes. Several different lipid transport proteins have been isolated and characterized (reviewed by Wirtz and Gadella, 1990). In addition to the cytosolic transport, energy-independent lateral diffusion of lipid molecules from one organelle to another is thought to occur at so-called membrane bridges.

The transfer of newly synthesized cholesterol from the site of synthesis (the endoplasmic reticulum) to the plasma membrane appears to occur by vesicle-mediated transport, since this transport is inhibited at 15°C (Kaplan

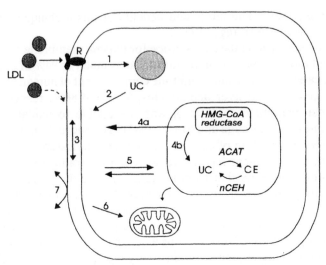

FIGURE 1 Schematic of cholesterol flow routes in peripheral cells. Peripheral cells obtain cholesterol through the low density lipoprotein (LDL) pathway or by endogenous synthesis. When the cellular requirement for cholesterol increases, LDL receptors (R) are synthesized and transported to the cell surface. LDL particles bind specifically to the receptors and subsequently are endocytosed (1) and degraded in lysosomes (Goldstein and Brown, 1977). The lysosomal cholesterol derived from LDL cholesteryl esters is transferred mainly to the plasma membrane (2), which serves as an expandable pool of free cholesterol (3). Uptake of cholesterol from LDL particles directly into the plasma membrane compartment, both in free and esterified form, also occurs (Slotte *et al.*, 1988). Newly synthesized cholesterol is transported rapidly from the endoplasmic reticulum, where the key regulatory enzyme HMG-CoA reductase resides, to the plasma membrane (4a). A fraction of the native cholesterol may become esterified immediately after synthesis (4b). Excess unesterified cholesterol (UC) is transported to the endoplasmic reticulum (5), where cholesteryl ester (CE) formation is catalyzed by the enzyme ACAT. Cholesteryl ester hydrolysis is catalyzed by neutral cholesteryl ester hydrolase (nCEH) which, in conjunction with ACAT, forms the cholesteryl ester cycle. In steroidogenic cells, in which cholesterol is required as a substrate for steroidogenesis, transfer of cholesterol to the mitochondria occurs from the plasma membrane compartment (6) and possibly also from intracellular sites upon cholesteryl ester hydrolysis. For peripheral nonsteroidogenic cells, the transfer of plasma membrane cholesterol to phospholipid-rich extracellular acceptors, such as high density lipoproteins (HDL), is the only way to dispose of excess cholesterol (7).

and Simoni, 1985) and is sensitive to energy poisons (DeGrella and Simoni, 1982). Cholesterol is transferred from the endoplasmic reticulum immediately after synthesis, with a reported $t_{1/2}$ of 10–20 min for transport to the plasma membrane at 37°C (DeGrella and Simoni, 1982; Lange *et al.*, 1991). Nascent cholesterol apparently is not trafficked by the secretory protein pathway through the Golgi apparatus, since inhibitors of exocytosis such as monensin and brefeldin A were shown not to affect this transport (Kaplan and Simoni, 1985; Urbani and Simoni, 1990).

The transport of low density lipoprotein (LDL)-derived cholesterol from the lysosomes appears to be directed mainly to the plasma membrane, before it is transferred to the regulatory pool of the esterification enzyme acyl CoA:cholesterol acyl transferase (ACAT) or to the mitochondria in steroidogenic cells (Johnson *et al.*, 1990; Nagy and Freeman, 1990). The transfer of lysosomal cholesterol to the plasma membrane compartment differs from the transfer of endogenous cholesterol in that this transport is insensitive to treatment with energy poisons (Liscum, 1990) but is inhibited by various pharmacological agents such as the hydrophobic amines U18666A and imipramine (Liscum and Faust, 1989; Rodriguez-Lafrasse *et al.*, 1990; Liscum and Collins, 1991). The transport of LDL-cholesterol to the plasma membrane has been reported to occur within 50 min of LDL degradation (Brasaemle and Attie, 1990; Johnson *et al.*, 1990). Liscum and Dahl (1992) suggested that the transfer of both endogenous and exogenous cholesterol could occur by vesicular transport, the difference being that the endogenous cholesterol might be subjected to an energy-requiring sorting step in the endoplasmic reticulum. Transport of LDL-derived cholesterol from lysosomes to plasma membranes is defective in Niemann–Pick Type C fibroblasts (Kruth *et al.*, 1986; Pentchev *et al.*, 1986). These cells, as well as other mutant cell lines with defective cholesterol transport (Cadigan *et al.*, 1990; Dahl *et al.*, 1992), should prove useful in future studies on intracellular cholesterol transport mechanisms.

Since excess free cholesterol is toxic to the cells, this lipid is stored in the form of cholesteryl ester droplets. ACAT, which is located in the endoplasmic reticulum, appears to be regulated primarily by substrate levels (reviewed by Suckling and Stange, 1985). Xu and Tabas (1991) showed that the LDL-induced activation of ACAT in macrophages occurs only after cellular cholesterol pools are expanded to a certain critical threshold level (approximately 25% above basal cholesterol level). The transport of cholesterol to the substrate pool of ACAT seems not to require mediation by sterol carrier protein 2 (SCP$_2$), since the cellular level of SCP$_2$ does not affect the kinetics of cholesterol esterification (van Heusden *et al.*, 1985; Liscum and Dahl, 1992). ACAT and neutral cholesteryl ester hydrolase (nCEH) form the cellular cholesteryl ester cycle, which regulates the level of intracellular free cholesterol (Brown and Goldstein, 1983). The degradation of cytoplasmic cholesteryl esters by nCEH is stimulated by cyclic AMP-dependent protein kinase (Trzeciak and Boyd, 1973; Khoo *et al.*, 1981), which also has been shown to stimulate cholesteryl ester clearance to high-density lipoproteins (HDL) in J774 macrophages

(Bernard *et al.*, 1991). Stimulatory effects of HDL on intracellular cholesteryl ester hydrolysis and cholesterol mobilization to the plasma membrane are well documented in the literature (Ho *et al.*, 1980; Slotte *et al.*, 1987; Aviram *et al.*, 1989), but the mechanisms underlying these effects have become a subject of investigation only recently. Data have been published to support the hypothesis of HDL-induced signal transduction, which appears to occur through protein kinase C activation (Theret *et al.*, 1990; Mendez *et al.*, 1991; Pörn *et al.*, 1991a).

In steroidogenic cells, which utilize cholesterol as a precursor for steroid hormones, the transport of cholesterol to mitochondria plays an essential role in cellular cholesterol homeostasis. However, the source of cholesterol used for steroidogenesis differs depending on the cell type. In Leydig tumor cells, the pool of unesterified cholesterol used for steroidogenesis has been shown to reside in the plasma membrane (Freeman, 1987) and, on stimulation with dibutyryl cyclic AMP, this cholesterol is transported to mitochondria (Freeman, 1989). In cells of the adrenal cortex, cholesteryl ester droplets have been implicated as the source of cholesterol for steroidogenesis (Bisgaier *et al.*, 1985). The transport of cholesterol to adrenal mitochondria apparently requires intact microfilaments and microtubules, since this movement is inhibited by vinblastine and cytochalasin B (Crivello and Jefcoate, 1980); SCP$_2$ has been shown to mediate transfer of cholesterol from lipid droplets to adrenal mitochondria (Chanderbhan *et al.*, 1982).

III. CHOLESTEROL–PHOSPHOLIPID INTERACTION IN MEMBRANES

A. Molecular Modes of Cholesterol–Phospholipid Interaction

The hydrophobic properties of cholesterol give it a very low water solubility (critical micellar concentration around 40 nM; Haberland and Reynolds, 1976). However, cholesterol is solubilized readily into phospholipid membranes. The planar cholesterol molecule is rigid with the exception of the isooctyl side chain at carbon 17, and is intercalated in the bilayer parallel to the acyl chains of the phospholipids (Taylor *et al.*, 1981; Dufourc *et al.*, 1984). The strength of the molecular interactions between cholesterol and various phospholipid classes is known to vary (Demel *et al.*, 1977; Wattenberg and Silbert, 1983). The strong interactions between cholesterol and sphingomyelin have been demonstrated with a variety of experimental approaches. In monolayer membranes, cholesterol is known to condense the lateral packing density of sphingomyelins more than it condenses that of other phospholipids (Lund-Katz *et al.*, 1988). The rate of cholesterol desorption from sphingomyelin-containing membranes also is known to be much slower than the rates observed for other phospholipid systems (Fugler *et al.*, 1985; Bhuvaneswaran and Mitropoulos, 1986; Yeagle and Young, 1986; Phillips *et al.*, 1987; Kan *et al.*, 1991a,b; Bittman, 1992). The oxidizability of membrane cholesterol by cholesterol oxidase has been shown to be reduced greatly in the presence of sphingomyelin (Grönberg and Slotte, 1990; Grönberg *et al.*, 1991), possibly reflecting the tight lateral packing density in such membranes.

The preferential interaction of cholesterol with sphingomyelin in membranes may result from the increased opportunities for van der Waals interactions between the molecules (Lund-Katz *et al.*, 1988). Sphingomyelin, with its asymmetric acyl chain lengths, also has been postulated to create packing defects (high-energy voids) in the hydrophobic core of a bilayer containing other phospholipids with symmetric acyl chain lengths. Cholesterol has been postulated to partition into such regions specifically to prevent the formation of the high-energy voids (McIntosh *et al.*, 1992a,b).

B. Lateral Heterogeneity in Lipid Distribution

The transbilayer as well as the lateral distribution of lipids in membranes is highly asymmetric (Severs and Robenek, 1983; Schroeder *et al.*, 1991; Rothblat *et al.*, 1992). Since the molecular interaction between cholesterol and phosphatidylcholine or sphingomyelin appears to vary, cholesterol may form distinct domains with phosphatidylcholine and with sphingomyelin when both these phospholipid classes are present in a membrane. In several studies, cholesterol was observed to associate with phosphatidylcholine with a stoichiometry of 1 : 1, whereas the corresponding stoichiometry was 2 : 1 with sphingomyelin (Slotte, 1992). These results suggest that sphingomyelin had a better capacity to solubilize cholesterol, and support the notion of distinct cholesterol–phospholipid domains.

With another experimental approach, in which the integral membrane protein cytochrome P450$_{scc}$ was used as a reporter of cholesterol availability with a membrane, researchers found that the cholesterol–sphingomyelin interaction was cooperative and differed in properties from the interactions between cholesterol and glycerolipids (Stevens *et al.*, 1986). These results, in conjunction with the results obtained in monolayer systems (Grönberg and Slotte, 1990; Grönberg *et al.*, 1991; Slotte 1992), clearly support the possibility of cholesterol–sphingomyelin microdomains in model and biological membranes. Next we discuss the consequences of manipulations of membrane phospholipid composition, which can affect these putative cholesterol–phospholipid microdomains directly.

IV. EFFECTS OF PHOSPHOLIPID DEGRADATION ON CHOLESTEROL FLOW IN THE CELL

A significant positive correlation is known to exist between the contents of cholesterol and sphingomyelin in the membranes of rat liver hepatocytes (Patton, 1970). The direct involvement of sphingomyelin in affecting the cellular distribution of cholesterol first was suggested by experiments in which sphingomyelin was introduced into the plasma membrane of cultured cells from liposomes (Gatt and Bierman, 1980; Kudchodkar et al., 1983). This treatment was shown to reduce the rate of endogenous cholesterol esterification dramatically, and to increase markedly the rate of cholesterol biosynthesis. Since both the esterification and the biosynthesis of cholesterol are under strict metabolic regulation, the results implied that sphingomyelin incorporation into the plasma membrane of cultured cells led to a net flow of cholesterol from intracellular sites to the plasma membrane compartment.

With these results in mind, testing whether a selective degradation of plasma membrane sphingomyelin in intact cells would lead to an opposite flow of cell cholesterol, from the cell surface into the cell, was logical. The availability of a phospholipase C-type sphingomyelinase (*Staphylococcus aureus*), specific for sphingomyelin made such studies feasible.

A. Effects of Sphingomyelin Degradation on Cellular Cholesterol Flow

The activation of ACAT frequently has been used to monitor the flow of cholesterol mass in cells. Slotte and Bierman (1988) were the first to show that degradation of sphingomyelin mass (using 0.1 U/ml sphingomyelinase) in cultured fibroblasts led to a dramatic activation of the endogenous esterification of cholesterol. This first study was followed by related studies with different cell types, in which similar effects on the activation of ACAT consistently have been observed (Gupta and Rudney, 1991; Stein et al., 1992). The sphingomyelinase-induced activation of ACAT was not seen when cells were exposed to degradation products of sphingomyelin such as phosphorylcholine or sphingosine (Slotte and Bierman, 1988; Gupta and Rudney, 1991). The activation of ACAT, following the degradation of sphingomyelin in native cells, was interpreted to be a consequence of an increased flow of cholesterol mass from the cell surface to the endoplasmic reticulum.

Usually, 50–80% of the prelabeled pool of [^{3}H]sphingomyelin is degraded in a cell by 0.1 U/ml sphingomyelinase in 30–60 min (Pörn and Slotte, 1990; Slotte et al., 1990a; Pörn et al., 1991b). Since the size of the degradable pool of [^{3}H]sphingomyelin is similar in native and glutaralde-hyde-fixed cells (J. P. Slotte, M. I., Pörn, and A. S. Härmälä, unpublished observation), this fraction is likely to represent the amount of sphingomyelin mass in the outer leaflet of the plasma membrane. Using plasma membrane isolation and sphingomyelin mass analysis, Lange and co-workers (1989) observed that fibroblast plasma membranes contain approximately 90% of the total sphingomyelin mass. This value is very close to the amount obtained using sphingomyelinase in fibroblasts (i.e., 80%; Pörn and Slotte, 1990).

Using a modified cholesterol oxidase assay (Slotte et al., 1989), we have been able to quantitate the amount of cholesterol mass present in plasma membranes of cultured cells, and to determine the extent of cholesterol translocation away from the cell surface in sphingomyelinase-treated cells (Slotte et al., 1989; Pörn and Slotte, 1990; Slotte et al., 1990a). The complete degradation of sphingomyelin mass from the external leaflet of the cellular plasma membrane invariably leads to a rapid and substantial translocation of cholesterol from the cell surface (Slotte et al., 1989; Pörn and Slotte, 1990; Slotte et al., 1990a). The extent of cholesterol translocation differs from one cell type to another (between 20 and 60% of the cellular free cholesterol), but also within different lines of a similar cell type (i.e., human fibroblasts; J. P. Slotte, M. I. Pörn, and A. S. Härmälä, unpublished observations).

The recovery of cells from the effects of sphingomyelin degradation were tested first with fibroblasts and transformed human neuroblastoma cells (Pörn and Slotte, 1990). A restoration of sphingomyelin mass (after removal of the initially added sphingomyelinase) was observed to tend to restore the "normal" distribution of cholesterol. However, since the resynthesis of sphingomyelin mass in these cells was a very slow process, another cell model was chosen (baby hamster kidney cells) that reportedly had a very rapid rate of sphingomyelin synthesis (Allan and Quinn, 1988). Results from experiments with BHK-21 cells are shown in Fig. 2. With this cell type, the initial loss of about 60% of sphingomyelin mass was corrected rapidly when the exogenously added sphingomyelinase was removed; the cells restored their sphingomyelin mass to the normal level within 3–4 hr (Fig. 2A). Plasma membrane cholesterol was translocated rapidly and extensively from the cell surface after the degradation of sphingomyelin. However, the resynthesis of sphingomyelin mass in the BHK-21 cells led to a rapid flow of cholesterol back to the cell surface, thus restoring the pretreatment state of cholesterol distribution (Fig. 2B). A similar pattern can be seen in the response to the endogenous esterification reaction. When cholesterol flowed into the cell, the ACAT reaction was activated markedly (Fig. 2C). However, when the flow of cholesterol

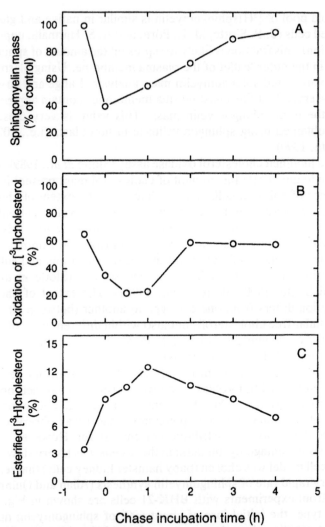

FIGURE 2 Effects of sphingomyelin degradation and resynthesis on cholesterol distribution and cholesterol esterification activity in cultured BHK-21 cells. Confluent cells, prelabeled with [³H]cholesterol, first were exposed for 30 min to 0.1 U/ml sphingomyelinase, and then rinsed and postincubated without sphingomyelinase for indicated periods of time. (A) Mass of sphingomyelin. (B) Distribution of [³H]cholesterol. (C) Amount of cellular [³H]cholesteryl esters during the incubation. Reproduced from Slotte *et al.* (1990a) with permission.

was turned in the direction of the plasma membrane (during the sphingomyelin restoration phase), the cellular level of [³H]cholesteryl esters decreased extensively.

In addition to affecting the rate of endogenous cholesterol esterification, the sphingomyelinase-induced cholesterol translocation also affected the rate of cholesterol biosynthesis. Slotte and Bierman (1988) originally demonstrated that the incorporation of sodium [¹⁴C]acetate into *de novo* sterols was reduced markedly in cells treated with sphingomyelinase compared with control fibroblasts. Later, Gupta and Rudney (1991) extended these experiments to include rat intestinal epithelial cells, human skin fibroblasts, and Hep G2 cells, and demonstrated that sphingomyelin degradation in conjunction with the resulting cholesterol flow directly downregulated the activity of hydroxy methylglutaryl (HMG) CoA reductase, the key regulatory enzyme in cholesterol biosynthesis.

Since a fraction of plasma membrane cholesterol was translocated to the endoplasmic reticulum on depletion of cell surface sphingomyelin, the question arose whether cholesterol also would flow to other intracellular organelles such as the mitochondria. To elucidate this question, mouse Leydig tumor cells were used in studies concerning the effects of sphingomyelin depletion on steroidogenesis. In addition to the decrease in cell cholesterol oxidizability, which was observed in all cell types studied, the cellular conversion of [³H]cholesterol to [³H]steroid hormones was increased on sphingomyelinase treatment of Leydig cells (Pörn *et al.*, 1991b). Thus, the sphingomyelinase–induced cholesterol flow from the plasma membrane also appears to be directed to the mitochondria. Unfortunately, marker signals for other intracellular organelles are not available, but the translocated plasma membrane cholesterol may be distributed to other intracellular organelles as well.

The first studies of the effects of sphingomyelin degradation on cellular cholesterol translocation were performed in culture media devoid of extracellular cholesterol acceptors (Slotte and Bierman, 1988). Therefore whether the sphingomyelinase-induced cholesterol translocation into intracellular compartments could be retarded with an extracellular cholesterol acceptor was not known. When this possibility was tested with HDL₃ as a cholesterol acceptor, sphingomyelin degradation was observed still to result in an activation of endogenous [³H]cholesteryl ester formation, the extent of which was not significantly different from that in control cells (sphingomyelinase treatment but no HDL₃; Slotte *et al.*, 1990b). However, Gupta and Rudney (1991) later performed similar experiments with a different cholesterol acceptor, and observed that the effects of sphingomyelin degradation on ACAT activation and HMG CoA reductase inhibition in IEC-6 cells were inhibited fully in the presence of 500 μM phospholipid vesicles.

B. Degradation of Phosphatidylcholine versus Cholesterol Flow

To test whether a selective degradation of plasma membrane phosphatidylcholine also leads to cholesterol translocation, a phosphatidylcholine-specific phospholipase C-type enzyme from *Bacillus cereus* was used. The basic experimental approach was similar to that used in the sphingomyelinase experiments. Surprisingly, a selective degradation of phosphatidylcholine did not cause a detectable cholesterol translocation in cells (Pörn *et al.*, 1993). Under the conditions used, phospholipase C degraded about 15% of the total cellular phosphatidylcholine, which corresponded to about 50 nmol phosphatidylcholine per mg cell protein. Sphingomyelinase treatment of the same cells degraded about 22 nmol sphingomyelin per mg cell protein. The partial degradation of phosphatidylcholine left the cells morphologically intact; the cells excluded Trypan Blue to a similar extent as control cells. Since the removal of a substantial amount of phosphatidylcholine from the outer leaflet of fibroblast plasma membrane did not alter the cholesterol balance in the membrane (assayed both with cholesterol oxidase and with the activity of ACAT), phosphatidylcholine–cholesterol interactions were concluded to be of lesser importance for the plasma membrane cholesterol balance, at least when sphingomyelin was present in the membrane.

C. Mechanisms of Sphingomyelinase-Induced Cholesterol Translocation

The mechanism by which cholesterol is translocated from the plasma membrane after the degradation of sphingomyelin is unknown. In red blood cells, sphingomyelin degradation by sphingomyelinase is known to result in endovesiculation and dramatic changes in cell morphology (Verkleij *et al.*, 1973; Allan and Walklin, 1988). In eukaryotic cells, the visible effects (as determined with phase-contrast microscopy) of shipngomyelin degradation are minor and transient (Slotte and Bierman, 1988). No measurable and extensive endovesiculation has been observed in sphingomyelinase-treated fibroblasts, using markers for fluid-phase endo- and pinocytosis (J. P. Slotte, unpublished observations).

The exposure of the exofacial leaflet to sphingomyelinase results in the degradation of sphingomyelin, leading to a transient domain containing cholesterol and ceramide. Such a domain would be expected to be thermo-dynamically unstable, since ceramide and cholesterol would be able to undergo transbilayer flip–flop (Pagano, 1989). Whereas the generation of ceramide in red blood cells is known to lead to the segregation of ceramide into a lateral phase of its own (Verkleij *et al.*, 1973), ceramide introduced into the plasma membranes of eukaryotic cells is known to transfer to intracellular organelles, especially the Golgi apparatus (Lipsky and Pagano, 1983; Van Meer *et al.*, 1987).

Because of a possible destabilization of the endoleaflet, or as a result of other effects, cholesterol is transported rapidly from the plasma membrane in the direction of endomembranes. This transfer step is rapid, since cholesterol has been observed in the substrate pool of ACAT within 5–10 min of the induction of the transfer process (Slotte and Bierman, 1988), and may involve vesicular transfer vehicles.

V. ARE SPHINGOMYELINASES ENDOGENOUS REGULATORS OF CHOLESTEROL FLOW?

The effects of sphingomyelin degradation on cellular cholesterol homeo-stasis have been achieved to date exclusively with the addition of exogenous sphingomyelinase. However, many cell types contain at least two distinct enzymes that degrade sphingomyelin; a lysosomal acid sphingo-myelinase (Pentchev *et al.*, 1977; Jones *et al.*, 1981,1982) and a membrane-associated sphingomyelinase with optimal activity at neutral pH (Schneider and Kennedy, 1967; Gatt, 1976; Rao and Spence, 1976). Hannun and co-workers have shown that endogenous neutral sphingomye-linase in leukemia cells can be activated by treatment with vitamin D_3, tumor necrosis factor α, and interferon γ (Okazaki *et al.*, 1989; Kim *et al.*, 1991). The degradation products of sphingomyelin, particularly ceramide and sphingosine, appear to function as second messengers (reviewed by Kolesnick, 1991). However, these second messengers apparently are not the cause of the sphingomyelinase-induced effect on cellular cholesterol homeostasis, as mentioned earlier. However, the degradation of sphingomyelin by activated endogenous sphingomyelinases (either acid or neutral) may result in a changed cholesterol concentration in the specific microdomain in which the hydrolysis takes place.

In another report, the effects of sphingomyelin hydrolysis on cholesterol absorption into human intestinal Caco-2 cells were examined (Chen *et al.*, 1992). This interesting study showed that sphingomyelin degradation in Caco-2 cells by exogenously added sphingomyelinase decreased the uptake of cholesterol from bile salt micelles by 50%. Since these authors

also showed that human pancreatic juice contains neutral sphingomyelinase activity, they concluded that endogenous sphingomyelinase (in pancreatic juice) also could affect the rate of cholesterol absorption in the
intestine, by regulating the level of sphingomyelin in the intestinal absorptive cells.

We believe that the results obtained to date clearly imply a major role
for endogenous sphingomyelinases in the regulation of cellular cholesterol
distribution. However, we acknowledge that a tremendous amount of
effort still must be devoted to this area before all the important facets of
the interrelationship between sphingomyelin and cholesterol have been
unraveled.

References

Allan, D., and Quinn, P. (1988). Resynthesis of sphingomyelin from plasma membrane
    phosphatidylcholine in BHK cells treated with *Staphylococcus aureus* sphingomyeli-
    nase. *Biochem. J.* **254**, 765–771.
Allan, D., and Walklin, C. M. (1988). Endovesiculation of human erythrocytes exposed to
    sphingomyelinase C—a possible explanation for the enzyme-resistant pool of sphingo-
    myelin. *Biochim. Biophys. Acta* **938**, 403–410.
Aviram, M., Bierman, E. L., and Oram, J. F. (1989). High density lipoprotein stimulates
    sterol translocation between intracellular and plasma membrane pools in human
    monocyte-derived macrophages. *J. Lipid Res.* **30**, 65–76.
Bernard, D. W., Rodriguez, A., Rothblat, G. H., and Glick, J. M. (1991). cAMP stimulates
    cholesteryl ester clearance to high density lipoproteins in J774 macrophages. *J. Biol.
    Chem.* **266**, 710–716.
Bhuvaneswaran, C., and Mitropoulos, K. A. (1986). Effect of liposomal phospholipid compo-
    sition on cholesterol transfer between microsomal and liposomal vesicles. *Biochem. J.*
    238, 647–652.
Bisgaier, C. L., Chanderbhan, R., Hinds, R. W., and Vahouny, G. V. (1985). Adrenal
    cholesterol esters as substrate source for steroidogenesis. *J. Steroid Biochem.* **23**,
    967–974.
Bittman, R. (1992). A review on the kinetics of cholesterol movement between donor and
    acceptor bilayer membranes. *In* "Cholesterol in Model Membranes" (L. X. Finegold,
    ed.). CRC Press, Boca Raton, Florida.
Brasaemle, D. L., and Attie, A. D. (1990). Rapid intracellular transport of LDL-derived
    cholesterol to the plasma membrane in cultured fibroblasts. *J. Lipid Res.* **31**, 103–112.
Brown, M. S., and Goldstein, J. L. (1983). Lipoprotein metabolism in the macrophage:
    Implications for cholesterol deposition in atherosclerosis. *Annu. Rev. Biochem.* **52**,
    223–261.
Cadigan, K. M., Spillane, D. M., and Chang, T.-Y. (1990). Isolation and characterization
    of Chinese hamster ovary cell mutants defective in intracellular low density lipoprotein-
    cholesterol trafficking. *J. Cell Biol.* **110**, 295–308.
Chanderbhan, R., Noland, B. J., Scallen, T. J., and Vahouny, G. V. (1982). Sterol carrier
    protein$_2$. *J. Biol. Chem.* **257**, 8928–8934.
Chapman, D., Owens, N. F., Phillips, M. C., and Walker, D. A. (1969). Mixed monolayers
    of phospholipids and cholesterol. *Biochim. Biophys. Acta* **183**, 458–465.
Chen, H., Born, E., Mathur, S. N., Johlin, F. C., and Field, F. J. (1992). Sphingomyelin

content of intestinal cell membranes regulated cholesterol absorption. *Biochem. J.* **286,** 771–777.

Colbeau, A., Nachbaur, J., and Vignais, P. M. (1971). Enzymic characterization and lipid composition of rat liver subcellular membranes. *Biochim. Biophys. Acta* **249,** 462–492.

Crivello, J. F., and Jefcoate, C. R. (1980). Intracellular movement of cholesterol in rat adrenal cells. *J. Biol. Chem.* **255,** 8144–8151.

Dahl, N. K., Reed, K. L., Daunais, M. A., Faust, J. R., and Liscum, L. (1992). Isolation and characterization of chinese hamster ovary cells defective in the intracellular metabolism of low density lipoprotein-derived cholesterol. *J. Biol. Chem.* **267,** 4889–4896.

Dawidowicz, E. A. (1987). Dynamics of membrane lipid metabolism and turnover. *Annu. Rev. Biochem.* **56,** 43–61.

DeGrella, R. F., and Simoni, R. D. (1982). Intracellular transport of cholesterol to the plasma membrane. *J. Biol. Chem.* **257,** 14256–14262.

Demel, R. A., Bruckdorfer, K. R., and Van Deenen, L. L. M. (1972a). Structure requirement of sterols for the interaction with lecithin at the air–water interface. *Biochim. Biophys. Acta* **255,** 311–320.

Demel, R. A., Bruckdorfer, K. R., and Van Deenen, L. L. M. (1972b). The effect of sterol structure on the permeability of liposomes to glucose, glycerol, and Rb^+. *Biochim. Biophys. Acta* **255,** 321–330.

Demel, R. A., Jansen, J. W. C. M., Van Dijck, P. W. M., and van Deenen, L. L. M. (1977). The preferential interaction of cholesterol with different classes of phospholipids. *Biochim. Biophys. Acta* **465,** 1–10.

Dufourc, E. J., Parish, E. J., Chitrakorn, S., and Smith, I. C. P. (1984). Structural and dynamical details of cholesterol-lipid interactions as revealed by deuterium NMR. *Biochemistry* **23,** 6062–6071.

Finean, J. B. (1990). Interaction between cholesterol and phospholipid in hydrated bilayers. *Chem. Phys. Lipids* **54,** 147–156.

Freeman, D. A. (1987). Cyclic AMP mediated modification of cholesterol traffic in Leydig tumor cells. *J. Biol. Chem.* **262,** 13061–13068.

Freeman, D. A. (1989). Plasma membrane cholesterol: Removal and insertion into the membrane and utilization as substrate for steroidogenesis. *Endocrinology* **124,** 2527–2534.

Fugler, L., Clejan, S., and Bittman, R. (1985). Movement of cholesterol between vesicles prepared with different phospholipids or sizes. *J. Biol. Chem.* **260,** 4098–4102.

Gally, H. U., Seelig, A., and Seelig, J. (1976). Cholesterol induced rod-like motion of fatty acyl chains in lipid bilayers, a deuterium magnetic resonance study. *Hoppe-Zeyler's Z. Physiol. Chem.* **357,** 1447–1450.

Gatt, S. (1976). Magnesium-dependent sphingomyelinase. *Biochem. Biophys. Res. Commun.* **68,** 235–241.

Gatt, S., and Bierman, E. L. (1980). Sphingomyelin suppresses the binding and utilization of LDL by skin fibroblasts. *J. Biol. Chem.* **255,** 3371–3376.

Goldstein, J. L., and Brown, M. S. (1977). The low-density lipoprotein pathway and its relation to atherosclerosis. *Annu. Rev. Biochem.* **46,** 897–930.

Gottlieb, M. H. (1977). The reactivity of human erythrocyte membrane cholesterol with a cholesterol oxidase. *Biochim. Biophys. Acta* **466,** 422–428.

Grönberg, L., and Slotte, J. P. (1990). Cholesterol oxidase catalyzed oxidation of cholesterol in mixed lipid monolayers—Effects of surface pressure and phospholipid composition on catalytic activity. *Biochemistry* **29,** 3173–3178.

Grönberg, L., Ruan, Z.-S., Bittman, R., and Slotte, J. P. (1991). Interaction of cholesterol with synthetic sphingomyelin derivatives in mixed monolayers. *Biochemistry* **30**, 10746–10754.

Gupta, A. K., and Rudney, H. (1991). Plasma membrane sphingomylein and the regulation of HMG CoA reductase activity and cholesterol biosynthesis in cell cultures. *J. Lipid Res.* **32**, 125–136.

Haberland, M. E., and Reynolds, J. A. (1976). Self-association of cholesterol in aqueous solution. *Proc. Natl. Acad. Sci. U.S.A.* **70**, 2313–2316.

Ho, Y. K., Brown, M. S., and Goldstein, J. L. (1980). Hydrolysis and excretion of cytoplasmic cholesteryl esters by macrophages: Stimulation by high density lipoprotein and other agents. *J. Lipid Res.* **21**, 391–398.

Johnson, W. J., Chacko, G. K., Phillips, M. C., and Rothblat, G. H. (1990). The efflux of lysosomal cholesterol from cells. *J. Biol. Chem.* **265**, 5546–5553.

Jones, C. S., Shankaran, P., and Callahan, J. W. (1981). Purification of sphingomyelinase to apparent homogeneity by using hydrophobic chromatography. *Biochem. J.* **195**, 373–382.

Jones, C. S., Shankaran, P., Davidson, D. J., Poulos, A., and Callahan, J. W. (1982). Studies on the structure of sphingomyelinase. *Biochem. J.* **209**, 291–297.

Kan, C.-C., Bittman, R., and Hajdu, J. (1991a). Phospholipids containing nitrogen- and sulphur-linked chains: Kinetics of cholesterol exchange between vesicles. *Biochim. Biophys. Acta* **1066**, 95–101.

Kan, C.-C., Ruan, Z.-S., and Bittman, R. (1991b). Interaction of cholesterol with sphingomyelin in bilayer membranes: Evidence that the hydroxy group of sphingomyelin does not modulate the rate of cholesterol exchange between vesicles. *Biochemistry* **30**, 7759–7766.

Kaplan, M. R., and Simoni, R. D. (1985). Transport of cholesterol from the endoplasmic reticulum to the plasma membrane. *J. Cell Biol.* **101**, 466–453.

Khoo, J. C., Mahoney, E. M., and Steinberg, D. (1981). Neutral cholesterol esterase activity in macrophages and its enhancement by cAMP-dependent protein kinase. *J. Biol. Chem.* **256**, 12659–12661.

Kim, M.-Y., Linardic, C., Obeid, L., and Hannun, Y. (1991). Identification of sphingomyelin turnover as an effector mechanism for the action of tumor necrosis factor α and γ-interferon. *J. Biol. Chem.* **266**, 484–489.

Kolesnick, R. N. (1991). Sphingomyelin and derivatives as cellular signals. *Progr. Lipid Res.* **30**, 1–38.

Kruth, H. S., Comly, M. E., Butler, J. D., Vanier, M. T., Fink, J. K., Wenger, D. A., Patel, S., and Pentchev, P. G. (19860. Type C Niemann–Pick disease. *J. Biol. Chem.* **261**, 16769–16774.

Kudchodkar, B. J., Albers, J. J., and Bierman, E. L. (1983). Effects of positively-charged sphingomyelin liposomes on cholesterol metabolism of cells in culture. *Atherosclerosis* **46**, 353–367.

Lange, Y. (1991). Disposition of intracellular cholesterol in human fibroblasts. *J. Lipid Res.* **32**, 329–339.

Lange, Y. (1992). Tracking cell cholesterol with cholesterol oxidase. *J. Lipid Res.* **33**, 315–321.

Lange, Y., and Matthies, H. J. G. (1984). Transfer of cholesterol from its site of synthesis to the plasma membrane. *J. Biol. Chem.* **259**, 14624–14630.

Lange, Y., and Ramos, B. V. (1983). Analysis of the distribution of cholesterol in the intact cell. *J. Biol. Chem.* **258**, 15130–15134.

Lange, Y., and Steck, T. L. (1985). Cholesterol-rich intracellular membranes: A precursor to the plasma membrane. *J. Biol. Chem.* **260**, 15592–15597.

Lange, Y., Matthies, H., and Steck, T. L. (1984). Cholesterol oxidase susceptibility of red cell membrane. *Biochim. Biophys. Acta* **769,** 551–562.

Lange, Y., Swaisgood, M. H., Ramos, B. V., and Steck, T. L. (1989). Plasma membranes contain half of the phospholipid and 90% of the cholesterol and sphingomyelin in cultured fibroblasts. *J. Biol. Chem.* **264,** 3786–3793.

Lange, Y., Echevarria, F., and Steck, T. L. (1991). Movement of zymosterol, a precursor of cholesterol, among three membranes in human fibroblasts. *J. Biol. Chem.* **266,** 21439–21443.

Lipsky, N. G., and Pagano, R. E. (1983). Intracellular translocation of fluorescent sphingolipids in cultured fibroblasts—Endogenously synthesized sphingomyelin and glucocerebroside analogues pass through the Golgi apparatus en route to the plasma membrane. *J. Cell Biol.* **100,** 27–34.

Liscum, L. (1990). Pharmacological inhibition of the intracellular transport of low-density lipoprotein-derived cholesterol in Chinese hamster ovary cells. *Biochim. Biophys. Acta* **1045,** 40–48.

Liscum, L., and Collins, G. J. (1991). Characterization of Chinese hamster ovary cells that are resistant to 3-β-[2-(diethylamino)ethoxy]androst-5-en-17-one inhibition of low density lipoprotein-derived cholesterol metabolism. *J. Biol. Chem.* **266,** 16599–16606.

Liscum, L., and Dahl, N. K. (1992). Intracellular cholesterol transport. *J. Lipid Res.* **33,** 1239–1254.

Liscum, L., and Faust, J. R. (1989). The intracellular transport of low density lipoprotein-derived cholesterol is inhibited in chinese hamster ovary cells cultured with 3-β-[2-(diethylamino)ethoxy]androst-5-en-17-one. *J. Biol. Chem.* **264,** 11796–11806.

Lund-Katz, S., Laboda, H. M., McLean, L. R., and Phillips, M. C. (1988). Influence of molecular packing and phospholipid type on rates of cholesterol exchange. *Biochemistry* **27,** 3416–3423.

McGookey, D. J., Fagerberg, K., and Anderson, R. G. W. (1983). Filipin–cholesterol complexes in uncoated vesicle membrane derived from coated vesicles during receptor-mediated endocytosis of low-density lipoprotein. *J. Cell Biol.* **96,** 1273–1278.

McIntosh, T. J., Simon, S. A., Needham, D., and Huang, C.-H. (1992a). Structure and cohesive properties of sphingomyelin/cholesterol bilayers. *Biochemistry* **31,** 2012–2020.

McIntosh, T. J., Simon, S. A., Needham, D., and Huang, C.-H. (1992b). Interbilayer interactions between sphingomyelin and sphingomyelin/cholesterol bilayers. *Biochemistry* **31,** 2020–2024.

Mendez, A. J., Oram, J. F., and Bierman, E. L. (1991). Protein kinase C as a mediator of high density lipoprotein receptor-dependent efflux of intracellular cholesterol. *J. Biol. Chem.* **266,** 10104–10111.

Nagy, L., and Freeman, D. A. (1990). Effect of cholesterol transport inhibitors on steroidogenesis and plasma membrane cholesterol transport in cultured MA-10 Leydig tumor cells. *Endocrinology* **126,** 2267–2276.

Okazaki, T., Bell, R. M., and Hannun, Y. A. (1989). Sphingomyelin turnover induced by vitamin D_3 in HL-60 cells. *J. Biol. Chem.* **264,** 19076–19080.

Orci, L., Montesano, R., Meda, P., Malaisse-Lagae, F., Perrelet, A., and Vasalli, P. (1981). Heterogenous distribution of filipin–cholesterol complexes across the cisternae of the Golgi apparatus. *Proc. Natl. Acad. Sci. U.S.A.* **78,** 293–297.

Pagano, R. E. (1989). A fluorescent derivative of ceramide: physical properties and use in studying the Golgi apparatus of animal cells. *Meth. Cell Biol.* **29,** 75–85.

Papahadjopoulos, D., Nir, S., and Oki, S. (1971). Permeability properties of phospholipid membranes: Effect of cholesterol and temperature. *Biochim. Biophys. Acta* **266**, 561–583.

Patton, S. (1970). Correlative relationship of cholesterol and sphingomyelin in cell membranes. *J. Theor. Biol.* **29**, 489–491.

Patzer, E. J., Wagner, R. R., and Barenholz, Y. (1978). Cholesterol oxidase as a probe for studying membrane organization. *Nature (London)* **274**, 394–395.

Pentchev, P. G., Brady, R. O., Gal, A. E., and Hibbert, S. R. (1977). The isolation and characterization of sphingomyelinase from human placental tissue. *Biochim. Biophys. Acta* **488**, 312–321.

Pentchev, P. G., Kruth, H. S., Comly, M. E., Butler, J. D., Vanier, M. T., Wenger, D. A., and Patel, S. (1986). Type C Niemann–Pick disease. *J. Biol. Chem.* **261**, 16775–16780.

Phillips, M. C., Johnson, W. J., and Rothblat, G. H. (1987). Mechanisms and consequences of cellular cholesterol exchange and transfer. *Biochim. Biophys. Acta* **906**, 223–276.

Pörn, M. I., and Slotte, J. P. (1990). Reversible effects of sphingomyelin degradation on cholesterol distribution and metabolism in fibroblasts and transformed neuroblastoma cells. *Biochem. J.* **271**, 121–126.

Pörn, M. I., Ares, M., and Slotte, J. P. (1993). Degradation of plasma membrane phosphatidylcholine appears not to affect the cellular cholesterol distribution. *J. Lipid Res.* **34**, 1385–1392.

Pörn, M. I., Åkerman, K. E. O., and Slotte, J. P. (1991a). High-density lipoproteins induce a rapid and transient release of Ca^{2+} in cultured fibroblasts. *Biochem. J.* **279**, 29–33.

Pörn, M. I., Tenhunen, J., and Slotte, J. P. (1991b). Increased steroid hormone secretion in mouse Leydig tumor cells after induction of cholesterol translocation by sphingomyelin degradation. *Biochim. Biophys. Acta* **1093**, 7–12.

Rao, B. G., and Spence, M. W. (1976). Sphingomyelinase activity at pH 7.4 in human brain and a comparison to activity at pH 5.0. *J. Lipid Res.* **17**, 506–515.

Renkonen, O., Gahmberg, C. G., Simons, K., and Kääriäinen, L. (1972). The lipids of the plasma membranes and endoplasmic reticulum from cultured baby hamster kidney cells (BHK-21). *Biochim. Biophys. Acta* **255**, 66–78.

Rodriguez-Lafrasse, C., Rousson, R., Bonnet, J., Pentchev, P. G., Louisot, P., and Vanier, M. T. (1990). Abnormal cholesterol metabolism in imipramine-treated fibroblast cultures. Similarities with Niemann–Pick type C disease. *Biochim. Biophys. Acta* **1043**, 123–128.

Rothblat, G. H., Mahlberg, F. H., Johnson, W. J., and Phillips, M. C. (1992). Apolipoproteins, membrane cholesterol domains, and the regulation of cholesterol efflux. *J. Lipid Res.* **33**, 1091–1097.

Schneider, P. B., and Kennedy, E. P. (1967). Sphingomyelinase activity in normal human spleen and in spleens from subjects with Niemann–Pick disease. *J. Lipid Res.* **8**, 202–210.

Schroeder, F., Jefferson, J. R., Kier, A. B., Knittel, J., Scallen, T. J., Wood, W. G., and Hapala, I. (1991). Membrane cholesterol dynamics: Cholesterol domains and kinetic pools. *Proc. Soc. Exp. Biol. Med.* **196**, 235–252.

Severs, N. J., and Robenek, H. (1983). Detection of microdomains in biomembranes—An appraisal of recent development in freeze-fracture cytochemistry. *Biochim. Biophys. Acta* **737**, 373–408.

Slotte, J. P. (1992). Enzyme-catalyzed oxidation of cholesterol in mixed phospholipid monolayers reveals the stoichiometry at which free cholesterol clusters disappear. *Biochemistry* **31**, 5472–5477.

Slotte, J. P., and Bierman, E. L. (1988). Depletion of plasma-membrane sphingomyelin rapidly alters the distribution of cholesterol between plasma membranes and intracellular cholesterol pools in cultured fibroblasts. *Biochem. J.* **250**, 653–658.

Slotte, J. P., Oram, J. F., and Bierman, E. L. (1987). Binding of high density lipoproteins to cell receptors promotes translocation of cholesterol from intracellular membranes to the cell surface. *J. Biol. Chem.* **262**, 12904–12907.

Slotte, J. P., Chait, A., and Bierman, E. L. (1988). Cholesterol accumulation in aortic smooth muscle cells exposed to LDL–Contribution of free cholesterol transfer. *Arteriosclerosis* **8**, 750–758.

Slotte, J. P., Hedström, G., Rannström, S., and Ekman, S. (1989). Effects of sphingomyelin degradation on cholesterol oxidizability and steady-state distribution between the cell surface and the cell interior. *Biochim. Biophys. Acta* **985**, 90–96.

Slotte, J. P., Härmälä, A.-S., Jansson, C., and Pörn, M. I. (1990a). Rapid turn-over of plasma membrane sphingomyelin and cholesterol in BHK cells after exposure to sphingomyelinase. *Biochim. Biophys. Acta* **1030**, 251–257.

Slotte, J. P., Tenhunen, J., and Pörn, M. I. (1990b). Effects of sphingoymelin degradation on cholesterol mobilization and efflux to high-density lipoproteins in cultured fibroblasts. *Biochim. Biophys. Acta* **1025**, 152–156.

Smutzer, G., and Yeagle, P. L. (1985). Phase behavior of DMPC-cholesterol mixtures—A fluorescence anisotropy study. *Biochim. Biophys. Acta* **814**, 274–280.

Stein, O., Ben-Naim, M., Dabach, Y., Hollander, G., and Stein, Y. (1992). Modulation of sphingomyelinase-induced cholesterol esterification in fibroblasts, Caco-2 cells, macrophages and smooth muscle cells. *Biochim. Biophys. Acta* **1126**, 291–298.

Stevens, V. L., Lambeth, J. D., and Merrill, A. H., Jr. (1986). Use of cytochrome P-450$_{scc}$ to measure cholesterol–lipid interactions. *Biochemistry* **25**, 4287–4292.

Stockton, B. W., and Smith, I. C. P. (1976). A deuterium NMR study of the condensing effect of cholesterol on egg phosphatidylcholine bilayer membranes. *Chem. Phys. Lipids* **17**, 251–263.

Straume, M., and Litman, B. J. (1987). Influence of cholesterol on equilibrium and dynamic bilayer structure of unsaturated acyl chain phosphatidylcholine vesicles as determined from high order analysis of fluorescence anisotropy decay. *Biochemistry* **26**, 5121–5126.

Suckling, K. E., and Stange, E. F. (1985). Role of acyl-CoA:cholesterol acyltransferase in cellular cholesterol metabolism. *J. Lipid Res.* **26**, 647–671.

Sweet, W. D., and Schroeder, F. (1988). Lipid domains and enzyme activity. *In* "Lipid Domains and the Relationship to Membrane Function" (R. C. Aloia, C. C. Curtain, and L. M. Gordon, eds.), pp. 17–42. Liss, New York.

Szabo, G. (1974). Dual mechanism for the action of cholesterol on membrane permeability. *Nature (London)* **252**, 47–49.

Taylor, M. G., Akiyama, T., and Smith, I. C. P. (1981). The molecular dynamics of cholesterol in bilayer membranes—A deuterium NMR study. *Chem. Phys. Lipids* **29**, 327–339.

Theret, N., Delbart, C., Aguie, G., Fruchart, J. C., Vassaux, G., and Ailhaud, G. (1990). Cholesterol efflux from adipose cells is coupled to diacylglycerol production and protein kinase C activation. *Biochem. Biophys. Res. Commun.* **173**, 1361–1368.

Trzeciak, W. H., and Boyd, G. S. (1973). The effect of stress induced by ether anaesthesia on cholesterol content and cholesteryl-esterase activity in rat-adrenal cortex. *Eur. J. Biochem.* **37**, 327–333.

Urbani, L., and Simoni, R. D. (1990). Cholesterol and vesicular stomatitis virus G protein take separate routes from the endoplasmic reticulum to the plasma membrane. *J. Biol. Chem.* **265**, 1919–1923.

Van Heusden, G. P. H., Souren, J., Geelen, M. J. H., and Wirtz, K. W. A. (1985). The synthesis and esterification of cholesterol by hepatocytes and H35 hepatoma cells are independent of the level of nonspecific lipid transfer protein. *Biochim. Biophys. Acta* **846,** 21–25.

Van Meer, G. (1987). Plasma membrane cholesterol pools. *Trends Biochem. Sci.* **12,** 375–376.

Van Meer, G., Stetzer, E. H. K., Wijnaendts-van-Resandt, R. W., and Simons, K. (1987). Sorting of sphingolipids in epithelial (Madin-Darby canine kidney) cells. *J. Cell Biol.* **105,** 1623–1635.

Verkleij, A. J., Zwaal, R. F., Roelofsen, B., Comfurius, P., Kastelijn, D., and Van Deenen, L. L. M. (1973). The asymmetric distribution of phospholipids in the human red cell membrane. A combined study using phospholipases and freeze-etch electron microscopy. *Biochim. Biophys. Acta* **323,** 178–193.

Wattenberg, B. W., and Silbert, D. F. (1983). Sterol partitioning among intracellular membranes—Testing a model for cellular sterol distribution. *J. Biol. Chem.* **258,** 2284–2289.

Wirtz, K. W. A., and Gadella, T. W. J., Jr. (1990). Properties and modes of action of specific and non-specific phospholipid transfer proteins. *Experientia* **46,** 592–599.

Xu, X.-X., and Tabas, I. (1991). Lipoproteins activate acyl-coenzyme A:cholesterol acyltransferase in macrophages only after cellular cholesterol pools are expanded to a critical threshold level. *J. Biol. Chem.* **266,** 17040–17048.

Yeagle, P. L. (1985). Cholesterol and the cell membrane. *Biochim. Biophys. Acta* **822,** 267–287.

Yeagle, P. L. (1989). Lipid regulation of cell membrane structure and function. *FASEB J.* **3,** 1833–1842.

Yeagle, P. L. (1991). Modulation of membrane function by cholesterol. *Biochimie* **73,** 1303–1310.

Yeagle, P. L., and Young, J. E. (1986). Factors contributing to the distribution of cholesterol among phospholipid vesicles. *J. Biol. Chem.* **261,** 8175–8181.

Yeagle, P. L., Martin, R. B., Lala, A. K., Lin, H. K., and Bloch, K. (1977). Differential effects of cholesterol and lanosterol on artificial membranes. *Proc. Natl. Acad. Sci. U.S.A.* **74,** 4924–4926.

CHAPTER 20

Glycosphingolipid Trafficking in the Endocytic Pathway

Jan Willem Kok and Dick Hoekstra
Department of Physiological Chemistry, University of Groningen, 9712 KZ Groningen, The Netherlands

I. INTRODUCTION

Glycosphingolipids are expressed exclusively in the outer leaflet of the plasma membrane, thus facing the extracellular environment (Thompson and Tillack, 1985; Wiegandt, 1985; and references therein). Furthermore, the expression in terms of composition is cell-type specific and may change during cell growth, differentiation, and oncogenic transformation (Hakomori, 1981,1984; Curatalo, 1987). In conjunction with this typical expression, these lipids are thought to play a role in various cellular processes that occur at the cell surface such as cell adhesion, membrane stabilization and curvature, and growth factor receptor modulation. In addition, these lipids are abused by viruses and toxins that use them as binding sites to gain intracellular access (Curatalo, 1987).

In addition to being expressed at the cell surface, glycosphingolipids may reside intracellularly as well. However, no consensus exists about the actual distribution of glycosphingolipids among the plasma membrane and intracellular organelles. The largest fraction often is assumed to be in the plasma membrane, but the opposite, that is, a predominantly intracellular localization, also has been claimed (Van Meer, 1989, and references therein). What could be the reason for their presence inside the cell, given that no distinct intracellular functions have been associated with glycosphingolipids to date? First, they are, of course, synthesized inside the cell. Obviously they will reside at least transiently in organelles of the biosynthetic route: at their site of synthesis in the endoplasmic reticulum (ER) and within the Golgi apparatus and vesicular structures involved in their consecutive transport to the plasma membrane. In addition to this outbound cellular trafficking pathway, which brings about a flow of (lipid) molecules from intracellular organelles to the plasma membrane, the inbound endocytic pathway exists. Along this pathway, molecules flow from the plasma membrane through various maturation stages of endosomes to the lysosomes. The latter pathway is of use to the cell to take up nutrients, to down-regulate receptors, and, in general, to balance the biosynthetic renewal of membrane components. Glycosphingolipids also are subject to endocytic internalization, as is described in a subsequent section in considerable detail. A priori, however, no reason seems to exist for glycosphingolipids to be internalized, since they presumably perform their functions at the cell surface. Obviously, when a cell internalizes a protein (receptor), lipid molecules are internalized concomitantly, since they are an inherent part of the vesicles that are involved in endocytic uptake. Still, this basic membrane-forming function of lipids could be performed by phospholipids alone. In this regard, one must consider that glycosphingolipids usually constitute only a minor (5%) fraction of the total lipid in cellular membranes (Sweeley, 1985; Wiegandt, 1985), leading to the conclusion that they are not simply used as membrane building units, but rather perform specific cell biological functions, as indicated earlier.

What could be the rationale for endocytic glycosphingolipid uptake? Uptake could be specific and may, for instance, be related to down-regulation or, in a more general context, to the regulation of expression of bioactive glycosphingolipids at the cell surface. Alternatively, uptake is nonspecific and simply occurs as a consequence of the pinching off of a plasma membrane domain that contains the lipids. After endocytic uptake, glycosphingolipids could be (partially) degraded in the lysosomes. Again, this event may be functional in down-regulation or it may occur as a consequence of the lipid reaching the lysosomes, where catabolic enzymes are present. No strong analogy exists to protein degradation and renewal in case of malfunctioning due to aging, since lipid molecules are much more simple in structure and probably are not as susceptible to damage and dysfunction as proteins are. In other words, cells have no need for a high turnover rate of lipids for renewal purposes. Moreover, during endocytosis, the rate and extent of membrane internalization are not likely to be matched by the rate of lipid synthesis needed to replenish the plasma membrane lipid pool. Therefore, cells must have recycling pathways at their disposal. Such pathways can prevent molecules from ending up in lysosomes by diverting them from endosomes back to the plasma membrane. Note that recycling is defined here as the interorganellar movement of a molecule followed by its return to the original site, without intervening metabolic processing (Kok *et al.*, 1992). This definition fundamentally differs from that of metabolic recycling (cf. Trinchera *et al.*, 1990). Recycling occurs, for instance, in the endocytic pathway of a receptor–ligand complex, involving internalization and return of the unmodified and intact receptor to the plasma membrane. For protein receptors such as the transferrin receptor, this process can occur very specifically and efficiently. For (glycosphingo)lipids, recycling may occur less specifically since lipids are inherent components of recycling carrier vesicles.

In this chapter, we first outline the events that take place in the endocytic trafficking pathways, followed by a discussion of the participation of glycosphingolipids in these routes. Finally, the concept of endocytic glycosphingolipid uptake will be placed in a broader context of the overall regulation of glycosphingolipid expression at the cell surface in relation to their cell biological functions.

II. ENDOCYTIC TRAFFICKING PATHWAYS

After decades of research, the endocytic trafficking pathways are fairly well documented. Several intracellular organelles have been identified that are involved in the uptake and processing of cell surface molecules (Evered, 1982; Pastan and Willingham, 1985; Mellman *et al.*, 1987; Hoekstra *et al.*, 1989, and references therein; Brown and Greene, 1991). However, some major issues are still to be resolved, several of which are discussed in this section since they bear relevance to the trafficking of (glycosphingo)lipids.

A. Endocytic Uptake Mechanisms

First, at the very beginning of the process, that is, the invagination of a membrane domain resulting in the formation of an endocytic vesicle, several mechanisms seem to operate. In the receptor-mediated endocytic pathway, the invagination takes place at a specific membrane domain, called a coated pit, which is characterized by the presence of a clathrin lattice on the cytoplasmic membrane leaflet. A clathrin-coated vesicle is formed that is (partly) uncoated consecutively by an uncoating ATPase (Rothman and Schmid, 1986; Eskelinen *et al.*, 1991). The uncoated vesicles then fuse with other primary endocytic vesicles or with pre-existing early endosomes. Ligands taken up via this pathway, such as the epidermal growth factor (EGF), low density lipoprotein (LDL), and iron-loaded transferrin, are sorted in early endosomes, followed by their ultimate degradation in the lysosomes or their recycling to the cell surface (Fig. 1).

In addition to this uptake mechanism, at least two others seem to exist (Morré *et al.*, 1983; Steinman *et al.*, 1983; Mellman, 1984)—phagocytosis, by which large particles are taken up by cells (mainly in specialized cells such as leukocytes and macrophages), and fluid-phase endocytosis (or pinocytosis). The latter mechanism involves fluid uptake in endocytic vesicles that invaginate from plasma membrane areas that may be clathrin-coated or non-(clathrin-)coated (see subsequent discussion). In the literature, ample discussion has been presented on whether fluid uptake via noncoated vesicles really occurs or whether all fluid taken up by cells can be accounted for by receptor-mediated endocytosis (Morré *et al.*, 1983; Steinman *et al.*, 1983; Mellman, 1984). The fluid-phase endocytic pathway can be monitored by fluid-phase markers such as horse radish peroxidase (HRP), [³H]sucrose, and Lucifer Yellow (Griffiths *et al.*, 1989). Evidently, these markers do not discriminate between a strict fluid-phase endocytic pathway and the receptor-mediated endocytic pathway, since the latter also involves fluid uptake in the endocytic vesicle content. Therefore, assessing the contribution of either pathway to the total cellular uptake is rather difficult. However, the relative contribution of the strict fluid-phase endocytic uptake to overall endocytosis appears to be cell-type dependent. In baby hamster kidney (BHK) fibroblasts this uptake is con-

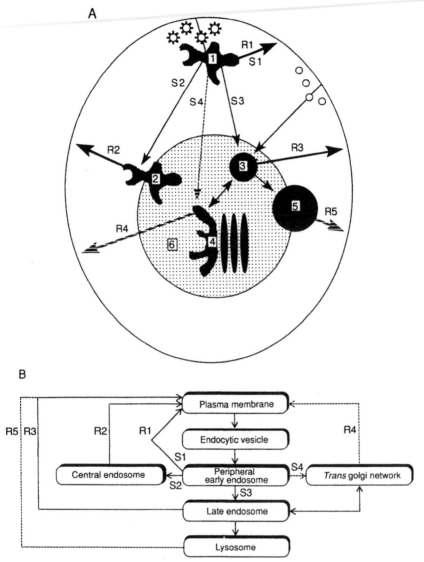

FIGURE 1 Endocytic trafficking pathways. Schematic cellular representation (A) and flow diagram (B) of the endocytic trafficking pathways and intracellular organelles involved in the flow of lipid and protein molecules. S1–S4 represent (hypothetical) sorting pathways from peripheral early endosomes; R1–R5 represent (hypothetical) recycling pathways from various intracellular organelles to the plasma membrane. (A) Molecules are taken up either through coated vesicles or non-(clathrin)-coated vesicles. In the receptor-mediated endocytic pathway, molecules reach peripheral early endosomes (1), from which they can be sorted to central endosomes (2), late endosomes (3), the trans-Golgi network (4), or back to the plasma membrane (recycling). From the late endosomes, molecules can move further to the lysosomes (5) or recycle to the plasma membrane (6; nucleus). (B) Flow diagram between the organelles involved in the receptor-mediated endocytic pathway, as depicted in (A). For details on the flow of specific lipid (Section IV,B) or protein (Section II,B) molecules, see text.

sidered to be of minor importance (Marsh and Helenius, 1980), whereas fluid-phase endocytosis appears substantial in, for instance, A431 epithelial cells (Hopkins *et al.*, 1985; Sandvig *et al.*, 1987).

Nowadays, the more appropriate terminology of "coated" and "non-coated" endocytosis is widely used to distinguish endocytic uptake mechanisms (Montesano *et al.*, 1982; Tran *et al.*, 1987; Hansen *et al.*, 1991). Distinct non-coated endocytic vesicles have been observed with electron microscopy (EM), next to and likely not derived from coated vesicles (Hansen *et al.*, 1991). One can discriminate physiologically between the two pathways either by manipulation of *intracellular* pH or potassium levels or by changing the *extracellular* osmolarity. A decrease of the intracellular pH or a depletion of intracellular K^+ as well as sucrose-induced hyperosmolarity of the extracellular environment results in specific inhibition of clathrin-dependent endocytic uptake, whereas the clathrin-independent pathway apparently is not affected (Sandvig *et al.*, 1987; Heuser and Anderson, 1989; Oka *et al.*, 1989).

Caveolae, which long have been recognized (Yamada, 1955), have received new attention with the identification of the coat protein caveolin (Rothberg *et al.*, 1992). Caveolae originally were described as noncoated membrane invaginations. However, they are in fact plasma membrane domains that are decorated at the cytoplasmic surface with striated coats. This non-clathrin coat has been proposed to function in an alternative coated endocytic pathway and may have a function similar to that of the clathrin coat in controlling membrane curvature and acting as a sink for certain membrane molecules. Although caveolae are implicated in uptake of folate and toxins (Montesano *et al.*, 1982; Tran *et al.*, 1987; Rothberg *et al.*, 1990a), no evidence exists as yet that they actually form endocytic vesicles (Rothberg *et al.*, 1990a,1992). Therefore, the relationship to the noncoated endocytic pathway, as discussed earlier, remains unclear.

In conclusion, no clear picture can be given currently of the involvement of various plasma membrane domains from which differently coated or non-coated endocytic mechanisms in endocytic uptake originate. Many aspects of the interrelation of these mechanisms, their specificity in capturing membrane molecules, and, above all, their physiological significance remain to be resolved. With respect to the latter issue, one may wonder why a cell actually possesses several uptake mechanisms since the endocytic pathways seem to merge rather quickly, resulting in delivery of endocytosed molecules to the same endosomal/lysosomal system (Tran *et al.*, 1987). However, the exact interrelation of the consecutive intracellular pathways and the exact point and extent of merging of the several pathways have not been established and require further study. In this respect, lipid transport studies, which have benefited greatly from results on protein trafficking, also may contribute to a better understanding of these pathways.

B. Endosomal Pathways

We now return to the discussion of the distinct steps of the receptor-mediated endocytic pathway. Just after the pinching off of endocytic vesicles, these vesicles, after (partial) uncoating, appear to become highly fusogenic (Braell, 1987; Gruenberg and Howell, 1987; Mayorga et al., 1988). They fuse either with each other or with pre-existing early endosomes. The early endosome (Fig. 1), located in the cell periphery, is a very important endocytic organelle (Helenius et al., 1983; Hopkins, 1983) and is involved in directing molecules to various destinations (sorting) such as late endosomes, the trans-Golgi network (TGN), or back to the plasma membrane (recycling).

The morphology of the endosomal system has been studied extensively (for example, see Hopkins et al., 1990; Kawai and Hatae, 1991; Tooze and Hollinshead, 1991), especially in BHK cells (Marsh et al., 1986; Griffiths et al., 1989). Inside the cell, molecules can be transported from peripherally located early endosomes to centrally located lysosomes. The transport route may involve several developmental stages of endosomes. Indeed, in BHK cells, endosomes have been localized in the peripheral cytoplasm as well as in the perinuclear region (Marsh et al., 1986). However, the actual transport mechanisms connecting endosomal compartments with each other (and with other organelles) have not been resolved. To date, three different hypotheses exist. Two of these, the so-called "maturation model" and "vesicle shuttle model" originally were proposed by Helenius et al. (1983). A novel hypothesis has been introduced by Hopkins et al. (1990) that can be referred to as the "continuous endosomal network model."

According to the maturation model, early endosomes gradually mature into late endosomes while moving from the cell periphery to the perinuclear region. Maturation involves changes in composition due, on the one hand, to budding of vesicles that causes removal of components from the endosome and, on the other, to fusion of incoming vesicles that results in insertion of new material. In this model, no pre-existing stable early endosomes exist, but they are continuously turning into late endosomes while new early endosomes are being formed from plasma membrane-derived vesicles. The vesicle shuttle model states that stable early and late endosomal structures exist that are connected by shuttling vesicles that transport the molecules to be processed.

This dispute has not been settled definitively to date, as can be inferred from more recent publications, including reviews, in favor of either the vesicle shuttle hypothesis (Gruenberg *et al.*, 1989; Griffiths and Gruenberg, 1991) or the maturation hypothesis (Murphy, 1991; Stoorvogel *et al.*, 1991; Dunn and Maxfield, 1992). To complicate the issue, a novel hypothesis (Hopkins *et al.*, 1990) states that the endosomal system consists of several large interconnected tubular networks, displaying distinct functional regions, through which molecules can flow from a peripheral intracellular location to the perinuclear area. From this perspective, neither separate vesicular compartments nor budding and fusing shuttle vesicles are of importance.

Note that the morphology and the mechanism of functioning of the endosomal system may depend, to a large extent, on the cell type studied. In agreement with this notion are observations by Tooze and Hollinshead (1991) indicating that tubular endosomes can be seen readily in Hep2 cells (cf. Hopkins *et al.*, 1990), PC12 cells, and HeLa cells, but are sparse in 3T3 and BHK-21 cells.

In addition to the different views on the transport mechanisms operating in the endosomal pathway, a widely variable nomenclature exists for the organelles involved. In this chapter, we propose the following terminology, based to a large extent on the transport routes of recycling ligands such as transferrin compared with those of lysosomally directed ligands, as described in the literature (Dickson *et al.*, 1983; Klausner *et al.*, 1983; Hanover *et al.*, 1984; Schmid *et al.*, 1988) and as we observed them in BHK cells (Eskelinen *et al.*, 1991; Kok *et al.*, 1992). The proposed flow pathways are quite consistent with the model of Dunn and Maxfield (Dunn *et al.*, 1989; Dunn and Maxfield, 1990,1992), but a different terminology is preferred to include a distinction in cytoplasmic localization (peripheral versus central) and to address the fact that different endosomal stages display sorting and/or recycling capacity. The early endosome is defined as the organelle that contains transferrin 2 min after triggering its endocytic uptake by raising the temperature from 4°C to 37°C. In BHK and Chinese hamster ovary (CHO) cells, this distinct structure is localized in the cell periphery so we designate it peripheral early endosome (Fig. 1). Compartments similar to the peripheral early endosome have been defined as early endosome (Schmid *et al.*, 1988), receptosome (Pastan and Willingham, 1983), CURL (Geuze *et al.*, 1983), CEV (Eskelinen *et al.*, 1991), and sorting endosome (Dunn and Maxfield, 1990). The peripheral early endosome receives input from plasma membrane-derived endocytic vesicles (Fig. 1).

Transferrin can be transported from these peripheral early endosomes to central endosomes (Fig. 1), located in the juxtanuclear centriolar region of the cell ("cell center"; Dunn and Maxfield, 1990; Eskelinen *et al.*, 1991). The central endosome also has been termed CEV (Eskelinen *et al.*, 1991) or recycling endosome (Dunn and Maxfield, 1990). The peripheral early endosomes are apparently capable of sorting molecules since, in addition to recycling to the plasma membrane, transferrin is directed to central endosomes whereas other ligands such as EGF (Dickson *et al.*, 1983), alpha$_2$-macroglobulin (Goldenthal *et al.*, 1988), asialoglycoprotein (Stoorvogel *et al.*, 1987), LDL (Dunn *et al.*, 1989), and ricin (Kok *et al.*, 1992) are transported to late endosomes and finally end up in lysosomes (Fig. 1; see also Schmid *et al.*, 1988).

The movement of molecules from peripheral early endosomes to central endosomes or late endosomes clearly depends on the microtubular system (Sakai *et al.*, 1991; Kok *et al.*, 1992) and seems to be independent of the intermediate filament and microfilament networks. The transfer of molecules may reflect the movement of the organelle as such along microtubules (Herman and Albertini, 1984; Matteoni and Kreis, 1987; Baba *et al.*, 1991). The input into the peripheral early endosome from the plasma membrane, as well as the recycling from this organelle to the plasma membrane, does not seem to be dependent on cytoskeletal elements at all (Sakai *et al.*, 1991; Kok *et al.*, 1992).

C. Involvement of the Golgi Apparatus

A third issue that is not resolved completely to date is the involvement of the Golgi apparatus in endocytic trafficking routes, in particular the role of the TGN (Griffiths and Simons, 1986). The latter organelle is a very important sorting compartment in the outbound biosynthetic pathway, but also has been proposed to function as a crossover point between the inbound endocytic and the outbound trafficking pathway. In addition to a well-established transport cycle of biosynthetic lysosomal enzymes between the TGN and late endosomes (Kornfeld and Mellman, 1989), an input from the early endocytic pathway into the TGN may exist, followed by a recycling to the plasma membrane (Fig. 1), thus constituting a transport cycle between the plasma membrane and the TGN. Studying the possible interaction between the endocytic and the biosynthetic pathways is complicated by the fact that endosomal and Golgi structures may be found in very close proximity in the cell center, but nevertheless could be functionally distinct (Griffiths and Simons, 1986). Thus, a central endosomal localization of molecules could be misinterpreted as a Golgi localization. However, several studies do report a Golgi-mediated transferrin recycling pathway (Stein and Sussman, 1986; Fishman and Fine, 1987;

Stoorvogel *et al.*, 1988). Especially those studies (Fishman and Fine, 1987; Stoorvogel *et al.*, 1988) showing the existence of TGN vesicles that contain both a newly synthesized protein (as a marker of the biosynthetic pathway) and endocytosed transferrin are convincing in this respect, although reports are not always concurrent (cf. Hedman *et al.*, 1987). Note, however, that the early endocytic trafficking pathway to the TGN is not likely to represent a major pathway, since only an estimated 5% of internalized ricin reaches the TGN via this route (Van Deurs *et al.*, 1988).

D. Recycling

Finally, we want to address the question of recycling in the endocytic pathway, as defined in the introduction (Section I). A priori, recycling pathways could exist starting from all known endocytic compartments to the plasma membrane. No doubt recycling can occur from peripheral early endosomes and central endosomes (Dunn *et al.*, 1989; Sakai *et al.*, 1991; Kok *et al.*, 1992). Transferrin efficiently recycles from peripheral early endosomes, whereas a smaller fraction continues to move inward toward central endosomes and possibly to the TGN, from which organelle(s) the protein also recycles. Thus, in this way transferrin is sorted efficiently from other ligands that move to late endosomes and lysosomes. However, very little is known about the recycling of molecules to the plasma membrane once they have reached late endosomes. Recycling has been proposed to occur even from lysosomes (Van Deurs and Nilausen, 1982; Buktenica *et al.*, 1987), by means of relatively small (<100 nm) vesicles. This suggestion is in contrast with the general idea that this endocytic organelle constitutes a dead end street for trafficking, in which molecules end up only to be degraded.

With respect to recycling from the TGN, this process obviously occurs once a molecule reaches this organelle from the endocytic pathway, since transport from the TGN to the plasma membrane is an inherent part of the biosynthetic pathway. In this "recycling" pathway, remodeling of molecules (repair) by biosynthetic TGN-localized enzymes may occur.

In conclusion, although several issues in endocytic trafficking pathways and underlying mechanisms remain to be resolved, the overall pathways taken by different proteinaceous ligands gradually are becoming understood. In turn, these ligands can be used as markers of the endocytic pathways in studies concerning lipid trafficking. On the other hand, research in lipid trafficking may yield new insights into the flow of membranes involved in endocytosis.

III. METHODOLOGY FOR STUDYING ENDOCYTIC LIPID TRAFFICKING

The methods used to study lipid trafficking are, in principle, very similar to those applied in research on protein (ligand) transport in the endocytic pathway.

A. Endogenous Unlabeled Lipids

First, one can analyze the molecular composition of isolated organelles. From these data, combined with information on the site of synthesis of a specific (protein or lipid) molecule, one can draw conclusions about the flow pathways that are thought to be necessary to achieve such a distribution over the cellular organelles. Evidently, this method does not yield any direct information. Furthermore, it has the disadvantage that it depends highly on the purity of the isolated organelle, since small contaminations of organelles that are enriched in the compound of interest can disturb the result severely (for review, see Van Meer, 1989).

One specific example of such studies (Reynier *et al.*, 1991) has revealed that the plasma membrane of differentiated HT29 cells, a human colon adenocarcinoma cell line, has a specific phospholipid composition compared with the cell homogenate. The choline phospholipids, sphingomyelin and phosphatidylcholine, are relatively enriched whereas the aminophospholipids, phosphatidylserine and phosphatidylethanolamine, are present in relatively lower abundance in the plasma membrane. Since undifferentiated HT29 cells do not display a specific plasma membrane phospholipid composition, these data suggest that a phospholipid sorting mechanism is operational in the differentiated cells.

Another method that allows the study of the flow of endogenous unlabeled glycolipids, in an indirect fashion, is the use of the binding capacity of glycolipids toward toxins, such as cholera or shiga toxin (Gonatas *et al.*, 1983; Sandvig *et al.*, 1991). By following the fate of the labeled toxin, either by fluorescence or by electron microscopy, one indirectly follows the fate of the glycolipid. Obviously, care should be taken in the interpretation of the results of such studies, since the toxin may not bind exclusively to glycolipids but also to glycoproteins, which may follow or even lead to different uptake pathways. Furthermore, the toxin may be released from the glycolipid during its intracellular transport.

Finally, one can perform immunofluorescence studies on the (intracellular) localization of glycolipids with the aid of indirectly fluorescently labeled (for instance, see Butor *et al.*, 1991) or gold-labeled (Van Genderen *et al.*, 1991) specific antibodies against these lipids. One disadvantage of this technique is the inability to monitor the flow of the lipid directly in living cells, which is possible when using fluorescently tagged lipids.

B. Radioactive Tracers

To study the flow of a lipid (or protein) molecule more directly, one must label the compound of interest. Labeling can be done with a radioactive tracer, as has been done extensively for protein ligands such as transferrin to study the kinetics of endocytic uptake and recycling to the plasma membrane (Klausner et al., 1983; Stein and Sussman, 1986). This method can be applied to (glyco)lipids as well. However, in contrast to protein ligands which are water soluble and therefore can be administered to cells in the medium, enabling their binding to specific receptors, (radioactively labeled) lipids generally are poorly water soluble and must be introduced into the lipid bilayer of the plasma membrane. The presence of the labeled lipid as an integral part of the cellular membrane is a prerequisite for studying its genuine flow pathways (Hoekstra and Kok, 1992). In this respect, the more complex glycosphingolipids (gangliosides) have physicochemical properties that allow them to insert spontaneously as monomers into the plasma membrane of cultured cells, where they mix with the pool of endogenous glycosphingolipids and thus function as genuine tracers, as has been shown using electron spin-labeled analogs of gangliosides (Schwarzmann and Sandhoff, 1990).

Following this approach of radiolabeled lipids, the study of their intracellular fate can be performed in several ways. A direct method is autoradiography of EM preparations (Gonatas et al., 1983; Knoll et al., 1988). Alternatively, EM analysis of lipid transport can be done using specific antibodies against various (glycosphingo)lipids (Van Genderen et al., 1991). These EM methods are very promising because they could, in principle, reveal the localization of lipids in great detail. However, the resolution may be limited by the large grain size in autoradiography and the build-up of a large antibody complex in immuno-EM, respectively.

Another method used to follow the fate of isotopically labeled lipids consists of allowing transport processes to take place during a certain time period, followed by lysis of the cells and separation of the organelles of interest. However, in addition to problems in obtaining homogeneous populations of organelles without significant contaminations (see Section III,A), another drawback is related specifically to the nature of lipids. During the homogenization and purification procedure, lipid exchange between organelles may take place, leading to scrambling of the original lipid distribution.

Finally, (glycosphingo)lipid transport routes have been inferred from metabolic studies on lipids, radiolabeled at different positions of the molecule, combined with knowledge of the intracellular localization of their catabolic and anabolic enzymes (Trinchera *et al.*, 1990,1991). The use of inhibitors of certain pathways, or interference with temperature-sensitive transport steps by reducing the incubation temperature, can aid in dissecting the (endocytic) transport route of a specific lipid. Even studying the localization of enzymes, such as glycosyltransferases, per se can lead to models of flow pathways for glycolipids, especially their processing through the Golgi apparatus, where most of these enzymes are located. Thus, ganglioside head group elongation is most likely to occur during flow of the molecule through the Golgi complex in an oriented fashion from *cis* to *trans* (Schwarzmann and Sandhoff, 1990). Further, it has been claimed, based on the processing of radioactive glycolipids, that intact glycolipids can escape from degradation in the endocytic pathway and move to organelles involved in the biosynthetic pathway to be used as precursors for more complex glycolipids (see Section IV).

Metabolic radioactive labeling of glycolipids has been combined elegantly with derivatization of the endogenous plasma membrane pool of these lipids to determine transport kinetics from intracellular biosynthetic organelles to the plasma membrane (Miller-Podraza and Fishman, 1982; for review, see Voelker, 1991). Unfortunately, these metabolic labeling studies are limited to biosynthetic transport studies.

C. Fluorescent Markers

The use of fluorescent groups as markers, covalently attached to lipids, has become a popular tool in lipid trafficking research (Pagano and Sleight, 1985; Hoekstra and Kok, 1992; Kok and Hoekstra, 1993). This type of labeling offers the great advantage that it allows direct visualization of the (endocytic) fate of the lipid by fluorescence microscopy. Furthermore, the flow of the lipid can be related directly to its metabolic fate, which also can be analyzed easily with these fluorescently tagged lipids. The most widely used fluorescent lipid analogs are those containing the fluorescent group 7-nitrobenz-2-oxa-1,3-diazol-4-yl (NBD) attached via a short-chain (C_6) fatty acid to the hydrophobic lipid backbone (C_6-NBD-lipid; Kok and Hoekstra, 1993), first introduced by Pagano and co-workers (Struck and Pagano, 1980; Pagano and Sleight, 1985). However, other fluorescent lipid analogs have been used in transport studies (for instance, see Spiegel *et al.*, 1984; Huang and Dietsch, 1991; Van't Hof *et al.*, 1992; Vos *et al.*, 1992). The C_6-NBD-lipids are particularly suitable for endocytic uptake studies, notwithstanding the importance of C_6-NBD-lipid precursors such as C_6-NBD-ceramide in the investigation of biosynthetic lipid transport (Lipsky and Pagano, 1985; Van Meer et al., 1987). The NBD analogs combine the characteristics of reasonable water solubility (in contrast to most natural lipids) and high hydrophobic partition coefficients. As a result, they readily insert into (cellular) membranes, but also can exchange easily between membranes. Thus, they can be inserted into the outer leaflet of the plasma membrane of the cell to become an integral part of this membrane (see discussion in Section III,B). When insertion is done at 2°C, when all intracellular transport processes are arrested, subsequent instant warming of the cells to 37°C allows controlled monitoring of the endocytic fate of the NBD-lipid. Furthermore, the C_6-NBD-lipid can be removed from the plasma membrane by a so-called back-exchange, using either liposomes (Struck and Pagano, 1980) or bovine serum albumin (BSA) (Kok et al., 1989) as lipid scavenger. This procedure distinguishes the plasma membrane NBD-lipid pool from the intracellular pool, which is very helpful in determining internalization kinetics, in better visualizing intracellular labeling patterns by fluorescence microscopy, and in determining the occurrence and extent of lipid recycling from intracellular organelles back to the plasma membrane (Kok et al., 1989,1992).

Finally, the use of these fluorescent lipids can be combined with the use of fluorescently tagged protein ligands to perform double-labeling studies to identify intracellular lipid transport pathways, based on the knowledge of previously identified protein ligand pathways (for instance, those of transferrin; Kok et al., 1989,1992; Koval and Pagano, 1989).

One obvious disadvantage of the C_6-NBD-lipid approach is the inability to study the possible effects of fatty acid heterogeneity on processes such as sorting and recycling. Some reports propose fatty acyl chain length as a lipid sorting signal (Bertho et al., 1991; Longmuir and Haynes, 1991; Zaal et al., 1994).

The discussion of endocytic lipid trafficking in Section IV is based largely on results obtained with C_6-NBD derivatives of the more simple neutral (glyco)sphingolipids such as glucosylceramide and sphingomyelin, since these studies offer the most detailed picture to date of lipid flow pathways in relation to labeled ligands as markers for these pathways.

D. Photoaffinity Labels and Cell-Free Systems

Although a reasonable amount of information is now available on the (endocytic) trafficking pathways of (glyco)sphingolipids (see Section IV), not much is known about the mechanisms behind processes such as sorting and recycling and about factors involved in the regulation of lipid flow, in general. Now that the framework of lipid trafficking is more or less set, the resolution of these basic questions forms the challenge of future research in this field. The experimental setup that is used in this respect may include the following approaches.

First, regulatory proteins may be involved in the transport of lipids. In the endocytic pathway, interactions between (glyco)sphingolipids on the one hand and catabolic enzymes and so-called activator proteins, which perform a helper function in the breakdown of lipids, on the other have been studied in considerable detail, especially in relation to lipid storage diseases (Koval and Pagano, 1991; O'Brien and Kisimoto, 1991; Sandhoff *et al.*, 1992). In analogy, one also can envision a role for specific proteins in processes such as sorting and recycling of lipids. A promising method by which to study potential functional interactions between lipids and proteins involves the use of protein-reactive photoaffinity labels (Sonnino *et al.*, 1989,1992; Rosenwald *et al.*, 1991). To this end, (glycosphingo)lipids are derivatized in such a way that they contain both a radioactive marker and a light-sensitive reactive group. These lipids can be inserted into the plasma membrane of cells, analogous to fluorescent lipids, followed by internalization along the endocytic pathway. After controlled lipid uptake in the dark, the photosensitive probe is activated by UV light, after which the activated intermediate will react with and covalently attach to neighboring molecules. Thus, protein molecules that perform putative regulatory functions in lipid trafficking and, therefore, must be in close proximity to these lipid molecules may become radioactively labeled. Obviously, other proteins such as metabolic enzymes also will become labeled.

A second approach to studying mechanisms of lipid transport and the factors involved includes the use of cell-free systems, analogous to the extensive research that has been undertaken in recent years in the field of intracellular fusion (Goda and Pfeffer, 1989; Gruenberg and Howell, 1989). These studies have shown that various fusion events in the endocytic pathway are ATP dependent and require cytosolic proteins. Sensitivity of fusion to GTPγS has led to the discovery of the involvement of various GTP-binding rab proteins in endocytic trafficking (Mayorga *et al.*, 1989; Chavrier *et al.*, 1990). Transferrin recycling has been shown to require ATP and cytosol (Podbilewicz and Mellman, 1990); the GTP-binding protein rab4 is involved in controlling the transferrin recycling pathway (Van der Sluijs *et al.*, 1992). Similar mechanisms may operate in the control of endocytic uptake and recycling of (glycosphingo)lipids (cf. Wattenberg, 1990).

Finally, in resolving the underlying mechanisms of lipid transport pathways, studies of cells from patients with inherited lipid storage diseases (cf. Koval and Pagano, 1990) or of mutant cell lines with a malfunctioning sorting or recycling machinery, as seems to be the case in cells overexpressing rab proteins (Van der Sluijs *et al.*, 1992), will be worthwhile.

IV. ENDOCYTIC GLYCOSPHINGOLIPID TRAFFICKING

As indicated in Section III, the description of the endocytic glycosphingolipid trafficking pathways is based largely on studies carried out with C_6-NBD derivatives of the more simple neutral (glyco)sphingolipids. (Glyco)sphingolipids have a hydrophobic backbone called ceramide that is embedded in the lipid monolayer of the membrane. To this backbone, a polar head group is attached that is located at the lipid–water interface of the membrane. In the case of sphingomyelin, the head group is phosphorylcholine, as in phosphatidylcholine. All other sphingolipids have one or more (neutral or acidic) sugar residues linked to the ceramide [glyco(sphingo)lipids]. Neutral glycolipids have only neutral sugar residues such as glucose, galactose, and *N*-acetyl galactosamine, whereas acidic glycolipids contain, in addition to neutral sugars, one or more *N*-acetyl neuraminic acid groups (gangliosides) or sulfate groups (sulfatides). One of the most simple glycolipids is glucosylceramide, which contains a glucose moiety as the head group. By the addition of an increasing number of sugar residues, the glycolipids become very complex; collectively about 300 different types have been identified to date. [For more information on the structure of these lipids the reader is referred to several reviews (Schwarzmann and Sandhoff, 1990; Koval and Pagano, 1991; Hoekstra and Kok, 1992) that deal with the metabolism and intracellular transport of sphingolipids, and to several textbooks (Sweeley, 1985; Wiegandt, 1985).]

The ceramide backbone contains a sphingosine moiety to which a fatty acid is attached via an amide linkage. The fatty acid can be replaced by a C_6-NBD fatty acid to yield a fluorescently labeled sphingolipid (for detailed methodology, see Kok and Hoekstra, 1993). This fluorescent labeling of the sphingolipid allows one to follow the fate of the lipid directly during endocytic uptake (see also Section III).

A. Endocytic Uptake Mechanisms

It has been concluded that glycolipids are internalized by non-coated endocytosis, based on observations that cholera toxin and tetanus toxin are internalized via this pathway while it is known that these toxins bind to cell surface G_{M1} and di- and trisialogangliosides, respectively (Montesano *et al.*, 1982; Critchley *et al.*, 1985; Curatalo, 1987; Tran *et al.*, 1987; Pacuszka and Fishman, 1992). Hence, these gangliosides seem to be internalized via noncoated membrane domains. Furthermore, several indirect lines of evidence may lead to the conclusion that glycolipids are involved in the caveolar uptake mechanism.

Glycosphingolipids are found to be enriched in specialized membrane domains in association with glycosyl-phosphatidylinositol (GPI)-anchored proteins (Brown and Rose, 1992) as well as with cholesterol (for review, see Van Meer, 1989). First, the folate receptor, which is a GPI-anchored membrane protein, is associated with caveolae (Rothberg *et al.*, 1990a). Second, cholesterol plays a critical role in maintaining the caveolar membrane domain (Rothberg *et al.*, 1990b). Thus, in addition to noncoated endocytic uptake, glycolipids may be implicated in the caveolar uptake mechanism. However, note that the interrelationship between noncoated membrane domains and caveolae, as well as actual endocytic uptake into the cell through caveolae, has not been established (see also Section II,A). These two mechanisms may be mutually exclusive or (partly) overlapping.

Studies on the internalization of another glycolipid-binding toxin, shiga toxin, have indicated that (other) glycolipids are endocytosed from coated pits and, thus, follow the receptor-mediated endocytic pathway, since the toxin itself follows this pathway (Sandvig *et al.*, 1989). In this case, the neutral glycolipid globotriaosylceramide, seems to function as the toxin binding site. However, more direct evidence exists for internalization of neutral glycolipids along the (clathrin-dependent) pathway of receptor-mediated endocytosis. Fluorescently labeled glucosylceramide has been shown to be internalized together with (fluorescent) transferrin, a marker of this pathway. After 2 min of uptake in BHK cells, both the lipid and the protein are located in peripheral early endosomes (Kok *et al.*, 1989). This uptake pathway for glucosylceramide in these cells is specific, since uptake does not occur via another non-receptor-mediated endocytic pathway that can be probed with *N*-(lissamine rhodamine B sulfonyl)phosphatidylethanolamine (*N*-Rh-PE; Kok *et al.*, 1990,1992).

Along the latter pathway, the uptake of the fluid-phase marker Lucifer Yellow also occurs. The initial invagination step may occur at a noncoated membrane domain or a caveola. *N*-Rh-PE appears to be present as small clusters in the plasma membrane (Kok *et al.*, 1990), which could explain its selective sorting to and internalization from non(clathrin-)coated membrane domains. This process is analogous to the fate of the GPI-anchored folate receptor (Rothberg *et al.*, 1990a), which clusters independent of its association with any membrane specialization while clusters become associated with caveolae, despite the absence of a membrane-spanning region and a cytoplasmic tail (cf. *N*-Rh-PE), which is thought to be relevant for the trapping of receptors in clathrin-coated pits.

The receptor-mediated endocytic pathway followed by glucosylceramide can be generalized to include other neutral glycolipids (Kok et al., 1991) as well as sphingomyelin (Koval and Pagano, 1989). Thus, from the foregoing observations, one may come to the conclusion that the more simple neutral (glyco)sphingolipids are internalized along the pathway of receptor-mediated endocytosis, whereas the more complex gangliosides are taken up through non-(clathrin-)coated membrane domains. However, such a conclusion relies on the limited work performed to date that supports such a direction. Some arguments are in contradiction with this simple conclusion. First, gangliosides have been shown to be an inherent part of isolated endocytic coated vesicles (Alfsen et al., 1984; Gravotta and Maccioni, 1985). Furthermore, using EM, exogenous derivatized gangliosides have been shown to be subject to endocytosis via coated pits after insertion into the outer leaflet of the plasma membrane (Schwarzmann and Sandhoff, 1990). These studies thus implicate the receptor-mediated endocytic pathway in the uptake of gangliosides. Another contention that is difficult to reconcile with the preceding conclusion is that the neutral glycolipids are thought to be segregated into small size microdomains in both model systems and biological membranes, whereas gangliosides are not (Thompson and Tillack, 1985).

In conclusion, more comparative studies on the internalization mechanisms involved in the uptake of various (glyco)sphingolipids, in combination with specific markers for the uptake mechanisms, are necessary, to establish rigorously the specificity of sphingolipid uptake pathways.

B. Endosomal Pathways

In fibroblasts, the intracellular trafficking of the neutral sphingolipids glucosylceramide and sphingomyelin through the endosomal pathway has been studied in detail, using C_6-NBD analogs (Kok et al., 1989,1992; Koval and Pagano, 1989,1990).

1. Glucosylceramide

Glucosylceramide is taken up into BHK cells along the receptor-mediated endocytic pathway, with transferrin, and is found in peripheral early endosomes after 2 min of incubation at 37°C (Fig. 1; Kok et al., 1989). The coated and uncoated endocytic vesicles that constitute the intermediate stages between the plasma membrane and peripheral early endosomes probably are too small to be discerned by fluorescence microscopy. The localization of C_6-NBD-glucosylceramide in distinct vesicular structures in the cell periphery indicates that, indeed, no large interconnected endosomal networks exist in these cells (cf. Section II,B), since the laterally mobile lipid would distribute immediately throughout such a system.

Beyond this early endosomal stage of endocytosis, the glycolipid follows several routes and partly segregates from the marker transferrin. A small percentage of the glucosylceramide is transported with transferrin to the Golgi area, which is a distinct juxtanuclear area in BHK cells where, besides the Golgi apparatus, the centrioles and central endosomes reside (cell center). At the level of fluorescence microscopy, resolution is too low to be able to localize the glycolipid in either the Golgi apparatus or the central endosomes. Colocalization with transferrin is, in this respect, not conclusive since this marker has been reported to traverse both compartments (see Sections II,B, II,C, and IV,D for further discussion on the Golgi involvement in endocytic glycosphingolipid trafficking). Furthermore, from peripheral early endosomes, another part of the glucosylceramide fraction is found to be transported to large perinuclear late endosomes (Fig. 1), in conjunction with the endocytic marker ricin. Interestingly, N-Rh-PE, which initially is taken up into the cell through a clathrin-independent mechanism separate from transferrin (Section IV,B), also is transported to the late endosomes. These studies on endocytic lipid trafficking (Kok et al., 1990,1992) indicate that the late endosomal compartment is the merging point of the clathrin-dependent and the clathrin-independent pathways (cf. Section II,A). N-Rh-PE and ricin move further from the late endosomes to the lysosomes. However, glucosylceramide is not found in the latter compartment and apparently can escape from this organelle. Indeed, prior to reaching the lysosomes, the lipid recycles to the plasma membrane.

The transport of glucosylceramide from peripheral early endosomes to both the Golgi area and the late endosomes depends on an intact microtubular system (cf. Section II,B). When this cytoskeletal system is perturbed with nocodazole, glucosylceramide does not move beyond the peripheral early endosomal level, even after prolonged incubation (30 min) at 37°C. These peripheral endosomes continuously receive new plasma membrane-derived material, since they still can become labeled by newly presented transferrin when the ligand is allowed to be taken up during only 2 min. The transport block is reversible and transport of glucosylceramide to the cell center is resumed after reestablishment of the microtubular system. In contrast to microtubules, the central movement of the glycolipid, or any other of its endocytic transport routes, does not seem to depend on actin filaments (Kok et al., 1992). However, associations of glycosphingolipids with different types of intermediate filaments in various cell types, including fibroblasts, have been reported (Gillard et al., 1991,1992), suggesting a role of intermediate filaments in intracellular transport and possibly sorting of glycosphingolipids; such a role, however, remains to be established. Unfortunately, no drugs are available to disrupt these cytoskeletal systems, making the study of their functional involvement difficult.

The endocytic fate of C_6-NBD-glucosylceramide in BHK cells involving peripheral early, central, and late endosomes cannot be generalized to all other cell types. In *undifferentiated* HT29 human colon epithelial tumor cells intact glucosylceramide has been shown to be directed to the Golgi apparatus after endocytic uptake (Kok *et al.*, 1991). This pathway appears to be specific for both this glycolipid and this cell type. Thus, in the same cell type, other neutral glycolipids and sphingomyelin are transported to the endosomal/lysosomal system. Moreover, in the *differentiated* counterpart of these HT29 cells, all sphingolipids including glucosylceramide follow the same pathway to the endosomal/lysosomal system. The specific sorting of glucosylceramide is the first "real" lipid sorting event (cf. Section IV,D) that has been reported to occur in the endocytic pathway. In the biosynthetic pathway in polarized Madin–Darby canine kidney (MDCK) cells, glucosylceramide also has been shown to be sorted from sphingomyelin, although in this case the glycolipid is directed specifically to the apical plasma membrane domain of the cells (Van Meer *et al.*, 1987). Note, however, that the endocytic trafficking pathways of sphingolipids in epithelial cells have not yet been established in as much detail as those in fibroblasts.

2. Sphingomyelin

In Chinese hamster V79 cells, sphingomyelin is enriched in endosomes relative to the plasma membrane, suggesting that the lipid is taken up abundantly by endocytosis (Urade *et al.*, 1988). In CHO cells, the endocytic fate of C_6-NBD-sphingomyelin has been studied in detail (Koval and Pagano, 1989); its fate is very similar to that of C_6-NBD-glucosylceramide in BHK cells. Also, colocalization with transferrin is observed in the cell center. Importantly, in these cells it could be shown that the sphingolipid and transferrin are confined mostly to central endosomes in the centriolar region, whereas the Golgi apparatus displays a distinct perinuclear localization. On treatment with nocodazole, sphingomyelin was found only in peripheral early endosomes, as inferred from an analogy with glucosylceramide in BHK cells. Appearance of C_6-NBD-sphingomyelin in late endosomes was not reported, but transport to the lysosomes was studied elegantly in Niemann–Pick Type A (NP-A) cells in a comparative study with normal human skin fibroblasts (Koval and Pagano, 1990). In these NP-A cells, which lack the lysosomal sphingomyelinase, sphingomyelin accumulated in the lysosomes. This result contrasts with the situation in normal cells, in which no lysosomal staining was observed. It was concluded that transport of C_6-NBD-sphingomyelin to the lysosomes also occurs in normal cells, but that fast degradation prevents accumulation to a degree that can be visualized. Obviously, sphingomyelin is not transported massively along this pathway to the lysosomes. Indeed, the rate of transfer to this compartment was calculated to be approximately 20-fold slower than the rate of sphingomyelin recycling. The results obtained with the endocytic trafficking of glucosylceramide in BHK cells are, again, in agreement with this finding since in these cells also the lysosomally directed pathway seems to be a very minor route compared with the recycling pathway.

A priori, the lysosome is not necessarily the only compartment in which sphingomyelin breakdown can take place. In addition to a soluble lysosomal acid sphingomyelinase (A-SMase), which is deficient in the Niemann-Pick Type A and B sphingomyelin storage diseases, several cell types contain a membrane bound Mg^{2+}-dependent neutral sphingomyelinase (N-SMase) that is located predominantly at the cell surface (Koval and Pagano, 1991). Thus, sphingomyelin hydrolysis already may take place at the plasma membrane, as is observed for C_6-NBD-sphingomyelin when incubated with cells at 7°C, at which temperature endocytosis is still arrested (Koval and Pagano, 1989,1990). In this respect, note that proper insertion of the sphingolipid into the plasma membrane (cf. Section III,B) is a prerequisite for involvement of the lipid in intracellular membrane flow and for its serving as a substrate for the membrane-bound N-SMase. Indeed, when sphingomyelin is fed to cells as part of a lipoprotein complex, it primarily reaches the lysosomes, where the lipid is degraded by the A-SMase (Levade et al., 1991a,b). An analogy may exist with the breakdown of glucosylceramide. In this case, in addition to the lysosomal glucocerebrosidase that is deficient in the glucosylceramide storage disease Gaucher, a distinct second glucocerebrosidase appears to exist that is more tightly membrane-associated and has a higher pH optimum (Van Weely et al., 1993). This enzyme does not seem to be localized in the plasma membrane, but resides in an endocytic compartment that is reached rapidly by the substrate glucosylceramide during endocytosis. In this context, note that we have observed some hydrolysis of C_6-NBD-glucosylceramide at 37°C in nocodazole-treated cells, in which the glycolipid does not move beyond the early peripheral endosomal compartment (J. W. Kok and D. Hoekstra, unpublished observations).

3. Gangliosides

The intracellular endocytic transport routes followed by gangliosides have not been documented in the literature. In one review (Schwarzmann and Sandhoff, 1990), the use of NBD derivatives of the gangliosides G_{M3},

G_{M2}, G_{M1}, and G_{D1a} is reported. All these gangliosides as well as NBD-lactosylceramide predominantly label lysosomes. Experimental evidence to support such a localization, or the transport pathways as such, were not described. However, as discussed in the same review and elsewhere (Van Meer, 1989), cellular gangliosides have very long half-lives (10–50 hr) and, therefore, do not seem to reach the lysosomes efficiently. Thus, also gangliosides may be prone to rather efficient recycling between endosomes and the plasma membrane and/or are exluded efficiently from internalization at the plasma membrane.

C. Involvement of the Golgi Apparatus

Several different types of study have revealed that the endocytic trafficking pathway through the Golgi apparatus in fibroblasts is quantitatively of minor importance and most likely involves only the TGN. First, with the NBD analogs of sphingomyelin and glucosylceramide, this has been shown directly; C_6-NBD-sphingomyelin traverses peripheral early endosomes and central endosomes, which in CHO cells are physically well separated from the Golgi complex. A large fraction recycles, whereas a minor fraction is transported further to lysosomes. Recycling is not affected by monensin, a drug that disrupts the Golgi apparatus and results in inhibition of transport from the Golgi apparatus to the plasma membrane. In contrast, the transport of newly synthesized C_6-NBD-sphingolipids (in the Golgi apparatus) to the plasma membrane is inhibited by monensin. During endocytosis, some C_6-NBD-sphingomyelin is degraded to C_6-NBD-ceramide, which accumulates in the Golgi apparatus where it is used as a sphingolipid precursor. The transport of the products, among which is newly synthesized C_6-NBD-sphingomyelin, to the plasma membrane again is susceptible to monensin inhibition (Koval and Pagano, 1989). In BHK cells, the central endosomes are not well separated from the Golgi apparatus in the Golgi area. However, again no extensive inhibition by monensin of the recycling of glucosylceramide is seen. Furthermore, brefeldin A, which is known to cause drastic morphological changes in the Golgi apparatus due to redistribution of the *cis*- to *trans*-Golgi compartments into the endoplasmic reticulum (Lippincott-Schwartz *et al.*, 1990), does not have any effect on the distribution pattern of glucosylceramide processed along the endocytic pathway. Thus, the *cis*- to *trans*-Golgi compartments do not seem to be involved in trafficking of endocytosed glucosylceramide.

Second, from experiments on quantitation of intracellular toxin distribution on endocytosis, one can infer the fate of the glycolipid to which the toxin is bound (Section III,A). Of the total fraction of internalized shiga toxin and ricin toxin, 10% and 5%, respectively, are found in the Golgi complex (Van Deurs *et al.,* 1988; Sandvig *et al.,* 1991). In the case of ricin, ~75% of the Golgi-associated fraction was present in the TGN. Thus, if these data are extrapolated to the lipids to which the toxins bind, only small fractions of internalized neutral glycolipids can reach the (*trans*) Golgi (network).

Third, indirect support for a limited participation of the Golgi in endocytic glycolipid flow comes from experiments in which radioactively labeled gangliosides were administered to cultured fibroblasts (Schwarzmann and Sandhoff, 1990). Only a small fraction of these (precursor) gangliosides was observed to be glycosylated directly, resulting in more complex gangliosides. This result indicates that the administered ganglioside probably is integrated into the plasma membrane and internalized; in addition to being degraded in the lysosomes or recycled intact to the cell surface, a small fraction of ganglioside is directed to the Golgi apparatus where it is modified further. Import into the Golgi probably occurs in a late Golgi compartment, since exogenous G_{M2} gives rise to direct formation of G_{D1a}, whereas the less glycosylated G_{M3} does not. Given the model that higher glycosylated gangliosides are synthesized and used as precursors in more distal Golgi compartments, G_{M3} entry into a late Golgi compartment would not lead to G_{M2} synthesis and, consequently, not to G_{D1a} formation. Furthermore, a small amount of internalized NBD-G_{M1} is glycosylated to NBD-G_{D1a} in the presence of brefeldin A, suggesting again that import occurs in a late Golgi compartment (TGN).

It is generally assumed that endocytic transport to the Golgi apparatus originates from a donor organelle at the endosomal level rather than involving a connection via the lysosome. The only catabolite that is thought to leave the lysosome, heading for the Golgi, is ceramide, the transport of which may not be vesicular (see Hoekstra and Kok, 1992, for further discussion).

Some additional comments about the Golgi involvement in glycosphingolipid trafficking are warranted. It has been inferred (Trinchera *et al.,* 1990,1991) that, when formed as a degradation product from administered more complex glycolipids (either G_{M1} or lactosylceramide) in rat liver cells, glucosylceramide can escape further catabolism and move from the lysosomes to the Golgi apparatus, where it can be used as a precursor for ganglioside biosynthesis. Thus, glucosylceramide not only leaves the lysosome intact, but also enters an early Golgi compartment, where glycosyltransferases for its anabolism reside. In (undifferentiated) HT29 cells, endocytosed C_6-NBD-glucosylceramide has directly been shown to be transported preferentially to the Golgi apparatus (Section IV,B,1). In cultured oligodendrocytes, a fluorescent analog of sulfatide (i.e., fatty acyl labeled with rhodamine) has been claimed to follow an endocytic pathway leading first, via endocytic vesicles, to the Golgi apparatus and subsequently to the lysosomal system. Here, breakdown to rhodamine-ceramide would occur, which again flows back to the Golgi apparatus, where it is used for biosynthesis of sphingomyelin (Vos *et al.*, 1992). In conclusion, the extent of the involvement of the Golgi apparatus in endocytic (glyco)sphingolipid trafficking may be highly variable among different cell types.

D. Recycling

Recycling in the endocytic pathway, as defined in the introduction (Section I)—involving internalization and return of the intact unmodified molecule to the plasma membrane—first was shown directly for the C_6-NBD derivatives of sphingomyelin and glucosylceramide (Kok *et al.*, 1989; Koval and Pagano, 1989). These sphingolipids follow intracellular endocytic pathways similar to those of transferrin, a protein that is known to be recycled efficiently to the plasma membrane from both peripheral early and central endosomes. In addition to participating in these recycling pathways, some of the C_6-NBD-glucosylceramide is transported to late endosomes, from which it also appears to be able to recycle to the plasma membrane, in contrast with, for instance, the endocytic markers ricin and *N*-Rh-PE, which are transported further to the lysosomes. In accordance with the more extensive trafficking pathways of glucosylceramide compared with transferrin, the half-time for recycling of the glycolipid in BHK cells is longer (12.5 min; Kok *et al.*, 1992) than that for the very efficient recycling of transferrin (5 min; Eskelinen *et al.*, 1991).

Small amounts of C_6-NBD-glucosylceramide may be transported to the lysosomes, in analogy to C_6-NBD-sphingomyelin (Section IV,B,2) and as has been reported to occur for the nondegradable NBD-glucosylthioceramide (Schwarzmann and Sandhoff, 1990). Whether recycling from the lysosomes to the plasma membrane occurs (cf. Section II,D), cannot be determined from the available data. However, this process seems unlikely given the apparent high rate of breakdown of the sphingolipid in normal cells. After degradation, an efficient efflux of the resulting C_6-NBD-ceramide probably occurs that precludes observation of lysosomal labeling. As noted earlier for NP-A cells, in breakdown-deficient cells accumulation of sphingolipid in the lysosomes occurs.

The picture that emerges from these studies is one of a very efficient recycling of C_6-NBD-glucosylceramide after internalization, with the involvement of several endosomal compartments (Fig. 1). The question arises whether "true" sorting phenomena are involved in these recycling events or, more generally, in the entire endocytic trafficking of (glyco)-sphingolipids. A discussion of this theme should involve bulk flow estimates of the endocytic (recycling) pathways. Thus, in the biosynthetic pathway, endogenous glycosphingolipids have been considered to move to the cell surface, after their synthesis, at a rate consistent with bulk flow (Young *et al.*, 1992). Koval and Pagano (1991) discuss their results in terms of sorting of sphingomyelin between the degradative pathway to the lysosomes and the recycling pathway back to the plasma membrane. No absolute data on bulk flow in these two pathways are available. These authors state, however, that efficient recycling of sphingomyelin is in accordance with models based on endosome morphology, which propose that the majority of the endosomal membrane is returned to the cell surface. In the case of C_6-NBD-glucosylceramide, internalization and recycling result in a distribution of the NBD-glycolipid between the plasma membrane and intracellular compartments that is in reasonable agreement with measurements of the surface area of the plasma membrane versus that of the endosomal apparatus in BHK cells, indicating that under steady state conditions the glycolipid is distributed approximately according to available surface area (Kok *et al.*, 1992). This result suggests that this sphingolipid also is transported according to bulk flow. On the other hand, at the level of the late endosome, two membrane-inserted fluorescent lipids—C_6-NBD-glucosylceramide and N-Rh-PE—apparently are sorted from each other, the first recycling and the second moving quantitatively to the lysosomes.

It has been reported that two topologically and metabolically separate pools of sphingomyelin exist in BHK cells (Quin and Allan, 1992). One of the pools would be the endocytic/recycling pool, residing in the plasma membrane and endosomes. The other pool probably resides in the endoplasmic reticulum (ER)/Golgi, where newly synthesized sphingomyelin molecules could be sorted from those transported to the plasma membrane to retain them in the Golgi or move them back to the ER. Whether similar distinct pools exist in other cell types and for other sphingolipids, for example, glucosylceramide in BHK cells, remains to be established.

Clearly, in (undifferentiated) HT29 cells a true sorting event occurs between C_6-NBD-sphingolipids, most of them (including sphingomyelin) being processed along pathways that are probably similar to those described for glucosylceramide in BHK cells and sphingomyelin in CHO cells. However, in undifferentiated HT29 cells, C_6-NBD-glucosylceramide is sorted preferentially to the Golgi apparatus from which it can recycle to the cell surface. This maybe a salvage pathway for this specific glycolipid, given the observation that glucosylceramide is synthesized more abundantly from the precursor C_6-NBD-ceramide in the undifferentiated cell type than in the differentiated counterpart.

In conclusion, in some cell types and possibly in fibroblasts in general, (glyco)sphingolipids move according to bulk flow in the endocytic pathway, whereas in other cell types (epithelial cells), specific sorting events between sphingolipids occur, both in the endocytic and in the biosynthetic pathway.

V. ENDOCYTIC GLYCOSPHINGOLIPID TRAFFICKING AND SIGNAL TRANSDUCTION

Internalization of (glyco)sphingolipids during endocytosis could be considered nonspecific uptake of lipid molecules that happen to be in an invaginating membrane domain. Alternatively, this event can be viewed in the context of the overall regulation of (glyco)sphingolipid expression and distribution. In addition to the biosynthetic activity of sphingolipid synthesizing enzymes, the regulation of transport to the plasma membrane, lipid turnover at the plasma membrane (in relation to signal transduction), and the removal of lipids from the plasma membrane (down-regulation) may serve as mechanisms by which to regulate the plasma membrane pool size. Regulation of this pool is obviously of great importance with respect to the cell biological functions of these lipids (see Section I). Many different sphingolipids have been proposed to function at various levels of signal transduction at the plasma membrane, either directly (as receptors or modifiers of growth factor receptors, protein kinases, and so on) or indirectly through the action of their breakdown products such as lysosphingolipids, ceramide, and sphingosine (Hannun and Bell, 1989; Merrill and Jones, 1990; Merrill, 1991; Ballou, 1992; Kolesnick, 1992).

The endocytic uptake of both sphingomyelin and glucosylceramide, which are described extensively in previous sections, also may be of functional significance as a mechanism of down-regulation, since both sphingolipids have been implicated in cellular signal transduction. Sphingomyelin can be degraded in the plasma membrane by the neutral sphingomyelinase to yield the bioactive molecule ceramide. Ceramide, in turn, can be degraded by a plasma membrane ceramidase to yield sphingosine which, among other functions, inhibits protein kinase C. The head group of phosphatidylcholine can be transferred to ceramide, resulting in the resynthesis of sphingomyelin (which was degraded originally) and diacyl-glycerol, a stimulator of protein kinase C. This system of sphingomyelin hydrolysis/synthesis has been suggested to function as a futile cycle for the generation of diacylglycerol (Merrill and Jones, 1990; Koval and Pagano, 1991).

Using inhibitors of either the synthesis or the breakdown of glucosylceramide, it was shown that the cellular glucosylceramide level is related to cell growth. These effects are mediated by modulation of protein kinase C activity and inositol 1,4,5-triphosphate (IP_3) formation (Shayman et al., 1991; Mahdiyoun et al., 1992). In accordance with these observations, we have found that undifferentiated, highly proliferative HT29 epithelial tumor cells display a high level of glucosylceramide in the plasma membrane relative to the differentiated, less proliferative counterpart (Babia et al., 1993). Furthermore, in the undifferentiated cells, glucosylceramide is sorted preferentially in the endocytic pathway to the Golgi apparatus, avoiding degradation in the lysosomes (Section IV,B,1). There are indications that the sphingolipid composition at the plasma membrane of HT29 cells is not determined only by synthesis and vesicular transport to the plasma membrane, since biosynthetic carrier vesicles have a sphingolipid composition that differs from that of the plasma membrane (Babia et al. unpublished observations). Therefore, the final plasma membrane levels of the various sphingolipids in these cells may be determined by endocytic uptake and degradation as well.

Another example of a proposed plasma membrane–cell interior cycle of glycolipids involves the ganglioside G_{M3} and its breakdown product lactosylceramide. G_{M3} is involved in density-dependent regulation of cell growth, through inhibition of EGF-induced phosphorylation. The following flow model is suggested (see also Merrill, 1991; Hoekstra and Kok, 1992). G_{M3} is converted to lactosylceramide by a sialidase at the cell surface, relieving the inhibitory effect on the EGF receptor. Lactosylceramide then is taken up by endocytosis, to be resialylated in the Golgi apparatus and returned to the cell surface. This model clearly supports an important role for the endocytic pathway in the regulation of cell surface levels of bioactive sphingolipids.

In conclusion, although direct evidence is still lacking, it is reasonable to assume that endocytosis of (glyco)sphingolipids functions as an additional mechanism for modulation of plasma membrane sphingolipid levels and, consequently, as a regulatory mechanism in signal transduction. Therefore, studying the possible involvement of endocytic uptake in the regulation of the flow of bioactive sphingolipid breakdown products such as ceramide and sphingosine may be interesting. Also, the existence of different endocytic uptake mechanisms (Sections II,A and IV,A) may be reappraised in light of their possible involvement in signal transduction pathways.

Acknowledgments

Many thanks are due to our collaborators Sinikka Eskelinen, Teresa Babia, Martin ter Beest, and Karin Hoekstra, who contributed to original work cited in this paper. Part of the authors' work was carried out under the auspices of the Netherlands Foundation for Chemical Research (SON) with financial support from the Netherlands Foundation for Scientific Research (NWO).

References

Alfsen, A., de Paillerets, C., Prasad, K., Nandi, P. K., Lippoldt, R. E., and Edelhoch, H. (1984). Organization and dynamics of lipids in bovine brain coated and uncoated vesicles. *Eur. Biophys. J.* **11,** 129–136.

Baba, T., Shiozawa, N., Hotchi, M., and Ohno, S. (1991). Three-dimensional identification of endosomes in macrophages by a replica-scraping method. *Eur. J. Cell Biol.* **56,** 147–150.

Babia, T., Kok, J. W., Hulstaert, C., de Weerd, H., and Hoekstra, D. (1993). Differential metabolism and trafficking of sphingolipids in differentiated versus undifferentiated HT2g cells. *Int. J. Cancer* **54,** 839–845.

Ballou, L. R. (1992). Sphingolipids and cell function. *Immunol. Today* **13,** 339–341.

Bertho, P., Moreau, P., Morre, D. J., and Cassagne, C. (1991). Monensin blocks the transfer of very long chain fatty acid containing lipids to the plasma membrane of leek seedlings. Evidence for lipid sorting based on fatty acyl chain length. *Biochim. Biophys. Acta* **1070,** 127–134.

Braell, W. A. (1987). Fusion between endocytic vesicles in a cell-free system. *Proc. Natl. Acad. Sci. U.S.A.* **84,** 1137–1141.

Brown, D. A., and Rose, J. K. (1992). Sorting of GPI-anchored proteins to glycolipid-enriched membrane subdomains during transport to the apical cell surface. *Cell* **68,** 533–544.

Brown, V. I., and Greene, M. I. (1991). Molecular and cellular mechanisms of receptor-mediated endocytosis. *DNA Cell Biol.* **10,** 399–409.

Buktenica, S., Olenick, S. J., Salgia, R., and Frankfater, A. (1987). Degradation and regurgitation of extracellular proteins by cultured mouse peritoneal macrophages and baby hamster kidney fibroblasts. Kinetic evidence that the transfer of proteins to lysosomes is not irreversible. *J. Biol. Chem.* **262,** 9469–9476.

Butor, C., Stelzer, E. H. K., Sonnenberg, A., and Davoust, J. (1991). Apical and basal Forsmann antigen in MDCK II Cells: A morphological and quantitative study. *Eur. J. Cell Biol.* **56,** 269–285.

Chavrier, P., Parton, R. G., Hauri, H. P., Simons, K., and Zerial, M. (1990). Localization of low molecular weight GTP binding proteins to exocytic and endocytic compartments. *Cell* **62,** 317–329.

Critchley, D. R., Nelson, P. G., Habig, W. H., and Fishman, P. H. (1985). Fate of tetanus toxin bound to the surface of primary neurons in culture. *J. Cell Biol.* **100,** 1499–1507.

Curatalo, W. (1987). Glycolipid function. *Biochim. Biophys. Acta* **906,** 137–160.

Dickson, R. B., Hanover, J. A., Willingham, M. C., and Pastan, I. (1983). Prelysosomal divergence of transferrin and epidermal growth factor during receptor-mediated endocytosis. *Biochemistry* 22, 5667–5674.

Dunn, K. W., and Maxfield, F. R. (1990). Use of fluorescence microscopy in the study of receptor-mediated endocytosis. *In* "Noninvasive Techniques in Cell Biology" (J. K. Foskett and S. Grinstein, eds.), pp. 153–176. Wiley-Liss, New York.

Dunn, K. W., and Maxfield, F. R. (1992). Delivery of ligands from sorting endosomes to late endosomes occurs by maturation of sorting endosomes. *J. Cell Biol.* 117, 301–310.

Dunn, K. W., McGraw, T. E., and Maxfield, F. R. (1989). Iterative fractionation of recycling receptors from lysosomally destined ligands in an early sorting endosome. *J. Cell Biol.* 109, 3303–3314.

Eskelinen, S., Kok, J. W., Sormunen, R., and Hoekstra, D. (1991). Coated endosomal vesicles: Sorting and recycling compartment for transferrin in BHK cells. *Eur. J. Cell Biol.* 56, 210–222.

Evered, D. (1982). "Membrane Recycling," Ciba Foundation Symposium 92. Pitman Press, London.

Fishman, J. B., and Fine, R. E. (1987). A *trans* Golgi-derived exocytic coated vesicle can contain both newly synthesized cholinesterase and internalized transferrin. *Cell* 48, 157–164.

Geuze, H. J., Slot, J. W., Strous, G. J. A. M., Lodish, H. F., and Schwartz, A. L. (1983). Intracellular site of asialoglycoprotein receptor-ligand uncoupling: Double label immunoelectron microscopy during receptor-mediated endocytosis. *Cell* 32, 277–287.

Gillard, B. K., Heath, J. P., Thurmon, L. T., and Marcus, D. M. (1991). Association of glycosphingolipids with intermediate filaments of human umbilical vein endothelial cells. *Exp. Cell Res.* 192, 433–444.

Gillard, B. K., Thurmon, L. T., and Marcus, D. M. (1992). Association of glycosphingolipids with intermediate filaments of mesenchymal, epithelial, glial, and muscle cells. *Cell Motil. Cytoskel.* 21, 255–272.

Goda, Y., and Pfeffer, S. R. (1989). Cell-free systems to study vesicular transport along the secretory and endocytic pathways. *FASEB J.* 3, 2488–2495.

Goldenthal, K. I., Hedman, K., Chen, J. W., August, J. T., Vihko, P., Pastan, I., and Willingham, M. C. (1988). Pre-lysosomal divergence of alpha$_2$-macroglobulin and transferrin: A kinetic study using a monoclonal antibody against a lysosomal membrane glycoprotein (LAMP-1). *J. Histochem. Cytochem.* 36, 391–400.

Gonatas, N. K., Stieber, A., Gonatas, J., Mommoi, T., and Fishman, P. H. (1983). Endocytosis of exogenous GM1 ganglioside and cholera toxin by neuroblastoma cells. *Mol. Cell. Biol.* 3, 91–101.

Gravotta, D., and Maccioni, H. J. F. (1985). Gangliosides and sialoglycoproteins in coated vesicles from bovine brain. *Biochem. J.* 225, 713–721.

Griffiths, G., and Gruenberg, J. (1991). The arguments for pre-existing early and late endosomes. *Trends Cell Biol.* 1, 5–9.

Griffiths, G., and Simons, K. (1986). The *trans* Golgi network: Sorting at the exit site of the Golgi complex. *Science* 234, 438–443.

Griffiths, G., Back, R., and Marsh, M. (1989). A quantitative analysis of the endocytic pathway in baby hamster kidney cells. *J. Cell Biol.* 109, 2703–2720.

Gruenberg, J., and Howell, K. E. (1987). An internalized transmembrane protein resides in a fusion-competent endosome for less than 5 minutes. *Proc. Natl. Acad. Sci. U.S.A.* 84, 5758–5762

Gruenberg, J., and Howell, K. E. (1989). Membrane traffic in endocytosis: Insights from cell-free assays. *Annu. Rev. Cell Biol.* 5, 453–481.

Gruenberg, J., Griffiths, G., and Howell, K. E. (1989). Characterization of the early endosome and putative endocytic carrier vesicles in vivo and with an assay of vesicle fusion
in vitro. *J. Cell Biol.* **108**, 1301–1316.

Hakomori, S.-I. (1981). Glycosphingolipids in cellular interaction, differentiation, and oncogenesis. *Annu. Rev. Biochem.* **50**, 733–764.

Hakomori, S.-I. (1984). Glycosphingolipids as differentiation dependent tumor-associated
markers and as regulators of cell proliferation. *Trends Biochem. Sci.* **9**, 453–458.

Hannun, Y. A., and Bell, R. M. (1989). Functions of sphingolipids and sphingolipid breakdown products in cellular regulation. *Science* **243**, 500–507.

Hanover, J. A., Willingham, M. C., and Pastan, I. (1984). Kinetics of transit of transferrin
and epidermal growth factor through clathrin-coated membranes. *Cell* **39**, 283–293.

Hansen, S. H., Sandvig, K., and Van Deurs, B. (1991). The preendosomal compartment
comprises distinct coated and noncoated endocytic vesicle populations. *J. Cell Biol.*
113, 731–741.

Hedman, K., Goldenthal, K. L., Rutherford, A. V., Pastan, I., and Willingham, M. C.
(1987). Comparison of the intracellular pathways of transferrin recycling and vesicular
stomatitis virus membrane glycoprotein exocytosis by ultrastructural double-label cytochemistry. *J. Histochem. Cytochem.* **35**, 233–243.

Helenius, A., Mellman, I., Wall, D., and Hubbard, A. (1983). Endosomes. *Trends Biochem.
Sci.* **8**, 245–250.

Herman, B., and Albertini, D. F. (1984). A time-lapse video image intensification analysis
of cytoplasmic organelle movements during endosome translocation. *J. Cell Biol.* **98**,
565–576.

Heuser, J. E., and Anderson, R. G. W. (1989). Hypertonic media inhibit receptor-mediated
endocytosis by blocking clathrin-coated pit formation. *J. Cell Biol.* **108**, 389–400.

Hoekstra, D., and Kok, J. W. (1992). Trafficking of glycosphingolipids in eukaryotic cells;
Sorting and recycling of lipids. *Biochim. Biophys. Acta* **1113**, 277–294.

Hoekstra, D., Eskelinen, S., and Kok, J. W. (1989). Transport of lipids and proteins during
membrane flow in eukaryotic cells. *In* "Organelles in Eukaryotic Cells: Molecular
Structure and Interactions" (J. M. Tager, A. Azzi, S. Papa, and F. Guerreri, eds.),
pp. 59–83. Plenum Press, New York.

Hopkins, C. R. (1983). The importance of the endosome in intracellular traffic. *Nature
(London)* **304**, 684–685.

Hopkins, C. R., Miller, K., and Beardmore, J. M. (1985). Receptor-mediated endocytosis
of transferrin and epidermal growth factor receptors: A comparison of constitutive and
ligand-induced uptake. *J. Cell Sci. Suppl.* **3**, 173–186.

Hopkins, C. R., Gibson, A., Shipman, M., and Miller, K. (1990). Movement of internalized
ligand-receptor complexes along a continuous endosomal reticulum. *Nature (London)*
346, 335–339.

Huang, R. T. C., and Dietsch, E. (1991). Cellular incorporation and localization of fluorescent derivatives of gangliosides, cerebroside and sphingomyelin. *FEBS Lett.* **281**,
39–42.

Kawai, Y., and Hatae, T. (1991). Three-dimensional representation and quantification of
endosomes in the rat kidney proximal tubule cell. *J. Electron Microsc.* **40**, 411–415.

Klausner, R. D., Van Renswoude, J., Ashwell, G., Kempf, C., Schechter, A. N., Dean,
A., and Bridges, K. R. (1983). Receptor-mediated endocytosis of transferrin in K562
cells. *J. Biol. Chem.* **258**, 4715–4724.

Knoll, G., Burger, K. N. J., Bron, R., Van Meer, G., and Verkleij, A. J. (1988). Fusion of
liposomes with the plasma membrane of epithelial cells: Fate of incorporated lipids as
followed by freeze fracture and autoradiography of plastic sections. *J. Cell Biol.* **107**,
2511–2521.

Kok, J. W., and Hoekstra, D. (1993). Fluorescent lipid analogues. Applications in cell- and membrane biology. *In* "Fluorescent Probes for Biological Function of Living Cells. A Practical Guide" (W. I. Mason, ed.), pp. 100–119. Academic Press, London.

Kok, J. W., Eskelinen, S., Hoekstra, K., and Hoekstra, D. (1989). Salvage of glucosylceramide by recycling after internalization along the pathway of receptor-mediated endocytosis. *Proc. Natl. Acad. Sci. U.S.A.* **86,** 9896–9900.

Kok, J. W., ter Beest, M., Scherphof, G., and Hoekstra, D. (1990). A non-exchangeable fluorescent phospholipid analog as a membrane traffic marker of the endocytic pathway. *Eur. J. Cell Biol.* **53,** 173–184.

Kok, J. W., Babia, T., and Hoekstra, D. (1991). Sorting of sphingolipids in the endocytic pathway of HT29 cells. *J. Cell Biol.* **114,** 231–239.

Kok, J. W., Hoekstra, K., Eskelinen, S., and Hoekstra, D. (1992). Recycling pathways of glucosylceramide in BHK cells: Distinct involvement of early and late endosomes. *J. Cell Sci.* **103,** 1139–1152.

Kolesnick, R. (1992). Ceramide: A novel second messenger. *Trends Cell Biol.* **2,** 232–236.

Kornfeld, S., and Mellman, I. (1989). The biogenesis of lysosomes. *Annu. Rev. Cell Biol.* **5,** 483–525.

Koval, M., and Pagano, R. E. (1989). Lipid recycling between the plasma membrane and intracellular compartments: Transport and metabolism of fluorescent sphingomyelin analogues in cultured fibroblasts. *J. Cell Biol.* **108,** 2169–2181.

Koval, M., and Pagano, R. E. (1990). Sorting of an internalized plasma membrane lipid between recycling and degradative pathways in normal and Niemann–Pick, Type A fibroblasts. *J. Cell Biol.* **111,** 429–442.

Koval, M., and Pagano, R. E. (1991). Intracellular transport and metabolism of sphingomyelin. *Biochim. Biophys. Acta* **1082,** 113–125.

Levade, T., Gatt, S., Maret, A., and Salvayre, R. (1991a). Different pathways of uptake and degradation of sphingomyelin by lymphoblastoid cells and the potential participation of the neutral sphingomyelinase. *J. Biol. Chem.* **266,** 13519–13529.

Levade, T., Gatt, S., and Salvayre, R. (1991b). Uptake and degradation of several pyrene-sphingomyelins by skin fibroblasts from control subjects and patients with Niemann–Pick disease. *Biochem. J.* **275,** 211–217.

Lippincott-Schwartz, J., Donaldson, J. G., Schweizer, A., Berger, E. G., Hauri, H.-P., Yuan, L. C., and Klausner, R. D. (1990). Microtubule-dependent retrograde transport of proteins into the ER in the presence of brefeldin A suggests an ER recycling pathway. *Cell* **60,** 821–836.

Lipsky, N. G., and Pagano, R. E. (1985). Intracellular translocation of fluorescent sphingolipids in cultured fibroblasts: Endogenously synthesized sphingomyelin and glucocerebroside analogues pass through the Golgi apparatus en route to the plasma membrane. *J. Cell Biol.* **100,** 27–34.

Longmuir, K. J., and Haynes, S. (1991). Evidence that fatty acid chain length is a type II cell lipid-sorting signal. *Am. J. Physiol.* **260,** L44–L51.

Mahdiyoun, S., Deshmukh, G. D., Abe, A., Radin, N. S., and Shayman, J. A. (1992). Decreased formation of inositol triphosphate in Madin-Darby canine kidney cells under conditions of β-glucosidase inhibition. *Arch. Biochem. Biophys.* **292,** 506–511.

Marsh, M., and Helenius, A. (1980). Adsorptive endocytosis of Semliki Forest virus. *J. Mol. Biol.* **142,** 439–454.

Marsh, M., Griffiths, G., Dean, G. E., Mellman, I., and Helenius, A. (1986). Three-dimensional structure of endosomes in BHK-21 cells. *Proc. Natl. Acad. Sci. U.S.A.* **83,** 2899–2903.

Matteoni, R., and Kreis, T. E. (1987). Translocation and clustering of endosomes and lysosomes depends on microtubules. *J. Cell Biol.* **105,** 1253–1265.

Mayorga, L. S., Diaz, R., and Stahl, P. D. (1988). Plasma membrane-derived vesicles containing receptor-ligand complexes are fusogenic with early endosomes in a cell-free system. *J. Biol. Chem.* **263**, 17213–17216.

Mayorga, L. S., Diaz, R., and Stahl, P. D. (1989). Regulatory role for GTP-binding proteins in endocytosis. *Science* **244**, 1475–1477.

Mellman, I. (1984). Membrane recycling during endocytosis. *In* "Molecular and Cellular Aspects of Lysosomes. Lysosomes in Biology and Pathology" (J. T. Dingle, R. T. Dean, and W. Sly, eds.), Vol. 7, pp. 201–229. Elsevier, Amsterdam.

Mellman, I., Howe, C., and Helenius, A. (1987). The control of membrane traffic on the endocytic pathway. *Curr. Top. Membr. Transp.* **29**, 255–288.

Merrill, A. H., Jr. (1991). Cell regulation by sphingosine and more complex sphingolipids. *J. Bioenerg. Biomembr.* **23**, 83–104.

Merrill, A. H., Jr., and Jones, D. D. (1990). An update of the enzymology and regulation of sphingomyelin metabolism. *Biochim. Biophys. Acta* **1044**, 1–12.

Miller-Podraza, H., and Fishman, P. H. (1982). Translocation of newly synthesized gangliosides to the cell surface. *Biochemistry* **21**, 3265–3270.

Montesano, R., Roth, J., Robert, A., and Orci, L. (1982). Non-coated membrane invaginations are involved in binding and internalization of cholera and tetanus toxin. *Nature (London)* **296**, 651–653.

Morre, D. J., Widnell, C. J., and Thilo, L. (1983). Membrane dynamics: Flow routes and quantitation of membrane transport and recycling. *Fed. Proc.* **43**, 2884–2887.

Murphy, R. (1991). Maturation models for endosome and lysosome biogenesis. *Trends Cell Biol.* **1**, 77–82.

O'Brien, J. S., and Kishimoto, Y. (1991). Saposin proteins: Structure, function, and role in human lysosomal storage disorders. *FASEB J.* **5**, 301–308.

Oka, J. A., Christensen, M. D., and Weigel, P. H. (1989). Hyperosmolarity inhibits galactosyl receptor-mediated but not fluid phase endocytosis in isolated rat hepatocytes. *J. Biol. Chem.* **20**, 12016–12024.

Pacuszka, T., and Fishman, P. H. (1992). Intoxification of cultured cells by cholera toxin: Evidence for different pathways when bound to ganglioside GM1 or neoglycoproteins. *Biochemistry* **31**, 4773–4778.

Pagano, R. E., and Sleight, R. G. (1985). Defining lipid transport pathways in animal cells. *Science* **229**, 1051–1057.

Pastan, I., and Willingham, M. C. (1983). Receptor-mediated endocytosis: Coated pits, receptosomes and the Golgi. *Trends Biochem. Sci.* **8**, 250–254.

Pastan, I., and Willingham, M. C. (1985). "Endocytosis." Plenum Press, New York.

Podbilewicz, B., and Mellman, I. (1990). ATP and cytosol requirements for transferrin recycling in intact and disrupted MDCK cells. *EMBO J.* **9**, 3477–3487.

Quin, P., and Allan, D. (1992). Two separate pools of sphingomyelin in BHK cells. *Biochim. Biophys. Acta* **1124**, 95–100.

Reynier, M., Sari, H., d'Anglebermes, M., Ah Kye, E., and Pasero, L. (1991). Differences in lipid characteristics of undifferentiated and enterocytic-differentiated HT29 human colonic cells. *Cancer Res.* **51**, 1270–1277.

Rosenwald, A. G., Pagano, R. E., and Raviv, Y. (1991). Activation of 5-[^{125}I]iodonaphthyl-1-azide via excitation of fluorescent [*N*-(7-nitrobenz-2-oxa-1,3-diazol-4-yl)] lipid analogs in living cells. *J. Biol. Chem.* **266**, 9814–9821.

Rothberg, K. G., Ying, Y.-S., Kolhouse, J. F., Kampen, B. A., and Anderson, R. G. W. (1990a). The glycophospholipid-linked folate receptor internalizes folate without entering the clathrin-coated pit endocytic pathway. *J. Cell Biol.* **110**, 637–649.

Rothberg, K. G., Ying, Y.-S., Kamen, B. A., and Anderson, G. W. (1990b). Cholesterol controls the clustering of the glycophospholipid-anchored membrane receptor for 5-methyltetrahydrofolate. *J. Cell Biol.* **111**, 2931–2938.

Rothberg, K. G., Heuser, J. E., Donzell, W. C., Ying, Y.-S., Glenney, J. R., and Anderson, R. G. W. (1992). Caveolin, a protein component of caveolae membrane coats. *Cell* **68**, 673–682.

Rothman, J. E., and Schmid, S. L. (1986). Enzymatic recycling of clathrin from coated vesicles. *Cell* **46**, 5–9.

Sakai, T., Yamashina, S., and Ohnishi, S.-I. (1991). Microtubule-disrupting drugs blocked delivery of endocytosed transferrin to the cytocenter, but did not affect return of transferrin to plasma membrane. *J. Biochem.* **109**, 528–533.

Sandhoff, K., Van Echten, G., Schroder, M., Schnabel, D., and Suzuki, K. (1992). Metabolism of glycolipids: The role of glycolipid-binding proteins in the function and pathobiochemistry of lysosomes. *Biochem. Soc. Trans.* **20**, 695–699.

Sandvig, K., Olsnes, S., Petersen, O. W., and Van Deurs, B. (1987). Acidification of the cytosol inhibits endocytosis from coated pits. *J. Cell Biol.* **105**, 670–689.

Sandvig, K., Olsnes, S., Brown, J. E., Peterson, O. W., and Van Deurs, B. (1989). Endocytosis from coated pits of shiga toxin: A glycolipid-binding protein from *Shigella dysenteriae* 1. *J. Cell Biol.* **108**, 1331–1343.

Sandvig, K., Prydz, K., Ryd, M., and Van Deurs, B. (1991). Endocytosis and intracellular transport of the glycolipid-binding ligand Shiga toxin in polarized MDCK cells. *J. Cell Biol.* **113**, 553–562.

Schmid, S. L., Fuchs, R., Male, P., and Mellman, I. (1988). Two distinct subpopulations of endosomes involved in membrane recycling and transport to lysosomes. *Cell* **52**, 73–83.

Schwarzmann, G., and Sandhoff, K. (1990). Metabolism and intracellular transport of glycosphingolipids. *Biochemistry* **29**, 10865–10871.

Shayman, J. A., Deshmukh, G. D., Mahdiyoun, S., Thomas, T. P., Wu, D., Barcelon, F. S., and Radin, N. S. (1991). Modulation of renal epithelial growth by glucosylceramide. Association with protein kinase C, sphingosine and diacylglycerol. *J. Biol. Chem.* **266**, 22968–22974.

Sonnino, S., Chigorno, V., Acquotti, D., Pitto, M., Kirshner, G., and Tettamanti, G. (1989). A photoreactive derivative of radiolabeled GM1 ganglioside: Preparation and use to establish the involvement of specific proteins in GM1 uptake by human fibroblasts in culture. *Biochemistry* **28**, 77–84.

Sonnino, S., Chigorno, V., Valsecchi, M., Pitto, M., and Tettamanti, G. (1992). Specific ganglioside-cell protein interactions: A study performed with GM1 ganglioside derivative containing photoactivable azide and rat cerebellar granule cells in culture. *Neurochem. Int.* **20**, 315–321.

Spiegel, S., Schlessinger, J., and Fishman, P. H. (1984). Incorporation of fluorescent gangliosides into human fibroblasts: Mobility, fate, and interaction with fibronectin. *J. Cell Biol.* **99**, 699–704.

Stein, B. S., and Sussman, H. H. (1986). Demonstration of two distinct transferrin receptor recycling pathways and transferrin-independent receptor internalization in K562 cells. *J. Biol. Chem.* **261**, 10319–10331.

Steinman, R. M., Mellman, I., Muller, W. A., and Cohn, Z. A. (1983). Endocytosis and the recycling of plasma membrane. *J. Cell Biol.* **96**, 1–27.

Stoorvogel, W., Geuze, H. J., and Strous, G. J. (1987). Sorting of endocytosed transferrin and asialoglycoprotein occurs immediately after internalization in HepG2 cells. *J. Cell Biol.* **104**, 1261–1268.

Stoorvogel, W., Geuze, H. J., Griffiths, J. M., and Strous, G. J. (1988). The pathways of endocytosed transferrin and secretory protein are connected in the *trans*-Golgi reticulum. *J. Cell Biol.* **106**, 1821–1829.

Stoorvogel, W., Strous, G. J., Geuze, H. J., Oorschot, V., and Schwartz, A. L. (1991). Late endosomes derive from early endosomes by maturation. *Cell* **65**, 417–427.

Struck, D. K., and Pagano, R. E. (1980). Insertion of fluorescent phospholipids into the plasma membrane of a mammalian cell. *J. Biol. Chem.* **255**, 5405–5410.

Sweeley, C. C. (1985). Sphingolipids. In "Biochemistry of Lipids and Membranes" (D. E. Vance and J. E. Vance, eds.), pp. 361–403. Benjamin/Cummings, Menlo Park, California.

Thompson, T. E., and Tillack, T. W. (1985). Organization of glycosphingolipids in bilayers and plasma membranes of mammalian cells. *Annu. Rev. Biophys. Bioeng.* **14**, 361–386.

Tooze, J., and Hollinshead, M. (1991). Tubular early endosomal networks in AT20 and other cells. *J. Cell Biol.* **115**, 635–653.

Tran, D., Carpentier, J.-L., Sawano, F., Gordon, P., and Orci, L. (1987). Ligands internalized through coated or non-coated invaginations follow a common intracellular pathway. *Proc. Natl. Acad. Sci. U.S.A.* **84**, 7957–7961.

Trinchera, M., Ghidoni, R., Sonnino, S., and Tettamanti, G. (1990). Recycling of glucosylceramide and sphingosine for the biosynthesis of gangliosides and sphingomyelin in rat liver. *Biochem J.* **270**, 815–820.

Trinchera, M., Carrettoni, D., and Ghidoni, R. (1991). A part of glucosylceramide formed from exogenous lactosylceramide is not degraded to ceramide but re-cycled and glycosylated in the Golgi apparatus. *J. Biol. Chem.* **266**, 9093–9099.

Urade, R., Hayashi, Y., and Kito, M. (1988). Endosomes differ from plasma membranes in the phospholipid molecular species composition. *Biochim. Biophys. Acta* **946**, 151–163.

Van der Sluijs, P., Hull, M., Webster, P., Male, P., Goud, B., and Mellman, I. (1992). The small GTP-binding protein rab4 controls an early sorting event on the endocytic pathway. *Cell* **70**, 729–740.

Van Deurs, B., and Nilausen, K. (1982). Pinocytosis in mouse L-fibroblasts: Ultrastructural evidence for a direct membrane shuttle between the plasma membrane and the lysosomal compartment. *J. Cell Biol.* **94**, 279–286.

Van Deurs, B., Sandvig, K., Petersen, O. W., Olsnes, S., Simons, K., and Griffiths, G. (1988). Estimation of the amount of internalized ricin that reaches the *trans*-Golgi network. *J. Cell Biol.* **106**, 253–267.

Van Genderen, I. L., Van Meer, G., Slot, J. W., Geuze, H. J., and Voorhout, W. (1991). Subcellular localization of Forssman glycolipid in epithelial MDCK cells by immunoelectronmicroscopy after freeze substitution. *J. Cell Biol.* **115**, 1009–1019.

Van Meer, G. (1989). Lipid traffic in animal cells. *Annu. Rev. Cell Biol.* **5**, 247–275.

Van Meer, G., Stelzer, E. H. K., Wijnaendts-Van Resandt, R. W., and Simons, K. (1987). Sorting of sphingolipids in epithelial (Madin-Darby canine kidney) cells. *J. Cell Biol.* **105**, 1623–1635.

Van't Hoff, W., Silvius, J., Wieland, F., and Van Meer, G. (1992). Epithelial sphingolipid sorting allows for extensive variation of the fatty acyl chain and the sphingosine backbone. *Biochem. J.* **283**, 913–917.

Van Weely, S., Brandsma, M., Strijland, A., Tager, J. M., and Aerts, J. M. F. G. (1992). Demonstration of the existence of a second, non-lysosomal glucocerebrosidase that is not deficient in Gaucher disease. *Biochim. Biophys. Acta* **1181**, 55–62.

Voelker, D. R. (1991). Organelle biogenesis and intracellular lipid transport in eukaryotes. *Microbiol. Rev.* **55**, 543–560.

Vos, J. P., Giudici, M. L., Van Golde, L. M. G., Preti, A., Marchesini, S., and Lopes-Cardozo, M. (1992). Cultured oligodendrocytes metabolize a fluorescent analogue of sulphatide; Inhibition by monensin. *Biochim. Biophys. Acta* **1126**, 269–276.

Wattenberg, B. W. (1990). Glycolipid and glycoprotein transport through the Golgi complex are similar biochemically and kinetically. Reconstitution of glycolipid transport in a cell-free system. *J. Cell Biol.* **111,** 421–428.

Wiegandt, H. (1985). "Glycolipids. New Comprehensive Biochemistry." Vol. 10. Elsevier, Amsterdam.

Yamada, E. (1955). The fine structure of the gall bladder epithelium of the mouse. *J. Biophys. Biochem. Cytol.* **1,** 445–458.

Young, W. W., Lutz, M. S., and Blackburn, W. A. (1992). Endogenous glycosphingolipids move to the cell surface at a rate consistent with bulk flow estimates. *J. Biol. Chem.* **267,** 12011–12015.

Zaal, K., Kok, J. W., Kuipers, F., and Hoekstra, D. (1994). Lipid trafficking in hepatocytes. Relevance to biliary lipid secretion. *Adv. Mol. Cell Biol.* (*in press*).

CHAPTER 21

Lipid Polarity and Sorting in Epithelial Cells

Wouter van 't Hof* and Gerrit van Meer
Department of Cell Biology, Medical School, University of Utrecht, 3584 CH Utrecht, The Netherlands

* Present address: Department of Cell Biology and Anatomy, Cornell University Medical College, New York, New York 10021.

I. LIPID ORGANIZATION IN CELLULAR MEMBRANES

A. Lipid Diversity

Mammalian cells are enveloped by their plasma membrane. The cytoplasm is filled with numerous intracellular membranes that confine the various cellular organelles. In conjunction with their specific functions, the individual membranes possess unique protein and lipid compositions. One general aim in cell biological research is to understand how this situation is created and how it relates to the functions of biomembranes. The plasma membrane is of vital importance to the cell since it protects the cytoplasmic space and because the surface proteins and lipids determine the interactions of the cell with the environment. Therefore, knowing how cells shape the lipid composition of their surface, that is, the outer exoplasmic leaflet of the plasma membrane bilayer, and how this composition is maintained and modulated in time is relevant. Epithelial cells are especially interesting in this sense, since their plasma membranes consist of two domains that face two different environments.

The basic structure of cellular membranes is the lipid bilayer, which also serves as a scaffold for the membrane-associated proteins (Gorter and Grendel, 1925; Singer and Nicolson, 1972). The bilayer configuration results from the amphipathic character of membrane lipids. The vast majority of membrane lipids belongs to one of three classes: phospholipids, glycosphingolipids (GSLs), and cholesterol, with a molar ratio in most cells on the order of 75% : 5% : 20%. This membrane lipid diversity is increased further by several factors. Each phospholipid possesses one of some 10 different polar head groups; in addition, the fatty acyl chains at the 1 and the 2 position of the glycerol backbone may be any of 10 or more different species. The variation in the GSLs is even larger because of multiple carbohydrate-containing head groups. However, not all theoretically possible lipids occur in any particular cell. Phospholipids prefer a saturated fatty acid in the 1 position and an unsaturated fatty acid in the 2 position. Further, each cell only expresses a limited number of GSLs. Still, approximately 500 different lipid species are estimated to be present in a particular cell. Obviously, having all these different lipid molecules in its membranes must be of functional significance for the cell. A first indication for a connection between lipid form and function is the observation that the membrane of each organelle, in connection with a specialized function, possesses not only a specific complement of proteins but also a unique lipid composition.

B. Intracellular Distribution

Analysis by cell fractionation studies has indicated that the various organelles have different lipid compositions, and that this lipid distribution is remarkably uniform among different cell types (see Chapter 1). However, compared to the differences in protein composition, the differences in lipid composition are far less absolute. Since generally the same types of lipids are present in each organelle, the lipid compositions differ in the ratio between the various lipid species. For instance, along the exocytic pathway from the endoplasmic reticulum (ER) via the Golgi complex to the plasma membrane (Fig. 1), a gradual shift is observed in lipid composition that consists of an increase in the phospholipids sphingomyelin (SM) and phosphatidylserine (PS), in cholesterol, and in GSLs that is compensated for by a loss in phosphatidylcholine (PC) and phosphatidylinositol (PI). The relative amount of phosphatidylethanolamine (PE) remains constant throughout. The membranes of the *trans*-Golgi network (TGN) endosomes, and lysosomes, connected to the plasma membrane by vesicular

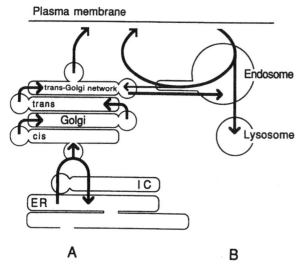

A   B

FIGURE 1 Vesicular transport routes in mammalian cells. (A) Biosynthetic exocytic pathway from the endoplasmic reticulum, via the Golgi complex to the plasma membrane. (B) Endocytic pathway from the plasma membrane via endosomes to lysosomes. The exocytic and the endocytic route are interconnected at the level of the *trans*-Golgi network and endosomes. ER, endoplasmic reticulum; IC, intermediate compartment.

transport routes (Fig. 1), have lipid compositions similar to that of the plasma membrane (see van Meer, 1989).

C. Lipid Asymmetry in Membranes

The vectorial nature of several cellular membrane functions is met by an asymmetric distribution of proteins and lipids across the lipid bilayer. Whereas protein asymmetry consists of a single orientation of a protein with respect to the exoplasmic–cytoplasmic vector across the membrane, lipids display asymmetry in the sense that the two leaflets of the lipid bilayer have different lipid compositions (see also Chapters 1 and 2), making the compositional differences between cellular lipid pools more pronounced. Lipid asymmetry has been well characterized for plasma membranes, with the erythrocyte as a model system (Op den Kamp, 1979; Zachowski and Devaux, 1990; Devaux, 1991), and for endocytic membranes (Sandra and Pagano, 1978). Higher GSLs, containing more than two sugar moieties, have been localized exclusively in the exoplasmic luminal leaflet of cellular membranes (Thompson and Tillack, 1985). Various lines of evidence suggest that SM is located in the exoplasmic leaflet exclusively (van Meer, 1989). Also, the larger part (70–80%) of the other choline-containing phospholipid, PC, resides in the exoplasmic leaflet. Interestingly, GSLs and SM possess the acyl-sphingosine backbone whereas PC, which is also present in the cytoplasmic leaflet, carries a diacylglycerol backbone. The aminophospholipids PE and PS are concentrated in the cytoplasmic leaflet of plasma membranes (see Chapter 1). For PS this enrichment is almost absolute. Data on the transbilayer organization of cholesterol are controversial (Brasaemle et al., 1988; Zachowski and Devaux, 1990; see also Chapter 19).

Therefore, the overall consensus on the composition of the external leaflet of the plasma membrane is that 90% or more of the lipids are accounted for by GSLs, SM, PC, and cholesterol. These lipids generally display low rates of transbilayer mobility (flip–flop) in plasma membranes. GSLs do not translocate at all, as also appears to be true for SM (Devaux, 1991). In the case of PC, two pools exist in the plasma membrane that intermix only very slowly (Pagano and Sleight, 1985; Devaux, 1991). In contrast, the aminophospholipids appear to be removed actively from the outside and translocated to the cytoplasmic leaflet by an energy-requiring aminophospholipid translocator (see Devaux, 1991; see also Chapter 1). As discussed later, flip–flop does not seem to be quantitatively important in lipid transport to the external leaflet of the plasma membrane when compared with vesicular transport.

A low flip–flop rate for PC is also characteristic of endocytic membranes (Sandra and Pagano, 1978). Moreover, fluorescent analogs of PC and SM do not translocate to the cytoplasmic leaflet when participating in the endocytic pathway (Pagano and Sleight, 1985; Kok *et al.*, 1989,1991; Koval and Pagano, 1989,1990,1991; cf. however, Sleight and Abanto, 1989). Finally, short-chain SM does not flip–flop from its luminal location in the Golgi complex (Lipsky and Pagano, 1985a; van Meer *et al.*, 1987; Helms *et al.*, 1990; Karrenbauer *et al.*, 1990; Jeckel *et al.*, 1992; Kobayashi *et al.*, 1992).

II. LIPID POLARITY IN EPITHELIAL CELLS

A. Lipid Composition of the Two Plasma Membrane Domains

In addition to the transbilayer heterogeneity, in epithelial cells the plasma membrane contains two specialized domains in the plane of the lipid bilayer. The apical plasma membrane domain faces the external environment whereas the opposing basolateral domain interacts with neighboring cells and the underlying tissue. To cope with the varying demands of the different environments and to deal with their specific functions, the two domains exhibit different compositions, a property termed epithelial surface polarity. The domains contain unique sets of proteins (Simons and Fuller, 1985; Rodriguez-Boulan and Powell, 1992), and the lipid compositions also have been found to be widely different, a phenomenon referred to as epithelial lipid polarity.

The most extensive studies on the lipid composition of apical and basolateral plasma membrane domains have been performed on rodent intestinal cells (Forstner *et al.*, 1968; Douglas *et al.*, 1972; Forstner and Wherrett, 1973; Kawai *et al.*, 1973; Brasitus and Schachter, 1980; Hauser *et al.*, 1980). These studies made use of cell fractionation and were facilitated by the organization of the apical membrane of intestinal cells in a highly differentiated brush border that is purified easily. The major difference between the lipid composition was found to be a 2- to 4-fold higher level of GSLs apically, concomitant with a 2- to 4-fold decrease in PC and SM. The amounts of the other phospholipids and cholesterol as percentages of total lipid were equivalent (Table I).

The concentration of GSLs in the brush border is unusually high. For example, the plasma membrane of the human erythrocyte has a molar ratio of GSLs to phospholipids to cholesterol of 1 : 51 : 48 (Cooper, 1978). For the brush border of mouse intestinal cells, these numbers were found to be 1 : 1 : 1 (Table I). GSLs have been localized to the exoplasmic bilayer

TABLE I

Lipid Composition of Apical and Basolateral Plasma Membrane
Domains of Mouse Intestinal Cells[a]

Lipid	Apical (mol%)	Basolateral (mol%)	Polarity (apical/basolateral)
Glycosphingolipid	33.3	7.9	4.22
Phospholipid	33.3	65.8	
PC	8.3	33.6	0.25
SM	2.8	6.5	0.43
Cholesterol	33.3	26.3	

[a] Calculated from cell fractionation data on the ICR strain of mouse, from Kawai *et al.* (1974).

leaflet exclusively (Thompson and Tillack, 1985; Thompson *et al.*, 1986). With the assumption of a 50:50 distribution of cholesterol across the membrane, the apical surface of the intestinal epithelium is covered completely by GSLs and cholesterol. This GSL cover may fulfill a general role in membrane stabilization (Curatolo, 1987a,b). In intestinal epithelia, it protects against phospholipase A_2 (PLA$_2$), abundant in the intestinal lumen, to which sphingolipids are resistant.

Apical and basolateral domains have been purified from a variety of epithelia and species (Bode *et al.*, 1976; Kremmer *et al.*, 1976; Stubbs *et al.*, 1979; Schwertz *et al.*, 1983; Hise *et al.*, 1984; Meier *et al.*, 1984; Carmel *et al.*, 1985; Molitoris and Simon, 1985; Jaffe *et al.*, 1987). When total lipid was analyzed, the situation was very similar to that in intestinal cells, with a 2-fold higher apical level of GSLs and a 2-fold apical depletion in phospholipids. When GSLs were not analyzed, clear cut differences were reported for the phospholipid composition of the two domains. Researchers also have concluded that cholesterol is enriched in the apical domain (Kremmer *et al.*, 1976; Chapelle and Gilles-Baillien, 1983; Meier *et al.*, 1984; Carmel *et al.*, 1985; Molitoris and Simon, 1985). However, this conclusion was based on a comparison of the cholesterol:phospholipid ratios of the respective domains. When the concentration of cholesterol is expressed as a molar percentage of the total membrane lipids (cholesterol plus phospholipids plus GSLs), the concentration of cholesterol in the two domains is identical (van Meer, 1988; see Table I). The typical polarized distribution of surface lipids also was found to be preserved in model systems of epithelial cell lines in culture, with Madin–Darby canine kidney (MDCK) cells as a paradigm (van Meer and Simons, 1982,1986; Nichols *et al.*, 1987,1988; van Genderen *et al.*, 1991).

B. Maintenance of Lipid Polarity by the Tight Junctions

The apical and basolateral surface domains of epithelial cells are parts of a continuous plasma membrane; lipids rapidly diffuse in the plane of the plasma membrane (Dragsten *et al.*, 1982; Yechiel and Edidin, 1987). What prevents apical and basolateral lipids from intermixing? Situated at the boundary between the apical and the basolateral plasma membrane, the tight junction is an obvious candidate (Cereijido, 1992; Farquhar and Palade, 1963). The tight junction, a specific zone of cell–cell contact that encircles the apex of each epithelial cell, connects neighboring epithelial cells into a closed monolayer, thereby forming a highly selective physiological barrier between the external environment and the internal milieu of the body. For membrane proteins, the barrier function has been deduced from the observation that opening of the tight junctions led to mixing of apical and basolateral proteins (for a discussion, see Simons and Fuller, 1985; Rabito, 1992).

The issue of whether tight junctions prevent lipid diffusion has been tested in the MDCK model system in various ways. Based on observations with a variety of fluorescent amphiphiles, Dragsten *et al.* (1981) proposed a model in which the tight junction acts as a diffusion barrier in the exoplasmic leaflet of the plasma membrane, whereas lipids are free to diffuse through a continuous cytoplasmic leaflet. Although in this first study the possibility of equilibration through the cytosol could not be excluded, the model later was substantiated by a number of independent observations on the behavior of an endogenous lipid (Spiegel *et al.*, 1985) and of fluorescent and radiolabeled lipids introduced into the plasma membrane via fusion techniques (van Meer and Simons, 1986; van Meer *et al.*, 1986,1987; Knoll *et al.*, 1988).

Models for the structure of the tight junction have been proposed based on its ultrastructural features (see Hirokawa, 1982; Pinto da Silva and Kachar, 1982). At the site of the tight junction, the intracellular space is occluded and the two apposed plasma membranes appear to fuse. In freeze-fracture replicas, a complementary pattern of anastomosing strands and furrows is observed; whether these consist of proteins (e.g., Hirokawa, 1982) or of an intramembranous hexagonal cylinder of lipids (Kachar and Reese, 1982; Pinto da Silva and Kachar, 1982) is unclear. In the lipid model, the cytoplasmic leaflet is continuous between the apical and basolateral plasma membrane. The exoplasmic leaflets are interrupted by the hexagonal lipid cylinder. In contrast, the exoplasmic leaflets would be in continuity with those of the plasma membrane of the adjacent cells. Morphological observations argue against this continuity (Nichols *et al.*, 1986; van Meer *et al.*, 1986; van Genderen *et al.*, 1991) and, thus, do not favor the hexagonal lipid model (van Meer *et al.*, 1992). Unfortunately, only two tight junction-associated proteins have been identified to date (Anderson and Stevenson, 1992). Still, the presence of proteins in the core of tight junctions would be more consistent with the fact that the tight junction acts as a highly selective barrier to paraepithelial ion diffusion (Reuss, 1992).

Free diffusion predicts an identical lipid composition of the cytoplasmic leaflet in the apical and basolateral domain, with interesting consequences for lipid asymmetry in the basolateral membrane. Calculations imply that 65–90% of the PE compared with only 10–25% of the PC is localized in the cytoplasmic leaflet (van Meer and Simons, 1982,1986). Studies with fluorescent SM indicate that most if not all SM is situated in the basolateral exoplasmic leaflet (van Meer *et al.*, 1987). These calculated transbilayer orientations in the basolateral domain are strikingly similar to those found for the plasma membrane of erythrocytes (Devaux, 1991). Cells seem to require a uniform lipid composition in the cytoplasmic leaflet of their plasma membrane. The lipid heterogeneity on the outer cell surface of epithelial cells suggests that the lipid composition of the exoplasmic leaflets can be adapted to meet the demands of the external environment.

III. GENERATION OF LIPID POLARITY

A. Intracellular Biosynthesis and Transport of Surface Lipids

The common precursor for sphingolipid biosynthesis, ceramide, is made in the ER. The biosynthesis of SM has been assigned to the *cis*-Golgi from colocalization with *cis*- and not with *trans*-Golgi markers after cell fractionation (Futerman *et al.*, 1990; Jeckel *et al.*, 1990). The active site of SM synthase and its products have been found to reside in the luminal exoplasmic leaflet (Futerman *et al.*, 1990; Helms *et al.*, 1990; Jeckel *et al.*, 1990,1992; Karrenbauer *et al.*, 1990). From there, SM travels in the luminal leaflet of transport vesicles and is delivered to the outer leaflet of the plasma membrane with a half-time of less than 30 min (van Meer *et al.*, 1987; Helms *et al.*, 1990; Karrenbauer *et al.*, 1990).

Until recently, biosynthesis of GSL was thought to occur on the luminal exoplasmic leaflet of Golgi membranes. In cell fractionation studies, UDP-glucose:*N*-acyl sphingosine glucosyltransferase (GlcT), the synthase of the simplest glycolipid glucosylceramide (GlcCer), displayed a bimodal distribution over *cis*-Golgi and a second unknown compartment (Futerman and Pagano, 1991; Jeckel *et al.*, 1992). However, the active site of GlcT was found to be located on the cytoplasmic surface of the Golgi complex

(Coste *et al.*, 1986; Futerman and Pagano, 1991; Trinchera *et al.*, 1991). In addition, newly synthesized GlcCer was observed to have a cytosolic orientation (Coste *et al.*, 1986; Jeckel *et al.*, 1992). Newly synthesized GlcCer has been proposed to reach the cell surface by cytoplasmic exchange, possibly mediated by lipid transfer proteins (Dowhan, 1991; Wirtz, 1991) and subsequent flip–flop in the plasma membrane (Sasaki, 1990). However, a variety of observations supports a vesicular mode of transport of GlcCer from the Golgi to the cell surface. GlcCer transport was inhibited in mitotic cells (Kobayashi and Pagano, 1989) at temperatures below 15°C (van Meer *et al.*, 1987; van 't Hof and van Meer, 1990), and was disturbed by treatment of cells with monensin or nocodazole (Lipsky and Pagano, 1985a; van Meer and van 't Hof, 1993). GlcCer was found on the luminal surface of transport vesicles originating from the TGN (Karrenbauer *et al.*, 1990; Kobayashi *et al.*, 1992). These data suggest that GlcCer flip–flops after biosynthesis in the Golgi complex.

Also, the transfer of galactose onto glucosylceramide to yield lactosylceramide (LacCer) seems to occur on the cytoplasmic aspect of the Golgi (Trinchera *et al.*, 1991). Intracellular transport of LacCer was studied using the cell-free system developed by Rothman and co-workers (Rothman and Orci, 1992) for the dissection of the molecular machinery involved in vesicular transport. LacCer transport was found to be biochemically and kinetically similar to that of glycoproteins, suggesting a vesicular mechanism for LacCer transport through the Golgi complex (Wattenberg, 1990). In addition to GlcCer, LacCer also most likely translocates in the Golgi, where it serves as the precursor for more complex GSLs for which the glycosyltransferases have been assigned to the lumen of the Golgi (Schwarzmann and Sandhoff, 1990; Trinchera *et al.*, 1991).

The biosynthesis of PC occurs at the cytoplasmic leaflet of the ER (Bell *et al.*, 1981; Pagano, 1990). Facilitated by a "PC flippase" (Bishop and Bell, 1985; Devaux, 1991), PC rapidly translocates across the ER membrane, explaining how unilateral lipid synthesis on the cytoplasmic ER surface can result in a lipid bilayer. Since rapid flip–flop of PC disappears along the exocytic route, possibly by retention of the PC flippase in the ER, the PC pool on the cell surface can be predicted to be derived from the PC that has entered the luminal membrane leaflet by flip–flop in the ER and subsequently has been transported by vesicular transport. Obviously, the cytoplasmically oriented PC pool is available for other transport mechanisms such as monomer exchange across the cytoplasmic space. This possibility may explain the observations by Kaplan and Simoni (1985) that PC transport to the plasma membrane displayed a half-time of 2 min, much faster than any half-time of vesicular transport measured between ER and plasma membrane. In addition, the existence of monomer exchange or any other mechanism would explain why the observed PC transport was inhibited only partially by low temperature and energy poisons. In conclusion, all evidence is consistent with intracellular synthesis and a vesicular mechanism for transport of PC, SM, and GlcCer from the Golgi complex to the outer cell surface.

B. Biosynthetic Sphingolipid Sorting in MDCK Cells

In a series of papers, Lipsky and Pagano (1983, 1985a,b) described that the exchangeable lipid N-6[7-nitro-2-benzoxa-1,3-diazol-4-yl]aminocaproyl sphingosine (C_6-NBD-ceramide) was incorporated into cells, was trapped in the Golgi complex, and was converted to C_6-NBD-GlcCer and C_6-NBD-SM. These products subsequently were transported to the plasma membrane by a vesicular mechanism. Since these molecules are fluorescent analogs of glycolipids and phospholipids, which are enriched on the apical and basolateral domains respectively (Table I), the comparison of the synthesis in the Golgi and the subsequent transport of C_6-NBD-GlcCer and C_6-NBD-SM to the plasma membrane in epithelial cells should serve as an excellent model in which to study where and how the two lipids obtain their different polarities, that is, how they are sorted from one another.

After insertion of C_6-NBD-ceramide into MDCK cells, at temperatures below 20°C the fluorescent products accumulated in the Golgi area. Transport to the plasma membrane subsequently could be monitored at 37°C by trapping the fluorescent molecules on arrival at the cell surface by selective depletion onto bovine serum albumin (BSA) in the medium. Polarized transport to the two surfaces could be assayed separately, since the tight junctions act as diffusion barriers for both NBD lipids and BSA. Equal amounts of C_6-NBD-SM appeared at the two plasma membrane domains. In contrast, nearly 3-fold more C_6-NBD-GlcCer was delivered to the apical than to the basolateral surface (see, for example, Table II). Thus, newly synthesized sphingolipids had been sorted. Because the assay procedure involved continuous depletion of products during the 37°C incubation, this lipid sorting event was concluded to have occurred intracellularly before the lipids reached the cell surface by vesicular transport (van Meer et al., 1987). Analogous conclusions have been drawn concerning the sorting of plasma membrane proteins in MDCK cells (Caplan and Matlin, 1989; Lisanti et al., 1989b; Simons and Wandinger-Ness, 1990; Bomsel and Mostov, 1991; Mostov et al., 1992; Nelson, 1992; Rodriguez-Boulan and Powell, 1992). The TGN, the last intracellular compartment before exit to the plasma membrane, has been proposed as the site at

TABLE II

Polarized Delivery and Sorting of C_6-NBD Lipids in MDCK, Caco-2. and FRT Cells

Epithelial cell line	Polarity GlcCer (apical/basolateral)	Polarity SM (apical/basolateral)	Relative polarity (Sorting)	[n]
Intestinal Caco-2	4.2 ± 1.7	0.6 ± 0.1	7.5 ± 2.5	[8][a]
Kidney MDCK	1.9 ± 0.2	0.9 ± 0.2	2.2 ± 0.3	[3][b]
Thyroid FRT	0.2 ± 0.06	0.5 ± 0.1	0.4 ± 0.03	[4][b]

[a] Data from van 't Hof *et al.* (1992) for fully differentiated Caco-2 cells (cf. Table I for *in vivo* polarities).
[b] See Zurzolo *et al.* (1993b). For experimental conditions and calculations see van 't Hof and van Meer (1990) and van 't Hof *et al.* (1992).

which newly synthesized apical and basolateral proteins (Griffiths and Simons, 1986) and lipids (Simons and van Meer, 1988) are sorted.

C. Involvement of Microdomains in Biosynthetic Lipid Sorting

Based on the results in MDCK cells, a model for biosynthetic lipid sorting has been proposed (van Meer *et al.*, 1987). According to the model, the key event in lipid sorting is the generation of GSL microdomains in the luminal leaflet of the TGN. The GSL microdomain subsequently buds into a transport vesicle with an apical destination. The luminal leaflet of the basolateral transport vesicles would, as a consequence, be depleted of the GSLs and enriched in phospholipids, namely PC and SM (Fig. 2). If this hypothesis were true, this system would be the first clear example in which lipid microdomains are of functional importance for the cell. One key feature of the process is that sphingolipids have a common structural characteristic, their sphingosine backbone, that is not present in glycero-phospholipids. Sphingolipids, especially the GSLs carrying carbohydrate head groups, have a tendency to associate by hydrogen bond formation to form separate domains in model membrane systems (Pascher, 1976; Thompson and Tillack, 1985; Thompson *et al.*, 1986; Boggs, 1987; Curatolo, 1987a). In cells, this microdomain formation could be facilitated or induced by the conditions prevailing in the lumen of the sorting compartment, for example, low pH and ionic conditions. Glycerophospholipids cannot form intermolecular hydrogen bonds in the same regions of the lipid bilayer since their ester or ether groups can act only as acceptors for the formation of hydrogen bonds (Curatolo, 1987a).

For this lipid sorting model to be correct, translocation of newly synthe-

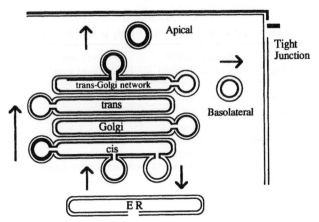

FIGURE 2 Epithelial lipid sorting model: segregation of apical glycosphingolipids from basolateral phospholipids in the luminal leaflet of the membrane of the *trans*-Golgi network. Transport of sorted lipids to the correct domain occurs by direct vesicular transport pathways. Maintenance of lipid polarity is assured by the tight junctions, barriers for lipid diffusion in the exoplasmic leaflet, selectively.

sized GlcCer (see preceding discussion) must occur in the Golgi complex. Indirect evidence from fluorescence microscopy experiments suggested that biosynthetic C_6-NBD-GlcCer indeed became trapped in a compartment colocalizing with the Golgi complex in Caco-2 cells (van't Hof and van Meer, 1990). Biosynthetic GlcCer analogs also have been found on the luminal surface of purified Golgi (Karrenbauer *et al.*, 1990) and of transport vesicles originating from the TGN (Kobayashi *et al.*, 1992). Still, the final elucidation of the mechanism by which GlcCer reaches the outer cell surface awaits the purification and immunolocalization of the putative "GlcCer translocase" activity.

We have obtained good evidence for a lipid sorting event in the lumen of the Golgi complex from the observations that different SM analogs reach the cell surface with a 2-fold difference in polarity (van't Hof *et al.*, 1992). The first indications are that C_6-NBD-SM and α-OH-C_6-NBD-SM also display considerable differences in their polarities of delivery (P. van der Bijl, M. Lopes-Cardozo, I. L. van Genderen, and G. van Meer, unpublished observations). A comparison of the various SM analogs within the very same cells, using double-label experiments with different fluorescent and radiolabeled ceramide precursors, will have to confirm independently the lipid sorting capacity in the lumen of the Golgi complex in epithelial cells.

IV. EPITHELIAL LIPID VERSUS PROTEIN SORTING

A. Sorting of Membrane Proteins and Lipids

Epithelial protein and lipid sorting have been proposed to be interlinked closely by partitioning of apical but not basolateral membrane proteins in GSL microdomains. The segregation step could be mediated by transmembrane sorting proteins. These proteins, or sorted transmembrane proteins themselves, could connect the luminal sorting information to cytoplasmic transport machinery for delivery to the correct plasma membrane domain (Simons and van Meer, 1988). To date, despite extensive efforts, no apical or basolateral sorting signals have been identified consistently on the luminal ectodomains of apical and basolateral proteins (see Rodriguez-Boulan et al., 1991; Rodriguez-Boulan and Powell, 1992).

In nonpolarized cells, the constitutive exocytic route to the plasma membrane is thought to operate by default, that is, delivery to the cell surface requires no specific signals (Wieland et al., 1987). Specific retention signals on proteins are thought to be essential for residence in the intracellular compartments along the biosynthetic route. The KDEL sequence on soluble resident ER proteins is an example (Pelham, 1991). In the past, the apical domain in epithelial cells has been regarded as a specialized part of the plasma membrane compared with nonpolarized cells based on functional, compositional, and topological properties. The apical pathway was proposed as the specific route requiring sorting signals, whereas the basolateral pathway, equivalent to the exocytic route in nonpolarized cells, was considered to work by default (Simons and Wandinger-Ness, 1990). According to this hypothesis, the one crucial event in epithelial protein sorting would be the active insertion of apical proteins into GSL microdomains. On the other hand, the identification of specific basolateral sorting signals on the cytoplasmic tails of various membrane proteins (Brewer and Roth, 1991; Casanova et al., 1991; Hunziker et al., 1991; Le Bivic et al., 1991; Matter et al., 1992) has suggested that the basolateral route is the specialized pathway (Hopkins, 1991; Mostov et al., 1992). In this model, basolateral proteins would be excluded actively from GSL domains.

No direct correlation exists between protein sorting and lipid sorting in different epithelial cell types. Some intestinal brush border enzymes, which in Caco-2 cells depend on transcytosis for final apical targeting (Le Bivic et al., 1990; Matter et al., 1990), are directly delivered apically after transfection into MDCK cells (Wessels et al., 1990; Low et al., 1991). Sphingolipid sorting, however, has been found to be identical in MDCK (van Meer et al., 1987) and Caco-2 cells (van 't Hof and van Meer, 1990).

Moreover, Caco-2 and Fischer rat thyroid (FRT) cells sort influenza hem-agglutinin (HA) to the apical and vesicular stomatitis virus (VSV) G pro-tein to the basolateral domain (Zurzolo et al., 1992), whereas sphingolipid sorting is reversed quantitatively in FRT cells (see Table II) relative to Caco-2 cells.

These results can be explained by the interesting speculation that, in epithelial cells, specific apical and basolateral protein sorting mechanisms both operate and that membrane proteins contain an apical or a basolateral targeting signal, or combinations of these. In the last case, the final site of delivery would rely on the difference in affinities of the respective signals for the two sorting mechanisms. Sorting, in other words, would consist of a competition between an apical and a basolateral mechanism. GSL microdomain formation, driven by hydrogen bonding and induced or facilitated by the conditions in the lumen of a sorting compartment, could provide the appropriate microenvironment for recognition of an apical sorting signal on a membrane protein. Basolateral sorting most likely would be driven by a cytoplasmic receptor. Structural differences in the mechanisms, for example, in the absolute amount or in the conforma-tion of a sorting component, could explain the different sorting patterns of a signal protein in different epithelial cell types.

B. Sorting of Glycosyl-Phosphatidylinositol Proteins and Lipids

Glycosyl-phosphatidylinositol (GPI)proteins constitute a novel class of proteins in which the C terminus is attached to the outer exoplasmic leaflet of cellular membranes via a GPI-lipid anchor (see Chapter 11). These proteins have been identified in a variety of mammalian cells and parasitic protozoans, with a host of different functions (Ferguson and Williams, 1988; Low and Saltiel, 1988; Low, 1989; Cross, 1990). To date, the precise function of the GPI linkage itself is not known. Biosynthetic studies have indicated that GPI proteins are synthesized with a membrane-linked C-terminal extension which, in the ER, is replaced by GPI within 1–5 min of completion of protein synthesis (Low, 1989; Cross, 1990). The attach-ment of GPI has been found to be directed by a cleavable signal at the C terminus (Caras and Weddell, 1989; Caras et al., 1989). The minimal requirements for GPI anchoring have been traced to a hydrophobic domain and a cleavage–attachment site consisting of a pair of small amino acids, positioned 10–12 residues N-terminally from the hydrophobic domain (Caras, 1991; Moran and Caras, 1991a,b). In all cases, the GPI glycolipid was observed to contain a conserved core structure consisting, in linear sequence, of PI, glucosamine, three mannoses, and a phosphorylethanolamine that is linked to the C terminus of the protein. The heterogeneity in the structure of GPI proteins among different cell types and species results from addition of galactoses, mannoses, or N-acetyl galactosamine to the core structure (Ferguson and Williams, 1988; Low and Saltiel, 1988; Low, 1989; Cross, 1990). Currently available evidence suggests the existence of preformed precursor GPI glycolipids that are transferred en bloc to the newly synthesized proteins, most likely by a transpeptidase mechanism that is independent of protein synthesis and energy (Mayor et al., 1991).

In epithelial cells, the intriguing observation was made that endogenous GPI proteins are localized exclusively on the apical surface (Lisanti et al., 1988,1990b). In hippocampal neurons, the GPI protein Thy-1 was localized exclusively on the axonal surface, the presumed neuronal equivalent of the epithelial apical domain (Dotti and Simons, 1990; Dotti et al., 1991; Rodriguez-Boulan and Powell, 1992), implying that the asymmetrical distribution of GPI proteins in polarized mammalian cells is a general phenomenon. The apical distribution in MDCK cells was found to be generated by direct apical delivery after biosynthesis and a subsequent slow rate of apical endocytosis (Lisanti et al., 1990a). Replacement of the transmembrane peptide of a normally basolateral protein with a GPI anchor resulted in redirection of the protein to the apical surface (Brown et al., 1989; Lisanti et al., 1989a; Powell et al., 1991), clearly showing that, in these cells, the GPI anchor is an apical sorting signal. The observations that, in MDCK cells, GSLs and GPI proteins are disposed apically has led to the suggestion of a common sorting mechanism (Rodriguez-Boulan and Nelson, 1989; Lisanti and Rodriguez-Boulan, 1990,1991). Additional support for this idea was obtained by studies in FRT cells, in which the polarity of delivery of both newly synthesized sphingolipids (see Table II) and GPI proteins (Zurzolo et al. (1993a); see also Chapter 11) is reversed. Brown and Rose (1992) tested the hypothesis that GPI anchors, being glycolipids, act as apical sorting signals in MDCK cells through partitioning into GSL microdomains. The fact that sphingolipids and biosynthetic GPI proteins displayed a selective resistance against Triton X-100 extraction from cells in the cold was considered evidence for such an event. However, at present the possibility cannot be ruled out that coresistance against the detergent treatment results from common biophysical properties of sphingolipids and GPI proteins, independent of clustering in microdomains. Nonetheless, application of the Triton X-100 assays to FRT cells and other epithelial cell lines with different sorting phenotypes could provide additional insight into the issue of whether sorting of GPI proteins depends on interactions with GSL microdomains. In the context of the model for membrane protein sorting proposed earlier, the GPI proteins are special in the sense that they lack a cytoplasmic tail. Therefore, the luminal signal for (apical) sorting would not be hindered by competition of the opposite cytoplasmic (basolateral) signal, resulting in strict asymmetrical cell surface delivery.

V. OTHER LIPID SORTING EVENTS

A. Sphingomyelin Sorting in the cis-Golgi

As described already, SM is synthesized in the cis-cisterna of the Golgi complex. In the past, the low level of SM in the ER (see van Meer, 1989) was considered a consequence of unidirectional transport from ER to the Golgi complex. However, evidence has been presented for return traffic of resident ER proteins that have slipped through to the cis-Golgi (Pelham, 1991). Also, under influence of the drug brefeldin A, the Golgi complex was observed to disassemble and mix with the ER (Klausner et al., 1992). Based on these observations, the existence of bidirectional transport routes between ER and Golgi has been proposed, leaving open the questions of whether the retrograde pathway has a vesicular nature (Klausner et al., 1992) and whether intermediate compartments are involved in transport between ER and Golgi (Schweizer et al., 1991). In any case, how SM is excluded from such a retrograde transport route must be explained. One interesting possibility is that SM is sorted in the cis-Golgi into the outgoing pathway by clustering into sphingolipid microdomains, separating the sphingolipids from the glycerolipids. This cis-Golgi sphingolipid sorting event could be the more general case of glycosphingolipid sorting in the TGN of epithelial cells, with possible important implications for the transport of newly synthesized cholesterol from the ER to the cell surface (van Meer, 1989).

B. Endocytic NBD-Sphingolipid Sorting

In a number of studies, researchers have concluded that sphingolipid sorting occurs in the endocytic pathway (Kok et al., 1989,1991; Koval and Pagano, 1989,1990). The vast majority of exogenously added C_6-NBD-GlcCer (Kok et al., 1989) and C_6-NBD-SM (Koval and Pagano 1990) has been described to be salvaged or sorted after internalization to the recycling pathway instead of to the degradative route to the lysosomes (Fig. 2). Sorting is this sense is defined as the preference of a single lipid to follow a particular transport pathway. However, if the membrane flux is larger in the recycling pathway than in the degradative route, C_6-NBD-GlcCer and C_6-NBD-SM simply would follow the bulk of transport. "Sorting," therefore, does not seem to be the proper term to describe such a process. This term is better reserved for an event by which the concentration of a lipid is increased in one pathway over another. This increase in concentration would have to be expressed as an enrichment of one lipid over total lipid or lipid surface area, that is, an increase in surface density. Since no lipid bulk flow markers are currently available, the comparison of the behavior of different lipid classes within the same experiment can be used as a relative measure of lipid sorting (van Meer and Burger, 1992). Using such an experimental approach, it was observed that, after insertion of C_6-NBD-GlcCer and C_6-NBD-SM into the plasma membrane of undifferentiated HT29 cells at low temperature and subsequent warming to 37°C, both products were internalized but their intracellular destination differed. By colocalization, C_6-NBD-GlcCer was found in the Golgi area whereas C_6-NBD-SM overlapped with endosomal and lysosomal compartments (Kok *et al.*, 1991). Additional biochemical analysis of the different compartments and more direct quantitation of the different transport routes will have to substantiate this endocytic lipid sorting event, which surprisingly seems to be absent in differentiated HT29 cells (Kok *et al.*, 1991).

VI. FINAL REMARKS

In this chapter, we have discussed the status of current research on the generation of epithelial lipid polarity. What developments are to be expected in the next years? One important theme will be the extension of the lipid sorting assay to all lipid classes present on the epithelial cell surface. First, this issue concerns PC, the major lipid on the basolateral cell surface. Using a double-label protocol with radiolabeled choline and a truncated analog of phosphatidic acid, an assay has been developed for the measurement of delivery of newly synthesized PC to the cell surface (W. van 't Hof and G. van Meer, unpublished results). In addition, investigators have discovered that the use of a ceramide carrying a short α-OH fatty acyl chain resulted in substantial production of both galactosyl- and glucosylceramide (P. van der Bijl, I. L. van Genderen, M. Lopes-Cardozo, and G. van Meer, unpublished observations). Comparison of the sorting of the two glycosphingolipids, which is being pursued, forms a new approach to study the involvement of hydrogen bonding in the sorting process. Obviously, the comparison of different epithelial cell types will be important in these studies, as exemplified by Table II.

A second focus of the research on epithelial lipid sorting will be the characterization of transcytosis, the vesicular pathway that connects the two plasma membrane domains of epithelial cells in both directions. Newly devised assays for quantitating lipid transcytosis (I. L., van Genderen, W. van 't Hof, and G. van Meer, unpublished observations) have demonstrated that, in MDCK cells, the equivalent of 20% of the total surface area of each plasma membrane domain is transported to the opposite surface per hour. The major challenge will be to test the prediction that lipid sorting along this pathway must occur to prevent lipid intermixing between the two cell surface domains.

The enzymes responsible for the synthesis of the simple sphingolipids are expected to be purified and cloned within the next 5 years. The availability of antibodies against these enzymes potentially will settle the controversies over the sites and sidedness of the individual events in lipid biosynthesis by biochemical and morphological means.

Even when these expectations are realized, studies on the interactions between the various lipid classes in model membranes still will be required to understand the behavior of the lipids at the molecular level. The hope seems justified that this broadened insight into the transport and sorting pattern of lipids in the various cell types will facilitate a more comprehensive view of the correlation between the epithelial sorting of lipids and that of the different classes of membrane proteins. Based on our experience over the past decade, it seems realistic to state that the elucidation of the molecular details of the epithelial sorting machinery will require yet another 10, hopefully exciting, years.

Acknowledgments

We are grateful to Alex Sandra for comments on the manuscript. During this work, G. van Meer was supported by a senior fellowship from the Royal Netherlands Academy of Arts and Sciences.

References

Anderson, J. M., and Stevenson, B. R. (1992). The molecular structure of the tight junction. *In* "Tight Junctions" (M. Cereijido, ed.), pp. 77–90. CRC Press, Boca Raton, Florida.

Bell, R. M., Ballas, L. M., and Coleman, R. A. (1981). Lipid topogenesis. *J. Lipid Res.* **22,** 391–403.

Bishop, W. R., and Bell, R. M. (1985). Assembly of the endoplasmic reticulum phospholipid bilayer: The phosphatidylcholine transporter. *Cell* **42,** 51–60.

Bode, F., Baumann, K., and Kinne, R. (1976). Analysis of the pinocytic process in rat kidney. II. Biochemical composition of pinocytic vesicles compared to brush-border microvilli, lysosomes and basolateral plasma membranes. *Biochim. Biophys. Acta* **433,** 294–310.

Boggs, J. M. (1987). Lipid intermolecular hydrogen bonding: Influence on structural organization and membrane function. *Biochim. Biophys. Acta* **906,** 353–404.

Bomsel, M., and Mostov, K. E. (1991). Sorting of plasma membrane proteins in epithelial cells. *Curr. Opin. Cell Biol.* **3**, 647–653.

Brasaemle, D. L., Robertson, A. D., and Attie, A. D. (1988). Transbilayer movement of cholesterol in the human erythrocyte membrane. *J. Lipid Res.* **29**, 481–489.

Brasitus, T. A., and Schachter, D. (1980). Lipid dynamics and lipid–protein interactions in the rat enterocyte basolateral and microvillus membranes. *Biochemistry* **19**, 2763–2769.

Brewer, C. B., and Roth, M. G. (1991). A single amino acid change in the cytoplasmic domain alters the polarized delivery of influenza virus hemagglutinin. *J. Cell Biol.* **114**, 413–421.

Brown, D. A., and Rose, J. K. (1992). Sorting of GPI-anchored proteins to glycolipid-enriched membrane subdomains during transport to the apical cell surface. *Cell* **68**, 533–544.

Brown, D. A., Crise, B., and Rose, J. K. (1989). Mechanism of membrane anchoring affects polarized expression of two proteins in MDCK cells. *Science* **245**, 1499–1501.

Caplan, M. J., and Matlin, K. S. (1989). Sorting of membrane and secretory proteins in polarized epithelial cells. *In* "Functional Epithelial Cells in Culture" (K. Matlin and J. Valentich, eds.), pp. 71–127. Liss, New York.

Caras, I. W. (1991). An internally positioned signal can direct attachment of a glycophospholipid membrane anchor. *J. Cell Biol.* **113**, 77–85.

Caras, I. W., and Weddell, G. N. (1989). Signal peptide for protein secretion directing glycophospholipid membrane anchor attachment. *Science* **243**, 1196–1198.

Caras, I. W., Weddell, G. N., and Williams, S. R. (1989). Analysis of the signal for attachment of a glycophospholipid membrane anchor. *J. Cell Biol.* **108**, 1387–1396.

Carmel, G., Rodrigue, F., Carrière, S., and Le Grimellec, C. (1985). Composition and physical properties of lipids from plasma membranes of dog kidney. *Biochim. Biophys. Acta* **818**, 149–157.

Casanova, J. E., Apodaca, G., and Mostov, K. E. (1991). An autonomous signal for basolateral sorting in the cytoplasmic domain of the polymeric immunoglobulin receptor. *Cell* **66**, 65–75.

Cereijido, M. (1992). Evolution of ideas on the tight junction. *In* "Tight Junctions" (M. Cereijido, ed.), pp 1–13. CRC Press, Boca Raton, Florida.

Chapelle, S., and Gilles-Baillien, M. (1983). Phospholipids and cholesterol in brush border and basolateral membranes from rat intestinal mucosa. *Biochim. Biophys. Acta* **753**, 269–271.

Cooper, R. A. (1978). Influence of increased membrane cholesterol on membrane fluidity and cell function in human red blood cells. *J. Supramol. Struct.* **8**, 413–430.

Coste, H., Martel, M. B., and Got, R. (1986). Topology of glucosylceramide synthesis in Golgi membranes from porcine submaxillary glands. *Biochim. Biophys. Acta* **858**, 6–12.

Cross, G. A. M. (1990). Glycolipid anchoring of plasma membrane proteins. *Annu. Rev. Cell Biol.* **6**, 1–39.

Curatolo, W. (1987a). The physical properties of glycolipids. *Biochim. Biophys. Acta* **906**, 111–136.

Curatolo, W. (1987b). Glycolipid function. *Biochim. Biophys. Acta* **906**, 137–160.

Devaux, P. F. (1991). Static and dynamic lipid asymmetry in cell membranes. *Biochemistry* **30**, 1163–1173.

Dotti, C., and Simons, K. (1990). Polarized sorting of viral glycoproteins to the axon and dendrites of hippocampal neurons in culture. *Cell* **62**, 63–72.

Dotti, C., Parton, R. G., and Simons, K. (1991). Polarized sorting of glypiated proteins in hippocampal neurons. *Nature (London)* **349**, 158–161.

Douglas, A. P., Kerley, R., and Isselbacher, K. J. (1972). Preparation and characterization

558   Wouter van 't Hof and Gerrit van Meer

of the lateral and basal plasma membranes of the rat intestinal epithelial cell. *Biochem. J.* **128**, 1329–1338.

Dowhan, W. (1991). Phospholipid-transfer proteins. *Curr. Opin. Cell Biol.* **3**, 621–625.

Dragsten, P. R., Blumenthal, R., and Handler, J. S. (1981). Membrane asymmetry in epithelia, is the tight junction a barrier to diffusion in the plasma membrane? *Nature (London)* **294**, 718–722.

Dragsten, P. R., Handler, J. S., and Blumenthal, R. (1982). Fluorescent membrane probes and the mechanism of maintenance of cellular asymmetry in epithelia. *Fed. Proc.* **41**, 48–53.

Farquhar, M. G., and Palade, G. E. (1963). Junctional complexes in various epithelia. *J. Cell Biol.* **17**, 375–412.

Ferguson, M. A. J., and Williams, A. F. (1988). Cell-surface anchoring of proteins via glycosylphosphatidylinositol structures. *Annu. Rev. Biochem.* **57**, 285–320.

Forstner, G. G., and Wherrett, J. R. (1973). Plasma membrane and mucosal glycosphingolipids in the rat intestine. *Biochim. Biophys. Acta* **306**, 446–459.

Forstner, G. G., Tanaka, K., and Isselbacher, K. J. (1968). Lipid composition of the isolated rat intestinal microvillus membrane. *Biochem. J.* **109**, 51–59.

Futerman, A. H., and Pagano, R. E. (1991). Determination of the intracellular sites and topology of glucosylceramide synthesis in rat liver. *Biochem. J.* **280**, 295–302.

Futerman, A. H., Stieger, B., Hubbard, A. L., and Pagano, R. E. (1990). Sphingomyelin synthesis in rat liver occurs predominantly at the cis and medial cisternae of the Golgi apparatus. *J. Biol. Chem.* **265**, 8650–8657.

Gorter, E., and Grendel, F. (1925). On bimolecular layers of lipids on the chromocytes of the blood. *J. Exp. Med.* **41**, 439–443.

Griffiths, G., and Simons, K. (1986). The trans Golgi network: Sorting at the exit site of the Golgi complex. *Science* **234**, 438–443.

Hauser, H., Howell, K., Dawson, R. M. C., and Bowyer, D. E. (1980). Rabbit small intestinal brush border membrane preparation and lipid composition. *Biochim. Biophys. Acta* **602**, 567–577.

Helms, J. B., Karrenbauer, A., Wirtz, K. W. A., Rothman, J. E., and Wieland, F. T. (1990). Reconstitution of steps in the constitutive secretory pathway in permeabilized cells. Secretion of glycosylated tripeptide and truncated sphingomyelin. *J. Biol. Chem.* **265**, 20027–20032.

Hirokawa, N. (1982). The intramembrane structure of tight junctions: An experimental analysis of the single-fibril and two-fibril models using the quick-freeze method. *J. Ultrastruct. Res.* **80**, 288–301.

Hise, M. K., Mantulin, W. W., and Weinman, E. J. (1984). Fluidity and composition of brush border and basolateral membranes from rat kidney. *Am. J. Physiol.* **247**, F434–F439.

Hopkins, C. R. (1991). Polarity signals. *Cell* **66**, 827–829.

Hunziker, W., Harter, C., Matter, K., and Mellman, I. (1991). Basolateral sorting in MDCK cells requires a distinct cytoplasmic domain determinant. *Cell* **66**, 907–920.

Jaffe, S., Oliver, P. D., Farooqui, S. M., Novak, P. L., Sorgente, N., and Kalra, V. K. (1987). Separation of luminal and abluminal membrane enriched domains from cultured bovine aortic endothelial cells: Monoclonal antibodies specific for endothelial cell plasma membranes. *Biochim. Biophys. Acta* **898**, 37–52.

Jeckel, D. J., Karrenbauer, A., Birk, R., Schmidt, R. R., and Wieland, F. T. (1990). Sphingomyelin is synthesized in the cis-Golgi. *FEBS Lett.* **261**, 155–157.

Jeckel, D. J., Karrenbauer, A., Burger, K. N. J., van Meer, G., and Wieland, F. T. (1992). Glucosylceramide is synthesized at the cytosolic surface of various Golgi subfractions. *J. Cell Biol.* **117**, 259–268.

Kachar, B., and Reese, T. S. (1982). Evidence for the lipidic nature of tight junction strands. *Nature (London)* **296**, 464–466.

Kaplan, M. R., and Simoni, R. D. (1985). Intracellular transport of phosphatidylcholine to the plasma membrane. *J. Cell Biol.* **101**, 441–445.

Karrenbauer, A., Jeckel, D. J., Just, W., Birk, R., Schmidt, R. R., Rothman, J. E., and Wieland, F. T. (1990). The rate of bulk flow from the Golgi to the plasma membrane. *Cell* **63**, 259–267.

Kawai, K., Fujita, M., and Nakao, M. (1974). Lipid components of two different regions of an intestinal epithelial cell membrane of mouse. *Biochim. Biophys. Acta* **369**, 222–233.

Klausner, R. D., Donaldson, J. G., and Lippincott-Schwartz, J. (1992). Brefeldin A: Insights into the control of membrane traffic and organelle structure. *J. Cell Biol.* **116**, 1071–1080.

Knoll, G., van Meer, G., Bron, R., Burger, K. N. J., and Verkleij, A. J. (1988). Fusion of liposomes with the plasma membrane of epithelial cells: Fate of implanted lipids as followed by freeze-fracture and autoradiography. *J. Cell Biol.* **107**, 2511–2521.

Kobayashi, T., and Pagano, R. E. (1989). Lipid transport during mitosis. Alternative pathways for delivery of newly synthesized lipids to the cell surface. *J. Biol. Chem.* **264**, 5966–5973.

Kobayashi, T., Pimplikar, S. W., Parton, R. G., Bhadki, S., and Simons, K. (1992). Sphingolipid transport from the trans Golgi network to the apical surface in permeabilized MDCK cells. *FEBS Lett.* **300**, 227–231.

Kok, J. W., Eskelinen, S., Hoekstra, K., and Hoekstra, D. (1989). Salvage of glucosylceramide by recycling after internalization along the pathway of receptor-mediated endocytosis. *Proc. Natl. Acad. Sci. U.S.A.* **86**, 9896–9900.

Kok, J. W., Babia, T., and Hoekstra, D. (1991). Sorting of sphingolipids in the endocytic pathway of HT29 cells. *J. Cell Biol.* **114**, 231–239.

Koval, M., and Pagano, R. E. (1989). Lipid recycling between the plasma membrane and intracellular compartments: Transport and metabolism of fluorescent sphingomyelin analogues in cultured fibroblasts. *J. Cell Biol.* **108**, 2169–2181.

Koval, M., and Pagano, R. E. (1990). Sorting of an internalized plasma membrane lipid between recycling and degradative pathways in normal and Niemann–Pick, Type A fibroblasts. *J. Cell Biol.* **111**, 429–442.

Koval, M., and Pagano, R. E. (1991). Intracellular transport and metabolism of sphingomyelin. *Biochim. Biophys. Acta* **1082**, 113–125.

Kremmer, T., Wisher, M. H., and Evans, W. H. (1976). The lipid composition of plasma membrane subfractions originating from the three major functional domains of the rat hepatocyte cell surface. *Biochim. Biophys. Acta* **455**, 655–664.

Le Bivic, A., Quaroni, A., Nichols, B., and Rodriguez-Boulan, E. (1990). Biogenetic pathways of plasma membrane proteins in Caco-2, a human intestinal epithelial cell line. *J. Cell Biol.* **111**, 1351–1362.

Le Bivic, A., Sambuy, Y., Patzak, N., Patill, M., Chao, M., and Rodriguez-Boulan, E. (1991). An internal deletion in the cytoplasmic tail reverses the apical localization of human NGF receptor in transfected MDCK cells. *J. Cell Biol.* **115**, 607–618.

Lipsky, N. G., and Pagano, R. E. (1983). Sphingolipid metabolism in cultured fibroblasts: Microscopic and biochemical studies employing a fluorescent ceramide analogue. *Proc. Natl. Acad. Sci. U.S.A.* **80**, 2608–2612.

Lipsky, N. G., and Pagano, R. E. (1985a). Intracellular translocation of fluorescent sphingolipids in cultured fibroblasts: Endogenously synthesized sphingomyelin and glucocerebroside analogues pass through the Golgi apparatus en route to the plasma membrane. *J. Cell Biol.* **100**, 27–34.

Lipsky, N. G., and Pagano, R. E. (1985b). A vital stain for the Golgi apparatus. *Science*, **228**, 745–747.

Lisanti, M. P., and Rodriguez-Boulan, E. (1990). Glycophospholipid membrane anchoring provides clues to the mechanism of protein sorting in polarized epithelial cells. *Trends Biochem. Sci.* **15**, 113–118.

Lisanti, M. P., and Rodriguez-Boulan, E. (1991). Polarized sorting of GPI-linked proteins in epithelia and membrane microdomains. *Cell Biol. Int. Rep.* **15**, 1023–1049.

Lisanti, M. P., Sargiacomo, M., Graeve, L., Saltiel, A., and Rodriguez-Boulan, E. (1988). Polarized apical distribution of glycosyl-phosphatidylinositol-anchored proteins in a renal epithelial cell line. *Proc. Natl. Acad. Sci. U.S.A.* **85**, 9557–9561.

Lisanti, M. P., Caras, I. W., Davitz, M. A., and Rodriguez-Boulan, E. (1989a). A glycophospholipid membrane anchore acts as an apical targeting signal in polarized epithelial cells. *J. Cell Biol.* **109**, 2145–2156.

Lisanti, M. P., Le Bivic, A., Sargiacomo, M., and Rodriguez-Boulan, E. (1989b). Steady state distribution and biogenesis of endogenous Madin-Darby Canine Kidney glycoproteins: Evidence for intracellular sorting and polarized cell surface delivery. *J. Cell Biol.* **109**, 2117–2127.

Lisanti, M. P., Caras, I. W., Gilbert, T., Hanzel, D., and Rodriguez-Boulan, E. (1990a). Vectorial apical delivery and slow endocytosis of a glycolipid-anchored fusion protein in transfected MDCK cells. *Proc. Natl. Acad. Sci. U.S.A.* **87**, 7419–7423.

Lisanti, M. P., Le Bivic, A., Saltiel, A. R., and Rodriguez-Boulan, E. (1990b). Preferred apical distribution of glucosyl-phosphatidylinositol (GPI) anchored proteins: A highly conserved feature of the polarized epithelial cell phenotype. *J. Membr. Biol.* **113**, 155–167.

Low, M. G. (1989). The glycosyl-phosphatidylinositol anchor of membrane proteins. *Biochim. Biophys. Acta* **988**, 427–454.

Low, M. G., and Saltiel, A. R. (1988). Structural and functional role of glycosylphosphatidylinositol in membranes. *Science* **239**, 268–275.

Low, S. H., Wong, S. H., Tang, B. L., Subramaniam, V. N., and Hong, W. (1991). Apical cell surface expression of rat dipeptidyl peptidase IV in transfected Madin-Darby canine kidney cells. *J. Biol. Chem.* **266**, 13391–13396.

Matter, K., Brauchbar, M., Bucher, K., and Hauri, H.-P. (1990). Sorting of endogenous plasma membrane proteins occurs from two sites in cultured human intestinal epithelial cells (Caco-2). *Cell* **60**, 429–437.

Matter, K., Hunziker, W., and Mellman, I. (1992). Basolateral sorting of LDL-receptor in MDCK cells: The cytoplasmic domain contains two tyrosine-dependent targeting determinants. *Cell* **71**, 741–753.

Mayor, S., Menon, A. K., and Cross, G. A. M. (1991). Transfer of glycosylphosphatidylinositol membrane anchors to polypeptide acceptores in a cell-free system. *J. Cell Biol.* **114**, 61–71.

Meier, P. J., Sztul, E. S., Reuben, A., and Boyer, J. L. (1984). Structural and functional polarity of canalicular and basolateral plasma membrane vesicles isolated in high yield from rat liver. *J. Cell Biol.* **98**, 991–1000.

Molitoris, B. A., and Simon, F. R. (1985). Renal cortical brush-border and basolateral membranes: Cholesterol and phospholipid composition and relative turnover. *J. Membr. Biol.* **83**, 207–215.

Moran, P., and Caras, I. W. (1991a). A nonfunctional sequence converted to a signal for glycophosphatidylinositol membrane anchor attachment. *J. Cell Biol.* **115**, 329–336.

Moran, P., and Caras, I. W. (1991b). Fusion of sequence elements from non-anchored proteins to generate a fully functional signal for glycophosphatidylinositol membrane anchor attachment. *J. Cell Biol.* **115**, 1595–1600.

Mostov, K. E., Apodaca, G., Aroeti, B., and Okamoto, C. (1992). Plasma membrane protein sorting in polarized epithelial cells. *J. Cell Biol.* **116**, 577–583.

Nelson, W. J. (1992). Regulation of cell surface polarity from bacteria to mammals. *Science* **258**, 948–955.

Nichols, G. E., Borgman, C. A., and Young, W. W., Jr. (1986). On tight junction structure: Forssman glycolipid does not flow between MDCK cells in an intact epithelial monolayer. *Biochem. Biophys. Res. Commun.* **138**, 1163–1169.

Nichols, G. E., Shiraishi, T., Allietta, M., Tillack, T. W., and Young, W. W., Jr. (1987). Polarity of the Forssman glycolipid in MDCK epithelial cells. *Biochim. Biophys. Acta* **930**, 154–166.

Nichols, G. E., Shiraishi, T., and Young, W. W., Jr. (1988). Polarity of neutral glycolipids, gangliosides, and sulfated lipids in MDCK epithelial cells. *J. Lipid Res.* **29**, 1205–1213.

Op den Kamp, J. A. F. (1979). Lipid asymmetry in membranes. *Annu. Rev. Biochem.* **48**, 47–71.

Pagano, R. E. (1990). Lipid traffic in eukaryotic cells: Mechanisms for intracellular transport and organelle-specific enrichment of lipids. *Curr. Opin. Cell Biol.* **2**, 652–663.

Pagano, R. E., and Sleight, R. G. (1985). Emerging problems in the cell biology of lipids. *Trends Biochem. Sci.* **10**, 421–425.

Pascher, I. (1976). Molecular arrangements in sphingolipids. Conformation and hydrogen bonding of ceramide and their implication on membrane stability and permeability. *Biochim. Biophys. Acta* **455**, 433–451.

Pelham, H. R. B. (1991). Recycling of proteins between the endoplasmic reticulum and Golgi complex. *Curr. Opin. Cell Biol.* **3**, 585–591.

Pinto da Silva, P., and Kachar, B. (1982). On tight-junction structure. *Cell* **28**, 441–450.

Powell, S. K., Cunningham, B. A., Edelman, G. M., and Rodriguez-Boulan, E. (1991). Targeting of transmembrane and GPI-anchored forms of N-CAM to opposite domains of a polarized epithelial cell. *Nature (London)* **353**, 76–77.

Rabito, C. A. (1992). Tight junctions and apical/basolateral polarity. *In* "Tight Junctions" (M. Cereijido, ed.), pp. 203–214. CRC Press, Boca Raton, Florida.

Reuss, L. (1992). Tight junction permeability to ions and water. *In* "Tight Junctions" (M. Cereijido, ed.), pp. 49–66. CRC press, Boca Raton, Florida.

Rodriguez-Boulan, E., and Nelson, W. J. (1989). Morphogenesis of the polarized epithelial cell phenotype. *Science* **245**, 718–725.

Rodriguez-Boulan, E., and Powell, S. K. (1992). Polarity of epithelial and neuronal cells. *Annu. Rev. Cell Biol.* **8**, 395–427.

Rodriguez-Boulan, E., Lisanti, M. P., Powell, S. K., Gilbert, T., and Le Bivic, A. (1991). Protein targeting pathways and sorting signals in epithelial cells. *In* "Tight Junctions" (M. Cereijido, ed.), pp. 231–241. CRC Press, Boca Raton, Florida.

Rothman, J. E., and Orci, L. (1992). Molecular dissection of the secretory pathway. *Nature (London)* **355**, 409–415.

Sandra, A., and Pagano, R. E. (1978). Phospholipid asymmetry in LM cell plasma membrane derivatives: Polar head group and acyl chain distributions. *Biochemistry* **17**, 332–338.

Sasaki, T. (1990). Glycolipid transfer protein and intracellular traffic of glucosylceramide. *Experientia* **46**, 611–616.

Schwarzmann, G., and Sandhoff, K. (1990). Metabolism and intracellular transport of glycosphingolipids. *Biochemistry* **29**, 10865–10871.

Schweizer, A., Matter, K., Ketcham, C. M., and Hauri, H.-P. (1991). The isolated ER-Golgi intermediate compartment exhibits properties that are different from ER and cis-Golgi. *J. Cell Biol.* **113**, 45–54.

Schwertz, D. W., Kreisberg, J. I., and Venkatachalam, M. A. (1983). Characterization of rat kidney proximal tubule brush border membrane-associated phosphatidylinositol phosphodiesterase. *Arch. Biochem. Biophys.* **224**, 555–567.

Simons, K., and Fuller, S. D. (1985). Cell surface polarity in epithelia. *Annu. Rev. Cell Biol.* **1**, 243–288.

Simons, K., and van Meer, G. (1988). Lipid sorting in epithelial cells. *Biochemistry* **27**, 6197–6202.

Simons, K., and Wandinger-Ness, A. (1990). Polarized sorting in epithelia. *Cell* **62**, 207–210.

Singer, S. J., and Nicolson, G. L. (1972). The fluid mosaic model of the structure of cell membranes. *Science* **175**, 720–731.

Sleight, R. G., and Abanto, M. N. (1989). Differences in intracellular transport of a fluorescent phosphatidylcholine analog in established cell lines. *J. Cell Sci.* **93**, 363–374.

Spiegel, S., Blumenthal, R., Fishman, P. H., and Handler, J. S. (1985). Gangliosides do not move from apical to basolateral plasma membrane in cultured epithelial cells. *Biochim. Biophys. Acta* **821**, 310–318.

Stubbs, C. D., Ketterer, B., and Hicks, R. M. (1979). The isolation and analysis of the luminal plasma membrane of calf urinary bladder epithelium. *Biochim. Biophys. Acta* **558**, 58–72.

Thompson, T. E., and Tillack, T. W. (1985). Organization of glycosphingolipids in bilayers and plasma membranes of mammalian cells. *Annu. Rev. Biophys. Biophys. Chem.* **14**, 361–386.

Thompson, T. E., Barenholz, Y., Brown, R. E., Correa-Freire, M., Young, W. W., Jr., and Tillack, T. W. (1986). Molecular organization of glycosphingolipids in phosphatidylcholine bilayers and biological membranes. *In* "Enzymes of Lipid Metabolism" (L. Freysz, ed.), pp. 387–396. Plenum Press, New York.

Trinchera, M., Fabbri, M., and Ghidoni, R. (1991). Topography of glycosyltransferases involved in the initial glycosylation of gangliosides. *J. Biol. Chem.* **266**, 20907–20912.

van Genderen, I. L., van Meer, G., Slot, J. W., Geuze, H. J., and Voorhout, W. (1991). Subcellular localization of Forssman glycolipid in epithelial MDCK cells by immunoelectronmicroscopy after freeze substitution. *J. Cell Biol.* **115**, 1009–1019.

van Meer, G. (1988). How epithelial cells grease their microvilli. *Trends Biochem. Sci.* **13**, 242–243.

van Meer, G. (1989). Lipid traffic in animal cells. *Annu. Rev. Cell Biol.* **5**, 247–275.

van Meer, G., and Burger, K. N. J. (1992). Sphingolipid trafficking—sorted out? *Trends Cell Biol.* **2**, 332–337.

van Meer, G., and Simons, K. (1982). Viruses budding from either the apical or the basolateral plasma membrane domain of MDCK cells have unique phospholipid compositions. *EMBO J.* **1**, 847–852.

van Meer, G., and Simons, K. (1986). The function of tight junctions in maintaining differences in lipid composition between the apical and basolateral cell surface domains of MDCK cells. *EMBO J.* **5**, 1455–1464.

van Meer, G., and van 't Hof, W. (1993). Epithelial sphingolipid sorting is insensitive to reorganization of the Golgi by nocodazole, but is abolished by monensin in MDCK cells and by brefeldin A in Caco-2 cells. *J. Cell Sci.* **104**, 833–842.

van Meer, G., Gumbiner, B., and Simons, K. (1986). The tight junction does not allow lipid molecules to diffuse from one epithelial cell to the next. *Nature (London)* **322**, 639–641.

van Meer, G., Stelzer, E. H. K., Wijnaendts-van-Resandt, R. W., and Simons, K. (1987). Sorting of sphingolipids in epithelial (Madin-Darby canine kidney) cells. *J. Cell Biol.* **105**, 1623–1635.

van Meer, G., van 't Hof, W., and van Genderen, I. (1992). Tight junctions and polarity of lipids. *In* "Tight Junctions" (M. Cereijido, ed.), pp. 187–201. CRC Press, Boca Raton, Florida.

van 't Hof, W., and van Meer, G. (1990). Generation of lipid polarity in intestinal epithelial

(Caco-2) cells, Sphingolipid synthesis in the Golgi complex and sorting before vesicular traffic to the plasma membrane. *J. Cell Biol.* **111**, 977–986.

van 't Hof, W., Silvius, J., Wieland, F., and van Meer, G. (1992). Epithelial sphingolipid sorting allows for extensive variation of the fatty acyl chain and the sphingosine backbone. *Biochem. J.* **283**, 913–917.

Voorhout, W. F., van Genderen, I. L., Yoshioka, T., Fukami, K., Geuze, H. J., and van Meer, G. (1992). Subcellular localization of glycolipids as revealed by immunoelectronmicroscopy. *Trends Glycosci. Glycotechnol.* **4**, 533–546.

Wattenberg, B. W. (1990). Glycolipid and glycoprotein transport through the Golgi complex are similar biochemically and kinetically. Reconstitution of glycolipid transport in a cell free system. *J. Cell Biol.* **111**, 421–428.

Wessels, H. P., Hansen, G. H., Fuhrer, C., Look, A. T., Sjöström, H., Noren, O., and Spiess, M. (1990). Aminopeptidase N is directly sorted to the apical domain in MDCK cells. *J. Cell Biol.* **111**, 2923–2930.

Wieland, F. T., Gleason, M. L., Serafini, T. A., and Rothman, J. E. (1987). The rate of bulk flow from the endoplasmic reticulum to the cell surface. *Cell* **50**, 289–300.

Wirtz, K. W. A. (1991). Phospholipid transfer proteins. *Annu. Rev. Biochem.* **60**, 73–99.

Yechiel, E., and Edidin, M. (1987). Micrometerscale domains in fibroblast plasma membranes. *J. Cell Biol.* **105**, 755–760.

Zachowski, A., and Devaux, P. F. (1990). Transmembrane movements of lipids. *Experientia* **46**, 644–656.

Zurzolo, C., Polistina, C., Saini, M., Gentile, R., Aloj, L., Migliaccio, G., Bonatti, S., and Nitsch, L. (1992). Opposite polarity of virus budding and of viral envelope glycoprotein distribution in epithelial cells derived from different tissues. *J. Cell Biol.* **117**, 551–564.

Zurzolo, C., Lisanti, M. P., Caras, I. W., Nitsch, L., and Rodriguez-Boulan, E. (1993a). Glycosylphosphatidylinositol-anchored proteins are preferentially targeted to the basolateral surface in Fischer rat thyroid epithelial cells. *J. Cell Biol.* **121**, 1031–1039.

Zurzolo, C., van 't Hof, W., van Meer, G., and Rodriguez-Boulan, E. (1993b). VIP21/caveolin, glycosphingolipid clusters and the sorting of glycosylphosphatidylinositol anchored proteins in epithelial cells. *EMBO J.* (*in press*).

CHAPTER 22

Reconstitution of Glycolipid Transport between Compartments of the Golgi in a Cell-Free System

Binks W. Wattenberg
Cell Biology Unit, The Upjohn Company, Kalamazoo, Michigan 49007

I. WHY RECONSTITUTE GLYCOLIPID TRANSPORT *IN VITRO*?

An impressive understanding of the genesis and movement of glyco-sphingolipids has been gained by the study of intact cells. The use of pulse–chase labeling techniques and the advent of fluorescent precursors has revealed many aspects of the pathways and requirements of trafficking of these essential molecules. What is to be gained by breaking open cells and studying this process with isolated organelles? Two goals, one immediate and one long term, that would be difficult or impossible to achieve through the study of intact cells can be attained using the broken cell technology. The short-term question, addressed in another publication from this laboratory (Wattenberg, 1990), is whether glycolipids and glyco-proteins are transported by the same machinery. Abundant evidence suggests that protein transport through the Golgi is accomplished by the sequential formation, targeting, and fusion of transport vesicles (Rothman and Orci, 1992). Several tests can be applied to determine if this is also true for glycolipid transport. A more comprehensive question pertains to the nature of the molecular mechanisms that govern the movement of glycosphingolipids through the Golgi and to the cell surface. The essence of the cell-free system is that it allows direct manipulation of the organelles and their environment without the intervening complication of the plasma membrane or cellular metabolism. This reconstitution allows investigation of the role of proteins *and* lipids in this process by the direct manipulation of each. The armamentarium of assays that can be applied to the problem includes fractionation of cytosolic components, use of chemical and immunochemical inhibitors, manipulation of concentrations of small molecules, and alteration of the lipid content of the Golgi. In addition, in this isolated system, kinetic analysis can be applied with considerable precision. Finally, advantage can be taken of the substantial background derived from the study of protein trafficking. This chapter focuses on the assay developed in this laboratory using isolated Golgi fractions. However the measurement of glycolipid transport to the cell surface also has been achieved in permeabilized cells (Helms *et al.*, 1990; Kobayashi *et al.*, 1992). These systems afford many of the advantages just mentioned of the assay using isolated Golgi fractions. Results from the permeabilized cell systems are cited where appropriate.

II. ASSAY FOR GLYCOLIPID TRANSPORT BETWEEN COMPARTMENTS OF THE GOLGI

The logic of this assay is borrowed directly from that used to reconstitute protein transport between Golgi compartments (Balch *et al.*, 1984a). This strategy uses two populations of Golgi membranes, each derived from different cells. The molecule whose transport will be measured initially resides in a "donor" Golgi population, derived from cells deficient in a relevant glycosylation reaction. Thus, the molecule of interest is underglycosylated in the donor membranes. Movement to the "acceptor" population, which is competent for the glycosylation reaction that is lacking in donor membranes, is marked by the appropriate carbohydrate modification. In the transport system discussed here (Fig. 1), the "donor" membranes are derived from the lec2 glycosylation mutant of Chinese hamster ovary (CHO) cells (Stanley and Siminovitch, 1977). Lec2 cells are deficient in the membrane transport system that delivers CMP-sialic acid from its site of synthesis in the cytosol to the lumen of the Golgi (Deutscher *et al.*, 1984). Because the sialyltransferases acting on both glycolipids and glycoproteins use CMP-sialic acid as a substrate, the consequence of this mutation is a deficiency of sialylation of both proteins and lipids. As will

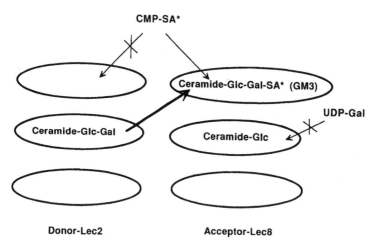

FIGURE 1   Assay for intercompartmental glycolipid transport. Transport of lactosylceramide is measured between donor and acceptor Golgi populations. The donor population is from *lec2*, a glycosylation mutant lacking the Golgi CMP-sialic acid transporter that is deficient in all sialylated glycoconjugates including the ganglioside G_{M3} (ceramide-Glc-Gal-SA). The predominant glycolipid in these cells is, therefore, lactosylceramide (ceramide-Glc-Gal). The acceptor population is from *lec8*, a mutant lacking the UDP-galactose transporter (Deutscher and Hirschberg, 1986). The major glycolipid in these cells is glucosylceramide, which is not a substrate for the sialyltransferase that converts lactosylceramide to G_{M3}. If lactosylceramide is transported from donor to acceptor, it will be sialylated. Transport is quantified in the assay by the inclusion of CMP-[^{3}H]sialic acid in the incubation and measurement of the incorporation of [^{3}H]sialic acid into G_{M3}.

become apparent, this consequence is a distinct advantage experimentally because the transport of lipids and proteins can be measured in the same system. As a result of this mutation, the major glycosphingolipid in lec2 cells and, therefore, in the donor Golgi membranes is lactosylceramide (LacCer) rather than the G_{M3} (sialic acid 2,3-LacCer) present in wild-type CHO cells. Acceptor membranes are derived from a different glycosylation mutant, lec8, which lacks the Golgi transporter for UDP-galactose (Deutscher and Hirschberg, 1986). The predominant glycolipid in these cells, therefore, is glucosylceramide because, although the cells contain a complete apparatus for sialylation, the sialyltransferase substrates that terminate in galactose cannot be generated. When these two membrane populations are mixed, the only way to generate G_{M3} is for the LacCer in the donor membranes to be exposed to the sialylation machinery present in the acceptor membranes. The generation of G_{M3} therefore is interpreted to indicate that LacCer has been transported from donor to acceptor membranes. The measurement itself is accomplished routinely by adding

CMP-[^{3}H]sialic acid to the transport reactions and measuring the levels of [^{3}H] found in G_{M3} by one-dimensional thin-layer chromatography (TLC) (Wattenberg, 1990). In Fig. 1, the transport is depicted to be from a medial compartment of the Golgi to the *trans* compartment. In practice, however, knowing exactly which Golgi compartment in the donor is the origin of LacCer is not possible. Originally the destination was thought to be a *trans*-Golgi compartment, since protein sialyltransferases had been localized to the *trans*-Golgi. However, evidence indicates that the LacCer sialyltransferase may reside in a more medial Golgi compartment (Trinchera and Ghidoni, 1989). For these reasons, we can say with certainty only that this assay measures transport between Golgi cisternae. Specifying exactly which cisternae are involved is not currently possible.

The same membranes used to measure glycolipid transport can be used to measure the transport of a membrane protein. Lec2 cells are infected with vesicular stomatitis virus (VSV), which produces a single membrane glycoprotein (VSV G), and donor membranes are prepared. This VSV G protein is not sialylated and will become so only if transport of VSV G from donor to acceptor occurs. In this way, protein and glycolipid transport can be measured under exactly the same conditions and a precise comparison can be made.

III. IS IT TOO GOOD TO BE TRUE? ASSESSING THE VALIDITY OF THE *IN VITRO* TRANSPORT ASSAY

Since the introduction by Rothman and colleagues of a cell-free system measuring transport of VSV G between compartments of the Golgi, spirited discussion has taken place about how well this system mimics events in the intact cell (Mellman and Simons, 1992). At its core, this reconstitution was simply an extension of the biochemist's approach to studying intracellular reactions by breaking open the cell and measuring a chemical transformation using partially or fully purified cellular components. However, in this case, not a single reaction but many are reconstituted and, in many cases, neither the substrates nor the products are simple chemical entities, but are molecular assemblies such as Golgi membranes and transport vesicles. Given this complexity, questioning the fidelity of the reconstruction is valid, indeed, essential. However, recognizing that the transport assay may reconstitute all the details of transport occurring in the intact cell imperfectly and still be of enormous value in uncovering essential components of the system is equally important.

The first issue that must be addressed is whether the sialylation of LacCer is simply the result of nonspecific fusion of Golgi cisternae from donor and acceptor, rather than of vesicular transport. Careful studies of the morphology of Golgi membranes, performed under identical transport conditions, indicate that the Golgi stacks do not change in size or shape (Braell *et al.*, 1984). Therefore no morphological evidence for fusion of donor and acceptor Golgi exists. In addition, the glycosylation of proteins in a similar system is extremely specific; only transported proteins, not resident Golgi proteins as would be expected if fusion were occurring, are glycosylated (B. Wattenberg, unpublished data).

Having ruled out nonspecific Golgi fusion, an additional test of the accuracy of the transport assay is whether it exhibits biochemical properties expected of the *in vivo* process. These features include a requirement for energy, elevated temperature, and intact membranes. Figure 2 shows the levels of G_{M3} generated under various incubation conditions to test these expectations. A complete incubation leads to the incorporation of almost 4000 cpm [^{3}H]sialic acid into G_{M3}. If donor or acceptor membranes are omitted from the assay, very little incorporation is found, demonstrating that neither the lec2-derived Golgi nor the lec8-derived Golgi have the ability individually to generate G_{M3}. If ATP is omitted from the incubation, no labeling of G_{M3} is seen. ATP is required in virtually all

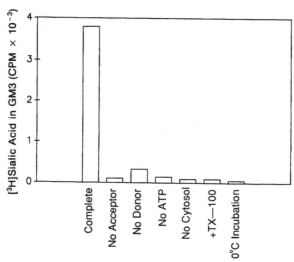

FIGURE 2 Conditions promoting glycolipid transport *in vitro*. Complete incubations included donor and acceptor membranes, Chinese hamster ovary (CHO) cell cytosol, and an ATP regenerating system. Incubations were performed for 2 hr at 37°C except as noted. Omission of any one of the components, addition of 0.1% Triton X-100, or incubation on ice instead of 37°C eliminates transport.

transport systems studied, both *in vivo* and *in vitro*, including permeabilized systems measuring sphingolipid transport (Kobayashi and Pagano, 1988; Helms *et al.*, 1990). However, the glycosylation reaction mediated by the LacCer sialyltransferase does not require ATP. The ATP dependence is a strong demonstration that G_{M3} is not generated by the access of LacCer to the sialyltransferase in fragmented or permeable lec2 and lec8 Golgi. This conclusion is supported by the finding that intentionally solubilizing the membranes with Triton X-100 completely destroys the ability of membranes to accomplish the sialylation of G_{M3}, indicating that the formation of G_{M3} noted under transport conditions requires that both substrates and glycosyltransferase be confined to the lumen of the Golgi.

IV. ARE GLYCOLIPIDS AND GLYCOPROTEINS TRANSPORTED BY THE SAME MECHANISM?

This transport reaction fulfills a number of criteria that indicate that it does indeed reflect the process by which glycolipid is transported between compartments of the Golgi. Another confirmation of this assertion would be documentation the participation of a protein that is known to be required for protein transport through the Golgi. The *N*-ethylmaleimide-sensitive factor (NSF) has been shown to be required for this transport (Block *et al.*, 1988), as well as for transport from the endoplasmic reticulum (ER) to the Golgi (Beckers *et al.*, 1989), both in a mammalian system and in yeast systems, and for fusion of endocytic vesicles (Diaz *et al.*, 1989). The case for NSF involvement in vesicular traffic is particularly compelling. NSF initially was discovered by analysis of the *in vitro* Golgi protein transport assay. Subsequently this factor was found to be encoded by a gene that, when mutated, produces a block in protein transport in living yeast cells (Wilson *et al.*, 1989). This result confirms the participation of NSF in vesicular transport *in vivo*. A blocking antibody against NSF was tested in the glycolipid transport assay and was found to inhibit transport (Fig. 3). To correlate the transport of glycolipid with that of glycoprotein, the very same membranes used to measure glycolipid transport were used to measure the effect of the anti-NSF antibody on the transport of VSV G, as outlined earlier (Fig. 3). The responses of the two transport systems to this antibody were nearly identical. The requirement for NSF was confirmed by inactivating NSF with *N*-ethylmaleimide and measuring the response of each system to the addition of purified NSF (not shown). Again, the response of the glycolipid and glycoprotein transport assays were very similar.

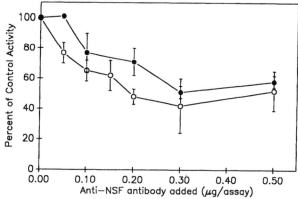

FIGURE 3 Glycolipid and glycoprotein transport are inhibited similarly by antibody against the N-ethylmaleimide sensitive factor (NSF). Glycoprotein (open circles) and glycolipid (filled circles) transport incubation mixtures contained acceptor and vesicular stomatitis virus (VSV) infected donor, Chinese hamster ovary (CHO) cell cytosol, the ATP regenerating system, and buffer. Glycoprotein transport assays contained 0.5 μCi [³H]CMP-sialic acid whereas glycolipid assays contained twice as much radioactivity. Assays were incubated with the indicated amount of antibody against NSF (6.7 mg/ml in 20 mM Tris/150 mM NaCl/1 mM EDTA) for 15 min on ice before initiating the assays for 2 hr at 37°C. The volumes of added antibody were compensated by the addition of buffer of the same composition. The averages of duplicate assays are shown.

Another group of proteins whose involvement in transport processes has become widely established consists of the GTP binding proteins. The small molecular weight GTP binding proteins of the rab and ARF (ADP-ribosylation factor) families, as well as (more tentatively) those of the trimeric GTP binding protein families, have been implicated in various transport processes (Balch, 1990). The nonhydrolyzable analog of GTP, GTPγS, will inhibit transport from the ER to the Golgi in both mammalian and yeast systems (Baker *et al.*, 1988; Ruohola *et al.*, 1988; Beckers and Balch, 1989), as well as protein transport through the Golgi (Melançon *et al.*, 1987). The effect of GTPγS therefore was tested on glycolipid transport and, in the same membranes, on glycoprotein transport (Fig. 4). Once again the responses of the transport of these two types of molecules were found to be identical. A similar inhibition by GTPγS has been noted in a permeabilized system measuring sphingolipid transport (Helms *et al.*, 1990).

This biochemical characterization was complemented by a kinetic comparison of glycolipid and glycoprotein transport. The kinetics of a process

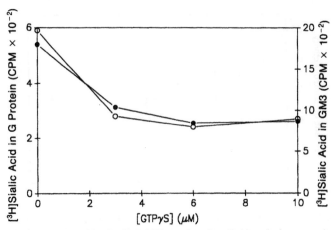

FIGURE 4 GTPγS is identically inhibitory for glycolipid and glycoprotein transport. Incubations for analysis of glycolipid (G_{M3}, closed circles) and glycoprotein (G protein, open circles) transport were performed identically. GTPγS was added to the indicated final concentrations. Assays were incubated for 2 hr at 37°C.

are a reflection of the underlying mechanisms, so differences in the machinery transporting glycolipids and glycoproteins should be apparent in their respective rates of transport. The time course of sialic acid addition to both LacCer and VSV G protein is shown in Fig. 5. Each shows a lag time of approximately 20–25 min, followed by a linear rate of sialic acid incorporation and a plateau after about 2 hr. In addition to the rate of glycosylation, the rate of entry into a transport intermediate known as the low cytosol-requiring intermediate (LCRI) was measured also (Wattenberg et al., 1986). The dependence of transport on cytosol concentration decreases as the reaction proceeds. The initial interpretation of this result was that a wave of transport has carried vesicles from the donor to the acceptor membranes, but these vesicles are slow to fuse with acceptor membranes (Wattenberg et al., 1986). In this model, fusion requires less cytosol than the steps that precede it, leading to the reduced cytosol dependence. As illustrated in Fig. 6, entry into this intermediate occurs substantially before glycosylation itself (compare Fig. 5 and Fig. 6) and exhibits no lag time. Using this kinetic measure, the rates of glycolipid and glycoprotein transport are again very similar. Note, however, that kinetic analysis of transport based on glycosylation rests on the assumption that glycosylation is very rapid relative to the rate of transport. This assumption does not hold in the system measuring protein transport to the medial Golgi, in which transport is marked by the addition of N-

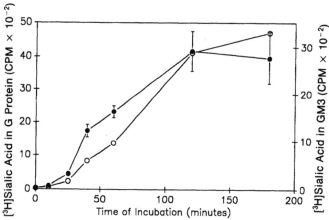

FIGURE 5 The time course of glycolipid and glycoprotein transport are similar. Incubations were initiated at $t = 0$ and 50-μl aliquots were removed at each indicated time point and placed on dry ice. At the conclusion of the time course, all samples were thawed and analyzed for transport of either glycolipid (G_{M3}, solid circles) or vesicular stomatitis virus (VSV) G protein (open circles).

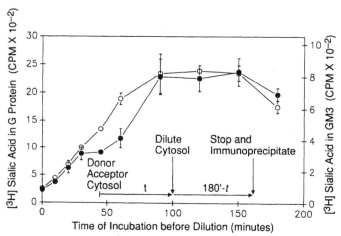

FIGURE 6 The kinetics of transport into the low cytosol-requiring transport intermediate is similar for glycoprotein and glycolipid transport. Measurement of entry into the low cytosol-requiring intermediate is accomplished in a two-step assay. In the first incubation, for each time point, a mixture was made of 5 μl each of acceptor and vesicular stomatitis virus (VSV) infected donor, buffer, and the ATP regeneration system; 1.25 μl Chinese hamster ovary (CHO) cell cytosol; and 0.5 μCi CMP-sialic acid in a volume of 25 μl. After incubation for the indicated times, 25-μl aliquots were removed and diluted into 100 μl incubation buffer with the ATP regenerating system, 0.5 μCi/25 μl CMP-sialic acid, and 10 μl/25 μl 35% sucrose to balance the sucrose contained in the membrane fractions. The diluted samples then were incubated further for $t - 180$ min. Samples then were analyzed for glycolipid transport (G_{M3}, solid circles) or VSV G protein transport (open circles).

acetyl glucosamine (GlcNAc) to VSV G protein (Hiebsch and Wattenberg, 1992). In that system, following the lag time, the kinetics of GlcNAc addition have been shown to be dominated by the glycosylation reaction itself, not by transport, that is, transport precedes glycosylation by an undetermined time. How does this feature affect the interpretation of the data just illustrated? The lag time preceding linear incorporation of label does not depend on glycosylation (Balch *et al.*, 1984b). Similarly, entry into the LCRI transport intermediate is measured in a way that is independent of glycosylation kinetics. Therefore these two measures are reliable indicators of underlying transport events. The similarity of lag time and of entry into the LCRI for glycolipid and glycoprotein transport is a strong indication that these two classes of molecules utilize similar, if not identical, transport mechanisms.

V. WHERE TO NEXT?

The evidence cited in this chapter, in conjunction with the abundance of evidence from intact cells, is a strong indication that glycolipids are transported through the cell by the same vesicular mechanism that transports glycoproteins. This conclusion bolsters the "bulk flow" concept (Pfeffer and Rothman, 1987). In its most simple form, the bulk flow model states that membrane is transported through the secretory pathway continuously and without regard to content. No special signal or physical property is required for a membrane component to be incorporated into a transport vesicle and be shuttled to the next destination on the route. Instead, in this model, special signals or mechanisms are required to retain a component at any particular site. Glycolipids have been suggested to be good bulk flow markers, lacking any amino-acid-based targeting signals. The similarity between glycoprotein and glycolipid transport therefore suggests that glycoproteins also progress by a bulk flow pathway. If so, why study glycolipid transport when the study of glycoprotein transport is further advanced and will lead to the same conclusions? As with most models, the bulk flow concept may be correct in outline but is an oversimplified view of the actual mechanism. Evidence suggests, for example, that proteins tethered to the membrane by a glycosyl-phosphatidylinositol-based lipid are incorporated into a sphingolipid-rich membrane domain (Brown and Rose, 1992). Whether this domain is treated differently from bulk lipid in the transport process remains to be seen. Such domains also have been implicated in the sorting of membrane components to the different membrane surfaces of polarized epithelia (van Meer and Simons,

1986). Discovering the effects of interaction with other membrane components on the transport of glycolipids will be important.

It is becoming clear that the influence of the lipid environment on the transport of glycolipids will be an important and approachable area of study. One of the most notable characteristics of sphingolipids is that they have very strong interactions with one another (Pascher, 1976). They also interact strongly with membrane cholesterol (de Kruyff et al., 1973), and sphingolipid–cholesterol interaction modulates sphingolipid–sphingolipid interaction (Rintoul et al., 1979). The effect of these interactions on transport should be easily approachable using the cell-free transport assay. Cholesterol levels could be modulated by incubation with sterol-containing lipid vesicles or sterol-free vesicles to increase or decrease sterol content, respectively. Alterations of sphingolipid or phospholipid levels might be achieved by fusing vesicles of the appropriate composition into the Golgi (Kobayashi and Pagano, 1988) or by manipulating sphingolipid levels in vivo using inhibitors of sphingolipid synthesis (Vunnam and Radin, 1980). Since this system can be used to measure protein and glycolipid transport simultaneously, differentiating the effects of these lipid changes on the transport of these two different types of molecules will be possible. Little attention has been paid to the involvement of the lipid matrix in vesicular transport; this system provides an ideal starting place for such studies.

References

Baker, D., Hicke, L., Rexach, M., Schleyer, M., and Schekman, R. (1988). Reconstitution of SEC gene product-dependent intercompartmental protein transport. Cell 54, 335–344.

Balch, W. E. (1990). Small GTP-binding proteins in vesicular transport. Trends Biochem. Sci. 15, 473–477.

Balch, W. E., Dunphy, W. G., Braell, W. A., and Rothman, J. E. (1984a). Reconstitution of the transport of protein between successive compartments of the Golgi measured by the coupled incorporation of N-acetylglucosamine. Cell 39, 405–416.

Balch, W. E., Glick, B. S., and Rothman, J. E. (1984b). Sequential intermediates in the pathway of intercompartmental transport in a cell-free system. Cell 39, 525–536.

Beckers, C. J. M., and Balch, W. E. (1989). Calcium and GTP: Essential components in vesicular trafficking between the endoplasmic reticulum and Golgi apparatus. J. Cell Biol. 108, 1245–1256.

Beckers, C. J., Block, M. R., Glick, B. S., Rothman, J. E., and Balch, W. E. (1989). Vesicular transport between the endoplasmic reticulum and the Golgi stack requires the NEM-sensitive fusion protein. Nature (London) 339, 397–398.

Block, M. R., Glick, B. S., Wilcox, C. A., Wieland, F. T., and Rothman, J. E. (1988). Purification of an N-ethylmaleimide-sensitive protein catalyzing vesicular transport. Proc. Natl. Acad. Sci. U.S.A. 85, 7852–7856.

Braell, W. A., Balch, W. E., Dobbertin, D. C., and Rothman, J. E. (1984). The glycoprotein that is transported between successive compartments of the Golgi in a cell-free system resides in stacks of cisternae. Cell 39, 511–524.

Brown, D. A., and Rose, J. K. (1992). Sorting of GPI-anchored proteins to glycolipid-enriched membrane subdomains during transport to the apical cell surface. *Cell* **68**, 533–544.

de Kruyff, B., Demel, R. A., Slotboom, A. J., van Deenen, L. L., and Rosenthal, A. F. (1973). The effect of the polar headgroup on the lipid-cholesterol interaction: A monolayer and differential scanning calorimetry study. *Biochim. Biophys. Acta* **307**, 1–19.

Deutscher, S. L., and Hirschberg, C. B. (1986). Mechanism of galactosylation in the Golgi apparatus. A Chinese hamster ovary cell mutant deficient in translocation of UDP-galactose across Golgi vesicle membranes. *J. Biol. Chem.* **261**, 96–100.

Deutscher, S. L., Nuwayhid, N., Stanley, P., Briles, E. I., and Hirschberg, C. B. (1984). Translocation across Golgi vesicle membranes: A CHO glycosylation mutant deficient in CMP-sialic acid transport. *Cell* **39**, 295–299.

Diaz, R., Mayorga, L. S., Weidman, P. J., Rothman, J. E., and Stahl, P. D. (1989). Vesicle function following receptor-mediacted endocytosis requires a protein active in Golgi transport. *Nature (London)* **339**, 398–400.

Helms, J. B., Karrenbauer, A., Wirtz, K. W. A., Rothman, J. E., and Wieland, F. T. (1990). Reconstitution of steps in the constitutive secretory pathway in permeabilized cells. *J. Biol. Chem.* **265(32)**, 20027–20032.

Hiebsch, R. R., and Wattenberg, B. W. (1992). Vesicle fusion in protein transport through the Golgi in vitro does not involve long-lived prefusion intermediates. A reassessment of the kinetics of transport as measured by glycosylation. *Biochemistry* **31**, 6111–6118.

Kobayashi, T., and Pagano, R. E. (1988). ATP-dependent fusion of liposomes with the Golgi apparatus of perforated cells. *Cell* **55**, 797–805.

Kobayashi, T., Pimplikar, S. W., Parton, R. G., Bhakdi, S., and Simons, K. (1992). Sphingolipid transport from the trans-Golgi network to the apical surface in permeabilized MDCK cells. *Eur. Biochem. Soc.* **300(3)**, 227–231.

Melançon, P., Glick, B. S., Malhotra, V., Weidman, P. J., Serafini, T., Gleason, M. L., Orci, L., and Rothman, J. E. (1987). Involvement of GTP-binding "G" proteins in transport through the Golgi stack. *Cell* **51**, 1053–1062.

Mellman, I., and Simons, K. (1992). The Golgi complex: In vitro veritas? *Cell* **68**, 829–840.

Pascher, I. (1976). Molecular arrangements in sphingolipids. Conformation and hydrogen bonding of ceramide and their implication on membrane stability and permeability. *Biochem. Biophys. Acta.* **455**, 433–451.

Pfeffer, S. R., and Rothman, J. E. (1987). Biosynthetic protein transport and sorting by the endoplasmic reticulum and Golgi. *Annu. Rev. Biochem.* **56**, 829–852.

Rintoul, D. A., Chou, S.-M., and Silbert, D. F. (1979). Physical characterization of sterol-depleted LM-cell plasma membranes. *J. Biol. Chem.* **254**, 10070–10077.

Rothman, J. E., and Orci, L. (1992). Molecular dissection of the secretory pathway. *Nature (London)* **335**, 409–415.

Ruohola, H., Kabcenell, A. K., and Ferro-Novick, S. (1988). Reconstitution of protein transport from the endoplasmic reticulum to the Golgi complex in yeast: The acceptor Golgi compartment is defective in the *sec23* mutant. *J. Cell Biol.* **107**, 1465–1476.

Stanley, P., and Siminovitch, L. (1977). Complementation between mutants of CHO cells resistant to a variety of plant lectins. *Somat. Cell Genet.* **3**, 391–405.

Trinchera, M., and Ghidoni, R. (1989). Two glycosphingolipid sialyltransferases are localized in different sub-Golgi compartments in rat liver. *J. Biol. Chem.* **264(27)**, 15766–15769.

van Meer, G., and Simons, K. (1986). The function of tight junctions in maintaining differences in lipid composition between the apical and the basolateral cell surface domains of MDCK cells. *EMBO J.* **5**, 1455–1464.

Vunnam, R. R., and Radin, N. S. (1980). Analogs of ceramide that inhibit glucocerebroside synthetase in mouse brain. *Chem. Phys. Lipids* **26(3),** 265–278.

Wattenberg, B. W. (1990). Glycolipid and glycoprotein transport through the Golgi complex are similar biochemically and kinetically. Reconstitution of glycolipid transport in a cell free system. *J. Cell Biol.* **111(2),** 421–428.

Wattenberg, B. W., Balch, W. E., and Rothman, J. E. (1986). A novel prefusion complex formed during protein transport between Golgi cisternae in a cell-free system. *J. Biol. Chem.* **261,** 2202–2207.

Wilson, D. W., Wilcox, C. A., Flynn, G. C., Chen, E., Kuang, W.-J., Henzel, W. J., Block, M. R., Ullrich, A., and Rothman, J. E. (1989). A fusion protein required for vesicle-mediated transport in both mammalian cells and yeast. *Nature (London)* **339,** 355–359.

CHAPTER 23

Lipid Transport from the Hepatocyte into the Bile

Attilio Rigotti, María Paz Marzolo, and Flavio Nervi
Departamento de Gastroenterología, Facultad de Medicina, Pontificia Universidad Católica, Santiago, Chile

I. INTRODUCTION

The hepatocyte and the enterocyte are the only two mammalian cell types that have the enzymatic machinery and the organellar structure for the packing and transport of highly hydrophobic lipids (e.g., triacylglycerols, cholesterol esters, and free cholesterol) into plasma lipoproteins and chylomicrons, respectively. The liver plays a major role in whole body cholesterol homeostasis, acquiring cholesterol from endogenous synthesis and by lipoprotein internalization. Hepatic cholesterol can be recruited for differents fates: incorporation into cell membranes, packing into very low density lipoproteins and secretion into sinusoids, conversion into bile acids to be secreted into bile, and, finally, direct secretion as free cholesterol into the biliary canaliculus. Intrahepatic cholesterol trafficking is probably much more complex than that in other cell types. This complexity is determined by the functional and structural organization of the hepatocyte as a highly polarized epithelial cell with distinctive sorting and transport systems directed both to the sinusoidal (basolateral) domain and to the canalicular (apical) domain. Some of these transport systems ultimately are responsible for the secretion of bile acid molecules and phospholipid–cholesterol polymolecular aggregates into bile (for review, see Coleman, 1987; Turley and Dietschy, 1988; Strasberg and Hofmann, 1990; Marzolo *et al.*, 1990; Nathanson and Boyer, 1991; Coleman and Rahman, 1992).

The biliary lipid secretory pathway of the liver is essentially unknown compared with the lipoprotein secretory pathway through the sinusoidal membrane. The principal reason for the lack of information about lipid transport from the hepatocyte into the bile is related to the inherent methodological difficulties of the experimental approaches designed to disclose the intracellular traffic and secretion of biliary cholesterol and phospholipid. To date, no firm evidence exists of specific lipoprotein associations in bile, making following the intrahepatocytic traffic of the putative precursor of biliary-type phospholipid–cholesterol aggregates extremely difficult. As reviewed in this volume, lipid molecules can move between intracellular membranes by vesicle budding and fusion, spontaneous or protein-mediated transport of lipid monomers through the cytosol, and lateral diffusion between organelles connected by membrane bridges (see Chapters 18, 19, and 21). In contrast to different methodological approaches to studying protein translocation and transport, no specific probes for biliary cholesterol and phospholipids exist. These difficulties have hampered the application of common techniques of cell biology and biochemistry to analyzing the intracellular–canalicular flux of cholesterol and phospholipids into the bile. In addition, technical problems in obtaining canalicular bile samples from which to isolate and characterize the primary carriers of biliary lipids have restricted the study of biliary lipid secretion to the analysis of bile specimens obtained from the common bile duct, both from humans and experimental animals. In these experiments, researchers have assumed that the biliary transit time and bile flow coming from bile duct epithelium do not modify the structure and composition of biliary lipid carriers. In fact, bile duct-dependent factors may induce significant changes in the different carriers of biliary lipids.

Despite these limitations, in the last years the use of an increasing number of biochemical and morphological techniques (e.g., gel filtration chromatography, quasi elastic light scattering, x-ray diffraction, transmission and freeze-fracture electron microscopy) has allowed the initial elucidation of the cellular and molecular mechanisms of biliary lipid secretion.

Understanding the secretion, solubilization, and transport of cholesterol in extracellular fluids, including plasma and bile, is critical not only in physiological conditions, but also in pathophysiological processes. Disturbances of cholesterol transport and solubilization may favor the development of two highly prevalent diseases in Western countries; the deposition of excess cholesterol within the artery walls, leading to atherosclerosis, and the precipitation of cholesterol crystals in gall bladder bile, initiating the formation of cholesterol gallstones.

The focus of this review is on the functional organization of the hepatocyte as related to lipid secretion with special emphasis in the biliary secretory pathway. We discuss the metabolic regulation, the intrahepatic transport, and the canalicular secretion of biliary lipids.

II. LIPID CARRIERS IN BILE

A. General Remarks

Major advances in this field occurred in the last several years with the demonstration and isolation of phospholipid unilamellar vesicles in native human supersaturated bile (Sömjen and Gilat, 1983; Pattinson, 1985; Ulloa et al., 1985; Lee et al., 1987) and in rat unsaturated bile (Ulloa et al., 1987). Biliary cholesterol is solubilized and transported, in fact, in phosphatidylcholine unilamellar vesicles, in bile acid–phosphatidylcholine mixed micelles, and in phosphatidylcholine multilamellae (Sömjen et al., 1990; Amigo et al., 1992; Rigotti et al., 1993). Another important observation during this period was that cholesterol crystal formation in bile, a critical step of the pathogenesis of cholesterol gallstone formation, occurs after aggregation of vesicles and not from mixed micelles (Halpern et al., 1986; Harvey et al., 1987; Peled et al., 1988; Schriever and Jüngst, 1989). Finally, the role of biliary proteins in biliary cholesterol transport, solubilization, and precipitation is currently a subject of intensive research.

Apparently the physical form adopted by the lipids in bile will depend on several factors, including the absolute and relative concentration of each lipid and the detergent nature of the bile acid pool secreted into bile (Cohen et al., 1989b; Cohen and Carey, 1990). The primary secretory unit of biliary cholesterol and phospholipids is the unilamellar vesicle. These vesicles will fuse in the canalicular lumen to form multilamellae under conditions of high lipid concentration. The role of biliary proteins in this process is not known. In the biliary tree, micellation of lipids will occur as a function of time and of the degree of cholesterol saturation and bile salt composition (Fig. 1).

B. Identification and Isolation of Lipid Carriers in Bile

Bile is a complex solution containing several amphipathic constituents (phosphatidylcholine, bile acids, and some proteins) and a highly insoluble constituent free cholesterol (Hay and Carey, 1990). The surface properties of these molecules explain their tendency to aggregate in different polymolecular aggregates and to interact with cell membranes (for review, see Cabral and Small, 1989; Hay and Carey, 1990; Mazer, 1990). Table I shows the principal constituents of native human and rat bile.

In vitro studies have demonstrated that high cholesterol concentrations

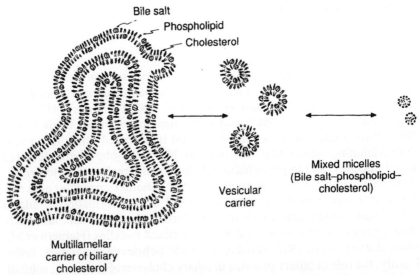

FIGURE 1 Hypothetical molecular interconversions between the different carriers of biliary lipids isolated from bile. The proportion of each polymolecular aggregates would depend on the absolute and relative concentrations of phospholipid, cholesterol, and bile acids. Dynamic and reversible changes, represented by the arrows, normally exist in the biliary tree. The thermodynamic equilibrium of the system in humans normally is achieved in the gall bladder where, after concentration, the major proportion of lamellar carriers (vesicles and multilamellae) will be solubilized in mixed micelles.

TABLE I

Composition of Native Human and Rat Bile[a]

	Human bile	Rat bile
Lipid concentration (mM)		
Bile acids	21	25
Phospholipids	7	7
Cholesterol	5	0.8
Cholesterol saturation (%)	265	39
Proteins (mg/ml)	1.3	1.1

[a] Values represent the mean of measurements of 7 patients with T-tube after operation for common bile duct stones and 7 bile fistula Wistar male rats. The major bile acids in human bile are glycoconjugates (~65%); in rat bile, these are tauroconjugates (~75%). The primary human bile acids, cholic and chenodeoxycholic acid, account for 40% each of the bile acids. The secondary bile acids, deoxycholic and lithocholic acid, account for approximately 20% and 1%, respectively. Bile acids in the rat are more heterogeneous with the presence of variable amounts of α and β muricholic acid and hyodeoxycholic acid (Heuman, 1989).

in dilute model solutions are solubilized in unilamellar phospholipid vesicles (Mazer and Carey, 1983; Mazer et al., 1990). These observations are consistent with the identification by quasielastic light scattering studies, gel filtration chromatography, ultracentrifugation, and electron microscopy of unilamellar vesicles as major carriers of biliary cholesterol in native hepatic and gall bladder supersaturated human bile (Sömjen and Gilat, 1983,1985; Mazer et al., 1984; Pattinson, 1985; Ulloa et al., 1985; Lee et al., 1987). Unilamellar vesicles were identified by electron microscopy in unsaturated rat bile and in the canaliculi of rats with partial depletion of the bile acid pool (Ulloa et al., 1987), as shown in Fig. 2. Previous failures to identify the vesicular carriers of biliary lipids were likely to be due in part to the potent detergent effect of bile acids, which easily would dissolve native canalicular vesicles into a mixed micellar phase, during sample preparation for electron microscopy. In addition, the artifactual effect induced by common ultramicroscopic techniques presumably was overemphasized by different authors, since lamellar carriers (unilamellar vesicles and multilamellae) previously were identified in bile but were not characterized further, or were considered artifacts (Howell et al., 1970; Oh and Holzbach, 1976).

In more recent studies using small-angle X-ray diffraction, electron microscopy after high resolution chromatography, and ultracentrifugation, phospholipid multilamellae were demonstrated as major cholesterol carri-

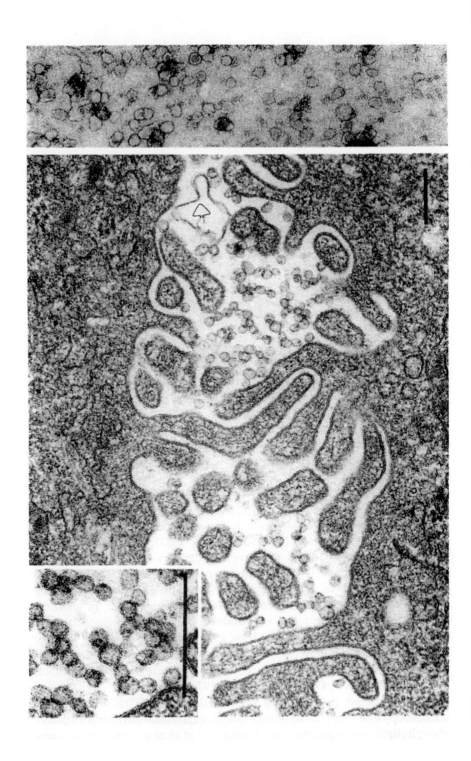

ers in human supersaturated gall bladder bile (Sömjen *et al.*, 1990a). Multi-lamellae have been found in the canaliculi of rats with massive secretion of biliary cholesterol (Amigo *et al.*, 1992; Rigotti *et al.*, 1993), as shown in Fig. 3. In this model, we intended to prepare native experimental bile with a composition similar to human native bile by feeding rats with 0.5% diosgenin, a sapogenin, plus 0.02% simvastatin, an inhibitor of hydroxy-methylglutaryl (HMG) CoA reductase, for 1 wk. Under these experimental conditions, biliary cholesterol saturation increased from 38% to 283%, a value similar to that found in human native bile (Table I) (Rigotti *et al.*, 1993). As a result of these studies, native biliary lipids were shown to be transported mainly in lamellar carriers and not in mixed micelles, as generally thought in the past.

Among the major problems in physicochemical and pathophysiological research of bile are accurate isolation and quantitation of biliary lipid carriers. Native bile is in a nonequilibrium state; separation procedures introduce important perturbing conditions that must be considered in fractionation experiments. Gel filtration chromatography and ultracentrifugation have been used extensively for separation of lamellar carriers from mixed micelles (Amigo *et al.*, 1990a,b; Donovan and Carey, 1990; Sömjen *et al.*, 1990). For analytical studies, high performance gel filtration chromatography using the intermicellar–intervesicular concentration of bile salts is the most accurate separation method (Donovan and Carey, 1990; Donovan *et al.*, 1991). For preparative purposes, density gradient ultracentrifugation allows the processing of large volumes of bile in a short period of time (Amigo *et al.*, 1990a,b). Vesicles and multilamellae are harvested at the top of the centrifuge tube at $d < 1.060$ g/ml (the major proportion of unilamellar vesicles is in fractions of $d < 1.030$ g/ml; Amigo *et al.*, 1990,1992; Rigotti *et al.*, 1993). Sample dilution and bile salt addition, two important pitfalls in gel filtration, are not required during isolation of biliary lamellar carriers by ultracentrifugation.

The relative distribution of unilamellar vesicles, multilamellae, and mixed micelles in native bile is a function of the absolute concentration of biliary lipids, the species, and their relative proportion (Cohen and Carey, 1990; Cohen *et al.*, 1990; Donovan and Carey, 1990; Donovan *et*

FIGURE 2 Electron micrograph of a bile canaliculus of a bile salt-depleted rat liver that shows the presence of unilamellar vesicles within the lumen (*left*) ($\times 32,800$). Higher magnification (*inset*) illustrates more clearly the structure of the vesicles and can be compared with the structure of the canalicular membrane. Similar unilamellar vesicles were found in the fresh bile specimens of the same animal (*right*). Bars, 0.25 μm. Reprinted from Ulloa *et al.*, (1987) with permission.

al., 1991). Newly secreted phospholipid–cholesterol aggregates may be dissolved rapidly by bile acids into a mixed micellar phase, depending on the detergency of the bile acid pool (Donovan *et al.*, 1991). In fact, native vesicles markedly decrease, or disappear when hepatic bile is concentrated and stored in the gall bladder (Pattinson and Chapman, 1986). In addition, the vesicular cholesterol–phospholipid ratio increases as a function of biliary transit time because phospholipids are solubilized preferentially and rapidly into bile acid mixed micelles relative to cholesterol (Cohen and Carey, 1990).

The discovery that biliary cholesterol and phospholipids are preferentially solubilized and transported in lamellar carriers (unilamellar and multilamellar vesicles) in native bile has important implications. This idea supports the concept that biliary cholesterol and phospholipids are secreted in unilamellar vesicles and subsequently may fuse, originating multilamellae, or may be dissolved into mixed micelles, according to total lipid concentration and the detergency of the bile acid pool. The ultrastructure and lipid composition of these carriers are similar to those of the lamellar bodies and surfactant-like material present in other epithelia, such as lung (Wright and Clemens, 1985; Tierney, 1989), tongue (Holland *et al.*, 1989), skin (Elias *et al.*, 1988; O'Guin *et al.*, 1989), stomach, and small intestine (De Schryver-Kecskemeti *et al.*, 1989; Kao and Lichtenberger, 1991). In addition, protein composition analysis has demonstrated that isolated and purified lamellar lipids from rat and human bile (Miquel *et al.*, 1993; Rigotti *et al.*, 1993) have a protein profile similar to that of alkaline phosphatase-rich small intestinal surfactant-like material (Eliakim *et al.*, 1989). Additional protein characterization and careful comparisons between lamellar complexes from different tissues might contribute to our understanding of the nature and origin of these widely distributed lamellar particles. The presence of these lamellar complexes involved in the intracellular storage and the secretion and solubilization of surface-active lipids in differents organs suggests that the organelles involved in their assembly, sorting, transport and secretion may be the same in pneumocytes type II, keratinocytes, intestinal mucous cells and hepatocytes.

FIGURE 3 Electron micrograph of a bile canaliculus of a diosgenin plus simvastatin-fed rat showing multilamellae and vesicles within the lumen. In this experimental model, bile acid secretion remain in the normal range but biliary cholesterol secretion increases by one order of magnitude and phosphatidylcholine secretion by 200% (Amigo *et al.*, 1992; Rigotti *et al.*, 1993). The micrograph was kindly provided by J. Garrido, Department of Cell and Molecular Biology, Pontificia Universidad Católica de Chile.

C. Protein–Lipid Association in Bile

Important biliary constituents are a series of different proteins, the majority of which are serum proteins (Mullock et al., 1978; LaRusso, 1984; Reuben, 1984). The search for specific biliary proteins associated to biliary lipid carriers has not been fruitful. This matter is extremely important since it may represent a key step in disclosing the cell biology of biliary lipid secretion.

Several groups have reported the presence of putative specific lipoprotein aggregates in bile, such as lipoprotein-X (Manzato et al., 1976) and the bile–lipoprotein complex (Nalbone et al., 1973,1985; Martigne et al., 1988; Domingo et al., 1990,1992), the roles of which in biliary lipid secretion and solubilization have not been elucidated. Plasma lipoprotein-X, usually found in cholestasis, showed the same ultrastructure as and similar lipid composition to biliary vesicles (Felker et al., 1978), but its specific protein profile is unknown. Lipoprotein-X found in plasma of patients with cholestatic diseases presumably corresponds to unilamellar biliary vesicles that are secreted in reverse form through the hepatic sinusoidal membrane into plasma. In the case of the 300–500 Å bile–lipoprotein complex (Nalbone et al., 1973), the major protein component has been reported to be a low molecular weight amphipathic anionic polypeptide with strong affinity for phospholipids and calcium (reviewed by Ostrow, 1993). More recently, this anionic polypeptide fraction has been associated preferentially with the vesicular carrier of biliary lipids (Ostrow, 1993). Other groups have identified different plasma apolipoproteins in bile, the functional significance of which is presently unknown (reviewed by LaRusso, 1984; Reuben, 1984). These biliary apolipoproteins, or their fragments, probably are derived from circulating lipoproteins. Biliary apolipoprotein B secretion has been found to be dependent on receptor-mediated uptake of low density lipoprotein at the hepatic sinusoidal membrane and to be correlated with biliary lipid secretion (Kawamoto et al., 1987), but no experimental data are available for a direct involvement of apolipoproteins in biliary cholesterol excretion and solubilization.

Studies from our laboratory have shown that highly purified biliary lamellar carriers harvested from native human and rat bile present a unique and constant protein profile, with a series of hydrophobic glycoproteins between 62–67 and 200 kDa (Miquel et al., 1992,1993; Rigotti et al., 1993). A 130-kDa hydrophobic glycoprotein was isolated from human biliary lamellae and used for preparation of a rabbit polyclonal antibody, which cross-reacted with a homologous rat protein in liver homogenates. By Western blotting, we established that the purified low-density fraction of bile–Metrizamide gradients, containing unilamellar vesicles and multilamellae, was enriched with this 130-kDa glycoprotein relative to bile acid–mixed micelles. Amino acid sequencing of the N terminus of this protein demonstrated a complete identity with aminopeptidase N, a canalicular transmembrane hydrophobic glycoprotein (Núñez *et al.*, 1993a,b; Rigotti *et al.*, 1993). The functional significance of the association of aminopeptidase N and of the others proteins with the lamellar carriers of biliary lipids is currently unknown. Specifically, the role of these proteins in the biogenesis, transport, and intracellular sorting of biliary lipids has not been elucidated (see subsequent discussion).

Different authors have studied the presence of canalicular and organellar enzymes in bile, looking for markers of the mechanism of biliary lipid secretion (reviewed by Coleman, 1987). In fact, canalicular membrane enzymes [e.g., alkaline phosphatase, 5'-nucleotidase, gammaglutamyl transpeptidase (GGTP); Evans *et al.*, 1976; Coleman *et al.*, 1979; Godfrey *et al.*, 1981; Barnwell *et al.*, 1983] and lysosomal enzyme activities (LaRusso and Fowler, 1979) can be detected in bile, but demonstrating a structural association of these proteins with biliary lipid carriers or a functional relationship between these enzyme markers and the process of biliary cholesterol and phospholipid secretion has not been possible. The prevailing view has been that canalicular enzymes are solubilized directly by bile acids into mixed micelles (Coleman, 1987). However, the biliary vesicle-associated aminopeptidase N is the first evidence for a canalicular plasma membrane enzyme interacting with biliary lipid carriers (Rigotti *et al.*, 1993). In addition, these findings have shown that aminopeptidase N is relatively enriched in the mixed vesicular and multilamellar fraction isolated by ultracentrifugation and purified by column chromatography from native human bile (Rigotti *et al.*, 1993). The major proportion of total biliary proteins, however, usually is isolated in the bile acid-rich high density fractions of bile–Metrizamide gradients, suggesting a specific association between aminopeptidase N and biliary lamellar carriers. The mechanisms involved in biliary protein–lipid associations have not been elucidated, but the hydrophobic properties of different proteins secreted into bile apparently facilitate (and explain) their association with the lipid bilayer of the lamellar carriers. This may be the case for some hydrophobic membrane glycoproteins, such as aminopeptidase N.

Another interesting aspect of protein–lipid interaction in bile is related to the discovery that some biliary proteins may influence cholesterol crystal formation and growth in bile (Harvey and Strasberg, 1993). Because biliary cholesterol crystallization occurs from biliary vesicles (Halpern *et al.*, 1986; Harvey *et al.*, 1987; Peled *et al.*, 1988; Schriever and Jüngst, 1989), the identification of vesicle-associated biliary proteins is very important, since they may play a critical role in modulating cholesterol solubilization and, subsequently, cholesterol precipitation in bile (Miquel *et al.*, 1992,1993). Interestingly, we found that biliary vesicle-associated aminopeptidase N has a strong *in vitro* cholesterol crystallization-promoting activity. We also have demonstrated a high significant correlation between the cholesterol nucleation time of gall bladder bile and the concentration of biliary aminopeptidase N, suggesting a major role in the pathogenesis of cholesterol gallstone disease (Núñez *et al.*, 1993a,b).

III. ORIGIN, ASSEMBLY, AND SORTING OF BILIARY LIPIDS

A. Functional Organization of the Hepatocyte as Related to Lipid Secretion

The hepatocyte is a highly polarized epithelial cell with a sinusoidal receptor-rich domain in contact with plasma, a lateral domain attached to adjacent cells, and the canalicular domain, or apical pole, delimited by the tight junctions. Lipids are secreted through the sinusoidal membrane in the form of lipoproteins and through the canalicular membrane as monomers (bile acids) and 40- to 90-nm vesicles (phosphatidylcholine and cholesterol; Ulloa *et al.*, 1985,1987; Cohen *et al.*, 1989a). The biliary lipid secretory pathway of the hepatocyte is essentially unknown compared with our understanding of the cell biology of lipoprotein assembly and secretion through the sinusoidal membrane. The presence of a specific lipoprotein, or proteoliposome, as a major cholesterol and phospholipid carrier in bile has been postulated (Nalbone *et al.*, 1973,1985; Manzato *et al.*, 1976); to date, no firm experimental evidence exists for these types of lipid carriers in bile.

The endocytic and exocytic pathways of hepatic lipoproteins have been fairly well characterized (Figure 4; for review, see Angelin, 1984; Marsh, 1984; Brown and Goldstein, 1986; Gibbons, 1990). Triglycerides and cholesterol are secreted, associated to apo B, as very low density lipoprotein (VLDL) particles. Nascent VLDL initiates its synthesis and maturation in the endoplasmic reticulum (ER). The assembly of apo B into the VLDL lipid core has not been elucidated completely. The particles are transported in vesicles to the Golgi, where they incorporate more lipid and are directed vectorially in a secretory granule to the sinusoidal membrane for exocytosis. Whether the transport of VLDL from the *trans*-Golgi cisternae across the sinusoidal membrane is constitutive or regulated is still not known. A fraction of VLDL also is transported directly from the ER to the sinusoidal membrane (Merre, 1981). The secretory pathway of VLDL is dependent on the functional integrity of the microtubular system (Stein *et al.*, 1974).

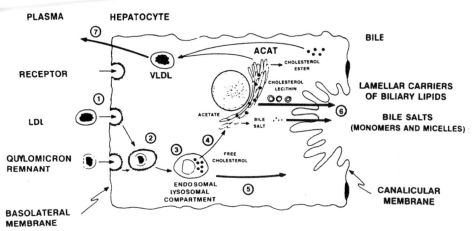

FIGURE 4   Endocytic and secretory pathways of hepatic lipids. (1) Low density lipoproteins (LDL) and chylomicrons are taken up through specific receptors located in coated pits. (2) Particles are internalized by endosomes that fuse with lysosomes. (3) Hydrolysis of different constituents occurs in lysosomes. (4) Free cholesterol presumably is transported bound to sterol carrier proteins to the smooth endoplasmic reticulum (SER) where several important metabolic events occur: the flux of cholesterol to the ER reciprocally regulates HMG-CoA reductase activity and excess cholesterol is esterified by the ACAT enzyme and stored in oil droplets. Part of the newly synthesized cholesterol in the ER and part of the internalized lipoprotein cholesterol will be used for bile acid synthesis. (5,6) Preformed and neosynthesized free cholesterol and phospholipids are sorted into specific vesicles and carried to the canalicular plasma membrane for biliary lipid secretion. (7) Nascent very low density lipoproteins (VLDL) are synthesized in the ER and further processed and assembled in the Golgi apparatus for final sinusoidal excretion by exocytosis. These sinusoidal and canalicular pathways of hepatic cholesterol secretion are interrelated functionally.

The intrahepatocytic transport of phospholipid–cholesterol vesicles to the canalicular membrane for secretion into bile has not been elucidated with certainty. Indirect evidence (see subsequent discussion) suggests that a fraction of biliary lipids (neosynthesized biliary cholesterol and phospholipids) originates in a specific region of the ER, where bile-destined lipids are sorted and packed into intracellular vesicles to be transported vectorially through the Golgi apparatus and secreted finally into bile (Marzolo et al., 1990; Coleman and Rahman, 1992). Note that monensin, a carboxylic ionophore that interferes with the sorting and transport of proteins from Golgi (Griffith et al., 1983), decreases both the sinusoidal secretion of VLDL (Rustan et al., 1985) and canalicular secretion of cholesterol and phospholipid (Casu and Camogliano, 1990). Another fraction of biliary cholesterol and phosphatidylcholine (preformed lipids) could be presumably transported to the canalicular region from the endosomal compartment after processing of lipoproteins internalized by receptor-mediated uptake. Multivesicular bodies, which are derived from lipoprotein internalization (Hornick *et al.*, 1985), may be an intracellular precursor of biliary lipid secretion.

The cytoskeleton appears to be involved in the biliary lipid secretory pathway. Microtubules are major structural components of the cytoskeleton that determine canalicular lipid traffic (Gregory *et al.*, 1975,1978; Dubin *et al.*, 1980; Barnwell *et al.*, 1984; Crawford *et al.*, 1988). The pericanalicular region is very rich in microfilaments that confer the distinctive morphology on the apical pole of the hepatocyte. This cytoskeletal component may play an important role in determining an active force for the extrusion of bile constituents and for driving bile flow. Using hepatocyte couplets in culture, researchers have shown that periodic variations occur in the volume and surface area of canaliculi formed between adjacent hepatocytes (Oshio and Phillips, 1981). Changes in volume as a function of time are accelerated by taurocholate (Miyairi *et al.*, 1984) and abolished by cytochalasin and phalloidin (Phillips and Satir, 1988). Although this motor activity of the biliary pole certainly may represent an important propulsive force for bile flow down the canaliculi (Phillips and Satir, 1988; Watanabe *et al.*, 1991), the activity of the microfilaments around the pericanalicular cytoplasm also may represent a mechanical force for biliary lipid secretion through the canalicular membrane.

B. Hepatic Transport of Bile Acids

Bile acids are taken up from sinusoidal blood mainly by carrier-mediated Na^+-dependent transport, but other potential Na^+-independent transporters have been described (Frimmer and Ziegler, 1988). Using photoaffinity labeling techniques, several studies have been performed to identify the sinusoidal bile acid transporters; at least two different membrane proteins may be involved in bile acid uptake (Kramer *et al.*, 1982; Wieland *et al.*, 1984). More recently, a complementary DNA encoding one of the rat liver bile acid uptake system has been cloned (Hagenbuch *et al.*, 1991). After uptake by the sinusoidal membrane, bile acids are bound to cytoplasmic binding proteins (Erlinger, 1993). A fraction of the monomers also is partitioned in intracellular membranes (Strange *et al.*, 1979) where these molecules may play a role in the sorting and intracellular transport of biliary lipids. Biliary secretion of bile acids is remarkably efficient, with transit times on the order of 1 min from sinusoidal pole to bile (Lowe *et al.*, 1984). Translocation of bile acids into bile across the canalicular membrane is a process carried out by a Na^+-independent transport system (Inoue *et al.*, 1984) and by an ATP-driven bile salt carrier (Meier *et* al., 1987; Adachi et al., 1991; Nathanson and Boyer, 1991). Canalicular secretion is the rate-limiting step in transhepatic bile acid flux (Hardison et al., 1981). This transport system is subjected to adaptive up- and down-regulation according to the flux of bile acids through the hepatocyte (Accatino et al., 1988,1993; Icarte et al., 1991).

The concentration of bile acids in the hepatocyte under basal conditions is in the range of 0.2 mM, less than one order of magnitude of the critical micellar concentration (Okishio and Nair, 1966; Strange et al., 1979; Simion et al., 1984). Therefore, that bile acids may solubilize intracellular membranes into mixed micelles as a mechanism for intracellular transport of biliary lipids is very unlikely. The major intrahepatic bile acid movement probably involves bile acid–protein complexes, which diffuse through the cytosol to the bile acid transporter of the canalicular membrane (Erlinger, 1993). The complex then would dissociate and the unbound bile acid would be secreted by facilitated and/or ATP-dependent transport into the canaliculus (Frimmer and Ziegler, 1988).

However, several pieces of morphological evidence suggest important functional relationships between bile acid monomers and the ER, the Golgi, and the pericanalicular area (Jones et al., 1979; Simion et al., 1984; Lamri et al., 1988; Crawford et al., 1991). These subcellular compartments may participate in the vectorial transport of bile acids to the canalicular membrane. The presence of specific binding sites for bile acid monomers in the ER and Golgi may have important implications in the sorting and intracellular transport of biliary lipids. At high bile acid flux through the hepatocyte, the monomers have been suggested to be taken up by a carrier of the Golgi (Simion et al., 1984), transferred into the Golgi lumen, and then into Golgi-derived vesicles. These vesicles would be transported vectorially toward the canalicular membrane by a mechanism dependent on the functional integrity of the microtubular system (Crawford et al., 1988; Erlinger, 1993).

The mechanism by which bile acids influence biliary secretion of phospholipids and cholesterol is poorly understood. Bile acids at concentrations below the critical micellar concentration have been suggested to change the behavior of cholesterol in model systems (Chijiiwa and Nagai, 1989) and to disrupt phospholipid–cholesterol lamellae (Schubert et al., 1986; Schubert and Smith, 1988). Some data support the possibility that bile acid monomers may promote bidirectional transfer of cholesterol between hepatic membranes and high density lipoproteins (HDL). This effect is proportional to the hydrophobicity of the bile acid molecule (Vlahcevic et al., 1990). Bile acids also have been found to induce vesiculation of model lipid systems with lipid composition similar to that of intracellular organelles (Cohen and Carey, 1990). However, whether the bile acids bound to their intracellular binding proteins may induce the budding of vesicles from specific microdomains of the ER and Golgi, from which biliary cholesterol and phospholipid may be delivered to the canalicular membrane remains to be elucidated (cf. Fig. 5).

C. Hepatic Transport of the Precursor of Biliary Cholesterol and Phospholipids

Approximately 3–20% of biliary phospholipid and cholesterol are newly synthesized in the ER prior to transport into bile (Gregory *et al.*, 1975; Long *et al.*, 1978; Turley and Dietschy, 1981; Robins and Brunengraber, 1988).

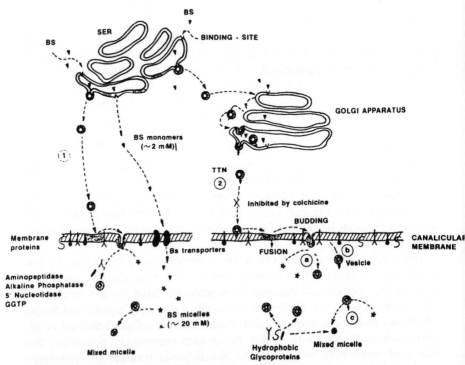

FIGURE 5 Shedding model of biliary lipid secretion. The precursors of biliary vesicles originate from the smooth endoplasmic reticulum (SER). They reach the canalicular membrane directly (1) or through the Golgi apparatus (2), after being concentrated in the *trans* Golgi network (TGN). Bile salt (BS) micelles are formed in the canaliculus, facilitating the budding of specific microdomains (a) of the canalicular membrane to form vesicles (b). These vesicles and canalicular membrane ectoenzymes are solubilized rapidly by BS micelles (c).

However, at least 80% of biliary lipids originate from a preformed pool of membranes that has not been localized clearly. Apparently, this pool must include components of the plasma membrane, the ER, the lysosomal–endosomal compartment, the Golgi, the transtubular network cytoskeleton, and the lipoprotein cholesterol–phospholipid after incorporation into the catabolic pathway of the hepatocyte. The subcellular organization of this transport system is still poorly understood. Whether the huge amounts of membranes involved in the traffic of different compounds through the endocytic pathways of the hepatocyte represent a source of biliary lipids is not known (Evans, 1981).

From the knowledge of the subcellular distribution of cholesterol (Colbeau et al., 1971), the comparison of normal and genetically defective cholesterol transport (Liscum et al., 1989; Cadigan et al., 1990), and the use of pharmacological inhibitors of intracellular cholesterol traffic (Liscum, 1990; Rodriguez-Lafrasse et al., 1990), researchers have postulated that intracellular cholesterol movement is not a random process but is highly controlled and directed. Moreover, in nonhepatic cells, the transport of endogenously synthesized cholesterol from ER to the plasma membrane seems to be distinct from the movement of lipoprotein-derived exogenous cholesterol to the plasma membrane (Liscum and Dahl, 1992). The relevance of these findings to the understanding of intrahepatic cholesterol transport is not known. Evidence suggests separate pathways of transport for newly synthesized and preformed cholesterol into bile (Robins et al., 1985).

Several studies have been performed to determine the role of specific plasma lipoproteins as preformed precursors of biliary lipids in bile using radioactive tracers, suggesting that biliary cholesterol may originate preferentially from plasma HDL (Schwartz et al., 1978; Portman et al., 1980). The functional significance of these studies, however, is limited because lipid tracers such as cholesterol or phosphatidylcholine freely interchange among plasma lipoproteins and with membranes, making the interpretation of kinetic studies almost impossible (Turley and Dietschy, 1988). Theoretically, lipid transported by specific lipoproteins could be channeled preferentially to the canalicular domain. Evidence supports the possibility of intracellular trafficking of apo E-rich VLDL remnants and LDLs to different endosomal compartments in macrophages (Tabas et al., 1990), a function that would depend on the presence of apo E. This differential distribution of lipoprotein contents within the cells makes a preferential channeling of specific lipoprotein cholesterol and phospholipid into the bile more likely. In Fu 5AH rat hepatoma cells, LDL cholesterol has been shown to be incorporated rapidly into the plasma membrane 40 min after uptake and lysosomal hydrolysis of cholesterol esters (Johnson et al., 1990).

However, this intracellular trafficking of free cholesterol from lysosomes into the plasma membrane of nonhepatic cells is not inhibited by colchicine, cytochalasin, or nocodazole (Liscum and Dahl, 1992). In contrast, biliary cholesterol and phospholipid are decreased markedly by inhibitors of the cytoskeleton (Gregory *et al.*, 1975,1978; Dubin *et al.*, 1980; Barnwell *et al.*, 1984; Crawford *et al.*, 1988). In addition, the contribution of lysosomal constituents to bile is very low; interference of chloroquine with lysosomal function does not influence biliary lipid output significantly (Coleman and Rahman, 1992). Based on these experiences, the mechanisms involved in lipoprotein-derived cholesterol movement for biliary secretion remain enigmatic.

The subcellular functional organization of intrahepatic cholesterol transport into bile is still poorly understood. A major problem in elucidating the mechanisms of intracellular lipid transport has been the lack of simple and appropriate methods for studying the process *in vivo*. The use of fluorescent probes and spin-labeled analogs of phospholipids has permitted researchers to examine intracellular transport and intramembrane translocation of lipids in living cells (Voelker, 1992).

Because of the hydrophobic properties of cholesterol and phospholipids, these molecules do not move by spontaneous intracellular aqueous diffusion and must be translocated within the cell, either associated to lipid binding proteins or in a vesicular form. Cholesterol and phospholipid transfer proteins have been described, purified, and identified in different eukaryotic cells including hepatocytes (reviewed by Reinhart, 1990; Wirtz *et al.*, 1990; Cleves *et al.*, 1991). These proteins are capable of exchanging cholesterol and phospholipids between isolated membranes *in vitro* rather than producing a net lipid transfer. A major issue with respect to the role of these lipid binding proteins is whether they function in net intracellular lipid transport or whether the lipid binding and exchanging properties have other physiological purposes. Analysis of spontaneous or induced mutant cell lines and transgenic experiments should provide new insights into this intriguing topic. To date, no experimental evidence supports a role for phospholipid and cholesterol carrier proteins in the process of intrahepatic biliary lipid transport and sorting.

The major fraction of cholesterol and phospholipid transfer to the canalicular domain of the hepatocyte presumably is determined by vesicular movement. Extensive accumulation of vesicles is detected in the pericanalicular area when biliary lipid secretion is stimulated by bile acid infusion (Jones *et al.*, 1979). Otherwise, secretion of cholesterol and phospholipid into bile is decreased significantly by microtubule inhibitors and other compounds that disrupt the subcellular processes related to vesicular movement (Gregory *et al.*, 1975,1978; Dubin *et al.*, 1980; Barnwell *et al.*, 1984; Crawford *et al.*, 1988). One of the major tasks in this field is the isolation of these intrahepatic transport vesicles. Putative vesicle intermediates in cholesterol movement have been identified in nonhepatic mammalian cells (Lange and Steck, 1985; Urbani and Simoni, 1990). The most important technical problems are the identification, acquisition, and preservation of sufficient amounts of intrahepatic vesicles for lipid and protein analysis and the methodological design for purifying specific canaliculus-destined vesicles from the total population of intrahepatic vesicles. If the hydrophobic glycoproteins that we have detected in native biliary vesicles (Miquel *et al.*, 1992,1993) are early associated to biliary-type lipids from their intrahepatic origin, these proteins can be used as specific markers to facilitate the identification and isolation of these precursor vesicles after hepatic homogenization and fractionation procedures.

IV. CANALICULAR SECRETION

A. General Remarks

For many years, the prevailing view about the mechanism of biliary lipid secretion was that bile acid micelles were able to solubilize biliary phospholipids and cholesterol from the intracellular and canalicular membranes (Hardison and Apter, 1972). This micellation process would take into account the positive correlation of the transhepatic flux of total bile acids and the detergency of the bile acid pool with the rates of biliary phospholipid and cholesterol outputs (Hofmann and Small, 1968; Small, 1970, Barnwell *et al.*, 1984; Carulli *et al.*, 1984; Yousef *et al.*, 1987). Kinetic analysis of the bile acid-dependent secretion of biliary lipids has shown that bile acids are secreted into bile prior to the secretion of phospholipid and cholesterol (Lowe *et al.*, 1984). The authors of this study suggested that bile acids are first secreted into bile, where they aggregate into simple micelles. These micelles then can solubilize biliary lipids and some proteins from microdomains of biliary-type lipid in the canalicular membrane by a process of vesiculation and micellation (Lowe *et al.*, 1984). The identification of aminopeptidase N as a constituent of native biliary vesicles provides additional experimental support for this theoretical model of biliary lipid secretion (Rigotti *et al.*, 1993).

Although this putative secretory mechanism may be responsible for some fraction of biliary lipid excretion, experimental evidence exists against this membrane-solubilizing mechanism as a major determinant of biliary lipid output. *In vitro* solubilization of cell membranes by bile acids demonstrates nonspecific micellation of membrane phospholipids at rates that may not explain the highly efficient and specific secretion of biliary phosphatidylcholines (Coleman *et al.*, 1979; Graham and Northfield, 1987). Phosphatidylcholine constitutes more than 90% of biliary phospholipids in different species. The biliary fatty acid pattern is rich in 1-palmitoyl, 2-linoleyl and 1-palmitoyl, 2-oleoyl phosphatidylcholine. In contrast, phosphatidylcholine constitutes approximately 50% of the total canalicular plasma membrane phospholipids, which have a different fatty acid pattern, with a higher proportion of stearoyl and arachidonyl species than biliary phosphatidylcholines (Gregory *et al.*, 1975; Evans *et al.*, 1976; Booker *et al.*, 1989; Chanussot *et al.*, 1990; Robins *et al.*, 1991). The phospholipid composition of lamellar carriers of biliary lipids isolated from rats with massive lipid secretion is shown in Table II, with the phospholipid composition of plasma membranes isolated from rat liver (Rosario *et al.*, 1988). These observations demonstrate that biliary lipids are not secreted into bile by just a nonspecific random bile acid-dependent solubilization process of the outer leaflet of the canalicular membrane. In fact, biliary sphingomyelin and phosphatidylethanolamine are minor constituents of total biliary

TABLE II

Comparison of the Relative Phospholipid Composition of Whole Rat Bile, Lamellar Carriers, and Hepatic Membranes [a]

Phospholipid	Composition (% of total phospholipids)			
	Whole bile	Lamellar carriers	Sinusoidal membranes[b]	Canalicular membranes[b]
Phosphatidylcholine	94.8	95.3	44.6	35.5
Lysophosphatidylcholine	0.5	0.3	0.8	1.6
Sphingomyelin	0.1	0.2	11.0	22.1
Phosphatidylethanolamine	4.5	3.7	28.4	23.8
Phosphatidylinositol	—	—	6.4	4.4
Phosphatidylserine	—	—	7.6	11.2

[a] Values represent the mean of 3 determinations from pooled bile collected from rats fed diosgenin plus simvastatin. Lamellar carriers were harvested from the top fraction ($d < 1.030$ g/ml) of a bile–metrizamide density gradient and were purified by two sequential gel filtration chromatography steps. Phospholipid classes were separated by thin layer chromatography in silica gel G plates and developed in chloroform–methanol–acetic acid–water (50 : 30 : 8 : 4). Phospholipids were visualized by 8-anilinenaphthalenesulfonic acid.

[b] Values for sinusoidal and canalicular membranes were obtained from the work of Rosario *et al.* (1988).

phospholipids, whereas they are quantitatively important in the canalicular membrane (Evans *et al.*, 1976). Their concentrations in bile only increase under high transhepatic fluxes of hydrophobic bile acids with membrane-damaging properties (Coleman *et al.*, 1979).

The quantitatively important flux of lipid secreted into bile per unit time strongly supports the existence of a highly specialized and efficient secretory system for biliary cholesterol and phosphatidylcholine at the canalicular membrane. According to the experimental data accumulated in the last decade, biliary lipids are, in fact, very likely to be secreted normally into bile through several mechanisms, including exocytosis, that remain to be elucidated.

B. Theories of Canalicular Lipid Transport

Two theories have been proposed to explain the mechanism of biliary lipid secretion at the canalicular membrane: the "fusion-budding" or "shedding" model (Lowe *et al.*, 1984; Coleman and Rahman, 1992) and exocytosis (Evans, 1981; Marzolo *et al.*, 1990).

In the shedding model represented in Fig. 5, specific microdomains in the canalicular membrane are postulated to result from the fusion of biliary lipid precursor vesicles. These microdomains would be much more fluid and would bud by the detergent effect of bile acid micelles already present in the canaliculus. Bile acids also would solubilize different enzymes from the outer leaflet of the canalicular membrane. The demonstration of a specific association of a canalicular plasma membrane ectoenzyme, such as aminopeptidase N, with native biliary vesicles supports this model (Rigotti *et al.*, 1993). Intracellular resupply of biliary-type lipids destined for specific canalicular microdomains would be accomplished through a continuous vesicular traffic from intracellular organelles, including SER, Golgi apparatus, and the TGN. This trafficking would be stimulated by bile acids and would depend on the functional integrity of the cytoskeleton. As a variant of this model, the normal turnover of some canalicular constituents also may facilitate the shedding of phospholipids and cholesterol into bile, as occurs with other epithelia. The shedding of vesicular material from the surface of eukaryotic cells is a common process that presumably occurs through different cellular mechanisms (reviewed by Beaudoin and Grondin, 1991).

The exocytic pathway is represented in Fig. 6. This secretory mechanism is universally present in cells and is highly efficient for the secretion of polymolecular aggregates, including proteins and lipoproteins (Palade, 1975). Vesicle-containing secretory granules, resembling hepatic multive-

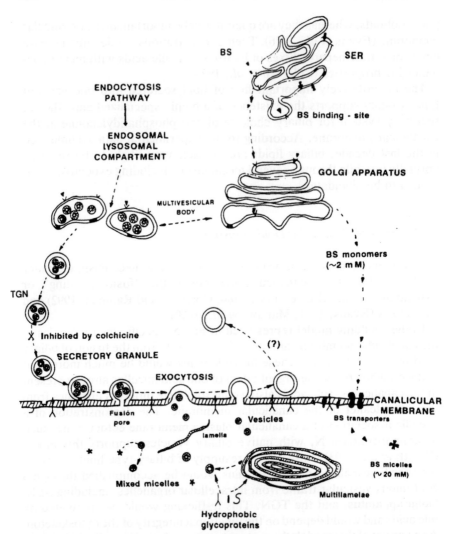

FIGURE 6 Exocytosis model of biliary lipid secretion. Biliary vesicles are concentrated in specific multivesicular bodies of the endosomal–lysosomal compartment. Phospholipid and cholesterol for biliary secretion originate in part from the endocytosis pathway and from specific membranes of the smooth endoplasmic reticulum (SER). Sorting and assembly occurs in the Golgi apparatus by a bile acid-dependent mechanism. The "biliary secretory granules" that contain vesicles are directed through the *trans* Golgi network (TGN), an organelle particularly abundant in the pericanalicular area. Vesicles, micelles, and multilamellae are the carriers of biliary cholesterol. Canalicular as well as intracellular hydrophobic glycoproteins may associate with lamellar carriers of biliary lipids from both inside the hepatocyte or within the canaliculus.

sicular granules, have been described in other cell types by quick-freezing methods (Ornberg *et al.*, 1986). These intragranular vesicles are released from adrenal chromaffin cells during secretion of catecholamines (Ornberg *et al.*, 1986). Biliary vesicles may be secreted through the canalicular membrane by exocytosis of putative homologous intrahepatic organelles. In this exocytic model, a precanalicular compartment of multivesicular bodies containing biliary vesicles would be formed by lipid constituents derived and sorted from both the ER and the endosomal–lysosomal system, with a subsequent intracellular transport coupled to the transhepatic flux of bile acids.

Note the analogy between biliary lipid secretion and lung surfactant secretion by the pneumocytes type II cells. The phospholipid composition of lung surfactant (Rooney, 1985) is similar to that of biliary phospholipids. Phosphatidylcholine represents approximately 80% of total lung surfactant phospholipids. This material is secreted into the alveoli by a highly efficient exocytic mechanism (Wright and Clemens, 1987). We believe that this mechanism also may be responsible for the secretion of the major proportion of biliary lipids, since multilamellae have been found in the biliary canaliculi (Rigotti *et al.*, 1993). Challenging this possibility is the apparent lack of ultramicroscopic evidence of multilamellar granules in the pericanalicular area, as the intracellular precursor compartment for the exocytic pathway of biliary lipids. Our interpretation is that, during sample preparation for conventional transmission electron microscopy, these multivesicular bodies may be dissolved by the progressive increase in bile acid concentration in the liver specimen. Nondisrupting ultrastructural techniques, such as X-ray diffraction and cryo-electron microscopy should be developed to overcome this problem. The identification of specific proteins associated with lamellar carriers of biliary cholesterol is critical to identifying and following this theoretical secretory pathway. Preliminary immunocytochemical evidence suggests that a biliary lamellae-associated 62- to 67-kDa hydrophobic glycoprotein is localized in pericanalicular organelles (Rigotti *et al.*, unpublished observations). Identification of this protein may prove useful in future experiments designed to unravel the secretory process of biliary lipids.

V. MODULATION OF BILIARY CHOLESTEROL SECRETION

A. Role of Bile Acids and Phospholipids

Biliary cholesterol and phospholipid are coupled tightly to bile acid secretion under physiological conditions. The effectiveness of the various

bile acid species in driving the co-secretory mechanism of biliary phospholipids and cholesterol outputs varies in relation to their hydrophobicity
(Carulli et al., 1984; Roda et al., 1988; Bilhartz and Dietschy, 1989; Hofmann, 1990). The quantitative significance of this factor is, however,
relatively minor (Marzolo et al., 1990). The hydrophobicity of bile acid
molecules may have a critical role in the process of recruiting and sorting
cholesterol–phospholipid vesicles from the biliary lipid-precursor compartment before extrusion into the canalicular lumen (Armstrong and
Carey, 1982; Vlahcevic et al., 1990).

Several organic anions, including bilirubin, iodipamide (Apstein and
Robins, 1982; Verkade et al., 1990), ampicillin (Apstein and Russo, 1985),
ceftriaxone (Xia et al., 1990), and cefmetazole (Cava et al., 1991) have
been shown to decrease biliary lipid output. These anions have been used
to identify the site and mechanism of the co-secretory coupling of bile
acids and biliary lipids. However, no definitive evidence exists for the
exact mechanism(s) of action of these drugs and their specificity. All these
anions possess hydrophobic characteristics; their uncoupling effect on the
bile acid co-secretory mechanism of biliary lipids has been postulated to
occur inside the hepatocyte or in the canalicular membrane (for review,
see Coleman and Rahman, 1992).

Although these various studies support the concept that the three biliary
lipids usually are coupled during bile secretion, several experiences suggest a major modulating role for bile acid-independent mechanisms in
biliary cholesterol secretion. At low bile acid secretion rates, relatively
more cholesterol than phospholipids is secreted into bile, suggesting that
a fraction of biliary cholesterol may be secreted independently of bile
acids (Dowling et al., 1971; Hardison and Apter, 1972; Wheeler and King,
1972; Wagner et al., 1976; Hofmann, 1990). In addition, under normal
enterohepatic circulation of bile acids, modification of biliary phospholipid
output by choline feeding is paralleled by cholesterol output (Robins and
Amstrong, 1976). Similarly, ethinylestradiol and diosgenin feeding stimulates biliary cholesterol and phospholipid output (Nervi et al., 1984,1988;
Berr et al., 1988) and bean intake has the same effect in the rat (Rigotti
et al., 1989), without major changes in bile acid output. Otherwise, pathological conditions such as obesity and cholesterol gallstone disease present
absolute higher rates of biliary cholesterol secretion, compared with normal states, with unchanged bile acid secretion (Shaffer and Small, 1977).
These observations indicate that, under a physiological transhepatic flux
of bile acids, the rate of biliary cholesterol output is determined significantly by the metabolic setting of intrahepatic cholesterol (see subsequent
discussion).

B. Metabolic Regulation of Biliary Cholesterol Output

Biochemical and morphological evidence is consistent with the existence of a functional precursor compartment within the hepatocyte for biliary lipid secretion (Gregory *et al.*, 1975; Nemchausky *et al.*, 1977; Nervi *et al.*, 1984,1988; Stone *et al.*, 1985,1992; Turley and Dietschy, 1988). Cholesterol input into the hepatic secretory pathways (i.e., synthesis of macromolecular lipid complexes for plasma and biliary lipid secretion) is brought about by *de novo* synthesis in the SER and by uptake of chylomicron and VLDL remnants, LDL, and HDL (Fig. 4). LDL and HDL deliver cholesterol into the hepatocyte through a receptor- or non-receptor-mediated pathway. Under conditions of relative intracellular excess, free cholesterol is esterified rapidly by acyl CoA: cholesterol acyl-transferase (ACAT) for cholesterol storage or for resecretion in VLDL particles by an exocytic process (Dixon and Ginsberg, 1993). Studies done in this and other laboratories have indicated that the pharmacological or dietary manipulation of ACAT activity may determine the hepatocyte availability of cholesterol for secretion into bile reciprocally (Del Pozo *et al.*, 1983; Nervi *et al.*, 1983,1984; Stone *et al.*, 1985; Rigotti *et al.*, 1989). Similarly, stimulation of hepatic VLDL production and secretion by feeding fructose to rats dramatically abolished the >500% increment of biliary cholesterol output induced by diosgenin as shown in Fig. 7 (Nervi *et al.*, 1988). The functional interrelationship between VLDL and the biliary cholesterol secretory pathways, supporting the concept of the existence of a common precursor compartment of hepatocytic cholesterol, has been confirmed in acute experiments (Stone and Evans, 1992). This correlation between the sinusoidal and canalicular secretory pathways of hepatic cholesterol is apparently independent of VLDL apo B secretion (Marzolo *et al.*, 1993).

This series of studies also has shown that bile acid synthesis measured *in vivo* remains constant despite major changes in biliary cholesterol channeling into bile (Nervi *et al.*, 1988; Marzolo *et al.*, 1990). This finding suggests that the free cholesterol pool used for bile acid synthesis is functionally unrelated to the pool from which VLDL cholesterol and biliary cholesterol originate. These data are consistent with previous isotopic analysis that demonstrated that bile acid synthesis preferentially originates from newly synthesized cholesterol, whereas biliary cholesterol originates from preformed stores (Norman and Norum, 1974; Schwartz *et al.*, 1977). However, human studies have suggested that biliary cholesterol and bile acids may share a common intrahepatic metabolic compartment (Einarson *et al.*, 1985). Although dietary cholesterol transported in

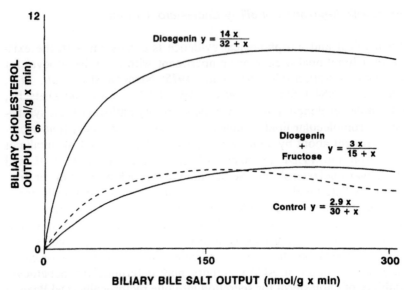

FIGURE 7 Biliary cholesterol secretion as a function of bile acid output in rats fed different diets that modify hepatic cholesterol metabolism. The rectangular hyperbola with the formula $y = \dfrac{ax}{b + x}$ is the best fitting of the experimental data. The calculated value of *a* corresponds to the maximal secretion rate of cholesterol; the term *b* refers to the rate of bile acid secretion at which half-maximal biliary cholesterol output is obtained. Diosgenin stimulates biliary cholesterol output. This effect is correlated with a decrease in the activity of the enzyme ACAT in the endoplasmic reticulum (ER). Fructose stimulates very low density lipoprotein (VLDL) production and sinusoidal cholesterol output. The maximal secretion rate of biliary cholesterol induced by diosgenin is normalized by adding fructose to the experimental diet. The curves of the figure were constructed from the experimental data reported by Nervi *et at.* (1988).

chylomicron remnants to the liver may increase biliary cholesterol output massively in prairie dogs and mice, this effect is not apparent in human and rat. Under physiological conditions, hepatic cholesterol neosynthesis is not a major determinant of biliary cholesterol secretion (Turley and Dietschy, 1988). However, the derepression of hepatic cholesterol synthesis in lovastatin- or simvastatin-fed rats may increase biliary cholesterol output (Bilhartz *et al.*, 1989; Yamauchi, 1991; Amigo *et al.*, 1992; Rigotti *et al.*, 1993) associated with a marked proliferation of SER membranes (A. Rigotti *et al.*, unpublished observations).

Collectively, these studies on the regulation of biliary cholesterol secretion provide evidence to support the concept that metabolic compartmentalization of hepatic cholesterol represents a major determinant of the amount of cholesterol available for biliary lipid secretion. As a result of these studies, we can conclude that, in humans and other species under basal conditions, the input of lipoprotein cholesterol or newly synthesized cholesterol into the hepatic precursor membranes of biliary cholesterol does not drive the canalicular secretion of cholesterol-rich vesicles directly. Anabolic processes such as hepatic cholesterol esterification and VLDL production are more likely to determine primarily the amount of intrahepatocytic cholesterol available for the bile acid-dependent co-secretory mechanism of biliary lipids.

VI. SUMMARY AND CONCLUSIONS

Evidently, our current knowledge of the detailed cellular process of biliary lipid secretion is very limited. At present, no sufficient data exist from which to elaborate a coherent scheme of the intrahepatocytic traffic and sorting of the precursor vesicles that carry biliary cholesterol and phospholipids. Major drawbacks to unravelling this secretory pathway are the technical difficulties in isolating the putative precursor carriers and the absence of a protein marker for these carriers. The identification of a specific protein associated with the lamellar carriers of biliary cholesterol may represent a key step for the discovery of the intracellular pathway of biliary lipid transport. Bile acids represent important modulators of biliary lipid secretion. These molecules presumably play a key role in the sorting mechanism of cholesterol and phospholipid from the hepatocyte into bile. Whether this role is played through bile acid-induced biliary lipid secretion from inside the hepatocyte or at the canalicular membrane is not known. Similarly, no data related to the potential role of lipid carrier proteins in the quantitative translocation and sorting of biliary lipid monomers in the secretory pathway are available. An important finding is that biliary cholesterol output can be increased by a number of dietary and pharmacological manipulations. Thee observations have demonstrated that the secretory pathways of hepatic cholesterol are interconnected functionally and reciprocally. Biliary lipids presumably are secreted by different mechanisms, including shedding of the canalicular membrane and specific exocytosis. The detergent effect of bile acids on the canalicular membrane, under physiological conditions, presumably determines a minor fraction of the total lipid secreted into bile. Most likely, the major proportion of specific phosphatidylcholine and cholesterol vesicles is secreted through a highly specialized exocytic process. Further analysis, including cellular and molecular approaches, is required to obtain new insights into cholesterol and phospholipid secretion into bile. This topic deserves much more attention from cellular and molecular biologists. In addition, a novel area of potential significance is related to the role of specific biliary proteins as promotors of cholesterol crystallization. These aspects may be critical for a better understanding of the initial steps of the molecular pathology of cholesterol gallstone formation.

Acknowledgments

The authors thank Aldo Greco and Geltrude Mingrone for helpful discussion and support (Istituto di Clinica Medica Universita Cattolica del Sacro Cuore, Roma, Italia). They also express their gratitude to Jorge Garrido, Miguel Bronfman, and Enrique Brandán for their advice and discussion (Departamento de Biología Celular y Molecular, Pontificia Universidad Católica de Chile). They also acknowledge Miriem Aguad for excellent assistance in the preparation of the manuscript. The original studies presented in this chapter were supported in part by grants from the following institutions: Instituto per la Cooperazione Universitaria and Direzzione Generale Cooperazione allo Svilupo, Roma, Italia; Volkswagenwerk Stiftung and Alexander von Humboldt Foundation, Germany; and Fondo Nacional de Desarrollo Científico y Tecnológico (FONDECYT), Chile. María Paz Marzolo was supported in part by a grant from FONDECYT; Attilio Rigotti was supported in part by Fundación Andes.

References

Accatino, L., Hono, J., Maldonado, M., Icarte, M., and Persico, R. (1988). Adaptative regulation of hepatic bile salt transport. Effect of prolonged bile salt depletion in the rat. *J. Hepatol.* **7,** 215–223.

Accatino, L., Hono, J., Koenig, C., Pizarro, M., and Rodríguez, L. (1993). Adaptative changes of hepatic bile salt transport in a model of reversible interruption of the entero-hepatic circulation in the rat. *J. Hepatol. (in press)*.

Adachi, Y., Kobayashi, H., Kurumi, Y., Shouji, M., Kitaro, M., and Yamamoto, T. (1991). ATP-dependent taurocholate transport by rat liver canalicular membrane vesicles. *Hepatology* **14,** 655–659.

Amigo, L., Covarrubias, C., and Nervi, F. (1990a). Rapid isolation of vesicular and micellar carriers of biliary lipids by ultracentrifugation. *J. Lipid Res.* **31,** 341–347.

Amigo, L., Covarrubias, C., and Nervi, F. (1990b). Separation and quantitation of cholesterol carriers in native bile by ultracentrifugation. *Hepatology* **12,** 130S–133S.

Amigo, L., Garrido, J., Jr., Garrido, J., Rigotti, A., and Nervi, F. (1992). Biliary surfactant-ant-like material. Morphological evidence of multilamellae in the canaliculi of rats with massive biliary lipid output. *Gastroenterology* **102,** A774.

Angelin, B. (1984). Regulation of hepatic lipoprotein receptor expression. *In* "Liver and Lipid Metabolism" (S. Calandra, N. Carulli, and G. Salvioli, eds.), pp. 187–202. Exerpta Medica, Amsterdam.

Apstein, M., and Robins, S. (1982). Effect of organic anions on biliary lipids in the rat. *Gastroenterology* **83,** 1120–1126.

Apstein, M., and Russo, A. (1985). Ampicillin inhibits biliary cholesterol secretion. *Dig. Dis. Sci.* **30,** 253–256.

Armstrong, M. J., and Carey, M. C. (1982). The hydrophilic/hydrophobic balance of bile salts: Inverse correlation between reverse-phase high performance liquid chromatography mobilites and micellar cholesterol solubilizing capacities. *J. Lipid Res.* **23,** 70–80.

Barnwell, S., Godfrey, P., Lowe, P., and Coleman, R. (1983). Biliary protein output by isolated perfused liver. Effects of bile salts. *Biochem. J.* **210,** 549–557.

Barnwell, S., Lowe, P., and Coleman, R. (1984). The effects of colchicine on secretion into bile of bile salts, phospholipids, cholesterol and plasma membrane enzymes: Bile salts are secreted unaccompanied by phospholipids and cholesterol. *Biochem. J.* **220,** 723–731.

Beaudoin, A., and Grondin, G. (1991). Shedding of vesicular material from the cell surface of eukaryotic cells: Different cellular phenomena. *Biochim. Biophys. Acta* **1071,** 203–219.

Berr, F., Stellaard, F., Goetz, A., Hammer, C., and Paumgartner, G. (1988). Ethinylestradiol stimulates a biliary cholesterol-phospholipid cosecretory mechanism in the hamster. *Hepatology* **8,** 619–624

Bilhartz, L., and Dietschy, J. M. (1989). Bile salt hydrophobicity influences cholesterol recruitment from rat liver *in vivo* when cholesterol synthesis and lipoprotein uptake are constant. *Gastroenterology* **95,** 771–779.

Bilhartz, L., Spady, D. K., and Dretschy, J. M. (1989). Inappropriate hepatic cholesterol synthesis expands the cellular pool of sterol availability for recruitment by bile acids in the rat. *J. Clin. Invest.* **84,** 1181–1187.

Booker, M., Scott, T., and La Morte, W. (1989). Effect of dietary cholesterol on phosphatidylcholines and phosphatidylethanolamines in bile and gallbladder mucosa in the prairie dog. *Gastroenterology* **97,** 1261–1267.

Brown, M., and Goldstein, J. (1986). A receptor-mediated pathway for cholesterol homestasis. *Science* **232,** 34–46.

Bush, N., and Holzbach, T. (1990). Crystal growth-inhibiting proteins in bile. *Hepatology* **12,** 195S–199S.

Cabral, D., and Small, D. (1989). Physical chemistry of bile. *In* "Handbook of Physiology. The Gastrointestinal System III" (S. Schultz, J. Forte, and B. Rauner, eds.), Section 6, pp. 621–662. Waverly Press, San Diego, California.

Cadigan, K. M., Sipillane, D. M., and Chang, T. Y. (1990). Isolation and characterization of Chinese hamster ovary cell mutants defective in intracellular low density lipoprotein cholesterol trafficking. *J. Cell Biol.* **110,** 295–308.

Carulli, N., Loria, P., Bertolotti, M., Ponz de Leon, M., Medici, G., Menozzi, D., Zironi, F., and Iori, R. (1984). Role of bile acid pool composition on biliary lipid secretion. *In* "Liver and Lipid Metabolism" (S. Calandra, N. Carulli, and G. Salvioli, eds.), pp. 93–105. Exerpta Medica, Amsterdam.

Casu, A., and Camogliano, L. (1990). Glycerophospholipids and cholesterol composition of bile in bile-fistula rats treated with monensin. *Biochim. Biophys. Acta* **1043,** 113–115.

Cava, F., González, J., González-Buitrago, J., Muriel, C., and Jiménez, R. (1991). Inhibition of biliary cholesterol and phospholipid secretion by cefmetazole. *Biochem. J.* **275,** 591–595.

Chanussot, F., Lafont, H., Hauton, J., Tuchweber, B., and Yousef, I. (1990). Studies on the origin of biliary phospholipid. Effect of dehydrocholic acid and cholic acid infusions on hepatic and biliary phospholipids. *Biochem. J.* **270,** 691–695.

Chijiiwa, K., and Nagai, M. (1989). Interaction of bile salt monomer and cholesterol in the aqueous phase. *Biochim. Biophys. Acta* **1001,** 111–114.

Cleves, A., McGee, T., and Bankaitis, V. (1991). Phospholipid transfer proteins: A biological debut. *Trends Biochem. Sci.* **1,** 30–34.

Cohen, D., and Carey, M. (1990). Physical chemistry of biliary lipids during bile formation. *Hepatology* **12,** 143S–148S.

Cohen, D., Angelico, M., and Carey, M. (1989a). Quasielastic light scattering evidence for vesicular secretion of biliary lipids. *Am. J. Physiol.* (*Gastrointest. Liver Physiol.*) **257,** G1–G8.

Cohen, D., Angelica, M., and Carey, M. (1989b). Structural alterations in lecithin-cholesterol vesicles following interactions with monomeric and micellar bile salts: Physical–chemical basis for subselection of biliary lecithin species and aggregative states of biliary lipid during bile formation. *J. Lipid Res.* **31**, 55–76.

Colbeau, A., Nachbaur, T., and Vignais, P. M. (1971). Enzymic characterization and lipid composition of rat liver subcellular membranes. *Biochim. Biophys. Acta* **249**, 962–989.

Coleman, R. (1987). Biochemistry of bile secretion. *Biochem. J.* **244**, 249–261.

Coleman, R., and Rahman, K. (1992). Lipid flow in bile formation. *Biochim. Biophys. Acta* **1125**, 113–133.

Coleman, R., Iqhal, S., Godfrey, P., and Billington, D. (1979). Composition of several mammalian biles and their membrane damaging properties. *Biochem. J.* **178**, 201–208.

Crawford, J., Berken, C., and Gollan, J. (1988). Role of the hepatocyte microtubular system in the excretion of bile salts and biliary lipid: Implications for intracellular vesicular transport. *J. Lipid Res.* **29**, 144–156.

Crawford, J., Vinter, D., and Gollan, J. (1991). Taurocholate induces pericanalicular localization of C_6-NBD-ceramide in isolated hepatocytes couplets. *Am. J. Physiol.* **260**, G119–G132.

Del Pozo, R., Nervi, F., Covarrubias, C., and Ronco, B. (1983). Reversal of progesterone-induced biliary cholesterol output by dietary cholesterol and ethynyl-estradiol. *Biochim. Biophys. Acta* **753**, 164–172.

De Schryver-Kecskemeti, Eliakim, R., Carroll, S., Stenson, W., Moxley, M., and Alpers, D. (1989). Intestinal surfactant-like material. A novel secretory product of the rat enterocyte. *J. Clin. Invest.* **84**, 1355–1361.

Dixon, J., and Ginsberg, H. N. (1993). Regulation of hepatic secretion of apolipoprotein B-containing lipoproteins: Information obtained from cultured liver cells (Review). *J. Lipid Res.* **34**, 167–179.

Domingo, N., Botta, D., Martigne-Cros, M., Lechene de la Porte, P., Park-Leung, P., Hauton, J., and Lafont, H. (1990). Evidence for the synthesis and secretion of APF—A bile lipid associated protein—by isolated rat hepatocytes. *Biochim. Biophys. Acta* **1044**, 243–248.

Domingo, N., Grosclaude, J., Bekaert, E., Mege, D., Chapman, M., Shimizu, S., and Ayrault-Jarrier, M. (1992). Epitope mapping of the human amphipatic anionic polypeptide (APF): Similarity with a calcium-binding protein (CBP) isolated from gallstones and bile, and immunological cross-reactivity with apolipoprotein A-I. *J. Lipid Res.* **33**, 1419–1430.

Donovan, J. M., and Carey, M. C. (1990). Separation and quantitation of cholesterol carriers in bile. *Hepatology* **12**, 94S–105S.

Donovan, J., Timofeyeva, N., and Carey, M. (1991). Influence of total lipid concentration, bile salt:lecithin ratio, and cholesterol content on inter-mixed micellar/vesicular (non-lecithin-associated) bile salt concentrations in model bile. *J. Lipid Res.* **32**, 1501–1512.

Dowling, H., Mack, J., and Small, D. (1971). Biliary lipid secretion and bile composition following acute and chronic interruption of the enterohepatic circulation in the Rhesus monkey. *J. Clin. Invest.* **50**, 1917–1926.

Dubin, M., Maurice, M., Feldman, G., and Erlinger, S. (1980). Influence of colchicine and phalloidin on bile secretion and hepatic ultrastructure in the rat: Possible interaction between microtubules and microfilaments. *Gastroenterology* **79**, 646–654.

Einarson, K., Nilssell, K., Leijd, B., and Angelin, B. (1985). Influence of age on secretion of cholesterol and synthesis of bile acids by the liver. *N. Engl. J. Med.* **313**, 277–282.

Eliakim, R., DeSchryver-Kecskemeti, K., Nogee, L., Stentson, W. F., and Alpers, H. H. (1989). Isolation and characterization of small intestinal surfactant-like particle containing alkaline phosphatase and other digestive enzymes. *J. Biol. Chem.* **264**, 20614–20619.

Elias, P., Menon, G., Grayson, S., and Brown, B. (1988). Membrane structural alterations in murine stratum corneum: Relationship to the localization of polar lipids and phospholipases. *J. Invest. Dermatol.* **91**, 3–10.

Erlinger, S. (1993). Intracellular events in bile acid transport by the liver. *In* "Hepatic Transport and Bile Secretion" (N. Tavoloni and P. D. Berk, eds.), pp. 467–475. Raven Press, New York.

Evans, W. (1981). Membrane traffic at the hepatocyte's sinusoidal and canalicular surface domains. *Hepatology* **1**, 452–457.

Evans, W., Kremmer, T., and Culvenor, J. (1976). Role of membranes in bile formation. Comparison of the composition of bile and a liver-canalicular plasma-membrane subfraction. *Biochem. J.* **154**, 589–595.

Felker, T., Hamilton, R., and Havel, R. (1978). Secretion of lipoprotein-X by perfused livers of rats with cholestasis. *Proc. Natl. Acad. Sci. U.S.A.* **75**, 3459–3463.

Frimmer, M., and Ziegler, K. (1988). The transport of bile acids in liver cells. *Biochim. Biophys. Acta* **947**, 75–99.

Gibbons, G. (1990). Assembly and secretion of hepatic very-low-density lipoprotein. *Biochem. J.* **268**, 1–13.

Godfrey, P., Warner, M., and Coleman, R. (1981). Enzymes and proteins in bile. Variations in outputs in rat cannula bile during and after depletion of the bile salt pool. *Biochem. J.* **196**, 11–16.

Graham, J., and Northfield, T. (1987). Solubilization of lipids from hamster bile-canalicular and contiguous membranes and from human erythrocytes membranes by conjugated bile salts. *Biochim. J.* **242**, 825–834.

Gregory, D., Vlahcevic, R., Schatzki, P., and Swell, L. (1975). Mechanism of secretion of biliary lipids. I. Role of bile canalicular and microsomal membranes in the synthesis and transport of biliary lecithin and cholesterol. *J. Clin. Invest.* **55**, 105–114.

Gregory, D., Vlahcevic, R., Prugh, M., and Swell, L. (1978). Mechanisms of secretion of biliary lipids: Role of a microtubular system in hepatocellular transport of biliary lipids in the rat. *Gastroenterology* **74**, 93–100.

Griffiths, G., and Simons, K. (1986). The trans Golgi network: Sorting at the exit site of the Golgi complex. *Science* **234**, 438–443.

Hagenbuch, B., Stieger, B., Foguet, M., Lubbert, H., and Meier, P. J. (1991). Functional expression cloning and characterization of hepatocyte Na^+–bile acid cotransport system. *Proc. Natl. Acad. Sci. U.S.A.* **88**, 10629–10633.

Halpern, Z., Dubley, M. A., Lynn, M. C., Nader, J. M., Breuer, A. C., and Holzbach, R. T. (1986). Vesicle aggregation in model systems of supersaturated bile. Relation to crystal nucleation and lipid composition of the vencular phase. *J. Lipid Res.* **27**, 295–306.

Hardison, W., and Apter, J. (1972). Micellar theory of biliary cholesterol excretion. *Am. J. Physiol.* **222**, 61–67.

Hardison, W., Hattoff, D., Miyai, K., and Weiner, R. (1981). Nature of bile acid maximum secretion rate in the rat. *Am. J. Physiol. (Gastrointest. Liver Physiol.)* **241**, 337G–343G.

Harvey, P. R. C., and Strasberg, S. M. (1993). Will the real cholesterol-nucleating and anti-nucleating proteins please stand up? *Gastroenterology* **104**, 646–650.

Harvey, P. R. C., Sömjen, G., Lichtenberger, M. S., Petrunka, C. N., Gilat., T., and Strasberg, S. M. (1987). Nucleation of cholesterol from vesicles isolated from bile of patients with and without cholesterol gallstones. *Biochim. Biophys. Acta* **921**, 198–204.

Hay, D., and Carey, M. (1990). Chemical species of lipids in bile. *Hepatology* **12**, 6S–16S.

Heuman, D. M. (1989). Quantitative estimation of the hydrophilic–hydrophobic balance of mixed bile salt solutions. *J. Lipid Res.* **30**, 719–730.

Hofmann, A. (1990). Bile acid secretion, bile flow and biliary lipid secretion in humans. *Hepatology* **12,** 17S–25S.

Hofmann, A., and Small, D. (1967). Detergent properties of bile salts: Correlation with physiological function. *Annu. Rev. Med.* **18,** 333–376.

Holland, V., Zampighi, G., and Simon, S. (1989). Morphology of fungiform papillae in canine lingual epithelia. *J. Comp. Neurol.* **279,** 13–27.

Hornick, C., Hamilton, R., Spaziani, E., Enders, G., and Havel, R. (1985). Isolation and characterization of multivesicular bodies from rat hepatocytes; An organelle distinct from secretory vesicles of the Golgi apparatus. *J. Cell Biol.* **100,** 1558–1569.

Howell, J., Lucy, J., Pirola, R., and Bouchier, I. (1970). Macromolecular assemblies of lipid in bile. *Biochim. Biophys. Acta* **210,** 1–6.

Icarte, M., Pizarro, M., and Accatino, L. (1991). Adaptive regulation of hepatic bile salt transport: effects of alloxan-diabetes in the rat. *Hepatology* **14,** 671–678.

Inoue, M., Kinne, R., Tran, T., and Arias, I. (1984). Taurocholate transport by rat liver canalicular membrane vesicles: Evidence for the presence of a Na-independent transport system. *J. Clin. Invest.* **73,** 659–663.

Johnson, W., Chacko, G., Phillips, M., and Rothblat, G. (1990). The efflux of lysosomal cholesterol from cells. *J. Biol. Chem.* **265,** 5546–5553.

Jones, A., Schumker, P., and Mooney, J. (1979). Alterations in hepatic pericanalicular cytoplasm during enhanced bile secretory activity. *Lab. Invest.* **40,** 512–517.

Kao, Y., and Lichtenberger, L. (1991). Phospholipid- and neutral lipid-containing organelles of rat gastroduodenal mucous cells. Possible origin of the hydrophobic mucosal lining. *Gastroenterology* **101,** 7–21.

Kawamoto, T., Mao, S., and La Russo, N. (1987). Biliary excretion of apolipoprotein B by the isolated perfused rat liver. *Gastroenterology* **92,** 1236–1242.

Kramer, W., Brickel, U., and Buscher, H. P. (1982). Bile salt binding polypeptide in plasma membranes of hepatocytes revealed by photoaffinity labeling. *Eur. J. Biochem.* **129,** 13–18.

Lamri, Y., Roda, A., Dumont, M., Feldmann, G., and Erlinger, S. (1988). Immunoperoxidase localization of bile salts in rat liver cells. *J. Clin. Invest.* **82,** 1173–1182.

Lange, Y., and Steck, T. (1985). Cholesterol-rich intracellular membranes: A precursor to plasma membrane. *J. Biol. Chem.* **260,** 15592–15594.

La Russo, N. (1984). Proteins in bile: How they get there and what they do. *Am. J.Physiol. (Gastrointest. Liver Physiol.)* **247,** G199–G205.

LaRusso, N. F., and Fowler, S. (1979). Coordinate secretion of acid hydrolases in rat bile: Hepatocyte exocytosis of lysosomal protein? *J. Clin. Invest.* **64,** 949–954.

Lee, S., Park, H., Madami, H., and Kaler, E. (1987). Partial characterization of a nonmicellar system of cholesterol solubilization in bile. *Am. J. Physiol. (Gastrointest. Liver Physiol.)* **252,** G374–G384.

Liscum, L. (1990). Pharmacological inhibition of the intracellular transport of low-density lipoprotein-derived cholesterol in Chinese hamster ovary cells. *Biochim. Biophys. Acta* **1045,** 40–48.

Liscum, L., and Dahl, N. (1992). Intracellular cholesterol transport. *J. Lipid Res.* **33,** 1239–1254.

Liscum, L., Ruggiero, R. M., and Faust, J. R. (1989). The intracellular transport of low density lipoprotein-derived cholesterol is defective in Niemann–Pick type C fibroblasts. *J. Cell Biol.* **108,** 1625–1636.

Little, T., Lee, S., Madani, H., Kaler, E., and Chinn, K. (1991). Interconversions of lipid aggregates in rat and model bile. *Am. J. Physiol. (Gastrointest. Liver Physiol.)* **260,** G70–G79.

Long, T., Jakoi, L., Stevens, R., and Quarfordt, S. (1978). The sources of rat biliary cholesterol and bile acid. *J. Lipid Res.* **19,** 872–878.

Lowe, P., Barnwell, G., and Coleman, R. (1984). Rapid kinetic analysis of the bile-salt-dependent secretion of phospholipid, cholesterol and a plasma-membrane enzyme into bile. *Biochem. J.* **222,** 631–637.

Manzato, E., Fellin, R., Baggio, G., Walch, S., Neubeck, W., and Seidel, D. (1976). Formation of lipoprotein-X. Its relationship to bile compounds. *J. Clin. Invest.* **57,** 1248–1260.

Marsh, J. (1984). Hepatic synthesis and secretion of lipoproteins and apolipoproteins. *In* "Liver and Lipid Metabolism" (S. Calandra, N. Carulli, and G. Sahioli, eds.), pp. 13–26. Excerpta Medica, Amsterdam.

Martigne, M., Domingo, N., Lechene de la Porte, P., Lafort, H., and Hauton, J. (1988). Identification and localization of apoprotein fraction of the bile lipoprotein complex in human gallstones. *Scand. J. Gastroenterol.* **23,** 731–737.

Marzolo, M. P., Rigotti, A., and Nervi, F. (1990). Secretion of biliary lipids from the hepatocyte. *Hepatology* **12,** 134S–142S.

Marzolo, M. P., Amigo, L., and Nervi, F. (1993). Hepatic production of lipoproteins, catabolism of low density lipoprotein, biliary lipid secretion and bile salt synthesis in rats fed a bean (*Phaseoulus vulgaris*) diet. *J. Lipid Res.* **34,** 807–814.

Mazer, N. (1990). Quasielastic light scattering studies of aqueous biliary lipid systems and native bile. *Hepatology* **12,** 39S–44S.

Mazer, N., and Carey, M. (1983). Quasi-elastic light-scattering studies of aqueous biliary lipid systems. Cholesterol solubilization and precipitation in model bile solutions. *Biochemistry* **22,** 426–442.

Mazer, N. A., Schurtenberger, P., Carey, M. C., Preisig, R., Weigand, K., and Kanzig, W. (1984). Quasi-elastic light-scattering studies of native bile from the dog: Comparison with aggregative behavior of biliary lipid systems. *Biochemistry* **23,** 1994–2005.

Meier, P., St. Meier-Abt, A., and Boyer, J. (1987). Properties of the canalicular bile acid transport system in the rat. *Biochem. J.* **242,** 465–449.

Merré, D. (1981). An alternative pathway for secretion of lipoprotein particles in rat liver. *Eur. J. Cell Biol.* **26,** 21–25.

Miquel, J., Rigotti, A., Rojas, E., Brandan, E., and Nervi, F. (1992). Isolation and purification of human biliary vesicles with potent cholesterol-nucleation-promoting activity. *Clin. Sci.* **82,** 175–180.

Miquel, J., Núñez, L., Rigotti, A., Amigo, L., Brandan, E., and Nervi, F. (1993). Isolation and partial characterization of cholesterol pronucleating hydrophobic glycoproteins associated with native biliary vesicles. *FEBS Lett.* **318,** 45–49.

Miyairi, M., Oshio, C., Wanatabe, S., Smith, C., Yousef, I., and Phillips, M. (1984). Taurocholate accelerates bile canalicular contractions in isolated rat hepatocytes. *Gastroenterology* **87,** 788–792.

Mullock, B., Dobrota, M., and Hinton, R. (1978). Sources of the proteins of rat bile. *Biochim. Biophys. Acta* **543,** 497–507.

Nalbone, G., Lafont, H., Domingo, N., Lairon G., Pautrat, G., and, Hauton, Y. (1973). Ultramicroscopic study of bile lipoprotein complex. *Biochimie* **55,** 1503–1506.

Nalbone, G., Lafont, H., Vigne, J., Domingo, N., Lairon, D., Chabert, C., and Lechene, P. (1985). The apoprotein fraction of the bile lipoprotein complex: Isolation, partial characterization and phospholipid binding properties. *Biochimie* **61,** 1029–1041.

Nathanson, M., and Boyer, J. (1991). Mechanisms and regulation of bile secretion. *Hepatology* **14,** 551–566.

Nemchausky, B., Layden, T., and Boyer, J. (1977). Effects of chronic cholertic infusion of bile acids on the bile canaliculus: A biochemical and morphological study. *Lab. Invest.* **36**, 259–267.

Nervi, F., Bronfman, M., Allalon, W., Deperieux, E., and Del Pozo, R. (1984). Regulation of biliary cholesterol secretion in the rat: Role of hepatic cholesterol esterification. *J. Clin. Invest.* **74**, 2226–2237.

Nervi, F., Marinovic, I., Rigotti, A., and Ulloa, N. (1988). Regulation of biliary cholesterol secretion: Functional relationship between the canalicular and sinusoidal secretory pathways in the rat. *J. Clin. Invest.* **82**, 1818–1827.

Norman, P. T., and Norum, K. R. (1974). Newly synthesized hepatic cholesterol as precursor for cholesterol and bile acid in rat bile. *Scand. J. Gastroenterol.* **11**, 427–432.

Núñez, L., Rigotti, A., Amigo, L., Puglielli, L., Ibáñez, L., Raddatz, A., Guzmán, S., Zúñiga, A., and Nervi, F. (1993a). Biliary cholesterol crystallization-promoting activity of aminopeptidase-N isolated from native biliary vesicles and its quantitative correlation with cholesterol nucleation time in gallbladder bile of gallstone patients. *Gastroenterology* **104**, A965.

Núñez, L., Amigo, L., Rigotti, A., Pughelli, L., Mingione, G., Greco, A. V., and Nervi, F. (1993b). Cholesterol crystallization-promoting activity of aminopeptidase-N isolated from the vesicular carrier of biliary lipids. *FEBS Lett.* **329**, 84–88.

O'Guin, W., Manabe, M., and Sun, T. (1989). Association of a 25-kDa protein with membrane coating granules of human epidermis. *J. Cell Biol.* **109**, 2313–2321.

Oh, S., and Holzbach, T. (1976). Transmission electron microscopy of biliary mixed lipid micelles. *Biochim. Biophys. Acta* **441**, 498–505.

Okishio, T., and Nair, P. (1966). Studies on bile acids. Some observations on the intracellular localization of major bile acids in rat liver. *Biochemistry* **5**, 3662–3668.

Ornberg, R. L., Duong, L. T., and Pollard, H. B. (1986). Intragranular vesicles: New organelles in the secretory granules of adrenal chromaffin cells. *Cell Tissue Res.* **245**, 547–553.

Oshio, C. and Phillips, J. (1981). Contractility of bile canaliculi: Implications for liver function. *Science* **212**, 1041–1042.

Ostrow, D. (1993). APF/CBP, an anionic polypeptide in bile and gallstones that may regulate calcium salt and cholesterol precipitation from bile. *Hepatology* **16**, 1494–1496.

Palade, G. (1975). Intracellular aspects of the process of protein secretion. *Science* **189**, 347–358.

Pattinson, N. (1985). Solubilization of cholesterol in human bile. *FEBS. Lett.* **181**, 339–342.

Pattinson, N., and Chapman, B. (1986). Distribution of biliary cholesterol between mixed micelles and non-micelles in relation to fasting and feeding in humans. *Gastroenterology* **91**, 697–702.

Peled, Y., Halpern, Z., Baruch, R., Goldman, G., and Gilat, T. (1988). Cholesterol nucleation form its carriers in human bile. *Hepatology* **8**, 914–918.

Phillips, M., and Satir, P. (1988). The cytoskeleton of the hepatocyte: Organization, relationships, and pathology. *In* "The Liver: Biology and Pathobiology" (I. Arias, W. Jakoby, H. Popper, D. Schachter, and D. Shafritz, eds.), pp. 11–27. Raven Press, New York.

Portman, O., Alexander, M., and O'Malley, J. (1980). Metabolism of free and esterified cholesterol and apolipoproteins of plasma low and high density lipoproteins. *Biochim. Biophys. Acta* **619**, 545–558.

Reinhart, M. P. (1990). Intracellular sterol trafficking. *Experientia* **46**, 569–579.

Reuben, A. (1984). Biliary proteins. *Hepatology* **5**, 46S–50S.

Rigotti, A., Marzolo, M. P., Ulloa, N., González, O., and Nervi, F. (1989). Effect of bean intake on biliary lipid secretion and on hepatic cholesterol metabolism in the rat. *J. Lipid Res.* **30**, 1041–1048.

Rigotti, A., Núñez, L., Amigo, L., Puglielli, L., Garrido, J., Santos, M., González, S., Mingrone, G., Greco, A., and Nervi, F. (1993). Biliary lipid secretion: Immunolocalization and identification of a protein associated with lamellar cholesterol carriers in supersaturated rat and human bile. *J. Lipid Res. (in press).*

Robins, S., and Amstrong, M. (1976). Biliary lecithin secretion. II. Effects of dietary choline and biliary lecithin synthesis. *Gastroenterology* **70**, 397–402.

Robins, S., and Brunengraber, H. (1988). Origin of biliary cholesterol and lecithin in the rat: Contribution of new synthesis and preformed stores. *J. Lipid Res.* **23**, 604–608.

Robins, S., Fasulo, J., Collins, M., and Patton, G. (1985). Evidence for separate pathways of transport of newly synthesized and preformed cholesterol into bile. *J. Biol. Chem.* **260**, 6511–6513.

Robins, S., Fasulo, J., Raymond, L., and Patton, G. (1989). The transport of lipoprotein cholesterol into bile: A reassessment of kinetic studies in the experimental animal. *Biochem. Biophys. Acta* **1004**, 327–331.

Robins, S., Fasulo, J., Robins, V., and Patton, G. (1991). Utilization of different fatty acids for hepatic and biliary phosphatidylcholine formation and the effect of changes in phosphatidylcholine molecular species on biliary lipid secretion. *J. Lipid Res.* **32**, 985–992.

Roda, A., Grigolo, B., Roda, E., Simoni, P., Pelliciari, R., Natalini, B., Fini, A., and Morselli Labate, A. (1988). Quantitative relationship between bile acid structure and biliary lipid secretion in rats. *J. Pharmacol. Sci.* **77**, 596–605.

Rodriguez-Lafrasse, C., Rousson, R., Bonnet, J., Pentchev, P. R., Louisot, P., and Vanier, M. T. (1990). Abnormal cholesterol metabolism in imipramine-treated fibroblasts culture. Similarities with Niemann-Pick type C disease. *Biochim. Biophys. Acta* **1043**, 123–128.

Rooney, S. A. (1985). The surfactant system and lung phospholipid biochemistry. *Am. Rev. Resp. Dis.* **131**, 439–460.

Rosario, J., Sutherland, E., Zaccaro, L., and Simon, F. (1988). Ethinyl-estradiol administration selectively alters liver sinusoidal membrane lipid fluidity and protein composition. *Biochemistry* **27**, 3939–3946.

Rustan, A., Nossen, J., Berg, T., and Drevon, C. (1985). The effects of monensin on secretion of very-low-density lipoprotein and metabolism of asialofetuin by cultured rat hepatocytes. *Biochem. J.* **227**, 529–536.

Schriever, C., and Jüngst, D. (1989). Association between cholesterol–phospholipid vesicles and cholesterol crystals in human gallbladder bile. *Hepatology* **9**, 541–546.

Schubert, R., and Schmidt, K. H. (1988). Structural changes in vesicle membranes and mixed micelles of various lipid compositions after binding of different bile salts. *Biochemistry* **27**, 8787–8794.

Schubert, R., Beyer, K., Wolburg, H., and Schmidt, K. H. (1986). Structural changes in membranes of large unilamellar vesicles after binding of sodium cholate. *Biochemistry* **25**, 5263–5269.

Schwartz, C. C., Berman, M., Vlhacevic, R., Halloran, L. C. Gregory, D. H., and Swell, L. (1977). Multicompartmental analysis of cholesterol metabolism in man: Characterization of the hepatic bile acid and biliary cholesterol precursor sites. *J. Clin. Invest.* **61**, 408–423.

Schwartz, C., Halloran, L., Vlahcevic, R., Gregory, D., and Swell, L. (1978). Preferential utilization of free cholesterol from high-density lipoproteins for biliary cholesterol secretion in man. *Science* **200**, 62–64.

Shaffer, E., and Small, D. (1977). Biliary lipid secretion in cholesterol gallstone disease: The effect of cholecystectomy and obesity. *J. Clin. Invest.* **59**, 828–840.

Shipley, G. (1990). Structural studies of the lipid components of bile. *Hepatology* **12**, 33S–38S.

Simion, F., Fleisher, B., and Fleisher, S. (1984). Subcellular distribution of bile acids, bile salts, and taurocholate binding in rat liver. *Biochemistry* **23**, 6459–6466.

Small, D. (1970). The formation of gallstones. *Adv. Intern. Med.* **16**, 243–264.

Sömjen, G., and Gilat, T. (1983). A non-micellar mode of cholesterol transport in human bile. *FEBS. Lett.* **156**, 265–268.

Sömjen, G., and Gilat, T. (1985). Contribution of vesicular and micellar carriers to cholesterol transport in human bile. *J. Lipid Res.* **26**, 699–704.

Sömjen, G., Marikovsky, Y., Lelkes, P., and Gilat, T. (1986). Cholesterol–phospholipid vesicles in human bile: An ultrastructural study. *Biochim. Biophys. Acta* **879**, 14–21.

Sömjen, G., Marikovsky, Y., Wachtel, E., Harvey, R., Rosemberg, R., Strasberg, S., and Gilat, T. (1990a). Phospholipid lamellae are cholesterol carriers in human bile. *Biochim. Biophys. Acta* **1042**, 28–35.

Sömjen, G. J., Rosenberg, R., and Gilat, T. (1990b). Gel filtration and quasielastic light scattering studies of human bile. *Hepatology* **12**, 123S–129S.

Stein, O., Sanger, L., and Stein, Y. (1974). Colchicine-induced inhibition of lipoprotein and protein secretion into the serum and lack of interference with secretion of biliary phospholipids and cholesterol by rat liver in vivo. *J. Cell Biol.* **62**, 90–99.

Stone, B., and Evans, D. (1992). Evidence for a common biliary cholesterol and VLDL cholesterol precursor pool in rat liver. *J. Lipid Res.* **33**, 1665–1675.

Stone, B., Erickson, S., Craig, W., and Cooper, A. (1985). Regulation of rat biliary cholesterol secretion by agents that alter intrahepatic cholesterol metabolism. *J. Clin. Invest.* **76**, 1773–1781.

Strange, R., Chapman, B., Johnston, J., Nimmo, I., and Percy-Robb, I. (1979). Partitioning of bile acids into subcellular organelles and in vivo distribution of bile acids in rat liver. *Biochim. Biophys. Acta* **573**, 535–545.

Strasberg, S., and Hofmann, A. (1990). Biliary cholesterol transport and precipitation: Introduction and overview of conference. *Hepatology* **12**, 1S–5S.

Tabas, I., Lim, S., Xu, X. X., and Maxfield, F. R. (1990). Endocytosed beta VLDL and LDL are delivered to different intracellular vesicles in mouse peritoneal macrophages. *J. Cell Biol.* **111**, 929–940.

Tierney, D. (1989). Lung surfactant: Some historical perspectives leading to its cellular and molecular biology. *Am. J. Physiol. (Lung)* **257**, L1–L12.

Turley, S., and Dietschy, J. (1981). The contribution of newly synthesized cholesterol to biliary cholesterol in the rat. *J. Biol. Chem.* **256**, 2438–2446.

Turley, S., and Dietschy, J. (1988). The metabolism and excretion of cholesterol by the liver. *In* "The Liver: Biology and Pathobiology" (I. Arias, W. Jakoby, H. Popper, D. Schachter, and D. Shafritz, eds.), pp. 617–641. Raven Press, New York.

Ulloa, N., Garrido, J., and Nervi, F. (1985). Vesicular carriers of biliary cholesterol: Ultracentrifugal isolation and morphological evidence in native bile. *Gastroenterology* **88**, A1701.

Ulloa, N., Garrido, J., and Nervi, F. (1987). Ultracentrifugal isolation of vesicular carriers of biliary cholesterol in native human and rat bile. *Hepatology* **7**, 235–244.

Urbani, L., and Simoni, R. D. (1990). Cholesterol and vesicular stomatitis virus G protein take separate routes from the endoplasmic reticulum to plasma membrane. *J. Biol. Chem.* **265**, 1919–1923.

Verkade, H., Wolbers, M., Havinga, R., Uges, D., Vonk, R., and Kuipers, F. (1990). The uncoupling of biliary lipid from bile acid secretion by organic anions in the rat. *Gastroenterology* **99**, 1485–1492.

Vlahcevic, R., Gurley, E., Heuman, D., and Hylemon, P. (1990). Bile salts in submicellar concentrations promote bidirectional cholesterol transfer (exchange) as a function of their hydrophobicity. *J. Lipid Res.* **31**, 1063–1071.

Voelker, D. R. (1992). Organelle biogenesis and intracellular lipid transport in eukaryotes. *Microbiol. Rev.* **55**, 543–560.

Wagner, C., Trotman, B., and Soloway, R. (1976). Kinetic analysis of biliary lipid excretion in man and dog. *J. Clin. Invest.* **57**, 473–477.

Watanabe, N., Tsukada, N., Smith, C., and Phillips, J. (1991). Motility of bile canaliculi in the living animal: Implications for bile flow. *J. Cell Biol.* **113**, 1069–1080.

Wheeler, H., and King, K. (1972). Biliary excretion of lecithin and cholesterol in the dog. *J. Clin. Invest.* **51**, 1337–1350.

Wieland, T., Nassel, M., and Kramer, W. (1984). Identity of hepatic transmembrane transport system for bile salts and phospholipid by photoaffinity labeling. *Proc. Natl. Acad. Sci. U.S.A.* **81**, 5232–5236.

Wirtz, K. W. A., and Gadella, T. W. J. (1990). Properties and mode of action of specific and non-specific phospholipid transfer proteins. *Experientia* **46**, 592–599.

Wright, J., and Clemens, J. (1985). Metabolism and turnover of lung surfactant. *Am. Rev. Resp. Dis.* **135**, 426–444.

Xia, Y., Lambert, K., Schteingart, C., Gu, J., and Hofmann, A. (1990). Concentration biliary secretion of ceftriaxone. Inhibition of lipid secretion and precipitation of calcium ceftriaxone in bile. *Gastroenterology* **99**, 454–464.

Yamauchi, S., Linscheer, W., and Beach, D. (1991). Increase in serum and bile cholesterol and HMG-CoA reductase by lovastatin in rats. *Am. J. Physiol. (Gastrointest. Liver Physiol.)* **260**, G625–G630.

Yousef, I., Barnwell, S., Gratton, F., Tuchweber, B., Weber, A., and Roy C. (1987). Liver cell membrane solubilization may control maximum secretory rate of cholic acid in the rat. *Am. J. Physiol. (Gastrointest. Liver Physiol.)* **252**, G84–G91.

PART V

Summary and Perspectives

Quo Vadis, Lipidologist?

Kai Simons
European Molecular Biology Laboratory, 6900 Heidelberg, Germany

Investigators studying biological membranes traditionally have been divided into two camps, one concentrating on the lipid constituents and the other analyzing the membrane proteins. The lipid biochemists led the way in the early days by developing powerful analytical methods to characterize the phospholipid, glycolipid, and cholesterol species that build up the lipid bilayer, the common structure of all biological membranes. Research on membrane proteins was hampered for a long time by the difficulty of solubilizing these hydrophobic molecules in a way that would enable their isolation for functional characterization.

In the 1970s, research on biological membranes made important advances. The mechanism of detergent solubilization of biological membranes became understood, facilitating a rational approach to membrane protein purification. Use of easily accessible membranes—such as the erythrocyte plasma membrane, *Halobacterium* purple membranes, and viral envelopes—contributed to understanding how proteins are integrated into the lipid bilayer. During the 1970s, the two camps joined forces, a crucial step for the advances that characterized the research during this period. Additional support came from the field of membrane biophysics. Not only was the bilayer structure firmly established, but the dynamics of the lipid and the protein constituents were revealed. Lateral and rotational diffusion and flip–flop rates from one leaflet to the other were measured. As a result, the present textbook views of membrane structure and dynamics were formulated.

Despite this promising start, during the 1980s a widening gulf between the lipid and the protein camps became evident, and research on proteins began to dominate the field. Strategies to identify membrane proteins were transformed completely by recombinant DNA method. The molecular guises of enzymes, ion pumps, channel proteins, and receptors that long had eluded investigators finally were being unraveled at an increasing pace. The machinery responsible for the translocation of proteins across membranes was being dissected with the help of elegant biochemical and genetic approaches. Molecular research on membrane trafficking was initiated. Today, the molecular mechanisms responsible for vesicular transport are being disclosed with considerable success. However, the research continues to be focused overwhelmingly on the protein components. The lipids have been left in the background and, at least from the perspective of protein biochemists, are viewed as merely filling the holes that proteins make in the two-dimensional fluid. The impressive advances in lipid research in the 1980s were concerned primarily with the involvement of inositol lipids in signal transduction. The understanding of the lipid matrices of the different cellular membranes has not progressed much over the last 10 years relative to the advances in protein research. This widening gulf is reflected in conferences and symposia on molecular membranology at which lipids are sadly neglected.

In my opinion, one reason for this development is the fact that lipid laboratories around the world generally have remained isolated from the advances that have changed the face of molecular cell biology. Too much effort has been spent on model lipid systems with little relevance to cellular membranes. Lipids are being studied in isolation and not in their cellular context. Of course, several encouraging exceptions exist. The introduction of fluorescent lipid analogs has contributed greatly to promoting research on lipid traffic in cells. The use of mutations affecting yeast lipids also is beginning to have an impact on the field. However, clearly dramatic changes in lipid research must be introduced to bridge the gulf to the protein camp. Such changes can be accomplished only by reorganization of university departments so that biochemistry, biophysics, and cell biology of cellular membranes are being pursued in proximity and with synergy. My optimistic view is that this movement is already well under way and will dominate the field in the 1990s. A marriage between all these areas involved in membrane research undoubtedly will result in many successful offspring.

Many important issues must be explored, among the most interesting of which is the dynamic lateral organization of the bilayer. The existence of microdomains long has been debated, but little new insight into this important aspect of membrane organization has been forthcoming. The methods available seem not to be powerful enough to reveal clusters that are being formed and reformed continuously with rapid kinetics. Would it be possible to detect a lipid cluster within a cellular membrane with the dynamic characteristics of a detergent micelle? The answer to date has been no. Glycosphingolipids are known to associate with each other, a feature that has been postulated to be a critical organizing principle in the sorting of lipids and proteins into the apical transport vesicles from the *trans*-Golgi network of epithelial cells. However, no direct evidence from biophysical measurements supports such a view. The glycolipids on the cell surface also seem to cluster, possibly with cholesterol. One such interesting microdomain of glycolipids and cholesterol is represented by caveolae, specialized invaginations of the plasma membrane. Glycophospholipid-anchored proteins seem to partition preferentially into such microdomains involved in apical sorting and in caveolar organization. Presently, no clues exist to the mechanisms involved. Indirect evidence suggests that microdomains of glycolipids and glycophospholipid-anchored proteins on the cell surface play a role in signal transduction. Several cytoplasmic tyrosine kinases seem to be associated with such membrane clusters. These microdomains could segregate signal transducing events along the cell surface to enhance the specificity of ligand–receptor interactions with the signal transducing machinery. The dynamic coupling of the ligand and receptors with caveolae could be brought about by adaptors such as the alpha subunits of trimeric G proteins. The most intriguing experimental finding is that some nonionic detergents such as Triton X-100 can be used at 0°C to extract the microdomains from the cell more or less intact. This idea is at least one interpretation of the results obtained in several laboratories. No information is available to date on the physicochemical basis of the detergent resistance of the glycolipid–cholesterol–protein complexes, but these complexes provide an elegant shortcut for isolating the proteins involved.

Another interesting feature of cellular membranes is that their lipid bilayers are asymmetric. The cell expends a substantial amount of energy to generate and maintain the asymmetry of its bilayers. Little is known of how lipid asymmetry contributes to membrane function. One possibility is that lipid asymmetry stabilizes membrane protein topology. However, lipid asymmetry also could play an important role in membrane trafficking. One prevailing view of the formation of membrane vesicles that mediate transport from one cellular compartment to another assumes that protein coats such as clathrin polygons provide a driving force for inducing membrane curvature in the donor membrane during vesiculation. The lipids would function only as fluid matrices with no specific involvement in the vesiculation process. However, the lipid molecules also may be reorganized during vesicle budding to facilitate the increase in curvature. Here lipid asymmetry could play an important role. Studies with a yeast mutant affecting phospholipid transfer suggest a specific function for lipids in vesicular transport in the secretory pathway. One possibility is that lipid asymmetry is disturbed in this yeast mutant, and that this disturbance leads to a "traffic jam" in biosynthetic transport.

Not only are the properties of the lipid bilayer with its multitude of lipid species poorly understood, but little is known about how the cellular lipids contribute to the organization of the different cell types in multicellular organisms. The problem of the elucidation of the molecular mechanisms underlying cellular morphogenesis includes the generation of the different lipid profiles of cellular membranes. For instance, epithelial cells differentially sort glycolipids to their apical cell surface. How is this done? Some glycolipid species such as galactosylceramides tend to be excluded from the apical route; this lipid is enriched in the basolateral plasma membrane instead. Could this comparmentalization be interpreted to mean that glucosylceramides and their glycosylated derivatives associate preferentially with each other, avoiding galactosylceramides? In other words, is glycolipid clustering dependent on the chemical make-up of the lipids? Neurons also are polarized, segregating their cell surfaces into axons and dendrites. Several experiments suggest similarities between the neuronal sorting of surface proteins and lipids and the epithelial sorting mechanisms. Is the axonal plasmalemma covered with an external layer of glycolipids? Does this associating glycolipid layer enhance nerve conduction by decreasing membrane permeability? How are different glycolipid species sorted in neurons? Can different glycolipid clusters be used as sorting platforms? Oligodendrocytes and Schwann cells are responsible for forming the myelin sheaths that wrap around the axons. Myelin formation is an unusually interesting problem of membrane biogenesis. The major lipid constituent of myelin is galactosylceramide. This glycolipid could play an essential role in the coordinated formation of myelin. Obviously myelin research will progress only if both the lipid and the protein constituents are included in the analysis. The many important diseases affecting myelin provide an added stimulus for an integrated research program.

Other areas that would benefit from new research strategies include the secretion of lipid–protein assemblies. The serum lipoproteins have been studied intensively. However, despite all efforts, little is known of the mechanisms that are responsible for assembling these particles during biosynthetic transport. Imaginative methodology will be required to reveal how their biogenesis is accomplished. Another interesting assembly is the lamellar body produced by lung alveolar cells. A mixture of lipids, mainly dipalmitoyl phosphatidylcholine, and an array of surfactant proteins are assembled into lamellar bodies and secreted in a regulated manner from the alveolar epithelial cells. After secretion, these bodies are disassembled and function as a surfactant to reduce surface tension along the lung alveolar epithelium. Almost nothing is known about how these lamellar bodies are assembled. Therapy of respiratory distress syndrome, a frequent cause of infant mortality, would benefit greatly from an increased understanding of how lung surfactant is assembled and disassembled.

One additional lipid–protein assembly deserving more attention is the Odland body, a type of a lamellar body produced and secreted by the cornified layer of the epidermis. These lamellar bodies have a completely different composition than those containing lung surfactant. The Odland bodies are greatly enriched in glycosphingolipids and cholesterol. After secretion, the lamellar bodies are integrated into the intercellular space of the cornified layer of the skin. The lipid components of these assemblies are thought to be responsible for the barrier properties of skin. Practically nothing is known of how these bodies are formed and how they contribute to the structure and function of the permeability barrier operating in the epidermis.

These few examples illustrate the exciting prospects ahead of us. There is no lack of interesting research problems requiring answers. Good luck—the choice is yours.

Index